Springer Collected Works in Mathematics

For further volumes:
http://www.springer.com/series/11104

伊藤清

Kiyosi Itô

# Selected Papers

*Editors*

D. W. Stroock · S. R. S. Varadhan

Reprint of the 1987 Edition

Springer

*Author*
Kiyosi Itô (1915 – 2008)
Kyoto University
Kyoto
Japan

*Editors*
D. W. Stroock
Massachusetts Institute of Technoloy
Cambridge, MA
USA

S. R. S. Varadhan
Courant Institute
New York University
New York
USA

AMS Classifications: 60G07, 60Gl5, 60H05, 60HI0, *60105*

ISSN 2194–9875
ISBN 978-1-4614-9630-4 (Softcover)
978-0-387-96326-6 (Hardcover)
DOI 10.1007/978-1-4612-5370-9
Springer New York Heidelberg Dordrecht London

Library of Congress Control Number: 2012954381



Printed on acid-free paper

Springer is part of Springer Science+Business Media (www.springer.com)

# Contents*



* Numbers in brackets refer to the Bibliography of Kiyosi Itô (see pages xix–xxi).

# Introduction

The central and distinguishing feature shared by all the contributions made by K. Itô is the extraordinary insight which they convey. Reading his papers, one should try to picture the intellectual setting in which he was working. At the time when he was a student in Tokyo during the late 1930s, probability theory had only recently entered the age of continuous-time stochastic processes: N. Wiener had accomplished his amazing construction little more than a decade earlier (Wiener, N., "Differential space," *J. Math. Phys.* **2,** (1923)), Lévy had hardly begun the mysterious web he was to eventually weave out of Wiener's paths, the generalizations started by Kolmogorov (Kolmogorov, A.N., "Über die analytische Methoden in der Wahrscheinlichkeitsrechnung," *Math Ann.* **104** (1931)) and continued by Feller (Feller, W., "Zur Theorie der stochastischen Prozesse," *Math Ann.* **113,** (1936)) appeared to have little if anything to do with probability theory, and the technical measure-theoretic *tours de force* of J.L. Doob (Doob, J.L., "Stochastic processes depending on a continuous parameter," *TAMS* **42** (1937)) still appeared impregnable to all but the most erudite. Thus, even at the established mathematical centers in Russia, Western Europe, and America, the theory of stochastic processes was still in its infancy and the student who was asked to learn the subject had better be one who was ready to test his mettle. In view of the state of affairs in Russia and the West, one can imagine how tremendously difficult it must have been for the student Itô sitting in Japan. Unlike most of us, it was not his task to simply learn a subject which was already well understood by his mentors. Instead, he was expected to first learn the subject essentially on his own and then explain it to everyone around him. That he was equal to this task, and more, is amply demonstrated by the papers contained in this volume. Indeed, Itô has spent his professional career explaining stochastic processes not only to the Japanese but also to the rest of us as well. By looking upon his work as an ongoing attempt to provide the insight with which to elucidate the essential structural underpinnings of the subject, one begins to appreciate the extent to which Itô has molded the way in which we all think about stochastic processes. The rest of this introduction is devoted to an expansion of this point.

Everyone who is likely to pick up this book has at least heard that there exists a subject called the theory of stochastic integration and that K. Itô is the Lebesgue of this branch of integration theory (Paley and Wiener were its Riemann). However, many of us have little or no idea why Itô got involved in the topic. Surely he must have had in mind something other than the integration of an arbitrary previsable *cadlag* process against an arbitrary *cadlag* semimartingale. With the translation of his 1942 article [2], those of us whose Japanese is less than it ought to be are at last treated to the pleasure of learning what was the insight behind the insight. What we learn in this article is that Itô had discovered what is the *tangent* to an *integral curve of probability measures*. (Of course, Kolmogorov's forward equation contains the same discovery, but it fails to provide the same insight.) That is, Itô views the transition function $P(t,x,\cdot)$ of a continuous-time Markov process as a flow of probability measures issuing from the point mass $\delta_x$ at time $t = 0$. In order to find the tangent to this flow, he reasons as follows: the tangent to a flow is the best linear approximation to that flow; in the context of probability measures, the role of lines is played by the infinitely divisible laws (a process which proceeds by independent identically distributed increments is not changing direction and is therefore going in a straight line); and the procedure for finding the best linear approximation is to magnify via a rescaling. Hence,

$$L(x,\cdot) = \lim_{n\to\infty} (P(1/n,x,\cdot))^{*n}$$

is the tangent to $t \to P(t,x,\cdot)$ at $t = 0$ in the sense that the best infinitely divisible approximation to the process governed by $P(t,x,\cdot)$ is the stationary independent increment process associated with $L(x,\cdot)$ (i.e., having distribution $L(x,\cdot)$ at time one). Thus, a Markov process leaves the point $x$ as if it were, instantaneously, this independent increment process.

Knowing that someone was sufficiently clever to reach the preceding conclusions, it should come as no more than a mild surprise that the same person was able to figure out how to integrate the *vector field* $x \to L(x,\cdot)$ and thereby recapture $P(t,x,\cdot)$ from a knowledge of the field $L$. Turning to the simplest case, namely, when the state space is the real line and $L(x,\cdot)$ is, for each $x$, a Gaussian with mean value $b(x)$ and variance $a(x)$, let us see where the reasoning given above leads. As a first guess, one is tempted to try an Euler approximation scheme in order to integrate $L(x,\cdot)$. That is, one might try constructing $P(t,x,\cdot)$ as the limit as $n \to \infty$ and $k/n \to t$ of $P_n(k/n,x,\cdot)$, where

$$P_n(0,x,\cdot) = \delta_x,$$

$$P_n((k+1)/n,x,\cdot) = \int G_{a(y),b(y)}(1/n,y,\cdot)P_n(k/n,x,dy)$$

and $G_{a(y),b(y)}(1/n,y,\cdot)$ denotes the Gaussian with mean $b(y)/n$ and variance $a(y)/n$. Such an approach would have worked. In fact, E.E. Lévy had successfully carried it out (at least when $a(\cdot)$ is uniformly positive and both $a(\cdot)$ and $b(\cdot)$ have bounded continuous first derivatives many years before (cf. (Lévy, E.E., *Rend. del Circ. Mat Palermo* **24**, pp. 275–317 (1907)). However, from Itô's point of view, there are two serious objections to this approach. First, although it was on probabilistic grounds that we wrote down this scheme, probability theory alone does not immediately suggest a method with which to complete the program. (Remember that the invariance principle was not available to Itô.) Secondly, even if one produces $P(t,x,\cdot)$ by this procedure, one is still a long way from realizing the stochastic process (i.e., the measure on function space) for which $P(t,x,\cdot)$ is the transition function. That is, one is at the same stage in the construction of the associated diffusion as one is in the construction of Brownian motion after writing down the heat kernel but before one knows Weiner's theorem. Of course, the existing work of Kolmogorov and Doob could have been used to complete this part of the program, but by the time one had finished, probability theory would have been relegated to a minor role.

The route adopted by Itô completely circumvents the objections mentioned above. What he did next in [2] was to transfer his intuitive picture to the paths themselves. That is, he decided to take his intuition about the way in which a Markov process leaves a point more literally. Thus, in the diffusion context, he reasoned that the paths themselves should leave $x$ just like Wiener paths having instantaneous mean $b(x)$ and instantaneous variance $a(x)$. With this in mind, it is clear that the differential $dx(t)$ of the diffusion path $x(t)$ satisfies

(1) $$dx(t) = a(x(t))^{1/2}dw(t) + b(x(t))dt, \qquad x(0) = x,$$

where $w(\cdot)$ denotes a Wiener path. (Notice that, in this context, there is no ambiguity about how the increment $dw(t)$ is to be taken: it must be taken in the future in order to fit the intuition.) Having written down equation (1), Itô still had two formidable tasks ahead of him. In the first place, he still had to make sense out the equation, and it was to resolve this knotty technical problem that he invented what is now called the theory of *Itô integration*. The second problem which confronted him was that of connecting the paths $x(\cdot)$ determined by (1) with the transition function $P(t,x,\cdot)$. Obviously, if his reasoning was correct, $P(t,x,\cdot)$ should be the distribution of $x(t)$ under Wiener measure. But how can one check that this is indeed the case. Presumably, the most satisfactory and direct check would be to show that the distribution of $x(t)$ is the solution to Kolmogorov's forward equation. The solution to this identification problem went through several stages, the initial and perhaps the most critical of which can be seen already in [2], and culminated with the justly famous *Itô formula* [13]:

(2) $$df(x(t)) = f'(x(t))dx(t) + 1/2f''(x(t))(dx(t))^2,$$

where $(dx(t))^2$ is interpreted as $a(x(t))dt$. Of course, all this is now ancient history. Nonetheless, one can still get a measure of the depth at which Itô was working by trying to explain to a novice, even a mathematically sophisticated one, just how it was that Itô solved these problems. Also, it is intriguing to see Itô's ideas come full circle in the recent work on the stochastic integration of forms on manifolds which has forced us to make a careful reexamination of the principles underlying the theory. In particular, the fine treatment of this theory given by P.A. Meyer [*Geometrie Differentielle Stochastique, Seminaire de Probabilités XVI, 1980–81* (Lecture Notes in Mathematics 921). Springer-Verlag, Berlin, 1981, pp. 165–207] hints of a return to Itô's original ideas about integral curves of probability measures.

Before moving on to some of Itô's other accomplishments, we must bring to the reader's attention a remarkable recent contribution which Itô made to the theory of stochastic integration. During the sixties, more and more of the applied stochastic process community had adopted the Stratonovich version of stochastic integration instead of Itô's. Their reason for doing so is easy to understand: Stratonovich's integral behaves much better under changes of variable and, for this reason, solutions to the associated integral equations are nicely approximated by solutions to ordinary integral equations. However, it is fair to say that until the appearance of [37] mathematicians had a difficult time reconciling their own theory with that of their applications oriented colleagues. With his typical incisiveness, Itô isolated and resolved all these difficulties in a few pages. Shortly thereafter, he went on, in [40], to explain Stratonovich's integral in entirely new terms and show how this new formulation gives rise to a completely satisfactory stochastic integral version of the fundamental theorem of calculus. In doing so, he introduced the idea of looking at martingales under different filtrations, an idea which has not as yet been fully exploited but has nonetheless proved to be powerful in the hands of people like M. Yor.

Although it is difficult for us to imagine now, the reception accorded Itô's discoveries by European and American probabilists was not immediately enthusiastic. Indeed, with the exception of the brilliant treatment of Itô's theory given by Doob in *Stochastic Processes* (New York, Wiley (1953)), the mathematicians (as opposed to engineers) in the West appear to have ignored Itô's stochastic calculus during the fifties and early sixties. During this same period, Itô himself turned his attention to other matters. In 1955, Itô was a visitor to the Princeton Institute for Advanced Study. Because of the presence of Solomon Bochner and William Feller on the faculty at Princeton University, Princeton in the mid-fifties was a very exciting place for someone with Itô's interests. In particular, Feller had just completed his soon to become famous analysis of one-dimensional diffusions generators (cf. *TAMS* **77,** pp. 1–31 (1954) and *Ann. Math.* **55,** pp. 417–436 (1954)) and was about to turn his student H.P. McKean, Jr., loose in the mathematical world. Feller's theory of one-dimensional diffusions clearly demonstrated that a complete description of diffusion theory cannot be achieved unless one allows more general generators than those which Kolmogorov and he himself had considered in the thirties. In particular, Feller's work forces one to look at generators which are not expressible as classical differential operators. Of course, as soon as the generator is no longer a classical differential operator, the associated diffusion process can no longer be constructed as the solution to a "classical" Itô stochastic differential equation and it becomes a non-trivial problem to describe in what sense (if any) these "generalized" diffusions are related to Brownian motion. Although other authors, in particular E.B. Dynkin (cf. *Theory Prob. and Appl.* **I**, pp. 34–54 (1956)), made important steps towards a probabilistic interpretation of the generators which arise in Feller's theory and V.A. Volkonskii (cf. *Theory Prob. and Appl.* **III**, pp. 310–326 (1958)) discovered a preliminary version of the relationship between Feller's generalized diffusions and Brownian motion, the first people to give a complete solution to the problem were Itô and McKean, who, at the time when they made their key discovery, were unaware of the work which Dynkin and Volkonskii were doing. Their key discovery was that of the role which Brownian *local time* plays and remains one of the essential tools in the analysis of one-dimensional diffusions. Because Itô and McKean's contributions to this subject are contained in their justly famous book (*Diffusion Processes and Their Sample Paths,* Springer-Verlag, Berlin (1965)), it is unnecessary to expand here on the ideas which they introduced. Suffice it to say that once again Itô had gone directly to the heart of the matter.

At the sixth (and last) Berkeley Symposium on Mathematical Statistics and Probability Theory, Itô presented a paper ([34] in this volume) that displayed the same startling originality and insight which had been the trademark of his earlier investigations. His purpose in this article was to show how a continuous-time Markov process may be decomposed into *excursions* away from a recurrent state. The idea of decomposing Markov processes into such excursions was already a familiar one in the discrete-time setting and had been exploited with great success there. Further, extensions of the technique to continuous-time but non-instantaneous state processes had been made. Itô himself, in a paper with McKean [24], had applied the idea to classify Brownain motions on a half line. However, prior to these articles by Itô and McKean, only Lévy had had the courage to consider excursions in the context of instantaneous state processes, and Lévy's treatment seemed to be inextricably tied to special properties of Brownian motion (in particular, to Brownian scaling). Moreover, although Lévy succeeded in describing Brownian excursions, it was not clear from his treatment that one could reconstitute the Brownian motion from his decomposition. Thus, even though the idea had antecedents, Itô's paper represents a major original advance.

Without going into the details it is difficult to fully appreciate Itô's accomplishment. Nonetheless, one can get a sense of the intrinsic difficulties if one considers a few of the basic facts and tries to imagine how one might overcome the problems which they impose. Suppose, for example, that one wants to describe the excursions which a one-dimensional Brownian motion makes from the origin. The set $M$ of times $t \geqslant 0$ when the path is at 0 is (almost surely) a closed set which (like the Cantor set) contains no intervals and no isolated points. Thus, already the problem of *counting* the excursions in some reasonable way is a highly non-trivial one. Further, the advantage which one hopes to gain from a decomposition into excursions comes from the idea (obviously true in the non-instantaneous context) that the successive excursions should be independent and identically distributed. But how is one going to achieve such a description when every pair of excursions is separated by infinitely many other excursions? The problem here has a familiar ring to anyone who has studied the problem of resolving an instantaneous-jump Lévy process into its Poisson components. As it turns out, not only are these problems similar to one another but so are their solutions. Indeed, after putting everything in the correct time scale (namely, the *local time scale*), what Itô did is describe the excursion process in terms of a path-valued Poisson point process. The picture thereby achieved is the precise analogue of the description which he [1] and Lévy gave of independent increment processes. Although his work on this subject is still relatively new, its power is already apparent from its applications to the construction of diffusions satisfying boundary conditions (cf. S. Watanabe, "Construction of diffusion processes by means of Poisson point processes of Brownian excursions," *Proc. Third Japan–USSR Symp. Prob. Theory,* Lec. Notes in Math., Springer-Verlag, p. 550 (1976)) and recent work by J. Pitman and M. Yor on the windings of Brownian motion.

Just as this volume is itself a *Selecta* and not a *Collected Works*, so this introduction is a mere sampling. Our hope is that it will whet the reader's appetite and encourage him to read some of the original papers by one of the truly original and influential contributors to our discipline. Our own professional careers having been as enriched as they have from our acquaintance with Itô and his work, we are inclined to believe that others cannot help but profit from learning directly from one of the masters.

We take this opportunity to thank J. Pitman for allowing us to consult with him about excursion theory and Itô's role in its development. The preceding brief account of that theory is based on a letter which we received from him.

S.R.S. Varadhan
D.W. Stroock

# Foreword

As my foreword to this book I would like to outline my research development and its background, which may be of some interest to the reader.

When I was a student in the Department of Mathematics at the University of Tokyo (1935–1938), I was fascinated by the rigorous arguments and the beautiful structures seen in pure mathematics, but also I was concerned with the fact that many mathematical concepts had their origins in mechanics. Fiddling around with mathematics and mechanics, I came close to stochastic processes through statistical mechanics. At that time when probability theory was not popular in Japan, I felt myself rather isolated. But I was able to continue my work thanks to the kind encouragement of Professor Shokichi Iyanaga in whose seminar I was participating.

Having read A. Kolmogorov's *Grundbegriffe der Wahrscheinlichkeitsrechnung* (1933) I became convinced that probability theory could be developed in terms of measure theory as rigorously as in other fields of mathematics. In P. Lévy's book *Théorie de l'addition des variables aléatoires* (1937) I saw a beautiful structure of the sample paths of stochastic processes deserving the name of mathematical theory. From this book I learned stochastic processes, Wiener's Brownian motion (Wiener process), Poisson process, and processes with independent increments (differential processes). I was particularly interested in the decomposition theorem for differential processes, the core of this book. But I had a hard time following Lévy's argument because of his unique intrinsic description. Fortunately I noticed that all ambiguous points could be clarified by means of J. L. Doob's idea of regular versions presented in his paper "Stochastic processes depending on a continuous parameter" [*Trans. Amer. Math. Soc.* **42**, 1938]. Checking Lévy's argument carefully from Doob's viewpoint, I was able to introduce Poisson random measures of jumps to really understand Lévy's spirit of the decomposition theorem. In accordance with Professor Iyanaga's suggestion I published this result [1] in the *Japanese Journal of Mathematics* in 1942.

As a natural extension of Lévy's book I read A. Kolmogorov's paper on a sample continuous Markov process (diffusion process), "Über die analystische Methoden in der Wahrscheinlichkeitsrechnung" [*Math. Ann.* **104,** 1931], and W. Feller's paper on a general Markov process with jumps, "Zur Theorie der stochastichen Prozesse (Existenz-und Eindeutigkeitssätze)" [*Math. Ann.* **113,** 1936]. In these papers I saw a powerful analytic method to study the transition probabilities of the process, namely Kolmogorov's parabolic equation and its extension by Feller. But I wanted to study the paths of Markov processes in the same way as Lévy observed differential processes. Observing the intuitive background in which Kolmogorov derived his equation (explained in the Introduction of the paper), I noticed that a Markovian particle would perform a time homogeneous differential process for the infinitesimal future at every instant, and arrived at the notion of a stochastic differential equation governing the paths of a Markov process that could be formulated in terms of the differentials of a single differential process. It took some years to carry out this idea. When I finished this job for a sample continuous Markov process (Kolmogorov's case), I wrote a paper on the result in Japanese (English translation [2]) and intended to write it in English after having been able to carry out my idea for a general Markov process (Feller's case).

I published this Japanese paper in a mimeographed journal (1942) of Osaka University. This journal was being published to encourage young mathematicians all over Japan to communicate their ideas at that time when there was no xeroxing.

I do not know anyone who read this paper thoroughly when it appeared except my friend G. Maruyama. As I heard from him later, he read it in the military camp where he had been drafted for World War II. Later he wrote a paper on the subject (*Cir. Mat. Palermo,* 1955). At that time Maruyama and I were the only probabilists in Japan who were really interested in the sample paths.

I wrote these two papers [1], [2] when I was working in the Statistical Bureau of the Government (1939–1943), where I was given sufficient time for my own study thanks to the kindness of Mr. T. Kawashima, Head of the Bureau.

In 1943 I obtained a teaching position at Nagoya University and stayed there until I moved to Kyoto in 1952. At Nagoya I was very happy to work with Professor Kôsaku Yosida, though this period was the dark age of World War II and its aftermath.

At that time Yosida had just completed his famous semigroup theory and was developing it further to apply it to various fields of analysis, including the Kolmogorov equation for time homogeneous Markov processes. Through discussions with him I obtained much knowledge about functional analysis that turned out to be very useful later.

I was also lucky to become acquainted with Professor S. Kakutani who returned from Princeton to Osaka because of the war breaking out. I listened with much interest to his lectures on ergodic theory, positive-definite functions on non-abelian groups, and the relation between Brownian motions and harmonic functions. I enjoyed discussions with him, because he was also interested in the sample paths and showed interest in my work.

Although I had heard much of N. Wiener's great contribution to probability theory, I had not read his work carefully until I went to Nagoya. Even his theory of Brownian motion I learned from Lévy's book and Doob's papers. Reading some of his papers I was impressed by the originality with which he initiated not only measure-theoretic probability theory but also path-theoretic process theory as early as the 1920's.

Being motivated by Kakutani's lecture on ergodic theory, Maruyama's work on ergodic properties of stationary Gaussian processes, and discussions with H. Anzai, Kakutani's student, I slightly modified Wiener's homogeneous chaos to define multiple Wiener integral [4] (1951). Later I learned that Wiener had made the same modification in a different way. In 1953 I introduced the complex multiple Wiener integral and used it to give another proof of Maruyama's result [17].

When I wrote my manuscript on stochastic differential equations for general Markov processes several years after I had planned to in 1942, I learned that there was no journal in Japan to publish such a long note because of the shortage of papers under the bad economic conditions after the war. So I sent it to Professor J. L. Doob and asked him about any possibility of its being published in the United States. He made a kind arrangement for publishing it in the *Memoirs of the American Mathematical Society* in 1951 [12].

In the later half of my Nagoya period manifold theory was beginning to draw the attention of young people. Being stimulated by such an atmosphere, I became interested in a diffusion on a compact manifold whose generator is a non-degenerate elliptic operator. The analytic theory had been initiated by Kolmogorov, and Yosida was treating it from his semi-group viewpoint. I wanted to construct the path of the diffusion by writing a stochastic differential equation in terms of local coordinates and wrote three papers [10], [11], [15]. Although my work was not complete at all, I obtained a chain rule of stochastic differentials [13] as a byproduct and noticed that this rule is useful to understand the probabilistic meaning of the generator.

In 1952 I moved to Kyoto University, from which I retired in 1979. During those 27 years I was at Kyoto for half of the period in total, because I was abroad for the rest, at Princeton (1954–1956), Stanford (1961–1964), Aarhus, Denmark (1966–1969), and Cornell (1969–1975). I am very grateful to Professors Y. Akizuki, A. Kobori, A. Komatsu, and M. Hukuhara for their kindness and generosity, thanks to which I was able to enrich my knowledge in mathematics as well as in other respects through my experience in the United States and in Europe. In particular I learned various aspects of probability theory through discussions with W. Feller, H. P. McKean, R. Getoor, J. L. Doob, S. Karlin, K. L. Chung, H. Hoffmann-Jørgensen, F. Spitzer, H. Kesten, L. Gross, H. H. Kuo, D. W. Stroock, S. R. S. Varadhan, P. Malliavin, P. Meyer, J. Neveu, H. Föllmer, E. B. Dynkin, and many others.

In spite of my frequent absence from Kyoto I was lucky to have had a number of students who contributed much to the field of my interest and are still working actively and bringing up young probabilists. N. Ikeda, T. Hida, M. Nisio, S. Watanabe, M. Fukushima and H. Kunita are among them.

In 1952 when I moved to Kyoto, Doob's famous book *Stochastic Processes* appeared. I was impressed by his beautiful theory of martingales and its applications to various problems. I was also glad to see the stochastic integral discussed in the framework of martingales.

## FOREWORD

My experience at Princeton (1954–1956) was most exciting because of the presence of Professor W. Feller and his student H. P. McKean, Jr., now a Professor at the Courant Institute. When I met Feller, I was impressed by his deep insight into the paths of stochastic processes in spite of all his papers being written from the analytic side. He suggested to me and to McKean as well the problem of constructing the paths of the most general one-dimensional diffusion whose analytic theory he had completed by that time. Since I did not really understand Feller's theory at that time, I learned it from McKean; I gave him the whole set of my reprints, all of which had been written from the path-theoretic viewpoint. He read my papers very quickly. Later he wrote a nice book on stochastic integrals (1964), compact but full of interesting materials, using the relation between stochastic integrals and martingales ingeniously. After lively and often exciting discussions we succeeded in constructing the Brownian path with an elastic barrier, the problem Feller proposed to us for the first step, by using P. Lévy's local time. Thus we obtained a clue for constructing the paths of Feller's diffusion. Thereafter our joint work went on rather smoothly, even though we sometimes had technical or more serious difficulties to overcome. When we got an almost complete picture of the one-dimensional diffusion processes, we decided to write a book, which was published as *Diffusion Processes and Their Sample Paths* by Springer-Verlag in 1965. For almost ten years McKean worked very hard to collect new materials and to organize them. Without his tremendous effort our joint work would have never appeared in book form.

In 1957–1958 McKean visited Kyoto and gave a series of lectures on diffusion processes. His lectures were so inspiring that they resulted in the emergence of quite a few active probabilists in Japan. We wrote two joint papers, one for random walks and potentials [20] and the other for the construction of Brownian motion on the half-line for every possible boundary condition [21] whose analytic theory had been established under some restrictions by Feller.

Since 1960 the general theory of Markov processes represented by E. B. Dynkin's work and the potential theoretic Markov process theory represented by G. Hunt's work have developed extensively and rapidly. Doob's martingale theory played an important role in this development. Also the theory of stochastic differential equations has been completely transformed by martingale integrals (initiated by Doob and completed by H. Kunita, S. Watanabe, and P. Meyer), the Stroock–Varadhan martingale problems, the stochastic differential geometry of J. Eells, K. D. Elworthy, and P. Malliavin, and Malliavin calculus. I wrote several papers related to this development, three of which I explain below.

When I was at Stanford (1961–1964), I had a plan to construct harmonic tensor fields on a Riemannian manifold by using Brownian motion. Though I was not able to solve this problem, I obtained the idea of stochastic parallel displacement as a byproduct [23] (1962). Since this idea drew no attention except that E. B. Dynkin introduced tensor diffusions, I gave up my plan. Later I was very glad to hear that J. Eells, K. D. Elworthy, and P. Malliavin took up my idea around 1970 when they established their theory of stochastic differential geometry, in which my problem was beautifully solved. Such a solution was far beyond my reach when I introduced stochastic parallel displacement.

When I heard of the Stratonovich integral from engineering mathematicians around 1970, I was too stubborn to accept it, because their reason to use it seemed to me to be just technical. But when I learned from D. W. Stroock his construction of Brownian motion on the sphere using the Stratonovich integral, I suddenly realized the mathematical meaning of this integral. I wrote a paper [37] (1975) to clarify the relation between my integral and the Stratonovich integral in terms of martingale integrals and to illustrate some mathematical applications.

In 1970, when I was at Cornell, I wrote a paper to determine all possible behaviors of a Markov process at a recurrent point [34]. This was a generalization of my joint paper with McKean [24], but the motivation was quite different. When I was at Aarhus (1966–1969), I became interested in infinite-dimensional stochastic differential equations to deal with stochastic dynamical systems of infinite degrees of freedom and tried for the first step to study basic facts and to check special examples. After several years it became my habit to observe even finite-dimensional facts from the infinite-dimensional viewpoint. This habit led me to reduce the problem above to a Poisson point process with values in the space of excursions.

When I became free from all official duties by retiring from Gakushuin University (1979–1985), I was given a chance to spend one year in the Institute for Mathematics and Its Applications at the University of Minnesota thanks to the kind consideration of Professors H. Weinberger, S. Orey, and D. W. Stroock. Since all workshops for this year are for stochastic differential equations and their applications, I am listening to the lectures with great interest and learning that there are many interesting new theories of which I have not had the slightest idea. It would be my greatest pleasure if I could study these new theories leisurely after I returned to Kyoto.

I would like to express my sincere thanks to Professors S. R. S. Varadhan and D. W. Stroock for their arrangement of publication of my selected papers and their extremely kind Introduction, and I am also grateful to the staff at Springer-Verlag for their enthusiastic cooperation.

K. Itô

# Bibliography of Kiyosi Itô

## Papers


[1] On stochastic processes (infinitely divisible laws of probability) (Doctoral thesis). *Japan. Journ. Math.* **XVIII,** 261–301 (1942).

[2] Differential equations determining a Markoff process (original Japanese: Zenkoku Sizyo Sugaku Danwakai-si). *Journ. Pan-Japan Math. Coll.* No. 1077 (1942).

[3] On the ergodicity of a certain stationary process. In: *Proc. Imp. Acad. Tokyo* **20,** 54–55 (1944).

[4] A kinematic theory of turbulence. In: *Proc. Imp. Acad. Tokyo* **20,** 120–122 (1944).

[5] On the normal stationary process with no hysteresis. In: *Proc. Imp. Acad. Tokyo* **20,** 199–202 (1944).

*[6] A screw line in Hilbert space and its application to the probability theory. *Proc. Imp. Acad. Tokyo* **20,** 203–209 (1944).

[7] Stochastic integral. In: *Proc. Imp. Acad. Tokyo* **20,** 519–524 (1944).

*[8] On Student's test. *Proc. Imp. Acad. Tokyo* **20,** 694–700 (1944).

[9] On a stochastic integral equation. In: *Proc. Japan Acad.* **22,** 32–35 (1946).

[10] Stochastic differential equations in a differentiable manifold. *Nagoya Math. Journ.* **1,** 35–47 (1950).

[11] Brownian motions in a Lie group. In: *Proc. Japan Acad.* **26,** 4–10 (1950).

[12] On stochastic differential equations. *Mem. Amer. Math. Soc.* **4,** 1–51 (1951).

[13] On a formula concerning stochastic differentials. *Nagoya Math. Journ.* **3,** 55–65 (1951).

[14] Multiple Wiener integral. *Journ. Math. Soc. Japan* **3,** 157–169 (1951).

[15] Stochastic differential equations in a differentiable manifold (2). *Mem. Coll. Science, Univ. Kyoto, Ser. A,* **28,** 81–85 (1953).

[16] Stationary random distributions. *Mem. Coll. Science, Univ. Kyoto, Ser. A.,* **28,** 209–223 (1953).

[17] Complex multiple Wiener integral. *Japan. Journ. Math.* **22,** 63–86 (1952).

[18] Isotropic random current. In: *Proc. Third Berkeley Symp. Math. Statist. Prob.* **II,** 125–132 (1955).

[19] Spectral type of the shift transformation of differential processes with stationary increments. *Trans. Amer. Math. Soc.* **81,** 253–263 (1956).

[20] Potentials and the random walk (with H. P. McKean, Jr.). *Illinois Journ. Math.* **4,** 119–132 (1960).

[21] Wiener integral and Feynman integral. In: *Proc. Fourth Berkeley Symp. Math. Statist. Prob.* **II,** 227–238 (1960).

[22] Construction of diffusions. *Ann. Fac. Sci. Univ. Clermont* **2,** 23–32 (1962).


* These do not appear in this "selection."

[23] The Brownian motion and tensor fields on Riemannian manifold. In: *Proc. Intern. Congr. Mathemat.* (Stockholm), 536–539 (1962).

[24] Brownian motions on a half line (with H. P. McKean, Jr.). *Illinois Journ. Math.* **7,** 181–231 (1963).

[25] The expected number of zeros of continuous stationary Gaussian processes. *Journ. Math. Kyoto Univ.* **3,** 207–216 (1964).

[26] On stationary solutions of a stochastic differential equation (with M. Nisio). *Journ. Math. Kyoto Univ.* **4,** 1–75 (1964).

[27] Transformation of Markov processes by multiplicative functionals (with S. Watanabe). *Ann. Inst. Fourier, Univ. Grenoble* **XV,** 13–30 (1965).

[28] The canonical modification of stochastic processes. *Journ. Math. Soc. Japan* **20,** 130–150 (1968).

[29] On the convergence of sums of independent Banach space valued random variables (with M. Nisio). *Osaka Journ. Math.* **5,** 35–48 (1968).

[30] Generalized uniform complex measures in the Hilbertian metric space with their application to the Feynman integral. In: *Proc. Fifth Berkeley Symp. Math. Statist. Prob.* **II,** 145–161 (1965).

[31] On the oscillation functions of Gaussian processes (with M. Nisio). *Math. Scand.* **22,** 209–223 (1968).

[32] Canonical measurable random functions. In: *Proc. Int. Conf. Funct. Anal. Rel. Topics* (Tokyo), 369–377 (1969).

[33] The topological support of a Gauss measure on Hilbert space. *Nagoya Math. Journ.* **38,** 181–183 (1970).

[34] Poisson point processes attached to Markov processes. In: *Proc. Sixth Berkeley Symp. Math. Statist. Prob.* **III,** 225–239 (1970).

*[35] Stochastic differentials of continuous local martingales. In: *Stability of Stochastic Dynamical Systems* (Lecture Notes in Mathematics **294**) Springer-Verlag, Berlin, 1–7 (1972).

*[36] Stochastic integration. In: *Vector and Operator Valued Measures and Applications,* Academic Press, New York, 141–148 (1973).

[37] Stochastic differentials. *Appl. Math. and Opt.* **1,** 374–381 (1974).

[38] Stochastic parallel displacement. In: *Probabilistic Methods in Differential Equations* (Lecture Notes in Mathematics **451**), Springer-Verlag, Berlin, 1–7 (1975).

*[39] Stochastic calculus. In: *Mathematical Problems in Physics* (Lecture Notes in Physics **39**) Springer-Verlag, Berlin, 218–223 (1975).

[40] Extension of stochastic integrals. In: *Proc. Int. Symp. Stochastic Differential Equations* (Kyoto), 95–109 (1976).

*[41] Introduction to stochastic differential equations (with S. Watanabe). In: *Proc. Int. Symp. Stochastic Differential Equations* (Kyoto), i–xxx (1976).

*[42] Continuous additive $\mathcal{S}'$-processes.

*[43] Stochastic analysis in infinite dimensions. In: *Stochastic Analysis* (A. Friedman and M. Pinksy, eds.), Academic Press, New York, 187–197 (1978).

*[44] Infinite dimensional Ornstein-Uhlenbeck processes. In: *Taniguchi Symp. SA,* Katata, 197–224 (1982).

[45] Regularization of linear random functionals (with M. Nowata). In: *Probability Theory and Mathematical Statistics, Fourth USSR–Japan Symposium Proceedings, 1982* (Lecture Notes in Mathematics **1021**), Springer-Verlag, Berlin, 257–267 (1983).

[46] Distribution-valued processes arising from independent Brownian motions. *Math. Zeit.* **182,** 17–33 (1983).

*[47] A stochastic differential equation in infinite dimensions. *Contemporary Math.* **26,** 163–169 (1984).

Books*

[a] *Foundation of Probability Theory* (in Japanese), Iwanami, Tokyo, 1943.

[b] *Probability Theory* (in Japanese), Iwanami, Tokyo, 1952.

[c] *Diffusion Processes and Their Sample Paths*, Springer, 1965.

[d] *Probability Theory* (in Japanese), Iwanami Series of Fundamental Mathematics, Analysis (I) vii, Iwanami, Tokyo, 1978, revised 1983.

[e] *Introduction to Probability Theory* (English translation of [d] Chaps. I–IV), Cambridge Univ. Press, 1984.

[f] *Foundations of Stochastic Differential Equations in Infinite Dimensional Spaces,* (CBMS-NSF Reg. Conf. Ser. in Appl. Math 4) SIAM, 1984.

# 5. *On stochastic processes*

## (*Infinitely divisible laws of probability*)

By Kiyosi ITÔ.

(Received 1 August, 1941.)

The main aim of this paper is to give a new rigorous proof of the known formula of P. Lévy(1) on the infinitely devisible law (of probability) in making use of the scheme of stochastic differential processes introduced recently by J. L. Doob.(2) Chapter I is devoted to preliminary explanations of concepts and notations used later on. In Chapter II, I shall prove several theorems concerning the topology of the probability distributions on the real number space; these theorems will be useful in the following chapters.

In Chapter III, I shall establish a relation between "an infinitely divisible law" and "a stochastic process." Logically to say, they constitute two different concepts; the former is a probability distribution on the real number space and the latter is a random variable which takes values in a functional space. It is the chief object of Chapter III to relate these two concepts (Cf. Theorem 3.1.). The "$\mathfrak{L}$-process" appearing there is not essentially different from what is called a stochastic differential process with no fixed discontinuity.

In Chapter IV, I shall determine a canonical form of $\mathfrak{L}$-processes (Cf. Theorem 4.4, and 4.5.), whence the formula of Lévy is easily deduced. The idea of this chapter is essentially due to the work of Lévy, but it may be not without interest to express it in a clearer form. In Chapter V, I shall prove two theorems concerning temporally homogeneous $\mathfrak{L}$-processes, by which we can easily prove the uniqueness theorem of Lévy(3) concerning his formula.

## I. Preliminaries.

*A probability distribution* on a space $\Omega$, as is well-known, is defined as a real-valued set-function $P$ on $\Omega$, defined for any set belonging to a

---

*) The cost of this research has been defrayed from the Scientific Research Expenditure of the Department of Education.

(1) See P. Lévy: Théorie de l'addition des variables aléatoires (1937), p. 180 Th. 54.1, and see also A. Khintchine: Déduction nouvelle d'une formule de M. Paul Lévy. Bull. Univ. Etat Moscou, Sér. Int. Sect. A. Math. et Mechan. 1 (1937).

(2) See J. L. Doob: Stochastic processes depending on a continuous parameter, Th. 3.8 (Transactions of the American Mathematical Society, Vol. 42).

(3) See P. Lévy: loc. cit. (1) p. 186 Th. 55.1.

certain completely additive class of subsets of $\Omega$ and satisfying the following conditions:

(1) $$P(E) \geqq 0$$

(2) $$P(\Omega) = 1$$

(3) for disjointed subsets $E_1, E_2, \ldots,$ of $\Omega$, we have

$$P(\sum_1^\infty E_k) = \sum_1^\infty P(E_k).$$

We shall often make use of the term "*a P-measurable set*" instead of "a set for which $P$ is defined."

Let $B$ be a certain completely additive class of subsets of a space $X$. When we consider the space together with $B$, we shall call it *a field* and denote it by $(X, B)$. We can consider any topological space $X$ as a field in taking as $B$ the least completely additive class of subsets of $X$ containing all open sets on $X$. A space $\Omega$ on which a probability distribution $P$ is difined can also be considered as a field $(\Omega, B_P)$, $B_P$ being the class of all $P$-measurable subsets. Such a field is called *a probability field* and is denoted by $(\Omega, P)$.

One of the most important concepts in the probability theory is that of the "*random variable*". *An $(X, B)$-valued random variable on a probability field $(\Omega, P)$* is defined as a function $x(\omega)$ on $\Omega$ which takes values in $X$ and such that $E(\omega\,;\, x(\omega) \in E)$ is $P$-measurable for any subsets $E$ in $B$. The image $x(\Omega)$ of $\Omega$ by $x$ is a subset of $X$. If we denote this subset by $X'$, we shall be able to define a completely additive class $B'$ on $X'$ as the class of the subsets such as $X' \cdot E$, $E$ running over $B$. The random variable $x$ may be considered also as an $(X', B')$-random variable. Such a consideration may be not important if $X$ is a simple space such as the real number space or the $n$-dimensional vector space, but it was shown by J. L. Doob(4) how important it is in case $X$ is a functional space.

A function defined on $\Omega$ except on a subset of $\Omega$ with probability *0* is called a random variable in the larger sense, if it satisfies the above condition of $P$-measurability.

In order to avoid ambiguities we shall make use of different notations to represent a random variable itself and the value which the variable takes for a certain element $\omega$ in $\Omega$; the former will be denoted by the symbol such as "$x$" or "$(x(\omega)\,;\, \omega \in \Omega)$", and the latter by the symbol such as $x(\omega)$.

(4) See J. L. Doob: loc. cit. (3).

We shall now define *the direct product of the fields* $(X_\alpha, B_\alpha)$ for $\alpha \in A$, $A$ being a parameter domain. The direct product $X$ of the space $X_\alpha$ for $\alpha$ in $A$, as is well-known, is the set of all correspondences such as

$$(x_\alpha\,;\ \alpha \in A)\,,$$

where $x_\alpha$ is an element in $X_\alpha$ for any value $\alpha$ in $A$. The direct product of $E_\alpha$ for $\alpha \in A$, $E_\alpha$ being respectively any element in $B_\alpha$, will be called a cylinder subset of $X$, if $E_\alpha \neq X_\alpha$ for a finite number of values of $\alpha$. Let $B$ denote the least completely additive class of subsets of $X$ which contains all cylinder subsets. The field $(X, B)$ is the required direct product.

**Definition 1.1.** Let $x_\alpha$ be any $(X_\alpha, B_\alpha)$-valued random variable on $(\Omega, P)$, where $\alpha$ runs over a parameter domain $A$. Let $(X, B)$ denote their product field. The $(X, B)$-valued random variable on $(\Omega, P)$:

$$\big((x_\alpha(\omega)\,;\ \alpha \in A)\,;\ \omega \in \Omega\big)$$

is called *the combination of* $x_\alpha$ for $\alpha \in A$ and is denoted by $(x_\alpha\,;\ \alpha \in A)$. Except in the case that $A$ is enumerable, this definition may perhaps lose its sense, if we replace here "randow variable in the larger sense" instead of "random variable".

**Definition 1.2.** Let $x$ be an $(X, B)$-valued random variable, and $f$ a transformation from $(X, B)$ to $(X', B')$ such that the set $f^{-1}(E')$ may belong to $B$ for any set $E'$ in $B'$. Then $(f(x(\omega))\,;\ \omega \in \Omega)$ will be an $(X', B')$-valued random variable. This variable is called *a B-measurable function of* $x$ and is denoted by $f(x)$.

**Definition 1.3.** Let $x_\alpha$ be any $(X_\alpha, B_\alpha)$-valued random variable on $(\Omega, P)$ where $\alpha$ runs over a parameter domain $A$. When, for any finite system of parameter-values $\alpha_1 \alpha_2 \ldots \alpha_n (\in A)$ and for any subset $E_{\alpha_i}$ of $X_{\alpha_i} (i = 1, 2, \ldots n)$, we have

$$P\big((x_{\alpha_1} \in E_1)(x_{\alpha_2} \in E_2)\cdots(x_{\alpha_n} \in E_n)\big) = P(x_{\alpha_1} \in E_1)P(x_{\alpha_2} \in E_2)\cdots P(x_{\alpha_n} \in E_n)\text{(5)},$$

we say that $x_\alpha (\alpha \in A)$ are *independent*.

**Corollary 1.1.** *Let $f_\alpha(x_\alpha)$ be a B-measurable function of $x_\alpha$ where $\alpha$ runs over $A$. If $x_\alpha (\alpha \in A)$ be independent, then $f_\alpha(x_\alpha)$ will be also independent.*

(5) Let $A(\omega)$ be a proposition depending on an element $\omega$ in a probability field $(\Omega, P)$. Then we shall denote the set of the $\omega$-elements for which $A(\omega)$ is true by $A$ itself, if we have no fear of confusions; this notation may be consistent with the conventional use.

If two random variables $x_1$, $x_2$ are independent, we say also that $x_1$ is *independent of* $x_2$.

**Corollary 1.2.** *If $x_1, x_2, \ldots, x_n$ be independent, then $x_i$ will be also independent of the combination of $x_{i+1}, x_{i+2} \ldots$, and $x_n$ for any suffix $i$, and conversely.*

**Definition 1.4.** Let $x$ be an $(X, B)$-valued random variable on $(\Omega, P)$. $P(x^{-1}(E))$ will be a probability distribution on $X$. We shall call this distribution *the probability law of* $x$, and denote it by $P_x$; we also say that $x$ is *subject to this law* $P_x$.

**Definition 1.5.** Let $x$ be a random variable on $(\Omega, P)$ which takes values in the $n$-dimensional vector space. In case $\int_\Omega \| x(\omega) \| P(d\omega) < \infty$[6], we shall define the mean value of $x$ as $\int_\Omega x(\omega) P(d\omega)$, and denote it by $\bar{x}$ or $m(x)$.

**Definition 1.6.** Let $P$ be a probability distribution on the $n$-dimensional vector space $R^n$. In case $\int_{R^n} \| \lambda \| P(d\lambda) < \infty$ we shall define the mean value of $P$ as $\int_{R^n} \lambda P(d\lambda)$, and denote it by $m(P)$.

**Corollary 1.3.** $\bar{x} = m(P_x)$

**Definition 1.7.** Let $x$ be a real-valued random variable on $(\Omega, P)$. *The standard deviation of* $x$ is defined as $\sqrt{\overline{(x-\bar{x})^2}}$ if this value exists and is finite. It will be denoted by $\sigma(x)$. *The concentration function of* $x$ with regard to a positive number $l$ is defined as $\underset{\infty > \lambda > -\infty}{\text{l.u.b.}} \overline{P(\lambda \leqq x \leqq \lambda + l)}$ and denoted by $Q(x, l)$. *The characteristic function of* $x$ is defined as $\overline{e^{izx}}$ and is denoted by $\varphi_x(z)$.

**Definition 1.8.** Let $P$ be a probability distribution on the real number space $R$. *The standard deviation of* $P$ is defined as $\sqrt{\int_{-\infty}^{+\infty} (\lambda - m(P))^2 P(d\lambda)}$ if this value exists and is finite, and is denoted by $\sigma(P)$. *The concentration function of* $P$ with regard to $l$ is defined as $\underset{\infty > \lambda > -\infty}{\text{l.u.b.}} P([\lambda, \lambda + l])$[7] and is denoted by $Q(P, l)$.

The characteristic function of $P$ is defined as $\int_{-\infty}^{+\infty} e^{iz\lambda} P(d\lambda)$, and is denoted by $\varphi_P(z)$.

**Corollary 1.4.** $\sigma(x) = \sigma(P_x)$, $Q(x, l) = Q(P_x, l)$, $\varphi_x(z) = \varphi_{P_x}(z)$.

---

(6) $\| \quad \|$ means the euclidian metric.

(7) By $[ab)$ we shall denote the interval closed on the left and open on the right. $[ab]$, $(ab]$ or $(ab)$ may also be understood similarly.

**Definition 1.9.** Let $x$ be an $(X, B)$-valued random variable on $(\Omega, P)$ and $y$ on $(\Omega', P')$. If $P_x(E) = P_y(E)$ for any set $E$ in $B$, we say that $x$ and $y$ are *equivalent in law as $(X, B)$-valued random variables*; we shall write " $x \underset{(X,B)}{\sim} y$ " or simply " $x \sim y$ ".

**Definition 1.10.** If $F$, $G$ be probability distributions on the real number space, the following set-function $H$ will be also a probability distribution:

$$H(E) = \int_{-\infty}^{+\infty} F(E(+)(-\lambda))G(d\lambda)\,(^8).$$

This distribution is called the convolution of $F$ and $G$, and is denoted by $F * G$.

**Corollary 1.5.** *If $x$, $y$ be independent real-valued random variables, then we have*

$$P_{x+y} = P_x * P_y$$

**Definition 1.11.** Let $(\Omega_\alpha, P_\alpha)$ be a probability field separable in $P_\alpha$ where $\alpha$ runs over $A$. Let $\Omega$ denote the product space of $\Omega_\alpha$. We can constitute(9) a probability distribution $P$ on $\Omega$ so that

$$P(E) = \prod_{\alpha \in A} P_\alpha(E_\alpha),$$

$E_\alpha$ being identified with $\Omega_\alpha$ except for a finite number of values of $\alpha$ and $E$ being the product space of $E_\alpha$ for $\alpha \in A$. Then $(\Omega, P)$ is called *the direct product of* $(\Omega_\alpha, P_\alpha)$ for $\alpha \in A$. The *representation* in the direct product $(\Omega, P)$ of a random variable $x_\beta$ on $(\Omega_\beta, P_\beta)$ will be defined as the random variable $x'_\beta$ on $(\Omega, P)$ satisfying $x'_\beta(\omega) = x_\beta(\omega_\beta)$ for $\omega = (\omega_\alpha;\ \alpha \in A)$.

**Corollary 1.6.** *Let $x_\alpha$ be a random variable on $(\Omega_\alpha, P_\alpha)$ where $\alpha$ runs over $A$. Let $x'_\alpha$ denote the representation of $x_\alpha$ on the direct product of $(\Omega_\alpha, P_\alpha)$ for $\alpha \in A$. Then $x'_\beta(\alpha \in A)$ are independent.*

## II. On probability distributions on the real number space.

In this chapter we shall treat probability distributions on the real number space and call them distributions simply. Let $F$ be any distribution. Then $F([-\infty, \lambda])$ will have the following properties, if we consider it as a function of $\lambda$:

---

(8) $E(+)a$ means the set $E(x+a;\ x \in E)$.

(9) See J. L. Doob: Stochastic processes with an integral-valued parameter, Th. 1.1 (Transactions of the American Mathematical Society, Vol. 44), and E. Hopf: Ergodentheorie, p. 4, Mass und Integral in Produkträumen (Ergebnisse der Mathematik, Vol. 5, No. 2). The word "separable" used here is explained at the end of this paper.

(1) monotone increasing[10]

(2) if $\lambda \to \infty$, then it will tend to *1*, and if $\lambda \to -\infty$, then it will tend to *0*,

(3) continuous on the right hand.

We shall denote this function by the same notation $F$. As Lévy[11] has remarked, it is convenient to consider that, at each discontinuity point $\lambda$ of $F$, $F(\lambda)$ may represent any values between $F(\lambda-0)$ and $F(\lambda+0)$. Then the curve $y=F(x)$ proves to be a continuous curve. By the distance $\rho$ between two distributions $F$, $G$ we shall understand the least upper bound of the distance between the points at which any straight line perpendicular to the bisector of the first quadrant in the $(xy)$-plane intersects the curves $y=F(x)$ and $y=G(x)$ respectively.

$q(F)$, as defined in the following expression, will be called *the mean concentration function of* $F$:

$$q(F)=\int_0^\infty Q(F,\,l)e^{-l}dl\,, \tag{4}$$

$Q(F,\,l)$ being the concentration function of $F$ with regard to $l$.

Let $F_1$ and $F_2$ be distributions. If there exists a real number $\lambda$ such that $F_1(E(+)\lambda)=F_2(E)$ for any $B$-measurable set $E$, we say that $F_1$ is equivalent to $F_2$ and write $F_1 \sim F_2$. This relation is evidently reflexive, symmetric, and transitive. We classify all distributions by means of this equivalence relation, and represent classes by small letters $f, g, \dots$, and especially by $e$ the *unit class* containing the unit distribution $E$ which attributes to *0* the probability *1*. *The concentration function* or *the mean concentration function of a class* is defined as that of some distribution belonging to the class, and *the convolution of two classes* is defined as the class which contains the convolution of distributions chosen from each of the classes; these definitions are clearly independent of the choice of distributions from the classes in question.

If there exists a class $h$ such that $f=g*h$, we shall write, by definition, $f \leqslant g$. This relation satisfies evidently the following conditions:

(5a) $f \leqslant f$, (5b) $f \leqslant g$ and $g \leqslant h$ imply $f \leqslant g$,

(5c) $f \leqslant g$ and $g \leqslant f$ imply $f=g$.

The set of all classes forms thus *a partially ordered set with regard to this order relation.*

(10) In this paper we shall make use of the term "monotone increasing" in the sense of "monotone non-decreasing".

(11) See P. Lévy: loc. cit. (1) p. 28, foot-note (1).

*The distance* $\rho$ between two classes is defined as the minimum of the distance between two distributions respectively chosen from the classes. The distance $\rho$ satisfies clearly the following conditions;

(6a) $\rho(f,g) \geqq 0$ (6b) $\rho(f,g) = \rho(g,f)$ (6c) $\rho(f,g)+\rho(g,h) \geqq \rho(f,h)$

(6d) $\rho(f,f) = 0$ (6e) $\rho(f,g) = 0$ implies $f = g$.

It is also easy to see that the set of all classes forms *a complete set with regard to this distance.*

We shall prove a theorem concerning the relation of this order and this distance $\rho$.

**Theorem 2.1.** *Any set* $\mathfrak{K}$ *of classes lowerly bounded with regard to the above order relation by a class* $f$ *is totally bounded with regard to the above distance* $\rho$. (We shall adopt the term "order-bounded" or "$\rho$-bounded" to signify "bounded with regard to the above order" or "to the above distance $\rho$".)

*Proof.* It is sufficient to prove that any sequence $\{f_k\}$ in $\mathfrak{K}$ may have a $\rho$-convergent subsequence. We will choose from each class of $f_1, f_2, \ldots$, and $f$ the distributions whose medians are equal to 0, and denote them by $F_1, F_2, \ldots$, and $F$ respectively, where the median of $F$ means one of the values $x$ such that $F(x) = \frac{1}{2}$. By *the principle of decreasing*(12) *of the concentration function we have*

$$Q(F_k, l) \geqq Q(F, l).$$

The median of $F_1, F_2, \ldots$, and $F$ being equal to *0*, we can deduce from this inequality that, as far as $l$ is so large that $Q(F, l) > \frac{1}{2}$, we have

$$1-F_k(l)+F_k(-l) \leqq 1-Q(F_k, l) \leqq 1-Q(F, l).$$

$1-Q(F, l)$ tending to 0 with $\frac{1}{l}$, a theorem of Lévy(13) informs us that the sequence $\{F_k\}$ has a $\rho$-convergent subsequence $\{G_k\}$. Let $g_1, g_2, \ldots$ denote respectively the classes containing $G_1, G_2, \ldots$. Then $g_k$ is a $\rho$-convergent subsequence of $\{f_k\}$.

We shall introduce now another distance into the set of classes; we define namely:

$$\rho_1(f,g) = |q(f)-q(g)| \tag{7}$$

(12) See P. Lévy: loc. cit. (1) p. 90.
(13) See P. Lévy: loc. cit. (1) p. 63.

This distance $\rho_1$ satisfies the above conditions (6a)—(6d) but not necessarily (6e).

**Theorem 2.2.** *The distance $\rho_1$ satisfies the condition (6e) on a linearly ordered set of classes.*

*Proof.* It is sufficient to show that $f \leqslant g$ and $q(f) = q(g)$ imply $f = g$. As $f \leqslant g$, we have $f = g * h$ where $h$ is a certain class. If we choose $G \in g$ and $H \in h$, then we have $G * H \in f$. As the principle of decreasing of the concentration function informs us $Q(G * H, l) \leqq Q(G, l)$, we have $Q(f, l) \leqq Q(g, l)$. But we have by the assumption $q(f) = q(g)$, namely,

$$\int_0^\infty Q(f, l)e^{-l}dl = \int_0^\infty Q(g, l)e^{-l}dl. \tag{8}$$

As $Q(f, l)$ and $Q(g, l)$ are monotone functions continuous on the right hand, we have $Q(f, l) = Q(g, l)$ i.e. $Q(G * H, l) = Q(G, l)$. By a theorem of Lévy(14) we can deduce, from this identity, $H \in e$, namely $f = g$.

**Theorem 2.3.** *The distances $\rho$ and $\rho_1$ induce equivalent topologies on any set $\mathfrak{R}$ of classes of distributions which is linearly ordered and lowerly order-bounded.*

*The proof* of this theorem consists of two parts. In the first part we shall show that there exists a positive number $\delta(\mathfrak{E})$ for any positive number $\mathfrak{E}$ such that $\rho(f, g) < \delta(\mathfrak{E})$ may imply $\rho_1(f, g) < \mathfrak{E}$, and in the second part we shall show conversely that there exists a positive number $\delta(\mathfrak{E})$ for any positive number $\mathfrak{E}$ such that $\rho_1(f, g) < \delta(\mathfrak{E})$ may imply $\rho(f, g) < \mathfrak{E}$.

The first part: Let $F_0$, $G_0$ denote distributions in $f$ and $g$ respectively such that $\rho(F_0, G_0) = \rho(f, g)$. On the curve $y = F_0(x)$ there exist now two points $(x_1 y_1)$ $(x_2 y_2)$ such that

$$x_2 - x_1 = l \tag{9}$$

$$y_2 - y_1 = Q(F_0, l). \tag{10}$$

Draw two straight lines perpendicular to the bisector of the first quadrant through $(x_1 y_1)$ and $(x_2 y_2)$; and let $(x_1' y_1')$ and $(x_2' y_2')$ denote respectively the points of intersection with the curve $y = G_0(x)$. Let $\delta$ denote $\rho(F_0, G_0)$. Then we have

$$|x_1 - x_1'| \leqq \delta, \quad |y_1 - y_1'| \leqq \delta, \quad |x_2 - x_2'| \leqq \delta, \quad \text{and} \quad |y_2 - y_2'| \leqq \delta,$$

and so

(14) See P. Lévy: loc. cit. (1) p. 19, Th. 29,1.

$$x_2'-x_1' \leqq x_2-x_1+2\delta = l+2\delta$$

$$y_2'-y_1' \geqq y_2-y_1-2\delta = Q(F_0,\, l)-2\delta\,,$$

from which we deduce

$$Q(G_0,\, l+2\delta) \geqq Q(F_0,\, l)-2\delta$$

i.e.

$$\int_0^\infty Q(G_0,\, l+2\delta)e^{-l}dl \geqq q(f)-2\delta\,. \tag{11}$$

But

$$\int_0^\infty Q(G_0,\, l+2\delta)e^{-l}dl = e^{2\delta}\int_{2\delta}^\infty Q(G_0,\, l)e^{-l}dl \leqq e^{2\delta}q(g) \leqq (1+4\delta)q(g)$$

holds for $\delta<\dfrac{1}{4}$. Therefore we have, as $q(g)\leqq 1$, for $\delta<\dfrac{1}{4}$,

$$\int_0^\infty Q(G_0,\, l+2\delta)e^{-l}dl \leqq q(g)+4\delta\,. \tag{12}$$

By (11) and (12) we have $q(g)+4\delta \geqq q(f)-2\delta$ or $q(g)\geqq q(f)-6\delta > q(f)-\mathfrak{E}$ for $\delta<min\left(\dfrac{\mathfrak{E}}{6},\ \dfrac{1}{4}\right)$. Similarly we have $q(f)>q(g)-\mathfrak{E}$, and so we have finally, for $\rho(f,g)<min\left(\dfrac{\mathfrak{E}}{6},\ \dfrac{1}{4}\right)$, $\rho_1(f,\, g) = |\,q(f)-q(g)\,| < \mathfrak{E}$.

For the second part of the proof, it is sufficient to prove that the existence of a positive number $\mathfrak{E}_0$ and three sequences $\{f_n\}$, $\{g_n\}$, $\{h_n\}$ in $\mathfrak{K}$, such that $f_n = g_n * h_n$, $\rho_1(f_n, g_n) \leqq \dfrac{1}{n}$ and $\rho(f_n, g_n) \geqq \mathfrak{E}_0$ for every natural number $n$, may be reduced to a contradiction. By Theorem 2.1 we can see that there exists a subsequence $\{n_k\}$ of $\{n\}$ such that

$$\rho\text{-}\lim_{k\to\infty} f_{n_k} = f\,,\quad \rho\text{-}\lim_{k\to\infty} g_{n_k} = g\,,\quad \text{and}\quad \rho\text{-}\lim_{k\to\infty} h_{n_k} = h\,.$$

(The limit distributions $f$, $g$, and $h$ may not exist in $\mathfrak{K}$, but it does not matter.) By the assumption we have $f_{n_k} = g_{n_k} * h_{n_k}$ and so

$$f = g*h \tag{13}$$

By the first part of the proof we can see that $\rho_1(f_{n_k}, g_{n_k}) \leqq \dfrac{1}{n_k}$ implies

$$\rho_1(f,g) = 0\,. \tag{14}$$

From (13) and (14) follows $f=g$ by Theorem 2.2, in cotradiction with the inequality $\rho(f,g)\geqq \mathfrak{E}_0$ which follows from $\rho(f_{n_k}, g_{n_k}) \geqq \mathfrak{E}_0$.

**Corollary 2.1.** *If* $f_1 \geqslant f_2 \geqslant f_3 \geqslant \cdots \geqslant f$, *then there exists* $\rho\text{-}\lim_{n\to\infty} f_n$.

*Proof.* We have $q(f_1) \geqq q(f_2) \geqq \cdots \geqq q(f)$ by the hypothesis, and so $q(f_n)$ is a convergent sequence. $\{f_n\}$ is therefore a $\rho_1$-fundamental sequence. As the set $\mathfrak{K}$ of $f_n$ $(n = 1, 2, \ldots)$ is linearly ordered and lowerly order-bounded, $\{f_n\}$ is also a $\rho$-fundamental sequence by Theorem 2.2. By the $\rho$-completeness of the set of all classes of distributions we obtain the existence of $\rho\text{-}\lim_{n\to\infty} f_n$.

**Corollary 2.2.** *If* $f_1 \leqslant f_2 \leqslant \cdots$, *then there exists* $\rho\text{-}\lim_{n\to\infty} f_n$.

**Corollary 2.3.** *Let* $\{f_t\}$ *be a set of classes depending on a continuous parameter* $t$ *such that* $t < s$ *may imply* $f_t \geqslant f_s$. *Then the set of the* $\rho$*-discontinuity* $t$*-points of* $f_t$ *forms an enumerable or finite set.*

*Proof.* By Theorem 2.3 the set of all $\rho$-discontinuity $t$-points of $f_t$ and the set of all $\rho_1$-discontinuity $t$-points of $f_t$ are coincident. But the latter forms, as is well-known, an enumerable or finite set, since $q(f_t)$ is a monotone function of $t$. So it is also the case with the former.

**Theorem 2.4.** *Let* $\mathfrak{K}$ *be a totally* $\rho$*-bounded set. Then there exists* $\delta(\mathcal{E})$ *for any positive number* $\mathcal{E}$ *such that* $\rho(h, e) < \delta(\mathcal{E})$ *may imply* $|q(f * h) - q(f)| < \mathcal{E}$ *for any class* $f$ *in* $\mathfrak{K}$.

*Proof.* It is sufficient to prove that the existence of a positive number $\mathcal{E}_0$ and two sequence $\{f_n\}$, $\{h_n\}$ such that $f_n \in \mathfrak{K}$, $\rho(h_n, e) < \frac{1}{n}$ and $|q(f_n * h_n) - q(f_n)| \geqq \mathcal{E}_0$ may be reduced to a contradiction. As $\mathfrak{K}$ is totally $\rho$-bounded, there exists a $\rho$-convergent subsequence $\{f_{n_k}\}$ of $\{f_n\}$. Let $f_0$ denote the limit of $\{f_{n_k}\}$. But as $\rho(h_n, e) < \frac{1}{n}$, $h_{n_k}$ will $\rho$-converge to $e$. By the $\rho$-continuity of the function $q$ and the convolution, we have $q(f * e) - q(f)| \geqq \mathcal{E}_0$, i.e. $0 \geqq \mathcal{E}_0$, which is contradictory.

## III. A relation between infinitely divisible laws and $\mathfrak{L}$-processes.

**Definition 3.1.** Let $R$ be a "*resolution*" of $F: F = F_1 * F_2 * F_3 * \cdots * F_n$. *By the degree of divisibility* $\Delta(F, R)$ *of a distribution* $F$ *in* $R$ we understand $(\max_{1 \leqq i \leqq n} \rho(F_i, E))^{-1}$. *The degree of divisivility of* $F$ is defined as the least upper bound of $\Delta(F, R)$, $R$ running over all resolutions of $F$. In the case $\Delta(F) = \infty$, $F$ will be called *an infinitely divisible law.*

**Corollary 3.1.** *Gaussian distributions and Poissonian distributions are infinitely divisible.*

Let $\mathfrak{F}_{ab}$ denote the set of all real-valued functions $\{f(t)\}$ defined on the $t$-interval $[ab]$ satisfying the following conditions:

(1) there exist $f(t-0)$ and $f(t+0)$ at any $t$-point,

(2) $$f(t+0) = f(t)\,,$$

(3) $$f(0) = 0\,.$$

Let $\mathfrak{B}_{ab}$ denote the least completely additive class of subsets on $\mathfrak{F}_{ab}$ containing all cylinder subsets[15] of $\mathfrak{F}_{ab}$. Let $x$ be an $(\mathfrak{F}_{ab}, \mathfrak{B}_{ab})$-valued random variable, and $y$ the value of $x$ at $t$. Then $y$ will be a real-valued random variable. We shall denote this variable by $x_t$. Then $x$ is clearly the combination of $x_t$.

**Definition 3.2.** *An $\mathfrak{L}$-process $x$ over* $[ab]$ is defined as an $(F_{ab}, \mathfrak{B}_{ab})$ valued random variable satisfying the following conditions:

1°. $x_t = x_{t-0}$ holds with probability 1[16] for any $t$ in $[ab]$.

2°. For any system of non-overlapping $t$-intervals $(t_1 s_1), (t_2 s_2), \ldots (t_n s_n)$, $x_{s_i} - x_{t_i}$ $(i = 1, 2, \ldots n)$ are independent, that is to say, $x_s - x_t$ are independent of the combination of $x_\tau$ for $0 \leqq \tau \leqq t$, where $a \leqq t \leqq s \leqq b$; we shall say in this case that $x_t$ is differential.

*Note.* If $x$ is an $\mathfrak{L}$-process over $[ab]$, then $(x_t - x_\alpha;\ \alpha \leqq t \leqq \beta)$ will also be an $\mathfrak{L}$-process over $[\alpha\beta]$ where $a \leqq \alpha \leqq \beta \leqq b$.

**Definition 3.3.** An $\mathfrak{L}$-process $x$ on $(\Omega, P)$ is called a *$\mathfrak{G}$-process,* if, $x_t(\omega)$ is continuous in $t$ for any element $\omega$ in $\Omega$.

**Definition 3.4.** An $\mathfrak{L}$-process $x$ on $(\Omega, P)$ is called a *$\mathfrak{P}$-process,* if, for any element $\omega$ in $\Omega$, $x_t(\omega)$ is a step-function of $t$ which increases only by jumps with the height 1.

As is well-known, the variation between two $t$-points of an $\mathfrak{L}$-process is subject to an infinitely divisible law, and that of a $\mathfrak{G}$-process to a Gaussian distribution, and that of a $\mathfrak{P}$-process to a Poissonian distribution. Conversely we shall obtain the following theorem.

**Theorem 3.1.** *Let $F$ be an infinitely divisible law (or a Gaussian distribution, or a Poissonian distribution). We can constitute an $\mathfrak{L}$-process $(x_t;\ 0 \leqq t \leqq 1)$ (or a $\mathfrak{G}$-process, or a $\mathfrak{P}$-process) such that $x_1$ may be subject to the given law $F$.*

---

(15) See A. Kolmogoroff: Grundbegriffe der Wahrscheinlichkeitsrechnung, p. 25 (Ergebnisse der Mathematik Vol. 2, No. 3).

(16) Let $M_\alpha(\omega)$ be a proposition where $\alpha$ runs over a parameter domain $A$, and $\omega$ over a probability field $(\Omega, P)$. If $P(\prod_{\alpha \in A} M_\alpha) = 1$, then we say that $M_\alpha$ holds for any $\alpha$ in $A$ with probability *1*, or that, with probability *1*, $M_\alpha$ holds for any $\alpha$ in $A$. If $P(M_\alpha) = 1$ for any $\alpha$ in $A$, then we say that $M_\alpha$ holds with probability *1* for any $\alpha$ in $A$, or that, for any $\alpha$ in $A$, $M_\alpha$ holds with probability *1*.

*Proof.* We shall discuss only the case of "$\mathfrak{L}$-*process*", as the other cases can be also treated analogously. First we shall show the existence of $F_{st}$ $(0 \leqq s \leqq t \leqq 1)$ satisfying the following conditions:

$$(4) \qquad F_{01} = F$$

$$(5) \qquad F_{st} * F_{tu} = F_{su}$$

$$(6) \qquad \rho\text{-}\lim_{s \to u} F_{su} = F_{uu} (= E) .$$

Let $\mathfrak{K}$ denote the set of all classes lowerly order-bounded by the class $f$ containing the distribution $F$. By theorem 2.1 we can see that $\mathfrak{K}$ is a $\rho$-compact set. $F$ being infinitely divisible, there exists a sequence of the resolutions of $f$:

$$(7) \qquad f = f_1^k * f_2^k * \cdots * f_{n_k}^k \ (k = 1, 2, 3, \ldots)$$

with the property:

$$(8) \qquad \rho(f_i^k, e) < \frac{1}{k} \ (i = 1, 2, \ldots n_k) .$$

Let $f_{ml}^k$ denote the convolution of all classes such as $f_\sigma^k$ in the $k$-th resolution (7) satisfying the condition:

$$(9) \quad 1-(1-q(f))\frac{l-1}{2^m} > q(f_1^k * f_2^k * \cdots * f_\sigma^k) \geqq 1-(1-q(f))\frac{l}{2^m},$$

where $k, m = 1, 2, 3, \ldots, l = 1, 2, 3, \ldots, 2^m$. If such class does not exist, we shall define $f_{ml}^k$ as $e$. All classes thus obtained will belong to $\mathfrak{K}$. Therefore any sequence such as $\{f_{ml}^1, f_{ml}^2, \ldots\}$ will have a $\rho$-convergent subsequence. By means of *the diagonal method* we can find a subsequence $\{k'\}$ of $\{k\}$ independent of the suffix $(ml)$ such that $\{f_{ml}^{1'}, f_{ml}^{2'}, \ldots\}$ may $\rho$-converge to a limit $f_{ml}$, since all the pairs $\{(ml)\}$ appearing in the suffix form an enumerable set.

For $t = \dfrac{l}{2^m}$ and $s = \dfrac{k}{2^m}$ we shall define $g_{ts}^m$ by

$$(10) \qquad g_{ts}^m = f_{m,l+1} * f_{m,l+2} * \cdots * f_{m,k} .$$

$g_{ts}^m$ is clearly independent of $m$ on account of the identity $f_{m+1,2\sigma-1} * f_{m+1,2\sigma} = f_{m\sigma}$ which follows at once from the definition; we can denote $g_{ts}^m$ by $g_{ts}$ with no confusion. Concerning the system $\{g_{ts}\}$ thus obtained, we shall verify the following properties:

$$(11) \qquad g_{ts} * g_{su} = g_{tu} \ (0 \leqq t \leqq s \leqq u \leqq 1) ,$$

(12) $$g_{01} = g$$

(13) $$q(g_{0t}) = 1-(1-q(f))t\,.$$

(11) and (12) are clear. In order to prove (13), we shall first remark

(14) $$q(g_{0t}) = \lim_{k\to\infty} q\,(f_{m1}^{k'} * f_{m2}^{k'} * \cdots * f_{ml}^{k'}) \left(t = \frac{l}{2^m}\right),$$

which is clear on account of the $\rho$-continuity of $q$(Cf. the first part of the proof of Theorem 2.3). But we have by the definition of $\{f_{ml}^k\}$

(15) $$q(f_{m1}^{k'} * f_{m2}^{k'} * \cdots * f_{ml}^{k'}) \geqq 1-(1-q(f))\frac{l}{2^m} = 1-(1-q(t))t$$

(15′) $$q(f_{m1}^{k'} * f_{m2}^{k'} * \cdots * f_{ml}^{k'} * f_{\sigma+1}^{k'}) < 1-(1-q(f))\frac{l}{2^m} = 1-(1-q(f))t\,,$$

where $\sigma$ is the number of the factors of the $k'$-th resolution (7) used to constitute $f_{m1}^{k'} * f_{m2}^{k'} * \cdots * f_{ml}^{k'}$. On account of $f_{m1}^{k'} * f_{m2}^{k'} * \cdots * f_{ml}^{k'} \in \mathfrak{R}$ and $\rho(f_{\sigma+1}^{k'}, e) < \frac{1}{k'}$, we have, by Theorem 2.4.

(16) $$|\,q(f_{m1}^{k'} * f_{m2}^{k'} * \cdots * f_{ml}^{k'}) - q(f_{m1}^{k'} * f_{m2}^{k'} * \cdots * f_{\sigma+1}^{k'})\,| < \mathfrak{E}(k)\,,$$

where $\mathfrak{E}(k)$ is a positive number tending to $0$ with $\frac{1}{k}$. From (14), (15), (15′), and (16) follows the identity (13).

We shall now define $g_{ts}$ for every pair of real numbers $(ts)$ with $0 \leqq t < s \leqq 1$ so that this definition may be compatible with the above definition. Let $\{t_i\}$ $\{s_i\}$ be sequences of rational numbers with the form: $\frac{l}{2^m}$ such that

(17) $$t_1 \geqq t_2 \geqq t_3 \geqq \cdots \to t\,, \text{ and}$$

(18) $$s_1 \leqq s_2 \leqq s_3 \leqq \cdots \to s\,.$$

The existence of $\rho\text{-}\lim_{n\to\infty} g_{t_n s_n}$ is clear by Corollary 2.1, and furthermore this limit is independent of the choice of the sequence $\{t_n\}$ $\{s_n\}$; we can denote it by $g_{t+0,\,s-0}$. Analogously we can define $g_{t-0,\,s+0}$ in the case $0 \leqq t \leqq s \leqq 1$. We shall prove

(19) $$g_{t+0,\,s-0} = g_{t-0,\,s+0}\ (t < s)\,, \text{ and}$$

(20) $$g_{t-0,\,t+0} = e\,.$$

From (13) follows

$$(21)\qquad q(g_{0,\,t-0}) = q(g_{0,\,t+0})\ \ (=1-(1-q(f))t)\,.$$

But we have, by the definition,

$$(22)\qquad g_{0,\,t-0} * g_{t-0,\,t+0} = g_{0,\,t+0}\,.$$

From (21) and (22) follows (20) by Theorem 2.2, and accordingly we have

$$g_{t-0,\,s+0} = g_{t-0,\,t+0} * g_{t+0,\,s-0} * g_{s-0,\,s+0} = e * g_{t+0\ s-0} * e = g_{t+0,\,s-0}\,.$$

If we define $g_{ts}$ as $g_{t-0,\,s+0}(=g_{t+0,\,s-0})$, the system $\{g_{ts}\}$ satisfies evidently the conditions:

$$(23)\qquad g_{ts} * g_{su} = g_{tu}\,,$$

$$(24)\qquad g_{01} = g\,,$$

$$(25)\qquad \rho\text{-}\lim_{\substack{t\to u\\ s\to u}} g_{ts} = g_{uu}(=e)\,.$$

Let $G_t$ be the Gaussian distribution with the mean value 0 and with the standard deviation $t$ where $t\in[0,1]$, and $h_t$ be the class containing $G_t$. Now choose from $g_{0t} * h_t$ the distribution $F_{0t} * G_t$ whose median is that of $F * G_1$ multiplied by $t$; we can see that the median of $F_{0t} * G_t$ is uniquely determined and that $F_{0t}$ is also uniquely determined. Both $F_{00} = E$ and $F_{01} = F$ are clear. As $g_{tu} = g_{ts} * g_{su}$, we have $F_{tu} \sim F_{ts} * F_{su}$ i.e.

$$(26)\qquad F_{tu} = F_{ts} * F_{su} * E_m\,,$$

$E_m$ being the distribution which attributes the probability *1* to $m$. Therefore we have

$$F_{0u} = F_{0t} * F_{tu} = F_{0t} * F_{ts} * F_{su} * E_m = F_{0u} * E_m,$$

and so $m=0$, by which (5) follows at once from (26). We can easily see

$$(27)\qquad \rho\text{-}\lim_{s\to t} F_{0s} * G_s = F_{0t} * G_t$$

and so

$$(28)\qquad \rho\text{-}\lim_{s\to t} F_{0s} * G_t = F_{0t} * G_t$$

But

$$(29)\qquad F_{0s} * G_t * F_{st} = F_{0t} * G_t$$

In considering (25), (28) and (29), we can easily prove (6) by reductio ad absurdum.

Next we shall define an $\mathfrak{L}$-process $(x_t\,;\ 0 \leqq t \leqq 1)$ so that $x_1 - x_0$ may be subject to $F$. Let $\Omega^*$ denote the space of all real-valued function of $t$ defined on the interval $(0, 1)$. We shall define $x_t^*(\omega^*)$ by

$$(30) \qquad x_t^*(\omega^*) = f(t) - f(0) \quad (\omega^* = f(t)) .$$

We shall introduce a probability distribution $P^*_{s_1 s_2 \ldots s_n}$ into $\Omega^*$ as follows:

$$(31) \qquad P^*_{s_1 s_2 \ldots s_n}\Big((x^*_{s_1} - x^*_0 \in E_1)(x^*_{s_2} - x^*_{s_1} \in E_2) \cdots (x^*_{s_n} - x^*_{s_{n-1}} \in E_n)\Big)[17]$$
$$= F_{0 s_1}(E_1) F_{s_1 s_2}(E_2) \cdots F_{s_{n-1} s_n}(E_n) .$$

Then $(x_t^*\,;\ t = s_1, s_2, \ldots, s_n)$ is a stochastic *differential* process on $(\Omega^*, P^*_{s_1 s_2 \ldots s_n})$ depending on a *discontinuous* parameter $t$, and $x^*_{s_i} - x^*_{s_{i-1}}$ is subject to $F_{s_{i-1} s_i}$. Accordingly we shall have, by Corollary 1.5,

$$(32) \qquad P^*_{s_1 s_2 \ldots s_n}(x^*_{s_{i+1}} - x^*_{s_{i-1}} \in E) = (F_{s_{i-1} s_i} * F_{s_i s_{i+1}})(E) = F_{s_{i-1} s_{i+1}}(E) .$$

In accordance with the choice of the system $(s_1, s_2, \ldots, s_n)$, we can obtain different probability distributions on $\Omega^*$, but we can prove, by virtue of (32), that these distributions are coherent to each other, namely that they attribute to the same subset of $\Omega^*$ always the same probability; we can thus omit the suffix and write $P^*$. $P^*$ is an additive (not necessarily completely additive) measure on $\Omega^*$, which is completely additive on the Borel system of the cylinder sets[18] over any assigned finite system of $t$-values. By a theorem of Kolmogoroff[19] we can extend this measure and constitute a probability distribution on $\Omega^*$; we shall rewrite it as $P^*$. The process $(x_t^*\,;\ 0 \leqq t \leqq 1)$ proves to be a stochastic differential process on $(\Omega^*, P^*)$. As $x_s^* - x_t^*$ is evidently subject to $F_{ts}$, $x_t^*$ is continuous in $t$ with regard to the topology of "*convergence in probability*" on account of (6). Let $\Omega$ denote the set of all elements of $\Omega$ satisfying (1), (2), and (3). By virtue of a theorem of Doob[20] we can define a probability distribution $P$ on $\Omega$ so that $P(E^*\Omega) = P^*(E^*)$ for any $P^*$-measurable set $E^*$. If we define a random variable $x$ by $x(\omega) = \omega$, $x$ is the required $\mathfrak{L}$-process.

---

(17) $(x^*_{s_{i+1}} - x^*_{s_i} \in E)$ means the $\omega^*$-set $E(\omega^*\,;\ x^*_{s_{i+1}} - x^*_{s_i} \in E)$. Cf. foot-note(5).

(18) See foot-note (15).

(19) See A. Kolmogoroff: loc. cit. (15) p. 25. Cf. J. L. Doob: loc. cit. (9) Th. 1.1.

(20) See J. L. Doob: loc. cit. (2) Th. 3.8.

**Theorem 3.2.** *Let $\{F_k\}$ be a $\rho$-convergent sequence of distributions, and $F$ denote its limit. If the degree of divisibility $\Delta(F_k)$ increases indefinitely with $k$, then $F$ is infinitely divisible.*

*Proof.* Let $f_k\,(k=1, 2, \ldots)$ and $f$ denote respectively the classes containing the distributions $F_k\,(k=1, 2, \ldots)$ and $F$. We shall need only prove that the set $\sum_{k=1}^{\infty} E(g\,;\, g \geqslant f_k)$ is totally $\rho$-bounded, namely that any sequence of classes in this set may have a $\rho$-convergent subsequence; if this is established, we can prove the existence of $\{F_{st}\}$ satisfying (4), (5), (6) in the same manner as in the proof of the preceeding theorem.

Let $\{g_k\}$ be any sequence in $\sum_{k=1}^{\infty} E(g\,;\, g \geqslant f_k)$. If there is one (infinite) subsequence of $\{g_k\}$ contained in anyone of the sets $E(g\,;\, g \geqslant f_k)$ $(k=1, 2, \ldots)$, this subsequence may have a $\rho$-convergent subsequence by Theorem 2.1. In the contrary case we can find a subsequence $\{h_k\}$ of $\{g_k\}$ such that $h_k \geqslant f_{k'}$ $(k=1, 2, \ldots)$. Choose from each of $\{h_k\}$ the distribution whose median is equal to $0$, and denote it by $H_k\,(k=1, 2, \ldots)$.

Let $\{\lambda_i\}$ be an enumerable set everywhere dense in $(-\infty, +\infty)$. As $H_k(\lambda_i)$ are real numbers between $0$ and $1$, we can find a convergent subsequence of $\{H_k(\lambda_i)\}$. By means of the diagonal method we can find a subsequence $\{\bar{k}\}$ of $\{k\}$ such that the sequence $\{H_{\bar{k}}(\lambda_i)\,;\, k=1, 2, \ldots\}$ may be convergent for any natural number $i$. Denote the limit by $H(\lambda_i)$. Then we shall see that $\lambda_i < \lambda_j$ implies $H_{\bar{k}}(\lambda_i) \leqq H_{\bar{k}}(\lambda_j)$ and so $H(\lambda_i) \leqq H(\lambda_j)$. Define $\overline{H}(\lambda)$ as $\operatorname{g.l.b.}_{\lambda_i<\lambda} H(\lambda_i)$. Then it is easy to see that $\overline{H}(\lambda)$ is continuous except for a finite or enumerable set of $\lambda$-values, and that the sequence $H_{\bar{k}}(\lambda)$ converges to $\overline{H}(\lambda)$ at any continuity $\lambda$-point of $\overline{H}(\lambda)$. We shall now prove

$$\overline{H}(+\infty) = 1 \qquad \overline{H}(-\infty) = 0\,, \tag{33}$$

or

$$\lim_{\lambda\to\infty} (\overline{H}(\lambda) - \overline{H}(-\lambda)) = 1\,. \tag{33'}$$

Assume that

$$\lim_{\lambda\to\infty} (\overline{H}(\lambda) - \overline{H}(-\lambda)) = \alpha < 1\,. \tag{34}$$

We shall now prove that there exists, for any positive number $l$, a natural number $K(l)$ such that $k > K(l)$ may imply

$$H_{\bar{k}}(l+0) - H_{\bar{k}}(-l-0) \leqq \frac{\alpha+1}{2} < 1 \tag{35}$$

The discontinuity points of $\overline{H}(\lambda)$ forming a finite or enumerable set, we can choose the continuity points of $\overline{H}(\lambda)$ from each of $(-\infty, -l)$ and

$(l, \infty)$ Let $m$, $M$ denote these points respectively. There exists a positive number $K(l)$ such that $k > K(l)$ may imply $|H_{\bar{k}}(m) - \bar{H}(m)| < \frac{1-a}{4}$ and $|H_{\bar{k}}(M) - \bar{H}(M)| < \frac{1-a}{4}$, and accordingly

$$(36) \quad H_{\bar{k}}(l+0) - H_{\bar{k}}(-l-0) \leq H_{\bar{k}}(M) - H_{\bar{k}}(m) \leq \bar{H}(M) - \bar{H}(m)$$
$$+ \frac{1-a}{2} < \lim_{\lambda\to\infty} (\bar{H}(\lambda) - \bar{H}(-\lambda)) + \frac{1-a}{2}.$$

From (34) and (36) follows (35) at once.

In the case $Q(H_{\bar{k}}, l) \geq \frac{1}{2}$, $k > K(l)$ implies

$$Q(H_{\bar{k}}, l) \leq H_{\bar{k}}(l+0) - H_{\bar{k}}(-l-0) < \frac{a+1}{2}$$

on account of $H_{\bar{k}}(0) = \frac{1}{2}$, and so $k > K(l)$ always implies

$$Q(H_{\bar{k}}, l) \leq \max\left(\frac{1}{2}, \frac{a+1}{2}\right) = \frac{a+1}{2} \text{ i.e. } Q(h_{\bar{k}}, l) \leq \frac{a+1}{2} < 1.$$

As we have $f_{(\bar{k})'} \leqslant h_{\bar{k}}$ by the hypothesis, $k > K(l)$ implies $Q(f_{(\bar{k})'}, l) \leq Q(h_{\bar{k}}, l) \leq \frac{a+1}{2} < 1$. $f_{(\bar{k})'}$ $\rho$-converging to $f$, we have finally, $Q(f, l-0) \leq \frac{a+1}{2}$, whatever value $l$ may take. This contradicts $Q(f, +\infty) = 1$. Therefore (33) must hold, that is to say, $\bar{H}$ must be a distribution. The class containing $\bar{H}$ is the limit class of the subsequence $\{h_{\bar{k}}\}$ of $\{h_k\}$ or of $\{g_k\}$.

**Definition 3.5.** Let $\lambda$ be a positive number and $F$ a distribution. We shall define $\frac{1}{\lambda}F$ by $\left(\frac{1}{\lambda}F\right)(E) = F(\lambda(\times)E)$ (21).

Let $\{F_k\}$ be a sequence of distributions and let $\{\lambda_k\}$ be a divergent sequence of positive numbers such that $\frac{\lambda_{n+1}}{\lambda_n}$ may tend to *1* as $n$ increases indefinitely.

If $\frac{1}{\lambda_n}(F_1 * F_2 * \cdots * F_n)$ be $\rho$-convergent, then the $\rho$-limit is called *a limit law*(22).

**Theorem 3.3.** *Limit laws are infinitely divisible*(23).

(21) $\lambda(\times)E$ means $E(\lambda a;\ a \in E)$.

(22), (23) See P. Lévy: loc. cit. (1) p. 192.

First we shall prove a lemma due to A. Wintner([24]).

**Lemma 3.1.** *If* $\{H_n * G_n\}$ *and* $\{H_n\}$ *ρ-converge to the same limit* $H$, $\{G_n\}$ *will ρ-converge to* $E$.

*Proof of the lemma.* We shall make use of *reductio ad absurdum.* Assume that $\{G_n\}$ should not converge to $E$. Then $\{G_n\}$ will have a subsequence $\{G_{n'}\}$ such that

$$\rho(G_{n'}, E) \geqq \varepsilon_0 \ (n = 1, 2, 3, \ldots). \tag{37}$$

Let $h$, $g_n$, $h_n$ denote respectively the classes containing $H$, $G_n$, $H_n$. Then $h_n * g_n$ will ρ-converge to $h$. On account of $g_{n'} \geqslant h_{n'} * g_{n'}$ we can see that $\{g_{n'}\}$ has a ρ-convergent subsequence $\{g_{n''}\}$, if we recall what was stated at the biginning of the proof of Theorem 3.2. Let $g$ denote the ρ-limit of $\{g_{n''}\}$. Then we have $h * g = h$ and so $g = e$, and so we can find a sequence $\{m_n\}$ of real numbers such that

$$G_{n''} * E_{-m_n} \to E \ (n \to \infty). \tag{38}$$

Accordingly we have

$$H_{n''} * G_{n''} * E_{-m_n} \to H * E = H \ (n \to \infty). \tag{39}$$

But by the hypothesis we have

$$H_{n''} * G_{n''} \to H. \tag{40}$$

By (39) and (40) we have $m_n \to 0$, and so finally by (38)

$$G_{n''} \to E \ (n \to \infty), \tag{41}$$

in contradiction with (37).

*Proof of the Theorem 3.3.* It is sufficient to deduce, from the following three conditions, that $F$ is infinitely divisible:

$$\left.\begin{aligned} &\frac{1}{\lambda_n}(F_1 * F_2 * \cdots * F_n) \to F && (42)\\ &\frac{\lambda_{n+1}}{\lambda_n} \to 1 && (43)\\ &\lambda_n \to \infty && (44) \end{aligned}\right\} (n \to \infty).$$

([24]) A. Wintner: Asymptotic distributions and infinite convolutions (1937-1938), Lemma 5.1.

(42) means

$$\frac{\lambda_{n-1}}{\lambda_n}\cdot\frac{1}{\lambda_{n-1}}(F_1*F_2*\cdots*F_{n-1})*\frac{1}{\lambda_n}F_n\to F. \tag{45}$$

By Lemma 3.1 we can deduce from (43) and (45) that $\frac{1}{\lambda_n}F_n$ converges to $E$, namely that there exists a natural number $K(\mathfrak{E})$ for any positive number $\mathfrak{E}$ such that $n>K(\mathfrak{E})$ may imply $\rho\left(\frac{1}{\lambda_n}F_n, E\right)<\mathfrak{E}$. Let $\{n'\}$ denote the subsequence of $\{n\}$ such that $m<n'$ may imply $\lambda_m<\lambda_{n'}$; the existence of $\{n'\}$ follows at once from (3). Then we can see that $n'\geqq k>K(\mathfrak{E})$ implies

$$\rho\left(\frac{1}{\lambda_{n'}}F_k, E\right)\leqq\rho\left(\frac{1}{\lambda_k}F_k, E\right)<\mathfrak{E}. \tag{46}$$

$\lambda_{n'}$ increasing indefinitely with $n$, we can find a natural number $N(\mathfrak{E})(>K(\mathfrak{E}))$ such that $n>N(\mathfrak{E})$ may imply

$$\rho\left(\frac{1}{\lambda_{n'}}F_k, E\right)<\mathfrak{E}\quad (k=1, 2, 3, \ldots, K(\mathfrak{E})). \tag{47}$$

From (46) and (47) we can deduce that $n>N(\mathfrak{E})$ implies

$$\Delta\left(\frac{1}{\lambda_{n'}}(F_1*F_2*\cdots*F_{n'})\right)>\frac{1}{\mathfrak{E}}. \tag{48}$$

According to Theorem 3.2, (48) shows that the $\rho$-limit $F$ of $\left\{\frac{1}{\lambda_{n'}}(F_1*\cdots*F_{n'})\right\}$ is infinitely divisible.

## IV. A canonical form of $\mathfrak{L}$-processes.

Let $\mathfrak{L}\equiv(\mathfrak{L}_t;\ 0\leqq t\leqq 1)$ be an $\mathfrak{L}$-process on $(\Omega, P)$. If $\mathfrak{L}_{t+0}(\omega)-\mathfrak{L}_t(\omega)\equiv u\neq 0$, then we say that $\mathfrak{L}(\omega)\equiv(\mathfrak{L}_t(\omega);\ 0\leqq t\leqq 1)$ has a *jump* with the *height* $u$ at $t$. We shall represent such jump of $\mathfrak{L}(\omega)$ by the point $(tu)$ in the $(tu)$-plane. By *jumps in a certain* $(tu)$*-set* $E$ we shall understand the jumps which may be represented by the points in $E$.

In this paper we shall often need to indicate the property of a set of real numbers that the set is not only bounded but also separated from $0$ by a positive distance. For this purpose we shall make use of the sigh $*$; for example, "*set**" means "set with the above property". Terms such as "*interval**" "*Borel set**" are to be understood similarly.

Let $\mathfrak{P}$ be the combination of $\mathfrak{P}$-processes $\mathfrak{P}^U = (\mathfrak{P}_t^U;\ 0 \leqq t \leqq 1)$ on $(\Omega, P)$, $U$ running over all Borel sets*. $\mathfrak{P}$ will be called a *differential system of $\mathfrak{P}$-processes,* if it satisfies the following two conditions:

(1) for any system of disjointed Borel sets* $U_k (k = 1, 2, \dots)$ such that their sum $U$ may be also a Borel set*, we have $\mathfrak{P}^U = \ = \sum_{k=1}^{\infty} \mathfrak{P}^{U_k}$, namely $\mathfrak{P}_t^U(\omega) = \sum_{k=1}^{\infty} \mathfrak{P}_t^{U_k}(\omega)$ for any $t$ in $[0\ 1]$ and for any $\omega$ in $\Omega$.

(2) for any system of disjointed Borel sets* $U_k (k = 1, 2, \dots)$, $\mathfrak{P}^{U_1}, \mathfrak{P}^{U_2}, \cdots, \mathfrak{P}^{U_n}$ are independent.

We can consider $\mathfrak{P}$ as *a measure defined on Borel sets* which takes values of $\mathfrak{P}$-processes.* We can consider *the integration based on this measure.*

**Theorem 4.1.** *Let $\mathfrak{P}$ be a differential system of $\mathfrak{P}$-processes $\mathfrak{P}^U \left(\equiv (\mathfrak{P}_t^U;\ 0 \leqq t \leqq 1)\right)$. Then $\int_a^b u \mathfrak{P}^{du}$*[25] *$(ab > 0)$ is an $\mathfrak{L}$-process the number of whose jumps in a $(tu)$-set $[0t] \times U$ is equal to $\mathfrak{P}_t^U$ for any number $t$ in $[01]$ and for any Borel set $U$ contained in $(ab]$.*

*Proof.* We shall prove this theorem in the case $a > 0$, $b > 0$.

Let $t_1, t_2, \dots, t_n$ denote the positions of the jumps of $\mathfrak{P}^{ab]}(\omega)$. Then we have

$$(3) \qquad \mathfrak{P}_{t_i+0}^{(ab]}(\omega) - \mathfrak{P}_{t_i-0}^{(ab]}(\omega) = 1 \quad (i = 1, 2, \dots n).$$

On account of (1), $\mathfrak{P}_{t_i+0}^{(au]}(\omega)$, as well as $\mathfrak{P}_{t_i-0}^{(au]}$, is a function of $u$ continuous on the right hand, and so it is the case with $\mathfrak{P}_{t_i+0}^{(au]}(\omega) - \mathfrak{P}_{t_i-0}^{(au]}$. By the definition this function is monotone increasing and takes $0$ or $1$. Therefore we can find a number $u_i$ in $[ab]$ for $i = 1, 2, \dots n$ such that

$$(4) \qquad \mathfrak{P}_{t_i+0}^{(au]}(\omega) - \mathfrak{P}_{t_i-0}^{(au]}(\omega) = 0 \quad (u < u_i)$$
$$= 1 \quad (u \geqq u_i).$$

Consequently we have

$$(5) \qquad \mathfrak{P}_t^U(\omega) = \sum_{\substack{u_i \in U \\ 0 \leqq t_i \leqq t}} 1,$$

$$(6) \qquad \int_a^b u \mathfrak{P}_t^{du} = \sum_{0 \leqq t_i \leqq t} u_i.$$

(25) $\int_a^b$ means $\int_{a+0}^{b+0}$.

For any fixed point $t$, $\int_a^b u\mathfrak{P}_t^{du}(\omega)$ is a $P$-measurable function of $\omega$, i.e. a random variable, since it is the limit of a bounded and monotone sequence of the random variables:

$$(7) \qquad \sum_{a<\frac{K}{2^n}<b} \frac{K}{2^n}\mathfrak{P}_t\left(\frac{K}{2^n}, \frac{K+1}{2^n}\right](\omega) \quad (n=1, 2, \ldots).$$

It follows at once from (2) that $\int_{a'}^b u\mathfrak{P}^{du}$ is a stochastic differential process. The above expression (6) shows that $\int_a^b u\mathfrak{P}_t^{du}(\omega)$ increases only by jumps when $t$ increases. We can see by (5) and (6) that

$$(8) \qquad E\left(\omega\,;\int_a^b u\mathfrak{P}_{t+0}^{du}(\omega)-\int_a^b u\mathfrak{P}_{t-0}^{du}(\omega)\neq 0\right)$$
$$= E\left(\omega\,;\mathfrak{P}_{t+0}^{(ab]}(\omega)-\mathfrak{P}_{t-0}^{(ab]}(\omega)\neq 0\right).$$

As the right hand side of (8) has $P$-measure $0$ on account of the definition of $\mathfrak{P}$-process, so does also the left hand side. Consequently we shall see on account of (5) and (6) that $\int_a^b u\mathfrak{P}^{du}$ is an $\mathfrak{L}$-process the number of whose jumps in the $(tu)$-set $[0t]\times U$ is equal to $\mathfrak{P}_t^U$.

Next we shall discuss the behaviour of the integral in the vicinity of $u=\infty$.

**Theorem 4.2.** *Let $\mathfrak{P}$ be a differential system of $\mathfrak{P}$-processes satisfying the condition:*

$$(9) \qquad \int_a^\infty \overline{\mathfrak{P}_1^{du}}<\infty \ \left(\text{or} \int_{-\infty}^{-a}\overline{\mathfrak{P}_1^{du}}<\infty\right) \quad (a>0)$$

*Then $\int_a^\infty u\mathfrak{P}^{du}$ (or $\int_{-\infty}^{-a} u\mathfrak{P}^{du}$)*[26] *is an $\mathfrak{L}$-process in the larger sense, the number of whose jumps in a $(tu)$-set $[0t]\times U$ is equal to $\mathfrak{P}_t^U$ for any number $t$ in $[01]$ and for any Borel set $U$ in $(a,\infty)$ (or $(-\infty,-a]$) with probability 1.*

*Proof.* On account of Theorem 4.1, it is sufficient to prove that $\mathfrak{P}_1^{(a\infty)}$ is finite with probability 1. It is clear that

$$(10) \qquad P(\mathfrak{P}_1^{(ab]}\leq k)=\sum_{i=0}^k e^{-\eta(b)}\frac{(\eta(b))^i}{i} \quad (k=1, 2, \ldots),$$

(26) $\int_a^\infty u\mathfrak{P}du = \left((\lim_{b\to\infty}\int_a^b u\mathfrak{P}_t^{du}(\omega)\,;\ \omega\in\Omega)\,;\ 0\leq t\leq 1\right)$

where $\eta(b)$ denotes the mean value of $\mathfrak{P}_1^{[ab]}$. $\mathfrak{P}_1^{[ab]}$ is monotone increasing in $b$ and so $\mathfrak{P}_1^{(a\infty)}(\omega)$ will exist, if it is permitted to take infinity. Accordingly we have

$$(11) \qquad P(\mathfrak{P}_1^{(a\infty)} \leqq k) = \lim_{b\to\infty} P(\mathfrak{P}_1^{[ab]} \leqq k) = \sum_{i=0}^{k} e^{-\eta(\infty)} \frac{(\eta(\infty))^i}{\underline{i}}$$

since $\eta(\infty) < \infty$ by virtue of (9). Consequently we have

$$(12) \qquad P(\mathfrak{P}_1^{(a\infty)} < \infty) = \sum_{i=0}^{\infty} e^{-\eta(\infty)} \frac{(\eta(\infty))^i}{\underline{i}} = 1\,.$$

We shall now discuss the behaviour of the integral in the vicinity of $u = 0$. In the case that $\int_0^a \overline{\mathfrak{P}_1^{du}}(a > 0)$ is finite, we can see, as in the above proof, that $\mathfrak{P}_1^{(0a]}$ is finite with probability *1*, and so we can easily define $\int_0^a u\mathfrak{P}^{du}$. In a more general case that $\int_0^a u\overline{\mathfrak{P}^{du}}(a > 0)$ is finite, we can also define $\int_0^a u\mathfrak{P}^{du}$, but we shall consider here the most general case:

$$(13) \qquad \int_{0<|u|\leqq 1} u^2\overline{\mathfrak{P}^{du}} < \infty\,.$$

In such a case $\int_0^a u\mathfrak{P}^{du}$ itself is not always finite, but we can establish the following

**Theorem 4.3.** *Let $\mathfrak{P}$ be a differential system of $\mathfrak{P}$-processes satifying (13). Then the sequence of $\mathfrak{L}$-processes:*

$$(14) \qquad \int_{\frac{1}{n}}^{1} (u\mathfrak{P}^{du} - u\overline{\mathfrak{P}^{du}}) \equiv \left(\int_{\frac{1}{n}}^{1} u\mathfrak{P}_t^{du} - \int_{\frac{1}{n}}^{1} u\overline{\mathfrak{P}_t^{du}}\,;\ 0 \leqq t \leqq 1\right) \qquad (n = 1, 2, \ldots)$$

*will be convergent uniformly in t with probability 1. The limit is an $\mathfrak{L}$-process in the larger sense, the number of whose jumps in a (tu)-set ([0t] × U) is equal to $\mathfrak{P}_t^U$ for any number t in [01] and for any Borel set* U contained in [01] with probability 1.*

*Proof.* If we put

$$(15) \qquad x_n(t) = \int_{\frac{1}{n+1}}^{\frac{1}{n}} u\mathfrak{P}_t^{du} - \int_{\frac{1}{n+1}}^{\frac{1}{n}} u\overline{\mathfrak{P}_t^{du}},$$

we can easily see by the hypothesis that $x_n \equiv (x_n(t)\,;\ 0 \leqq t \leqq 1)\ (n = 1, 2, \ldots)$ are independent $\mathfrak{L}$-processes and satisfy

(16) $$\overline{x_n(t)} = 0\,, \quad \text{and}$$

(17) $$\sum_{n=1}^{\infty}\left(\sigma(x_n(1))\right)^2 < \infty.$$

In order to prove the first part of the theorem, it is sufficient to show

(18) $$P(\lim_{n\to\infty} \underset{l>m\geqq n}{\text{l.u.b.}}\ \underset{0\leq t\leq 1}{\text{l.u.b.}} \,|\sum_{m}^{l} x_i(t)| > 0) = 0.$$

On account of (17) there exists a natural number $N(\mathfrak{E})$ for any assigned positive number $\mathfrak{E}$ such that $m > n \geqq N(\mathfrak{E})$ may imply

(19) $$\sum_{n}^{m}\left(\sigma(x_i(1))\right)^2 < \mathfrak{E}^3.$$

$x_i$ ($i = 1, 2, \ldots$) being $\mathfrak{L}$-processes, we have, by 2° of Definition 3.2,

(20) $$\left(\sigma(x_i(t))\right)^2 + \left(\sigma(x_i(1)-x_i(t))\right)^2 = \left(\sigma(x_i(1))\right)^2.$$

From (19), (20) follows that $m > n \geqq N(\mathfrak{E})$ implies

(21) $$\sum_{n}^{m}\left(\sigma(x_i(t))\right)^2 < \mathfrak{E}^3$$

and

(22) $$\sum_{n}^{m}\left(\sigma(x_i(1)-x_i(t))\right)^2 < \mathfrak{E}^3$$

for any number $t$ in $[01]$. $x_i$ ($i = 1, 2, \ldots$) being independent, we can deduce from (21) and (22) respectively

(21′) $$\left(\sigma(\sum_{n}^{m} x_i(t))\right)^2 < \mathfrak{E}^3$$

(22′) $$\left(\sigma(\sum_{n}^{m} (x_i(1)-x_i(t)))\right)^2 < \mathfrak{E}^3.$$

By the inequality of Bienaymé (21′) and (22′) will imply respectively

(21″) $$P\left(|\sum_{n}^{m} x_i(t)| \geqq \mathfrak{E}\right) \leqq \mathfrak{E}$$

(22″) $$P\left(|\sum_{n}^{m} (x_i(1)-x_i(t))| \geqq \mathfrak{E}\right) \leqq \mathfrak{E}.$$

We shall now prove that $m > n > N(\mathfrak{E})$ may imply

(23) $$P(\underset{0\leq t\leq 1}{\text{l.u.b.}}\,|\sum_{n}^{m} x_i(t)| \geqq 2\,\mathfrak{E}) \leqq \frac{\mathfrak{E}}{1-\mathfrak{E}}.$$

For brevity we shall denote $\sum_n^m x_i(t)$ by $y(t)$. Then $y \equiv (y(t)\ ;\ 0 \leqq t \leqq 1)$ is an $\mathfrak{L}$-process and satisfies

$$(21''') \qquad P(|y(t)| \geqq \mathfrak{E}) < \mathfrak{E}, \quad \text{and}$$

$$(22''') \qquad P(|y(1)-y(t)| \geqq \mathfrak{E}) < \mathfrak{E}.$$

But (23) may be written also as

$$(23') \qquad P(\underset{0\leq t\leq 1}{\text{l.u.b.}}\, |y(t)| \geqq 2\,\mathfrak{E}) < \frac{\mathfrak{E}}{1-\mathfrak{E}}.$$

Let $\{t_i\}$ be an enumerable set dense in the interval $[01]$. $y$ being an $\mathfrak{L}$-process, we can see, by $1°$ of Definition 3.2, that

$$(24) \qquad \underset{0\leq t\leq 1}{\text{l.u.b.}}\, |y(t)| = \underset{i=1,2,\cdots}{\text{l.u.b.}}\, |y(t_i)| = \lim_{n\to\infty} \max_{1\leq i\leq n} |y(t_i)|$$

It is sufficient to prove, instead of (23′), that

$$(25) \qquad P(\max_{1\leq i\leq n} |y(t_i)| \geqq 2\,\mathfrak{E}) \leqq \frac{\mathfrak{E}}{1-\mathfrak{E}},$$

since the $\omega$-set $(\max_{1\leq i\leq n} |y(t_i)| \geqq 2\,\mathfrak{E})$ increases with $n$.

We shall arrange the sequence $t_1, t_2, \ldots, t_n$, according to the order of magnitude and write as $s_1, s_2, \ldots, s_n$. Then $y(1)-y(s_i)$ is independent of $y(s_1), y(s_2), \ldots, y(s_i)$. Therefore, if we put

$$(26) \quad A_i = (|y(s_i)| \geqq 2\,\mathfrak{E}) \prod_{\sigma=1}^{i-1} (|y(s_\sigma)| < 2\,\mathfrak{E}) \quad (i = 1, 2, \ldots, n)$$

$$(27) \qquad B_i = (|y(1)-y(s_i)| \leqq \mathfrak{E}) \quad (i = 1, 2, \ldots, n),$$

we have

$$(28) \qquad P(A_iB_i) = P(A_i)P(B_i) \qquad (i = 1, 2, \ldots, n).$$

As $(A_1, A_2, \ldots, A_n)$ is a disjointed system of $\omega$-sets on account of (26,) so is $(A_1B_1, A_2B_2, \ldots, A_nB_n)$. Therefore we have

$$\begin{aligned}
(29) \quad & P(A_1B_1 + A_2B_2 + \cdots + A_nB_n) \\
& = P(A_1B_1) + P(A_2B_2) + \cdots + P(A_nB_n) \\
& = P(A_1)P(B_1) + P(A_2)P(B_2) + \cdots + P(A_n)P(B_n) \quad \text{(by (28))} \\
& > (1-\mathfrak{E})(P(A_1) + P(A_2) + \cdots + P(A_n)) \quad \text{(by (22'''))} \\
& = (1-\mathfrak{E})P(A_1 + A_2 + \cdots + A_n).
\end{aligned}$$

But we have, by (26) and (27),

(30) $$A_1B_1+A_2B_2+\cdots+A_nB_n \subset (|y(1)|\geqq \mathfrak{E})$$

(31) $$A_1+A_2+\cdots+A_n = (\max_{1\leq i\leq n}|y(s_i)|\geqq 2\,\mathfrak{E}) = (\max_{1\leq i\leq n}|y(t_i)|\geqq 2\,\mathfrak{E})$$

From (21‴), (29), (30) and (31) follows (25).

As $x_1, x_2, \ldots$ are independent $\mathfrak{L}$-procees, we can analogously deduce from (23) that $n > N(\mathfrak{E})$ implies

(32) $$P(\underset{m>n}{\text{l.u.b.}}\ \underset{0\leq t\leq 1}{\text{l.u.b.}}\ |\sum_n^m x_i(t)|\geqq 4\,\mathfrak{E}) \leqq \frac{\dfrac{\mathfrak{E}}{1-\mathfrak{E}}}{1-\dfrac{\mathfrak{E}}{1-\mathfrak{E}}} = \frac{\mathfrak{E}}{1-2\,\mathfrak{E}},$$

if we recall the well-known inequality:

$$\underset{t}{\text{l.u.b.}}\,|a_t+b_t|\geqq \underset{t}{\text{l.u.b.}}\,|a_t| - \underset{t}{\text{l.u.b.}}\,|b_t|\,.$$

From (32) follows

(33) $$P(\underset{l>m>n}{\text{l.u.b.}}\ \underset{0\leq t\leq 1}{\text{l.u.b.}}\ |\sum_m^l x_i(t)|\geqq 8\,\mathfrak{E}) \leqq \frac{2\,\mathfrak{E}}{1-2\,\mathfrak{E}},$$

and accordingly

(34) $$P(\lim_{n\to\infty}\ \underset{l>m>n}{\text{l.u.b.}}\ \underset{0\leq t\leq 1}{\text{l.u.b.}}\ |\sum_m^l x_i(t)|\geqq 8\,\mathfrak{E}) \leqq \frac{2\,\mathfrak{E}}{1-2\,\mathfrak{E}},$$

from which follows (18).

In order to prove the second part of the theorem it will be sufficient to show that the number of the jumps of the process $\int_{\frac{1}{n}}^{1} u\mathfrak{P}^{du} - \int_{\frac{1}{n}}^{1} u\overline{\mathfrak{P}^{dn}}$ in the $(tu)$-set $[0\,t]\times U$ is equal to $\mathfrak{P}_t^U$ whenever the integer $n$ is so great that $\left[\frac{1}{n}, 1\right]$ may contain $U$; if this fact is established for such $n$, it is clear that it holds also with regard to the limit process $\lim_{n\to\infty}\left(\int_{\frac{1}{n}}^{1} u\mathfrak{P}^{du} - \int_{\frac{1}{n}}^{1} u\overline{\mathfrak{P}^{du}}\right)$ on account of the uniform convergency just proved above. The above statement being verified for the first term of the above process $\int_{\frac{1}{n}}^{1} u\mathfrak{P}^{du} - \int_{\frac{1}{n}}^{1} u\overline{\mathfrak{P}^{du}}$ by virtue of Theorem 4.1, we need only prove that the second term $-\int_{\frac{1}{n}}^{1} u\overline{\mathfrak{P}^{du}}$ is continuous in $t$. But this follows at once from the fact that $\overline{\mathfrak{P}_t^{[\frac{1}{n},1]}}$ is continuous in $t$ because of the condition 1° stated in Definition 3.2. Thus the theorem is proved.

**Theorem 4.4.** *Let $\mathfrak{L}$ be a differential system of $\mathfrak{P}$-processes ($\mathfrak{P}_t^U$; $0 \leqq t \leqq 1$) on ($\Omega$, $P$) satisfying the conditions (9), (13), and $\mathfrak{G} \equiv (\mathfrak{G}_t;\ 0 \leqq t \leqq 1)$ a $\mathfrak{G}$-process on ($\Omega$, $P$) independent of the system $\mathfrak{P}$. Then the process:*

$$\left(\mathfrak{G}_t + \lim_{n\to\infty} \int_{n>|u|>\frac{1}{n}} \left(u\mathfrak{P}_t^{du} - \frac{u}{1+u^2}\overline{\mathfrak{P}_t^{du}}\right);\ 0 \leqq t \leqq 1\right)$$

*is an $\mathfrak{L}$-process in the larger sense, the number of whose jumps in the (tu)-set $[0t] \times U$ is equal to $\mathfrak{P}_t^U$ for any number $t$ in (01) and for any Borel set* $U$ with probability 1.*

*Proof.* Since the sequence of $\int_{n>|u|>1} \frac{u}{1+u^2}\overline{\mathfrak{P}_t^{du}}$ ($n = 2, 3, \ldots$) converges uniformly in $t$ on account of (9), the limit $\int_{|u|>1} \frac{u}{1+u^2}\overline{\mathfrak{P}_t^{du}}$ is continuous in $t$, and $\int_{1\geqq|n|>\frac{1}{n}} \left(u - \frac{u}{1+u^2}\right)\overline{\mathfrak{P}_t^{du}}$ is also continuous on account of (13). Therefore we shall obtain this theorem by Theorem 4.1, 4.2, and 4.3.

Next we shall prove the converse of theorem 4.4:

**Theorem 4.5.** *Let $\mathfrak{L} \equiv (\mathfrak{L}_t;\ 0 \leqq t \leqq 1)$ be an $\mathfrak{L}$-process on ($\Omega$, $P$). Then there exists a differential system $\mathfrak{P}$ of $\mathfrak{P}$-processes ($\mathfrak{P}_t^U$; $0 \leqq t \leqq 1$) with the properties (9), (13), and a $\mathfrak{G}$-process $\mathfrak{G}$ independent of the system $\mathfrak{P}$ such that*

$$\mathfrak{L}_t = \mathfrak{G}_t + \lim_{n\to\infty} \int_{n>|u|>\frac{1}{n}} \left(u\mathfrak{P}_t^{du} - \frac{u}{1+u^2}\overline{\mathfrak{P}_t^{du}}\right) \tag{35}$$

*may hold for any number t with probability 1.*

*Note.* According to this theorem it may be justified to call the $\mathfrak{G}$-process appearing here *the continuous part of* $\mathfrak{L}$, and the differential system $\mathfrak{P}$ *the discontinuous part of* $\mathfrak{L}$.

*Proof.* Let $\mathfrak{P}_t^U$ denote the number of the jumps of $\mathfrak{L}$ in the (*tu*)-set $[0t] \times U$, where $U$ is a Borel set*. Then it is clear that $\mathfrak{P}_t^U(\omega)$ is always finite. First we shall prove that $\mathfrak{P}_{t_2}^U - \mathfrak{P}_{t_1}^U$ is a $B$-measurable function([27]) of the $\mathfrak{L}$-process ($\mathfrak{L}_s - \mathfrak{L}_{t_1}$; $t_1 < s \leqq t_2$). For this purpose it is sufficient to consider the special case that $U$ is a set such as $[\lambda\infty)$ ($\lambda > 0$). By the definition we have

$$(\mathfrak{P}_{t_2}^{[\lambda\infty)} - \mathfrak{P}_{t_1}^{[\lambda\infty)} \geqq 1) = \prod_n \sum_{\substack{r_i - r_j < \frac{1}{n} \\ t_1 < r_j < r_i < t_2}} \left(\mathfrak{L}_{r_i} - \mathfrak{L}_{r_j} > \lambda - \frac{1}{n}\right) + N_1, \tag{36}$$

([27]) See Definition 1.2.

where $(r_1, r_2, \ldots)$ is an enumerable set dense in (01), and $N_1$ denotes the set of all $\omega$-elements for which $\mathfrak{L}_t$ has a jump at $t_2$. $P(N_1)=0$ is clear on account of the condition 2° of Definition 3.2. Similarly we have

$$(37)\quad (\mathfrak{P}_{t_2}^{[\lambda\infty)}-\mathfrak{P}_{t_1}^{[\lambda\infty)}\geqq 2)=\sum_k \prod_n \sum_{\substack{t_1<r_j<r_i<r_{j'}<r_{i'}<t_2 \\ r_i-r_j<\frac{1}{n} \\ r_{i'}-r_{j'}<\frac{1}{n} \\ r_{j'}-r_i>\frac{1}{k}}} \left(\mathfrak{L}_{r_i}-\mathfrak{L}_{r_j}>\lambda-\frac{1}{n}\right) \frown \left(\mathfrak{L}_{r_{i'}}-\mathfrak{L}_{r_{j'}}>\lambda-\frac{1}{n}\right)+N_2,$$

where $P(N_2)=0$. The analogous identity will hold also for $(\mathfrak{P}_{t_2}^{[\lambda\infty)}-\mathfrak{P}_{t_1}^{[\lambda\infty)}\geqq k)$ $(k=3, 4, \ldots)$. As $\mathfrak{P}_{t_2}^{[\lambda\infty)}-\mathfrak{P}_{t_1}^{[\lambda\infty)}$ takes only 0 or natural numbers, those identities show that $\mathfrak{P}_{t_2}^{[t\infty)}-\mathfrak{P}_{t_1}^{[\lambda\infty)}$ is a B-measurable function of $(\mathfrak{L}_s-\mathfrak{L}_{t_1};\ t_1<s\leqq t_2)$. As $\mathfrak{L}$ is differential, the $\mathfrak{L}$-processes $(\mathfrak{L}_s-\mathfrak{L}_{t_i};\ t_i<s\leqq s_i)$ $(i=1, 2, \ldots, n)$ will be independent for any system of non-overlapping intervals $(t_i s_i)$ $(i=1, 2, \ldots, n)$, and accordingly $\mathfrak{P}_{s_i}^U-\mathfrak{P}_{t_i}^U$ $(i=1, 2, \ldots, n)$ will also be independent, that is to say, $(\mathfrak{P}_t^U;\ 0\leqq t\leqq 1)$ is a differential process. It is clear that $\mathfrak{P}^U=(\mathfrak{P}_t^U;\ 0\leqq t\leqq 1)$ satisfies the condition stated in Definition 3.4. The condition 2° in Definition 3.2 will hold for $\mathfrak{P}^U$ as it holds for $\mathfrak{L}$. Therefore $\mathfrak{L}^U$ is a $\mathfrak{P}$-process.

Next we shall prove that $x\equiv(\mathfrak{L}_t-\int_U u\mathfrak{P}_t^{du};\ 0\leqq t\leqq 1)$ and $\mathfrak{P}^U$ are independent for any Borel set* $U$. $x$ and $\mathfrak{P}^U$ being both differential, we need only prove that

$$(38)\quad P\Big((x_s-x_t\in B)(\mathfrak{P}_s^U-\mathfrak{P}_t^U=k)\Big)=P(x_s-x_t\in B)P(\mathfrak{P}_s^U-\mathfrak{P}_t^U=k)$$

holds for $t<s$, for any Borel set $B$, and for $k=0, 1, 2, \ldots$.

Now, $\mathfrak{P}_s^U-\mathfrak{P}_t^U$ being subject to a Poissonian distribution, we have

$$(39)\quad P(\mathfrak{P}_s^U-\mathfrak{P}_t^U=k)=e^{-\eta}\frac{\eta^k}{\underline{|k}}\qquad (\eta=\overline{\mathfrak{P}_s^U}-\overline{\mathfrak{P}_t^U})$$

$$=1\quad (\text{if } \eta=0,\ k=0)$$

In case $\eta=0$, (38) is clear. In case $\eta\neq 0$ we have

$$(40)\quad P(\mathfrak{P}_s^U-\mathfrak{P}_t^U=k)\neq 0\quad (k=0, 1, 2, \ldots).$$

In order to prove (38) it is sufficient to show that the conditional probability:

$$(41)\quad P_k(x_s - x_t \in B) \equiv P(x_s - x_t \in B \,/\, \mathfrak{P}_s^U - \mathfrak{P}_t^U = k)$$
$$= \frac{P((x_s - x_t \in B)(\mathfrak{P}_s^U - \mathfrak{P}_t^U = k))}{P(\mathfrak{P}_s^U - \mathfrak{P}_t^U = k)}$$

is independent of $k$, namely, that

$$(42)\quad P_0(x_s - x_t \in B) = P_k(x_s - x_t \in B)$$

holds for $k = 1, 2, \dots .$ We shall prove (42) in case $k = 1$, as we can do analogously in the other cases. We can assume $t = 0$ and $s = 1$ with no loss of generality. For brevity we shall adopt the following notations:

$$(43)\quad \Omega^0 \equiv (\mathfrak{P}_1^U = 0)$$

$$(44)\quad \Omega^1 \equiv (\mathfrak{P}_1^U = 1)$$

$$(45)\quad \Omega^0_{ni} \equiv \left(\mathfrak{P}^U_{\frac{i}{n}} - \mathfrak{P}^U_{\frac{i-1}{n}} = 0\right) \quad (i = 1, 2, \dots, n)$$

$$(46)\quad \Omega^1_{ni} \equiv \left(\mathfrak{P}^U_{\frac{i}{n}} - \mathfrak{P}^U_{\frac{i-1}{n}} = 1\right) \quad (i = 1, 2, \dots, n),$$

$$(47)\quad \Omega^0_{ni} \equiv \Omega^0_{n1}\Omega^0_{n2} \dots \Omega^0_{n,i-1}\,\Omega^0_{n,i+1} \dots Q^0_{nn} \quad (i = 1, 2, \dots, n).$$

Then we have

$$(48)\quad \Omega^0 = \Omega^0_{n1}\Omega^0_{n2} \dots \Omega^0_{nn} = \Omega^0_{ni}\Omega^0_{ni} \quad (i = 1, 2, \dots, n),$$

and

$$(49)\quad \Omega^1 = \sum_{i=1}^{n} \Omega^1_{ni}\bar{\Omega}^0_{ni}.$$

If we put for $\omega \in \Omega^1_{ni}\Omega^0_{ni}$

$$(50)\quad y^{(n)}(\omega) = \mathfrak{L}_1(\omega) - \mathfrak{L}_{\frac{i}{n}}(\omega) + \mathfrak{L}_{\frac{i-1}{n}}(\omega) \quad (i = 1, 2, \dots, n),$$

then we have, for $\omega \in \Omega^1$,

$$(51)\quad \lim_{n\to\infty} y^{(n)}(\omega) = x_1(\omega) \left(\equiv \mathfrak{L}_1 - \int_U u\,\mathfrak{P}_1^{du}\right),$$

and so

$$(52)\quad F^1_{x_1} = \lim_{n\to\infty} F^1_{y^{(n)}}$$

$F^1_{x_1}$ and $F^1_{y^{(n)}}$ being respectively the probability law of $x_1$ and $y^{(n)}$ considered as a random variable on $(\Omega^1 P_1)$.

Let $A_i$ denote the $\omega$-set $\left(\mathfrak{L}_1-\mathfrak{L}_{\frac{i}{n}}+\mathfrak{L}_{\frac{i-1}{n}}\in B\right)$ $(i=1, 2, \ldots, n)$. Then we have

(53) $$P_1(y^{(n)}\in B)=\frac{1}{P(\Omega^1)}\sum_{i=1}^{n}P(\Omega^1_{ni}\overline{\Omega}^0_{ni})\frac{P(A_i\Omega^1_{ni}\overline{\Omega}^0_{ni})}{P(\Omega^1_{ni}\overline{\Omega}^0_{ni})},$$

and

(54) $$\frac{P(A_i\,\Omega^1_{ni}\overline{\Omega}^0_{ni})}{P(\Omega^1_{ni}\,\overline{\Omega}^0_{ni})}=\frac{P(A_i\,\overline{\Omega}^0_{ni})P(\Omega^1_{ni})}{P(\Omega^1_{ni})P(\overline{\Omega}^0_{ni})}=\frac{P(A_i\overline{\Omega}^0_{ni})}{P(\overline{\Omega}^0_{ni})};$$

the first half of (54) follows from the hypothesis that $\mathfrak{L}$ is differential. Similarly we shall obtain

(55) $$P_0(A_i)=\frac{P(A_i\,\overline{\Omega}^0_{ni})}{P(\overline{\Omega}^0_{ni})}.$$

From (53), (54) and (55) follows

(56) $$P_1(y^{(n)}\in B)=\frac{1}{P(\Omega^1)}\sum_{i=1}^{n}P(\Omega^1_{ni}\,\overline{\Omega}^0_{ni})P_0(A_i)$$

i.e.

(57) $$F^1_{y^{(n)}}=\frac{1}{P(\Omega^1)}\sum_{i=1}^{n}P(\Omega^1_{ni}\Omega^0_{ni})F^0_{\mathfrak{L}_1-\mathfrak{L}_{\frac{i}{n}}+\mathfrak{L}_{\frac{i-1}{n}}},$$

where $F^0_{\mathfrak{L}_1-\mathfrak{L}_{\frac{i}{n}}+\mathfrak{L}_{\frac{i-1}{n}}}$ is the probability law of $\mathfrak{L}_1-\mathfrak{L}_{\frac{i}{n}}+\mathfrak{L}_{\frac{i-1}{n}}$ considered as a random variable on $(\Omega^0, P_0)$.

Now, we shall prove that $\mathfrak{L}$ is an $\mathfrak{L}$-process on $(\Omega^0, P_0)$. If we put

(58) $$\Omega_{\tau\sigma}=(\mathfrak{P}^U_\sigma-\mathfrak{P}^U_\tau=0),$$

we shall have

(59) $$\Omega^0=\Omega_{0t_1}\,\Omega_{t_1s_1}\,\Omega_{s_1t_2}\,\Omega_{t_2s_2}\,\Omega_{s_21}.$$

Let $c_i$ denote the $\omega$-set $(\mathfrak{L}_{s_i}-\mathfrak{L}_{t_i}\in B_i)$ $(i=1, 2)$. Then we can obtain

(60) $$P_0(c_1c_2)=\frac{P(c_1c_2\Omega^0)}{P(\Omega^0)}=\frac{P(c_1c_2\,\Omega_{0t_1}\Omega_{t_1s_1}\Omega_{s_1t_2}\Omega_{t_2s_2}\Omega_{s_21})}{P(\Omega_{0t_1}\Omega_{t_1s_1}\Omega_{s_1t_2}\Omega_{t_2s_2}\Omega_{s_21})}$$

$$=\frac{P(c_1\Omega_{t_1s_1})P(c_2\Omega_{t_2s_2})P(\Omega_{0t_1})P(\Omega_{s_1t_2})P(\Omega_{s_21})}{P(\Omega_{0t_1})P(\Omega_{t_1s_1})P(\Omega_{s_1t_2})P(\Omega_{t_2s_2})P(\Omega_{s_21})}$$

(as $\mathfrak{L}$ is differential).

$$= \frac{P(c_1\Omega_{t_1s_1})}{P(\Omega_{t_1s_1})}\frac{P(c_2\Omega_{t_2s_2})}{P(\Omega_{t_2s_2})}$$

$$= \frac{P(\Omega_{0t_1})P(c_1\Omega_{t_1s_1})P(\Omega_{s_11})}{P(\Omega_{0t_1})P(\Omega_{t_1s_1})P(\Omega_{s_11})} \cdot \frac{P(\Omega_{0t_2})P(c\Omega_{t_2s_2})P(\Omega_{s_21})}{P(\Omega_{0t_2})P(\Omega_{t_2s_2})P(\Omega_{s_21})}$$

$$= \frac{P(\Omega^0 c_1)}{P(\Omega^0)}\frac{P(\Omega^0 c_2)}{P(\Omega^0)} \quad \text{(as } \mathfrak{L} \text{ is differential)}$$

$$= P_0(c_1)P_0(c_2)\,.$$

Generally we shall have, for non-overlapping $t$-intervals $(t_is_i)$ $(i=1, 2, \ldots n)$,

$$P_0(c_1c_2\ldots c_n) = P_0(c_1)P_0(c_2)\ldots P_0(c_n)\,, \tag{61}$$

where $c_i \equiv (\mathfrak{L}_{s_i} - \mathfrak{L}_{t_i} \in B_i)$ $(i = 1, 2, \ldots, n)$. This shows that $\mathfrak{L}$ is differential on $(\Omega^0, P_0)$. $\mathfrak{L}$ is clearly an $(\mathfrak{F}_{01}, \mathfrak{B}_{01})$-valued random variable[28]. The condition 1° in Definition 3.2 can be verified as follows:

$$P_0(|\mathfrak{L}_s - \mathfrak{L}_t| > \varepsilon) \leqq \frac{P(|\mathfrak{L}_s - \mathfrak{L}_t| > \varepsilon)}{P(\Omega_0)} \to 0 \quad (\text{if } s \to t)\,.$$

We have thus established that $\mathfrak{L}$ is an $\mathfrak{L}$-process on $(\Omega^0, P_0)$. Therefore we can see that

$$F^0_{\mathfrak{L}_1 - \mathfrak{L}_s + \mathfrak{L}_t} = F^0_{\mathfrak{L}_1 - \mathfrak{L}_s} * F^0_{\mathfrak{L}_t} \ (t < s)\,, \tag{62}$$

and that $F_{\mathfrak{L}_t}$ is $\rho$-continuous in $t$ and consequently that $F^0_{\mathfrak{L}_1 - \mathfrak{L}_s + \mathfrak{L}_t}$ is also continuous in $t$ in the $(ts)$-region: $0 \leqq t \leqq s \leqq 1$. By the covering theorem of Heine-Borel we have, in the $(ts)$-region:

$$0 \leqq s - t \leqq \frac{1}{n} \quad 0 \leqq s \leqq t \leqq 1\,,$$

$$\rho(F^0_{\mathfrak{L}_1 - \mathfrak{L}_s + \mathfrak{L}_t},\ F^0_{\mathfrak{L}_1}) < \varepsilon_n\,, \tag{63}$$

where $\varepsilon_n$ is independent of $t$, $s$ and converges to $0$ with $\frac{1}{n}$. From (57) and (63) follows

$$\rho(F^1_{y(n)},\ F^0_{\mathfrak{L}_1}) < \varepsilon_n\,. \tag{64}$$

As $x_1 = \mathfrak{L}_1$ on $\Omega_0$, we have, when $n$ tends to infinity,

(28) See the biginning of Chapter III.

(65) $$\rho(F^1_{x_1}, F^0_{x_1}) = 0 \qquad \text{i.e.} \quad F^1_{x_1} = F^0_{x_1}$$

by virtue of (52).

Now, we shall prove that $\mathfrak{P}^{U_1}, \mathfrak{P}^{U_2}, \ldots, \mathfrak{P}^{U_n}$ and

$$\mathfrak{L} - \int_U u\mathfrak{P}^{du} \qquad (U = U_1 + U_2 + \cdots + U_n)$$

are independent if $(U_1, U_2, \ldots, U_n)$ is a system of disjointed Borel sets*. By the above statement we can see that $\mathfrak{P}^{U_1}$ and $\mathfrak{P} - \int_{U_1} u\mathfrak{P}^{du}$ are independent and that $\mathfrak{P}^{U_2}, \mathfrak{P}^{U_3}, \ldots, \mathfrak{P}^{U_n}$ and $\mathfrak{L} - \int_U u\mathfrak{P}^{du}$ are $B$-measurable functions of $\mathfrak{L} - \int_{U_1} u\mathfrak{P}^{du}$. And so $\mathfrak{P}^{U_1}$ and the combination $(\mathfrak{P}^{U_2}, \mathfrak{P}^{U_3}, \ldots, \mathfrak{P}^{U_n}, \mathfrak{L} - \int u\mathfrak{P}^{du})$ are independent. Similarly we shall see that $\mathfrak{P}^{U_i}$ and $(\mathfrak{P}^{U_{i+1}}, \mathfrak{P}^{U_{i+2}}, \ldots, \mathfrak{P}^{U_n}, \mathfrak{L} - \int_U u\mathfrak{P}^{du})$ are independent $(i = 2, 3, \ldots, n)$. Therefore $\mathfrak{P}^{U_1}, \mathfrak{P}^{U_2}, \ldots, \mathfrak{P}^{U_n}, \mathfrak{L} - \int_U u\mathfrak{P}^{du}$ are independent on account of Corollary 1.2. Consequently we can conclude that the combination of $\mathfrak{P}^U$ for $U$ running over all Borel sets* is a differential system of $\mathfrak{P}$-processes. This system satisfies (9) on account of the fact that $\mathfrak{P}^{[\lambda\infty)}$ and $\mathfrak{P}^{(-\infty, -\lambda)} (\lambda > 0)$ are also $\mathfrak{P}$-processes. We shall prove that (13) holds for this differential system of $\mathfrak{P}$-processes. The following notations will be adopted here for brevity:

$F_n$: the probability law of $\int_{1 \geq |u| > \frac{1}{n+1}} u\mathfrak{P}_1^{du} \quad (n = 1, 2, 3, \ldots)$

$f_n$; the class containing $F_n \ (n = 1, 2, \ldots)$

$g$: the class containing the probability low of $\mathfrak{L}_1$.

As $\int_{\frac{1}{i} \geq |u| > \frac{1}{i+1}} u\mathfrak{P}_1^{du} \ (i = 1, 2, \ldots, n-1)$ and $\mathfrak{L} - \int_{1 \geq |u| > \frac{1}{n+1}} u\mathfrak{P}_1^{du}$ are independent, we have $f_1 \geqslant f_2 \geqslant \cdots \geqslant g$. By Corallary 2.1 we can easily show the existence of $\rho\text{-}\lim_{n\to\infty} f_n$ and accordingly that of $\rho\text{-}\lim_{n\to\infty} F_n * E_{-m_n}$, where $\{m_n\}$ is a convenient sequence of real numbers. We shall denote this limit by $F$. Then we have

(66) $$\lim_{n\to\infty} \varphi_{F_n * E_{-m_n}}(z) = \varphi_F(z).$$

$\mathfrak{P}_1^U$ being subject to a Poissonian distribution, we have

(67) $$\varphi_{F_n * E_{-m_n}}(z) = \exp\left( \int_{\frac{1}{n+1}}^1 (e^{izu} - 1)\overline{\mathfrak{P}_1^{du}} - im_n z \right),$$

and so

$$(68)\qquad \left|\varphi_{F_n * E_{-m_n}}(z)\right| = \exp\left(\Re\int_{\frac{1}{n}}^{1}(e^{izu}-1)\overline{\mathfrak{P}^{du}}\right)^{(29)}$$

$$\leqq \exp\left(-\frac{z^2}{4}\int_{\frac{1}{n}}^{1} u^2\overline{\mathfrak{P}_1^{du}}\right)$$

Assume that $\int_{1\geqq|u|>0} u^2\overline{\mathfrak{P}_1^{du}} = \infty$. Then by virtue of (66) and (68) we have

$$(69)\qquad |\varphi_F(z)| = 0 \qquad \text{for } |z|>0\,.$$

$\varphi_F(z)$ being continuous in $z$ on account of the definition, we shall have

$$(70)\qquad \varphi_F(0) = 0\,.$$

This is contradictory, for $\varphi_F(0) = 1$ follows at once from the definition. Thus we have proved that $\mathfrak{P}$ satisfies the conditions (9) and (13). By Theorem 4.4 we can see that $\left(\lim_{n\to\infty}\int_{n>|u|>\frac{1}{n}}\left(u\mathfrak{P}_t^{du} - \frac{u}{1+u^2}\overline{\mathfrak{P}_t^{du}}\right);\ 0\leqq t\leqq 1\right)$ is an $\mathfrak{L}$-process in the larger sense, the number of whose jumps in a $(tu)$-set $[0t]\times U$ is equal to $\mathfrak{P}_t^U$ for any number $t$ in $[01]$ and for any Borel set* $U$ on the $\omega$-set $\Omega'$ of $P$-measure 1. Furthermore on $\Omega'$, the set of the $(tu)$-points corresponding to the jumps of $\mathfrak{L}(\omega)$ is coincident with that of $\lim_{n\to\infty}\int_{n>|u|>\frac{1}{n}}\left(u\mathfrak{P}^{du}(\omega) - \frac{u}{1+u^2}\overline{\mathfrak{P}^{du}(\omega)}\right)$. We shall define a $\mathfrak{G}$-process $\mathfrak{G}$ by

$$(71)\qquad \mathfrak{G}(\omega) = \mathfrak{L}(\omega) - \lim_{n\to\infty}\int_{n>|u|>\frac{1}{n}}\left((u\mathfrak{P}^{du}(\omega) - \frac{u}{1+u^2}\overline{\mathfrak{P}^{du}}\right) \qquad (\text{for } \omega\in\Omega')$$

$$= 0 \qquad (\text{for } \omega\,\overline{\in}\,\Omega')$$

The independence of $\mathfrak{G}$ and $\mathfrak{P}$ can be easily seen by virtue of the fact that, for any system of disjointed Borel sets* $U_1, U_2, \ldots, \mathfrak{P}^{U_1}, \mathfrak{P}^{U_2}, \ldots, \mathfrak{P}^{U_n}$ and $\mathfrak{L} - \int_U\left(u\,\mathfrak{P}^{du} - \frac{u}{1+u^2}\overline{\mathfrak{P}^{du}}\right)$ $(U = U_1 + U_2 + \cdots + U_n)$ are independent. Thus Theorem 4.5 is proved.

**Theorem of Lévy.** *Let $F$ be an infinitely divisible law. Then the characteristic function $\varphi_F(z)$ of $F$ can be expressed by*

$$(72)\qquad \varphi_F(z) = \exp\left(imz - \frac{\sigma^2}{2}z^2 + \lim_{n\to\infty}\int_{|u|>\frac{1}{n}}\left(e^{izu} - 1 - \frac{izu}{1+u^2}\right)n(du)\right)$$

(29) $\Re$ means the real part.

*where*

(73) *$m$ and $\sigma$ are real numbers, and $\sigma$ is positive.*

(74) *$n$ is a completely additive non-negative set-function defined for Borel sets*, and satisfies*

$$\int_{|u|>1} n(du) < \infty , \quad \int_{0<|u|<1} u^2 n(du) < \infty .$$

*Proof.* By Theorem 3.1 we can define an $\mathfrak{L}$-process $\mathfrak{L} \equiv (\mathfrak{L}_t\,;\, 0 \leqq t \leqq 1)$ such that $\mathfrak{L}_1$ may be subject to $F$. As $\varphi_F(z) = e^{\overline{\psi_1 z}}$ is clear by virtue of Corollary 1.4, we shall obtain at once this theorem from Theorem 4.5.

## V. Homogeneous $\mathfrak{L}$-processes.

**Definition 5.1.** A stochastic differential process $x \equiv (x_t\,;\, a \leqq t \leqq b)$ is termed *temporally homogeneous* (or simply homogeneous) if and only if $s-t = s'-t'$ implies $x_s - x_t \sim x_{s'} - x_{t'}$.

**Corollary 5.1.** *If $x$ is a homogeneous stochastic differential process, we have*

$$(x_{s_i} - x_{t_i}\,;\, i = 1, 2, \dots, n) \sim (x_{s_i'} - x_{t_i'}\,;\, i = 1, 2, \dots, n)$$

*for any system of pairs $(t_1 s_1), (t_2 s_2), \dots, (t_n s_n), (t_1' s_1'), \dots, (t_n' s_n')$ such that the set $(t_1, s_1, t_2, s_2, \dots, t_n, s_n,)$ may be transformed to the set $(t_1', s_1', t_2', s_2', \dots, t_n', s_n')$ by translation.*

**Theorem 5.1.** *Let $\mathfrak{L}(\mathfrak{L}')$ be a temporally homogeneous $\mathfrak{L}$-process over $(01)$ on $(\Omega, P)$ $((\Omega', P'))$. Then $\mathfrak{L}_1 \sim \mathfrak{L}_1'$ implies $\mathfrak{L} \sim \mathfrak{L}'$.*

*Proof.* It is sufficient to prove that, for any pair $t, s (t < s)$, $\mathfrak{L}_s - \mathfrak{L}_t$ and $\mathfrak{L}_s' - \mathfrak{L}_t'$ are equivalent in law. Let $F_{ts}(F_{ts}')$ denote the probability law of $\mathfrak{L}_s - \mathfrak{L}_t (\mathfrak{L}_s' - \mathfrak{L}_t')$ and $F_n(F_n')$ denote $F_{0\frac{1}{2^n}}(F'_{0\frac{1}{2^n}})$. From the hypothesis that $\mathfrak{L}$ and $\mathfrak{L}'$ are temporally homogeneous, we can deduce

$$(1) \qquad F_{st} = \rho\text{-}\lim_{n\to\infty} \mathop{\text{☆}}_{s<\frac{k}{2^n}<t} F_{\frac{k}{2^n}\frac{k+1}{2^n}} = \rho\text{-}\lim_{n\to\infty} \mathop{\text{☆}}_{s<\frac{k}{2^n}<t} F_n, \quad \text{and}$$

$$(2) \qquad F_{st}' = \rho\text{-}\lim_{n\to\infty} \mathop{\text{☆}}_{s<\frac{k}{2^n}<t} F'_{\frac{k}{2^n}\frac{k+1}{2^n}} = \rho\text{-}\lim_{n\to\infty} \mathop{\text{☆}}_{s<\frac{k}{2^n}<t} F_n'.$$

Therefore we need only prove

$$(3) \qquad F_n = F_n' \; (n = 1, 2, \dots).$$

By the hypothesis we have

$$(4) \qquad F = F_1 * F_1 = F_1' * F_1',$$

and so

$$(5) \qquad \varphi_F(z) = (\varphi_{F_1}(z))^2 = (\varphi_{F_1'}(z))^2 .$$

As $F$ is infinitely divisible, we have

$$(6) \qquad \varphi_F(z) \neq 0 ,$$

according to the theorem of Lévy cited in the preceeding chapter. From (5) and (6) follows

$$(7) \qquad \varphi_{F_1}(z) \neq 0 .$$

By virtue of (5) and (7) we can see that $\dfrac{\varphi_{F_1'}(z)}{\varphi_{F_1}(z)}$ is either *1* or $-1$. As $\varphi_F(z)$ and $\varphi_{F_1'}(z)$ are continuous by the definitions, $\dfrac{\varphi_{F_1'}(z)}{\varphi_{F_1}(z)}$ is also continuous, namely $\dfrac{\varphi_{F_1'}(z)}{\varphi_{F_1}(z)}$ is identically equal to *1* or to $-1$. Therefore we have $\dfrac{\varphi_{F_1'}(z)}{\varphi_{F_1}(z)} = 1$ i.e. $F_1 = F_1'$, as $\dfrac{\varphi_{F_1'}(0)}{\varphi_{F_1}(0)}$ is clearly equal to *1*. Similarly we can obtain $F_2 = F_2'$, $F_3 = F_3'$, ..., and finally $F_n = F_n'$.

**Theorem 5.2.** *Let $m$, $\sigma$, $n$ satisfy the conditions (73) (74) stated in the preceeding chapter. Then there exists a temporally homogeneous $\mathfrak{L}$-process whose continuous part $\mathfrak{G}$ satisfies*

$$(8) \qquad \sigma(\mathfrak{G}_1) = \sigma , \qquad m(\mathfrak{G}_1) = m$$

*and whose discontinuous part $\mathfrak{P}$ satisfies*

$$(9) \qquad \overline{\mathfrak{P}_1^U} = n(U) .$$

*Proof.* Let $\Omega_0$ denote the set of all continuous functions of $t$. By a theorem of Doob([30]) we can introduce a $P$-measure $P_0$ on $\Omega_0$ such that

$$P_0\Big(\prod_{i=1}^{n} E(f(t)\,;\,f(s_i)-f(t_i)\in B_i)\Big) = \prod_{i=1}^{n}\frac{1}{\sqrt{2\pi(s_i-t_i)}\sigma} \times \int_{B_i}\exp\Big(-\frac{(\lambda-ms_i+mt_i)^2}{2\sigma^2(s_i-t_i)}\Big)d\lambda$$

([30]) See J. L. Doob: loc. cit. (2) Th. 3.9.

may hold for $0 \leq t_1 < s_1 \leq t_2 < s_2 \leq \cdots \leq t_n < s_n \leq 1$ and for any Borel sets $B_1, B_2, \ldots, B_n$. If we put

$$\mathfrak{G}_t(\omega_0) = f(t) - f(0) \quad \text{for} \quad \omega_0 = f(t) \in \Omega_0,$$

then $\mathfrak{G} \equiv (\mathfrak{G}_t ; 0 \leq t \leq 1)$ is evidently a $\mathfrak{G}$-process on $(\Omega_0, P_0)$.

Let $D$ denote the set $E((tu); 0 \leq t \leq 1)$ of the $(tu)$-plane. We shall introduce *a measure N on D* as follows:

$$(10) \qquad N(E) = \iint_E n(du)dt$$

$E$ being a Borel subset of $D$. We can easily see that $N(E)$ is finite for any Borel set $E$ separated from the $t$-axis by a positive distance.

Let $C^{\alpha\beta}$ denote the $(tu)$-set $(01] \times (\alpha\beta]$ $(\alpha < \beta, \alpha\beta > 0)$. We shall now define a sequence of partitions of $C^{\alpha\beta}$. First, let $C_1, C_2, C_3, C_4$ be defined respectively as $\left(0, \frac{1}{2}\right] \times \left(\alpha, \frac{\alpha+\beta}{2}\right]$, $\left(0, \frac{1}{2}\right] \times \left(\frac{\alpha+\beta}{2}, \beta\right]$, $\left(\frac{1}{2}, 1\right] \times \left(\alpha, \frac{\alpha+\beta}{2}\right]$, $\left(\frac{1}{2}, 1\right] \times \left(\frac{\alpha+\beta}{2}, \beta\right)$. Then we have

$$(11) \qquad C^{\alpha\beta} = C_1 + C_2 + C_3 + C_4.$$

In the exactly same manner $C_i$ may be decomposed into $C_{i1} C_{i2} C_{i3} C_{i4}$ $(i = 1, 2, 3, 4)$. Continuing this decomposition we can define $C_{i_1 i_2 \ldots i_n}$ $(i_1, i_2, \ldots, i_n = 1, 2, 3, 4)$. In this paper this sequence of partitions will be called *the chain of* $C^{\alpha\beta}$, and each set such as $C_{i_1 i_2 \ldots i_n}$ a *chain-element*. For brevity we shall adopt here the following notations:

$$(12) \qquad \nu = N(c^{\alpha\beta})$$
$$\nu_{i_1 i_2 \ldots i_n} = N(C_{i_1 i_2 \ldots i_n}).$$

An integral-valued set function defined for any chain-element of $C^{\alpha\beta}$ will be called a *distribution-chain on,* $C^{\alpha\beta}$ if it satisfies

$$(13) \qquad m(C^{\alpha\beta}) = m(C_1) + m(C_2) + m(C_3) + m(C_4)$$
$$m(C_{i_1 i_2 \ldots i_n}) = \sum_{i=1}^{4} m(C_{i_1 i_2 \ldots i_n i}).$$

Let $\Omega_{\alpha\beta}$ denote the set of all distribution-chains on $C^{\alpha\beta}$. We shall now introduce a sequence of $P$-measures $\{P^{(n)}\}$ of $\Omega_{\alpha\beta}$ as follows:

$$(14) \qquad P^{(0)}\left(E(m; m(C^{\alpha\beta}) = k)\right) = e^{-\nu} \frac{\nu^k}{\underline{|k}} \quad (k = 0, 1, 2, \ldots),$$

(15) $P^{(n)}\Big(E(m\,;\,m(C_{i_1 i_2 \dots i_n}) = k_{i_1 i_2 \dots i_n} \text{ for } i_1, i_2, \dots, i_n = 1, 2, 3, 4)\Big)$

$$= \prod_{i_1=1}^{4} \prod_{i_2=1}^{4} \cdots \prod_{i_n=1}^{4} e^{-\nu_{i_1 i_2 \dots i_n}} \frac{\nu_{i_1 i_2 \dots i_n}^{k_{i_1 i_2 \dots i_n}}}{|k_{i_1 i_2 \dots i_n}} .$$

It can be easily verified by simple calculations that $P^{(0)}$, $P^{(1)}$, $P^{(2)}$, ... are coherent, namely that these measures are coincident for the same subset of $\Omega_{\alpha\beta}$. We can thus obtain an additive (not necessarily completely additive) measure $P$ on $\Omega_{\alpha\beta}$ which is coincident with $P^{(n)}$ for any $P^{(n)}$-measurable subset of $\Omega_{\alpha\beta}$. We shall prove that we can generalized $P$ and constitute a completely additive measure (probability) $P_{\alpha\beta}$. We need only prove that a monotone decreasing sequence $(M_1, M_2, \dots)$ of subsets of $\Omega_{\alpha\beta}$ may have at least one common point, if $P(M_1)$, $P(M_2)$, ... are bounded lowerly by a positive number $\delta$. Let $E_k$ denote the set $(m(C^{\alpha\beta})\,;\, m \in M_k)$ $(k = 1, 2, \dots)$. Then $(E_1, E_2, \dots)$ is a monotone decreasing sequence of the sets of non-negative integers. The following two cases may be possible.

(16) $E_1, E_2, \dots$ have at least one common element.

(17) the minimum $\mu_k$ of the numbers in $E_k$ increases indefinitely with $k$. In the latter case $P_{\alpha\beta}(M_k)$, as it is smaller than $\sum_{i=\mu_k}^{\infty} e^{-\nu} \frac{\nu^i}{|i}$ by the definition, should tend to $0$ with $\frac{1}{k}$ in contradiction with the assumption. Therefore (16) should hold. Let $\sigma$ be an element in $\prod_{i=1}^{\infty} E_i$, and $E'_k$ the set $E\Big((m(C_1), m(C_2), m(C_3), m(C_4))\,;\, m \in M_k, m(C^{\alpha\beta}) = \sigma\Big)$. Then $\{E'_k\}$ will be a monotone decreasing sequence of bounded sets of the points in the 4-dimensional space whose coordinates are all integers. It is clear that none of $E'_k$ is empty. Therefore $\prod_{k=1}^{\infty} E'_k$ will have at least one element. Denote this element by $(\sigma_1, \sigma_2, \sigma_3, \sigma_4)$. Continuing this operation, we can now define $\sigma_{i_1 i_2 \dots i_n}$ $(i_1, i_2, \dots, i_n = 1, 2, 3, 4)$. If we define a distribution-chain $m$ by

(18) $$m(C_{i_1 i_2 \dots i_n}) = \sigma_{i_1 i_2 \dots i_n},$$

then $m$ will belong to each of $M_1, M_2, \dots$ We have thus obtained a probability field $(\Omega_{\alpha\beta}, P_{\alpha\beta})$.

We shall now define a system of random variables $x^C$ by

(19) $$x^C(m) = m(C) \quad \text{(for any distribution chain } m),$$

$C$ running over all chain-elements on $C^{\alpha\beta}$. Then $x^C$ is subject to a Poisonian distribution with the mean value $N(C)$, and for any system of disjointed chain-elements $C_1, C_2, \dots, C_n$, the random variables $x^{C_1}$, $x^{C_2}, \dots, x^{C_n}$ are independent, and $x^C(m)$ is additive (not necessarily completely additive) in $C$, that is to say, for any two systems of disjointed chain-elements $(C_1, C_2, \dots, C_n)$ $(C_1', C_2', \dots, C_n')$ such that $C_1+C_2+\cdots+C_n = C_1'+C_2'+\cdots+C_n'$, we have always

$$x^{C_1}(m)+x^{C_2}(m)+x^{C_3}(m)+\cdots+x^{C_n}(m) \\ = x^{C_1'}(m)+x^{C_2'}(m)+x^{C_3'}(m)+\cdots+x^{C_n'}(m).$$

Now, let $(\Omega, P)$ denote the direct product[31] of the probability fields $(\Omega_0, P_0)$, $(\Omega_{2^i 2^{i+1}}, P_{2^i 2^{i+1}})$ and $(\Omega_{(-2^{i+1})(-2^i)}, P_{(-2^{i+1})(-2^i)})$ for $i$ running over all integers. We shall denote here the representations[32] of the processes $\mathfrak{G}$ (defined on $(\Omega_0, P_0)$) and $x^C$ (defined above) in $(\Omega, P)$ respectively by the same notations $\mathfrak{G}$ and $x^C$.

We shall now classify the chain-elements appearing in the course of defining $(\Omega, P)$. Chain-elements such as $C^{2^i, 2^{i+1}}$, $C^{(-2^{i+1}), (-2^i)}$ will be called *chain-elements of the first kind.* The chain-elements which were constructed from each of the chain-elements of the $(n-1)$-th kind in the same manner as $C_1, C_2, C_3, C_4$ were constructed from $C^{\alpha\beta}$ will be called *chain-elements of the n-th kind.*

For brevity we shall make use of "*(tu)-set°*" to indicate a bounded subset of the $(tu)$-set $E((tu);\ 0 \leqq t \leqq 1)$ separated from the $t$-axis by a positive distance. Terms such as *interval°*, *Borel (tu)-set°*, etc. may be understood similarly.

Let $G$ be an open $(tu)$-set°. Let $C_1^n, C_2^n, \dots, C_k^n$ be the chain-elements of the $n$-th kind whose closure is contained in $G$. Then $\sum_{i=1}^{k} C_i^n$ will increase with $n$ and tend to $G$. We shall define $y^G(\omega)$ as $\lim_{n\to\infty} \sum_{i=1}^{k} x^{C_i^n}(\omega)$.

Let $p$ be a point in the $(tu)$-set $E((tu);\ 0 \leqq t \leqq 1,\ u \neq 0)$, and $U\left(p, \frac{1}{n}\right)$ the $\frac{1}{n}$-neighbourhood of $p$. We shall define $y^{(p)}(\omega)$ by $\lim_{n\to\infty} y^{U(p, \frac{1}{n})}(\omega)$. Then it is clear that $y^{(p)}(\omega)$ is a non-negative integer. Therefore we shall obtain

$$y^G(\omega) = \sum_{p \in G} y^{(p)}(\omega). \tag{20}$$

We shall define $y^E(\omega)$ as $\sum_{p \in E} y^{(p)}(E)$ for any $(tu)$-set°$E$.

---

[31], [32] See Definition 1.11.

We shall now prove that $y^E(\omega)$ is $P$-measurable in $\omega$, namely that $y^E=(y^E(\omega)\,;\ \omega\in\Omega)$ is a random variable on $(\Omega, P)$ for any Borel $(tu)$-set° $E$. By the definition $y^G$ is evidently $P$-measurable in $G$ for any open $(tu)$-set° $G$. Let $F$ any closed $(tu)$-set°, and $U_n$ the $\frac{1}{n}$-neighbourhood of $F$, Then we have

$$y^F(\omega)=\lim_{n\to\infty} y^{U_n}(\omega)\,. \tag{21}$$

As $y^{U_n}(\omega)$ $(n=1, 2, 3, \ldots)$ are $P$-measurable in $\omega$, it is the case with $y^F(\omega)$. From the definition we can easily deduce that $y^E=(y^E(\omega)\,;\ \omega\in\Omega)$ is a random variable subject to a Poissonian distribution with the mean value $N(E)$, if $E$ is an open $(tu)$-set° or a closed $(tu)$-set°.

Let $E$ be any Borel $(tu)$-set°. Then we have

$$N(E)=\underset{F\subset E}{l.u.b.}\ N(F)=\underset{G\supset E}{g.l.b.}\ N(G)\quad (F:\ \text{closed},\ G:\ \text{open}). \tag{22}$$

This shows that there exists a sequence $\{F_k\}$ of closed $(tu)$-sets° and a sequence $\{G_k\}$ of open $(tu)$-sets° such that

$$F_1\subset F_2\subset F_3\subset\cdots\subset E\quad \lim_{n\to\infty} N(F_n)=N(E)\,, \tag{23}$$

$$G_1\supset G_2\supset G_3\supset\cdots\supset E\quad \lim_{n\to\infty} N(G_n)=N(E) \tag{24}$$

Let $G$ and $F$ denote respectively $\prod_{n=1}^{\infty} G_n$ and $\sum_{n=1}^{\infty} F_n$. Then we have

$$F\subset E\subset G\,,\ \text{and so}\ \ y^F(\omega)\leqq y^E(\omega)\leqq y^G(\omega)\,,\ \text{and} \tag{25}$$

$$N(F)=N(E)=N(G)\,. \tag{26}$$

$y^F(y^G)$ is evidently a random variable subject to a Poissonian distribution with the mean value $N(F)$ $(N(G))$ on account of the definition, and $y^{G-F}$ is also a random varirble subject to a Poissonian distribution with the mean value $N(G-F)$. As $N(G-F)=0$ by virtue of (25) and (26), we have, with probability $1$,

$$y^{G-F}=0 \qquad \text{i.e.}\ \ y^G=y^F \tag{27}$$

and by (25), (26), $y^F=y^E=y^G$, which shows that $y^E$ is a random variable subject to a Poissonian distribution with the mean value $N(E)$. It is clear that $y^E(\omega)$ is completely additive in $E$. We shall next prove that $y^{E_1}, y^{E_2}, \ldots, y^{E_n}$ are independent for any system of disjointed Borel $(tu)$-sets° $E_1, E_2, \ldots, E_n$. In the case that $E_1, E_2, \ldots, E_n$ are all open $(tu)$-sets° we can deduce this fact at once from the definition, if we recall that the analogous fact holds with regard to $x^E$. Let $F_1, F_2, \dots, F_n$ be any disjointed closed $(tu)$-sets°, and $U_1^{(m)}, U_2^{(m)}, \dots, U_n^{(m)}$ respectively the $\frac{1}{m}$-neighbourhood of $F_1, F_2, \dots, F_n$ $(m = 1, 2, \dots)$. Then there exists a natural number $M$ such that $m > M$ may imply that $U_1^{(m)}, U_2^{(m)}, \dots, U_n^{(m)}$ are disjointed. We have, for $m > M$,

$$P\Big(\prod_{i=1}^{n} (y^{U_i^{(m)}} \in B_i)\Big) = \prod_{i=1}^{n} P(y^{U_i^{(m)}} \in B_i) \tag{28}$$

where $B_1, B_2, \dots, B_n$ are arbitrary Borel sets of real numbers. Therefore we shall obtain

$$P\Big(\prod_{i=1}^{n} (y^{F_i} \in B_i)\Big) = \lim_{m\to\infty} P\Big(\prod_{i=1}^{n} (y^{U_i^{(m)}} \in B_i)\Big) = \prod_{i=1}^{n} \lim_{m\to\infty} P(y^{U_i^{(m)}} \in B_i)$$
$$= \prod_{i=1}^{n} P(y^{F_i} \in B_i).$$

Let $E_1, E_2, \dots, E_n$ be any system of disjointed Borel $(tu)$-sets°. Then we have a sequence of system of closed $(tu)$-sets° $\{F_{m1}, F_{m2}, \dots F_{mn}\}$ $(m = 1, 2, \dots)$ such that

$$F_{1i} \subset F_{2i} \subset F_{3i} \subset \cdots \subset E_i \quad (i = 1, 2, \dots, n) \tag{29}$$
$$\lim_{m\to\infty} y^{F_{mi}} = y^{E_i} \ (i = 1, 2, \dots, n) \quad \text{(with } P\text{-measure } 1).$$

As $(E_1, E_2, \dots, E_n)$ are a disjointed system, so is $(F_{m1}, F_{m2}, \dots, F_{mn})$. By virtue of the above discussion we can see that $y^{F_{m1}}, y^{F_{m2}}, \dots, y^{F_{mn}}$ are independent and so by (29) that $y^{E_1}, y^{E_2}, \dots, y^{E_n}$ are so independent.

Let $L_t^n$ denote the $(tu)$-set $E\Big((\tau u); \tau = t, |u| \geq \frac{1}{n}\Big)$. We shall now prove that the set $\Omega_n$ of the $\omega$-elements, for which there exists at least one $t$-value satisfying

$$y^{L_t^n}(\omega) \geq 2, \tag{30}$$

has $P$-measure $0$. Let $L_{ts}^n$ denote the $(tu)$-set $E\Big((\tau u); t \leq \tau \leq s, |u| \geq \frac{1}{n}\Big)$. Then the $\omega$-set $\Omega_n$ will be contained in the $\omega$-set:

$$\sum_{k=1}^{m} \Big(y^{L^n_{\frac{k-1}{m}\frac{k}{m}}} \geq 2\Big) \tag{31}$$

whose $P$-measure is smaller than

$$m\left\{1-e^{-\eta\frac{1}{m}}\left(1+\eta\frac{1}{m}\right)\right\} \tag{32}$$

where $\eta = N(L_{01}^n) = \int_0^1 dt \int_{|u|>\frac{1}{n}} n(du)$. As (32) tends to $0$ with $\frac{1}{m}$, $\Omega_n$ should have $P$-mearure $0$. Consequently the set $\Omega_0$ of the $\omega$-elements for which there exists at least one $t$-value satisfying

$$y^{L_t}(\omega) \geqq 2 \quad (L_t = E((\tau u)\,;\ \tau = t\,,\ u \neq 0)) \tag{33}$$

will have also $P$-measure $0$, since $\Omega_0$ is equal to $\sum_{n=1}^{\infty} \Omega_n$. We shall now subtract $\Omega_0$ from $\Omega$; this subtraction may cause no essential change on the consequences obtained above.

For any Borel set* $U$ we shall define $\mathfrak{P}_t^U(\omega)$ as $y^{[0t]\times U}$. Then $\mathfrak{P}^U \equiv (\mathfrak{P}_t^U(\omega)\,;\ 0 \leqq t \leqq 1)$ will be a $\mathfrak{P}$-process, and the combination $\mathfrak{P}$ of $\mathfrak{P}^U$ for $U$ running over all Borel sets* will be a differential system of $\mathfrak{P}$-processes. The $\mathfrak{G}$-process $\mathfrak{G}$ obtaind before and the system $\mathfrak{P}$ will satisfy the conditions stated in Theorem 4.4. Therefore we can see that the process $\mathfrak{G} + \lim_{n\to\infty} \int_{n>|u|>\frac{1}{n}} \left( u\mathfrak{P}^{du} - \frac{u}{1+u^2}\bar{\mathfrak{P}}^{du} \right)$ is precisely the $\mathfrak{P}$-process whose existence was to be proved.

**Theorem of Lévy.** *Let $m$, $\sigma$, $n$ satisfy the conditions (73) (74) in Chapter IV. Then (72) is the characteristic function of a certain infinitely divisible law.*

*Proof.* We shall construct an $\mathfrak{P}$-process $\mathfrak{P}$ as in Theorem 5.2. Then (72) is equal to $e^{iz\mathfrak{L}_1}$ i.e. $\varphi_{F_{\mathfrak{L}_1}}(z)$. $\mathfrak{L}$ being an $\mathfrak{L}$-process, $F_{\mathfrak{L}_1}$ is an infinitely divisible law.

**Uniqueness theorem of Lévy.** *An infinitely divisible law $F$ is expressed in the canonical form in only one way.*

*Proof.* Assume that $F$ be expressed in the following two ways:

$$\varphi_F(z) = \exp\left(imz - \frac{\sigma^2}{2}z^2 + \lim_{n\to\infty} \int_{n>|u|>\frac{1}{n}} \left(e^{izu} - 1 - \frac{izu}{1+u^2}\right) n(du)\right), \tag{34}$$

$$\varphi_F(z) = \exp\left(im'z - \frac{\sigma'^2}{2}z^2 + \lim_{n\to\infty} \int_{n>|u|>\frac{1}{n}} \left(e^{izu} - 1 - \frac{izu}{1+u^2}\right) n'(du)\right). \tag{35}$$

Then we can construct temporally homogeneous $\mathfrak{L}$-processes $\mathfrak{L}$, $\mathfrak{L}'$ by Theorem 5.2 such that

$$\mathfrak{L} = \mathfrak{G} + \lim_{n\to\infty} \int_{n>|u|>\frac{1}{n}} \left( u\mathfrak{P}^{du} - \frac{u}{1+u^2}\overline{\mathfrak{P}^{du}} \right) \tag{36}$$

$$\left( \mathfrak{G}_1 = m\,, \quad \sigma(\mathfrak{G}_1) = \sigma\,, \quad \mathfrak{P}_1^U = n(U) \right)$$

$$\mathfrak{L}' = \mathfrak{G}' + \lim_{n\to\infty} \int_{n>|u|>\frac{1}{n}} \left( u\mathfrak{P}'^{du} - \frac{u}{1+u^2}\overline{\mathfrak{P}'^{du}} \right) \tag{37}$$

$$\left( \mathfrak{G}_1' = m'\,, \ \sigma(\mathfrak{G}_1') = \sigma'\,, \ \mathfrak{P}_1'^U = n'(U) \right)$$

On account of Theorem 5.1 we can deduce $\mathfrak{L} \sim \mathfrak{L}'$ from the hypothesis: $\mathfrak{L}_1 \sim \mathfrak{L}_1'$. Since $\mathfrak{G}(\mathfrak{G}')$ and $\mathfrak{G}(\mathfrak{G}')$ are respectively the continuous part and the discontinuous part of $\mathfrak{L}(\mathfrak{L}')$, we have $\mathfrak{G} \sim \mathfrak{G}'$, $\mathfrak{P} \sim \mathfrak{P}'$, a fortiori $m = m'$, $\sigma = \sigma'$, $n = n'$.

Mathematical Institute,
Faculty of Science, Imperial University of Tokyo.

---

(Added in proof)

As to the phrase "a probability field separable in $P_a$" that I have used in Definition 1.11, some supplementary explanations seem to be necessary.

Let $(\Omega, P)$ be a probability field. It is well-known that the distance between two $P$-measurable sets $E_1$ and $E_2$ means $P(E_1 - E_2) + P(E_2 - E_1)$. If the system of all $P$-measurable subsets of $\Omega$ be separable with respect to this distance, we shall call this probability field $(\Omega, P)$ to be separable in $P$.

In the proof of Theorem 5.2 also I have constructed a probability field with the aid of Definition 1.11 without examining whether this separability condition is satisfied or not. But it can be easily verified that the condition is well satisfied there.

---

**Reprinted from**
*Japanese J. Math.* **XVIII, 261-301 (1942)**

# Differential Equations Determining a Markoff Process*

Kiyosi Itô

## Introduction

(I) For a simple Markoff process $x_1, x_2, \cdots$ parametrized by the set of natural numbers and having a finite number of possible states $a_1, a_2, \ldots, a_m$, one can consider many kinds of transition probabilities. For example, the probability of the event $x_l = a_j$ given the condition that $x_k = a_i$, or the probability of $x_{n+1} = a_{n+1}$ given that $x_1 = a_{i_1}$, $x_2 = a_{i_2}, \ldots, x_n = a_{i_n}$, and so forth. However, the computation of these probabilities can be reduced eventually to the consideration of the probabilities $\mathfrak{p}_{ij}^{(k)}$ $(k = 1, 2, \cdots; i, j = 1, 2, \cdots m)$ for the event $x_{k+1} = a_j$ given that $x_k = a_i$. This fact is explained, for example, in the book by Kolmogoroff[1]. Let us call $\mathfrak{p}_{ij}^{(k)}$'s the basic transition probabilities from now on.

Even when the number of possible states is not finite, the situation remains the same if, for instance, the states are represented by the set of real numbers.

However, if the Markoff process is parametrized not by the natural numbers but by the reals, namely, if the process has continuous parameter, then the situation becomes more complicated[2].

More generally, for a simple Markoff process with its states being represented by the real numbers and having continuous parameter, the problem of determining quantities corresponding to $\mathfrak{p}_{ij}^{(k)}$ mentioned above and of constructing the corresponding Markoff process once these quantities are given has been investigated systematically by Kolmogoroff[3], who reduced the problem to the study of differential equations or integro-differential equations satisfied by the transition probability function.

W. Feller[4] has proved under fairly strong assumptions that these equations possess a unique solution and furthermore that the solution exhibits the properties of transition probability function.

However, if we adopt more strict point of view such as the one J. Doob[5] has applied toward his investigation of stochastic processes, it seems to us that the aforementioned work done by Feller is not quite adequate. For example, even though the differential equation determining the transition probabil-

---

* Translated from the original Japanese. First published in *Journ. Pan-Japan Math. Coll.* No. 1077 (1942).

ity function of a continuous stochastic process was solved in §3 of that paper, no proof was given of the fact that it is possible to introduce by means of this solution a probability measure on some "continuous" function space.

The objective of this article, then, is:

1) to formulate the problem precisely, and
2) to give a rigorous proof, à la Doob, for the existence of continuous parameter stochastic processes.

**(II) Remark:** When $x$ is a real valued random variable, a proposition concerning $x$ can be represented in the form $x \in E$, where $E$ is a subset of $R^1$. Therefore, if $x^{-1}(E)$ is $P$-measurable, it makes sense to consider a proposition of the form $x \in E$. Furthermore if $x \sim x'$, namely, if $P(x \neq x') = 0$, then the $P$-measure of the symmetric difference $(x \in E)\ \Delta\ (x' \in E)$ is 0, and therefore, the proposition $x \in E$ is a "permissible concept."[6]

Next, when $x$ and $y$ are a pair of real-valued random variables, then propositions concerning $x$ and $y$ can be represented in the form "the pair $(x,y)$ belongs to some subset of $R^2$." For instance the proposition $x < y$ can be represented as $(x,y) \in E\ \{(\lambda,\mu);\ \lambda < \mu\}$. Consequently, it is clear that a proposition concerning $x$ and $y$ can be discussed if the corresponding subset of $R^2$ is a Borel set.

The situation remains the same for propositions concerning countable number of random variables, but it is different for the case where uncountably many random variables are involved. This is because the notion of the joint random variable for the case of uncountably many random variables is, in general, a "non-permissible concept"[7].

For example, when $x_t$ is a real-valued stochastic process with continuous parameter $t$, the consideration of propositions such as "$x_t$ is continuous in $t$" or "the least upper bound of $x_t$ is $M$" necessitates the consideration of $x_t$ jointly for all $t$. This is precisely where the problem arises. (Of course, if one is concerned with the proposition that $x_t$ is continuous at $t = t_0$ with respect to the topology of convergence in probability, one can express it as $\lim_{t\to t_0} P(\ |x_t - x_{t_0}| > \epsilon) = 0$. Namely, what is involved here is a proposition $|x_t - x_{t_0}| > \epsilon$ which concerns only with a pair of random variables $x_t$ and $x_{t_0}$, and accordingly no difficulty of the type mentioned above occurs).

However, it is not correct to say that it is impossible to consider propositions such as "$x_t$ is continuous in $t$."

**Definition.** We say that $x_t$ is continuous in $t$ $(\in[a,b])$ if there exists a random variable $y$ taking values in the space of continuous functions on $[a,b]$ such that for each $t$ $P(x_t = y_t) = 1$, where $y_t$ denotes the value of $y$ at $t$.

**Remark 1.** $y$ as above is determined uniquely by $x_t$. Hence, we shall call this $y$ the joint random variable for $x_t$ and denote it by $(x_\tau\ ;\ a \leqslant \tau \leqslant b)$ or more simply by $x_{ab}$. Furthermore, if $\alpha < \beta$ are real numbers in $[a,b]$, a random variable taking its values in the space of continuous functions on the interval $[\alpha,\beta]$ is obtained if we restrict $y$ to $[\alpha,\beta]$. We denote this random variable by $x_{\alpha\beta}$.

**Remark 2.** $(y_\tau;\ a \leqslant \tau \leqslant b)$ equals $y$ itself. This fact is obvious, but this is the reason which justifies the statement "$y$ is $x_{ab}$."

**Remark 3.** The proposition "$x_t$ is continuous at each fixed $t$ with respect to the topology of convergence in probability" and the proposition "$x_t$ is continuous in the sense of definition above" do not coincide. P. Lévy has distinguished the two concepts by saying "$x_t$ has no fixed discontinuity points" in the case of the former and "$x_t$ has no moving discontinuity points" for the latter[8]. (Of course, Lévy was concerned only with the case of differential processes.) Therefore, if $\{y_t\}$ has no moving discontinuity points, then with $(y_\tau;\ a \leqslant \tau \leqslant b)$ interpreted as above, it is possible to consider the proposition "$|y_\tau|$ is smaller than some finite number $M$".

# I. Differentiation

## §1. Definition of Differentiation of a Markoff Process

Let $\{y_t\}$ be a (simple) Markoff process and denote by $F_{t_0 t}$ the conditional probability distribution[1] of $y_t - y_{t_0}$ given that "$y_{t_0}$ is determined". $F_{t_0 t}$ is clearly a $P_{y_t}$-measurable $(\rho)$ function[2] of $y_t$, where $\rho$ denotes the Lévy distance among probability distributions.

**Definition 1.1.**[3]
If

$$F_{t_0 t}^{*[1/t-t_0]} \tag{1}$$

(here $[a]$ is the integer part of the number $a$, and "$*k$" denotes the $k$-fold convolution) converges in probability with respect to the Lévy distance $\rho$ as $t \to t_0+0$, then we call the limit random variable (taking values in the space of probability distributions) the derivative of $\{y_t\}$ at $t_0$ and denote it by

$$D_{t_0}\{y_t\} \text{ or } Dy_{t_0}. \tag{2}$$

**Corollary 1.1.** *$Dy_{t_0}$ is an infinitely divisible probability distribution.*[4]

*Proof.* It is possible to choose a sequence $t_1 \geqslant t_2 \geqslant \cdots \to t_0$ suitably so that

---

[1] This journal, vol **234**, Article #1033.

[2] This journal, vol **234**, Article #1033, *Ibid*. vol **235** Article #1043.

[3] This definition differs somewhat from the definition given by the author in Article #1033, volume **234** of this journal. However, the remarks made there are, of course, relevant here also.

[4] To be more precise, Corollary 1.1 is valid "with probability 1." However, as we remarked earlier, we shall omit the expression "with probability 1" unless it has to be emphasized.

$$F_{t_0 t_n}^{*[1/t_n - t_0]} \qquad (n = 1, 2, \cdots) \tag{3}$$

converges with probability 1. When $F_{t_0 t_n}^{*[1/t_n - t_0]}$ converges, its limit would be a so-called "limit law" in the sense of Khintchine and, therefore, it is infinitely divisible. Consequently, with probability 1, $Dy_{t_0}$ is an infinitely divisible probability distribution.

$Dy_{t_0}$ obtained above is a function of $t_0$ as well as of $y_{t_0}$, and so, we denote it by $L(t_0, y_{t_0})$ corresponds precisely to the "basic transition probability" discussed in the Introduction.

The precise formulation of the problem of Kolmogoroff, then, is to solve the equation

$$Dy_t = L(t, y_t) \tag{4}$$

when the quantity $L(t, y)$ is given.

## §2. A Comparison Theorem

Let us prove a comparison theorem for $Dy_{t_0}$ which we shall make use of later.

**Theorem 2.1.** *Let* $\{y_t\}$, $\{z_t\}$ *be simple Markoff processes satisfying the following conditions:*

(1) $y_{t_0} = z_{t_0}$.
(2) $E(y_t - z_t \mid y_{t_0}) = o(t - t_0)$, *where o is the Landau symbol.*
(3) $\sigma(y_t - z_t \mid y_{t_0}) = o(\sqrt{t - t_0})$.

*(Here* $E(x \mid y)$ *denotes the conditional expectation of x given y and* $\sigma(x \mid y)$ *denotes the conditional standard deviation of x given y. Also, the quantity o may depend on* $t_0$ *or* $y_{t_0}$*). Then, whenever* $Dz_{t_0}$ *exists,* $Dy_{t_0}$ *exists also, and* $Dy_{t_0} = Dz_{t_0}$ *holds.*

*Proof.* For given $\epsilon$ and $\eta$, we can choose $\delta(\epsilon, \eta)$ sufficiently small so that if $|t - t_0| < \delta(\epsilon, \eta)$ then with probability bigger than $1 - \eta$ the following are satisfied:

$$\rho(F_{z_t - z_0 | z_0}^{*[1/t - t_0]}, Dz_{t_0}) < \epsilon \tag{4}$$ [5]

(here by $F_{x|y}$ we mean the conditional probability distribution of $x$ given $y$)

$$|E(y_t - z_t \mid y_{t_0})| < \epsilon(t - t_0) \tag{5}$$

$$|\sigma(y_t - z_t \mid y_{t_0})|^2 < \epsilon(t - t_0). \tag{6}$$

[5] $F_{x|y}$ denotes the conditional probability distribution of $x$ given $y$.

Since $y_{t_0} = z_{t_0}$, we have

$$(7) \qquad y_t - y_{t_0} = (z_t - z_{t_0}) + (y_t - z_t).$$

Let us denote points of the $2n$ dimensional Euclidean space $R^{2n}$ by $(\zeta_1, \xi_1, \zeta_2, \xi_2, \cdots \zeta_n, \xi_n)$, and introduce a probability measure of $R^{2n}$ by

$$(8) \qquad P'(d\zeta_1 d\xi_1 d\zeta_2 d\xi_2 \cdots d\zeta_n d\xi_n) = \prod_{i=1}^{n} F(d\zeta_i d\xi_i),$$

where $F$ denotes the conditional probability distribution

$$(9) \qquad F_{(z_t - z_{t_0},\, y_t - z_t) | z_{t_0}}.$$

The mapping

$$(10) \qquad (\zeta_1, \xi_1, \zeta_2, \xi_2 \cdots \zeta_n, \xi_n) \to \zeta_i \quad (\text{or } \xi_i)$$

may be regarded as a random variable defined on the probability space $(R^{2n}, P')$. We denote this random variable again by $\zeta_i$ (or $\xi_i$). If we define

$$(11) \qquad \eta_i = \zeta_i + \xi_i, \qquad \eta = \sum \eta_i, \qquad \zeta = \sum \zeta_i, \qquad \xi = \sum \xi_i$$

then $\{\eta_i\}$ gives a set of mutually independent random variables, and so do $\{\zeta_i\}$ and $\{\xi_i\}$. Furthermore, the probability distribution functions of $\zeta_i$, $\xi_i$, $\eta_i$ are given, respectively, by

$$F_{z_t - z_{t_0} | y_{t_0}} \qquad (\text{that is, } F_{z_t - z_{t_0} | z_{t_0}})$$

$$F_{y_t - z_t | y_{t_0}} \quad \text{and} \quad F_{y_t - y_{t_0} | z_{t_0}}.$$

Therefore, if we take $n = [1/t - t_0]$, then

$$(12) \qquad F^{*[1/t-t_0]}_{z_t - z_{t_0} | z_{t_0}} = F'_\zeta$$

$$F^{*[1/t-t_0]}_{y_t - y_{t_0} | y_{t_0}} = F'_\eta$$

where $F'_\zeta$ and $F'_\eta$ are probability distribution functions for $\zeta$ and $\eta$, respectively. Therefore, by (4)

$$(13) \qquad \rho(F'_\zeta, Dz_{t_0}) < \epsilon,$$

and since $\xi_1, \xi_2, \cdots \xi_n$ are independent, we obtain

$$(14) \qquad (\sigma(\xi))^2 = \sum_{i=1}^{n} (\sigma(\xi_1))^2 < \left[\frac{1}{t-t_0}\right] (t - t_0)\, \epsilon \leqslant \epsilon$$

and

$$(15)\qquad |E(\xi)| \leqslant \sum_{i=1}^{n} |E(\xi_i)| < \left[\frac{1}{t - t_0}\right] (t - t_0)\epsilon \leqslant \epsilon.$$

Consequently, $E(\xi^2) = (E(\xi))^2 + (\sigma(\xi))^2 < \epsilon + \epsilon^2 < 2\epsilon$ $(\epsilon < 1)$, and

$$P(|\xi| > \epsilon^{1/4}) < 2\epsilon^{1/2},$$

that is, $P(|\eta - \zeta| > \epsilon^{1/4}) < 2\epsilon^{1/2}$, which in turn implies that

$$d(\eta,\zeta) < \epsilon^{1/4} + 2\epsilon^{1/2} < 3\epsilon^{1/4}.$$ [6]

According to P. Lévy[7], we also have

$$(16)\qquad \rho(F_\eta{}', F_\zeta{}') < 3\sqrt{2}\epsilon^{1/4}.$$

From (13) and (16) it follows that

$$(17)\qquad \rho(F_\eta{}', Dz_{t_0}) < \epsilon + 3\sqrt{2}\epsilon^{1/4} < (3\sqrt{2} + 1)\epsilon^{1/4},$$

and from (12) and (17) it follows further that

$$(18)\qquad \rho(F^{*[1/t - t_0]}_{y_t - y_{t_0} | y_{t_0}}, Dz_{t_0}) < (3\sqrt{2} + 1)\epsilon^{1/4}.$$

Thus, we see that the inequality (18) is valid for an arbitrary pair $\epsilon$, $\eta$ with probability greater than $1 - \eta$ as long as $|t - t_0| < \delta(\epsilon,\eta)$ is satisfied. We therefore conclude that $Dy_{t_0} = Dz_{t_0}$.

**Theorem 2.2.** *Let $\{y_t\}$ and $\{z_t\}$ be simple Markoff processes satisfying the following:*

$$(19)\qquad y_{t_0} = z_{t_0}$$

$$(20)\qquad d(y_t, z_t) = o(t - t_0)$$

*(here the quantity o may depend on $t_0$ and $y_{t_0}$).*

*Then, if $Dz_{t_0}$ exists, so does $Dy_{t_0}$ and, furthermore, $Dy_{t_0} = Dz_{t_0}$ is satisfied.*

**Remark.** $d(y_t, z_t) = \inf_{\alpha>0}\{P(|y_t - y_t| > \alpha) + \alpha\}$.

*Proof.* In view of hypothesis (20), we have, if $\delta(\epsilon)$ is chosen sufficiently small,

$$P\{|y_t - z_t| > \epsilon(t - t_0)\} < \epsilon(t - t_0)$$

---

[6,7] Theorie de l'addition des variables aléatorires, p. 51.
$d(x,y) = \inf_{\eta>0}\{P\{|x - y| > \eta\} + \eta\}$.

whenever $|t - t_0| < \delta(\epsilon)$.

With the same notations as in the proof of the preceeding theorem, we obtain

$$P(|\xi_i| > \epsilon(t - t_0)) < \epsilon(t - t_0) \qquad (i = 1, 2, \cdots n)$$

$$P(|\xi| > \epsilon \left[\frac{1}{t-t_0}\right](t-t_0)) < \sum_{i=1}^{n} P(|\xi_i| > \epsilon(t-t_0)) \quad (\text{where } n = \left[\frac{1}{t-t_0}\right])$$

$$< n\epsilon(t - t_0) = \epsilon(t - t_0)\left[\frac{1}{t-t_0}\right].$$

Therefore, $P(|\xi| > 2\epsilon) < 2\epsilon$.

We can then conclude the proof of this theorem just as in the proof of the preceding theorem.

## §3. Examples

**Example 1.** When $\{x_t\}$ is a (temporally) homogeneous differential process, $Dx_t = F_{x_1 - x_0}$ (independent of $(t, x_t)$).

**Example 2.** It happens quite frequently that $\{x_t\}$ is a differential process and the characteristic function $\varphi_{t,s}(z)$ of the difference $x_s - x_t$ is given in the following form:

$$\log \varphi_{t,s}(z) = \int_s^t \psi_\tau(z) d\tau,$$

where

$$\psi_\tau(z) = \text{i}m_\tau z - \frac{\rho_\tau^2}{2} z^2 + \left(\int_{-\infty}^{-0} + \int_{+0}^{+\infty}\right)(e^{izu} - 1 - \frac{izu}{1+u^2})n_\tau(du).$$

Let us assume that $\psi_\tau(z)$ is continuous in $\tau$ when $z$ is fixed and is equicontinuous in some neighborhood of $z = 0$. Then, we have $\log\varphi_{Dx_t}(z) = \psi_t(z)$ (independent of $x_t$). Here $\varphi_{Dx_t}(z)$ denotes the characteristic function of the probability distribution $Dx_t$.

Conversely, one can construct a corresponding differential process when $\psi_t(z)$ is given.

**Example 3.** Let $\{x_t\}$ be a Brownian motion. Namely, $\{x_t\}$ is a temporally homogeneous differential process without moving discontinuity points and $x_1 - x_0$ has the normal distribution. Then, by example 1, we see that

(1) $$Dx_t = \text{normal distribution}.$$

In the sequel, we shall denote for the sake of simplicity, by $G(a,b)$ the Gaussian distribution with mean $a$ and standard deviation $b$. Then, the normal distribution is written as $G(0,1)$. If we consider $\{y_t\}$ given by

(2) $$y_t = (x_t - x_0)^2,$$

then since

(3) $$y_t = (x_{t_0} - x_0)^2 + (x_t - x_{t_0})^2 + 2(x_{t_0} - x_0)(x_t - x_{t_0}),$$

we see that $y_t$ is a simple Markoff process. From

(4) $$\begin{aligned} y_t &= y_{t_0} + (x_t - x_{t_0})^2 + 2(x_{t_0} - x_0)(x_t - x_{t_0}) \\ &= \{y_{t_0} + (t - t_0) + 2(x_{t_0} - x_0)(x_t - x_{t_0})\} \\ &\qquad + \{(x_t - x_{t_0})^2 - (t - t_0)\} \end{aligned}$$

it follows that if we denote by $z_t$ the quantity inside of the first { } on the right-hand side above, then

(5) $$y_{t_0} = z_{t_0}$$

(6) $$\begin{aligned} E(y_t - z_t) &= E\{(x_t - x_{t_0})^2 - (t - t_0)\} \\ &= t - t_0 - (t - t_0) = 0 \end{aligned}$$

(7) $$\begin{aligned} (\sigma\{y_t - z_t\})^2 &= E\{(y_t - z_t)^2\} \\ &= E\{(x_t - x_{t_0})^4\} - (t - t_0)^2 \end{aligned}$$

$$\begin{aligned} E\{(x_t - x_{t_0})^4\} &= \int_{-\infty}^{\infty} \frac{\xi^4}{\sqrt{2\pi(t-t_0)}} \exp\left\{-\frac{\xi^2}{2(t-t_0)}\right\} d\xi \\ &= \left[\int_{-\infty}^{\infty} \frac{\lambda^4}{\sqrt{2\pi}} \exp\left\{-\frac{\lambda^2}{2}\right\} d\lambda\right] (t-t_0)^2 \\ &= 3(t - t_0)^2. \end{aligned}$$

Therefore, we have

(8) $$(\sigma\{y_t - z_t\})^2 = 2(t - t_0)^2 = o(t - t_0).$$

On the other hand, from the definition of $z_t$ it follows that

$$F_{z_t - z_{t_0} | x_{t_0} - x_0} = G(t - t_0, 2|x_{t_0} - x_0|\sqrt{t - t_0}).$$

Since the right-hand side above depends only on $|x_{t_0} - x_0|$, we conclude that

$$F_{z_t - z_{t_0} | (x_{t_0} - x_0)^2} = G(t - t_0, 2|x_{t_0} - x_0|\sqrt{t - t_0}),$$

which implies that

$$F_{z_t - z_0 | z_0} = G(t - t_0, 2 \mid x_{t_0} - x_0 | \sqrt{t - t_0}).$$

Consequently, we have $Dz_{t_0} = G(1, 2\sqrt{z_{t_0}})$. Therefore, if we use Theorem 2.1 we can conclude from the equation above and from (5), (6), and (8) that

$$Dy_{t_0} = G(1, 2\sqrt{z_{t_0}}).$$

# II. Integration

## 4. Definite Integral

Let $x_t$ be a Brownian motion satisfying $x_0 = 0$.[8] Assume that $b_t$ is a function of $x_{0t}$ (that is, of $(x_\tau: 0 \leqslant \tau \leqslant t)$)[9] having no moving discontinuity points. Denote by $\Delta$ a $2n+1$-tuple of real numbers $t_0, t_1, .., t_n, \tau_0, \tau_1, \cdots \tau_{n-1}$ satisfying the following conditions:

(1) $$0 = t_0 \leqslant t_1 \leqslant t_2 \leqslant \cdots \leqslant t_n = 1$$

(2) $$0 \leqslant \tau_0, \tau_1, \tau_2, \cdots \tau_{n-1} \leqslant 1$$

(3) $$\tau_i \leqslant t_i \ (i = 0, 1, 2, \cdots n-1).$$

Define $d(\Delta) = \max_{1 \leqslant i \leqslant n}(t_i - \tau_{i-1})$. Then, clearly, we have

(4) $$t_i - \tau_{i-1} \leqslant d(\Delta) \qquad (i = 1, 2, \cdots n).$$

Let us call

(5) $$y_\Delta = \sum_{i=1}^{n} b_{\tau_{i-1}}(x_{t_i} - x_{t_{i-1}})$$

a $\delta$-sum over the partition $(t_0, t_1, \cdots t_n)$ when $d(\Delta) < \delta$.

**Theorem 4.1.** *$y_\Delta$ converges in probability as $d(\Delta) \to 0$.*

*Proof.* Let $s_1, s_2, \ldots, s_m$ be a refinement of $t_1, t_2, \cdots t_n$ and let $\{\sigma_i\}$ be given by

(6) $$\sigma_{i-1} = \tau_{k-1} \text{ if } (s_{i-1}, s_i) \subset (t_{k-1}, t_k).$$

Then

---

[8] Cf. Example 3 of § 3.

[9] Cf. Introduction (II) of this article.

(7) $$\sum_{i=1}^{n} b_{\sigma_{i-1}}(x_{s_i} - x_{s_{i-1}})$$

is also a $\delta$-sum, and it equals $y_\Delta$. We call (7) the refined representation of $y_\Delta$ over the partitions $s_1, s_2, \cdots s_m$. When two $\delta$-sums, $y_\Delta$ and $y_{\Delta'}$ are given, we can consider the common refinement of partitions for $\Delta$ and $\Delta'$ by taking the union of sub-division points for each, and by considering the corresponding refined representations of $y_\Delta$ and $y_{\Delta'}$, we can represent $y_\Delta$ and $y_{\Delta'}$ as $\delta$-sums over a single partition.

(8) $$y_\Delta = \sum_{i=1}^{n} b_{\tau_{i-1}}(x_{t_i} - x_{t_{i-1}})$$

(9) $$y_{\Delta'} = \sum_{i=1}^{n} b_{\tau'_{i-1}}(x_{t_i} - x_{t_{i-1}})$$

Therefore,

(10) $$y_\Delta - y_{\Delta'} = \sum (b_{\tau_{i-1}} - b_{\tau'_{i-1}})(x_{t_i} - x_{t_{i-1}}).$$

Since $b_t$ has no moving discontinuity points, we can, for any given $\epsilon,\eta$, choose $\delta(\epsilon,\eta)$ sufficiently small so that

(11) $$P\left[\bigcap_{|t-s|<\delta(\epsilon,\eta)} (|b_t - b_s| < \epsilon)\right] > 1 - \eta.$$

If we now choose both $d(\Delta)$ and $d(\Delta')$ to be smaller than $\delta(\epsilon,\eta)$, and define $c_i$ $(i=1,2, \cdots n)$ by

(12) $$c_i = b_{\tau_{i-1}} - b_{\tau'_{i-1}} \quad \text{if } |b_{\tau_{i-1}} - b_{\tau'_{i-1}}| < \epsilon$$

and

$$= 0 \quad \text{if } |b_{\tau_{i-1}} - b_{\tau'}| \geqslant \epsilon,$$

then by (11) we obtain

(13) $$P(y_\Delta - y_{\Delta'} \neq \sum_i c_i(x_{t_i} - x_{t_{i-1}})) < \eta,$$

and since $c_i$ is clearly a function of $x_{0t_{i-1}}$ we see that $c_i$ and $x_{t_i} - x_{t_{i-1}}$ are mutually independent.

Furthermore,

(14) $$E\{(\sum_i c_i(x_{t_i} - x_{t_{i-1}}))^2\}$$
$$= \sum_i E\{c_i^2(x_{t_i} - x_{t_{i-1}})^2\} + 2\sum_{i<j} E\{c_i c_j(x_{t_i} - x_{t_{i-1}})(x_{t_j} - x_{t_{j-1}})\}$$

$$= \sum_i E(c_i^2)\, E\{(x_{t_i}-x_{t_{i-1}})^2\} + 2\sum_{i<j} E\{c_i c_j (x_{t_i}-x_{t_{i-1}})\} E\{x_{t_j}-x_{t_{j-1}}\} \quad [10]$$

$$\leqslant \sum_i \epsilon^2 (t_i - t_{i-1}) = \epsilon^2.$$

Consequently,

$$P\{|\sum_i c_i (x_{t_i}-x_{t_{i-1}})| > \sqrt{\epsilon}\} < \epsilon. \tag{15}$$

From (13) and (14) we obtain

$$P\{|y_\Delta - y_{\Delta'}| > \sqrt{\epsilon}\} < \epsilon + \eta \qquad \text{q.e.d}$$

We note that in the argument above, the interval [0, 1] may be replaced by any interval $[t, s]$.

**Definition 4.1.** By $\int_0^1 b_\tau dx_\tau$ we shall mean the limit of $y_\Delta$ whose existence was proved in Theorem 4.1. We interpret $\int_t^s b_\tau dx_\tau$ similarly.

## 5. Theorems on Definite Integrals

Throughout this section, we assume that $\{x_t\}$ is a Brownian motion and that integrands ($b_t$, $c_t$, etc.) considered are functions of $x_{0t}$ having no moving discontinuity points.

The integral defined in the preceding section satisfies the following properties familiar for the ordinary integral:

**Theorem 5.1.** $\int_t^s dx_\tau = x_s - x_t$.

**Theorem 5.2.** $\int_t^s (\lambda b_\tau + \mu c_\tau) dx_\tau = \lambda \int_t^s b_\tau dx_\tau + \mu \int_t^s c_\tau dx_\tau$.

**Theorem 5.3.** *If* $t < s < u$,

$$\int_t^s b_\tau dx_\tau + \int_s^u b_\tau dx_\tau = \int_t^u b_\tau dx_\tau.$$

**Theorem 5.4.** *Let* $y = \int_t^s b_\tau dx_\tau$. *If there exists a continuous function* $M(\tau)$ *such that*

$$E(b_\tau^2) \leqslant M(\tau), \tag{1}$$

[10] Since $|c_i| \leqslant \epsilon$, $E(c_i^2)$ exists and since $c_i$ and $x_{t_i} - x_{t_{i-1}}$ are independent,

$$E\{c_i^2 (x_{t_i} - x_{t_{i-1}})^2\} = E(c_i^2) E\{(x_{t_i} - x_{t_{i-1}})^2\}$$

$c_i c_j (x_{t_i} - x_{t_{i-1}})$ is a function of $x_{0t_{j-1}}$, and is independent of $x_{t_j} - x_{t_{j-1}}$ since $i < j$. Furthermore, $E(c_i c_j (x_{t_i} - x_{t_{i-1}}))$ exists because $|c_i c_j (x_{t_i} - x_{t_{i-1}})| \leqslant \epsilon^2 |x_{t_i} - x_{t_{i-1}}|$. Consequently,

$$E\{c_i c_j (x_{t_i} - x_{t_{i-1}})(x_{t_j} - x_{t_{j-1}})\} = E\{c_i c_j (x_{t_i} - x_{t_{i-1}})\}\, E\{x_{t_j} - x_{t_{j-1}}\}.$$

*then*

(2) $$E(y^2) \leqslant \int_t^s M(\tau)d\tau.$$

**Theorem 5.5.** *If a sequence* $\{b_\tau^{(n)}\}$ $(n=1,2,3,\cdots)$ *of stochastic processes converges to* $\{b_\tau\}$ *in the sense of strong topology, namely, if*

(3) $$P\{\sup_{s\leqslant\tau\leqslant t} | b_\tau^{(n)} - b_\tau | > \epsilon\}\to 0 \quad as\ n\to\infty,$$

*then* $\int_t^s b_\tau^{(n)}\, dx_\tau$ *converges in probability to* $\int_t^s b_\tau dx_\tau$.

Theorems 5.1–5.3 are obvious. In order to prove Theorems 5.4 and 5.5, we shall first prove the following lemma.

**Lemma 5.1.** *Let* $x, x_1, x_2, \cdots$ *be real-valued random variables satisfying the following properties:*

(4) $$x_1, x_2, \cdots \text{ converges with probability 1 to } x,$$

(5) $$x_1, x_2 \cdots \geqslant 0,$$

(6) $$E(x_n) \leqslant e_n \qquad (n = 1, 2, \cdots),$$

(7) $$e_n \to e.$$

*Then, we have*

(8) $$E(x) \leqslant e.$$

*Proof.* Let $y_n = \inf\{x_n, x_{n+1}, \cdots\}$. Then,

(9) $$0 = y_1 \leqslant y_2 \leqslant \cdots \to x$$

(10) $$0 \leqslant E(y_n) \leqslant E(x_n) \leqslant e_n \quad (n = 1, 2, \cdots).$$

Since $\{y_n\}$ is monotone increasing by (9),

$$E(x) = E(\lim_{n\to\infty} y_n) = \lim E(y_n).$$

On the other hand, from (10) it follows that

$$\lim_{n\to\infty} E(y_n) \leqslant \overline{\lim}\, e_n = \lim e_n = e.$$

Therefore, we obtain $E(x) \leqslant e$.

*Proof of Theorem 5.4.* We can choose a sequence $\{\Delta_n\}$ suitably with $d(\Delta_n)\to 0$ so that

$$y = \int_t^s b_\tau dx_\tau = \lim_{n\to\infty} \sum_{\Delta_n} b_{\tau_{i-1}}(x_{t_i} - x_{t_{i-1}}) \tag{11}$$

holds with probability 1.[11] Therefore

$$y_2 = \left(\int_t^s b_\tau dx_\tau\right)^2 = \lim_{n\to\infty} \left[\sum_{\Delta_n} b_{\tau_{i-1}}(x_{t_i} - x_{t_{i-1}})\right]^2.$$

If we let $y_n = \sum_{\Delta_n} b_{\tau_{i-1}}(x_{t_i} - x_{t_{i-1}})$, then

$$E(y_n^2) = \sum_i E(b_{\tau_{i-1}}^2)\, E\{(x_{t_i} - x_{t_{i-1}})^2\}$$
$$+ 2\sum_{i<j} E\{b_{\tau_{i-1}}(x_{t_i} - x_{t_{i-1}})b_{\tau_{j-1}}\}E(x_{t_j} - x_{t_{j-1}}).^{12}$$

Since $E(x_{t_i} - x_{t_{i-1}}) = 0$, we obtain $E(y_n^2) \leqslant \sum_{\Delta_n} M(\tau_{i-1})(t_i - t_{i-1})$.
Since the right-hand side above tends to $\int_t^s M(\tau)d\tau$ as $n\to\infty$, we obtain

$$E(y^2) \leqslant \int_t^s M(\tau)d\tau \quad \text{in view of Lemma 5.1.}$$

*Proof of Theorem 5.5.* Because of the assumption of convergence in probability, we can choose, for given $\epsilon$ and $\eta$, $n$ sufficiently large so that

$$P\{\sup_{t\leqslant\tau\leqslant s} |b_\tau^{(n)} - b_\tau| \geqslant \epsilon\} < \eta. \tag{12}$$

Let us now define $c_\tau$ as follows:

$$\begin{aligned} c_\tau &= \epsilon && \text{if } b_\tau^{(n)} - b_\tau > \epsilon \\ &= b_\tau^{(n)} - b_\tau && \text{if } -\epsilon \leqslant b_\tau^{(n)} - b_\tau \leqslant \epsilon \\ &= -\epsilon && \text{if } b_\tau^{(n)} - b_\tau < -\epsilon. \end{aligned}$$

Then, $c_\tau$ is a function of $x_{0\tau}$ and the stochastic process $\{c_\tau\}$ has no moving discontinuity points. Therefore, one can consider the integral $\int_t^s c_\tau dx_\tau$. According to (12) we have

---

[11] P. Lévy: *Ibid.*, pp. 55 and 56.

[12]
$$\begin{aligned} &E(|b_{\tau_{i-1}}(x_{t_i} - x_{t_{i-1}})b_{\tau_{j-1}}|) \\ &\leqslant \sqrt{E\{b_{\tau_{i-1}}^2(x_{t_i} - x_{t_{i-1}})^2\}\, E\{b_{\tau_{j-1}}^2\}} \\ &= \sqrt{E(b_{\tau_{i-1}}^2)E\{(x_{t_i} - x_{t_{i-1}})^2\}\, E(b_{\tau_{j-1}}^2)} \quad (\text{since } b_{\tau_{i-1}} \text{ is independent of } x_{t_i} - x_{t_{i-1}}) \\ &\leqslant \sqrt{M(\tau_{i-1})M(\tau_{j-1})(t_i - t_{i-1})}. \end{aligned}$$

Therefore, $b_{\tau_{i-1}}(x_{t_i} - x_{t_{i-1}})b_{\tau_{j-1}}$ is integrable.

$$(13)\qquad P\left\{\int_t^s c_\tau dx_\tau \neq \int_t^s (b_\tau^{(n)} - b_\tau)dx_\tau\right\} < \eta.$$

and furthermore, $|c_\tau| \leqslant \epsilon$ and hence $E(c_\tau^2) \leqslant \epsilon^2$. Therefore we have

$$E\left\{\left(\int_t^s c_\tau dx_\tau\right)^2\right\} \leqslant \epsilon^2(s-t),$$

from which it follows that

$$(14)\qquad P\left\{\left|\int_t^s c_\tau dx_\tau\right| > \sqrt{\epsilon}\right\} \leqslant \epsilon(s-t).$$

From (13) and (14) we obtain that

$$P\left\{|\int_t^s b_\tau^{(n)}\, dx_\tau - \int_t^s b_\tau dx_\tau| > \sqrt{\epsilon}\right\} \leqslant \epsilon(s-t) + \eta,$$

from which we conclude that $\int_t^s b_\tau^{(n)}\, dx_\tau$ converges in probability to $\int_t^s b_\tau dx_\tau$.

## 6. Indefinite Integrals

**Theorem 6.1.** $\{\int_0^t b_\tau dx_\tau\}$ $(0 \leqslant t \leqslant 1)$ *has no moving discontinuity points.*

More precisely, there exists a random variable $y$ taking values in the space of continuous functions on [0,1] such that for arbitrary $t$ $(0 \leqslant t \leqslant 1)$

$$P\left\{\int_0^t b_\tau dx_\tau = y_t\right\} = 1 \quad \text{is satisfied,}$$

where $y_t$ denotes the real-valued random variable taking the value of $y$ at $t$. (The fact that such $y$ is determined uniquely up to equivalence was explained in Introduction (II).)

**Definition 6.1.** We call $y$ in Theorem 6.1 the indefinite integral of $b_\tau$ with respect to $x_\tau$.

*Proof of Theorem 6.1.* $\{y_\Delta(t)\}$ defined by

$$(1)\qquad y_\Delta(t) = \sum_{i=1}^{k-1} b_{\tau_{i-1}}(x_{t_i} - x_{t_{i-1}}) + b_{\tau_{k-1}}(x_t - x_{t_{k-1}}) \quad \text{for } t \in [t_{k-1}, t_k]$$

obviously has no moving discontinuity points. Therefore, we may regard $y_\Delta(t)$ as the value at $t$ of some random variable—which we denote by $y_\Delta$ also—taking values in the space of continuous functions. It suffices, therefore, to show that, as $d(\Delta)\to 0$, $y_\Delta = (y_\Delta(t);\ 0 \leqslant t \leqslant 1)$ converges in probability with respect to the strong topology. This is because it would then follow that we can choose a sequence $\{\Delta_n\}$ so that the sequence $\{y_{\Delta_n}\}$ $(n=1, 2, \cdots)$ converges with probability 1, and if we donate the limit by $y$, then $y$ would be a continuous function with probability 1 since it is with probability 1 the limit of the uniformly convergent sequence of continuous functions $\{y_{\Delta_n}\}$.

Let us first state a lemma.

**Lemma 6.1.** *Let $x_1, x_2, \cdots, x_n$ be mutually independent real-valued random variables and let for each $i = 1, 2, \cdots, n$ $y_i$ be a real-valued random variable independent of $(x_i, x_{i+1}, \cdots, x_n)$. If*

(2) $$E(x_i) = 0, \quad E(y_i^2) < \infty \qquad (i = 1, 2, \ldots, n)$$

*then*

(3) $$P\left\{\max_{1\leqslant p\leqslant n} |y_1x_1 + y_2x_2 + \cdots + y_kx_k| \geqslant l\right\} \leqslant \frac{E\{(y_1x_1 + y_2x_2 + \cdots + y_nx_n)^2\}}{l^2}.$$

This lemma is an extension of the Kolmogoroff inequality (which corresponds to the case where $y_1 = y_2 = \cdots = y_n = 1$), and its proof is identical with that for the latter. Hence, we shall omit its proof.

Going back to the proof of Theorem 6.1, let us suppose that $s_1, s_2, \cdots$ is a sequence of points dense in (0,1), and represent by $0 = t_0 < t_1 < \cdots < t_n = 1$ a new partition of (0,1) obtained by joining $s_1, s_2, \cdots s_m$ to the points of partitions $\Delta'$ and $\Delta$. Then we have

(4) $$y_\Delta(t_k) = \sum_{i=1}^{k} b_{\tau_{i-1}}(x_{t_i} - x_{t_{i-1}})$$

(5) $$y_{\Delta'}(t_k) = \sum_{i=1}^{k} b_{\tau'_{i-1}}(x_{t_i} - x_{t_{i-1}}).$$

Here we choose $\delta(\epsilon,\eta)$ as in (11) of §4, and $d(\Delta)$, $d(\Delta')$ are $< \delta(\epsilon,\eta)$. Then, the quantities on the right hand side of (4) and (5) both represent a $\delta(\epsilon,\eta)$-sum. Now,

(6) $$y_\Delta(t_k) - y_\Delta'(t_k) = \sum_{i=1}^{k} (b_{\tau_{i-1}} - b_{\tau'_{i-1}})(x_{t_i} - x_{t_{i-1}})$$

and by defining $c_i$'s as in (12) of §4 we obtain

(7) $$P\left\{\bigcup_{k=1}^{n} (y_\Delta(t_k) - y_\Delta'(t_k) \neq \sum_{i=1}^{k} c_i(x_{t_i} - x_{t_{i-1}}))\right\} < \eta.$$

We also have

$$E\left\{\left[\sum_{i=1}^{n} c_i(x_{t_i} - x_{t_{i-1}})\right]^2\right\} < \epsilon^2. \tag{8}$$

Applying Lemma 6.1 by taking $c_i\ (x_{t_i} - x_{t_{i-1}})$ and $\sqrt{\epsilon}$ to be $y_i$, $x_i$ and $l$, respectively, of that lemma, we obtain

$$P\left\{\max_{1 \leqslant k \leqslant n} \left| \sum_{i=1}^{k} c_i\ (x_{t_i} - x_{t_{i-1}}) \right| \geqslant \sqrt{\epsilon}\right\} \leqslant \epsilon. \tag{9}$$

From (7) and (9) it then follows that

$$P\left\{\max_{1 \leqslant k \leqslant n} | y_\Delta(t_k) - y_{\Delta'}(t_k) | \geqslant \sqrt{\epsilon}\right\} \leqslant \epsilon + \eta, \tag{10}$$

and therefore that

$$P\left\{\max_{1 \leqslant i \leqslant m} | y_\Delta(s_i) - y_{\Delta'}(s_i) | \geqslant \sqrt{\epsilon}\right\} \leqslant \epsilon + \eta. \tag{11}$$

As $m \to \infty$, the set within { } of the left-hand side above increases to the set

$$\left\{\sup_{k=1}^{\infty} | y_\Delta(s_k) - y_{\Delta'}(s_k) | \geqslant \sqrt{\epsilon}\right\}.$$

Since $y_\Delta(\tau)$ and $y_{\Delta'}(\tau)$ are both continuous functions of $\tau$, and since $\{s_i\}$ is dense in (0,1), we conclude that

$$\sup_{k=1}^{\infty} | y_\Delta(s_k) - y_{\Delta'}(s_k) | = \sup_{0 \leqslant \tau \leqslant 1} | y_\Delta(\tau) - y_{\Delta'}(\tau) |,$$

and therefore we obtain

$$P\left\{\sup_{0 \leqslant \tau \leqslant 1} | y_\Delta(\tau) - y_{\Delta'}(\tau) | \geqslant \sqrt{\epsilon}\right\} \leqslant \epsilon + \eta. \qquad \text{q.e.d.}$$

## §7. Examples of Indefinite Integrals

**Example 1.**

$$\int_0^t x_\tau dx_\tau = \frac{1}{2}x_t^2 - \frac{1}{2}t.$$

**Example 2.**

$$\int_0^t x_\tau^2 dx_\tau = \frac{1}{3}x_t^3 - \int_0^t x_\tau d\tau.$$

**Example 3.**

$$\int_0^t a(x_\tau)dx_\tau = \int_0^{x_t} a(\lambda)d\lambda - \int_0^t \frac{a'(x_\tau)}{2}d\tau,$$

where we assume that $a'(\lambda)$ is continuous in $\lambda$.

Since Examples 1 and 2 are special cases of Example 3, we shall prove Example 3 only. Let us first of all define

$$(1) \qquad b(\xi) = \int_0^\xi a(\lambda)\, d\lambda.$$

Since $\{x_t\}$ is a process without moving discontinuity points, $x_t$ is bounded in [0,1] with probability 1 (see the remark in Introduction (II)). Therefore, if we choose, for a given $\eta$, $M$ sufficiently large, we have

$$(2) \qquad P\left\{ \sup_{0\leqslant\tau\leqslant 1} |x_t| \leqslant M \right\} > 1 - \eta.$$

Since $a'(\xi)$ is continuous and hence is uniformly continuous on $|\xi| \leqslant M$, we have, if we choose $\delta$ sufficiently small for a given $\epsilon$,

$$(3) \qquad |a'(\xi') - a'(\xi)| < \epsilon \quad \text{whenever } |\xi' - \xi| < \delta.$$

The fact that with probability 1 $x_t$ is continuous implies also that for the $\delta$ as above, we have, if we choose $\gamma$ sufficiently small,

$$(4) \qquad P\left( \bigcap_{\substack{|t-s|<\gamma \\ 0\leqslant t,s\leqslant 1}} (|x_t - x_s| < \delta) \right) > 1 - \eta.$$

Now, let $\Omega' = (\bigcap_{|t-s|<\gamma}(|x_t - x_s| < \delta))\cdot(\sup_{0\leqslant\tau\leqslant 1}|x_\tau| < M)$, then we see from (2) and (4) that

$$(4') \qquad P(\Omega') > 1 - 2\eta.$$

The continuity of $a'(\lambda)$ implies that the function $b(\xi)$ has continuous derivatives up to the second order. Therefore,

$$(5) \quad b(\xi') - b(\xi)$$

$$= b'(\xi)(\xi'-\xi) + \frac{b''(x)}{2}(\xi'-\xi)^2 + \frac{b''(\xi + \theta(\xi'-\xi)) - b''(\xi)}{2}(\xi'-\xi)^2$$

$$= a(\xi)(\xi'-\xi) + \frac{a'(\xi)}{2}(\xi'-\xi)^2 + \frac{a'(\xi+\theta(\xi'-\xi)) - a'(\xi)}{2}(\xi' - \xi)^2,$$

where $\theta$ lies in the interval [0,1]. Therefore,

$$(6)\quad b(x_s) - b(x_t)$$

$$= a(x_t)(x_s - x_t) + \frac{a'(x_t)}{2}(x_s - x_t)^2 + \frac{a'(x_t + \theta(x_s - x_t)) - a'(x_t)}{2}(x_s - x_t)^2.$$

Whenever $|t-s| < \gamma$, we have on $\Omega'$

$$|x_t + \theta(x_s - x_t)| \leqslant \max(|x_t|, |x_s|) < M,$$

and

$$|x_t + \theta(x_s - x_t) - x_t| \leqslant |x_s - x_t| < \delta.$$

Therefore, by (5) we get

$$(7)\qquad |a'(x_t + \theta(x_s - x_t)) - a'(x_t)| \leqslant \epsilon.$$

Let us now define $c_{t,s}$ by

$$(8)\qquad c_{t,s} \equiv \tfrac{1}{2}(a'(x_t + \theta(x_s - x_t)) - a'(x_t))$$

if the inequality (7) is satisfied, and

$$(9)\qquad c_{t,s} \equiv 0$$

otherwise. $a'(\xi)$, being continuous in $\xi$, is bounded on $|\xi| \leqslant M$; so, let $\sup_{|\xi| \leqslant M} |a'(\xi)| = R$.

Define $e_t$ by

$$e_t = \begin{cases} a'(x_t) & \text{if } |a'(x_t)| \leqslant R \\ 0 & \text{otherwise.} \end{cases}$$

On the set $\Omega'$, $c_{t,s}$ is given by (8) and $e_t = a'(x_t)$; consequently, on $\Omega'$ we have, as long as $|t-s| < \gamma$,

$$(10)\quad b(x_s) - b(x_t)$$

$$= a(x_t)(x_s - x_t) + \frac{a'(x_t)}{2}(s - t) + c_{t,s}(x_s - x_t)^2$$

$$+ \frac{1}{2} e_t((x_s - x_t)^2 - (s - t)).$$

Let us choose points $0 = t_0 < t_1 < t_2 < \cdots < t_n = t$ such that $\max_{1 \leqslant i \leqslant n} |t_i - t_{i-1}| < \gamma$, then we have on $\Omega'$

$$(11)\quad b(x_t) - b(x_0)$$

$$= \sum_{i=1}^{n} (b(x_{t_i}) - b(x_{t_{i-1}}))$$

$$= \sum_{i=1}^{n} a(x_{t_{i-1}})(x_{t_i} - x_{t_{i-1}}) + \sum_{i=1}^{n} \frac{a'(x_{t_{i-1}})}{2}(t_i - t_{i-1})$$

$$+ \sum_{i=1}^{n} c_{t_{i-1}t_i}(x_{t_i} - x_{t_{i-1}})^2 + \sum_{i=1}^{n} \frac{1}{2} e_{t_{i-1}}((x_{t_i} - x_{t_{i-1}})^2 - (t_i - t_{i-1})).$$

If we choose each $|t_i - t_{i-1}|$ sufficiently small (less than $\beta$, for example), we obtain

$$P\left\{ \left| \sum_{i=1}^{n} a(x_{t_{i-1}})(x_{t_i} - x_{t_{i-1}}) - \int_0^t a(x_\tau)dx_\tau \right| > \epsilon \right\} < \eta \tag{12}$$

$$P\left\{ \left| \sum_{i=1}^{n} \frac{a'(x_{t_{i-1}})}{2}(t_i - t_{i-1}) - \int_0^t \frac{a'(x_\tau)}{2} d\tau \right| > \epsilon \right\} < \eta, \tag{13}$$

$$E\left( \left| \sum_{i=1}^{n} c_{t_{i-1}t_i}(x_{t_i} - x_{t_{i-1}})^2 \right| \right) \leqslant E\left( \sum_{i=1}^{n} \epsilon (x_{t_i} - x_{t_{i-1}})^2 \right) \leqslant \epsilon t \leqslant \epsilon$$

$$P\left\{ \left| \sum_{i=1}^{n} c_{t_{i-1}} t_i (x_{t_i} - x_{t_{i-1}})^2 \right| > \sqrt{\epsilon} \right\} < \sqrt{\epsilon}. \tag{14}$$

If we take each $|t_i - t_{i-1}|$ to be smaller than $\epsilon/R^2$, we also obtain

$$E\left\{ \left( \sum_{i=1}^{n} \frac{e_{t_{i-1}}}{2}((x_{t_i} - x_{t_{i-1}})^2 - (t_i - t_{i-1}) \right)^2 \right\} \leqslant \sum_{i=1}^{n} \frac{R^2}{4} \left( 2(t_i - t_{i-1})^2 \right)$$

$$\leqslant \frac{R^2}{4} \frac{\epsilon}{R^2} 2 \sum_{i=1}^{n} (t_i - t_{i-1}) \leqslant \frac{\epsilon}{2} t \leqslant \frac{\epsilon}{2}$$

and so,

$$P\left\{ \left| \sum_{i=1}^{n} \frac{e_{t_{i-1}}}{2}((x_{t_i} - x_{t_{i-1}})^2 - (t_i - t_{i-1})) \right| > \sqrt[4]{\epsilon} \right\} \leqslant \frac{\sqrt{\epsilon}}{2}. \tag{15}$$

From (4′) and (11)–(15), it follows that

$$P\left\{ \left| \int_0^{x_t} a(\lambda)d\lambda - \int_0^t a(x_\tau)dx_\tau - \int_0^t \frac{a'(x_\tau)}{2} d\tau \right| > 2\epsilon + \sqrt{\epsilon} + \sqrt[4]{\epsilon} \right\}$$

$$< 2\eta + 2\eta + \epsilon + \frac{\sqrt{\epsilon}}{2}.$$

Since $\epsilon$ and $\eta$ are arbitrary, we have, for each fixed $t$,

$$\int_0^{x_t} a(\lambda)d\lambda = \int_0^t a(x_\tau)dx_\tau + \int_0^t \frac{a'(x_\tau)}{2} d\tau$$

and hence

(16) $$\int_0^t a(x_\tau)dx_\tau = \int_0^{x_t} a(\lambda)d\lambda - \int_0^t \frac{a'(x_\tau)}{2}d\tau$$

holds with probability 1. The fact that both sides of (16) are stochastic processes having no moving discontinuity points implies that (16) is in fact valid for every $t$.

**Example 4.** When $a(\tau)$ and $b(\tau)$ are continuous functions of $\tau$ and $x_\tau$ is a Brownian motion,

$$y_t = \int_0^t a(\tau)d\tau + \int_0^t b(\tau)dx_\tau$$

is a differential process and $y_s - y_t$ has

$$G\left(\int_t^s a(\tau)d\tau, \sqrt{\int_t^s (b(\tau))^2 d\tau}\right)$$

distribution. Here $G(\alpha,\beta)$ denotes the Gaussian distribution with mean $\alpha$ and standard deviation $\beta$.

*Proof.* The fact that $y_t$ is a differential process is clear. If we choose $\{\Delta_n\}$ suitably, we have

$$y_s - y_t = \lim_{n\to\infty}\left\{\int_s^t a(\tau)d\tau + \sum_{\Delta_n} b(\tau_{i-1})(x_{t_i} - x_{t_{i-1}})\right\}.$$

But the quantity within { } has a Gaussian distribution with its mean equal to $\int_s^t a(\tau)d\tau$ and its standard deviation equal to

$$\sqrt{\sum_i (b(\tau_{i-1}))^2(t_i - t_{i-1})} \to \sqrt{\int_s^t (b(\tau))^2 d\tau}\cdot \qquad \text{q.e.d.}$$

## §8. An Inequality Concerning the Absolute Value of an Indefinite Integral

**Theorem 8.1.** *Let $\{x_t\}$ and $\{b_t\}$ be as in §4. We assume further that $E(b_t^2)$ is continuous in $t$. Then we have*

(1) $$P\left(\sup_{0\leqslant t\leqslant 1}\left|\int_0^t b_\tau dx_\tau\right| \geqslant l\right) \leqslant \frac{\int_0^t E(b_\tau^2)d\tau}{l^2}.$$

**Remark.** This theorem is nothing but Lemma 6.1 with the sum being replaced by the integral.

*Proof of Theorem 8.1.* Let us consider $y_\Delta(t)$ which was defined by (1) in the proof of Theorem 6.1. If we choose suitably $\{\Delta_n\}$ with $d(\Delta_n)\to 0$, $y_{\Delta_n}(t)$, which we denote simply by $y_n(t)$ in the sequel, converges uniformly to $\int_0^t b_\tau dx_\tau$ with probability 1.

Let $\alpha_1, a_2, \cdots$ be a sequence of points dense in (0,1) and let us consider the partition $0 = s_0 < s_1 < s_2 \cdots < s_n = 1$ obtained by joining the points $\alpha_1, \alpha_2, \cdots, \alpha_k$ to the points of partition $\Delta_m$. Then the refined representation of $y_m = y_{\Delta_m}$ with respect to $s_0 < s_1 < \cdots < s_n$ is given by

$$y_m(s_j) = \sum_{i=1}^{j} b_{\sigma_{i-1}}(x_{s_i} - x_{s_{i-1}}).$$

Now,

$$E\left\{\left[\sum_{i=1}^{n} b_{\sigma_{i-1}}(x_{s_i} - x_{s_{i-1}})\right]^2\right\} \leqslant \sum_{i=1}^{n} E(b^2_{\sigma_{i-1}})(s_i - s_{i-1})$$

$$= \sum_{\Delta_m} E(b^2_{\tau_{i-1}})(t_i - t_{i-1}).$$

By Lemma 6.1

$$P\left\{\max_{1\leqslant j\leqslant n} |y_m(s_j)| \geqslant l\right\} \leqslant \frac{1}{l^2} \sum_{\Delta_m} E(b^2_{\tau_{i-1}})(t_i - t_{i-1}).$$

Consequently,

$$P\left\{\max_{1\leqslant i\leqslant k} |y_m(\alpha_i)| \geqslant l\right\} \leqslant \frac{1}{l^2} \sum_{\Delta_m} E(b^2_{\tau_{i-1}})(t_i - t_{i-1}).$$

Noting that $y_m(\tau)$ has no moving discontinuity points, we obtain, by letting $k \to \infty$,

$$P\left\{\sup_{0\leqslant\tau\leqslant 1} |y_m(\tau)| \geqslant l\right\} \leqslant \frac{1}{l^2} \sum_{\Delta_m} E(b^2_{\tau_{i-1}})(t_i - t_{i-1}).$$

Finally, by letting $m \to \infty$, we conclude that

$$P\left\{\sup_{0\leqslant t\leqslant 1} \left|\int_0^t b_\tau dx_\tau\right| \geqslant l\right\} \leqslant \frac{1}{l^2} \int_0^t E(b_\tau^2) d\tau.$$

## III. Differential Equations and Integral Equations

**§9.** We intend in this chapter to solve the differential equation

(1) $$Dy_t = G(a(t, y_t), b(t, y_t))$$

under the initial condition

(2) $$y_0 = c.$$

Here $G(\alpha,\beta)$ represents the Gaussian distribution with mean $\alpha$ and standard deviation $\beta$.

Let us first state the theorem.

**Theorem 9.1.** *Suppose $a(t,y)$ and $b(t,y)$ are both continuous in $(t,y)$ and suppose further that there exist constants $A$ and $B$ such that*

(3) $$|a(t,y) - a(t,y')| \leqslant A\,|y - y'|$$
$$|b(t,y) - b(t,y')| \leqslant B\,|y - y'|$$
$$\text{for } 0 \leqslant t \leqslant 1,\ -\infty < y,\ y' < \infty.$$

*Then, for a Brownian motion $x_t$, the equation*

(4) $$y_t = c + \int_0^t a(\tau,y_\tau)d\tau + \int_0^t b(\tau,y_\tau)dx_\tau$$

*has one and only one solution $y_t$ and it satisfies the equation* (1).

## §10. Proof of the Existence of a Solution to the Integral Equation Given Above (the Method of Successive Approximations!)

**Lemma 10.1.** *Let $\{a_t\}$ be a stochastic process having no moving discontinuity points, and suppose that there exists a continuous function $M(t)$ satisfying $E(a_t^2) \leqslant M(t)$. Then, we have*

(1) $$E\left\{\left(\int_t^s a_\tau d\tau\right)^2\right\} \leqslant (s-t)\int_t^s M(\tau)d\tau.$$

*Proof.* We have

(2) $$\left(\int_t^s a_\tau d\tau\right)^2 \leqslant (s-t)\int_t^s a_\tau^2 d\tau$$

and

$$E\left\{\left(\int_t^s a_\tau d\tau\right)^2\right\} \leqslant (s-t)\,E\left\{\int_t^s a_\tau^2 d\tau\right\} \leqslant (s-t)\int_t^s M(\tau)d\tau,$$

where the last inequality can be proved by using Lemma 5.1.

Let us define $y_t^{(k)}$ $(k=1,2,\cdots)$ successively as follows:

(3) $$y_t^{(0)} = c$$

$$(4)\quad y_t^{(k)} = c + \int_0^t a(\tau, y_\tau^{(k-1)})d\tau + \int_0^t b(\tau, y_\tau^{(k-1)})dx_\tau \qquad (k=1,2,\cdots).$$

First, we shall show that for each fixed $t$ $(0 \leqslant t \leqslant 1)$ $y_t^{(k)}(k=1,2,\cdots)$ converges in the mean square.[13] We note that

$$(5)\qquad y_t^{(1)} - y_t^{(0)} = \int_0^t a(\tau,c)d\tau + \int_0^t b(\tau,c)dx_\tau.$$

Since $a(\tau,c)$ and $b(\tau,c)$ are continuous in $\tau$, their absolute values are dominated over the closed interval $0 \leqslant \tau \leqslant 1$ by some finite constant $M$, namely,

$$(6)\qquad |a(\tau,c)| \leqslant M, \quad |b(\tau,c)| \leqslant M \qquad (0 \leqslant \tau \leqslant 1).$$

Consequently, by Lemma 10.1, we obtain from the first inequality of (6)

$$(7)\quad E\left\{ \left(\int_0^t a(\tau,c)d\tau\right)^2 \right\} \leqslant t\int_0^t M^2 d\tau \leqslant M^2t^2 \leqslant M^2 t \qquad (0 \leqslant t \leqslant 1).$$

By using Theorem 5.4, we obtain from the second inequality of (6)

$$(8)\qquad E\left\{ \left(\int_0^t b(\tau,c)dx_\tau\right)^2 \right\} \leqslant \int_0^t M^2 d\tau = M^2 t.$$

From (5), (7), and (8), and utilizing

$$(9)\qquad E\{(x+y)^2\} \leqslant \left(\sqrt{E(x^2)} + \sqrt{E(y^2)}\right)^2, \text{[14]}$$

we conclude that

$$(10)\qquad E\{(y_t^{(1)} - y_t^{(0)})^2\} \leqslant 4M^2 t.$$

Next, let us prove that for each $n$ the following inequalities are valid:

$$(7')\quad E\left\{ \left(\int_0^t (a(\tau,y_\tau^{(n)}) - a(\tau,y_\tau^{(n-1)}))d\tau\right)^2 \right\} \leqslant 4M^2R^{2(n-1)} \frac{t^{n+1}}{(n+1)!}A^2,$$

$$(8')\quad E\left\{ \left(\int_0^t (b(\tau,y_\tau^{(n)}) - b(\tau,y_\tau^{n-1})dx_\tau\right)^2 \right\}$$

[13] Cf. P. Lévy: *Ibid.* p. 52. 3⁰ La convergence en moyenne. We consider the case of $d=2$ here. Namely, the set of all real-valued random variables $x$ with $E(x^2) < \infty$ comprises a complete metric space with respect to the metric $\rho_m(x,y) = \sqrt{E((x-y)^2)}$. The convergence with respect to this metric is called the convergence in the mean square.

[14] This inequality means that the distance function $\rho_m$ described in footnote (13) satisfies the triangle inequality.

$$\leqslant 4M^2R^{2(n-1)}\frac{t^{n+1}}{(n+1)!}B^2,$$

$$(10') \quad E\left\{(y_t^{(n+1)} - y_t^{(n)})^2\right\} \leqslant 4M^2R^{2n}\frac{t^{n+1}}{(n+1)!}, \quad \text{where } R \equiv A + B.$$

If we define $a(\tau, y_\tau^{(-1)}) = 0$, $b(\tau, y_\tau^{(-1)}) = 0$, then (7′), (8′), and (10′) are all valid for $n = 0$ in view of (7), (8), and (10). The proof for the case of general $n$ is done by induction.

So, let us suppose (7′), (8′), and (10′) are valid for $n-1$. Then, from (3) of §9 we obtain

$$|a(\tau,y_\tau^{(n)}) - a(\tau,y_\tau^{(n-1)})| \leqslant A\,|y_\tau^{(n)} - y_\tau^{(n-1)}|$$

$$E\left\{\left[a(\tau,y_\tau^{(n)}) - a(\tau,y_\tau^{(n-1)})\right]^2\right\} \leqslant A^2E\left\{\left[y_\tau^{(n)} - y_\tau^{(n-1)}\right]^2\right\}$$

$$\leqslant 4M^2R^{2(n-1)}\frac{t^n}{n!}A^2.$$

By Lemma 10.1 we have (noting that $0 \leqslant t \leqslant 1$)

$$E\left\{\left(\int_0^t(a(\tau,y_\tau^{(n)}) - a(\tau,y_\tau^{(n-1)}))d\tau\right)^2\right\}$$

$$\leqslant \int_0^t 4M^2R^{2(n-1)}\frac{\tau^n}{n!}A^2d\tau = 4M^2R^{2(n-1)}\frac{t^{n+1}}{(n+1)!}A^2,$$

which shows that (7′) is valid for $n$.

In the same way (8′) can be proved for $n$ by using Theorem 5.4 in place of Lemma 10.1.

From (7′), (8′), and the following identity (∗) we can deduce (10′) by using the inequality (9):

$$(*) \quad y_t^{(n+1)} - y_t^{(n)}$$

$$= \int_0^t(a(\tau,y_\tau^{(n)}) - a(\tau,y_\tau^{(n-1)}))d\tau + \int_0^t(b(\tau,y_\tau^{(n)}) - b(\tau,y_\tau^{(n-1)}))dx_\tau.$$

Now, since it follows from (10′) that

$$\rho_m(y_t^{(n+1)}, y_t^{(n)}) \leqslant \sqrt{4M^2R^{2n}\frac{t^{n+1}}{(n+1)!}}, \quad \text{where } \rho_m(x,y) = \sqrt{E((x-y)^2)},$$[15]

[15] Cf. footnote (13).

and since

$$\sum_{n=0}^{\infty}\sqrt{4M^2R^{2n}\frac{t^{n+1}}{(n+1)!}} < \infty,$$

the sequence

$$(11)\qquad y_t^{(n)} = y_t^{(0)} + (y_t^{(1)} - y_t^{(0)}) + \cdots + (y_t^{(n)} - y_t^{(n-1)})$$

converges in the mean square. If we denote the limit by $y_t$, then $E(y_t^2) < \infty$ and $y_t$ has a well-defined mean value and standard deviation.

Next we shall show that $\{y_t\}$ is a stochastic process without moving discontinuity points. Since $\{y_t^{(n)}\}$ $(n = 1,2, \cdots)$ is, by Lemma 6.1, a sequence of stochastic processes without moving discontinuity points, it suffices to show that the sequence $\{y_t^{(n)}\}$ converges in probability with respect to the strong topology (cf. the proof of Theorem 6.1 where similar arguments were used). First, letting $t = 1$ in (8′) and using Theorem 8.1, we obtain

$$(12)\quad P\left\{\sup_{0\leqslant t\leqslant 1}\left|\int_0^t b(\tau,y_\tau^{(n)}) - b(\tau,y_\tau^{(n-1)})dx_\tau\right| > \frac{1}{2^n}\right\}$$

$$\leqslant 4M^2R^{2(n-1)}\frac{1}{(n+1)!}B^2 2^{2n} \qquad (n = 1, 2, \cdots).$$

Since the infinite series whose general term is given by the right hand side of (12) converges, the following statement holds with probability 1 because of the Borel-Cantelli Lemma: For all sufficiently large $n$

$$(13)\qquad \sup_{0\leqslant t\leqslant 1}\left|\int_0^t(b(\tau,y_\tau^{(n)}) - b(\tau,y_\tau^{(n-1)}))dx_\tau\right| \leqslant \frac{1}{2^n}.$$

Consequently, for any $\eta > 0$, we can choose $n_0(\eta)$ sufficiently large so that

(14) the inequality (13) is valid for all $n \geqslant n_0(\eta)$

everywhere except on a set of probability $\eta$. Furthermore, since $a(\tau,y_\tau^{(n_0)}) - a(\tau,y_\tau^{(n_0-1)})$ has no moving discontinuity points, we have for a sufficiently large $K = K(\eta)$,

$$(15)\qquad |a(\tau,y_\tau^{(n_0)}) - a(\tau,y_\tau^{(n_0-1)})| \leqslant K \qquad (0 \leqslant t \leqslant 1)$$

everywhere except on a set of probability $\eta$. If we denote by $\Omega'$ the set where (14) and (15) are valid simultaneously, we have

$$(16)\qquad P(\Omega') > 1 - 2\eta.$$

If we let $n = n_0$ in the identity $(*)$ above, and use (14) and (15), we can conclude that

$$\text{(17)} \qquad |y_t^{(n_0+1)} - y_t^{(n_0)}| \leqslant Kt + \left(\frac{1}{2}\right)^{n_0} \qquad (0 \leqslant t \leqslant 1)$$

is valid on $\Omega'$.

Next, we let $n = n_0 + 1$ in the identity $(*)$ and use (14), (17), and (3) of §9 to obtain

$$\begin{aligned} |y_t^{(n_0+2)} - y_t^{(n_0+1)}| &\leqslant \int_0^t A|y_t^{(n_0+1)} - y_t^{(n_0)}|\, dt + (1/2)^{n_0+1} \\ &\leqslant AK\frac{t^2}{2} + (1/2)^{n_0}(tA) + (1/2)^{n_0+1}. \end{aligned}$$

If we repeat the same procedure, we obtain

$$\begin{aligned} |y_t^{(n_0+m)} - y_t^{(n_0+m-1)}| &\leqslant A^{m-1}K\frac{t^m}{m!} + \sum_{r=0}^{m-1}\left(\frac{1}{2}\right)^{n_0+r}\frac{(tA)^{m-1-r}}{(m-1-r)!} \\ &= A^{m-1}K\frac{t^m}{m!} + \sum_{r=0}^{m-1}\left(\frac{1}{2}\right)^{n_0+m-1-r}\frac{(tA)^r}{r!} \\ &= A^{m-1}K\frac{t^m}{m!} + \left(\frac{1}{2}\right)^{n_0+m-1}\sum_{r=0}^{m-1}\frac{(2tA)^r}{r!} \\ &\leqslant A^{m-1}K\frac{1}{m!} + \left(\frac{1}{2}\right)^{n_0+m-1} e^{2A} \quad (m=1,2,3,\cdots). \end{aligned}$$

Since the infinite series whose general term is given by the right-hand side above converges, we conclude that

$$\sum_{m=1}^{\infty}(y_t^{(n_0+m)} - y_t^{(n_0+m-1)})$$

converges uniformly in $t$ on the set $\Omega'$. Therefore, the sequence $\{y_t^{(n)}\}$ of stochastic processes converges in probability with respect to the strong topology, and thus its limit $\{y_t\}$ has no moving discontinuity points.

From the definition of $y_t^{(n)}$ it follows also that $y_t^{(n)}$ is a function of $x_{0t}$, and consequently, so are $y_t$, $a(t, y_t)$, and $b(t, y_t)$.

Equation (3) of §9 implies also that since $y_t$ does not have any moving discontinuity points, $a(t, y_t)$ and $b(t, y_t)$ do not have them either.

Furthermore, from the fact that $\{y_t^{(n)}\}$ converges to $\{y_t\}$ in probability with respect to the strong topology it follows that $\int_0^t a(\tau, y_\tau^{(n)})d\tau$ converges in probability to $\int_0^t a(\tau, y_\tau)d\tau$, and by virtue of Theorem 5.5, $\int_0^t b(\tau, y_\tau^{(n)})dx_\tau$ converges in probability to $\int_0^t (b\tau, y_\tau)dx_\tau$. Consequently, $y_t$ satisfies the integral equation (4) of §9 in view of equation (4).

## §11. Proof of the Uniqueness of the Solution of the Integral Equation of §9

$1^0$. Let us suppose that there are two solutions $y'_t$ and $y''_t$ for the case when $|a(\tau,y)|$ and $|b(\tau,y)|$ are bounded (say by $M$) in $0 \leqslant \tau \leqslant 1$, $-\infty < y < \infty$. Then,

(1) $y'_t - y''_t$

$$= \int_0^t (a(\tau,y'_\tau) - a(\tau,y''_\tau))d\tau + \int_0^t (b(\tau,y'_\tau) - b(\tau,y''_\tau))dx_\tau$$

$$E\left\{ \left[\int_0^t (a(\tau,y'_\tau) - a(\tau,y''_\tau))d\tau\right]^2\right\} \leqslant \left(\int_0^t M d\tau\right)^2 \leqslant M^2t^2 < \infty.$$

By Theorem 5.4, we also have

$$E\left\{ \left[\int_0^t (b(\tau,y'_\tau) - b(\tau,y''_\tau))dx_\tau\right]^2\right\} \leqslant \int_0^t M^2 d\tau = M^2 t < \infty.$$

From (1) and the two preceding inequalities we obtain, in view of $E((x+y)^2) \leqslant (\sqrt{E(x^2)} + \sqrt{E(y^2)})^2$,

$$E\{(y'_t - y''_t)^2\} \leqslant (Mt + M\sqrt{t})^2 \leqslant 4M^2 \qquad (0 \leqslant t \leqslant 1) \tag{2}$$

Now, if $E\{(y'_t - y''_t)^2\} \leqslant K(t)$, it then follows from (3) of §9, Lemma 10.1, and Theorem 5.4 that

$$E\left\{ \left[\int_0^t (a(\tau,y'_\tau) - a(\tau,y''_\tau))d\tau\right]^2\right\} \leqslant t\int_0^t A^2K(\tau)d\tau$$

$$\leqslant A^2\int_0^t K(\tau)d\tau,$$

$$E\left\{ \left[\int_0^t (b(\tau,y'_\tau) - b(\tau,y''_\tau))dx_\tau\right]^2\right\} \leqslant \int_0^t B^2K(\tau)d\tau$$

$$= B^2\int_0^t K(\tau)d\tau.$$

From (1) and the preceding two inequalities we obtain

$$E\{(y'_t - y''_t)^2\} \leqslant (A + B)^2\int_0^t K(\tau)d\tau, \tag{3}$$

and from (2) and (3) we conclude that

$$E\{(y'_t - y''_t)^2\} \leqslant (A + B)^2 4M^2 t. \tag{4}$$

By using (4) and (5) again, we obtain

$$E\{(y'_t - y''_t)^2\} \leqslant (A + B)^4 4M^2 \frac{t^2}{2},$$

and repeating, we deduce that

$$(5) \qquad E\{(y'_t - y''_t)^2\} \leqslant 4M^2 \frac{((A+B)^2 t)^n}{n!} \to 0 \quad \text{as } n \to \infty.$$

Therefore, $E\{(y'_t - y''_t)^2\} = 0$ and hence

$$P(y'_t = y''_t) = 1.$$

**2⁰. The General Case.** Since both $|a(t,c)|$ and $|b(t,c)|$ are continuous functions of $t$ they are bounded by some finite number $\alpha$ on $0 \leqslant t \leqslant 1$. Therefore,

$$(6) \qquad \begin{aligned} |a(t,y)| &\leqslant |a(t,c)| + |a(t,y) - a(t,c)| \\ &\leqslant \alpha + A|y - c| \end{aligned}$$

$$(7) \qquad |b(t,y)| \leqslant \alpha + B|y - c|.$$

Let us define $\Omega_K$ by

$$(8) \qquad \Omega_K = \left(\sup_{0 \leqslant \tau \leqslant 1} |y'_\tau - c| < K\right) \cdot \left(\sup_{0 \leqslant \tau \leqslant 1} |y''_\tau - c| < K\right).$$

Then, $\Omega_K$ increases as $K$ increases, and

$$(9) \qquad P(\Omega_K) \to 1 \quad \text{as } K \to \infty.$$

Furthermore, on $\Omega_K$ we have

$$(10) \qquad |a(t,y_t)|,\ |b(t,y_t)| \leqslant 2\alpha + (A+B)K \qquad (0 \leqslant t \leqslant 1).$$

Denote the quantity on the right-hand side above by $M$, and define $b_M(t,y)$ as follows:

$$(11) \qquad \begin{aligned} b_M(t,y) &= M && \text{if } b(t,y) > M \\ &= b(t,y) && \text{if } -M \leqslant b(t,y) \leqslant M \\ &= -M && \text{if } b(t,y) < -M. \end{aligned}$$

Let us now suppose that there are two distinct solutions, and call them $y'_t$ and $y''_t$. Then,

$$(12) \qquad P(y'_t \neq y''_t) > 0.$$

If we can show that (12) leads to a contradiction, then we shall have the uniqueness of the solution. Now,

$$P(\Omega_K \cdot (y'_t \neq y''_t)) \geqslant P(y'_t \neq y''_t) - (1 - P(\Omega_K)). \tag{13}$$

Hence, from (9), (12), and (13) it follows that

$$P(\Omega_K \cdot (y'_t \neq y''_t)) > 0 \tag{14}$$

holds for sufficiently large $K$. But on the set $\Omega_K$ we have

$$b_M(t,y_t) = b(t,y_t),\ a_M(t,y_t) = a(t,y_t) \qquad (0 \leqslant t \leqslant 1),$$

and therefore, both $y'_t$ and $y''_t$ satisfy on $\Omega_K$ the equation

$$y_t = c + \int_0^t a_M(\tau,y_\tau)d\tau + \int_0^t b_M(\tau,y_\tau)dx_\tau. \tag{15}$$

By definition, $a_M(\tau,y_\tau)$ and $b_M(\tau,y_\tau)$ satisfy condition (3) of §9 and their absolute values are dominated by $M$. Therefore, by what was already proved in $1^0$, we see that the solution of (15) is unique. If we call this solution $y_t^{(M)}$, then on $\Omega_K$ we have $y_t^{(M)} = y'_t = y''_t$ except possibly on a subset of probability 0. This says that $P\{\Omega_K \cdot (y'_t \neq y''_t)\} = 0$, which contradicts (14).

## §12. Proof That the Solution to the Integral Equation of § 9 Is a Markoff Process

Denote by $z_s = f(t,\ s,\ \eta,\ \tilde{x}_{ts})$ the solution of

$$z_s = \eta + \int_t^s a(\tau,z_\tau)d\tau + \int_t^s b(\tau,z_\tau)dx_\tau. \tag{1}$$

Here, $\tilde{x}_{ts} = (x_\tau - x_t;\ t \leqslant \tau \leqslant s)$.

Denote by $\tilde{N}_{ts}(\eta)$ the set of $\omega$ for which $f$ above is not well-defined. This set is described by some conditions on $\tilde{x}_{ts}$. Let $y_s$ be the solution obtained in the preceding section, and let

$$\begin{aligned} y'_s &= y_s && \text{for } 0 \leqslant s \leqslant t \\ &= f(t,\ s,\ y_t,\ \tilde{x}_{ts}) && \text{for } s > t\,. \end{aligned} \tag{2}$$

Now, $f(t,\ s,\ y_t,\ \tilde{x}_{ts})$ is not well-defined either when

$$y_t \text{ is not well-defined} \tag{3}$$

or when

$$\omega \text{ belongs to } \bigcup_\eta (y_t = \eta)(\tilde{x}_{ts} \in \tilde{N}_{ts}(\eta)). \tag{4}$$

But the probability of the set described in (3) is obviously 0, while the probability of the set in (4) is given by

(5) $$\int_{-\infty}^{\infty} F_{y_t}(d\eta)P(\tilde{N}_{ts}(y_t) \mid y_t = \eta),$$

where $F_y$ denotes the probability distribution function of $y$. Since $y_t$ is a function of $x_{0t}$, $\tilde{x}_{ts}$ and $y_t$ are independent. Consequently, $P(\tilde{N}_{ts}(y_t) \mid y_t = \eta) = P(\tilde{N}_{ts}(\eta)) = 0$. This implies that the integral (5) is also equal to 0, and hence the probability of the set in (4) is 0.

If $s \leqslant t$, we have

(6) $$\begin{aligned} y'_t = y_s &= c + \int_0^s a(\tau, y_\tau)d\tau + \int_0^s b(\tau, y_\tau)dx_\tau \\ &= c + \int_0^s a(\tau, y'_\tau)d\tau + \int_0^s b(\tau, y'_\tau)dx_\tau, \end{aligned}$$

while if $s > t$, we have

(7) $$\begin{aligned} y'_s &= f(t, s, y_t, \tilde{x}_{ts}) \\ &= y_t + \int_t^s a(\tau, y'_\tau)d\tau + \int_t^s b(\tau, y'_\tau)dx_\tau \quad \text{(Definition of } f!). \end{aligned}$$

Substituting into (7) the equation (6) with $s = t$, we obtain

(8) $$y'_s = c + \int_0^s a(\tau, y'_\tau)d\tau + \int_0^s b(\tau, y'_\tau)dx_\tau.$$

Therefore, $\{y'_s\}$ coincides with the solution $\{y_s\}$ of the integral equation of §9, and this implies that $y_s = f(t, s, y_t, \tilde{x}_{ts})$, so

$$F_{y_s \mid x_{0t} = \xi_{0t}} = F_{f(t,s,y_t,\tilde{x}_{ts}) \mid x_{0t} = \xi_{0t}} = F_{f(t,s,y_t(\xi_{0t}),\tilde{x}_{ts})}$$

(we obtain the last identity since $f(t,s,y_t(\xi_{0t}), \tilde{x}_{ts})$ is a function of $\tilde{x}_{ts}$ and hence is independent of $x_{0t}$). This says that $F_{y_s \mid x_{0t} = \xi_{0t}}$ depends only on $y_t(\xi_{0t})$. Since $y_{0t}$ is a function of $x_{0t}$, $F_{y_s \mid y_{0t} = \eta_{0t}}$ also depends only on $\eta_t$.

## §13. Proof That the Solution of the Integral Equation of §9 Satisfies the Differential Equation (1) of §9

Let

(1) $$y_t = c + \int_0^t a(\tau, y_\tau)d\tau + \int_0^t b(\tau, y_\tau)dx_\tau.$$

If we consider the conditional probability distribution of the solution $y_t$ of equation (1) for $t \geqslant t_0$ given that $y_{t_0} = \eta$, then from what was discussed in the beginning of the preceding section it follows that this conditional probability distribution coincides with the (unconditional) probability distribution of the solution for the equation

(2) $$y_t = \eta + \int_{t_0}^t a(\tau, y_\tau)d\tau + \int_{t_0}^t b(\tau, y_\tau)dx_\tau \qquad (t \geqslant t_0).$$

Now,

$$\begin{aligned} y_t - y_{t_0} &= y_t - \eta \\ &= \int_{t_0}^t a(\tau, y_\tau)d\tau + \int_{t_0}^t b(\tau, y_\tau)dx_\tau \\ &= \int_{t_0}^t a(\tau, y_{t_0})d\tau + \int_{t_0}^t b(\tau, y_{t_0})dx_\tau \\ &\quad + \int_{t_0}^t (a(\tau, y_\tau) - a(\tau, y_{t_0}))d\tau \\ &\quad + \int_{t_0}^t (b(\tau, y_\tau) - b(\tau, y_{t_0}))dx_\tau \end{aligned}$$

and hence

$$(2') \qquad y_t - y_{t_0} = \int_{t_0}^t a(\tau, y_{t_0})d\tau + \int_{t_0}^t b(\tau, y_{t_0})dx_\tau$$

$$+ \int_{t_0}^t (a(\tau, y_\tau) - a(\tau, y_{t_0}))d\tau + \sum_{\Delta}^{t}(b(\tau_{i-1}, y_{\tau_{i-1}}) - b(\tau_{i-1}, y_{t_0}))(x_{t_i} - x_{t_{i-1}})$$

$$+ \left\{ \int_{t_0}^t (b(\tau, y_\tau) - b(\tau, y_{t_0}))dx_\tau - \sum_{\Delta}^{t}(b(\tau_{i-1}, y_{\tau_{i-1}}) - b(\tau_{i-1}, y_{t_0}))(x_{t_i} - x_{t_{i-1}}) \right\}$$

where the sum $\sum_{\Delta}^{t}$ is defined just as $y_\Delta(t)$ was defined in (1) of §6. If we denote by $\gamma_t$ the quantity appearing inside of { } in (2)′, and if we choose $d(\Delta)$ sufficiently small, then

$$(3) \qquad P\{|\gamma_t| \geqslant \epsilon(t - t_0)\} < \epsilon(t - t_0).$$

Now, since the solution of (1) should be obtained by the method of successive approximations of §10, letting $M$ be the maximum value for $s_0 \leqslant \tau \leqslant 1$ of the quantities $|a(\tau,\eta)| \cdot |b(\tau,\eta)|$, we see that

$$E\{(y_t - \eta)^2\} \leqslant \left( \sum_{n=0}^{\infty} \sqrt{4M^2R^{2n}\frac{(t-t_0)^{n+1}}{(n+1)!}} \right)^2, \quad \text{where } R = A+B$$

$$\leqslant K(t - t_0), \quad \text{where } K \equiv \sum_{n=0}^{\infty} \sqrt{4M^2R^{2n}\frac{1}{(n+1)!}} < \infty,$$

namely, $E\{(y_t - y_{t_0})^2\} \leqslant K(t-t_0)$. Therefore

$$\begin{aligned} E\{(a(\tau, y_\tau) - a(\tau, y_{t_0}))^2\} &\leqslant A^2E\{(y_t - y_{t_0})^2\} \\ &\leqslant A^2K\,(t - t_0), \end{aligned}$$

from which it follows that

$$(4)\quad E\left\{\left[\int_{t_0}^{t}(a(\tau,y_\tau)-a(\tau,y_{t_0}))d\tau\right]^2\right\}$$

$$\leqslant (t-t_0)\int_{t_0}^{t}KA^2(t-t_0)dt=\frac{A^2K}{2}(t-t_0)^3=o(t-t_0).$$

We also have

$$(5)\quad \left|E\left\{\int_{t_0}^{t}(a(\tau,y_\tau)-a(\tau,y_{t_0}))d\tau\right\}\right|$$

$$\leqslant \sqrt{E\left\{\int_{t_0}^{t}(a(\tau,y_\tau)-a(\tau,y_{t_0}))d\tau)^2\right\}}$$

$$\leqslant \frac{\sqrt{A^2K}}{2}(t-t_0)^{3/2}=o(t-t_0).$$

From (4) and (5) we obtain

$$(5')\quad \sigma\left\{\int_{t_0}^{t}(a(\tau,y_\tau)-a(\tau,y_{t_0}))d\tau\right\}=o(\sqrt{t-t_0}).$$

Next, from $E\{(y_t-y_{t_0})^2\}\leqslant K(t-t_0)$ we obtain

$$E\{(b(t,y_t)-b(t,y_{t_0}))^2\}\leqslant B^2K(t-t_0).$$

Therefore, the mean of $b(t,y_t)-b(t,y_{t_0})$ exists also, and

$$(6)\quad E\left\{\sum_{\Delta}^{t}(b(\tau_{i-1},y_{\tau_{i-1}})-b(\tau_{i-1},y_{t_0}))(x_{t_i}-x_{t_{i-1}})\right\}$$

$$=\sum_{\Delta}^{t}E(b(\tau_{i-1},y_{\tau_{i-1}})-b(\tau_{i-1},y_{t_0}))E(x_{t_i}-x_{t_{i-1}})$$

$$=0;$$

$$(7)\quad E\left\{\left[\sum_{\Delta}^{t}(b(\tau_{i-1},y_{\tau_{i-1}})-b(\tau_{i-1},y_{t_0}))(x_{t_i}-x_{t_{i-1}})\right]^2\right\}$$

$$\leqslant \sum_{\Delta}^{t}B^2K(\tau_i-t_0)(t_i-t_{i-1})$$

$$\leqslant \int_{t_0}^{t} B^2K(\tau - t_0)d\tau \quad \text{(note that } t_0 \leqslant \tau_i \leqslant t_{i-1} \leqslant t_i)$$

$$= \frac{B^2K}{2}(t - t_0)^2 = o(t - t_0).$$

From (6) and (7) we obtain that

$$(7') \qquad \sigma\left[\sum_{\Delta}^{t}(b(\tau_{i-1}, y_{\tau_{i-1}}) - b(\tau_{i-1}, y_{t_0}))(x_{t_i} - x_{t_{i-1}})\right] = o(\sqrt{t-t_0}).$$

From (2′), (3), (4), (5′), (6), and (7′) it follows that $Dy_{t_0}$ equals the derivative $Dz_{t_0}$ of

$$(8) \qquad z_t = y_{t_0} + \int_{t_0}^{t} a(\tau, y_{t_0})d\tau + \int_{t_0}^{t} b(\tau, y_{t_0})dx_\tau.$$

But

$$F_{z_t - z_{t_0}} = G\left(\int_{t_0}^{t} a(\tau, y_{t_0})d\tau, \sqrt{\int_{t_0}^{t}(b(\tau, y_{t_0}))^2 d\tau}\right).$$

Therefore,

$$F_{z_t - z_{t_0}}^{*[1/t-t_0]} = G\left(\left[\frac{1}{t-t_0}\right]\int_{t_0}^{t} a(\tau, y_{t_0})d\tau, \sqrt{\left[\frac{1}{t-t_0}\right]\int_{t_0}^{t}(b(\tau, y_{t_0}))^2 d\tau}\right)$$

Now,

$$\lim_{t\to t_0}\left[\frac{1}{t-t_0}\right]\int_{t_0}^{t} a(\tau, y_{t_0})d\tau$$

$$= \lim_{t\to t_0}\frac{1}{t-t_0}\int_{t_0}^{t} a(\tau, y_{t_0})d\tau = a(t_0, y_{t_0}).$$

Similarly,

$$\lim_{t\to t_0}\left[\frac{1}{t-t_0}\right]\cdot\int_{t_0}^{t}(b(\tau, y_{t_0}))^2 d\tau = (b(t_0, y_{t_0}))^2.$$

Therefore,

$$Dy_{t_0} = Dz_{t_0} = G(a(t_0, y_{t_0}), b(t_0, y_{t_0})). \qquad \text{q.e.d.}$$

## References

[1] A. Kolmogoroff: Grundbegriffe der Wahrscheinlichkeitsrechnung, p. 12.

[2] M. Fréchet: Recherches théoriques modernes sur la théorie des probabilités, tom 2, p. 215.

[3] A. Kolmogoroff: Über die analytischen Methoden in der Wahrscheinlichkeitsrechnung (*Math. Ann.* **104**, p. 415).

[4] W. Feller: Zur Theorie der stochastischen Prozesse (Existenz und Eindeutigkeitssätze) (*Math. Ann.* **113**, p. 113).

[5] J. L. Doob: Stochastic processes depending on a continuous parameter (*Trans. Am. Math. Soc.* **42** (1937)); Stochastic processes with an integral-valued parameter (*ibid* **44** (1938)).

[6] Article #1033 (this journal, vol **234**) §2.

[7] Article #1033 (this journal, vol **234**) §4.

# 13. *On the Ergodicity of a Certain Stationary Process**.

By Kiyosi ITÔ.

Mathematical Institute, Nagoya Imperial University.

(Comm. by S. KAKEYA, M.I.A., Feb. 12, 1944.)

Let $x_t(\omega)$ be any strictly stationary process[1]. The probability law of the process $x_t(\omega)$ is a probability distribution on $R^R$ which is invariant by the mapping $T_\tau$ that transforms $f(t) \in R^R$ into $f(t+\tau) \in R^R$ for any $\tau$. We shall say that the process $x_t(\omega)$ is ergodic in the (strongly) mixing type if it is the case with the group of the measure-preserving mappings $\{T_\tau\}$[2]. We shall establish the

*Theorem. Let $x_t(\omega)$ be any strictly*[3] *stationary process of Gaussian type*[4] *with the correlation function $\rho(\tau) \equiv \int_{-\infty}^{\infty} e^{i\lambda\tau} F(d\lambda)$*[5]. *The sufficient condition that $x_t(\omega)$ should be ergodic in the (strongly) mixing type is that the spectral measure $F$ is absolutely continuous.*

*Proof.* It is sufficient to show the identity:

$$(1) \qquad \lim_{\tau\to\infty} P\{(x_{s_1}, x_{s_2}, \ldots, x_{s_m}) \in E_m, (x_{t_1+\tau}, x_{t_2+\tau}, \ldots, x_{t_n+\tau}) \in E_n\}$$

or

$$\lim_{\tau\to\infty} P\{(x_{s_1}, x_{s_2}, \ldots, x_{s_m}, x_{t_1+\tau}, x_{t_2+\tau}, \ldots, x_{t_n+\tau}) \in E_m \otimes E_n\}$$

$$= P\{(x_{s_1}, x_{s_2}, \ldots, x_{s_m}) \in E_m\} P\{(x_{t_1}, x_{t_2}, \ldots, x_{t_n}) \in E_n\}$$

where $E_m$ and $E_n$ are any bounded Borel sets respectively in $R^m$ and in $R^n$ and $s_1 < s_2 < \cdots < s_m$, $t_1 < t_2 < \cdots < t_n$. We may assume $\mathcal{E}(x_t)=0$ and $\mathcal{E}(x_t^2)=1$ with no loss of generality.

If $u_i$, $i=1, 2, \ldots, p$, are all different, the matrix $\{\rho(u_i-u_j)\,;\, i, j = 1, 2, \ldots, p\}$ is strictly positive definite, that is $\sum_{i,j} \rho(u_i-u_j)\xi_i\bar{\xi}_j > 0$ for any system $\xi_i$, $i=1, 2, \ldots, p$, such that $\sum_i |\xi_i|^2 \neq 0$. In fact we have

$$(2) \qquad \sum_{i,j} \rho(u_i-u_j)\xi_i\bar{\xi}_j = \int_{-\infty}^{\infty} |\sum_k e^{i\lambda u_k}\xi_k|^2 F(d\lambda) \geqq 0 .$$

If the last equality holds, we shall have $\sum_k e^{i\lambda u_k}\xi_k = 0$ for any spectrum of $F$. Since $F$ is absolutely continuous, the set of all the spectra of $F$ has accumulation points $\neq \infty$. Therefore $\sum_k e^{i\lambda u_k}\xi_k$, as an integral

---

* The cost of this research has been defrayed from the Scientific Research Expenditure of the Department of Education.

1) Cf. A. Khintchine: Korrelationstheorie der stationären stochastischen Prozesse (Math. Ann. 109).

2) Cf. E. Hopf: Ergodentheorie (Erg. d. Math.) 1937, p. 36, Def. 11.1.

3) The condition "strictly" can be omitted since any weakly stationary process of Gaussian type is strictly stationary.

4) Cf. A. Khintchine, loc. cit. 1), the remark at the end of § 2.

5) The correlation function of any stationary process can be always expressible in this form. Cf. A. Khintchine, loc. cit. 1).

function of $\lambda$, is identically equal to 0. Thus we should have $\xi_k=0$, $k=1, 2, \dots, p$, contrary to the assumption. Therefore we have $\sum_{i,j} \rho(u_i-u_j)\xi_i\bar{\xi}_j > 0$.

The probability law of $(x_{s_1}, x_{s_2}, \dots, x_{s_m}, x_{t_1+\tau}, x_{t_2+\tau}, \dots, x_{t_n+\tau})$ is a normalized $(m+n)$-dimensional Gaussian distribution with the correlation matrix:

$$M(\tau) \equiv \begin{pmatrix} S & R^*(\tau) \\ R(\tau) & T \end{pmatrix}, \tag{3}$$

where $S$ and $T$ are respectively the correlation matrices of $(x_{s_1}, x_{s_2}, \dots, x_{s_m})$ and of $(x_{t_1}, x_{t_2}, \dots, x_{t_n})$ and the elements $r_{ij}(\tau)$, $i=1, 2, \dots, n$, $j=1, 2, \dots, m$ of $R(\tau)$ are equal to $\rho(t_i-s_j+\tau)$, and $R^*(\tau)$ is the transposed matrix of $R(\tau)$.

We may suppose that $s_1, s_2, \dots, s_m$, $t_1+\tau, t_2+\tau, \dots, t_n+\tau$ are all different for a sufficiently large $\tau$. So $M(\tau)$ is a strictly positive definite matrix. The probability law of $(x_{s_1}, x_{s_2}, \dots, x_{s_m}, x_{t_1+\tau}, x_{t_2+\tau}, \dots, x_{t_n+\tau})$ is

$$\frac{1}{(2\pi)^{\frac{m+n}{2}}\sqrt{\text{Det.}\,M(\tau)}} e^{-\frac{1}{2}(M(\tau)^{-1}\xi,\,\xi)} \qquad (\xi \in R^{m+n}). \tag{4}$$

As $F$ is absolutely continuous, $r_{ij}(\tau) \equiv \int_{-\infty}^{\infty} e^{i\lambda(t_i-s_j+\tau)}F(d\lambda)$ tends to 0 on account of the Riemann-Lebesgue theorem. Therefore we have

$$M(\tau) \to \begin{pmatrix} S & 0 \\ 0 & T \end{pmatrix}, \qquad M(\tau)^{-1} \to \begin{pmatrix} S^{-1} & 0 \\ 0 & T^{-1} \end{pmatrix}. \tag{5}$$

since Det. $T$ and Det. $S$ do not vanish. Therefore the expression (4) converges to

$$\frac{1}{(2\pi)^{\frac{m}{2}}\sqrt{\text{Det.}\,S}} e^{-\frac{1}{2}(S^{-1}\eta,\,\eta)} \frac{1}{(2\pi)^{\frac{n}{2}}\sqrt{\text{Det.}\,T}} e^{-\frac{1}{2}(T^{-1}\zeta,\,\zeta)} \tag{6}$$

uniformly as far as $(\eta, \zeta)$ runs over a bounded region in $R^{m+n}$. The factors in (6) are clearly the probability laws of $(x_{s_1}, x_{s_2}, \dots, x_{s_m})$ and of $(x_{t_1}, x_{t_2}, \dots, x_{t_n})$ respectively. Thus the identity (1) can be deduced at once.

# 28. *A Kinematic Theory of Turbulence**.

By Kiyosi ITÔ.
Mathematical Institute, Nagoya Imperial University.
(Comm. by S. KAKEYA, M.I.A., March 13, 1944.)

1. *Generalities.* In the theory of turbulence[1] the deviation of the velocity from its mean may be considered as a system of random vectors $u_\lambda(t, \mathfrak{x}, \omega)$, $\lambda=1,2,3$, where $t(\in R)$ is the time parameter and $\mathfrak{x}(\in R^3)$ denotes the position and $\omega\big(\in(\Omega, P)\big)$ is the elementary event. Then we have

$$(1) \qquad \mathcal{E}_\omega\big(u_\lambda(,\mathfrak{x},\omega)\big)=0\,.$$

When the system $\{u_\lambda(t,\mathfrak{x},\omega)\}$ is of Gaussian type[2], we say that the turbulence is of Gaussian type.

Now we define the moment tensor of the turbulence by

$$(2) \qquad R_{\lambda\mu}(t, \mathfrak{x}\,;\, s, \mathfrak{y})=\mathcal{E}_\omega\{u_\lambda(t, \mathfrak{x}, \omega)u_\mu(s, \mathfrak{y}, \omega)\}\,.$$

Then $R_{\lambda\mu}(t, \mathfrak{x}\,;\, s, \mathfrak{y})$ is a positive-definite function of $(\lambda, t, \mathfrak{x})$ and $(\mu, s, \mathfrak{y})$ in the sense of Bochner, namely we have

$$(3) \qquad R_{\lambda\mu}(t, \mathfrak{x}\,;\, s, \mathfrak{y})=R_{\mu\lambda}(s, \mathfrak{y}\,;\, t, \mathfrak{x}) \quad \text{and}$$

$$(4) \qquad \textstyle\sum_{ij}\xi_i\xi_j R_{\lambda_i\lambda_j}(t_i, \mathfrak{x}_i\,;\, t_j, \mathfrak{x}_j)\geqq 0\,;$$

in fact (3) is evident by (2) and the left side of (4) is equal to $\mathcal{E}_\omega\Big\{\big(\sum_i\xi_i u_{\lambda_i}(t_i, \mathfrak{x}_i, \omega)\big)^2\Big\}$. Conversely the function $R_{\lambda\mu}(t, \mathfrak{x}\,;\, s, \mathfrak{y})$ satisfying (3) and (4) may be considered as the moment tensor of a turbulence of Gaussian type[3].

A turbulence is defined as temporally homogeneous, if its moment tensor satisfies

$$(5) \qquad R_{\lambda\mu}(t+\tau, \mathfrak{x}\,;\, s+\tau, \mathfrak{y})=R_{\lambda\mu}(t, \mathfrak{x}\,;\, s, \mathfrak{y})\,.$$

It is defined as spatially homogeneous, if we have

$$(6) \qquad R_{\lambda\mu}(t, \mathfrak{x}+\mathfrak{a}\,;\, s, \mathfrak{y}+\mathfrak{a})=R_{\lambda\mu}(t, \mathfrak{x}\,;\, s, \mathfrak{y})\,.$$

We say that it is isotopic if we have always

$$(7) \qquad \sum_{\lambda'\mu'} k_{\lambda'\lambda}k_{\mu'\mu}R_{\lambda'\mu'}\big(t, \mathfrak{x}\,;\, s, \mathfrak{x}+K(\mathfrak{y}+\mathfrak{x})\big)=R_{\lambda\mu}(t, \mathfrak{x}\,;\, s, \mathfrak{y})$$

for any orthogonal transformation $K\equiv\{k_{\lambda\mu}\,;\, \lambda, \mu=1,2,3\}$. We can easily prove by (3) that the isotropism implies the homogenuity.

* The cost of this research has been defrayed from the Scientific Expenditure of the Department of Eduction.

1) H. P. Robertson: The invariant theory of isotropic turbulence, Proc. Cambr. Phil. Soc. 36, 1940.

2) Cf. K. Itô: ガウス型確率變數系ニツイテ (全國紙上數學談話會第 261 號).

3) See Theorem 3 in my above-cited note (2).

It seems to be an important and perhaps difficult problem to determine the canonical form of $R_{\lambda\mu}(t, \mathfrak{X}; s, \mathfrak{Y})$ which satisfies (3), (4), (5) and (7).

2. *The temporally homogeneous and isotropic turbulence at a point.* For the investigation of this subject we can consider $u_\lambda(t, \omega)$ and $R_{\lambda\mu}(t, s)$ respectively instead of $u_\lambda(t, \mathfrak{X}, \omega)$ and $R_{\lambda\mu}(t, \mathfrak{X}; s, \mathfrak{Y})$. By (3) and (4) we have

$$(3') \quad R_{\lambda\mu}(t, s) = R_{\mu\lambda}(s, t) \quad \text{and} \quad (4') \quad \sum_{ij} \xi_i \xi_j R_{\lambda_i' \lambda_j'}(t_i, t_j) \geqq 0 .$$

The conditions (5) and (7) may be written in the forms:

$$(5') \quad R_{\lambda\mu}(t+\tau, s+\tau) = R_{\lambda\mu}(t, s) \quad \text{and} \quad (6') \quad \sum_{\lambda'\mu'} k_{\lambda'\lambda} k_{\mu'\mu} R_{\lambda'\mu'}(t, s) = R_{\lambda\mu}(t, s) .$$

Theorem 1. A necessary and sufficient condition that $R_{\lambda\mu}(t, s)$ should be the moment tensor of a temporally homogeneous and isotropic turbulence at a point is that $R_{\lambda\mu}(t, s)$ is expressible by the form:

$$(8) \qquad R_{\lambda\mu}(t, s) = \delta_{\lambda\mu} \int_{-0}^{\infty} \cos\big(\xi(t-s)\big) F(d\xi) ,$$

where $\delta_{\lambda\mu}$ is the Kronecker's delta and $F$ is a measure distribution on $[0, \infty)$ with the finite total measure.

Proof. Necessity. The isotropism (6′) implies that $R_{\lambda\mu}(t, s)$ is an invariant tensor. Therefore we obtain $R_{\lambda\mu}(t, s) = \delta_{\lambda\mu} C(t, s)$. From the temporal homogenuity (5′) and the symmetric character (3′) follows that $C(t, s)$ is a function of $|t-s|$ only, say $C_1(|s-t|)$. Now we see by (4′) that $C(|\tau|)$ is a positive-definite function of $\tau$. Making use of the Bochner's theorem we obtain (8). The sufficiency is evident.

According to this theorem $u_\lambda(t, \omega)$ and $u_\mu(s, \omega)$ $(\lambda \neq \mu)$ are non-correlated in this turbulence. Therefore, if we assume further that the turbulence be of Gaussian type, the three stochastic processes $\big(u_\lambda(t, \omega);\ -\infty < t < \infty\big)$, $\lambda = 1, 2, 3$, will become independent. Nevertheless each process is clearly a stationary process of Gaussian type with the correlation function $\rho(\tau) \equiv \int_{-0}^{\infty} \cos \tau\xi F(d\xi) / F\big([0 \infty)\big)$. In this case the problem may be reduced to the investigation of such a process.

Next we mention a theorem concerning the ergodicity of this process, which includes the result[1] obtained before by the author; the proof can be achieved by the same idea and so will be omitted.

Theorem 2. A necessarity and sufficient condition that a normalized continnous (in mean) stationary process $u(t, \omega)$ of Gaussian type should be ergodic in the strongly mixing type is that its correlation function $\rho(\tau)$ satisfies

$$(9) \qquad \lim_{\tau \to \infty} \rho(\tau) = 0 .$$

The condition (9) means that the correlation coefficient tends to 0

1) See K. Itô: On the ergodicity of a certain stationary process, Proc. **20** (1944), 54–55.

as the time interval increases indefinitely. In practice we may assume that it is well satisfied. Then by Theorem 2 we can see that

$$(10)\qquad P\left\{\omega\,;\ \lim_{T\to\infty}\frac{1}{T}\int_0^T u(t,\omega)u(t+\tau,\omega)\,dt=\rho(\tau)\right\}=1\,.$$

This identity justifies the practical method in which we make use of the time-mean of $u(t,\omega)u(t+\tau,\omega)$ at a certain (realized) value of $\omega$ instead of its mathematical expectation $\rho(\tau)$. It is also the case with turbulence.

東東 3404
株式會社ヘラルド社

# 43. *On the Normal Stationary Process with no Hysteresis.*

By Kiyosi ITÔ.
Mathematical Institute, Nagoya Imperial University.
(Comm. by S. KAKEYA, M.I.A., April 12, 1944.)

§ 1. Let $(\Omega, P)$ be a probability field and $\mathfrak{M}$ a system of real valued random variables. $\mathfrak{M}$ is called to be *normal* or *of the Gaussian type*, if, for any $x_i(\omega) \in \mathfrak{M}$, $i=1, 2, \dots, n$, the random variable $\big(x_1(\omega), x_2(\omega), \dots, x_n(\omega)\big)$ is subjected to an $n$-dimensional (sometimes perhaps degenerated) Gaussian distribution. *This condition is equivalent to the property that, for any $x_i(\omega) \in \mathfrak{M}$, and for any real $a_i$, $\sum_{i=1}^{n} a_i x_i(\omega)$ is normally distributed.* Let $x_i(\omega)$, $i=1, 2, \dots, m$, $y_j(\omega)$, $j=1, 2, \dots, n$, be elements in a normal system $\mathfrak{M}$. *Then the non-correlatedness of $x_i(\omega)$ and $y_j(\omega)$ for any, $i, j$, $1 \leq i \leq m$, $1 \leq j \leq n$, implies the independence of $\big(x_1(\omega), x_2(\omega), \dots, x_m(\omega)\big)$ and $\big(y_1(\omega), y_2(\omega), \dots, y_n(\omega)\big)$.*

Let $x(t, \omega)$, $-\infty < t < \infty$, be a stochastic process. If the system of $x(t, \omega)$, $-\infty < t < \infty$, is normal, then the process will be said to be *normal*. If the (conditional) probability law of $x(t, \omega)$ under the condition that $x(t_1, \omega), x(t_2, \omega), \dots, x(t_n, \omega)$ should be given depends only on the value $x(t_n, \omega)$ for any $t_1 < t_2 < \dots < t_n < t$, we say that $x(t, \omega)$ *has no hysteresis* or *is a simple Markoff process.* This terminology is applied to the case of a stochastic sequence $x(k, \omega)$, $k=0, \pm 1, \pm 2, \dots$.

In this investigation the author owes much to the suggestions given by Mr. T. Kitagawa and Mr. M. Ogawara.

§ 2. *The form of the correlation function.*

*Theorem 1. Let $x(k, \omega)$ be a normal stationary (in the sense of A. Khintchine) sequence. A necessary and sufficient condition that $x(k, \omega)$ should have no hysteresis is that its correlation function $\rho(k)$ is of the form $a^k$, $-1 \leq a \leq 1$.*

*Proof.* In order to avoid trivial complications we assume $E_\omega\big(x(k, \omega)\big)=0$, and $E_\omega\big(x(k, \omega)^2\big)=1$. If we define $\big(f(\omega), g(\omega)\big)$ by $E_\omega\big(f(\omega)g(\omega)\big)$, the closed linear subspace determined by the set $x(k, \omega)$, $k=0, \pm 1, \pm 2, \dots$, is considered as a Hilbert space, where *orthogonality implies (stochastic) independence* (Cf. § 1).

*Sufficiency.* For the proof it is sufficient to show that the conditional probability law of $x(k, \omega)$ under the condition that $x(k-i, \omega) = \xi_i$, $i=1, 2, \dots, n$, depends only on $\xi_1$.

We put $y(k, \omega)=x(k, \omega)-ax(k-1, \omega)$. Then we have $E_\omega\big(y(k, \omega)x(k-i, \omega)\big)=a^i-aa^{i-1}=0$, $i=1, 2, \dots$. Since the sequence is normal, $y(k, \omega)$ is independent of $\big(x(k-1, \omega), x(k-2, \omega), \dots, x(k-n, \omega)\big)$. Therefore the probability law of $y(k, \omega)$ is invariant even if we add the condition: $x(k-i, \omega)=\xi_i$, $i=1, 2, \dots, n$. Therefore the probability law of $x(k,\omega)$ i.e. of $ax(k-1,\omega)+y(k,\omega)$ under the same condition depends only on $\xi_1$.

*Necessity.* We fix a natural number $n$ arbitrarily. Let $L$ denote the linear manifold determined by the set $x(-i,\omega)$, $i=1,2,\dots,n$ and $\sum_{i=1}^{n} a_i x(-i,\omega)$ the orthogonal projection of $x(o,\omega)$ into $L$. Now we put

$$(1) \qquad x(o,\omega)=\sum_{i=1}^{n} a_i x(-i,\omega)+y(\omega).$$

$y(\omega)$, being orthogonal to $L$, is independent of $\big(x(-1,\omega), x(-2,\omega),\dots, x(-n,\omega)\big)$. We consider two conditions (A) $x(-i,\omega)=a_i$, $i=1,2,\dots,n$, (B) $x(-1,\omega)=a_1$, $x(-i,\omega)=0$, $i=2,3,\dots,n$. The expectation of $x(0\,\omega)$ is equal to $\sum_{i=1}^{n} a_i^2$ under (A), while it is $a_1^2$ under (B). Since the sequence has no hysteresis the two values must be coincident, from which follows that $a_i=0$, $i=2,3,\dots,n$. Thus the identity (1) becomes $x(o,\omega)=a_1x(-1,\omega)+y(\omega)$. Therefore we have

$$(2) \qquad \begin{aligned}\rho(1)&=E_\omega\big(x(o,\omega)x(-1,\omega)\big)\\ &=a_1E_\omega\big(x(-1,\omega)^2\big)+E_\omega\big(y(\omega)x(-1,\omega)\big)=a_1\end{aligned}$$

Consequently we can see that $x(o,\omega)-\rho(1)x(-1,\omega)$, being equal to $y(\omega)$, is independent of $x(-n,\omega)$, i.e. that $E_\omega\Big(x(-n,\omega)\big(x(o,\omega)-\rho(1)x(-1,\omega)\big)\Big)=0$, and so that $\rho(n)-\rho(1)\rho(n-1)=0$. Since $\rho(0)=1$, we have $\rho(n)=\alpha^n$, $n=0,1,2,\dots$, where $\alpha=\rho(1)$, and so $-1\leqq\alpha\leqq 1$. $\rho(n)$, as an even function, is equal to $\alpha^{|n|}$ for any integer $n$.

*Theorem 2. Let $x(t,\omega)$ be any normal stationary continuous (in the strong topology in $L^2(\Omega,P)$) process. The necessary and sufficient condition that it should have no hysteresis is that its correlation function $\rho(\tau)$ is of the form $e^{-a|\tau|}$, $a\geqq 0$.*

*Proof.* We prove only the necessity, since the sufficiency can be shown in the same manner as before. By Theorem 1 we have clearly $\rho(\tau)=\rho\left(\frac{\tau}{2}\right)^2>0$ and $\rho\left(\frac{m}{n}\right)=\rho(1)^{\left|\frac{m}{n}\right|}=e^{-a\left|\frac{m}{n}\right|}$ $\big(a=-\log\rho(1)\geqq 0\big)$. By the continuity of the process we obtain that of $\rho(\tau)$ and so $\rho(\tau)=e^{-a|\tau|}$.

§ 3. *The form of the sequence (or process).*

*Theorem 3. Any normal stationary sequence with no hysteresis is expressible in one of the following three forms and its converse is also true.*

$$(3) \qquad x(k,\omega)=x(\omega), \qquad k=0,\pm1,\pm2,\dots$$

$$(4) \qquad x(k,\omega)=(-1)^k x(\omega), \qquad k=0,\pm1,\pm2,\dots$$

*In these cases $x(\omega)$ denotes a normally distributed random variable.*

$$(5) \qquad x(k,\omega)=\sum_{n=-\infty}^{k} \alpha^{k-n}y(n,\omega), \qquad k=0,\pm1,\pm2,\dots,$$

*where $-1<\alpha<1$ and $y(k,\omega)$, $k=0,\pm1,\pm2$, is an independent* *sequence of random variables subjected to the same Gaussian distribution and is expressible by* $x(k, \omega)$ *as follows:*

$$(5') \qquad y(k, \omega) = x(k, \omega) - \alpha x(k-1, \omega).$$

*Proof.* We may assume $E_\omega\big(x(k, \omega)\big)=0$ and $E_\omega\big(x(k, \omega)^2\big)=1$ with no loss of generality. By Theorem 1 the correlation function $\rho(k)$ is equal to $\alpha^k$, $-1 \leqq \alpha \leqq 1$. In case $\alpha=1$, $x(k, \omega)$ is of the form (3), since $E_\omega\Big(\big(x(k, \omega)-x(0, \omega)\big)^2\Big)=2-2\rho(k)=0$. Similarly we obtain (4) in case $\alpha=-1$.

Suppose that $-1<\alpha<1$. We define $y(k, \omega)$ by (5′). The probability law of $y(k, \omega)$ is a Gaussian distribution. But we have, for $k>h$,

$$\begin{aligned} E_\omega\big(y(k, \omega)y(h, \omega)\big) &= \rho(k-h) - \alpha\rho(k-h-1) - \alpha\rho(k-h+1) + \alpha^2\rho(k-h) \\ &= \alpha^{k-h} - \alpha\cdot\alpha^{k-h-1} - \alpha\cdot\alpha^{k-h+1} + \alpha^2\alpha^{k-h} \\ &= 0. \end{aligned}$$

Therefore $y(k, \omega)$, $k=0, \pm1, \pm2, \ldots$, is an independent sequence. From (5′) we deduce, by the principle of recursion,

$$\begin{aligned} x(k, \omega) &= y(k, \omega) + \alpha x(k-1, \omega) \\ &= \sum_{n=-N}^{k} \alpha^{k-n} y(n, \omega) + \alpha^{k+N+1} x(-N-1, \omega). \end{aligned}$$

But $E_\omega\Big(\big(\alpha^{k+N+1}x(-N-1, \omega)\big)^2\Big)=\alpha^{2(k+N+1)}$ tends to 0, as $N\to\infty$. Thus we obtain (5).

The converse proposition is evident.

*Theorem 4. Any continuous stationary process with no hysteresis is expressible in one of the following two forms, and its converse is also true.*

$$(6) \qquad x(t, \omega) = x(\omega), \qquad -\infty < t < \infty,$$

*where* $x(\omega)$ *is a normally distributed random variable.*

$$(7) \qquad x(t, \omega) = \int_{-\infty}^{t} e^{-\alpha(t-\tau)} d_\tau y(\tau, \omega),$$

*where* $y(t, \omega)$ *is a brownian motion, i.e. a temporally and spatially homogeneous process with no moving discontinuity, and is expressible by* $x(t, \omega)$ *as follows:*

$$(7') \qquad y(t, \omega) - y(s, \omega) = x(t, \omega) - x(s, \omega) + \alpha\int_s^t x(\tau, \omega)d\tau,$$

*and the integral in* (7) *is to be understood as a Riemann-Stieltjes integral in the sense of the strong topology in* $L^2(\Omega, P)$.

*Proof.* We may assume $E_\omega\big(x(t, \omega)\big)=0$ and $E_\omega\big(x(t, \omega)^2\big)=1$ with no loss of generality. By Theorem 1 the correlation function is equal to $e^{-\alpha|\tau|}$, $\alpha\geqq 0$. If $\alpha=0$, then the process is clearly expressible in (6).

Suppose that $\alpha>0$. We define $y(t, \omega)$ by (7′). Making use of the identity $E_\omega\big(x(t,\omega)x(s,\omega)\big)=\rho(t-s)=e^{-a|t-s|}$, we can prove, by easy calculations,

$$(8)\qquad E_\omega\big(y(t,\omega)-y(s,\omega)\big)=0 \qquad (s<t),$$

$$(9)\qquad E_\omega\big(\big(y(t,\omega)-y(s,\omega)\big)^2\big)=2a(t-s) \qquad (s<t), \qquad \text{and}$$

$$(10)\qquad E_\omega\big(\big(y(t,\omega)-y(s,\omega)\big)\big(y(s,\omega)-y(u,\omega)\big)\big)=0 \qquad (u<s<t).$$

Since the process $x(t,\omega)$ is normal, (10) implies that $y(t,\omega)-y(s,\omega)$ and $y(s,\omega)-y(u,\omega)$ are independent and normally distributed. Therefore the process $y(t,\omega)$ is a brownian motion.

The right side of (7) is clearly equal to

$$\int_{-\infty}^{t} e^{-a(t-\tau)}d_\tau x(\tau,\omega)+a\int_{-\infty}^{t} e^{-a(t-\tau)}x(\tau,\omega)d\tau$$
$$=[e^{-a(t-\tau)}x(\tau,\omega)]_{-\infty}^{t}-a\int_{-\infty}^{t} e^{-a(t-\tau)}x(\tau,\omega)d\tau+a\int_{-\infty}^{t} e^{-a(t-\tau)}x(\tau,\omega)d\tau$$
$$=x(t,\omega).$$

Thus we obtain (7).

The converse proposition is evident*.

* The cost of this research has been defrayed from the Scientific Expenditure of the Department of Education.

# 109. *Stochastic Integral.**

By Kiyosi ITÔ.

Mathematical Institute, Nagoya Imperial University.

(Comm. by S. KAKEYA, M.I.A., Oct. 12, 1944.)

**1.** *Introduction.* Let $(\Omega, P)$ be any probability field, and $g(t, \omega)$, $0 \leqq t \leqq 1$, $\omega \in \Omega$, be any *brownian motion*[1] on $(\Omega, P)$ i. e. a (real) stochastic differential process with no moving discontinuity such that $\mathcal{E}\big(g(s, \omega)-g(t, \omega)\big)=0$[2] and $\mathcal{E}\big(g(s, \omega)-g(t, \omega)\big)^2=|s-t|$. In this note we shall investigate an integral $\int_0^t f(\tau, \omega)\, d_\tau g(\tau, \omega)$ for any element $f(t, \omega)$ in a functional class $S^*$ which will be defined in § 2; the particular case in which $f(t, \omega)$ does not depend upon $\omega$ has already been treated by Paley and Wiener[3].

In § 2 we shall give the definition and prove fundamental properties concerning this integral. In § 3 we shall establish three theorems which give sufficient conditions for integrability. In § 4 we give an example, which will show a somewhat singular property of our integral.

**2.** *Definition and Properties.* For brevity we define the classes of measurable functions defined on $[0, 1] \times \Omega$: $G$, $S(t_0, t_1, \dots, t_n)$, $S$ and $S^*$ respectively as the classes of $f(t, \omega)$ satisfying the corresponding conditions, as follows,

$G$: $f(\tau, \omega)$, $g(\tau, \omega)$, $0 \leqq \tau \leqq t$, are independent of $g(\sigma, \omega)-g(t, \omega)$, $t \leqq \sigma \leqq 1$, for any $t$, $g(\tau, \omega)$ being the above mentioned brownian motion,

$S(t_0, t_1, \dots, t_n)$, $0=t_0<t_1<\dots<t_n=1 : f(t, \omega) \in G \wedge L_2([0, 1] \times \Omega)$ and $f(t, \omega)=f(t_{i-1}, \omega)$, $t_{i-1} \leqq t < t_i$, $i=1, 2, \dots, n$,

$S$: $f(t, \omega)$ belongs to $S(t_0, \dots, t_n)$ for a system $t_0, t_1, \dots, t_n$ which may depend upon $f(t, \omega)$; in other words $S \equiv \cup S(t_0, t_1, \dots, t_n)$,

$S^*$: $f(t, \omega) \in G$ and for any $\varepsilon$ there exists $h(t, \omega) \in \bar{S}$[4] such that

$$P\{\omega ; f(t, \omega)=h(t, \omega) \text{ for any } t\} > 1-\varepsilon .$$

At first for $f(t, \omega) \in S$ we define the stochastic integral $\int_0^t f(\tau, \omega)\, d_\tau g(\tau, \omega)$ (for brevity denote it by $I(t, \omega ; f)$) as follows:

---

* The cost of this research has been defrayed from the Scientific Expenditure of the Department of Education.

1) C. P. Lévy: Théorie de l'addition des variable aléatoire, P. 167, 1937, and also J. L. Doob: Stochastic processes depending on a continuous parameter, Trans., Amer. Math. Soc. vol. 42, Theorem 3.9.

2) $\mathcal{E}$ denotes the mathematical expectation, viz. $\mathcal{E} f(\omega)=\int_\Omega f(\omega) P(d\omega)$.

3) R. E. A. G. Paley and N. Wiener, Fourier transforms in the complex domain, Amer. Math. Soc. Coll. Publ. (1934), Chap. IX.

4) $\bar{S}$ means the closure of $S$ with respect to the norm in $L_2([0, 1] \times \Omega)$.

$$(2.1) \quad I(t, \omega; f) = \sum_{i=1}^{k} f(t_{i-1}, \omega).\big(g(t_i, \omega) - g(t_{i-1}, \omega)\big) + f(t_k, \omega)\big(g(t, \omega) - g(t_k, \omega)\big)$$

for $t_k \leqq t \leqq t_{k+1}$, if $f(t, \omega) \in S(t_0, t_1, \dots, t_n)$; this definition is independent of the special choice of $S(t_0, t_1, \dots, t_n)$. We have the

*Theorem 2.1.*

(*L*) $I(t, \omega; af+bg) = aI(t, \omega; f) + bI(t, \omega; g)$

(*N*) $I(t, \omega; 1) = g(t, \omega) - g(0, \omega)$,

(*C*) *$I(t, \omega; f)$ is a continuous function of $t$ with P-measure 1,*

(*I*) $\| I(t, \omega; f) \|_{\Omega}^{2}$[1] $= \| f(\tau, \omega) \|_{[0,t] \times \Omega}^{2}$ *for any* $t$, $0 \leqq t \leqq 1$,

(*B*) $P\{\omega; \sup_{0 \leqq t \leqq 1} | I(t, \omega; f) | \geqq b\} \leqq \frac{1}{b^2} \| f(\tau, \omega) \|_{[0,1] \times \Omega}^{2}$,

and

(*J*) *if $f(t, \omega) = h(t, \omega)$, $0 \leqq t \leqq 1$ for any $\omega \in \Omega_1$, $\Omega_1$ being any P-measurable subset of $\Omega$, then $I(t, \omega; f) = I(t, \omega; h)$, $0 \leqq t \leqq 1$, almost everywhere in $\Omega_1$.*

*Proof.* (*L*), (*N*), (*C*) and (*J*) are evident by the definition. In order to show (*I*) we may assume $t = t_k$ with no loss of generality. The left side of (*I*) is the expectation of $I(t, \omega, f)^2$, say $\mathcal{E} I(t, \omega; f)^2$.

$$(2.2) \quad \mathcal{E} I(t, \omega; 1)^2 = \sum_{i=1}^{k} \mathcal{E} f(t_{i-1}, \omega)^2 \big(g(t_i, \omega) - g(t_{i-1}, \omega)\big)^2 + 2 \sum_{i<j \leqq k} \mathcal{E} f(t_{i-1}, \omega) f(t_{j-1}, \omega) \big(g(t_i, \omega) - g(t_{i-1}, \omega)\big) \big(g(t_j, \omega) - g(t_{j-1}, \omega)\big).$$

In order to calculate this right side we shall achieve preliminary calculations. For brevity write $f_t$ and $g_t$ for $f(t, \omega)$ and $g(t, \omega)$ respectively. Since $f(t, \omega) \in G \frown L_2([0, 1] \times \Omega)$, we have, for $t < s < u < v$,

$$\mathcal{E} | f_t f_u (g_s - g_t) | \leqq \sqrt{\mathcal{E} f_u^2 \mathcal{E} f_t^2 (g_s - g_t)^2} = \sqrt{\mathcal{E} f_u^2 \mathcal{E} f_t^2 \mathcal{E} (g_s - g_t)^2} < \infty,$$

$$\mathcal{E} | f_t f_u (g_s - g_t)(g_v - g_u) | = \mathcal{E} | f_t f_u (g_s - g_t) | \mathcal{E} | g_v - g_u | < \infty,$$

$$\mathcal{E} f_t f_u (g_s - g_t)(g_v - g_u) = \mathcal{E} f_t f_u (g_s - g_t) \mathcal{E} (g_v - g_u) = 0.$$

Therefore we obtain

$$\mathcal{E} I(t, \omega; f)^2 = \sum_{i=1}^{k} \mathcal{E} f(t_{i-1}, \omega)^2 (t_i - t_{i-1}) = \int_0^{t_k} \int_{\Omega} f(t, \omega)^2 P(d\omega) d\tau,$$

i.e. $\| I(t, \omega; f) \|_{\Omega}^2 = \| f(t, \omega) \|_{[0,t] \times \Omega}^2$

For the proof of (*B*) we state the

*Lemma 2.1. Let $y_i(\omega)$, $x_i(\omega)$, $i = 1, 2, \dots\dots, n$, be any random variables. We assume, for $i = 1, 2, \dots, n$, that $y_1(\omega)$, $x_1(\omega)$, $\dots$, $y_{i-1}(\omega)$, $x_{i-1}(\omega)$, $y_i(\omega)$ are independent of $x_i(\omega)$, $x_{i+1}(\omega)$, $\dots$, $x_n(\omega)$. Then we have*

$$(2.3) \quad P\{\omega; \max_{1 \leqq i \leqq n} | y_1(\omega) x_1(\omega) + \cdots + y_i(\omega) x_i(\omega) | \geqq b\} \leqq \frac{1}{b^2} \mathcal{E} \big(y_1(\omega) x_1(\omega) + \cdots + y_n(\omega) x_n(\omega)\big)^2.$$

1) $\| \ \|_{\Omega}$ means the norm in $L_2(\Omega)$.

This lemma is an extension of Kolmogoroff's inequality, and its proof can be achieved in the same way and so will be omitted.

Let $s_i$, $i=0, 1, 2, \ldots,$. be any sequence dense in [0,1]. Any function $f(t, \omega)\in S(t_0, t_1, \ldots, t_n)$ may be considered as an element in $S(t_0^{(m)}, t_1^{(m)}, \ldots, t_{m+n+1}^{(m)})$, $t_0^{(m)}, t_1^{(m)}, \ldots, t_{m+n+1}^{(m)}$ being the sequence $s_0, s_1, \ldots, s_m, t_0, t_1, \ldots t_n,$ rearranged in the order of magnitude.
We obtain by the above lemma 2.1

$$P\{\omega ; \max_{0\leq i\leq m+n+1} |I(t_i^{(m)}, \omega ; f)| \geq b\} \leq \frac{1}{b^2}\mathcal{E} I(1, \omega ; f)^2 = \frac{1}{b^2}\|f(\tau, \omega)\|^2_{[0,1]\times\Omega},$$

a fortiori

$$P\{\omega ; \max_{0\leq i\leq m} | I(s_i, \omega ; f)| \geq b\} \leq \frac{1}{b^2}\|f(\tau, \omega)\|^2_{[0,1]\times\Omega}.$$

and so, as $m\to\infty$, we have

$$P\{\omega ; \sup_i | I(s_i, \omega, f)| \geq b\} \leq \frac{1}{b^2}\|f(\tau, \omega)\|^2_{[0,1]\times\Omega},$$

which implies $(B)$ on account of $(C)$.

*Theorem 2.2. There exists an extension of $I(t, \omega ; f)$ defined for any $f\in\bar{S}$ which satisfy $(L)$, $(N)$, $(C)$, $(I)$, $(B)$ and $(J)$. If $I_1(t, \omega ; f)$, $I_2(t, \omega ; f)$ be such extensions, then $I_1(t, \omega ; f)=I_2(t, \omega ; f)$ for any $t$ with $P$-measure 1 for any $f\in\bar{S}$.*

*Definition. The function $I(t, \omega ; f)$ determined up to $P$-measure 0 in the above theorem is called the (stochastic) integral of $f$ with respect to $g(t, \omega)$ and is denoted by $\int_0^t f(\tau, \omega)dg(\tau, \omega)$.*

*Proof of Theorem 2.2. Existence.* $I(1, \omega ; f)$ is a linear operation from $S\big(\leq L_2([0, 1]\times\Omega)\big)$ to $L_2(\Omega)$, which is isometric on account of $(I)$. We can extend $I(1, \omega ; f)$ and define a linear isometric operation from $\bar{S}$ to $L_2(\Omega)$. The extension is determined up to $P$-measure 0 for each $f(t, \omega)\in\bar{S}$. We denote it by $\tilde{I}(1, \omega ; f)$. Similarly for any $t$ we can define $\tilde{I}(t, \omega ; f)$ which satisfy $(L)$, $(N)$ and $(I)$.

Let $f_n(t, w)$ be a sequence in $S$ such that

$$\|f_{n+1}-f_n\|^2_{[0,1]\times\Omega} \leq \frac{1}{8^n}. \tag{2.4}$$

By $(B)$ we obtain

$$P\Big\{\omega ; \sup_{0\leq t\leq 1} | I(t, \omega ; f_{n+1})-I(t, \omega ; f_n)| \geq \frac{1}{2^n}\Big\} \leq \frac{1}{2^n} \tag{2.5}$$

By Borel-Contelli's theorem we have

$$\sup_{0\leq t\leq 1} | I(t, \omega ; f_{n+1})-I(t, \omega ; f_n)| < \frac{1}{2^n} \tag{2.6}$$

for a sufficiently large number $n$ with $P$-measure 1. Therefore $\{I(t, f_n ; \omega)\}$ will be convergent uniformly in $t$ with $P$-measure 1. Denote the limit by $I(t, \omega ; f)$.

$\tilde{I}(t, \omega ; f)$, as the $\|\ \|_\Omega$-limit of the sequence $I(t, \omega ; f_n)$, is also the limit of a subsequence of $\{I(t, \omega; f_n)\}$, (in the truth "of the sequence itself") with $P$-measure 1 for any $t$. Thus we have

$$P\{\omega;\ I(t, \omega; f)=\tilde{I}(t, \omega; f)\}=1. \tag{2.7}$$

Now we shall verify the properties $(L)$, $(N)$, $(C)$, $(I)$, $(B)$ and $(J)$ for this extension $I(t, \omega; f)$. $(N)$ is clear. $I(t, \omega; f_n)$ being continuous with $P$-measure 1 by $(C)$, $I(t, \omega; f)$ will also satisfy $(C)$. Since $(L)$ and $(I)$ hold for $\tilde{I}(t, \omega; f)$, it is also the case with $I(t, \omega; f)$. For the proof of $(B)$ we make use of the above-cited sequence $\{f_n\}$. We have clearly by $(B)$ $P\{\omega; \sup_{0\leq t\leq 1} |I(t, \omega; f_n)| \geqq b\} \leqq \frac{1}{b^2} \|f_n\|^2_{[0,1]\times\Omega}$. As $n\to\infty$, we obtain $(B)$, for $\{I(t, \omega; f_n)\}$ converges to $I(t, \omega; f)$ uniformly in $t$ with $P$-measure 1, while $\{f_n(t, \omega)\}$ $\|\ \|_{[0,1]\times\Omega}$- converges to $f(t, \omega)$.

In order to prove $(J)$ we need only prove that $I(t, \omega; f)=I(t, \omega; h)$ almost everywhere in $\Omega_1$ for each value of $t$, because $I(t, \omega; f)$ and $I(t, \omega; g)$ are continuous in $t$ with $P$-measure 1 on account of $(C)$. Let $f_n(t, \omega)$, $h_n(t, \omega)$, $n=1, 2, \ldots$ be sequences in $S$ such that

$$\|f_n-f\|^2_{[0,1]\times\Omega} \leqq \frac{1}{8^n}, \quad \|h_n-h\|^2_{[0,1]\times\Omega} \leqq \frac{1}{8^n}. \tag{2.8}$$

Define $k_n(t, \omega)$, by

$$\begin{aligned} k_n(t, \omega) &= f_n(t, \omega) \text{ for } \omega\in\Omega_1, \\ &= h_n(t, \omega) \text{ for } \omega\in\Omega-\Omega_1. \end{aligned} \tag{2.9}$$

Then we have

$$\|k_n-h\|^2_{[0,1]\times\Omega} \leqq \frac{2}{8^n}. \tag{2.10}$$

By (2.8) and (2.10) we obtain

$$\|I(t, \omega; f_n)-I(t, \omega; f)\|^2_\Omega \leqq \frac{1}{8^n},$$

$$\|I(t, \omega; k_n)-I(t, \omega; h)\|^2_\Omega \leqq \frac{2}{8^n}.$$

By the use of Bienaymé's inequality and Borel-Cantelli's theorem we see that $\{I(t, \omega; f_n)\}$ and $\{I(t, \omega; k_n)\}$ converge to $I(t, \omega; f)$ and to $I(t, \omega; h)$ respectively with $P$-measure 1. Since $(J)$ holds in $S$, we have $I(t, \omega; f_n)=I(t, \omega; k_n)$ almost everywhere in $\Omega_1$, and so $I(t, \omega; f)=I(t, \omega; k)$ almost everywhere in $\Omega_1$.

*Uniqueness.* Let $\{f_n(t, \omega)\}$ be any seqnence in $S$, $\|\ \|_{[0,1]\times\Omega}$-convergent to $f(t, \omega)$ $(\in S)$. Let $I_1(t, \omega; f)$ and $I_2(t, \omega; f)$ be two extensions. By $(I)$, $I_1(t, \omega; f_n)$ and $I_2(t, \omega; f_n)$ $\|\ \|_\Omega$-converge to $I_1(t, \omega; f)$ and to $I_2(t, \omega; f)$ respectively. Therefore we have $I_1(t, \omega; f)=I_2(t, \omega; f)$ with $P$-measure 1 for any $t$, and so $I_1(t, \omega; f)=I_2(t, \omega; f)$ for any $t$ with $P$-measure 1 on account of $(C)$.

*At last we shall define* $I(t, \omega; f)$ *for* $f(t, \omega)\in S^*$. We choose $f_n(t, \omega)\in S$, $n=1, 2, \ldots,$ such that $P\{\omega;\ f_n(t, \omega)=f(t, \omega)\} > 1-\epsilon$.

Write $\Omega_n$ for the $\omega$-set in { }. We define $I(t, \omega; f)$ as $I(t, \omega; f_n)$ on $\Omega_n - \bigcup_{k=1}^{n-1} \Omega_k$. Thus we can define $I(t, \omega; f)$ on the set $\bigcup_n \Omega_n$ of $P$-measure 1. This definition is independent of its procedure on account of $(J)$, and we can easily verify the properties $(L)$, $(N)$, $(C)$, and $(J)$ for this integral.

**3.** We shall show important subclasses of $S$ or of $S^*$.

*Theorem 3.1.* $L_2([0, 1]) \leq \bar{S}$[1]

The proof is brief and so will be omitted. By this theorem we see that our integral is an extension of that of Paley and Wiener.

*Theorem 3.2. Any bounded function $f(t, \omega)$ in $G$ belongs to $\bar{S}$.*

*Proof.* Let $M$ denote an upper bound of $|f(\tau, \omega)|$. We shall define $f(t, \omega)=0$ in the case: $t<0$. Then it holds that $f(\tau, \omega)$, $\tau \leq t$ and $g(\tau, \omega)$, $0 \leq \tau \leq t$ are independent of $g(\sigma, \omega)-g(t, \omega)$, $t \leq \sigma \leq 1$ for $0 \leq t \leq 1$. Define $\psi_n(t)$ by $\psi_n(t)=(k-1)2^{-n}$ if $(k-1)2^{-n} \leq t < k2^{-n}$. By (the slight modification of) Doob's Lemma[2] there exist a number $c$ and a sequence of integers $a_n$ such that $\lim_{n\to\infty} f\big(\psi_{a_n}(t-c)+c, \omega\big)=f(t, \omega)$ almost everywhere in $\Omega$. Put $f_n(t, \omega)=f\big(\psi_{a_n}(t-c)+c, \omega\big)$. Since we have $\psi_{a_n}(t-c)+c \leq t$ by the definition, $f_n(t, \omega)$ belongs to $G$, and since $|f_n(t, \omega)| \leq M$, we have $f_n(t, \omega) \in L^2_{[0,1]\times\Omega}$. Therefore we have $f_n(t, \omega) \in S$ by the definition. Since $|f_n(t, \omega)| \leq M$, $n=1, 2, \ldots$, and $\{f_n(t, \omega)\}$ converges to $f(t, \omega)$ almost everywhere, $\{f_n(t, \omega)\}$ will $\| \ \|_{[0,1]\times\Omega}$-converge to $f(t, \omega)$.

*Theorem 3.3. If any function $f(t, \omega) \in G$ is $P$-measurable in $\omega$ for any $t$ and is a function of $t$ continuous except possibly for discontinuities of the first kind*[3] *with $P$-measure 1, then $f(t, \omega)$ belongs to $S^*$.*

*Proof.* It is clear that $\sup_{0\leq t\leq 1} f(t, \omega)$ is equal to $\sup f(t, \omega)$ for $t$ running over all rational numbers in $[0, 1]$ with $P$-measure 1. Denote it by $M(\omega)$. Then $M(\omega)$ is measurable in $\omega$ and is finite with $P$-measure 1.

For any $\varepsilon$ we determine $N$ such that $P\{\omega; M(\omega)<N\}>1-\varepsilon$. Define $f_N(t, \omega)$ as $f(t, \omega)$ on this $\omega$-set in { } and as 0 otherwise. Then $f_N(t, \omega)$ is a measurable (in $t, \omega$) bounded function $\in G$, and so we have $f_N(t, \omega) \in \bar{S}$. On the other hand we have

$$P\{\omega; f(t, \omega)=f_N(t, \omega)\}=P\{\omega; M(\omega)<N\}>1-\varepsilon. \quad \text{q. e. d.}$$

**4.** *Example.* Let $F(x)$ be a function of $x$ such that $F''(x)$ may be continuous. By Theorem 3.3 we see $F'\big(g(t, \omega)\big) \in S^*$ The author has proved the equality[4]:

---

1) Any function of $t$ can also be considered as a function of $(t, \omega)$. In this sense $L_2(\Omega)$ will be considered as a subset of $L_2([0, 1]\times\Omega)$.

2) J, L. Doob. Loc. cit. p. 512 (*1*) Lemma *2.1.*

3) $f(t)$ is called to have a discontinuity of the first kind at $t$, if $f(t+0)$ and $f(t-0)$ exist and $f(t+0)=f(t)\neq f(t-0)$.

4) Cf. K. Itô: Markoff 過程ヲ定メル微分方程式 §7. 全國紙上數學談話會 第244號.

$$\int_0^t F'\big(g(\tau,\omega)\big)d_\tau g(\tau,\omega)=F\big(g(t,\omega)\big)-F\big(g(0,\omega)\big)$$

$$-\frac{1}{2}\int_0^t F''\big(g(\tau,\omega)\big)d\tau\,.$$

**In the last term we may see a characteristic property by which we distinguish " stochastic integral " from " ordinary integral."**

---

Reprinted from
*Proc. Imp. Acad. Tokyo* **20**, 519–524 (1944)

# 5. On a Stochastic Integral Equation.

By Kiyosi ITÔ.
Mathematical Institute, Nagoya Imperial University.
(Comm. by S. KAKEYA, M.I.A., Feb. 12, 1946.)

In his note "Stochastic Integral"[1] the author has discussed an integral of the type $\int_0^t f(\tau, \omega) d_\tau g(\tau, \omega)$, where $\omega$ is a variable taking values in a probability field $(\Omega, P)$ and $g(t, \omega)$ is a normalized brownian motion on $(\Omega, P)$. This note is devoted to the investigation of a stochastic integral equation:

$$(1) \qquad x(t, \omega) = c + \int_0^t a(\tau, x(\tau, \omega)) d\tau + \int_0^t b(\tau, x(\tau, \omega)) d_\tau g(\tau, \omega),$$

which is closely related to the researches of Markoff process by many authors, especially by S. Bernstein,[2] A. Kolmogoroff,[3] and W. Feller.[4]

*Theorem. Let* $a(t, x)$ *and* $b(t, x)$ *be continuous in* $(t, x)$ *and satisfy*

(2) $|a(t,x) - a(t,y)| \leqq A|x-y|$, (3) $|b(t,x) - b(t,y)| \leqq B|x-y|$, *where* $0 \leqq t \leqq 1$ *and* $-\infty < x, y < \infty$. *Then the integral equation (1) has one and only one continuous (in t with P-measure 1) solution.*

*Proof.* Firstly we shall find a solution by the method of successive approximation. We define $x_k(t, \omega)$ for $k = 0, 1, 2, \ldots$ as follows,

$$(4) \qquad x_0(t, \omega) \equiv c,$$

$$(5) \qquad x_k(t, \omega) = c + \int_0^t a(\tau, x_{k-1}(\tau, \omega)) d\tau + \int_0^t b(\tau, x_{k-1}(\tau, \omega)) d_\tau g(\tau, \omega);$$

the possibility of these definitions can be verified recursively if we make use of the properties of the stochastic integral shown in S.I..

By (5) we have, for $k = 0, 1, 2, \ldots$.

$$(6) \qquad x_{k+1}(t, \omega) - x_k(t, \omega) = \int_0^t (a(\tau, x_k(\tau, \omega)) - a(\tau, x_{k-1}(\tau, \omega))) d\tau$$
$$+ \int_0^t (b(\tau, x_k(\tau, \omega)) - b(\tau, x_{k-1}(\tau, \omega))) d_\tau g(\tau, \omega).$$

Since $a(t, x)$ and $b(t, x)$ are continuous, $|a(t, c)|$ and $|b(t, c)|$ are bounded in $0 \leqq t \leqq 1$ by a finite upper bound, say $M$. Then we have

1) These proceedings Vol. XX. No. 8. p. 519. This paper will be cited as S.I. in the following.

2) S. Bernstein: Equations différrentielles stochastiques, Actuarités Scientifiques 738.

3) A. Kolmogoroff: Über die analytischen Methoden in der Wahrscheinlichkeitsrechnung, Math. Ann. 104, p. 415.

4) W. Feller: Zur Theorie der stochastischen Prozesse. (Existenz und Eindeutigkeitssätze.), Math. Ann. 113, p. 113.

(7) $$E(x_1(t,\omega)-x_0(t,\omega))^2 \leqq \Big\{\Big(E\Big(\int_0^t a(\tau,c)d\tau\Big)^2\Big)^{\frac{1}{2}}+\Big(E\Big(\int_0^t b(\tau,c)d_\tau g(\tau,\omega)\Big)^2\Big)^{\frac{1}{2}}\Big\}^2$$
$$=\Big\{\Big|\int_0^t a(\tau,c)d\tau\Big|+\Big(\int_\Omega\int_0^t b(\tau,c)^2 d\tau P(d\omega)\Big)^{\frac{1}{2}}\Big\}^2$$
$$\leqq (Mt+Mt^{\frac{1}{2}})^2 \leqq 4M^2t.$$

In general we can show the following inequalities by mathematical induction for $k=1,2,3,\ldots,$

(8) $$E\Big(\int_0^t |a(\tau,x_k(\tau,\omega))-a(\tau,x_{k-1}(\tau,\omega))|d\tau\Big)^2 \leqq 4A^2(A+B)^{2(k-1)}M^2\frac{t^{k-1}}{\lfloor k+1},$$

(9) $$E\Big(\int_0^t (b(\tau,x_k(\tau,\omega))-b(\tau,x_{k-1}(\tau,\omega))d_\tau g(\tau,\omega)\Big)^2$$
$$\leqq 4B^2(A+B)^{2(k-1)}M^2\frac{t^{k-1}}{\lfloor k-1},$$

(10) $$E(x_k(t,\omega)-x_{k-1}(t,\omega))^2 \leqq 4(A+B)^{2(k-1)}M^2\frac{t^k}{\lfloor k}$$

By Bienaymé's inequality we can deduce from (8)

(11) $$P\Big\{\omega;\int_0^1|a(\tau,x_k(\tau,\omega))-a(\tau,x_{k-1}(\tau,\omega))|d\tau>\frac{1}{2^{k+1}}\Big\}$$
$$\leqq 4A^2(A+B)^{2(k-1)}M^2\frac{2^{k+1}}{\lfloor k+1}.$$

Since the series whose $k$-th term in the above right side is evidently convergent, we can conclude by Borel-Cantelli's theorem with $P$-measure 1 that

$$\int_0^1|a(\tau,x_k(\tau,\omega))-a(\tau,x_{k-1}(\tau,\omega))|d\tau \leqq \frac{1}{2^{k+1}},$$

a fortiori

(12) $$\sup_{0\leqq t\leqq 1}\Big|\int_0^t(a(\tau,x_k(\tau,\omega))-(a\,\tau,x_{k-1}(\tau,\omega)))d\tau\Big|\leqq\frac{1}{2^{k+1}}$$

but for finite exceptional values of $k$.

By a property of stochastic integral we obtain from (9)

(13) $$P\Big\{\omega;\sup_{0\leqq t\leqq 1}\Big|\int_0^t(b(\tau,x_k(\tau,\omega))-b(\tau,x_{k-1}(\tau,\omega)))\,d_\tau g(\tau,\omega)\Big|>\frac{1}{2^{k+1}}\Big\}$$
$$\leqq 4B^2(A+B)^{2(k-1)}M^2\frac{2^{k+1}}{\lfloor k+1}$$

and so we can see, by making use of Borel-Cantelli's theorem again, with $P$-measure 1, that

(14) $$\sup_{0\leqq t\leqq 1}\Big|\int_0^t(b(\tau,x_k(\tau,\omega))-b(\tau,x_{k-1}(\tau,\omega)))\,d_\tau g(\tau,\omega)\Big|\leqq\frac{1}{2^{k+1}}$$

but for finite exceptional values of $k$.

From (6), (12) and (14) we see, with $P$-measure 1, that

(15) $$\sup_{0\leqq t\leqq 1}\Big|x_k(t,\omega)-x_{k-1}(t,\omega)\Big|\leqq\frac{1}{2^{k+1}}+\frac{1}{2^{k+1}}=\frac{1}{2^k}$$

but for finite exceptional values of $k$. Thus the sequence $x_k(t,\omega)$ is uniformly (in $0\leqq t\leqq 1$) convergent with $P$-measure 1. We shall denote the limit by

$x(t, \omega)$, which will be shown to be a solution of (1) in the following.

Integrating both sides of (10) from $t = 0$ to $t = 1$, we obtain

$$(16)\quad \int_0^1\int_\Omega (x_k(t,\omega) - x_{k-1}(t,\omega))^2 P(d\omega)\,dt \leqq 4(A+B)^{2(k-1)} M^2 \frac{1}{|k+1}.$$

Therefore $x_k(t, \omega)$ is convergent in the norm of $L_2([0,1] \times \Omega)$ and so the limit $x(t, \omega)$ belongs to $L_2([0,1] \times \Omega)$. Since $x(t, \omega)$ belongs to **G**[5] by the definition and it is continuous (in $t$) as the uniform (in $t$) limit with $P$-measure 1, $x(t, \omega)$ belongs to **S***.[6] Consequently we have

$$(17)\quad x(t,\omega)\,\varepsilon\,\bar{S}.^{7)}$$

Therefore we can apply Theorem 2.2 in S.I. to $x(t, \omega)$.

Now we have

$$(18)\quad E\Big(\int_0^t a(\tau, x(\tau,\omega))\,d\tau - \int_0^t a(\tau, x_k(\tau,\omega))\Big)^2 \leqq t\int_0^t\int_\Omega A^2 (x,(\tau,\omega)$$
$$- x_k(\tau,\omega))^2 P(d\omega)\,d\tau \leqq A^2\int_0^1\int_\Omega (x(\tau,\omega) - x_k(\tau,\omega))^2 P(d\omega)\,d\tau$$

$$(19)\quad E\Big(\int_0^t b(\tau, x(\tau,\omega))\,d_\tau g(\tau,\omega) - \int_0^t b(\tau, x_k(\tau,\omega))\,d_\tau g(\tau,\omega)\Big)^2$$
$$\leqq t\int_0^t\int_\Omega B^2 (x(\tau,\omega) - x_k(\tau,\omega))^2 P(d\omega)\,d\tau$$
$$\leqq B^2\int_0^1\int_\Omega (x(\tau,\omega) - x_k(\tau,\omega))^2 P(d\omega)\,d\tau,$$

Taking the $L_2([0,1] \times \Omega)$-limits of both sides of (5) for $k \to \infty$, we obtain

$$(20)\quad x(t,\omega) = c + \int_0^t a(\tau, x(\tau,\omega))\,d\tau + \int_0^t b(\tau, x(\tau,\omega))\,d_\tau g(\tau,\omega)$$

with $P$-measure 1 for any assigned $t$. But the above both sides are continuous in $t$ with $P$-measure 1. Therefore (20) holds for any $t$ with $P$-measure 1. Thus we have obtained a solution $x(t, \omega)$ of (1) which is continuous in $t$ with $P$-measure 1.

Next we shall the uniqueness of the solution of (1). Let $y(t, \omega)$ and $z(t, \omega)$ be a continuous (in $t$ with $P$-measure 1) solution of (1):

$$(21)\quad y(t,\omega) = c + \int_0^t a(\tau, y(\tau,\omega))\,d\tau + \int_0^t b(\tau, y(\tau,\omega))\,d_\tau g(\tau,\omega),$$

$$(22)\quad z(t,\omega) = c + \int_0^t a(\tau, z(\tau,\omega))\,d\tau + \int_0^t b(\tau, z(\tau,\omega))\,d_\tau g(\tau,\omega).$$

In the case that $|a(t, x)|$ and $|b(t, x)|$ are bounded by an upper bound $G$, we have

$$(23)\quad E(y(t,\omega) - z(t,\omega))^2 \leqq \Big(\Big(E\Big(\int_0^t G\,d\tau\Big)^2\Big)^{\frac12} + \Big(E\Big(\int_0^t G^2\,d\tau\Big)^2\Big)^{\frac12}\Big)^2$$

But we have

$$(24)\quad E(y(t,\omega) - z(t,\omega))^2 \leqq (A+B)^2 \int_0^t E(y(t,\omega) - z(t,\omega))^2\,d\tau$$

5) Cf. S.I. 2.

6) Cf. S.I. 2.

7) Cf. S.I. Foot Note 4).

Therefore we have

(25) $E(y(t,\omega)-z(t,\omega))^2 \leqq 4\,G^2(A+B)^{2k}\dfrac{t^k}{\lfloor k}$.

Thus we obtain, as $k$ tends to $\infty$, $E(y(t,\omega)-z(t,\omega))^2=0$, and so $y(t,\omega) = z(t,\omega)$ with $P$-measure 1 for any $t$. By the continuity (in $t$) of $y(t,\omega)$ and $z(t,\omega)$, $y(t,\omega)=z(t,\omega)$ holds for any $t$ with $P$-measure 1.

In the general case we obtain, by the assumption,

(26) $|a(t,y)| \leqq |a(t,c)| + |a(t,y)-a(t,c)| \leqq M + A\,|y-c|$,

(27) $|b(t,y)| \leqq |b(t,c)| + |b(t,y)-b(t,c)| \leqq M + B\,|y-c|$.

Put

(28) $\Omega_k = \{\omega;\ \sup_{0\leqq t\leqq 1} |y(t,\omega)-c| < K\} \wedge \{\omega;\ \sup_{0\leqq t\leqq 1} |z(t,\omega)-c| < K\}$

$\Omega_k$ increases with $K$ and tends to a set $\Omega^*$ of $P$-measure 1 on account of the continuity of $y(t,\omega)$ and $z(t,\omega)$.

Now we have on $\Omega_k$

(29) $|a(t,y(t,\omega))|,\ \ |b(t,y(t,\omega))|,\ \ |a(t,z(t,\omega))|,\ \ |b(t,z(t,\omega))|$
$< M + (A+B)K$.

Denote the right side by $G$ and define $a_G(t,x)$ and $b_G(t,x)$ as follows.

(30)
$$\begin{aligned}
a_G(t,x) &= G, && \text{when } a(t,x) \geqq G,\\
a_G(t,x) &= a(t,x), && \text{when } |a(t,x)| < G,\\
a_G(t,x) &= -G, && \text{when } a(t,x) \leqq -G,\\
b_G(t,x) &= G, && \text{when } b(t,x) \geqq G,\\
b_G(t,x) &= b(t,x), && \text{when } |b(t,x)| < G,\\
b_G(t,x) &= -G, && \text{when } b(t,x) \leqq -G.
\end{aligned}$$

By (29) we have on $\Omega_k$

(31)
$$\begin{aligned}
a_G(t,y(t,\omega)) &= a(t,y(t,\omega)), & a_G(t,z(t,\omega)) &= a(t,z(t,\omega)),\\
b_G(t,y(t,\omega)) &= b(t,y(t,\omega)), & b_G(t,z(t,\omega)) &= b(t,z(t,\omega)).
\end{aligned}$$

and so on $\Omega_k$ both $y(t,\omega)$ and $z(t,\omega)$ satisfy a stochastic integral equation:

(32) $x(t,\omega) = c + \int_0^t a_G(\tau, x(\tau,\omega))\,d\tau + \int_0^t b_G(\tau, x(\tau,\omega))\,d_\tau g(\tau,\omega)$,

whicn has a unique continuous (in $t$ with $P$-measure 1) solution by the argument in the above special case. Thus we have $y(t,\omega)=z(t,\omega) = x_k(t,\omega)$ for any $t$ and for any $K$ and so $y(t,\omega)=z(t,\omega)$ for any $t$ on $\Omega^*$. Q.E.D.

Added in proof: The author has published a detailed investigation of the same subject in a more general case. Cf. On ~~stochastic processes (II)~~ ~~A~~ stochastic differential equations forth-coming ~~to the Japanese Journal of Mathematics.~~ in the Memoires of the Aemerican Mathematical Society.

Reprinted from
*Proc. Japan Academy* **22**, 32–35 (1946)

# STOCHASTIC DIFFERENTIAL EQUATIONS IN A DIFFERENTIABLE MANIFOLD

KIYOSI ITÔ

The theory of stochastic differential equations in a differentiable manifold has been established by many authors from different view-points, especially by P. Lévy [2] [1], F. Perrin [1], A. Kolmogoroff [1] [2] and K. Yosida [1] [2]. It is the purpose of the present paper to discuss it by making use of stochastic integrals.[2]

In § 1 we shall state some properties of stochastic integrals for the later use. We shall discuss stochastic differential equations in the $r$-dimensional Euclidean space in § 2 and in a differentiable manifold in § 3.

**1. Some properties of stochastic integrals.** Throughout this note we fix an *r-dimensional Brownian motion* [3]:

(1.1) $$\beta(t,\omega) = (\beta^1(t,\omega),\ \beta^2(t,\omega), \ldots, \beta^r(t,\omega)), \quad -\infty < t < \infty,$$

$\omega$ ($\in \Omega$) being the probability parameter with the probability law $P$ and $t$ being the time parameter. We assume that any function of $t$ and $\omega$ appearing in this note satisfies the following two conditions:

(1.2) it is measurable in $(t,\omega)$,

(1.3) the value it takes at $t = t_0$ is a *B-measurable function* [4] of the joint variable $(\beta(t,\omega),\ \tau \leqq t_0)$ for any $t_0$.

If it holds

(1.4) $$\xi(s,\omega) - \xi(t,\omega) = \int_t^s a(\tau,\omega)d\tau + \sum_{i=1}^{r}\int_t^s b_i(\tau,\omega)d\beta^i(\tau,\omega),\ ^{5)}$$
$$u \leqq s \leqq t \leqq v,\ \omega \in \Omega_1 (\subseteqq \Omega),$$

Received March 10, 1950.

1) The numbers in [ ] denote those of the references at the end of this paper.

2) K. Itô [1], [3].

3) By an $r$-dimensional Brownian motion we understand an $r$-dimensional random process whose components are all one dimensional Brownian motion (Cf. P. Lévy [1] p. 166, § 52, J. L. Doob [1] Theorem 3.9) independent of each other.

4) A mapping $f$ from $R^A$ into $R$ is called to be *B-measurable* if the inverse image of any Borel subset of $R$ by $f$ is also a Borel subset of $R^A$, that is an element of the least completely additive class that contains all rectangular subsets of $R^A$.
A random variable $\xi(\omega)$ is called to be a *B-measurable function of the joint variable* $(\xi_\alpha(\omega),\ \alpha \in A)$ if and only if there exists a $B$-measurable mapping $f$ from $R^A$ into $R$ such that $\xi(\omega) = f(\xi_\alpha(\omega),\ \alpha \in A)$ for every $\omega$. Cf. K. Itô [3] § 1.

5) The sense of this integral is to be understood as a *stochastic integral* introduced by the author. Cf. K. Itô [1], [3] § 7, § 8.

then we shall express this relation in the *differential form* as follows:

$$(1.5)\qquad d\xi(t,\omega) = a(t,\omega)dt + \sum_{j=1}^{r} b_j(t,\omega)d\beta^j(t,\omega)\,,\quad u \leqq t \leqq v,\ \omega \in \Omega_1.$$

THEOREM 1.1. *If*

$$(1.6)\quad d\xi^i(t,\omega) = a^i(t,\omega)dt + \sum_{j=1}^{r} b_j{}^i(t,\omega)d\beta^j(t,\omega)\,,\quad i = 1,2,\ldots,m,\ u \leqq t \leqq v\,,$$

$\omega \in \Omega_1$, *and if*

$$(1.7)\qquad \eta(t,\omega) = f(\xi(t,\omega)),\quad \xi(t,\omega) = (\xi^1(t,\omega), \xi^2(t,\omega), \ldots, \xi^m(t,\omega))\,,$$

*$f$ being a real-valued function of $C_2$-class* [6] *defined on an open subset of $R^m$ which contains all the points $\xi(t,\omega)$, $u \leqq t \leqq v$, $\omega \in \Omega_1$, then we have*

$$(1.8)\quad d\eta(t,\omega) = \Big\{\sum_i f_i(\xi(t,\omega))a^i(t,\omega) + \frac{1}{2}\sum_{ijk} f_{ij}(\xi(t,\omega))b_k{}^i(t,\omega)b_k{}^j(t,\omega)\Big\}dt$$
$$+ \sum_j\Big(\sum_i f_i(\xi(t,\omega))b_j{}^i(t,\omega)\Big)d\beta^j(t,\omega)\,,$$

*where*

$$f_i(x^1,\ldots,x^m) = \frac{\partial f}{\partial x^i}(x^1,\ldots,x^m),\quad f_{ij}(x^1,\ldots,x^m) = \frac{\partial^2 f}{\partial x^i \partial x^j}(x^1,\ldots,x^m)$$

We shall here mention only the outline of the proof.[7] First we shall state a lemma.

LEMMA. $$\int_s^t b(\tau,\omega)d\beta^i(\tau,\omega)\int_s^t c(\tau,\omega)d\beta^j(\tau,\omega)$$
$$= \delta^{ij}\int_s^t b(\tau,\omega)c(\tau,\omega)d\tau + \int_s^t b(\tau,\omega)\int_s^\tau c(\sigma,\omega)d\beta^j(\sigma,\omega)d\beta^i(\tau,\omega)$$
$$+ \int_s^t c(\sigma,\omega)\int_s^\sigma b(\tau,\omega)d\beta^i(\tau,\omega)d\beta^j(\sigma,\omega)\quad (\delta^{ij} = \textit{Kronecker's delta}).$$

We can prove this lemma first by considering the special case that both $b(\tau,\omega)$ and $c(\tau,\omega)$ are uniformly stepwise [8] in $(s,t)$ and next by taking the limit in the general case.

In order to prove (1.8) we need only to compute the following expression:

$$\eta(s,\omega) - \eta(t,\omega) = \sum_{\nu=1}^{n}(\eta(t_\nu^{(n)},\omega) - \eta(t_{\nu-1}^{(n)},\omega)),\quad t_\nu^{(n)} = t + \frac{\nu}{n}(s-t)\,,$$
$$= \sum_{\nu i} f_i(\xi(t_{\nu-1}^{(n)},\omega))(\xi^i(t_\nu^{(n)},\omega) - \xi^i(t_{\nu-1}^{(n)},\omega))$$
$$+ \frac{1}{2}\sum_{\nu ij}(f_{ij}(\xi(t_{\nu-1}^{(n)},\omega)) + \varepsilon_{ij\nu}(\omega))(\xi^i(t_\nu^{(n)},\omega) - \xi^i(t_{\nu-1}^{(n)},\omega))(\xi^j(t_\nu^{(n)},\omega) - \xi^j(t_{\nu-1}^{(n)},\omega))$$

---

6) A function is called to be *of $C_2$-class* if its partial derivatives of the second order are all continuous.

7) The author will publish the proof in details in another note.

8) $b(\tau,\omega)$ is called to be ***uniformly stepwise*** in $(s,t)$ if there exist a division of the interval $(s,t)$: $s = t_0 < t_1 < \ldots < t_n = t$ independent of $\omega$ such that $b(\tau,\omega)$ $b(t_{i-1},\omega)$, $t_{i-1} \leqq \tau \leqq t_i$, $i = 1,2,\ldots,n$. Cf. K. Itô [1], [3] §7.

by making use of the above lemma and to take the limit as $n \to \infty$.

THEOREM 1.2. *The functions* $a(t,\omega)$, $b_i(t,\omega)$, $i = 1,2,\ldots,r$, *are uniquely determined by* $\xi(t,\omega)$ *in the sense that*

$$\begin{aligned} d\xi(t,\omega) &= a(t,\omega)dt + \sum_i b_i(t,\omega)\, d\beta^i(t,\omega) \\ &= \tilde{a}(t,\omega)dt + \sum_i \tilde{b}_i(t,\omega)d\beta^i(t,\omega), \quad u \leqq t \leqq v, \quad \omega \in \Omega_1, \end{aligned}$$

*implies*

$$a(t,\omega) = \tilde{a}(t,\omega), \quad b_i(t,\omega) = \tilde{b}_i(t,\omega), \quad i = 1,2,\ldots,r,$$

*for almost all* $(t,\omega)$, $u \leqq t \leqq v$, $\omega \in \Omega_1$.

*Proof.* It suffices to show that

$$\xi(t,\omega) \equiv 0, \quad u \leqq t \leqq v, \quad \omega \in \Omega_1,$$

implies

$$a(t,\omega) = 0, \quad b_i(t,\omega) = 0, \quad i = 1,2,\ldots,r,$$

for almost all $(t,\omega)$, $u \leqq t \leqq v$, $\omega \in \Omega_1$.

By Theorem 1 we have

$$\begin{aligned} (\xi(s,\omega) - \xi(t,\omega))^2 &= \int_s^t \{2(\xi(\tau,\omega) - \xi(s,\omega))a(\tau,\omega) + \sum_i b_i(\tau,\omega)^2\}d\tau \\ &\quad + \sum_i \int_s^t 2(\xi(\tau,\omega) - \xi(s,\omega))b_i(\tau,\omega)d\beta^i(\tau,\omega), \end{aligned}$$

from which it follows that

$$\sum_{\nu=1}^{n} (\xi(u + \frac{\nu}{n}(v-u),\omega) - \xi(u + \frac{\nu-1}{n}(v-u),\omega))^2 \longrightarrow \int_u^v \sum_i b_i(\tau,\omega)^2 d\tau$$

in probability. But the left side is always equal to 0 by the assumption and so we obtain

$$\int_u^v \sum_i b_i(\tau,\omega)^2 d\tau = 0$$

for almost all $\omega$ in $\Omega_1$, which completes the proof.

We have already obtained the following inequality [9)]

$$(1.9) \qquad P_r\{\sup_{u \leqq t \leqq v} |\int_u^t b(\tau,\omega)d\beta^j(\tau,\omega)| \geqq \alpha\} \leqq \frac{1}{\alpha^2}\int_u^v E(b(\tau,\omega)^2)dt = O(v-u) \quad (\text{as } v \to u).$$

The following theorem gives us a more precise evaluation in the case that $b(t,\omega)$ is uniformly bounded.

THEOREM 1.3. *In the case:*

$$(1.10) \qquad |b(t,\omega)| \leqq K \quad (u \leqq t \leqq v,\ \omega \in \Omega), \quad 0 \leqq v-u < \alpha^2/(2K^2),$$

*we have*

$$(1.11) \qquad P_r\{\sup_{u \leqq t \leqq v} |\int_u^t b(\tau,\omega)d\beta^j(\tau,\omega)| \geqq \alpha\} \leqq \frac{8K^4(v-u)^2}{\alpha^4} = O(v-u)^2 \quad (\text{as } v \to u).$$

9) Cf. K. Itô [1] Th. 8, [3]. $P_r(\mathfrak{A}(\omega)) = P(\{\omega;\ \mathfrak{A}(\omega)\})$, $E(\xi(\omega)) = \int_\Omega \xi(\omega)P(d\omega)$.

*Proof.* By Theorem 1.1 we have

$$\left(\int_u^v b(t,\omega)d\beta^j(t,\omega)\right)^2 = \int_u^v b(t,\omega)^2 dt + 2\int_u^v b(t,\omega)\int_u^t b(\tau,\omega)d\beta^j(\tau,\omega)d\beta^j(t,\omega).$$

But (1.10) implies

$$\int_u^v b(t,\omega)^2 dt \leqq \frac{\alpha^2}{2}.$$

Therefore the left side of (1.11) is less than

$$P_r\{\sup_{u\leqq s\leqq v} |\int_u^s b(t,\omega)\int_u^t b(\tau,\omega)d\beta^j(\tau,\omega)d\beta^j(t,\omega)| \geqq \frac{\alpha^2}{4}\}$$

$$\leqq \left(\frac{4}{\alpha^2}\right)^2\int_u^v E\{(b(t,\omega)\int_u^t b(\tau,\omega)d\beta^j(\tau,\omega))^2\}dt \leqq \left(\frac{4}{\alpha^2}\right)^2 K^4\frac{(v-u)^2}{2} = \frac{8K^4(v-u)^2}{\alpha^4}$$

on account of (1.9).

**2. Stochastic differential equation on the $r$-dimensional Euclidean space.** In this paragraph we shall treat a stochastic differential equation:

(2.1) $$d\xi^i(t,\omega) = a^i(t,\xi(t,\omega))dt + \sum_j b_j^i(t,\xi(t,\omega))d\beta^j(t,\omega),$$
$$\xi(t,\omega) = (\xi^1(t,\omega),\ldots,\xi^r(t,\omega)),^{10)} \quad i = 1,2,\ldots,r, \quad u\leqq t\leqq v,$$

with the initial condition

(2.2) $$\xi^i(u,\omega) = c^i(\omega), \quad i = 1,2,\ldots,r,$$

where $c^i(\omega)$, $i = 1,2,\ldots,r$ are $B$-measurable functions of $(\beta(\tau,\omega), \tau\leqq u)$. This stochastic differential equation is equivalent to a stochastic integral equation:

(2.3) $$\xi^i(t,\omega) = c^i(\omega) + \int_u^t a^i(\tau,\xi(\tau,\omega))d\tau + \sum_j\int_u^t b_j^i(\tau,\xi(\tau,\omega))d\beta^j(\tau,\omega), \quad u\leqq t\leqq v.$$

THEOREM 2.1. *Under the following four assumptions* [11]:

(2.4) $$\sum_i |a^i(t,x) - a^i(t,y)|^2 \leqq A\|x-y\|^2,$$
$$\sum_{ij}|b_j^i(t,x) - b_j^i(t,x)|^2 \leqq B\|x-y\|^2,$$

*where*

$$\|x-y\|^2 = \sum_i |x^i - y^i|^2,$$

(2.5) $a^i(t,x)$, $b_j^i(t,x)$ *are all continuous in t for any x.*

(2.6) $$\sum_i |a^i(t,x)|^2 \leqq A_1, \quad \sum_{ij}|b_j^i(t,x)|^2 \leqq B_1,$$

(2.7) $$E(|c^i(\omega)|^2) \leqq C, \quad i = 1,2,\ldots,r,$$

*($A, B, A_1, B_1, C$ being constants independent of $(t,x)$), there exists one and only one solution of the equation* (2.3).

---

10) The $i$-component of $r$-dimensional vectors $x$, $y$, $c(\omega)$, $\xi(t,\omega)$, $\eta(t,\omega)$ etc. are denoted by $x^i$, $y^i$, $c^i(\omega)$, $\xi^i(t,\omega)$, $\eta^i(t,\omega)$ etc. respectively.

11) It is possible to show this theorem without any use of the assumption (2.6), (2.7) by the method the author has used in one-dimensional case (K. Itô [1], [3] Th. 11) but it is unnecessary for our present purpose to do so.

*Proof of the existence.* We shall make use of the successive approximation method to find a solution of (2.3). We define $\xi_n(t,\omega)$, $n=1,2,\ldots$, recurively as follows:

$$\xi_0{}^i(t,\omega)=c^i(\omega),\quad i=1,2,\ldots,r, \tag{2.8}$$

$$\xi_n{}^i(t,\omega)=c^i(\omega)+\int_u^t a^i(\tau,\xi_{n-1}(\tau,\omega))d\tau+\sum_j\int_u^t b_j{}^i(\tau,\xi_{n-1}(\tau,\omega))d\beta^j(\tau,\omega). \tag{2.9}$$

Then we have

$$\begin{aligned}\sum_i|\xi_1{}^i(t,\omega)-\xi_0{}^i(t,\omega)|^2&=\sum_i\left|\int_u^t a^i(\tau,\xi_0(\tau,\omega))d\tau+\sum_j\int_u^t b_j{}^i(\tau,\xi_0(\tau,\omega))d\beta^j(\tau,\omega)\right|^2\\&\leqq(r+1)\sum_i\left\{\left|\int_u^t a^i(\tau,\xi_0(\tau,\omega))d\tau\right|^2+\sum_j\left|\int_u^t b_j{}^i(\tau,\xi_0(\tau,\omega))d\beta^j(\tau,\omega)\right|^2\right\},\end{aligned}$$

$$\begin{aligned}&E(\|\xi_1(t,\omega)-\xi_0(t,\omega)\|^2)\\&\leqq(r+1)\left\{\sum_i(t-u)\int_u^t E(a^i(\tau,\xi_0(\tau,\omega)))^2)d\tau+\sum_{ij}\int_u^t E(b_j{}^i(\tau,\xi_0(\tau,\omega)))^2)d\tau\right\}\\&\leqq(r+1)\left\{(v-u)\int_u^t E(\sum_i a^i((\tau,\xi_0(\tau,\omega))^2)d\tau+\int_u^t E(\sum_{ij}b_j{}^i(\tau,\xi_0(\tau,\omega))^2)d\tau\right\}\\&\leqq(r+1)\{(v-u)A_1+B_1\}(t-u),\end{aligned} \tag{2.10}$$

in making use of (2.6). In the same manner we obtain

$$\begin{aligned}&E(\|\xi_n(t,\omega)-\xi_{n-1}(t,\omega)\|^2)\\&\leqq(r+1)\{(v-u)A+B\}\int_u^t E\{\|\xi_{n-1}(\tau,\omega)-\xi_{n-2}(\tau,\omega)\|^2\}d\tau\end{aligned} \tag{2.11}$$

by virtue of (2.4). From (2.10) and (2.11) it follows that

$$E(\|\xi_n(t,\omega)-\xi_{n-1}(t,\omega)\|^2)\leqq\frac{K^n(t-u)^n}{n!}\leqq\frac{K^n(v-u)^n}{n!}, \tag{2.12}$$

$K$ being a constant determined by $A_1, B_1, A, B, u$ and $v$.

Since

$$\sum_{n=1}^\infty\sqrt[3]{\frac{K^n(v-u)^n}{n!}}<\infty,$$

we see, by Borel-Cantelli's theorem, that there exists a function $\xi(t,\omega)$ such that

$$E(\|\xi(t,\omega)-\xi_n(t,\omega)\|^2)\to0\text{ for each }t.$$

$\xi^i(t,\omega)$, $i=1,2,\ldots,r$, are clearly $B$-measurable functions of $(\beta(\tau,\omega),\tau\leqq t)$ for each $t$. But we have further

$$\int_u^v E(\|\xi_m(t,\omega)-\xi_n(t,\omega)\|^2)dt\to0\quad(m,n\to\infty),$$

from which we see that $\xi^i(t,\omega)$, $i=1,2,\ldots,r$, are all measurable in $(t,\omega)$.

Taking the limit of (2.9) as $n\to\infty$, we see that $\xi(t,\omega)$ satisfies (2.3).

*Proof of the uniqueness.* Let $\xi(t,\omega)$ and $\tilde{\xi}(t,\omega)$ satisfies (2.3). In the same manner as (2.12) we have

$$E(\|(\xi(t,\omega) - \tilde{\xi}(t,\omega)\|^2) \leqq \frac{K^n(v-u)^n}{n!} \to 0,$$

which completes the proof.

THEOREM 2.2. *Assume* (2.4), (2.5), (2.6), (2.7) *and*

(2.13) $$a^i(t,x),\ b_j^i(t,x) \equiv 0 \ \textit{for}\ x \notin U$$

*and*

(2.14) $$c(\omega) \in U,$$

*U being an open subset of* $R^r$. *Then the solution* $\xi(t,\omega)$ *of* (2.3) *satisfies*

(2.15) $$\xi(t,\omega) \in U, \quad u \leqq t \leqq v,$$

*for almost all* $\omega$.

*Proof.* It suffices to prove that the totality $\Omega_1$ of $\omega$ for which $\xi(t,\omega) \notin U$ for some $t$ has $P$-measure 0. Assume that $P(\Omega_1) > 0$. Since $\xi(t,\omega)$ is continuous in $t$ for each $\omega$, there exist $t$ and $s$ $(u \leqq t \leqq s \leqq v)$ such that

(2.16) $$\xi(\tau,\omega) \notin U \text{ for } t \leqq \tau \leqq s, \quad \xi(t,\omega) \neq \xi(s,\omega)$$

on an $\omega$-set $\Omega_2 (\subseteqq \Omega_1)$ with positive $P$-measure. From (2.13), (2.14), and (2.16) it follows that

$$a^i(\tau, \xi(\tau,\omega)) = 0, \quad b_j^i(\tau, \xi(\tau,\omega)) = 0, \quad s \leqq \tau \leqq t, \quad \omega \in \Omega_2,$$

and so we have

$$\xi^i(s,\omega) = \xi^i(t,\omega) + \int_s^t a^i(\tau, \xi(\tau,\omega))d\tau + \int_s^t b_j^i(\tau, \xi(\tau,\omega))d\beta^j(\tau,\omega) = \xi^i(t,\omega).$$

almost everywhere on $\Omega_2$, which contradicts with (2.16).

**3. Stochastic differential equation in a differentiable manifold.**

Given any $r$-dimensional differentiable manifold of $C_2$-class [12] $M$. By a continuous random motion in $M$ we understand an $M$-valued function $\pi(t,\omega)$ of $t$ and $\omega$ which is measurable in $\omega$ for each $t$ and continuous in $t$ for each $\omega$. Hereafter we assume that

(3.1) $$\pi(t,\omega) \text{ is a } B\text{-measurable function of } (\beta(\tau,\omega),\ \tau \leqq t).$$

Let $\xi(t,\omega) \equiv (\xi^1(t,\omega), \xi^2(t,\omega), \ldots, \xi^r(t,\omega))$ be any local coordinate of $\pi(t,\omega)$. If we define $\xi(t,\omega) \equiv 0$ in case $\pi(t,\omega)$ is outsides of the coordinate neighbourhood, $\xi^i(t,\omega)$ proves to be a $B$-measurable function of $\beta(\tau,\omega)$, $\tau \leqq t$, on account of (3.1). We can easily see that $\xi^i(t,\omega)$ is measurable in $(t,\omega)$.

We shall consider a stochastic differential equation:

---

[12] By definition an $r$-dimensional differentiable manifold of $C_2$-class is a Hausdorff space with the second countability axiom and with coordinate neighbourhoods, each homeomorphic to the interior of a sphere of $r$-dimensional Euclidean space and such that the coordinate relationships between the coordinates of the two intersecting neighbourhoods are of $C_2$-class.

(3.2) $$d\xi^i(t,\omega) = a^i(t,\xi(t,\omega))dt + \sum_j b_j^i(t,\xi(t,\omega))d\beta^j(t,\omega),$$

which means that

(3.2) $$\xi^i(t,\omega) - \xi^i(s,\omega) = \int_s^t a^i(\tau,\xi(\tau,\omega))d\tau + \sum_j \int_s^t b_j^i(\tau,\xi(\tau,\omega))d\beta^j(\tau,\omega)$$

for any $\omega$ such that $\pi(\tau,\omega)$ is contained in the coordinate neighbourhood for $s \leqq \tau \leqq t$.

The functions $a^i(t,x)$ and $b_j^i(t,x)$, $(x = (x^1, x^2, \ldots, x^r))$ depend upon the special choice of the local coordinate. We assume that they are transformed between two local coordinates $x$ and $\bar{x}$ in the following manner:

(3.3) $$\begin{cases} \bar{a}^i(t,\bar{x}) = \sum_k \dfrac{\partial \bar{x}^i}{\partial x^k} a^k(t,x) + \dfrac{1}{2} \sum_{jkl} \dfrac{\partial^2 \bar{x}^i}{\partial x^k \partial x^l} b_j^k(t,x) b_j^l(t,x), \\ \bar{b}_j^i(t,\bar{x}) = \sum_k \dfrac{\partial \bar{x}^i}{\partial x^k} b_j^k(t,x). \end{cases}$$

Now we shall explain the reason why we choose such a transformation law. Let $\bar{x} = (\bar{x}^1, \bar{x}^2, \ldots, \bar{x}^r)$ and $x = (x^1, x^2, \ldots, x^r)$ be two local coordinates defined on a neighbourhood $U$ in $M$ such that

(3.4) $$\bar{x}^i = f^i(x^1, x^2, \ldots, x^r), \quad i = 1, 2, \ldots, r,$$

and let $\xi(t,\omega)(\bar{\xi}(t,\omega))$ denote the $x$ $(\bar{x})$-coordinate of $\pi(t,\omega)$. Then we have (3.2) and

(3.2′) $$d\bar{\xi}^i(t,\omega) = \bar{a}^i(t,\bar{\xi}(t,\omega))dt + \sum_j \bar{b}_j^i(t,\bar{\xi}(t,\omega))d\beta^j(t,\omega),$$

(3.5) $$\bar{\xi}^i(t,\omega) = f^i(\xi^1(t,\omega), \xi^2(t,\omega), \ldots, \xi^r(t,\omega)), \quad i = 1, 2, \ldots, r.$$

In making use of Theorem 1.1 we deduce from (3.2) and (3.5)

(3.6) $$\begin{aligned} d\bar{\xi}^i(t,\omega) = \Big\{ & \sum_k f_k^i(\xi(t,\omega)) a^k(t,\xi(t,\omega)) \\ & + \frac{1}{2} \sum_{klj} f_{kl}^i(\xi(t,\omega)) b_j^k(t,\xi(t,\omega)) b_j^l(t,\xi(t,\omega)) \Big\} dt \\ & + \sum_j \sum_k f_k^i(\xi(t,\omega)) b_j^k(t,\xi(t,\omega)) d\beta^j(t,\omega), \end{aligned}$$

whenever $\pi(t,\omega)$ is contained in $U$. By Theorem 1.2 it follows from (3.2′) and (3.6) that

$$\begin{aligned} \bar{a}^i(t,\xi(t,\omega)) = & \sum_k f_k^i(\xi(t,\omega)) a^k(t,\xi(t,\omega)) \\ & + \frac{1}{2} \sum_{klj} f_{kl}^i(\xi(t,\omega)) b_j^k(t,\xi(t,\omega)) b_j^l(t,\xi(t,\omega)), \end{aligned}$$

$$\bar{b}_j^i(t,\xi(t,\omega)) = \sum_k f_k^i(\xi(t,\omega)) b_j^k(t,\xi(t,\omega)).$$

In order to express a sufficient condition for the equation (3.2) to have a unique solution, we define the *boundedness* of $a^i(t,x)$ and $b_j^i(t,x)$ as follows.

DEFINITION. By a canonical coordinate around $p$ we understand any local coordinate which maps a neighbourhood of $p$ onto the interior of the unit sphere in the $r$-dimensional Euclidean space $R^r$ and especially transforms $p$ to the centre of the sphere.

By *a canonical coordinate system* on $M$ we understand a collection of canonical coordinates such that, for any point of $M$, there exists a canonical coordinate around the point in the collection.

$a^i(t,x)$, $b_j^i(t,x)$, $i,j=1,2,\ldots,r$ are called to be *bounded* if and only if there exists a canonical coordinate system and a constant $K$ such that we have always

(3.7) $\quad |a(t,x)|,\ \ |b(t,x)|<K,\ \ i,j=1,2,\ldots,r,\ \ u\leqq t\leqq v,\ \ x\in S,$

for any coordinate of the system.

THEOREM 3.1. *We assume that* $a^i(t,x)$, $b_j^i(t,x)$, $i,j=1,2,\ldots,r$, $0\leqq t\leqq 1$, *are all*

(3.8) *bounded in the sense of the above definition,*

(3.9) *continuous in* $t$ *for each* $x$,

*and*

(3.10) *of* $C_1$*-class in* $x$ *for each* $t$.

*Then there exists one and only one solution of the stochastic differential equation* (3.2) *with the initial condition*:

$$\pi(u,\omega)=p(\omega),$$

*where* $p(\omega)$ *is an* $M$*-valued* $B$*-measurable function of* $(\beta(\tau,\omega),\ \tau\leqq t)$.

*Proof of the existence.* We shall denote by $\{x_p=(x_p^1\ldots x_p^r);\ p\in M\}$ a canonical coordinate system which satisfies (3.7) and by $S(S_1,S_2)$ the interior of the sphere with the centre at the origin 0 and with the redius $1(1/3,2/3)$, and by $U_p(U_p')$ the totality of all the points of $M$ whose $x_p$-coordinate lies in $S(S_1)$. Since $M$ satisfies the second countability axiom, we obtain a sequence in $M$: $\{p_n\}$ for which $\{U_{p_n}'\}$ covers the whole space $M$. We denote $x_{p_n}$, $U_{p_n}$ and $U_{p_n}'$ respectively by $x_n\equiv(x_n^1,\ldots,x_n^r)$, $U_n$ and $U_n'$. We define $V_n$ by

$$V_n=U_n'-\bigcup_{i=1}^{n-1}U_i'\quad(n\geqq 2),\quad V_1=U_1'.$$

It is clear that

$$M=\sum_{n=1}^{\infty}V_n.^{13)}$$

Now we shall define a random motion $\pi_m(t,\omega)$ which satisfies the equation (3.2) on an $\omega$-set $(\Omega^{(m)})$ whose $P$-measure tends to 1 as $m\to\infty$.

We define a function $\lambda(x)$ on $R^r$ by

[13] In this paper we denote the sum of disjoint sets by $\Sigma$.

$$\lambda(x) = \begin{cases} 1 & (\|x\| \leqq 2/3) \\ 0 & (\|x\| \geqq 5/6) \\ 5 - 6\|x\| & (2/3 < \|x\| < 5/6) \end{cases}$$

and $\tilde{a}^i(t,x)$ and $\tilde{b}_j^i(t,x)$ by

$$\tilde{a}^i(t,x) = a(t,x)\lambda(x), \quad \tilde{b}_j^i(t,x) = b_j^i(t,x)\lambda(x).$$

Then $\tilde{a}^i(t,x)$ and $\tilde{b}_j^i(t,x)$ satisfies the assumptions in Theorem 2.1 and

$$\tilde{a}^i(t,x) = 0, \quad \tilde{b}_j^i(t,x) = 0, \quad i,j = 1,2,\ldots,r, \quad x \notin S.$$

We devide the time-interval $(u,v)$ into $m$ equal subintervals with

$$u = u_0 < u_1 < u_2 < \ldots < u_m = v.$$

We define $C_{n1}(\omega) = (C^1_{n1}(\omega),\ldots,C^r_{n1}(\omega))$ by

$$C_{n1}(\omega) = \begin{cases} \text{the } x_n\text{-coordinate of } p(\omega) & \text{if } p(\omega) \in V_n \\ 0 \equiv (0,0,\ldots,0) & \text{if } p(\omega) \notin V_n, \end{cases}$$

and consider a stochastic integral equation:

$$\xi_n{}^i(t,\omega) = C^i_{n1}(\omega) + \int_{u_0}^t \tilde{a}^i(\tau,\xi_n(\tau,\omega))d\tau + \sum_j \int_{u_0}^t \tilde{b}^i_{jn}(\tau,\xi_n(\tau,\omega))d\beta^j(\tau,\omega),$$
$$i = 1,2,\ldots,r, \quad u_0 \leqq t \leqq u_1.$$

By Theorem 2.1 and Theorem 2.2 we can see that this equation has a unique solution such that

$$\xi_n(t,\omega) \in S, \quad u_0 \leqq t \leqq u_1,$$

for almost all $\omega$.

We define $\pi_m(t,\omega)$, $u_0 \leqq t \leqq u_1$, in the following way:

(3.11) $\pi_m(t,\omega)$ = the point in $M$ with the $x_n$-coordinate $\xi_n(t,\omega)$ when $p(\omega) \in V_n$;

$\pi_m(t,\omega)$, $u_0 \leqq t \leqq u_1$, is clearly determined up to $P$-measure 0.

Next we consider a stochastic differential equation:

$$\xi_n{}^i(t,\omega) = C^i_{n1}(\omega) + \int_{u_1}^t \tilde{a}_n{}^i(\tau,\xi_n(\tau,\omega))d\tau + \sum_j \int_{u_1}^t \tilde{b}^i_{jn}(\tau,\xi_n(\tau,\omega))d\beta^j(\tau,\omega);$$
$$i = 1,2,\ldots,r, \quad u_1 \leqq t \leqq u_2,$$

where

$$C_{n2}(\omega) = \begin{cases} \text{the } x_n\text{-coordinate of } \pi_n(u,\omega) & \text{if } \pi_m(u_1,\omega) \in V_n \\ 0 \equiv (0,0,\ldots,0) & \text{if } \pi_m(u_1,\omega) \notin V_n. \end{cases}$$

In making use of Theorem 2.1 and Theorem 2.2 again we see that this equation has a unique solution such that

$$\xi_n(t,\omega) \in S, \quad u_1 \leqq t \leqq u_2,$$

for almost all $\omega$. We define $\pi_m(t,\omega)$, $u_1 \leqq t \leqq u_2$ by (3.11) again. By continuing this procedure we define $\pi_m(t,\omega)$, $u \leqq t \leqq v$. Now we define $\Omega^k_{nl}$ by

$$\Omega^k_{nl} = \{\omega;\, \pi_m(u_{k-1},\omega) \in V_n,\, \pi_m(u_k,\omega) \in V_l,\, \xi_n(\tau,\omega) \in S_2 \text{ for } u_{k-1} \leqq \tau \leqq u_k\},$$

and put

$$\Omega^{(m)} = \sum_{n_0 n_1 \ldots n_m} \Omega^1_{n_0 n_1} \Omega^2_{n_1 n_2} \ldots \Omega^m_{n_{m-1} n_m}.$$

We shall prove that

(3.12) $\pi_m(t,\omega)$ satisfies (3.2) almost everywhere on $\Omega^{(m)}$

and that

(3.13) $$P(\Omega^{(m)}) \to 1 \quad \text{as } m \to \infty.$$

Let $x$ be any local coordinate defined on $U$, and $\pi_m(t,\omega)$ lie in $U$ for $t_1 \leqq \tau \leqq t_2$ and $\omega \in \Omega^* (\subseteqq \Omega^{(m)})$. In order to show (3.12) it is sufficient to prove that the $x$-coordinate $\xi(t,\omega)$ of $\pi_m(t,\omega)$ satisfies

(3.14) $$d\xi^i(t,\omega) = a^i(t,\xi(t,\omega))dt + \sum_j b^i_j(t,\xi(t,\omega))d\beta^j(t,\omega), \quad t_1 \leqq t \leqq t_2, \quad \omega \in \Omega^*.$$

In case $\omega \in \Omega^1_{n_0 n_1}\Omega^2_{n_1 n_2} \ldots \Omega^m_{n_{m-1} n_m}\Omega^*$ we have

$$d\xi^i_{n_{k-1}}(t,\omega) = \tilde{a}^i_{n_{k-1}}(t,\xi_{n_{k-1}}(t,\omega))dt + \sum_j \tilde{b}^i_{n_{k-1}j}(t,\xi_{n_{k-1}}(t,\omega))d\beta^j(t,\omega),$$

$$u_{k-1} \leqq t \leqq u_k,$$

and so

(3.15) $$d\xi^i_{n_{k-1}}(t,\omega) = a^i_{n_{k-1}}(t,\xi_{n_{k-1}}(t,\omega))dt + \sum_j b^i_{n_{k-1}j}(t,\xi_{n_{k-1}}(t,\omega))d\beta^j(t,\omega),$$

$$u_{k-1} \leqq t \leqq u_k,$$

since $\xi_{n_{k-1}}(t,\omega) \in S_2$ by the definition of $\Omega_{n_{k-1}n_k}$.

By Theorem 1.1 and the transformation law (3.3) we can deduce (3.14) from (3.15), which proves (3.12).

In order to prove (3.3) it suffices to show

(3.16) $$P\{\Omega^1_{n_0 n_1} \ldots \Omega^k_{n_{k-1} n_k} \tilde{\Omega}_{n_k}\} \geqq (1 - G(1/m)^2) P(\Omega^1_{n_0 n_1} \ldots \Omega^k_{n_{k-1} n_k}),$$

$$k = 1, 2, \ldots,$$

where $\tilde{\Omega}_{n_k} = \sum_{n=1}^{\infty} \Omega_{n_k n}$ and $G$ is a constant which depends neither on $m$ nor on $k$ but only on $K$, $u$ and $v$; in fact it follows from (3.16) that

$$P(\Omega^m) \geqq (1 - G(1/m)^2) \sum P(\Omega^1_{n_0 n_1}\Omega^2_{n_1 n_2} \ldots \Omega^{m-1}_{n_{m-2} n_{m-1}}) \geqq (1 - G(1/m)^2)^m \to 1$$

$$(\text{as } m \to \infty).$$

Let $\omega_1$ and $\omega_2$ denote the joint variables

$$(\beta(\tau,\omega),\ \tau \leqq u_k) \text{ and } (\beta(\tau,\omega) - \beta(u_k,\omega),\ u_k \leqq \tau \leqq v)$$

respectively. Then $\xi_{n_k}(t,\omega)$ and $C^i_{n_k k+1}(\omega)$ are expressible as $\xi_{n_k}(t,\omega_1,\omega_2)$ and $C^i_{n_k k+1}(\omega_1)$. By the above procedure by which we have defined $\pi_m(t,\omega)$ we have

$$\xi_{nk}(t,\omega_1,\omega_2) = C^i_{n_k k+1}(\omega_1) + \int_{u_k}^t \tilde{a}^i_{n_k}(\tau, \xi_{nk}(t,\omega_1,\omega_2))dt$$
$$+ \sum_j \int_{u_k}^t \tilde{b}^i_{n_k j}(\tau, \xi_{nk}(\tau,\omega_1,\omega_2))d\beta^j(t,\omega).$$

Since the stochastic integral appearing in the above expression is a $B$-mesurable function of $\tilde{b}^i_{n_k j}(t, \xi_{nk}(t,\omega_1,\omega_2))$ and $\omega_1$ is independent of $\omega_2$, we see, by Fubini's theorem in Lebesgue integral, that

$$\xi_{nk}(t,\omega_1^0,\omega_2) = C^i_{n_k k+1}(\omega_1^0) + \int_{u_k}^t \tilde{a}^i_{n_k}(\tau, \xi_{nk}(\tau,\omega_1^0,\omega_2))d\tau$$
$$+ \sum_j \int_{u_k}^t \tilde{b}_{n_k j}(\tau, \xi_{nk}(\tau,\omega_1^0,\omega_2))d\beta^j(\tau,\omega)$$

for almost all $\omega$ except possibly for $\omega_1^0$ of $P_{\omega_1}$-measure 0, $P_{\omega_1}$ being the probability law of $\omega_1$.

$\Omega^1_{n_0 n_1}\Omega^2_{n_1 n_2} \ldots \Omega^k_{n_{k-1} n_p}$ and $\tilde{\Omega}_{nk}$ are expressed in the form:

$$\{\omega;\ \omega_1 \in E_1\},\ \{\omega;\ \pi_m(t,\omega) \in V_{nk},\ \sup_{u_k \leqq t \leqq u_{k+1}} \|\xi_{nk}(t,\omega)\| < 2/3\},$$

$E_1$ being a subset of $R^{[-\infty, u_k]}$, and the latter $\omega$-set includes the $\omega$-set

$$\{\omega;\ \sup_{u_{k-1} \leqq t \leqq u_k} \|\xi_{nk}(t,\omega) - C_{n_k k+1}(\omega)\| < 1/3\},$$

as $C_{n_k k+1}(\omega) \in S_1$. Therefore the left side of (3.16) is greater than

$$P_r\{\omega \in E_1,\ \sup_{u_{k-1} \leqq t \leqq u_k} \|\xi_{nk}(t,\omega) - C_{n_k k+1}(\omega)\| < 1/3\}$$
$$= P_r\{\omega_1 \in E_1,\ \sup_{u_{k-1} \leqq t \leqq u_k} \|\xi_{nk}(t,\omega_1,\omega_2) - C_{n_k k+1}(\omega_1)\| < 1/3\}$$
$$= \int_{E_1} P_r\{\sup_{u_{k-1} \leqq t \leqq u_k} \|\xi_{nk}(t,\omega_1^0,\omega_2) - C_{n_k k+1}(\omega_1^0)\| < 1/3\} P_{\omega_1}(d\omega_1^0)$$

(by the independence of $\omega_1$ and $\omega_2$)

$$\geqq \int_{E_1} (1 - G(1/m)^2) P_{\omega_1}(d\omega_1^0) \quad \text{(by Theorem 1.3)}$$
$$= (1 - G(1/m)^2) P_r(\omega_1 \in E_1),$$

which proves (3.16).

In order to obtain a solution of (3.2) we need only to put

$$\pi(t,\omega) = \begin{cases} \pi_1(t,\omega) \text{ for } \omega \in \Omega^{(1)} \\ \pi_2(t,\omega) \text{ for } \omega \in \Omega^{(m)} - \Omega^{(m-1)}, \quad m = 2,3,\ldots \end{cases}$$

As to the *uniqueness*, by making use of the uniqueness of the solution of (2.3), we can easily prove that any solution of (3.2) is coincident with the solution above obtained, which completes the proof of our theorem.

*Example* 1. If $M$ is a closed manifold, the condition (3.8) follows from (3.10) and so we can do without this condition (3.8).

*Example* 2. Let $M$ be a Lie group. If $a^i(t, x)$ and $b_j^i(t, x)$ are left (or right)-invariant with regard to the group operation, then the condition (3. 8) follows from (3. 10). In considering this fact we can define Brownian motions on the Lie group.

A relation between the process we have defined above by the stochastic differential equation (3. 2) and the Fokker-Planck equation is given by the following theorem.

THEOREM 3. 2. *The solution of* (3. 2) *is a simple Markoff process on* $M$ *and the transition probability law* $F(t, p, s, E)$—*the conditional probability law of* $\pi(s, \omega)$ *under the condition that* $\pi(t, \omega) = p$—*is coincident with the probability law of the solution* $\pi(s, \omega)$ $(s \geqq t)$ *of* (3. 2) *with the initial condition:*

$$\pi(t, \omega) \equiv p \tag{3.17}$$

*If* $f(p)$ *is a bounded function of* $C_2$*-class defined on* $M$, *then we have, as* $s \to t$,

$$\frac{1}{s-t}\left\{\int_M f(q)F(t, p, s, dq) - f(p)\right\} \to \left(\sum_i a^i(t, x)\frac{\partial}{\partial x^i} + \frac{1}{2}\sum_{ijk} b_k^i(t, x)b_k^j(t, x)\frac{\partial^2}{\partial x^i \partial x^j}\right)f(x), \tag{3.18}$$

$x$ *being a local coordinate of* $p$.

*Proof.* The first part can be easily proved in the same way as the proof of (3. 16). We shall prove the second part. We fix a local coordinate $x$ around $p$ for which $a^i(t, x)$, $b_j^i(t, x)$ are all bounded and we denote by $U$ the coordinate neighbourhood of $x$. We shall denote the solution of (3. 2) with the initial condition: $\pi(t, \omega) = p$ by the same notation $\pi(t, \omega)$ and the $x$-coordinate of $\pi(t, \omega)$ with $\xi(t, \omega)$, where we define $\xi(t, \omega) = 0$ if $\pi(t, \omega) \notin U$.

Whenever $\pi(s, \omega)$ lies in $U$, we have

$$\xi^i(s, \omega) = x^i + \int_t^s a^i(\tau, \xi(\tau, \omega))d\tau + \sum_j \int_t^s b_j^i(\tau, \xi(\tau, \omega))d\beta^j(\tau, \omega) \tag{3.19}$$

and so, by Theorem 1. 1,

$$\begin{aligned} f(\xi(s, \omega)) - f(x) = \int_t^s \Big\{ & \sum_i a^i(\tau, \xi(\tau, \omega))f_i(\xi(\tau, \omega)) \\ & + \frac{1}{2}\sum_{ijk} b_k^i(\tau, \xi(\tau, \omega))b_k^j(\tau, \xi(\tau, \omega))\Big\}d\tau \\ & + \sum_j \int_t^s \sum_i b_j^i(\tau, \xi(\tau, \omega))f_i(\xi(\tau, \omega))\,d\beta^j(\tau, \omega). \end{aligned} \tag{3.20}$$

In making use of Theorem 1. 3 we can easily see that

$$P_r\{\pi(\tau, \omega) \notin U \text{ for some } \tau,\ t \leqq \tau \leqq s\} = O(s-t)^2.$$

Thus we see that $f(\pi(s, \omega)) - f(p)$ coincides with the right side of (3. 20) except possibly on an $\omega$-set of $P$-measure $O(s-t)^2$. Since $f(p)$ is bounded on $M$ by the assumption, we have

$$\frac{1}{s-t}\left[\int_M f(q)F(t,p,s,dq)-f(p)\right]=\frac{E(f(\pi(s,\omega))-f(p))}{s-t}$$

$$=\frac{1}{s-t}\left[\int_t^s E\{\sum_i a^i(\tau,\xi(\tau,\omega))f_i(\xi(\tau,\omega))\right.$$

$$\left.+\frac{1}{2}\sum_{ijk}b_k{}^i(\tau,\xi(\tau,\omega))b_k{}^j(\tau,\omega))f_{ij}(\xi(\tau,\omega))\}d\tau+O(s-t)^2\right]$$

$$\to\sum_i a^i(t,x)\frac{\partial f}{\partial x^i}(x)+\frac{1}{2}\sum_{ijk}b_k{}^i(t,x)b_k{}^j(t,x)\frac{\partial^2 f}{\partial x^i\partial x^j}(x)\quad(s\to t),$$

which proves (3.18).

## Reference

J. L. Doob:

[1] Stochastic processes depending on a continuous parameter, Trans. Amer. Math. Soc. **42**, 1937.

K. Itô:

[1] Stochastic integral, Proc. Imp. Acad. Tokyo, Vol. **20**, No. 8 (1944).

[2] On a stochastic integral equation, Proc. Imp. Acad. Tokyo, Vol. **22**, No. 2 (1946).

[3] Stochastic differential equations, forthcoming in the Memoirs of Amer. Math. Soc.

A. Kolmogoroff:

[1] Zur Theorie der stetigen zufälligen Prozesse, Math. Ann. **108** (1933).

[2] Umkehrbarkeit der stetigen Naturgestze, Math. Ann. **113** (1937).

P. Lévy:

[1] Theorie de l'addition des variables aléatoires, Paris (1937).

[2] Processus stochastiques et mouvement brownien, Paris (1948).

F. Perrin:

[1] Étude mathematique du mouvement brownien de rotation, Ann. Éc. Norm., (3), **45** (1928).

K. Yosida:

[1] Brownian motion on the surface of the 3-sphere, Ann. of Math. Statistics, **20**, 2 (1949).

[2] Integration of Fokker-Planck's equation in a compact Riemannian space, Arkiv för Mathematik, **1**, 9 (1949).

*Mathematical Institute, Nagoya University*

[From the Proceedings of the Japan Academy, Vol. 26 (1950), No. 8]



# 47. *Brownian Motions in a Lie Group.*

By Kiyosi ITÔ.
Mathematical Institute, Nagoya University.
(Comm. by T. TAKAGI, M.J.A., Oct. 12, 1950.)

The notion of Brownian motions has been introduced by N. Wiener [1] [2][1] in the case of the real number space (or more generally the n-space) and by P. Lévy [3] in the case of the circle. We shall here extend this notion in the case of a general Lie group.[2]

§ *1. Definition and fundamental theorems.* Let $G$ be an n-dimensional Lie group. A random process $\pi(t)$ in $G$ is called to be a *right (left) invariant Brownian motion* in $G$, if it satisfies the following five conditions M, C, T, S and C*.

**M.** $\pi(t)$ is a simple Markoff process; we denote the transition probability law of $\pi(t)$ with $F(t, p, s. E)$ i.e.

$$F(t, p, s, E) = P_r\{\pi(s)\in E/\pi(t) = p\}.$$

**C.** Kolmogoroff-Feller's continuity condition [4] [5]. For any neighbourhood $U$ of $p$ it holds that

$$\lim_{s\to t+0} \frac{1}{s-t} F(t, p, s, G-U) = 0$$

and the following limits exist ($1 \leq i, j \leq n$)

$$a^i(t, p) \equiv \lim_{s\to t+0} \frac{1}{s-t} \int_U (x^i - x_0^i) F(t, x_0, s, dx),$$

$$B^{ij}(t, p) \equiv \lim_{s\to t+0} \frac{1}{s-t} \int_U (x^i - x_0^i)(x^j - x_0^j) F(t, x_0, s, dx),$$

where $(x^i)$ is a local coordinate defined on $U$ and $(x_0^i)$ is the coordinate of p.

**T.** temporal homogeneity. $F(t, p, s, E) = F(t+\tau, p, s+\tau, E)$.

**S.** spatial homogeneity.
right invariance $F(t, p, s, E) = F(t, pr, s, Er)$.
(left invariance $F(t, p, s, E) = F(t, lp, s, lE)$.)

**C*** continuity. Almost all sample motions[3] are continuous.

1) The numbers in [ ] correspond to those in the the references at the end of this paper.

2) Prof. K. Yosida has obtained a similar result in making use of his operator-theoretical method. See the preceding article.

3) In the analytical theory of probability a random motion is represented by a motion depending on a probability parameter. Any motion for each parameter value is called to be a sample motion.

*By a Brownian motion* in $G$ we understand a right invariant one or a left invariant one. A *both-sides invariant Brownian motion* is defined as a Brownian motion which is right invariant as well as left invariant.

Now, put

$$(1.1)\qquad D_t f(p) = \lim_{s\to t} \frac{1}{s-t}\int_G (f(q)-f(p))F(t, p, s, dq) .$$

Then we see by **C** and **T** that $D_t f(p)$ is written as

$$(1.1.\text{ a})\qquad D_t f(p) = a^i(p)\frac{\partial f}{\partial x^i}(p) + \frac{1}{2}B^{ij}(p)\frac{\partial^2 f}{\partial x^i \partial x^j}(p)^{4)}$$

for any bounded function $f(p)$ of class $C_2$, where $\| B^{ij}(p) \|$ is a symmetric non-negative-definite matrix by virtue of

$$(1.1.\text{ b})\qquad \xi_i \xi_j B^{ij}(p) = \lim_{s\to t+0} \frac{1}{s-t}\int (\xi_i(x^i - x_0^i))^2 F(t, x_0, s, dx) \geqq 0 ,$$

namely that $D$ is an elliptic differential operator in $G$ independent of $t$. Therefore we may eliminate $t$ and write simply as $D$. $D$ is called to be the *generating operator* of the Brownian motion $\pi(t)$.

We shall here state several fundamental theorems.

*Theorem 1.* (Characterization of generating operators). Let $D$ be any elliptic differential operator defined for any bounded function of class $C_2$. Then the following three conditions are equivalent to each other.

(G. 1) $D$ is a generating operator of a right (left) invariant Brownian motion.

(G. 2) $D$ commutes with any right (left) translation operator $R_r(L_l)$, where $R_r f(p) = f(pr)$ $(L_l f(p) = f(lp))$.

(G. 3) $D$ is expressible in the form :

$$(1.2)\qquad D = A^i X_i + \frac{1}{2} B^{ij} X_i X_j$$

where $\{X_i\}$ is a basis of the infinitesimal operators of left (right) translations and $A^i$, $B^{ij}$ are all real constants such that the matrix $\| B^{ij} \|$ is a symmetric non-negative-definite one.

*Theorem 2.* (A generalization of the Fokker-Planck equation [6]). If we put

$$f(s, p) = \int f(q) F(t, p, s, dq) \quad (t \leqq s)$$

$f(q)$ being a function of class $C_2$ which vanishes outsides of a

4) We shall eliminate the summation sign $\sum$ according to the usual rule of tensor caluclus.

compact set, then $f(s, p)$ satisfies the following partial differential equation:

$$\frac{\partial}{\partial s} f(s, p) = Df(s, p) \tag{1.3}$$

with the initial condition

$$f(t, p) = f(p) . \tag{1.4}$$

*Theorem 3.* (Uniqueness theorem). The transition probability law of a Brownian motion is uniquely determined by its generating operator.

*Theorem 4.* (A condition for the both-sides invariance). A necessary and sufficient condition that $D$ be the generating operator of a both-sides invariant Brownian motion is that $D$ is expressible in the form (2) such that $\{A^i\}$ and $\{B^{ij}\}$ satisfies, besides the above-stated conditions,

$$A^j C^l_{kj} = 0, \quad B^{ij} C^l_{kj} + B^{jl} C^i_{kj} = 0 \quad (1 \leq i, k, l \leq n) . \tag{1.5}$$

*Theorem 5.* (A generalization of "*differential*" property). Let $\pi(t)$ be a right (left) invariant Brownian motion in $G$. Then

$$\pi(s_i)\,\pi(t_i)^{-1}\,(\pi(t_i)^{-1}\,\pi(s_i)\,), \; i = 1, 2, \dots, m,$$

are independent G-valued random variables for $t_1 < s_1 \leq t_2 < s_2 \leq \dots \leq t_m < s_m$.

§ 2. *Proof of the theorems.*

*Proof of Th. 1.* We shall consider only the case of right invariant Brownian motions. It is clear by the definition that (G. 1) implies (G. 2). We shall prove that (G. 2) implies (G. 3)[5]. By (G. 2) we have

$$Df(r) = R_r Df(e) = DR_r f(e) = Df_r(e), \text{ where } f_r(p) \equiv f(pr).$$

By taking an adequate coordinate $(x^i)$ around $e$ we may assume that $X_i$ is expressed as

$$X_i g(x) = c^k_i(x)\,\frac{\partial g}{\partial x^k}(x),\; c^k_i(e) = \delta^k_i,\; x^i(e) = 0 .$$

Then we have

$$Dg(e) = (A^i X_i + \frac{1}{2} B^{ij} X_i X_j) g(e) ,$$

where

$$A^i = a^i(e) - \frac{1}{2} B^{jk}(e) \frac{\partial c^i_k}{\partial x^j}(e),\; B^{ij} = B^{ij}(e) .$$

5) The author's original proof was the same in essential as that stated here but more complicated. He owes much to M. Kuranishi for the simplification of the proof.

$\| B^{ij} \|$ is clearly a symmetric non-negative-definite matrix by the definition.

Therefore $Df_r(e)$ is written as

$$Df_r(e) = (A^i X_i + \frac{1}{2} B^{ij} X_i X_j) f_r(e),$$

where $A^i$, $B^{ij}$ satisfy the conditions stated in (G. 3). In considering that $X_i$ is an infinitesimal operators of left translations and so commutative with $R_r$, we obtain

$$Df(r) = (A^i X_i + \frac{1}{2} B^{ij} X_i X_j) f_r(e) = (A^i X_i + \frac{1}{2} B^{ij} X_i X_j) f(r).$$

Next, we shall prove that (G. 3) implies (G. 1). K. Yosida has shown, in making use of his operator theoretical method, that (G. 3) implies that $D$ is the generating operator of a simple Markoff process which satisfies **M**, **C**, **T** and **S**. By the use of a stochastic differential equation [7] we shall here show that $D$ is the generating operator of a right invariant Brownian motion, which satisfies **C*** besides the above four conditions; this will mean that (G. 3) implies (G. 1). We fix a *canonical coordinate* [7] $(x^i)$ around $e$, and define a canonical coordinate around $p$ by

$$x_p^i(qp) = x^i(q), \qquad 1 \leqq i \leqq n. \tag{2.1}$$

Then $\{(x_p^i),\ p \in G\}$ is a *canonical coordinate system* [7]. By the above argument we see by (G. 2) that

$$Df(p) = \left(a^i \frac{\partial}{\partial x_p^i} + \frac{1}{2} B^{ij} \frac{\partial^2}{\partial x_p^i \partial x_p^j}\right) f(p), \tag{2.2}$$

where $a^i$, $B^{ij}$ are all independent of $p$ and $\| B^{ij} \|$ is a nonnegative definite matrix. We fix a real matrix $\| b_j^i \|$, such that

$$b_k^i b_k^j = B^{ij}. \tag{2.3}$$

Now we shall consider an arbitrary local coordinate $(x^i)$ whose coordinate neighbourhood contains $p$. For this coordinate we define

$$a^i(x) = a^j \frac{\partial x^i}{\partial x_p^j} + \frac{1}{2} b_k^j b_k^l \frac{\partial^2 x^i}{\partial x_p^j \partial x_p^l}, \quad b_k^i(x) = b_k^j \frac{\partial x^i}{\partial x_p^j}. \tag{2.4}$$

Then we have

$$a^i(p) \frac{\partial f}{\partial x^i}(p) + \frac{1}{2} b_k^i(p) b_k^j(p) \frac{\partial^2 f}{\partial x^i \partial x^j}(p)$$

$$= a^i(p) \frac{\partial f}{\partial x_p^i}(p) + \frac{1}{2} b_k^i(p) b_k^j(p) \frac{\partial^2 f}{\partial x_p^i \partial x_p^j}(p) = Df(p).$$

Since $Df(p) \equiv (A^i X_i + \frac{1}{2} B^{ij} X_i X_j) f(p)$ is independent of the special choice of the local coordinate, it is so with the left side of the above equation. This implies that $a^i(x)$ is transformed in the following manner:

(2.5. a) $$\bar{a}^i(x) = a^j(x)\frac{\partial \bar{x}^i}{\partial x^j} + \frac{1}{2} b^j_l(x) b^k_l(x) \frac{\partial^2 \bar{x}^i}{\partial x^j \partial x^k}.$$

$b^i_j(x)$ is clearly transformed as follows by (2.4):

(2.5. b) $$\bar{b}^i_k(x) = b^j_k(x)\frac{\partial \bar{x}^i}{\partial x^j}.$$

Thus we may consider the following *stochastic differential equation* [7]:

(2.6) $$d\xi^i(t) = a^i_j(\xi(t))\, dt + b^i_j(\xi(t))\, d\beta^i(t),$$

$(\xi^i(t))$ being a local coordinate of a random motion $\pi(t)$ in $G$.

In order to show the existence of the solution of this equation we shall verify the conditions (3.8), (3.9) and (3.10) in Theorem 3.1 in [7]. (3.9) and (3.10) are evident. We shall easily verify (3.8) in considering that $a^i(x)$, $b^i_j(x)$ is determined by the same expression around every point with respect to the above canonical coordinate system by virtue of the definitions.

By Theorem 3.2 in [7] we see that the solution $\pi(t)$ is a continuous simple Markoff process whose transition probability law $F(t, p, s, E)$ satifies

(2.7) $$\lim_{s \to t+0} \frac{1}{s-t}\int (f(q)-f(p))\, F(t, p, s, dq)$$
$$= a^i(p)\frac{\partial f}{\partial x^i}(p) + \frac{1}{2} b^i_k(p) b^j_k(p) \frac{\partial^2 f}{\partial x^i \partial x^j}(p) = Df(p).$$

Thus we see that $\pi(t)$ satisfies the conditions **M**, **C**, **C*** in § 1. By comparing the solution of (2.6) with the initial condition:

(2.8. a) $$\pi(t) = p$$

with the solution of the same equation with the initial condition:

(2.8. b) $$\pi(t+\sigma) = p$$

and in remembering the temporal homogeneity of $(\beta^i(t))$, we can easily verify that $\pi(t)$ is temporally homogeneous. In order to show the spatial homogeneity we need only to remember that, if $\pi(\tau)$ is the solution of (2.6) with the initial condition: $\pi(t) = p$, then $\pi^*(\tau) \equiv \pi(\tau)r$ is the solution of (2.6) with $\pi^*(t) = pr$.

*Proof of Th. 2.* By the right-invariance we see that

$$f(s, p) = \int f(q\cdot p) F(t, p, s, dq\cdot p) = \int f(q\cdot p) F(t, e, s, dq),$$

which implies that $f(s, p)$ is a bounded function of class $C_2$ in $p$.

By the temporal homogeneity we have

$$\begin{aligned} f(s+\Delta, p) &= \int f(q)F(t, p, s+\Delta, dq) \\ &= \iint f(q)F(t, p, t+\Delta, dr)F(t+\Delta, r, s+\Delta, dq) \\ &= \iint f(q)F(t, p, t+\Delta, dr)F(t, r, s, dq) \\ &= \int f(s, r)F(t, p, t+\Delta, dr) \end{aligned}$$

and so

$$f(s+\Delta, p)-f(s, p) = \int (f(s, r)-f(s, p))F(t, p, t+\Delta, dr)$$

and accordingly

$$\lim_{\Delta\to+0} \frac{f(s+\Delta, p)-f(s, p)}{\Delta} = Df(s, p)$$

$Df(s, p)$ being continuous in $s$ as is easily verified, we obtain (1.3) from the above identity.

*Proof of Th. 3.* Let $F_1(t, p, s, E)$ and $F_2(t, p, s, E)$ be the transition probability law of the Brownian motions with the same generating operator $D$. We shall here prove that $F_1 = F_2$. For this it is sufficient to show that the functions

$$f_i(s,p) = \int f(q)F_i(t, p, s, dq), \quad i = 1, 2,$$

coincide with one another for any function $f$ of class $C_2$ which vanishes outsides of a compact set. Put

$$g(s, p) = e^{-s}(f_1(s, p)-f_2(s,p)). \tag{2.9}$$

Then $g(s, p)$ is the solution of the equation

$$\frac{\partial}{\partial s} g(s, p) = -g(s, p) + Dg(s, p) \tag{2.10}$$

with the initial condition:

$$g(t, p) = 0. \tag{2.11}$$

Since $|f_1(s, p) - f_2(s, p)| \leqq 2\max|f(p)|$, $g(s, p)$ tends to 0 uniformly in $p$ *as* $s\to\infty$. When $G$ is compact, $g(s, p)$ takes the maximum in $s \geqq t$, $p \in G$. When $G$ is not compact (but locally compact as a Lie group), we also see that $g(s, p)$ takes the maximum, in considering that

$$g(s, p) = e^{-s}\left(\int f(q\cdot p)F_1(t, e, s, dq) - \int f(q\cdot p)F_2(t, e, s, dq)\right)$$

tends to 0 uniformly in $t \leqq s \leqq t'$ ($t'$ being any assigned constant) as $p$ tends to the point at infinity of $G$. Let $g(s_0, p_0)$ be the maximum. When $s_0 = t$, we have $g(s_0, p_0) = 0$ by (2.11). When $s_0 > t$, we have

$$\frac{\partial}{\partial s} g(s_0, p_0) = 0 \, , \quad Dg(s_0, p_0) \leqq 0$$

in remembering the expression (2.2) of $D$. Therefore we see, by virtue of (2.10), that $g(s_0, p_0) \leqq 0$. Thus we see that $g(s, p) \leqq g(s_0, p_0) \leqq 0$. Similarly we obtain $g(s, p) \geqq 0$ in considering the minimum of $g(s, p)$. Consequently we have $g(s, p) \equiv 0$, i.e. $f_1(s,p) \equiv f_2(s, p)$.

*Proof of Th. 4.* Let $D$ be the generating operator of a both-sides invariant Brownian motion. Then we see, by Theorem 1, that $D$ is expressible in the form (1.2) and commutative with each $X_i$. Therefore we have ($C^l_{jk}$ = structural constants)

$$\begin{aligned} 0 &= DX_k - X_k D \\ &= A^i[X_i, X_k] + \frac{1}{2} B^{ij} X_i [X_j, X_k] + \frac{1}{2} B^{ij} [X_i, X_k] X_j \\ &= A^j C^l_{kj} X_l + \frac{1}{2} (B^{ij} C^l_{kj} + B^{jl} C^i_{kj}) X_i X_l \end{aligned}$$

and so we obtain (1.5). Thus the *necessity* is proved. By the above argument, the *sufficiency* is also evident.

We may easily show *Th. 5* by making use of the spatial homogeineity of $\pi(t)$.

## References

1) N. Wiener: Differential space, Jour. math. phys. Mass. Inst. Techn. 2, (1923).

2) Paley and Wiener: Fourier transforms in the complex domain, 1934, Chap. IX, X.

3) P. Lévy: L'addition des variables aléatoires définies sur une circonférence, Bull. Soc. math. France **67**, (1939).

4) A. Kolmogoroff: Zur Theorie der stetigen zufälligen Prozesse, Math. Ann. **108**, (1933).

5) W. Feller: Zur Theorie der stochastischen Prozesse. (Existenz und Eindeutigkeitssätze), Math. Ann. **113**.

6) A. Khintchine: Asymptotische Gesetze der Wahrscheinlichkeitsrechnung, Erg. Math. **2**, 4, (1933).

7) K. Itô: Stochastic differential equations in a differentiable manifold, Nagoya Math. Jour. **1**, (1950).

# ON STOCHASTIC DIFFERENTIAL EQUATIONS

By

KIYOSI ITO

Let $x_t$ be a simple Markoff process with a continuous parameter t, and $F(t,\xi;s,E)$ be the transition probability law of the process:

(1) $F(t,\xi;s,E) = \Pr\{x_s \in E/x_t = \xi\}$,

where the right side means the probability of $x_s \in E$ under the condition: $x_t = \xi$. The differential of $x_t$ at $t = s$ is given by the transition probability law of $x_t$ in an infinitesimal neighborhood of $t = s$:

(2) $F(s-\Delta_2, \xi; s+\Delta_1, E)$.

W. Feller[1] has discussed the case in which it has the following form:

(3) $F(s-\Delta_2,\xi;s+\Delta_1,E) = (1-p(s,\xi)(\Delta_1+\Delta_2))G(s-\Delta_2,\xi;s+\Delta_1,E)$
$+(\Delta_1+\Delta_2)p(s,\xi)P(s,\xi,E) + o(\Delta_1+\Delta_2)$,

where $G(s-\Delta_2,\xi;s+\Delta_1,E)$ is a probability distribution as a function of E and satisfies

(4) $\frac{1}{\Delta_1+\Delta_2}\int_{|\eta-\xi|>\delta} G(s-\Delta_2,\xi;s+\Delta_1,d\eta) \longrightarrow 0$,

(5) $\frac{1}{\Delta_1+\Delta_2}\int_{|\eta-\xi|\leq\delta} (\eta-\xi)^2\, G(s-\Delta_2,\xi;s+\Delta_1,d\eta) \longrightarrow 2a(t,\xi)$,

(6) $\frac{1}{\Delta_1+\Delta_2}\int_{|\eta-\xi|\leq\delta} (\eta-\xi)G(s-\Delta_2,\xi;s+\Delta_1,d\eta) \longrightarrow b(t,\xi)$,

for $\Delta_1+\Delta_2 \longrightarrow 0$ and $p(s,\xi) \geq 0$ and $P(s,\xi,E)$ is a probability distribution in E. The special case of "$p(s,\xi) = 0$" has already been treated by A. Kolmogoroff[2] and S. Bernstein.[3]

We shall introduce a somewhat general definition of the differential of the process $x_t$ (Cf. §5). Let $P_{s,\xi,\Delta_1,\Delta_2}$ denote the conditional probability law:

$\Pr\{x_{s+\Delta_1} - x_{s-\Delta_2} \in E/x_{s-\Delta_2} = \xi\}$, $\Delta_1, \Delta_2 \geq 0$.

If the $[1/\Delta_1+\Delta_2]$- times[4] convolution of $P_{s,\xi,\Delta_1,\Delta_2}$ tends to a probability law $L_{s,\xi}$ with regard to Lévy's law-distance as $\Delta_1+\Delta_2 \longrightarrow 0$, then $L_{s,\xi}$ is called the stochastic differential coefficient at s. $L_{s,\xi}$ is clearly an infinitely divisible law. In the above Feller case the logarithmic characteristic function[5]

Received by the editors March 29, 1949.

$\psi(z, L_{s,\xi})$ of $L_{s,\xi}$ is given by

$$(7)\quad \psi(z,L_{s,\xi}) = ib(s,\xi)z - a(s,\xi)z^2 + p(s,\xi)\int_{-\infty}^{\infty} (e^{iuz}-1)P(s,\xi,du(+)\xi).$$ [6]

A problem of stochastic differential equations is to construct a Markoff process whose stochastic differential coefficient $L_{t,\xi}$ is given as a function of $(t,\xi)$. W. Feller has deduced the following integro-differential equation from (3), (4), (5) and (6):

$$(8)\quad \frac{\partial}{\partial t}F(t,\xi;s,E) + a(t,\xi)\frac{\partial^2}{\partial \xi^2}F(t,\xi;s,E) + b(t,\xi)\frac{\partial}{\partial \xi}F(t,\xi;s,E)$$
$$-p(t,\xi)F(t,\xi;s,E) + p(t,\xi)\int_{-\infty}^{\infty} F(t,\eta;s,E)P(t,\xi,d\eta) = 0.$$

He has proved the existence and uniqueness of the solution of this equation under some conditions and has shown that the solution becomes a transition probability law, and satisfies (3),(4),(5) (6). He has termed the case: $p(t,\xi) \equiv 0$ as continuous case and the case: $a(t,\xi) \equiv 0$ and $b(t,\xi) \equiv 0$ as purely discontinuous case.

It is true that we can construct a simple Markoff process from the transition probability law by introducing a probability distribution into the functional space $R^R$ by Kolmogoroff's theorem,[7] but it is impossible to discuss the regularity of the obtained process, for example measurability, continuity, discontinuity of the first kind etc., as was pointed out by J. L. Doob.[8] To discuss the measurability of the process for example, J. L. Doob has introduced a probability distribution on a subspace of $R^R$ and E. Slutsky has introduced a new concept "measurable kernel".[9] We shall investigate the sense of the term "continuous case" and "purely discontinuous case" used by W. Feller from the rigorous view-point of J. L. Doob and E. Slutsky. A recent research of J. L. Doob[10] concerning a simple Markoff process taking values in an enumerable set has been achieved from this view-point. A research of R. Fortet[11] concerning the above continuous case seems also to stand on the same idea but the author is not yet informed of the details.

In his paper "ON STOCHASTIC PROCESSES (I)" [12] the author has deduced Lévy's canonical form of differential processes with no fixed discontinuities by making use of the rigorous scheme of J. L. Doob. Using the results of the above paper, we shall here construct the solution of the above stochastic differential equation in such a way that we may be able to discuss the regularity of the solution. For this purpose we transform the stochastic differential equation into a stochastic integral equation.

The first and most simple form of stochastic integral is Wiener's integral[13] which is an integral of a function $\sigma(t) \in L_2$ based on a brownian motion $g(t)$: $\int \sigma(t)dg(t)$. In this integral $\sigma(t)$ is not a random function. The author has ex-

extended this notion and defined an integral in case $\sigma(t)$ is a random function satissying some conditions.[14] A brownian motion is a temporally homogeneous and differential (i.e. spatially homogeneous) process with no moving discontinuity. The process $x(t) = \int_a^t \sigma(t)dg(t)$ obtained by Wiener's integral is not temporally homogeneous but spatially homogeneous. In order to obtain a simple Markoff process--which is in general neither temporally nor spatially homogeneous--we shall have to solve a stochastic integral equation:

$$x(t) = \int_a^t \sigma(\tau, x(\tau))dg(\tau)$$

or more generally

$$x(t) = c + \int_a^t m(\tau, x(\tau))d\tau + \int_a^t \sigma(\tau, x(\tau))dg(\tau).$$

The author has published a note[15] on this stochastic integral equation, which concerns the continuous case above mentioned.

In order to discuss the general case we shall have to consider a stochastic integral equation where the integral is based not on a brownian motion but on a more general temporally homogeneous differential process, which will be called a fundamental differential process (Cf. § 6) in this paper.

Chapter I is devoted to the explanation of the fundamental concepts. Some of them are well-known but we shall explain them in a rigorous form for the later use. In Chapter II we shall introduce a stochastic integral of a general type. The results of the author's previous paper[16] will be contained here in an improved form. The aim of this paper will be attained in Chapter III, where we shall investigate a stochastic differential equation and a stochastic integral equation.

The author expresses his hearty thanks to Professor S. Iyanaga, Professor K. Yosida, Professor S. Kakutani and Mr. H. Anzai who have encouraged him with their kind discussions and to Professor J. L. Doob who has given him valuable suggestions to improve the manuscript and friendly aid to publish it.

## I. Fundamental concepts.

§1. Function of random variables. Let $X$ be any set and $B_X$ be a completely additive class of subsets of $X$. When we consider $X$ together with $B_X$ we call it a Borel field $(X, B_X)$. It is evident that $B_X$ may be arbitrarily taken, but in case $X$ is the real number space $R^1$, then we usually take the system $B^1$ of all Borel subsets of $R^1$ as $B_X$, and in case $X$ is $R^A$, $B_X$ is usually the least completely additive class that contains all Borel cyclinder subsets of $R^A$, which we denote by $B^A$. If $B_X$ and $B_Y$ are associated respectively with X and with Y, then we usually associate with the product space the least completely additive class that contains all the sets of the form: $E_X \otimes Y$, $X \otimes E_Y$, $E_X \in B_X$, $E_Y \in B_Y$; this class will be denoted by $B_X \otimes B_Y$. The product of many Borel fields can be similarly defined.

Let $(X,B_X)$ and $(Y,B_Y)$ be Borel fields. A mapping $f(x)$ from X into Y is said to be B-measurable if $f^{-1}(E_Y) \in B_X$ for any $E_Y \in B_Y$. If $f(x)$ is a B-measurable mapping from $(X,B_X)$ into $(Y,B_Y)$ and if $g(x)$ is a B-measurable mapping from $(Y,B_Y)$ into $(Z,B_Z)$, then $g(f(x))$ will be a B-measurable mapping from $(X,B_X)$ into $(Z,B_Z)$.

Let $(\Omega,B_\Omega,P)$ be a probability field, where $\Omega$ is a set, $B_\Omega$ is a completely additive class of subsets of $\Omega$, and P is a probability distribution (p.d.) on $(\Omega,B_\Omega)$. An $(X,B_X)$-valued function $x(\omega)$ defined on $\Omega$ is called an $(X,B_X)$-valued random variable, if it is B-measurable i.e. $x^{-1}(E_X) \in B_\Omega$ for any $E_X \in B_X$. If we put $P_x(E_X) = P(x^{-1}(E_X))$ for $E_X \in B_X$. $P_x$ is a p.d. on a Borel field $(X,B_X)$ which is called the probability law (P.$\ell$.) of x; we also say that x is governed by $P_x$.

Let $x(\omega)$ be an $(X,B_X)$-valued random variable and $f(.)$ be a B-measurable mapping from $(X,B_X)$ into $(Y,B_Y)$. Put $y(\omega) = f(x(\omega))$. Then $y(\omega)$ will be a $(Y,B_Y)$-valued random variable. $y(\omega)$ is called a B-measurable function of $x(\omega)$.

Theorem 1. Let $y_n(\omega)$, n=1,2,..., be real-valued B-measurable functions of an $(X,B_X)$-valued random variable. If $y_n(\omega)$ be convergent in probability, then the limit variable $y(\omega)$ is also coincident with a B-measurable function of $x(\omega)$ up to P-measure 0.

Proof. By taking a subsequence if necessary, we may assume that $y_n(\omega)$ be convergent with P-measure 1. Put $y_n(\omega) = f_n(x(\omega))$. Then

$$P_x(\bigcap_p \bigcup_q \bigcap_{m,n>q} \{\xi; |f_m(\xi) - f_n(\xi)| < 1/p\}) = P(\bigcap_p \bigcup_q \bigcap_{m,n>q} \{\omega; |f_m(x(\omega)) - f_n(x(\omega))| < 1/p\}) = 1.$$

Put $f(\xi) = \lim f_n(\xi)$ in the above $\xi$-set and $= 0$ elsewhere. Then $f(\xi)$ is a B-measurable function of $\xi \in (X,B)$, since the above $\xi$-set belongs to $B_X$. We have clearly, with probability 1,

$$f(x(\omega)) = \lim_n f_n(x(\omega)) = \lim_n y_n(\omega) = y(\omega),$$

which completes the proof.

§2. Conditional probability law. Let $x(\omega)$ and $y(\omega)$ be random variables taking values in $(X, B_X)$ and $(Y, B_Y)$ respectively. A function $P_y(E_Y/\xi)$ of $E_Y \in B_Y$ and $\xi \in X$ will be called the (conditional) probability law of $y(\omega)$ under the condition that $x(\omega) = \xi$ and will be denoted by $P_y(E_Y/x(\omega) = \xi)$ or $\Pr\{y \in E_Y/x = \xi\}$, if and only if

(2.1) $P_y(E_Y/\xi)$ is a p.d. on $(Y, B_Y)$ for any $\xi$,

(2.2) $P_y(E_Y/\xi)$ is a B-measurable function of $\xi \in (X, B)$ for any $E_Y \in B_Y$, and

(2.3) $\int_{E_X} P_y(E_Y/\xi) P_x(d\xi) = \Pr\{x \in E_X \,\&\, y \in E_Y\} = P(x^{-1}(E_X) \cap y^{-1}(E_Y))$.

The existence and uniqueness (up to P-measure 0) of $P_y(E_Y/\xi)$ was proved by J. L. Doob[17] in the case that $(Y, B_Y)$ is the n-dimensional space $(R^n, B^n)$.

$P_y(E_Y/x(\omega))$ i.e. the function of $\omega$ obtained by replacing $\xi$ with $x(\omega)$ in $P_y(E/\xi)$ will be called the conditional p.$\ell$. of $y(\omega)$ under the condition that $x(\omega)$ is determined and it will be also denoted by $\Pr\{y \in E_Y/x(\omega)\}$: this is clearly a real-valued random variable for any assigned $E_Y$. By (2.3) we have

(2.4) $\mathcal{E}\, P_y(E_Y/x(\omega)) = \Pr\{y \in E_Y\}$ ($\mathcal{E}$ = expectation.).

If the p.$\ell$. of the combined random variable $(x(\omega), y(\omega))$, which clearly takes values in $(X \otimes Y, P_X \otimes B_Y)$, is coincident with the direct product measure of $P_x$ and $P_y$: $P_x \otimes P_y$ on $(X \otimes Y, B_X \otimes B_Y)$ then $x(\omega)$ and $y(\omega)$ are called to be independent. The independence of many random variables can be similarly defined. Clearly we have

Theorem 2.1. $x(\omega)$ and $y(\omega)$ be independent. Then

$P_y(E_Y/x(\omega) = \xi) = P_y(E_Y)$ for almost all $(P_x)$ $\xi$,

i.e.

$P_y(E_Y/x(\omega)) = P_y(E_Y)$ for almost all (P) $\omega$.

Theorem 2.2. $x(\omega)$ and $y(\omega)$ be independent. $G(\xi, \eta)$ be a B-measurable mapping from $(X \otimes Y, B_X \otimes B_Y)$ into $(R^1, B^1)$. Put $z(\omega) = G(x(\omega), y(\omega))$. Then we have

$P_z(E/x(\omega) = \xi) = \Pr\{G(\xi, y(\omega)) \in E\}$

for almost all $(P_x)$ $\xi$.

Proof. Since $x(\omega)$ and $y(\omega)$ are independent, we can make use of Fubini's theorem.

$$\Pr\{z \in E, x \in E_X\} = (P_x \otimes P_y)(\{(\xi, \eta); f(\xi, \eta) \in E, \xi \in E_X\})$$

$$= \int_{E_X} P_y(\{\eta; f(\xi, \eta) \in E\}) P_x(d\xi) = \int_{E_X} \Pr\{f(\xi, y(\omega)) \in E\} P_x(d\xi),$$

which completes the proof.

Theorem 2.3. Let $x(\omega)$ and $y(\omega)$ be independent and $G(\xi, \eta)$ any real-valued B-measurable function in $(\xi, \eta)$. If $G(x(\omega), y(\omega)) = 0$ with P-measure 1, then $G(\xi, y(\omega)) = 0$ with P-measure 1 for almost all $(P_x)$ $\xi$.

Proof. By Theorem 2.2 we have $\int_X \Pr\{G(\xi, y(\omega)) = 0\} P_x(d\xi)$ $\Pr\{G(x(\omega), y(\omega)) = 0\} = 1$ and so $\Pr\{G(\xi, y(\omega)) = 0\} = 1$ for almost all $(P_x)$ $\xi$.

§3. Transition probability law. $x(\tau, \omega)$ be a real random variable for any $\tau$, $a \le \tau \le b$. The system $x(\tau, \omega)$, $a \le \tau \le b$, is called a stochastic process, which is also considered as an $(R^I, B^I)$-valued random variable, I being the interval $[a,b]$[18]. The p.l. of $x(s, \omega)$ under the condition that $(x(\tau, \omega), a \le \tau \le t)$[19] is determined:

(3.1) $\Pr\{x(s,\omega) \in E / x(\tau, \omega), a \le \tau \le t\}$ $(t < s)$

is called the transition probability law of this process. If this is equal to

(3.2) $\Pr\{x(s, \omega) \in E / x(t, \omega)\}$

for almost all (P) $\omega$, the process is called a simple Markoff process. In such a process we put

(3.3) $F(t, \xi; s, E) = \Pr\{x(s, \omega) \in E / x(t, \omega) = \xi\}$.

Then we can easily prove, for almost all $(P_{x(t,\omega)})$ $\xi$,

(3.4) $F(t, \xi; s, E) = \int_{-\infty}^{\infty} F(t, \xi; u, d\eta) F(u, \eta; s, E)$, $\quad (t < u < s)$,

which is well-known as Chapman's equation.

If $x(s_\nu, \omega) - x(t_\nu, \omega)$, $\nu = 1, 2, \ldots, n$, are independent random variables for any system of non-overlapping intervals $(t_\nu, s_\nu)$, $\nu = 1, 2, \ldots, n$, then we call $x(\tau, \omega)$, $a \le \tau \le b$, a differential process. This is evidently a simple Markoff process whose transition p.l. is given by

(3.5) $F(t, \xi; s, E) = F_{t,s}(E(-)\xi)$,

where $F_{t,s}$ is the p.l. of $x(s, \omega) - x(t, \omega)$ and $E(-)\xi$ is the set $\{\eta - \xi; \eta \in E\}$; (3.5) will be obtained at once if we substitute $(x(\tau, \omega), a \le \tau \le t)$, $x(s, \omega) - x(t, \omega)$ and $x(s, \omega)$ respectively for $x(\omega)$, $y(\omega)$ and $z(\omega)$ in Theorem 2.2.

## §4. THREE ELEMENTS OF AN INFINITELY DIVISIBLE LAW OF PROBABILITY.

The logarithmic characteristic function (l.c.f.) of an infinitely divisible law of probability (i.d.l.) can be expressed in the form:

$$(4.1)\quad imz - \frac{\sigma^2}{2} z^2 + \int_{|u|>1} (e^{if(u)z} - 1)\frac{du}{u^2} + \int_{|u|\le 1} (e^{if(u)z} - 1 - if(u)z)\frac{du}{u^2}$$

in one and only one way, where $m$ is real, $\sigma \geq 0$, and $f(u)$ is monotone non-decreasing and right-continuous and

$$\int_{|u|\le 1} f(u)^2 \frac{du}{u^2} < \infty;$$

this formula is deduced at once from Levy's formula.[20] These $m, \sigma, f(u)$ will be called the <u>three elements</u> of this i.d.l. . The i.d.l. whose l.c.f. is

$$(4.2)\quad \psi_0(z) = iz - \frac{z^2}{2} + \int_{|u|>1} (e^{iuz} - 1)\frac{du}{u^2} + \int_{|u|\le 1} (e^{iuz} - 1 - izu)\frac{du}{u^2},$$

i.e. $m = \sigma = 1$, $f(u) \equiv u$,

will be called the fundamental i.d.l. in this note.

Theorem 4.1. Let $m(L), \sigma(L)$ and $f(u,L)$ be the three elements of an i.d.l. $L$. Then $m(L), \sigma(L)$ and $f(u,L)$(for any fixed $u$) are all B-measurable in $L=(L(E);E\in B^1)\in(R^{B^1},B^{B^1})$

Remark. By the expression " $m(L)$ is B-measurable in $L\equiv(L(E),E\in B^1)\in(R^{B^1},B^{B^1})$ we mean that there exists at least one B-measurable function $M(L)$ defined on the whole space $(R^{B^1},B^{B^1})$ such that we have $M(L) = m(L)$ for any $L \equiv (L(E), E\in B^1)$ that is an i.d.l. as a function of $E$.

Proof. Let $\phi(z,L)$ be the characteristic function of any i.d.l. $L$. For any $z$, $\phi(z,L)$ is B-measurable in $L\in(R^{B^1},B^{B^1})$, because, if we define $\Phi_z(L)$ by

$$\Phi_z(L) = \lim_{n\to\infty} \sum_{k=-n^2}^{n^2} \exp(ikz/n)L((k-1/n,k/n)) \text{ (if this limit exists)}$$

$$= 0 \text{ (if otherwise)},$$

then $\Phi_z(L)$ (for each $z$) is B-measurable function defined on the whole space $(R^{B^1},B^{B^1})$ and $\Phi_z(L) = \phi(z,L)$ for any i.d.l. $L$.

Let $\psi(z,L)$ be the logarithmic characteristic function of any i.d.l. Since $\psi(z,L)$ is the branch of $\log\phi(z,L)$ which is obtained from $\log\phi(0,L)=1$ by the analytic prolongation along the curve:

$$\phi(\lambda,L),\quad o \leq \lambda \leq z \ (\text{or } o \geq \lambda \geq z)$$

and so it is expressible as

$$\psi(z,L) = \lim_{n\to\infty} \sum_{k=1}^{n} \int_{o}^{1} \frac{\phi(\frac{k}{n}z,L) - \phi(\frac{k-1}{n}z,L)}{(\phi(\frac{k}{n}z,L) - \phi(\frac{k-1}{n}z,L))t+1}\,dt$$

we see that $\psi(z,L)$ is also B-measurable in L for any z. By virtue of the Levy's formula $\psi(z,L)$ is written in the form

$$\psi(z,L) = i\bar{m}(L)z - \frac{\sigma^2(L)}{2}z^2 + \int_{-\infty}^{\infty}\left(e^{izu} - 1 - \frac{izu}{1+u^2}\right)n(du,L),$$

where the measure n is determined by the following procedure (Cf. A. Khintchine: Déduction nouvelle d'une formule de P. Lévy, Bull. d. l'univ. d'etat a Moscou, Serie International, Sect. A, Vol. 1, Fasc. 1, 1937),

$$\Delta(t,L) = \int_{t-1}^{t+1}\psi(z,L)dz - 2\psi(t,L),$$

$$K(u,L) = \frac{1}{2\pi}\lim_{c\to\infty}\int_{-c}^{c}\frac{1-e^{itu}}{it}\Delta(t,L)dt,$$

$$G(v,L) = -\int_{-\infty}^{v}\frac{1}{2\left(1-\frac{\sin u}{u}\right)}dK(u,L),$$

$$n((a,\infty),L) = \int_{a-o}^{\infty}\frac{1+v^2}{v^2}dG(v,L) \qquad (a > 0),$$

$$n((-\infty,a),L) = \int_{-\infty}^{a+o}\frac{1+v^2}{v^2}dG(v,L) \quad (a < 0).$$

Therefore we can prove recursively the B-measurability of the above functions of L. Thus we obtain, for each $a > 0$, a B-measurable functions $N_a(L)$ defined on the whole space $(R^{B^1}, B^{B^1})$ such that $N_a(L) = n((a,\infty),L)$ for any i.d.l. L. We may assume that $N_a(L)$ is monotone-decreasing and left-continuous in a for each L, by taking the supremum of

$N_r(L)$, $r$ running over all rational numbers $r < a$, instead of $N_a(L)$, if necessary.

Now we shall prove, for each $u > 0$, that $f(u,L)$ is B-measurable in L. $f(u,L)$ is written in the following form by the definition.

$f(u,L) = \inf \{a; n((a,\infty),L) < \frac{1}{u}\} \quad (u > 0)$.

Therefore, if we put

$F_u(L) = \inf\{a; N_a(L) < \frac{1}{u}\}$,

$F_u(L)$ (for each $u > 0$) is a function defined on the whole space $RB^1$ and $F_u(L)=f(u,L)$ for any i.d.l. L. The B-measurability of $F_u(L)$ is clear on account of the fact that "$F_u(L) < a$" is equivalent to "$N_a(L) < \frac{1}{u}$", which follows from the definition of $F_u(L)$ and the monotone-property (in a) of $N_a(L)$. Thus we see that $f(u,L)$ (for each $u > 0$) is B-measurable in L. Similarly we can show that $f(u,L)$ $(u < 0)$ is B-measurable in L. It is clear that $f(0,L)$ $(\equiv 0)$ is B-measurable in L.

Now we put

$$\Phi(z,L) \equiv im(L)z - \frac{\sigma^2(L)}{2}z^2 \equiv \psi(z,L) - \int_{|u|>1} (\exp(izf(u,L))-1)\, \frac{du}{u^2}$$

$$- \int_{|u|\leq 1} (\exp(izf(u))-1-izf(u,L))\, \frac{du}{u^2}.$$

Then $\Phi(z,L)$ (for each z) is B-measurable in L, since $\psi(z,L)$ and $f(u,L)$ are B-measurable. But we have

$$m(L) = \frac{1}{2i}(4\Phi(1,L) - \Phi(2,L))$$

and

$$\sigma^2(L) = 2\Phi(1,L) - 4\Phi(2,L),$$

from which follows the B-measurability of $m(L)$ and $\sigma(L)$.

Theorem 4.2. Let $L_\alpha$, $\alpha \in A$, be any system of i.d.l. depending on $\alpha \in A$ and $m_\alpha$, $\sigma_\alpha$ and $f_\alpha(u)$ be the three elements of $L_\alpha$. In order that $L_\alpha$, $\alpha \in A$, be totally bounded in the sense of Levy's law-distance,[21] it is necessary and sufficient that each of $|m_\alpha|$, $\sigma_\alpha$ and $||f_\alpha||_n$, n=1,2,..., is bounded, where

$$||f_\alpha||_n^2 = \int_{|u|\leq n} f_\alpha(u)^2\, \frac{du}{u^2}.$$

Proof. $L_\alpha$ is decomposed as

$$L_\alpha = L_\alpha^{(1)} * L_\alpha^{(2)} * L_{\alpha n}^{(3)} * L_{\alpha n}^{(4)} * L_{\alpha n}^{(5)},$$

where the l.c.f. of the factors are respectively

$$im_\alpha z, \; -\frac{\sigma_\alpha^2}{2} z^2, \; iz\int_{1<|u|<n} f_\alpha(u)\frac{du}{u^2}, \; \int_{0<|u|<n} (e^{if_\alpha(u)z} - 1 - if_\alpha(u)z)\frac{du}{u^2}$$

and $\int_{|u|\geq n} (e^{if_\alpha(u)z} - 1)\frac{du}{u^2}$.

Sufficiency. If the condition is satisfied, $\{L_\alpha^{(1)}\}$, $\{L_\alpha^{(2)}\}$ are clearly totally bounded and $\{L_{\alpha n}^{(3)}\}$ is also totally bounded for any fixed n, since we have, by Schwarz' inequality,

$$\left|\int_{1<|u|<n} f_\alpha(u)\frac{du}{u^2}\right|^2 \leq 2\int_{1<|u|<n} f_\alpha(u)^2\frac{du}{u^2}.$$

$L_{\alpha n}^{(4)}$ has the expectation 0 and the standard deviation $\|f_\alpha\|$ and so $\{L_{\alpha n}^{(4)}, \alpha \in A\}$ is totally bounded. Therefore

$$\{L_{\alpha n}^* \equiv L_\alpha^{(1)} * L_\alpha^{(2)} * L_{\alpha n}^{(3)} * L_{\alpha n}^{(4)}, \alpha \in A\}$$

is totally bounded, and so we have

$$\lim_{c\to\infty} \inf_\alpha L_{\alpha n}^*((-c,c)) = 1.$$

But

$$L_\alpha((-c,c)) \geq L_{\alpha n}^*((-c,c))\, L_\alpha^{(5)}(\{0\}) = L_{\alpha n}^*((-c,c))\exp(-2/n).$$

Consequently we have

$$\lim_{c\to\infty} \inf_\alpha L_\alpha((-c,c)) \geq \exp(-2/n) \text{ and so } \lim_{c\to\infty} \inf_\alpha L_\alpha((-c,c)) = 1,$$

which completes the proof.

Necessity. Let Q(L,c) be Levy's concentration function[22] of the p.d.L. Suppose that $\{L_\alpha\}$ is totally bounded. Then

$$\inf_\alpha Q(L_\alpha^{(2)}, c) \to 1 \qquad \text{as } c \to \infty.$$

But we have $Q(L_\alpha^{(2)}, c) \geq Q(L_\alpha, c)$ by Levy's theorem concerning the non-decreasing of concentration function. Therefore

$$\inf_\alpha Q(L_\alpha^{(2)}, c) \longrightarrow 1 \quad \text{as} \quad c \longrightarrow \infty,$$

and so $\sigma_\alpha$ will be bounded since $L_\alpha^{(2)}$ is a Gaussian distribution with the mean 0 and the standard deviation $\sigma_\alpha$.

$L^{(5)}_{\alpha\, n}$ is decomposed as $L^{(5)}_{\alpha\, n} = L^{(5)}_{\alpha\, n+} * L^{(5)}_{\alpha\, n-}$, where the factors has the l.c.f.

$$\int_n^\infty (\exp(if_\alpha(u)z) - 1)du/u^2 \qquad \text{and} \qquad \int_{-\infty}^{-n} (\exp(if_\alpha(u)z) - 1)du/u^2$$

respectively. By the above-cited Lévy's theorem we see

$$(4.3) \quad \inf_\alpha Q(L^{(5)}_{\alpha\, n+}, c) \geq \inf_\alpha Q(L_\alpha, c) \longrightarrow 1 \qquad \text{as } c \longrightarrow \infty.$$

But $c < f_\alpha(n)$ implies $Q(L^{(5)}_{\alpha\, n+}, c) = \exp(-1/n)$, i.e.

$$(4.4) \quad Q(L^{(5)}_{\alpha\, n}, c) > \exp(-1/n) \text{ implies } \quad c \geq f_\alpha(u).$$

By (4.3) there exists c such that $Q(L^{(5)}_{\alpha\, n+}, c) > \exp(-1/n)$ and so that $c \geq f_\alpha(n)$ for $\alpha \in A$. Thus we see that $f_\alpha(n)$ is bounded for any assigned n. This is also the case for $f_\alpha(-n)$. Consequently we see that $f_\alpha(u)$ is bounded whenever $\alpha \in A$ and $|u| \leq n$, for any fixed n.

If $L_{\alpha(p)}$, p=1,2,..., be chosen from $\{L_\alpha\}$ such that $||f_{\alpha(p)}||_n$ increases indefinitely with p, $L^{(4)}_{\alpha(p)n}$ is approximately a Gaussian distribution with the mean 0 and the standard deviation $||f_{\alpha(p)}||$ as $p \longrightarrow \infty$ by the central limit theorem. Thus we have

$$Q(L^{(4)}_{\alpha(p)n}, ||f_{\alpha(p)}||_n) \longrightarrow \int_{-1}^{1} 1/\sqrt{2\pi} \exp(-t^2/2)dt < 1$$

as $p \longrightarrow \infty$, which contradicts with the fact that

$$\inf_\alpha Q(L^{(4)}_{\alpha\, n}, c) \geq \inf_\alpha Q(L_\alpha, c) \longrightarrow 1 \text{ (as } p \longrightarrow \infty).$$

Thus $||f_\alpha||_n$ proves to be bounded for any fixed n. Therefore

$\{L^{(3)}_{\alpha\, n} * L^{(4)}_{\alpha\, n} * L^{(5)}_{\alpha\, n}\}$ is totally bounded. Therefore $L_\alpha^{(1)}$ must be totally bounded and so $\{m_\alpha\}$ will be bounded.

§5. STOCHASTIC DIFFERENTIATION. $x(\tau,\omega)$, $a \le \tau \le b$, be a stochastic process on $(\Omega, B, P)$ and $F(E,\omega;\Delta_1,\Delta_2)$ be the conditional p.l. of $x(t+\Delta_1,\omega)-x(t-\Delta_2,\omega)$ $(\Delta_1,\Delta_2 \ge 0)$ under the condition that $x(\tau,\omega)$, $a \le \tau \le t-\Delta_2$, be determined. If the $[1/\Delta_1+\Delta_2]$-times convolution of $F(E,\omega;\Delta_1,\Delta_2)$ $\rho$-converges to a distribution $L(E,\omega)$ in probability as $\Delta_1+\Delta_2 \longrightarrow 0$, i.e. for any $\varepsilon > 0$, there exists $\delta = \delta(\varepsilon)$ such that $0 < \Delta_1+\Delta_2 < \delta$ implies

$$\Pr\{\rho(H(E,\omega;\Delta_1,\Delta_2), L(E,\omega)) > \varepsilon\} < \varepsilon,$$

$\rho$ being Lévy's law-distance, then we say that $x(\tau,\omega)$ is differentiable at t and we call $L(E,\omega)$ the differential coefficient of $x(\tau,\omega)$ at t, and we denote it with $D_t x(\tau,\omega)$ or briefly with $Dx(t,\omega)$. This is considered as an $(RB^1, BB^1)$-valued random variable. By taking a convenient sequence $\Delta_1'+\Delta_2' > \Delta_1''+\Delta_2'' > \dots \longrightarrow 0$, we see that $L(E,\omega)$ is the $\rho$-limit of the $[1/\Delta_1^{(n)}+\Delta_2^{(n)}]$-times convolution of $F(E,\omega;\Delta_1^{(n)},\Delta_2^{(n)})$ with P-measure 1, and so we obtain

Theorem 5.1. $Dx(t,\omega)$ is an i.d.l. with P-measure 1.

From the definition we obtain, by making use of Theorem 1,

Theorem 5.2. $Dx(t,\omega)$ is a B-measurable function of $(x(\tau,\omega), a \le \tau \le t)$. If $x(\tau,\omega)$, $a \le \tau \le b$, is a simple Markoff process, then $Dx(t,\omega)$, if it exists, is a B-measurable function of $x(t,\omega)$; the form of the function clearly depends on t. If $x(\tau,\omega)$, $a \le \tau \le b$, is a differential process, then $Dx(t,\omega)$, if it exists, does not depend on $\omega$ but on t. If $x(\tau,\omega)$, $a \le \tau \le b$, is a temporally homogeneous differential process, then $Dx(t,\omega)$ exists and depends neither on $\omega$ nor on t; the l.c.f. of $Dx(t,\omega)$ is equal to the l.c.f. of the p.l. of $x(b,\omega) - x(a,\omega)$ divided by b-a.

We can easily see that, if $F_n^{*n}$ $\rho$-converges to a probability law, then $F_n$ $\rho$-converges to the unit distribution, and so we have

Theorem 5.3. If $x(\tau,\omega)$ is differentiable at t, then it is continuous at t in probability i.e. t is not a fixed discontinuity of this process.

## II. Stochastic Integral.

The integral of the form:

$$\int \sigma(\tau)\,dg(\tau,\omega),$$

where $\sigma(\tau) \in L_2$ and $g(\tau, \omega)$ is a brownian motion, is well-known as Wiener's integral.[23] The author has extended this integral to the case in which $\sigma$ depends not only on $\tau$ but also on $\omega$ and called it a stochastic integral.[24] In this Chapter we treat a more general stochastic integral for the later use.

§6. FUNDAMENTAL DIFFERENTIAL PROCESS. Let $l(t, \omega)$, $a \le t \le b$, be a temporally homogeneous differential process such that both $l(t+0, \omega)$ and $l(t-0, \omega)$ exist and $l(t+0, \omega) = l(t, \omega)$, i.e. $l(t, \omega)$ is continuous in t except possibly for discontinuities of the first kind (hereafter we term this property with " belong to $d_1$-class" ). Further we require that the p.l. of $l(s, \omega) - l(t, \omega)$ has the l.c.f. $(s-t)\psi_o(z)$, where $\psi_o(z)$ is the l.c.f. of the fundamental i.d.l. . Then $l(t, \omega)$, $a \le t \le b$, is defined to be a fundamental differential process. Such a process can be realized on a conveniently defined probability field $(\Omega, B_\Omega, P)$, where the p.l. of $l(a, \omega)$ can be arbitrarily assigned.

Any jump of $l(t, \omega)$ is expressed by a point $(t, u) \in [a,b] \otimes R^1$, t being its position and u being its height: $l(t, \omega) - l(t-0, \omega)$. The number $p(E, \omega)$ of the jumps in E, E being a Borel subset of $[a,b] \otimes R^1$, can be considered a real random variable, which proves to be governed by the Poisson distribution with the mean:

$$\pi(E) = \int_E d\tau\, du/u^2.$$

$p(E, \omega)$ is evidently a function of $l(t, \omega)$, $a \le t \le b$. The system $\{p(E, \omega)\}$ is called the discontinuous part of $l(t, \omega)$, $a \le t \le b$, $l(t, \omega)$ can be expressed as

$$l(t, \omega) = l(a, \omega) + t + g(t, \omega) + \int_a^t \int_{|u| > 1} u p(d\tau\, du, \omega) + \int_a^t \int_{|u|<1} u q(d\tau du, \omega)$$

for any t, $a \le t \le b$, for almost all (P) $\omega$, where $q(E, \omega) = p(E, \omega) - \pi(E)$ and $g(t, \omega)$ is a brownian motion which is also a function of $l(\tau, \omega)$, $a \le \tau \le t$, and is called the continuous part of $l(\tau, \omega)$.

For any disjoint system $E_1, E_2, \ldots, E_n$, $p(E_1, \omega), p(E_2, \omega), \ldots, p(E_n, \omega)$ and $(g(\tau, \omega), a \le \tau \le b)$ are independent.

All these properties can be immediately deduced from the results in the above-cited paper.[25]

§7. STOCHASTIC INTEGRAL BASED ON g. We shall define here an integral of the form:

(7.1) $\int_E \sigma(\tau, \omega)dg(\tau, \omega)$, E being a Borel subset of (a,b], in such a way that it may be a natural extension of Wiener's integral.

First we shall consider the case in which E is an interval: $I_1=(\alpha,\beta]$. By $S(I_1)$ we denote the class of all functions $\sigma(\tau,\omega)$, $\alpha \leq \tau \leq \beta$, $\omega \in \Omega$, satisfying the following three conditions:

(S.1) $\sigma(t,\omega)$ is measurable in $(t,\omega)$,

(S.2) $\int_\alpha^\beta \sigma(\tau,\omega)^2 d\tau < \infty$ for almost all $\omega$, and

(S.3) for any t, $\alpha \leq t \leq \beta$, the system $(\sigma(\tau,\omega), \alpha \leq \tau \leq t;$ $g(\tau,\omega) - g(\alpha,\omega), \alpha \leq \tau \leq t)$ is independent of $(g(\tau,\omega) - g(t,\omega), t \leq \tau \leq \beta)$. As is easily verified, $S(I_1)$ is conditionally complete; if $\sigma_n \in S(I_1)$ tends to $\sigma_\infty$ for almost all $(t,\omega)$ and if $|\sigma_n| \leq \sigma_0 \in S(I_1)$, then $\sigma_\infty \in S(I_1)$.

Theorem 7.1. We can determine, for $\sigma \in S(I_1)$,

(7.1') $\int_\alpha^\beta \sigma(\tau,\omega)dg(\tau,\omega)$ or $\int_{I_1}(\tau,\omega)dg(\tau,\omega)$ or briefly $\int(\sigma,\omega)$ in one and only one way so that it may satisfy (G.1) and (G.2). Furthermore it satisfies (G.3), (G.4), (G.5) and (G.6).

(G.1) When $\sigma(t,\omega)$ is a uniformly stepwise function, i.e., when there exist $\alpha = t_0 < t_1 < \ldots < t_k = \beta$ independent of $\omega$ such that $\sigma(t,\omega) = \sigma(t_{\nu-1},\omega), t_{\nu-1} \leq t < t_\nu$, we have

$$\int(\sigma,\omega) = \sum_{\nu=1}^{k} \sigma(t_{\nu-1},\omega)(g(t_\nu,\omega) - g(t_{\nu-1},\omega)).$$

(G.2) If $\sigma_n \in S(I_1)$ tends to $\sigma_\infty$ for almost all $(\tau,\omega)$, and if $|\sigma_n| \leq \sigma_0 \in S(I_1)$ and further if every B-measurable function $\sigma(t,\omega)$ of $(\sigma_1,\sigma_2,\ldots)$

satisfies (S.3), then $\int(\sigma_n, \omega)$ converges to $\int(\sigma_\infty, \omega)$ in probability.

(G.3) $$\int(c_1\sigma_1+c_2\sigma_2,\omega) = c_1 \int(\sigma_1,\omega) + c_2 \int(\sigma_2,\omega)$$

if $\sigma_1, \sigma_2, c_1\sigma_1 + c_2\sigma_2 \in S(I_1)$.

(G.4) $$\mathcal{E}((\int_{I_1}(\sigma,\omega))^2) = \int_{I_1}\mathcal{E}(\sigma^2(\tau,\omega))d\tau$$

if the right side is finite.

(G.5) If $\sigma_1(\tau,\omega) = \sigma_2(\tau,\omega)$ for $\tau \in I_1, \omega \in \Omega_1$, $\Omega_1$ being a P-measurable set, then $\int(\sigma_1,\omega) = \int(\sigma_2,\omega)$ for almost all (P) $\omega \in \Omega_1$.

(G.6) If $\int_{I_1}\mathcal{E}(\sigma^2(\tau,\omega))d\tau < \infty$, then $\mathcal{E}(\int(\sigma,\omega)) = 0$.

Proof of the existence. In case $\sigma$ is a uniformly stepwise function we define by (G.1). It is evident that this definition satisfies (G.3), (G.4), (G.5) and (G.6). The condition (S.3) will be used in the proof of (G.4) and (G.5) and (G.6).

In order to define $\int(\sigma,\omega)$ for $\sigma \in S(I_1)$ such that

(7.2) $$\|\sigma\|^2 = \int_{I_1}\mathcal{E}(\sigma^2(\tau,\omega))d\tau < \infty,$$

we shall establish

Lemma 7.1. For any $\sigma \in S(I)$ satisfying (7.2) we can find a sequence of uniformly stepwise functions $\sigma_n \in S(I)$ such that

(7.3) $$\|\sigma_n - \sigma\|^2 = \int_{I_1}\mathcal{E}((\sigma_n(\tau,\omega) - \sigma(\tau,\omega))^2)d\tau$$

may tend to 0.

The proof can be achieved by the method[26)] J. L. Doob has used in his research of measurable stochastic processes. By defining $\sigma(\tau,\omega) = 0$ for $\tau \le \alpha$ or $\tau > \beta$, we may assume that $\sigma(\tau,\omega) \in L_2(R^1 \times \Omega)$, and so, for almost all $\omega$, $\sigma(\tau,\omega) \in L_2(R^1)$.

Now put

$$\phi_n(t) = (k-1)/n \quad \text{for } (k-1)/n \leq t < k/n, \quad k=0, \pm 1, \pm 2, \cdots, n=1,2,\cdots.$$

Then $\phi_n(t) \longrightarrow t$ as $n \longrightarrow \infty$, and so we have for all $t$ and for almost all (P) $\omega$,

$$\int_{-\infty}^{\infty} (\sigma(\phi_n(t) + s, \omega) - \sigma(t+s, \omega))^2 ds \longrightarrow 0 \ (n \longrightarrow \infty),$$

since $\sigma(t+s, \omega)$ belongs to $L_2(R^1)$ as a function of $s$ for almost all (P) $\omega$. The left side is always less than $4 \int_{-\infty}^{\infty} \sigma(s, \omega)^2 ds$, since $(a-b)^2 \leq 2a^2 + 2b^2$, and so

$$\int_{\Omega} \int_{t=\ \ -1}^{\beta} \int_{s=-\infty}^{\infty} (\sigma(\phi_n(t)+s, \omega) - \sigma(t+s, \omega))^2 ds dt P(d\omega) \longrightarrow 0.$$

Therefore there exists a sequence $a_1 < a_2 < \ldots$ for almost all $s$ such that

$$\int_{\Omega} \int_{\alpha}^{\beta} |\sigma(\phi_{a_n}(t)+s, \omega) - \sigma(t+s, \omega)|^2 \, dt P(d\omega)$$

$$= \int_{\Omega} \int_{\alpha-s}^{\beta-s} |\sigma(\phi_{a_n}(t)+s, \omega) - \sigma(t+s, \omega)|^2 \, dt \, P(d\omega)$$

$$\leq \int_{\Omega} \int_{\alpha-1}^{\beta} |\sigma(\phi_{a_n}(t)+s, \omega) - \sigma(t+s, \omega)|^2 \, dt \, P(d\omega) \longrightarrow 0.$$

Put

$$\sigma_n(t, \omega) = \sigma(\phi_{a_n}(t-s)+s, \omega), \quad n=1,2,\cdots .$$

Then $\{\sigma_n\}$ is the required sequence. Thus the lemma is proved.

Since $\sigma_n$, $n=1,2,\cdots$, are all uniformly stepwise, we have already defined $\int(\sigma_n, \omega)$ and by (G.4)

$$\mathcal{E}(|\int(\sigma_n,\omega) - \int(\sigma_m,\omega)|^2) = \mathcal{E}(|\int(\sigma_n - \sigma_m,\omega)|^2)$$

$$= \int_{I_1} \mathcal{E}((\sigma_n(\tau,\omega) - \sigma_m(\tau,\omega))^2) d\tau \longrightarrow 0,$$

as $m,n \longrightarrow \infty$. Thus we have $f(\omega)$ such that $\mathcal{E}((\int(\sigma_n,\omega) - f(\omega))^2) \longrightarrow 0$. This $f(\omega)$ does not depend on the special choice of the sequence $\{\sigma_n\}$, but it is determined by $\sigma$ only. We denote it by $\int(\sigma,\omega)$. For this extended definition we can easily verify the properties: (G.3), (G.4) and (G.5). The proof of (G.6) is following. Let $\{\sigma_n\}$ be the sequence obtained above from $\sigma$. Then we have

$$|\mathcal{E}\int(\sigma,\omega)| = |\mathcal{E}\int(\sigma,\omega) - \mathcal{E}\int(\sigma_n,\omega)| = |\mathcal{E}\int(\sigma - \sigma_n,\omega)|$$

$$\leq \sqrt{\mathcal{E}((\int(\sigma - \sigma_n,\omega))^2)} = \sqrt{\int \mathcal{E}((\sigma - \sigma_n)^2)\, dt} \longrightarrow 0.$$

For any $\sigma \in S(I_1)$ we define $\sigma_n$ by

$$\sigma_n(t,\omega) = \phi_n\left(\int_\alpha^t \sigma^2(\tau,\omega) d\tau\right)\sigma(t,\omega),$$

where $\phi_n(\lambda) = 1$ for $|\lambda| \leq n$, and $= 0$ for $|\lambda| > n$.

We shall prove that $\sigma_n(t,\omega)$ satisfies (S.1) and (S.3). It is sufficient to show that $\int_\alpha^t \sigma^2(\tau,\omega) d\tau$ is B-measurable in $(\sigma(\tau,\omega),\ a \leq \tau \leq t)$. Since

$$\int_\alpha^t \sigma^2(\tau,\omega) d\tau = \lim_n \int_\alpha^t \phi_n(\sigma(\tau,\omega))\sigma^2(\tau,\omega) d\tau$$

for any $\omega$, it is sufficient to show the B-measurability of

$$\int_\alpha^t \phi_n(\sigma(\tau,\omega))\sigma^2(\tau,\omega) d\tau, \quad n=1,2,\cdots$$

by Theorem 1, which is evident since we have, by Lemma 7.1,

$$\int_\alpha^t \phi_n(\sigma(\tau,\omega))\sigma^2(\tau,\omega) d\tau = \lim_{p\to\infty} \int_\alpha^t \phi_n(\sigma(\phi_{a_p}(\tau - s) + s,\omega))\sigma^2(\phi_{a_p}(t-s)+s,\omega) d\tau$$

$$= \lim_{p \to \infty} \sum_{a_p(t-s) < \nu < 1+a_p(t-s)} \phi_n\left(\sigma\left(\frac{\nu-1}{a_p}+s,\omega\right)\right)\sigma^2\left(\frac{\nu-1}{a_p}+s,\omega\right)\frac{1}{a_p}$$

for some $s$ and some sequence $a_1 < a_2 < \ldots$ .

Let $\Omega_n$ be the set of all $\omega$ such that $\int_{I_1} \sigma^2(\tau,\omega)d\tau \leq n$. Then we have $\Omega_1 \subseteq \Omega_2 \subseteq \ldots$ and $\bigcup \Omega_n = \Omega - N$, $P(N) = 0$ by (S.2). Furthermore $\sigma_n(t,\omega) = \sigma(t,\omega)$, $t \in I_1$, $\omega \in \Omega_n$. Since $\|\sigma_n\| < \infty$, we have already defined $\int(\sigma_n,\omega)$. We define $\int(\sigma,\omega)$ as $\int(\sigma_n,\omega)$ on $\Omega_n - \Omega_{n-1}$, and 0 on N.

This extended definition satisfies (G.3), (G.4), (G.5) and (G.6), as is easily verified. We shall prove (G.2). If $\|\sigma_o\| < \infty$, then $\|\sigma_n - \sigma_o\| \to 0$. Therefore we have $\mathcal{E}((\int(\sigma_n,\omega) - \int(\sigma_\infty,\omega))^2) \to 0$ by (G.4) and so $\int(\sigma_n,\omega)$ will tend to $\int(\sigma_\infty,\omega)$ in probability. In the general case we consider

$$\sigma_n^{(m)}(t,\omega) = \phi_m\left(\int_\alpha^t \sigma_o^2(\tau,\omega)d\tau\right)\sigma_n(t,\omega),\ n=1,2,\ldots,\infty.$$

By virtue of (S.2) we have

$$P(\{\omega;\sigma_n^{(m)}(t,\omega) = \sigma_n(t,\omega),\ n=1,2,\ldots,\infty\})$$

$$\geq P(\{\omega;\int_\alpha^\beta \sigma_o^2(\tau,\omega)d\tau < m\}) > 1 - \frac{\varepsilon}{2}$$

for a sufficiently large m. Since $\sigma_n^{(m)}(t,\omega) \to \sigma_\infty^{(m)}(t,\omega)$, $|\sigma_n^{(m)}| \leq \sigma_o^{(m)}$ and $\|\sigma_o^{(m)}\| < \infty$, we have, for a sufficiently large n,

$$P(\{\omega;\ |\int(\sigma_n^{(m)},\omega) - \int(\sigma_\infty^{(m)},\omega)| > \varepsilon\}) < \frac{\varepsilon}{2}.$$

We obtain from the above two conditions

$$P(\{\omega;\ |\int(\sigma_n,\omega) - \int(\sigma_\infty,\omega)| > \varepsilon\}) < \varepsilon .$$

Proof of the uniqueness. Let $\int_1(\sigma,\omega)$ and $\int_2(\sigma,\omega)$ satisfy (G.1) and (G.2). We see by (G.1) that $\int_1(\sigma,\omega) = \int_2(\sigma,\omega)$ for any uniformly stepwise function $\sigma \in S(I_1)$. If $\sigma \in S(I_1)$ is bounded ($|\sigma| \leq m$), we have $\|\sigma\| \leq m(\beta-\alpha)$. The sequence $\{\sigma_n\}$ obtained by Lemma 7.1 from $\sigma$ is uniformly bounded ($|\sigma_n| \leq m$). By taking a subsequence we may assume that $\sigma_n \longrightarrow \sigma$ for almost all $(\tau,\omega)$. $\sigma_n$ being uniformly stepwise, we have $\int_1(\sigma_n,\omega) = \int_2(\sigma_n,\omega)$, $n = 1,2,\ldots$, and so $\int_1(\sigma,\omega) = \int_2(\sigma,\omega)$ by (G.2).

By (G.2) and (G.3) we have

Theorem 7.2. Let $\sigma_n \in S(I_1)$, $n=1,2,\ldots$ . If every measurable function of the joint variable $(\sigma_1, \sigma_2, \ldots)$ satisfies (S.3) and if $\sum|\sigma_n| \in S(I_1)$, then we have

$$\int_{I_1} \sum_{n=1}^{\infty} \sigma_n(\tau,\omega)dg(\tau,\omega) = \sum_{n=1}^{\infty} \int_{I_1} \sigma_n(\tau,\omega)dg(\tau,\omega)$$

in the sense of 'limit in probability'.

Let $E$ be any Borel subset of $(\alpha,\beta]$. For $\sigma \in S(I_1)$ we define as follows:

$$\int_E \sigma(\tau,\omega)dg(\tau,\omega) = \int_I \sigma(\tau,\omega)C_E(\tau)dg(\tau,\omega),$$

where $C_E(\tau)$ is the characteristic function of the set $E$. This definition is clearly independent of the choice of $(\alpha,\beta]$ containing $E$, and so it is an extension of (7.1').

Theorem 7.3. Let $\sigma \in S(I)$. Then we can define $\int_E \sigma\, dg$ for $E \subsetneq I$. If $\{E_n\}$ be a disjoint sequence of Borel subsets of $I$. Then we have

$$\int_{\sum_n E_n} \sigma \, dg = \sum_n \int_{E_n} \sigma \, dg$$

in the sense of 'limit in probability'.

This is clear by the previous theorem.

§8. THE CONTINUOUS KERNEL OF THE INTEGRAL. Let $\sigma \in S(I)$, $I = (\alpha,\beta]$. Then $\sigma \in S(I_t)$, $I_t = (\alpha,t]$ for any $t$, $\alpha \leq t \leq \beta$ , and so we can define

$$(8.1)\quad \int_\alpha^t \sigma(\tau,\omega)dg(\tau,\omega),$$

which is uniquely (up to P-measure 0) determined for any assigned t.

Theorem 8. We can determine, for $\sigma \in S(I)$, a stochastic process:

$$\left( *\int_\alpha^t \sigma(\tau,\omega)dg(\tau,\omega),\ \alpha \le t \le \beta \right)$$

in one and only one way (up to P-measure 0) so that the process may be continuous in t with P-measure 1 and that

$$(G') \qquad *\int_\alpha^t \sigma(\tau,\omega)dg(\tau,\omega) = \int_\alpha^t \sigma(\tau,\omega)dg(\tau,\omega)$$

with P-measure 1 for any assigned t.

For this integral we have

$$(G'.4) \qquad c^2P(\{\omega;\sup_{\alpha\le t\le\beta}|*\int_\alpha^t \sigma(\tau,\omega)dg(\tau,\omega)| \ge c\})$$
$$\le \mathcal{E}((\int_\alpha^\beta \sigma(\tau,\omega)dg(\tau,\omega))^2) = \int_\alpha^\beta \mathcal{E}\sigma^2(\tau,\omega)d\tau,$$

if the right side is finite.

Proof of the existence. If $\sigma(\tau,\omega)$ is uniformly stepwise in $\alpha \le \tau \le \beta$, then it is so in $\alpha \le \tau \le t$. In this case we shall define

$$*\int_\alpha^t \sigma(\tau,\omega)dg(\tau,\omega)$$

by (G') i.e. by (G.1), which is clearly continuous on account of the continuity of $g(\tau,\omega)$. In order to show (G'.4) we need

Lemma 8. Let $y_\nu(\omega), x_\nu(\omega)$, $\nu = 1,2,\dots,m$, be random variables satisfying the following conditions:

(8.2) $\mathcal{E}(x_\nu) = 0$, $\mathcal{E}(x_\nu^2)$, $\mathcal{E}(y_\nu^2) < \infty$, $\nu = 1,2,\dots,m$,

(8.3) $(x_k(\omega), x_{k+1}(\omega),\dots,x_m(\omega))$ is independent of $(y_1(\omega), x_1(\omega), y_2(\omega), x_2(\omega),\dots, y_{k-1}(\omega), x_{k-1}(\omega), y_k(\omega))$ for $k=1,2,\dots,m$.

Then we have

$$(8.4) \qquad c^2P(\{\omega;\max_{k=1}^n |\sum_{\nu=1}^k y_\nu(\omega)x_\nu(\omega)| > c\})$$
$$\le \mathcal{E}((\sum_{\nu=1}^m y_\nu(\omega)x_\nu(\omega))^2) = \sum_{\nu=1}^m \mathcal{E}(y_\nu(\omega)^2)\mathcal{E}(x_\nu(\omega)^2).$$

In case $y_\nu(\omega) \equiv 1$, $\nu=1,2,3,\dots,m$, this lemma is nothing but the so-called Kolmogoroff's inequality, whose proof will be available for our lemma, if we give it a slight modification.

Now, let $\{s_n\}$ be a sequence dense in $(\alpha,\beta]$. By the continuity of $*\int_\alpha^t \sigma\, dg$, we have

$$\sup_{t\in I} \left| *\int_\alpha^t \sigma\, dg\right| = \sup_{\nu=1}^{\infty} \left| *\int_\alpha^{s_\nu} \sigma\, dg\right|.$$

Since $\{\omega\,;\ \max_{\nu=1}^m \left|\int_\alpha^{s_\nu}\sigma\, dg\right| > c\}$, $m=1,2,\dots$, is a monotone increasing sequence tending to $\{\omega\,;\ \sup_{\nu=1}^{\infty} \left|\int_\alpha^{s_\nu}\sigma\, dg\right| > c\}$, we need only prove

$$c^2 P\{\omega\,;\ \max_{\nu=1}^{m} \left|\int_\alpha^{s_\nu}\sigma\, dg\right| > c\} \le \mathcal{E}((\int_\alpha^\beta \sigma\, dg)^2) = \int_\alpha^\beta \mathcal{E}(\sigma^2)d\tau$$

for the proof of (G'.4). Let $\alpha = t_0 < t_1 < \dots < t_n = \beta$ be chosen so that $\sigma(\tau,\omega)$ may be constant in $(t_{\nu-1}, t_\nu)$ for each $\omega$, $\nu=1,2,\dots$. If we rearrange $s_1, s_2, \dots, s_m$, $t_0, t_1, \dots, t_n$ in the order of magnitude and denote it with $u_0, u_1, \dots, u_{m+n}$, it is sufficient to prove

$$c^2 P(\{\omega,;\max_{\nu=0}^{m+n} \left|*\int_\alpha^{s_\nu}\sigma\, dg\right| > c\}) \le \mathcal{E}((\int_\alpha^\beta \sigma\, dg)^2) = \int_\alpha^\beta \mathcal{E}(\sigma^2)d\tau$$, which we obtain at once by putting $y_\nu = \sigma(u_{\nu-1},\omega)$, $x_\nu = g(u_\nu,\omega) - g(u_{\nu-1},\omega)$ in the above Lemma 8.

In order to define $*\int_\alpha^t \sigma\, dg$ for $\sigma \in S(I)$ such that

$$(8.5)\qquad \|\sigma(\tau,\omega)\|^2 = \int_\alpha^\beta \mathcal{E}\sigma^2(\tau,\omega)d\tau < \infty,$$

we define $\sigma_n$, $n=1,2,\dots$, from $\sigma$ by Lemma 7.1. By choosing a convenient subsequence, we may assume

$$(8.6)\qquad \|\sigma_{n+1} - \sigma_n\|^2 \le 1/8^n,\ n=1,2,\dots .$$

$\sigma_{n+1} - \sigma_n$, $n=1,2,\dots$, being uniformly stepwise, we can apply (G'.4) to it and we have

$$P(\{\omega\,;\sup_{\alpha\le t<\beta} \left|*\int_\alpha^t \sigma_{n+1} - *\int_\alpha^t \sigma_n\right| \ge 1/2^n\}) \le 1/2^n.$$

Therefore $*\int_\alpha^t \sigma_n dg$ is uniformly (in t) convergent with P-measure 1 by Borel-Cantelli's theorem; the limit depends only on $\sigma(t,\omega)$ and it is independent of $\{\sigma_n\}$, if we choose them so that $\sigma_n(t,\omega)$ may be a function of $\sigma(\tau,\omega)$, $a \le \tau \le t$, for any t. We define $*\int_\alpha^t \sigma\, dg$ as this limit. But $*\int_\alpha^t \sigma_n dg = \int_\alpha^t \sigma_n dg$ and so $*\int_\alpha^t \sigma\, dg = \int_\alpha^t \sigma\, dg$ with P-measure 1 for any assigned t.

By the uniformity of the convergence, $*\int_\alpha^t \sigma\, dg$ is continuous in t and it satisfies (G'.4). For any function $\sigma$ for which (8.5) does not hold, we can define $\int_\alpha^t \sigma\, dg$ in the same way as the previous § 7.

<u>Proof of the uniqueness.</u> Let $*\int_1$ and $*\int_2$ satisfy the assigned conditions. By (G') we have

$$*\int_{1,\alpha}^{\gamma} \sigma\, dg = *\int_{2,\alpha}^{\gamma} \sigma\, dg = \int_\alpha^{\gamma} \sigma\, dg$$

for any rational number $\gamma$ with P-measure 1, and by the continuity of $*\int_1$ and $*\int_2$ we obtain

$$*\int_{1,\alpha}^{t} \sigma\, dg = *\int_{2,\alpha}^{t} \sigma\, dg$$

for any t with P-measure 1.

Definition 8. The above $*\int_\alpha^t \sigma\, dg$ is called the continuous kernel of $\int_\alpha^t \sigma\, dg$.

## §9. STOCHASTIC INTEGRALS BASED ON p AND q AND THEIR REGULAR KERNEL.

We consider p and q appearing in the resolution of a fundamental differential process (§6). Let I be a half-open 2-dimensional interval $(\alpha,\beta] \times (\gamma,\delta]$, $0 < \gamma, \delta < \infty$, $a \le \alpha \le \beta \le b$. By F(I) we denote the class of all functions $f(t,u,\omega)$, $\alpha \le t \le \beta$, $\gamma \le u \le \delta$, such that

(F.1) $f(t,u,\omega)$ is measurable in $(t,u,\omega)$,

(F.2) $\int_I |f(t,u,\omega)|\, d\tau\, du/u^2 < \infty$ with P-measure 1, and

(F.3) $(f(t,u,\omega), \alpha \le \tau < t; p(E,\omega), E \subseteq R^2(t))$ is independent of

$(p(E,\omega),\ E \subseteq I \frown R^2_+(t)),\ R^2_+(t), R^2_-(t)$ being respectively the half-plane $\{(\tau,u);\ \tau \geq t\}$,

In the same way as in §7 we obtain

Theorem 9.1. We can determine, for $f \in F(I)$,

$$(9.1) \qquad \int_\alpha^\beta\int_\gamma^\delta f(\tau,u,\omega)dp(\tau,u,\omega) \text{ or } \int_I f(\tau,u,\omega)dp(\tau,u,\omega)$$

or briefly $\int(f,\omega)$

in one and only one (up to P-measure 1) way so that it may satisfy (P.1) and (P.2). Furthermore it satisfies (P.3), (P.4), (P.5) and (P.6).

(P.1) When $f(\tau,u,\omega)$ is a uniformly stepwise function, i.e. when there exists $\alpha = t_0 < t_1 < \dots < t_n = \beta$, $\gamma = u_0 < u_1 < \dots < u_n = \delta$ independent of $\omega$ such that $f(\tau,u,\omega) = f(t_{\mu-1},u_{\nu-1},\omega)$, $t_{\mu-1} \leq t < t_\mu$, $u_{\nu-1} \leq u < u_\nu$, $\mu = 1,2,\dots,m$, $\nu = 1,2,\dots,n$, we have

$$\int(f,\omega) = \sum_{\mu=1}^{m}\sum_{\nu=1}^{n} f(t_{\mu-1},u_{\nu-1},\omega)p((t_{\mu-1},t_\mu)\times(u_{\nu-1},u_\nu),\omega).$$

(P.2) If $f_n(\tau,u,\omega) \in F(I)$ tends to $f_\infty(\tau,u,\omega)$ for almost all $(\tau,u,\omega)$, and if $|f_n| \leq f_0 \in F(I)$ and furthermore if every B-measurable function $f(\tau,u,\omega)$ of $(f_1,f_2,\dots)$ satisfies (F.3), then $\int(f_n,\omega)$ converges to $\int(f_\infty,\omega)$ in probability.

(P.3) $\int(c_1f_1+c_2f_2,\omega) = c_1\int(f_1,\omega) + c_2\int(f_2,\omega)$ if $f_1,f_2,c_1f_1+c_2f_2 \in F(I)$.

(P.4) $\mathcal{E}(\int(f,\omega)) = \int_I \mathcal{E}f(\tau,u,\omega)d\tau\, du/u^2$ if $\int_I \mathcal{E}|f(\tau,u,\omega)|d\tau\, du/u^2 < \infty$.

(P.5) If $f_1 = f_2$ for $(\tau,u) \in I$, $\omega \in \Omega_1$, $\Omega_1$ being a P-measurable set, then $\int(f_1,\omega) = \int(f_2,\omega)$ for almost all $\omega$ on $\Omega_1$.

Let $f(\tau,u,\omega)$, $\alpha \leq t \leq \beta$, $\gamma \leq u$, $\omega \in \Omega (\gamma > 0)$ belong to $F((\alpha,\beta]\times(\gamma,n])$ for any integer $n > \gamma$. Then we can define

$$\int_\alpha^\beta\int_\gamma^n f(\tau,u,\omega)dp(t,u,\omega), \qquad n > \delta .$$

Let $\Omega_n$ denote the set $\{\omega;\ p((\alpha,\beta]\times(n,\infty),\omega) = 0$. Then we have, for $\omega \in \Omega_n$ and $m > n$,

$$\int_\alpha^\beta\int_\gamma^m \text{fdp} = \int_\alpha^\beta\int_\gamma^n \text{fdp}.$$

But $\Omega_n \subseteq \Omega_{n+1}$, and $P(\Omega_n) = \exp\{-\int_{\tau=\alpha}^\beta\int_{u=n}^\infty d\tau\, du/u^2\} \longrightarrow 1$ as $n \longrightarrow \infty$.

Therefore $\lim_{n\longrightarrow\infty}\int_\alpha^\beta\int_\gamma^n$ fdp exists with P-measure 1 and it has the above properties: (F.1), (F.2), etc., and so we denote it with

$$\int_\alpha^\beta\int_\gamma^\infty f(\tau,u,\omega)dp(\tau,u,\omega).$$

Similarly we can define, for $\delta < 0$,

$$\int_\alpha^\beta\int_{-\infty}^\delta f(\tau,u,\omega)dp(\tau,u,\omega).$$

Let E be any Borel subsets of $I = (\alpha,\beta] \times (\gamma,\delta]$, $\gamma,\delta > 0$, $a < \alpha \leq \beta \leq b$. Then we define, for $f \in F(I)$, as follows,

$$\int_E f(t,u,\omega)dp(t,u,\omega) = \int_I f(t,u,\omega)C_E(t,u)dp(t,u,\omega),$$

where $C_E(t,u)$ is the characteristic function of E. This definition is clearly independent of the choice of the interval $I \supseteq E$.

From (P.2) and (P.3) we obtain

Theorem 9.2. If $E_n$, n=1,2,..., are disjoint Borel subsets of I, and if $f \in F(I)$, then we have

$$\int_{\sum E_n} \text{fdp} = \sum_{\nu=1}^n \int_{E_n} \text{fdp}$$

in the sense of limit in probability.

From (P.4) we obtain

Theorem 9.3. If $f \in F(I)$, and if $\int_E |f(\tau,u,\omega)|d\tau du/u^2 < \infty$, then

$$\mathcal{E}(\int_E \text{fdp}) = \int_E \mathcal{E}(f(\tau,u,\omega))dt\, du/u^2.$$

Now we define the regular kernel of $\int_E fdp$, $f \in F(I)$, which will be denoted by $*\int fdp$. Firstly we consider the case in which $f \geq 0$ and

$$(9.2) \qquad \int_I \mathcal{E}(f(t,u,\omega))dtdu/u^2 < \infty.$$

Let $\{\alpha_\nu\}$, $\{\gamma_\nu\}$ be dense in $(\alpha, \beta]$ and in $(\gamma, \delta]$ respectively. We assume that $\{\alpha_\nu\} \ni \alpha, \beta$ and $\{\gamma_\nu\} \ni \gamma, \delta$. We shall here call any finite sum of intervals of the form $(\alpha_\mu, \alpha_\nu] \times (\gamma_\lambda, \gamma_\kappa]$ an elementary set. Then the system of all elementary sets is enumerable and forms a finitely additive class. Let $E_1, E_2, \ldots, E_n$ disjoint elementary sets and E their sum. Then we have

$$\int_E fdp = \sum_{\nu=1}^n \int_{E_\nu} fdp$$

with P-measure 1. Since this system of all these equalities is enumerable, we see that they hold simultaneously with P-measure 1. Since $f \geq 0$, we obtain $\int_E fdp \geq 0$ for any elementary set E with P-measure 1. Thus $\int_E fdp$ is a finitely additive measure for almost all $\omega$.

Let G be any set open in I and B any Borel subset of I. We define

$$*\int_G fdp = \sup\{\int_E fdp;\ \text{E is any elementary set whose closure} \leq G\}$$

$$*\int_B fdp = \inf\{\int_G fdp;\ \text{G is any open (in I) set that contains B}\}.$$

As is easily proved, we have

$$*\int_B fdp = \sum_{n=1}^\infty *\int_{B_n} fdp, \qquad B = \sum_{n=1}^\infty B_n,$$

for any disjoint system of Borel sets $\{B_n\}$.

Now we shall prove that

$$(9.3) \qquad *\int_B fdp = \int_B fdp$$

with P-measure 1 for any Borel set B. For any open set G we can choose a monotone-increasing sequence of elementary sets $\{E_n\}$ so that the closure of $E_n$ is contained in G, n=1,2,... and that $G = \bigcup E_n$. By Theorem 9.2 we see that

$$\int_G fdp = \text{l.i.P.} \int_{E_n} fdp \quad (\text{l.i.P.} = \text{limit in probability})$$

and so that $\int_G fdp \leq * \int_G fdp$ with P-measure 1. But $\int_G fdp \geq * \int_G fdp$, since $\int_E fdp$ is monotone in E on account of $f \geq 0$. Thus we see that (9.3) holds for any open set. By (9.2) and (P.4), $\int_I fdp$ is finite for almost all. Therefore by Theorem 9.3 we have

$$\int_B fdp = \text{l.i.P.} \int_{B_n} fdp,$$

if $B_1 \supseteq B_2 \supseteq \ldots \longrightarrow B$. Consequently (9.3) holds also for any closed (in I) set B.

Let B be any Borel set. By (9.2) we can find a sequence of open sets G and a sequence of closed sets $\{F_n\}$ such that

$$F_1 \subseteq F_2 \subseteq \ldots . \subseteq B \subseteq \ldots \subseteq G_2 \subseteq G_1$$

and that

$$\int_{G_n - F_n} \mathcal{E}(f(t,u,\omega))dtdu/u^2 \leq 1/n, \quad n=1,2,\ldots .$$

It is clear that

$$\int_{F_1} fdp \leq \int_{F_2} fdp \leq \ldots \leq \int_B fdp \leq \ldots \leq \int_{G_2} fdp \leq \int_{G_1} fdp$$

with P-measure 1. Furthermore we have

$$\mathcal{E}\left( \int_{G_n} fdp - \int_{F_n} fdp\right) = \int_{G_n - F_n} \mathcal{E}(f)dtdu/u^2 \leq 1/n \longrightarrow 0.$$

Therefore

$$(9.4) \quad \int_B fdp = \text{l.i.P.} \int_{G_n} fdp = \text{l.i.P.} \int_{F_n} fdp.$$

But we have

$$* \int_{F_1} f dp \leq * \int_{F_2} f dp \leq \dots \leq * \int_B f dp \leq \dots \leq * \int_{G_2} f dp \leq * \int_{G_1} f dp$$

by the definition, and so

$$(9.5) \qquad \lim_n * \int_{F_n} f dp \leq * \int_B f dp \leq \lim_n * \int_{G_n} f dp.$$

As has been already proved,

$$* \int_{F_n} f dp = \int_{F_n} f dp, \quad * \int_{G_n} f dp = \int_{G_n} f dp, \; n=1,2,\dots,$$

with P-measure 1 and so we obtain (9.3) at once from (9.4) and (9.5).

For $f \in F(I)$ for which $f \geq 0$ but (9.2) does not hold, we define $* \int_B f dp$ by

$$* \int_B f dp = \lim_{N \to \infty} * \int_B \phi_N\left( \int_{I \cap R_-^2(t)} f(\tau, \lambda, \omega) d\tau d\lambda / \lambda^2 \right) f(t,u,\omega) dp(t,u,\omega),$$

where $\phi_N(\lambda)$ is the characteristic function of the interval $[-N,N]$.

For any general $f$ we shall define

$$* \int_B f dp = * \int_B \frac{|f|+f}{2} dp - * \int_B \frac{|f|-f}{2} dp.$$

Now, we shall define $\int_I f dq$. For $f \in F(I)$, $I = (\alpha, \beta] \times (\gamma, \delta]$, $\gamma, \delta > 0$, we put

$$(9.6) \qquad \int_I f dq = \int_I f dp - \int_I f(\tau, u, \omega) d\tau du/u^2.$$

Then we can easily prove that

$$(9.7) \qquad \mathcal{E}\left(\left(\int_I f dq\right)^2\right) = \int_I \mathcal{E}(f^2) d\tau du/u^2, \quad \mathcal{E}\left(\int_I f dq\right) = 0,$$

for $f$ such that

$$\int_I \mathcal{E}(f^2) d\tau du/u^2 < \infty.$$

For $I = (\alpha, \beta] \times (0, \delta]$, $a \leq \alpha \leq \beta \leq b$, $0 < \delta < \infty$, we define $F_1(I)$ as the class of all functions $f(t,u,\omega)$ satisfying (F.1), (F.3) and

(F'.2) $\int_I (f(\tau,u,\omega))^2 d\tau du/u^2 < \infty.$

We shall define $\int_I fdq$ for $f \in F_1(I)$. Firstly we consider the case:

(F".2) $\int_I \mathcal{E}(f(\tau,u,\omega)^2)\, d\tau du|u^2 < \infty$

Let $I_n$ denote $(\alpha,\beta] \times (1/n, \delta]$. Then

$$\mathcal{E}((\int_{I_n} |f(\tau,u,\omega)| d\tau du/u^2)^2) \leq \int_{I_n} d\tau du/u^2 \int_{I_n} \mathcal{E}(f(\tau,u,\omega)^2) d\tau du/u^2$$

with P-measure 1. Therefore $f \in F(I_n)$ and so we can define

$$\int_{I_n} fdq$$

by (9.6). By (9.7) and (F" .2) we have

$$\mathcal{E}((\int_{I_n} fdq - \int_{I_m} fdq)^2) = \mathcal{E}((\int_{I_n - I_m} fdq)^2)$$

$$= \int_{I_n - I_m} \mathcal{E}(f^2) d\tau du/u^2 = \int_\alpha^\beta \int_{\frac{1}{n}}^{\frac{1}{m}} \mathcal{E}(f^2) d\tau du/u^2 \longrightarrow 0.$$

We shall define

$$\int_I fdq = \text{l.i.m.} \int_{I_n} fdq \ (\text{l.i.m.} = \text{limit in mean}).$$

This extended definition satisfies (9.7) evidently.

In order to define the integral in the general c ase we put

$$f_n(t,u,\omega) = \phi_n(\int_{\tau=\alpha}^t \int_{u=0}^\delta (f(\tau,u,\omega)^2) d\tau du/u^2) f(t,u,\omega),$$

where $\phi_n$ is the characteristic function of the interval $(-n,n)$. By (F'.2) we have

$$P(\Omega_n) \uparrow 1, \text{ where } \Omega_n = \{\omega; f_n(\tau,u,\omega) = f(\tau,u,\omega) \text{ for } (\tau,u) \in I\}.$$

We shall define

$$\int_I fdq \text{ as } \int_I f_n dq \text{ on } \Omega_n.$$

Similarly we can define the integral in case $I = (\alpha,\beta] x (\gamma, 0]$. In case $I = (\alpha,\beta] x (\gamma,0]$. In case $I = (\alpha,\beta]\ x(0,\delta]$, we define

$$\int_I fdq = \int_{I_1} fdq + \int_{I_2} fdq,\ I_1 = (\alpha,\beta]\ x\ (\gamma,0],\ I_2 = (\alpha,\beta]\ x(0,\delta].$$

As in the previous integrals we can define $\int_E fdq$ as $\int fC_E dq$, $C_E(t,u)$ being the characteristic function of E.

For the regular kernel of this integral we establish

Theorem 9.4. Let $I = (\alpha,\beta] x(\gamma,\delta]$ and $E \subseteq (\gamma,\delta]$. For $f \in F_1(I)$ we can determine $*\int_\alpha^t \int_E fdq$, $\alpha \leq t \leq \beta$, which belongs to $d_1$-class as a function of t with P-measure 1 and satisfies

$$*\int_\alpha^t \int_E fdq = \int_{[\alpha,t]xE} fdq$$

with P-measure 1. We have, for this regular kernel,

$$c^2 \Pr\{ \sup_{\alpha \leq t \leq \beta} |* \int_\alpha^t \int_E fdq\ | > c\} \leq \int_\alpha^\beta \int_E \mathcal{E}((f(t,u,\omega))^2) dtdu/u^2.$$

The proof can be achieved in the same way as that of Theorem 8.

III. STOCHASTIC DIFFERENTIAL EQUATION AND STOCHASTIC INTEGRAL EQUATION.

§10. Stochastic differential equation. We shall solve a stochastic differential equation:

(10.1) $Dx(t,\omega) = L(t,x(t,\omega)),\ a \leq t \leq b,$

under the condition:

(10.2) $P_{x(a,\omega)} = L,$

where L is a given distribution on $R^1$.

Theorem 10. We can construct a simple Markoff process $x(t,\omega)$ on a convenient probability field $(\Omega, B_\Omega, P)$ so that $x(t,\omega)$ may satisfy (10.1) and (10.2) and that $x(t,\omega)$ may belong to $d_1$-class with P-measure 1, if the three elements $m(\tau,\xi)$, $\sigma(\tau,\xi)$ and $f(\tau,u,\xi)$ of $L(\tau,\xi)$ satisfies the following conditions (A) and (B):

(A) $|m(\tau,\xi) - m(\tau,\eta)| \leq M|\xi - \eta|,$

$$|\sigma(\tau,\xi)-\sigma(\tau,\eta)| \leq S|\xi-\eta|,$$

$$\| f(\tau,u,\xi) - f(\tau,u,\eta)\|_n \equiv \left( \int_{|u|\leq n} (f(\tau,u,\xi)-f(\tau,u,\eta))^2 du/u^2\right)^{\frac{1}{2}}$$

$$\leq F_n|\xi-\eta|,$$

where $M,S,F_n, n=1,2,\ldots,$ are independent of $(\tau,\xi,\eta)$,

(B) $L(\tau,\xi)$ is continuous in $\tau$ with regard to Lévy's distance for any fixed $\xi$.

Proof. We construct a fundamental differential process $\ell(\tau,\omega)$, $a \leq \tau \leq b$, on a convenient probability field so that $\ell(a,\omega)$ may be governed by L, and we consider a stochastic integral equation:

$$(10.3) \quad x(t,\omega) = \ell(a,\omega) + \int_a^t m(\tau,x(\tau,\omega))d\tau + \int_a^t \sigma(\tau,x(\tau,\omega))dg$$

$$+ \int_a^t \int_{|u|>1} f(\tau,u,x(\tau,\omega))dp + \int_a^t \int_{|u|<1} f(\tau,u,x(\tau,\omega))dq,$$

whose solution is the required stochastic process by virtue of the following §11 (Hereafter the stochastic integral means the continuous or regular kernel, even if the notation ' * ' be omitted.

§11. Stochastic integral equation.

Theorem 11. Let $L(\tau,\xi)$ satisfy the condition (A) and (B) in Theorem 10. Let $c(\omega)$ be independent of $\ell(\tau,\omega) - \ell(a,\omega)$, $a \leq \tau \leq b$. Then there exists one and only one (up to P-measure 0) stochastic process satisfying a stochastic integral equation:

$$(11.1) \quad x(t,\omega) = c(\omega) + \int_a^t m(\tau,x(\tau,\omega))d\tau + \int_a^t \sigma(\tau,x(\tau,\omega))dg$$

$$+ \int_a^t \int_{|u|>1} f(\tau,u,x(\tau,\omega))dp + \int_a^t \int_{|u|<1} f(\tau,u,x(\tau,\omega))dq$$

for $a \leq t \leq b$ with P-measure 1 and fulfilling the following property:

(11.2) $(x(\tau,\omega), \ell(\tau,\omega) - \ell(a,\omega); a < \tau \leq t)$ is independent of $(\ell(\tau,\omega) - \ell(t,\omega), t \leq \tau \leq b)$ for $a \leq t \leq b$. This solution is a simple Markoff process, which belongs to $d_1$-class with P-measure 1 and satisfies a stochastic differential equation:

(11.3) $Dx(t,\omega) = L(t,x(t,\omega))$.

Proof: We shall firstly remark that the condition (B) implies that each of $|m(t,\xi)|$, $|\sigma(t,\xi)|$, $\| f(\tau,u,\xi)\|$, n=1,2,..., are bounded in $a \le t \le b$ for any assigned $\xi$, which is deduced from Theorem 4.2, since $\{L(\tau,\xi),\ a \le \tau \le b\}$ is compact and so totally bounded as a continuous image of a compact set [a,b] for any assigned $\xi$.

Next we shall remark that (A) and (B) imply that $m(\tau,\xi)$, $\sigma(\tau,\xi)$ and $f(\tau,u,\xi)$ are all B-measurable. Since $L(t,\xi)$ is continuous in t for any $\xi$ by (B) and $m(t,\xi)$ is B-measurable in $L(t,\xi)$ by virtue of Theorem 4.1, $m(t,\xi)$ is also B-measurable in t for any $\xi$. Besides $m(t,\xi)$ is continuous in $\xi$ for any fixed t by (A). Thus $m(t,\xi)$ is B-measurable in $(t,\xi)$. Similarly $\sigma(t,\xi)$ and $f(t,u,\xi)$ (for any fixed u) are B-measurable in $(t,\xi)$. Therefore $f(t,u,\xi)$ is B-measurable in $(t,u,\xi)$ since $f(t,u,\xi)$ is right-continuous in u.

For brevity we introduce the following notations. We put

$$c_1K_1+c_2K_2 = (c_1m_1+c_2m_2,\ c_1\sigma_1+c_2\sigma_2,\ c_1f_1+c_2f_2)$$

for $K_1=(m_1,\sigma_1,f_1)$ and $K_2=(m_2,\sigma_2,f_2)$. When $K(\tau,\omega) = (m(\tau,\omega),\sigma(\tau,\omega),f(\tau,u,\omega))$, we define

$$\int_\alpha^\beta K(\tau,\omega)d\ell = \int_\alpha^\beta m(\tau,\omega)d\tau + \int_\alpha^\beta \sigma(\tau,\omega)dg + \int_\alpha^\beta\int_{|u|>1} f(\tau,u,\omega)dp$$
$$+ \int_\alpha^\beta\int_{|u|<1} f(\tau,u,\omega)dq.$$

The triple of three elements of an i.d.l. L is also denoted by the same notion L. We put

(11.4) $c_N(\omega) = \phi_N(c(\omega))c(\omega)$,

(11.5) $f_N(\tau,u,\xi) = \phi_N(u)f(\tau,u,\xi)$,

(11.6) $L_N(\tau,\xi)$ = the i.d.l. whose elements are $m(\tau,\xi)$, $\sigma(\tau,\xi)$ and $f_N(\tau,u,\xi)$,

where $\phi_N$ is the characteristic function of the interval (-N,N).

Firstly we shall prove the existence and uniqueness of the solution of the stochastic integral equation:

$$(11.7) \qquad x(t,\omega) = c_N(\omega) + \int_a^t L_N(\tau, x(\tau,\omega))d\ell$$

such that

(11.8) $(x(t,\omega), \ell(\tau,\omega) - \ell(a,\omega), a \le \tau \le t)$ is independent of $(\ell(\tau,\omega) - \ell(t,\omega), t \le \tau \le b)$.

In order to find a solution we make use of the method of successive approximations; we define $x_n(t,\omega)$, n=1,2,..., recursively by

(11.9.1) $x_o(t,\omega)$ = any measurable process such that $x_o(t,\omega)$ is a function of $(c_N(\omega); \ell(\tau,\omega) - \ell(a,\omega), a \le \tau \le t)$ for $a \le t \le b$ and that $\mathcal{E}(x_o(t,\omega)^2)$ is bounded,

$$(11.9.2) \qquad x_n(t,\omega) = c_N(\omega) + \int_a^t L_N(\tau, x_{n-1}(\tau,\omega))d\ell \ , \ a \le t \le b, \ n=1,2,\dots \ .$$

From (11.9.2), n=1, we have, for any fixed $\xi_o$,

$$x_1(t,\omega) = c_N(\omega) + \int_a^t L_N(\tau, \xi_o)d + \int_a^t (L_N(\tau, x_o(\tau,\omega)) - L_N(\tau, \xi_o))d\ell \ ,$$

$$\int_a^t L_N(\tau, \xi_o)d\ell = \int_a^t m(\tau, \xi_o)d\tau + \int_a^t \sigma(\tau, \xi_o)dg + \int_a^t \int_{1<|u|<N} f(\tau, u, \xi_o)du/u^2dt$$

$$+ \int_a^t \int_{|u|<N} f(\tau, u, \xi_o)dq = I_1 + I_2 + I_3 + I_4,$$

$$\mathcal{E}((\int_a^t L_N(\tau, \xi_o)d\ell)^2) = 4I_1^2 + 4\mathcal{E}(I_2^2) + 4I_3^2 + 4\mathcal{E}(I_4^2),$$

$$I_1^2 \le (t-a)\int_a^t m(\tau, \xi_o)^2 d\tau \le (b-a)\int_a^b m(\tau, \xi_o)^2 d\tau \ ,$$

$$\mathcal{E}(I_2^2) = \int_a^t \sigma(\tau, \xi_o)^2 \, d\tau \le \int_a^b \sigma(\tau, \xi_o)^2 \, d\tau \ ,$$

$$I_3^2 \leq \int_a^t \int_{1<|u|<N} d\tau\, du/u^2 \int_a^t \int_{1<|u|<N} f(\tau,u,\xi_o)^2\, d\tau\, du/u^2 \leq 2 \int_a^b \| f(\tau,u,\xi_o)\|_N^2\, d\tau$$

$$\mathcal{E}(I_4^2) = \int_a^t \int_{|u|<N} f(\tau,u,\xi_o)^2\, d\tau\, du/u^2 \leq \int_a^b \| f(\tau,u,\xi_o)\|_N^2\, d\tau\ .$$

Thus $\mathcal{E}((\int_a^t L_N(\tau,\xi_o)d\ell)^2)$, $a \leq t \leq b$, is bounded by Theorem 4.2 and the condition (B). Similarly we can prove the boundedness of

$$\mathcal{E}((\int_a^t (L_N(\tau,x_o(\tau,\omega)) - L_N(\tau,\xi_o))d\ell\ )^2),\ a \leq t \leq b,$$

by making use of the condition (A). Consequently $\mathcal{E}(x_1(t,\omega)^2)$ is bounded.

Furthermore $x_1(t,\omega)$ belongs to $d_1$-class with P-measure 1 and so measurable in $(t,\omega)$. Besides $(x(\tau,\omega), \ell(\tau,\omega) - \ell(a,\omega);\ a \leq \tau \leq t)$ is independent of $(\ell(\tau,\omega) - \ell(t,\omega);\ t \leq \tau \leq b)$ for any t, as is easily verified.

Thus we can define $x_2(t,\omega)$ by (11.9.2) and so recursively $x_n(t,\omega)$, n=3,4,..., and we have

$$x_{n+1}(t,\omega) - x_n(t,\omega) = \int_a^t (m(\tau,x_n(\tau,\omega)) - m(\tau,x_{n-1}(\tau,\omega)))d\tau$$

$$+ \int_a^t (\sigma(\tau,x_n(\tau,\omega)) - \sigma(\tau,x_{n-1}(\tau,\omega)))dg + \int_a^t \int_{1<|u|<N} (f(\tau,u,x_n(\tau,\omega))$$

$$- f(\tau,u,x_{n-1}(\tau,\omega)))\frac{du}{u^2}$$

$$+ \int_a^t \int_{1<|u|<N} (f(\tau,u,x_n(t,\omega)) - f(\tau,u,x_{n-1}(\tau,\omega)))du|u^2 = I_1 + I_2 + I_3 + I_4,$$

$$\mathcal{E}(I_1^2) \leq \mathcal{E}((\int_a^t |m(\tau, x_n(\tau,\omega)) - m(\tau, x_{n-1}(\tau,\omega))|d\tau)^2)$$

$$\leq M^2(b-a)\int_a^t \mathcal{E}((x_n(t,\omega) - x_{n-1}(t,\omega))^2)d\tau ,$$

$$\mathcal{E}(I_2^2) \leq S^2 \int_a^t \mathcal{E}((x_n(\tau,\omega) - x_{n-1}(\tau,\omega))^2)d\tau ,$$

$$\mathcal{E}(I_3^2) \leq \mathcal{E}((\int_a^t \int_{1<|u|<N} |f(\tau,u,x_n(\tau,\omega)) - f(\tau,u,x_{n-1}(\tau,\omega))|d\tau du/u^2)^2)$$

$$\leq 2(b-a)F_N^2 \int_a^t \mathcal{E}((x_n(\tau,\omega) - x_{n-1}(\tau,\omega))^2)d\tau ,$$

$$\mathcal{E}(I_4^2) \leq F_N^2 \int_a^t \mathcal{E}((x_n(\tau,\omega) - x_{n-1}(\tau,\omega))^2)d\tau ,$$

$$\mathcal{E}((x_{n+1}(t,\omega) - x_n(t,\omega))^2) \leq 4(M^2(b-a) + S^2+2(b-a)F_N^2+ F_N^2)$$

$$\times \int_a^t \mathcal{E}((x_n(\tau,\omega) - x_{n-1}(\tau,\omega))^2)d\tau .$$

But $\mathcal{E}((x_1(t,\omega) - x_0(t,\omega))^2)$ has a finite upper bound (G), as is above proved. We obtain recursively ($\alpha = 4(M^2(b-a)+S^2+2(b-a)F_N^2+F_N^2)$)

$$\text{(11.10)} \qquad \mathcal{E}((x_n(t,\omega) - x_{n-1}(t,\omega))^2) \leq \alpha^{n-1}G(t-a)^{n-1} / \lfloor n-1 ,$$

$$\text{(11.11)} \qquad \mathcal{E}((\int_a^t |m(\tau, x_n(\tau,\omega)) - m(\tau, x_{n-1}(\tau,\omega))d\tau)^2)$$

$$\leq \alpha^{n-1} GM(b-a)(t-a)^n / \lfloor n ,$$

(11.12) $$\mathcal{E}\left(\left(\int_a^t (\sigma(\tau,x_n(\tau,\omega)) - \sigma(\tau,x_{n-1}(\tau,\omega))dg\right)^2\right)$$

$$\leq \alpha^{n-1} GS^2 (t-a)^n / \lfloor n,$$

(11.13) $$\mathcal{E}\left(\left(\int_a^t \int_{1<|u|<N} |f(\tau,x_n(\tau,\omega)) - f(\tau,x_{n-1}(\tau,\omega)))dt\, du/u^2\right)\right)$$

$$\leq 2\alpha^{n-1} GS (F_N^2 (b-a)(t-a)^n / \lfloor n,$$

(11.14) $$\mathcal{E}\left(\left(\int_a^t \int_{|u|>N} (f(\tau,x_n(\tau,\omega)) - f(\tau,x_{n-1}(\tau,\omega)))\, dq\right)^2\right) \leq \alpha^{n-1} GF_N^2 (t-a)/\lfloor n.$$

Now, putting t=b in (11.11) and using Bienaymé-Tschebycheff's inequality, we obtain

$$P\left(\left\{\omega ;\ \sup_{a \leq t \leq b} \left| \int_a^t (m(\tau,x_n(\tau,\omega)) - m(\tau,x_{n-1}(\tau,\omega)))d\tau \right| \geq \lambda_n^{\frac{1}{4}} \right\}\right)$$

$$\leq P\left(\left\{\omega ;\ \int_a^b |m(\tau,x_n(\tau,\omega)) - m(\tau,x_{n-1}(\tau,\omega))| d\tau \geq \lambda_n^{\frac{1}{4}} \right\}\right) \leq \lambda_n^{\frac{1}{2}},$$

where $\lambda_n = \alpha^{n-1} GM^2 (b-a)/\lfloor n-1$. Since $\sum \lambda_n^{\frac{1}{4}}, \sum \lambda_n^{\frac{1}{2}} < \infty$,

$$\sum_n \int_a^t (m(\tau,x_n(\tau,\omega)) - m(\tau,x_{n-1}(\tau,\omega)))d\tau, \qquad a \leq t \leq b,$$

is uniformly convergent in $a \leq t \leq b$ with P-measure 1 by Borel-Cantelli's theorem. If we make use of (G'.4) (§8) and (11.12), we can prove in the same way that

$$\sum_n \int_a^t (\sigma(\tau,x_n(\tau,\omega)) - \sigma(\tau,x_{n-1}(\tau,\omega)))dg$$

is uniformly convergent with P-measure 1. Similarly

$$\sum_n \int_a^t \int_{N>|u|>1} (f(\tau,u,x_n(\tau,\omega)) - f(\tau,u,x_{n-1}(\tau,\omega)))d\tau du/u^2 \text{ and}$$

$$\sum_n \int_a^t \int_{N > |u|} (f(\tau,u,x_n(\tau,\omega)) - f(\tau,u,x_{n-1}(\tau,\omega)))dq$$

are uniformly convergent in $a \leq t \leq b$ with P-measure 1. Consequently $x_n(t,\omega)$ is also uniformly convergent in $a \leq t \leq b$ with P-measure 1. We denote this limit with $x(t,\omega)$. Then $x(t,\omega)$ belongs to $d_1$-class with P-measure 1, and so measurable in $(t,\omega)$, since it is so the case with $x_n(t,\omega)$, as is recursively proved. By letting $n \longrightarrow \infty$ in (11.9.2), we can easily see that $x(t,\omega)$ satisfies (11.7).

Let $y(t,\omega)$ and $z(t,\omega)$ be any two solutions of (11.7) such that $\mathcal{E}(y(t,\omega)^2)$ and $\mathcal{E}(z(t,\omega)^2)$ are bounded ($\leq G_1$). Then in the same way as above we see

$$\mathcal{E}((y(t,\omega) - z(t,\omega))^2) \leq 4(M^2(b-a) + S^2 + 2F_N^2(b-a) + F_N^2)$$

$$\times \int_a^t \mathcal{E}((y(\tau,\omega) - z(\tau,\omega))^2)dt \leq \frac{\alpha^n}{\lfloor n} (4G_1)(t-a)^n \longrightarrow 0.$$

Therefore $y(t,\omega) = z(t,\omega)$ with P-measure 1 for t. Since $y(t,\omega)$ and $z(t,\omega)$ belong to $d_1$-class as the solution of (11.7), we have $y(t,\omega) = z(t,\omega)$, $a \leq t \leq b$, with P-measure 1.

For the solution $x(t,\omega)$ obtained above by the successive approximations, $\mathcal{E}(x(t,\omega)^2)$ is bounded and so it does not depend on $x_o(t,\omega)$. Let $y(t,\omega)$ be any solution of (11.7). Starting from $x_o(t,\omega) = \phi_M(y(t,\omega))y(t,\omega)$, $\phi_M$ being the characteristic function of $(-M,M)$, we define $x_n(t,\omega)$, $n=1,2,\ldots$, by (11.9.2), and obtain the solution $x(t,\omega)$. Now we put

$$\Omega_M = \{\omega; |y(t,\omega)| \leq M,\ a \leq t \leq b\}.$$

Then $\Omega_M$ increases with M and $P(\Omega_M) \longrightarrow 1$ since $y(t,\omega)$ belongs to $d_1$-class with P-measure 1. $y(t,\omega)$ satisfying (11.7), we have, by (G.5) and (P.5). $y(t,\omega) = x_o(t,\omega)$

$=x_1(t,\omega)$ ... and so

$$x(t,\omega) = \lim_{n} x_n(t,\omega) = y(t,\omega)$$

for almost all $\omega$ on $\Omega_M$ and so with P-measure $1 (M \longrightarrow \infty)$. Thus we have proved the existence and uniqueness of the solution of (11.7), say $x(t,\omega;N)$.

We put

$$E_M = \{\omega; |c(\omega)| \leq M, \int_a^b \int_{|u|\geq M} dp(\tau,u,\omega) = 0\}.$$

Then $E_M$ increases with M and $P(E_M) \longrightarrow 1$ as $M \longrightarrow \infty$. Starting from $x_0(t,\omega) = x(t,\omega;M)$, we define $x_n(t,\omega)$ by (11.9.2) and so we obtain the solution of (11.7) $x(t,\omega;N)$ as the limit. For $\omega \in E_M (M > N)$ we have $x(t,\omega;M) = x_0(t,\omega) = x_1(t,\omega) = \ldots = x(t,\omega;N)$. Therefore $x(t,\omega;M)$, $a \leq t \leq b$, does not depend on M for a sufficiently large M with P-measure 1 and so $\lim_{M\longrightarrow\infty} x(t,\omega;M)$ exists and satisfies (11.1).

Let $y(t,\omega)$ be any solution of (11.1). We put

$F_N = E_N \cap \{\omega; |y(t,\omega)| \leq N, a \leq t \leq b\}$.

Then $y(t,\omega)$ satisfies (11.7) in $F_N$. Starting from

$$x_0(t,\omega) = \phi_N(y(t,\omega))y(t,\omega)$$

we obtain $x_n(t,\omega)$, $n=1,2,\ldots$, by (11.9.2). For $\omega \in F_N$ we have, by (G.5) and (P.5),

$y(t,\omega) = x_0(t,\omega) = x_1(t,\omega) = \ldots \longrightarrow x(t,\omega;N)$. Since $y(t,\omega)$ belongs to $d_1$-class with P-measure 1, we have $P(F_N) \longrightarrow 1$. Therefore $y(t,\omega)$ coincides with the above obtained solution. Thus we have proved the existence and uniqueness of the solution.

Let $x(t,\omega)$ be the solution. Then we have

$$(11.15) \qquad x(t,\omega) = x(s,\omega) + \int_s^t L(\tau, x(\tau,\omega))d\ell, \quad s \leq t \leq b.$$

This is also considered as a stochastic integral equation of the above type concerning $x(\tau,\omega)$, $s \leq \tau \leq b$. By the uniqueness of the solution, $x(t,\omega)$ is obtained by the above procedure and so $x(t,\omega)$ is a B-measurable function of $x(s,\omega)$ and $(\ell(\tau,\omega) - \ell(s,\omega),\ s \leq \tau \leq b)$, as is easily verified if we use Theorem 1. We put

$$x(t,\omega) = f_t(x(s,\omega);\ \ell(\tau,\omega) - \ell(s,\omega),\ s \leq \tau \leq b)$$

and

$$x(t,\omega;s,\xi) = f_t(\xi;\ell(\tau,\omega) - \ell(s,\omega),\ s \leq \tau \leq b).$$

From (11.15) we obtain

$$(11.16) \qquad x(t,\omega;s,\xi) = \xi + \int_s^t L(\tau,x(\tau,\omega;s,\xi))d\ell$$

with P-measure 1 for almost all $(P_{x(s,\omega)})$ $\xi$, by replacing, in Theorem 2.3, $x(\omega),y(\omega)$ and $G(x(\omega),y(\omega))$ respectively with $x(s,\omega),(\ell(\tau,\omega)-\ell(s,\omega),s\leq \tau\leq b)$ and $x(t,\omega)-x(s,\omega) - \int_s^t L(\tau,x(\tau,\omega))d\ell$ ; the measurability condition can be easily verified by the definition of the stochastic integral, if we use Theorem 1.

By Theorem 2.3 we have, for almost all (with regard to the probability law of $(x(\tau,\omega),\ a \leq \tau \leq s)$) $(\xi_\tau,\ a \leq \tau \leq s)$,

$$\Pr\{x(t,\omega) \in E/x(\tau,\omega) = \xi_t, a \leq \tau \leq s\} = \Pr\{x(t,\omega;s,\xi_s) \in E\},$$

$$\Pr\{x(t,\omega) \in E/x(s,\omega) = \xi_s\} = \Pr\{x(t,\omega;s,\xi_s) \in E\},$$

and so

$$\Pr\{x(t,\omega) \in E/x(\tau,\omega) = \xi_\tau,\ a \leq \tau \leq s\} = \Pr\{x(t,\omega) \in E/x(s,\omega) = \xi_s\},$$

which implies that $x(t,\omega)$ is a simple Markoff process.

It remains only to prove (11.3). By the above discussion we need only prove that, as $\Delta_1+\Delta_2 \longrightarrow 0$,

(11.17) the $[1/\Delta_1+\Delta_2]$ - times convolution of the p.l. of $x(s+\Delta_1,\omega;s-\Delta_2,\xi)$ tends to $L(s,\xi)$.

Let $x_N(t,\omega;s-\Delta_2,\xi)$ be the solution of the stochastic integral equation:

$$x_N(t,\omega;s-\Delta_2,\xi) = \xi + \int_{s-\Delta_2}^t L_N(\tau,x_N(\tau,\omega;s-\Delta_2,\xi))d\ell .$$

Then we have

(11.18) $\qquad x_N(t,\omega;s-\Delta_2,\xi) = x(t,\omega;s-\Delta_2,\xi)$ for $s-\Delta_2 \leq t \leq s+\Delta_1$ and for $\omega$ such that

$$(11.19) \qquad \int_{s-\Delta_2}^{s-\Delta_1} \int_{|u|>N} dp = 0.$$

Now we have

$$(11.20) \qquad x(s+\Delta_1, \omega; s-\Delta_2, \xi) = x(\omega) + y(\omega) + z(\omega),$$

where

$$x(\omega) = \int_{s-\Delta_2}^{s+\Delta_1} L(\tau, \xi) d\ell ,$$

$$y(\omega) = \int_{s-\Delta_2}^{s+\Delta_1} (L_N(\tau, x_N(\tau, \omega; s-\Delta_2, \xi)) - L_N(\tau, \xi)) d\ell ,$$

$$z(\omega) = \int_{s-\Delta_2}^{s+\Delta_1} (L(\tau, x(\tau, \omega; s-\Delta_2, \xi)) - L_N(\tau, x_N(\tau, \omega; s-\Delta_2, \xi))) d\ell$$

$$+ \int_{s-\Delta_2}^{s+\Delta_1} (L_N(\tau, \xi) - L(\tau, \xi)) d\ell \quad ).$$

$L(\tau, \xi)$ being continuous in $\tau$ for any $\xi$ by (B), the logarithmic characteristic function $\phi(z; \tau, \xi)$ of $L(\tau, \xi)$ will be uniformly continuous in t, whenever z runs over any assigned bounded region, and so the l.c.f. of the $[1/\Delta_1+\Delta_2]$-times convolution of $P_x$:

$$[1/\Delta_1+\Delta_2] \int_{s-\Delta_2}^{s+\Delta_1} \phi(z; \tau, \omega) d\tau$$

will tend to $\phi(z; \tau, \omega)$ and so

$$p_x^{*[1/\Delta_1+\Delta_2]} \qquad (\text{* means convolution})$$

is arbitrarily near $L(s, \xi)$. By the property of stochastic integrals we have

$$\mathcal{E}(y(\omega)) = o(\Delta_1+\Delta_2), \quad \mathcal{E}(y(\omega)^2) = o(\Delta_1+\Delta_2),$$

$$\Pr\{z(\omega) \neq 0\} \ ( = (\Delta_1+\Delta_2)(1-e^{-\frac{2}{N}})) = o(\Delta_1+\Delta_2).$$

Thus it is sufficient to prove the following lemma.

Lemma 11. $x(\omega)$, $y(\omega)$ and $z(\omega)$ be real random variables on $(\Omega, B, P)$ such that

$$(11.21) \qquad |\mathcal{E}y(\omega)| < \frac{\alpha}{n}, \ \mathcal{E}(y(\omega)^2) < \frac{\alpha}{n}, \ \Pr\{z(\omega) \neq 0\} < \frac{\alpha}{n} \ (\alpha < 1).$$

Then we have

$$\rho(P_x^{*^n}, P_\phi^{*^n}) < 4\sqrt{2}\alpha^{\frac{1}{3}},$$

where $\phi = x + y + z$ and $\rho$ is the Levy's distance.

Proof. Let $(\Omega^*, B^*, P^*)$ be the product probability field $(\Omega, B, P)^n$. For $\omega^* = (\omega_1, \omega_2, \ldots, \omega_n)$ we define $x^*_\nu(\omega^*) = x(\omega_\nu)$, $y^*_\nu(\omega^*) = y(\omega_\nu)$, $z^*_\nu(\omega^*) = z(\omega_\nu)$, $\nu = 1,2,\ldots,n$. Then $(x^*(\omega^*), y^*(\omega^*), z^*(\omega^*))$, $\nu = 1,2,\ldots,n$, are independent random vectors.

$P_x^{*n}$ and $P_y^{*n}$ are respectively the p.l. of $X^*(\omega^*) = \sum x^*_\nu(\omega^*)$ and $\Phi^*(\omega^*) = \sum x^*_\nu(\omega^*) + \sum y^*_\nu(\omega^*) + \sum z^*_\nu(\omega^*)$. But we have

$$\mathcal{E}((\sum y^*_\nu(\omega^*))) \le (\sum_\nu \mathcal{E}(y^*_\nu(\omega^*)))^2 + \sum_\nu \mathcal{E}(y^*_\nu(\omega^*)) \le (n\frac{\alpha}{n})^2 + n\frac{\alpha}{n} = \alpha^2 + \alpha < 2\alpha.$$

Therefore we have

$$\Pr\{|\sum_\nu y^*_\nu(\omega^*)| > \alpha^{\frac{1}{3}}\} \le 2\alpha^{\frac{1}{3}}.$$

But

$$\Pr\{\sum_\nu z^*_\nu(\omega^*) \neq 0\} \le \sum_\nu \Pr\{z^*_\nu(\omega^*) \neq 0\} \le n\frac{\alpha}{n} = \alpha.$$

Thus we have

$$\Pr\{|\Phi^*(\omega^*) - X^*(\omega^*)| > \alpha^{\frac{1}{3}}\} < 2\alpha^{\frac{1}{3}} + \alpha < 3\alpha^{\frac{1}{3}},$$

from which we obtain

$$\rho(\text{p.l. of } X^*(\omega^*), \text{p.l. of } \Phi^*(\omega^*)) < 4\sqrt{2}\alpha^{\frac{1}{3}}, \text{ q.e.d.}$$

Remark. In case $L(t, \xi)$ is a Gaussian distribution, the above obtained process is continuous with P-measure 1(continuous case). In the case when the l.c.f. of $L(t, \xi)$ is of the form:

$$\psi(z;t,\xi) = \int_{-\infty}^{\infty} (e^{i\lambda u} - 1) n_{t,\xi}(du),$$

the above process increases only with jumps with P-measure 1 (purely discontinuous case). The most simple case of $\sigma = 0$, $f \equiv 0$ is reduced to that of an ordinary differential equation:

$$\frac{dx}{dt} = m(t,x).$$

## IV. Appendix I.

We shall show an interesting property of the stochastic integral which does not appear in the ordinary integral.

Theorem.[27] Let $\lambda(\xi)$ be any function of $\xi$ with the continuous second derivative. Then we have

$$(1) \quad \int_a^b \lambda'(g(t,\omega))dg(t,\omega) = \lambda(g(b,\omega)) - \lambda(g(a,\omega)) - \frac{1}{2}\int_a^b \lambda''(g(t,\omega))dt.$$

Proof. For the brevity of the notation we may assume a = 0,b=1. We have clearly

$$(2) \quad \lambda(g(1,\omega)) - \lambda(g(0,\omega)) = I_1 + \frac{1}{2} I_2 + I_3 + I_4,$$

where

$$I_1 = \sum_{k=1}^{n} \lambda'(g(\tfrac{k-1}{n},\omega))(g(\tfrac{k}{n},\omega) - g(\tfrac{k-1}{n},\omega))$$

$$I_2 = \sum_{k=1}^{n} \lambda''(g(\tfrac{k-1}{n},\omega))\frac{1}{n}$$

$$I_3 = \sum_{k=1}^{n} \lambda''(g(\tfrac{k-1}{n},\omega))((g(\tfrac{k}{n},\omega) - g(\tfrac{k-1}{n},\omega))^2 - \tfrac{1}{n})$$

$$I_4 = \sum_{k=1}^{n} \delta(n,k,\omega)(g(\tfrac{k}{n},\omega) - g(\tfrac{k-1}{n},\omega))^2;$$

$\delta(n,k,\omega)$ tends to 0 uniformly in k as $n \longrightarrow \infty$ for almost all $\omega$ on account of the continuity of $\lambda''(\xi)$ and $g(t,\omega)$.

We can choose a sufficiently large N for any $\delta > 0$ such that $n > N$ implies

$$(3)\quad \Pr\{|\delta(n,k,\omega)| < \delta,\ k=1,2,\dots,n\} > 1-\delta .$$

If we define

$\delta^*(n,k,\omega) = \delta(n,k,\omega)$ for $|\delta(n,k,\omega)| < \delta$ and $= 0$ elsewhere, and if we put

$$I_4^* = \sum_{k=1}^{n} \delta^*(k,n,\omega)\left(g\left(\tfrac{k}{n},\omega\right) - g\left(\tfrac{k-1}{n},\omega\right)\right),$$

then we have

$$(4)\quad \Pr\{I_4 = I_4^*\} > 1-\delta .$$

$$(5)\quad \mathcal{E}((I^*)^2) \le \sum_k \delta^2 \mathcal{E}\left(\left(g\left(\tfrac{k}{n},\omega\right) - g\left(\tfrac{k-1}{n},\omega\right)\right)^2\right) = n\,\delta^2\,\frac{1}{n} = \delta^2.$$

By (4) and (5) we have, as $n \longrightarrow \infty$, $I_4 \longrightarrow 0$ in probability.

By the continuity (in t), of $\lambda''(g(t,\omega))$ we can choose a sufficiently large M for any $\delta > 0$ such that

$$(6)\quad \Pr\left\{\left|\lambda''\left(g\left(\tfrac{k-1}{n},\omega\right)\right)\right| < M, k=1,2,3,\dots,n\right\} > 1-\delta .$$

If we define

$\mu_M(\xi) = \lambda''(\xi)$ for $|\lambda''(\xi)| \le M$ and $= 0$ elsewhere, and if we put

$$I_{3,M} = \sum_{k=1}^{n} \mu_M\left(g\left(\tfrac{k-1}{n},\omega\right)\right)\left(\left(g\left(\tfrac{k}{n},\omega\right) - g\left(\tfrac{k-1}{n},\omega\right)\right)^2 - \tfrac{1}{n}\right),$$

then we have, by (6),

$$(7)\quad \Pr\{I_3 = I_{3,M}\} > 1-\delta ,$$

and by making use of the fact that $g(\tau,\omega)$ is a differential process,

$$(8)\ \mathcal{E}(I_{3,M}^2) \le M^2 \sum_{k=1}^{n} \mathcal{E}\left(\left(\left(g\left(\tfrac{k}{n},\omega\right) - g\left(\tfrac{k-1}{n},\omega\right)\right)^2 - \tfrac{1}{n}\right)^2\right)$$

$$= M^2 n \int_{-\infty}^{\infty} \sqrt{\frac{n}{2\pi}}\, e^{-n\xi^2/2} \left(\xi^2 - \tfrac{1}{n}\right)^2 d\xi = M^2/n \int_{-\infty}^{\infty} \frac{1}{\sqrt{2\pi}}\, e^{-\frac{\xi^2}{2}} (\xi^2-1)^2 d\xi \longrightarrow 0.$$

Therefore we have, as $n \longrightarrow \infty$, $I_3 \longrightarrow 0$ in probability.

Similarly we can prove that

$$I_1 \longrightarrow \int_0^1 \lambda'(g(t,\omega))\,dg(t,\omega)$$

in probability. By the continuity (in t) of $\lambda''(g(t,\omega))$ we have

$$I_2 = \sum \lambda''(g(\frac{k-1}{n}, \omega))\frac{1}{n} \longrightarrow \int_0^1 \lambda''(g(t,\omega))dt.$$

Thus our theorem is completely proved.

Mathematical Institute, Faculty of Science, Nagoya University.

## FOOTNOTES

1) W. Feller: Zur Theorie der stochastischen Prozesse (Existenz-und Eindeutigkeitssätze), Math. Ann. 113.

2) A. Kolmogoroff: Über die analytische Methoden in der Wahrscheinlichkeitsrechnung, Math. Ann. 104.

3) S. Bernstein: Equations differentielles stochastiques, Act. Sci. et Ind., 738,1938.

4) $[\alpha]$ is the greatest integer n such that $n \leq \alpha$.

5) The characteristic function of an infinitely divisible law is expressible in the form $\exp \psi(z)$ by Levy's theorem. This $\psi(z)$ will be called hereafter the logarithmic characteristic function of the infinitely divisible law. Cf. P. Lévy: Théorie de l'addition des variable aléatoires (1937), Chap. VII, and also my previous paper: On stochastic processes (I), Japanese Jour. of Math. Vol. 18, 1942.

6) $E(+)\xi$ is the set $\{\eta + \xi ; \eta \in E\}$.

7) A. Kolmogoroff: Grundbegriffe der Wahrscheinlichkeitsrechnung, Ergeb. der Math. Vol. 2, No. 6.

8) J. L. Doob: Stochastic processes depending on a continuous parameter, Trans. Amer. Math. Soc. Vol. 42, 1937.

9) E. Slutsky: Sur les fonctions aléatoires presque périodiques et sur la décomposition des fonctions aléatoires stationaires en composantes, Act. Sci. et Ind. 738, 1938.

10) J. L. Doob: Markoff chains -- denumerable case, Trans. Amer. Math. Soc. 58, 1945.

11) R. Fortet: Les fonctions aléatoires du type de Markoff associées à certaines équations linéaires aux derivees partielles du type parabolique, Journ. de Math. 22, 1943.

12) K. Ito: loc. cit. 5).

13) R. E. A. Paley and N. Wiener: Fourier transforms in the complex domain, Amer. Math. Soc. Coll. Publ. 1934, Chap. IX.

14) K. Ito: Stochastic integral, Proc. Imp. Acad. Tokyo, Vol. 20, No. 8.

15) K. Ito: On a stochastic integral equation, forthcoming in Proc. Imp. Acad. Tokyo, Vol. 22.

16) loc. cit. 14)

17) J. L. Doob: Stochastic processes with an integral-valued parameter, Trans. Amer. Math. Soc. Vol. 44, 1938. The proof of the theorem concerning the conditional probability law seems to be somewhat incomplete, but it is available for our special case.

18) By $[a,b]$ we understand the closed interval: $a \leq x \leq b$, and by $(a,b]$ the half-open interval: $a < x \leq b$. $[a,b)$ and $(a,b)$ are to be understood similarly.

19) By $(x(\tau,\omega),\ a \leq \tau \leq t)$ we understand the joint variable of $x(\tau,\omega)$ for $\tau$ such that $a \leq \tau \leq t$.

20) loc. cit. 5).

21) P. Lévy: loc. cit. 5) p. 55.

22) P. Lévy: loc. cit. 5) p. 90.

23) loc. cit. 13).

24) loc. cit. 14).

25) loc. cit. 5).

26) J. L. Doob: loc. cit. 8) Lemma 2.1.

27) This theorem has been reported by the author without the proof, loc. cit. 14).

## Appendix 2. A generalized Fokker-Plank equation.

When I sent this paper to Professor J. L. Doob, he suggested to me to show that the process of Theorem 11 satisfies the Fokker-Plank equation. Though I cannot yet prove it in its complete generality, I have been able to solve it to a certain extent. It seems to be of some use to add it as an appendix.

Let $x(t,\omega)$ be the solution of the stochastic integral equation in Theorem 11, and $x(t,\omega;s,\xi)$ be the solution of the stochastic integral equation:

$$(1)\quad x(t,\omega;s,\xi) = \xi + \int_s^t L(\tau, x(\tau,\omega;s,\xi))d\ell\,, \quad s \leq t \leq b.$$

By putting $a=s$ and $c(\omega) \equiv \xi$ in Theorem 11, we see that $x(t,\omega;s,\xi)$ is uniquely determined for each $\xi$ and obtained by the procedure stated in Theorem 11, so that $x(t,\omega;s,\xi)$ is B-measurable in $\xi$. Denote the probability law of $x(t,\omega;s,\xi)$ with $F(s,\xi;t,E)$. By the argument in the proof of Theorem 11 we have

(2) $F(s,\xi;t,E) = \Pr\{x(t,\omega) \in E / x(s,\omega) = \xi\}$
for almost all $(P_{x(t,\omega)})$ $\xi$.

**Theorem.** $F(s,\xi;t,E)$ has the following properties:

I. Chapman's equation.

$$(3)\quad F(s,\xi;t,E) = \int_{-\infty}^{\infty} F(s,\xi;u,d\eta)F(u,\eta;t,E)\quad (a \leq s \leq u \leq t \leq b).$$

II. Let $\Phi(\xi)$ any bounded function with the second derivative $\Phi''(\xi)$. Then we have

$$(4)\quad \lim_{\substack{\Delta_1+\Delta_2 \downarrow 0 \\ \Delta_1,\Delta_2 \geq 0}} \frac{1}{\Delta_1+\Delta_2}\left[\int \Phi(\eta)F(s-\Delta_1,\xi;s+\Delta_2,d\eta) - \Phi(\xi)\right]$$

$$= m(s,\xi)\frac{d\Phi}{d\xi}(\xi) + \frac{\sigma^2(s,\xi)}{2}\frac{d^2\Phi}{d\xi^2}(\xi) + \int\Big(\Phi(\xi+u) - \Phi(\xi)$$

$$- \frac{u}{1+u^2}\Phi'(\xi)\Big)n(du,s,\xi),$$

where $m(s,\xi)$, $(s,\xi)$ and $n(E,s,\xi)$ are determined by

(5) the logarithmic characteristic function of $L(s,\xi)$

$$= im(s,\xi)z - \frac{\sigma^2(s,\xi)}{2}z^2 + \int_{-\infty}^{\infty}\left(e^{izu} - 1 - \frac{izu}{1+u^2}\right)n(du,s,\xi).$$

III. Generalized Fokker-Plank equation.

If $\frac{\partial^2}{\partial\xi^2}F(s,\xi;t,E)$ exists, then

$$(6)\quad D_s^- F(s,\xi;t,E) = \lim_{\Delta \downarrow 0}\frac{1}{\Delta}\big(F(s,\xi;t,E) - F(s-\Delta,\xi;t,E)\big)$$

exists and we have, in using $m$, $\sigma$ and $n$ in II

$$(7)\quad D_s^- F(s,\xi;t,E) + m(s,\xi)\frac{\partial}{\partial\xi}F(s,\xi;t,E) + \frac{\sigma^2(s,\xi)}{2}\frac{\partial^2}{\partial\xi^2}F(s,\xi;t,E)$$

$$+\int_{-\infty}^{\infty}\left(F(s,\xi+u;t,E) - F(s,\xi;t,E) - \frac{u}{1+u^2}\frac{\partial}{\partial\xi}F(s,\xi;t,E)\right)n(du,s,\xi) = 0.$$

Remark 1. By specializing $m, \sigma, n$, we obtain the Fokker-Plank equations that have already been known.

Case 1. $m(s,\xi) = m(\xi), \sigma(s,\xi) = \sigma(\xi)$, $n(E,s,\xi) \equiv 0$... Fokker-Plank's original case.

Case 2. $n(E,s,\xi) \equiv 0$ ... Kolmogoroff's continuous case.

Case 3. $p(s,\xi) \equiv n((-\infty,\infty), s,\xi) < \infty$ . . . Feller's mixing case.

In this case, by putting

$$n(E,s,\xi) = p(s,\xi)P(s,\xi,E(+)\xi)$$

$$m(s,\xi) - \int \frac{u}{1+u^e} n(du,s,\xi) = b(s,\xi),$$

and

$$\frac{\sigma^2(s,\xi)}{2} = a(s,\xi),$$

we obtain Feller's equation:

$$\frac{\partial}{\partial s} F(s,\xi;t,E) + a(s,\xi)\frac{\partial^2}{\partial \xi^2} F(s,\xi;t,E) + b(s,\xi)\frac{\partial}{\partial \xi}F(s,\xi;t,E)$$
$$+ p(s,\xi)\int (F(s,\eta;t,E) - F(s,\xi;t,E))\, P(s,\xi,d\eta) = 0.$$

Remark 2. It is desirable to prove the existence of $\frac{\partial}{\partial \xi^2} F(s,\xi;t,E)$ under some adequate restrictions concerning $m, \sigma$ and $n$, but it is an open problem for the author.

Proof 1. By the proof of Theorem 11 we see that $x(t,\omega;s,\xi)$ is a B-measurable function of $\xi$ and $\ell(\tau,\omega) - \ell(s,\omega)$, $s \leq \tau \leq t$. Put

$$x(t,\omega,s,\xi) = f_{st}(\xi, \ell(\tau,\omega) - \ell(s,\omega), s \leq \tau \leq t).$$

Since

$$x(t,\omega;s,\xi) = \xi + \int_s^t L(\tau, x(\tau,\omega;s,\xi))\, d\ell$$

$$= \xi + \int_s^u + \int_u^t$$

$$= x(u,\omega;s,\xi) + \int_u^t L(\tau, x(\tau,\omega;s,\xi))\, d\ell, \ (s \leq u \leq t),$$

we have, by the uniqueness of the solution in Theorem 11,

$$x(t,\omega;s,\xi) = f_{ut}(x(u,\omega;s,\xi), \ell(\tau,\omega) - \ell(s,\omega),\ u \leq \tau \leq t)$$

for almost all $\omega$. $x(u,\omega;s,\xi)$ being a function of $\ell(\tau,\omega) - \ell(s,\omega)$, $s \leq \tau \leq u$, and so independent of $\ell(t,\omega) - (s,\omega)$, $u \leq \tau \leq t$, we have, by Theorem 2.2,

$$\Pr\{x(t,\omega;s,\xi) \in E\} = \Pr\{f_{ut}(\eta, \ell(\tau,\omega) - \ell(s,\omega), u \leq \tau \leq t) \in E\} P_x(u,\omega;s,\xi)(d\eta)$$

$$= \int \Pr\{x(t,\omega;u,\eta) \in E\}\ P_x(u,\omega;s,\xi)(d\eta),$$

which proves Chapman's equation.

Next we shall prove III, assuming II. Since we have, by I,

$F(s-\Delta,\xi;t,E) = \int F(s-\Delta,\xi;s,d\eta)\ F(s,\eta;t,E)$, we obtain III at once, by putting $\phi(\xi) = F(s,\xi;t,E)$, $\Delta_1 = \Delta$ and $\Delta_2 = 0$ in II.

We have only to prove II. In the proof of Theorem 11 we have shown that

$$(8)\quad F(s-\Delta_1,\xi;s+\Delta_2,E(+)\xi)^{*}\left[\frac{1}{\Delta_1+\Delta_2}\right] \longrightarrow L(s,\xi).$$

Therefore it is sufficient to show that (8) implies (4). For this we state some preliminary lemmas.

Lemma 1. Let the logarithmic characteristic function (l.c.f.) of an infinitely divisible law (i.d.l.) P be given by

$$(9)\quad \psi(z) = imz + \int_{-\infty}^{\infty} \left(e^{izu} - 1 - \frac{izu}{1+u^2}\right) \frac{1+u^2}{u^2}\, d\,G(u),\quad G(-\infty) = 0$$

and $\{P_k\}$ be a sequence of probability laws such that

$$P_k^{*k} \longrightarrow P.$$

Then

$$(10)\qquad G_k(u) \equiv k\int_{-\infty}^{u} \frac{v^2}{1+v^2}\, P_k(dv)$$

is bounded for $k=1,2,\ldots$ and $-\infty < u < \infty$, and

$$(11.a)\qquad G_k(u) \longrightarrow G(u)$$

for any continuity point $u$ of $G(u)$, and

$$(11.b) \qquad \int \frac{d\,G_k(u)}{u} \longrightarrow m.$$

This lemma can be proved in the same way as A. Khintchine's proof of Lévy's formule (A. Khintchine: Déduction nouvelle d'une formule de P. Lévy, Bull. d. l'univ. d'état à Moscou, Sér. inter. Sect. A. Math. et Meca. Vol. 1, Fasc. 1, 1937).

Lemma 2. Let P be an i. d. l. with the l. c. f. $\psi(z)$ in (9). Let $\{P_h\}$ be a system of probability laws, h running over (o,c), (c=positive constant) such that

$$(12) \qquad P_h^{*[\frac{1}{h}]} \longrightarrow P.$$

Then

$$(13) \qquad G_h(u) = \frac{1}{h}\int_{-\infty}^{u} \frac{v^2}{1+v^2} P_h(dv)$$

is bounded for $o < h < c$ and $-\infty < u < \infty$, and

$$(14.a) \qquad G_h(u) \longrightarrow G(u)$$

for any continuity point u of G(u), and

$$(14.b) \qquad \int \frac{dG_h(u)}{u} \longrightarrow m.$$

This lemma follows immediately from Lemma 1.

Fundamental Lemma. Let P be an i.d.l. with the l.c.f.:

$$(15) \qquad \psi(z) = imz - \frac{\sigma^2}{2} z^2 + \int_{-\infty}^{\infty} (e^{izu} - 1 - \frac{i z u}{1+u^2}) n(du),$$

and $\{P_h\}$ be a system of probability laws, h running over (o,c). Then (12) implies that we have, for any bounded function $\Phi(\xi)$ $(-\infty < \xi < \infty)$ with the second derivatives $\Phi''(\xi)$,

$$(16) \qquad \lim_{h \downarrow o} \frac{1}{h}\left[\int \Phi(\xi+u) P_h(du) - \Phi(\xi)\right] = m\,\Phi'(\xi) + \frac{\sigma^2}{2}\Phi''(\xi)$$

$$+ \int_{-\infty}^{\infty} (\Phi(\xi+u) - \Phi(\xi) - \frac{u}{1+u^2}\Phi'(\xi))\, n\,(du).$$

Proof. (Concerning the following proof the author has obtained many suggestions in his discussions with Professor K. Yosida, whose research on a generalization of Fokker-Plank equation will soon be published in some journal.)

(15) is written in the form (9) if we put

$$G(u) = \sigma^2 + \int_{-\infty}^{u} \frac{v^2}{1+v^2}\, n(dv) \quad (u > 0) \tag{17}$$

$$G(u) = \int_{-\infty}^{u} \frac{v^2}{1+v^2}\, n(dv) \quad (u < 0).$$

Defining $G_h(u)$ by

$$G_h(u) = \frac{1}{h}\int_{-\infty}^{u} \frac{v^2}{1+v^2}\, P_h(dv),$$

we obtain

$$\frac{1}{h}\left[\int \Phi(\xi + u) P_h(du) - \Phi(\xi)\right]$$

$$= \frac{1}{h}\left[\int (\Phi(\xi + u) - \Phi(\xi))\, P_h(du)\right.$$

$$= \int (\Phi(\xi + u) - \Phi(\xi))\, \frac{1+u^2}{u^2}\, d\, G_h(u)$$

$$= \Phi'(\xi) \int \frac{d\, G_h(u)}{u} + \int \left(\Phi(\xi + u) - \Phi(\xi) - \frac{u}{1+u^2}\, \Phi'(\xi)\right) \frac{1+u^2}{u^2}\, d\, G_h(u),$$

$$= \Phi'(\xi) \int \frac{d\, G_h(u)}{u} + \int \psi(u, \xi)\, d\, G_h(u),$$

where $\psi(u, \xi)$ is defined by

$$\psi(u, \xi) = \left(\Phi(\xi + u) - \Phi(\xi) - \frac{u}{1+u^2}\, \Phi'(\xi)\right) \frac{1+u^2}{u^2} \quad (u \neq 0) \tag{18}$$

$$\psi(0, \xi) = \Phi''(\xi).$$

Then $\psi(u,\xi)$ is bounded and continuous outsides of any neighborhood of u=o for any fixed $\xi$ by the boundedness of $\Phi$, while $\psi(u,\xi)$ is also bounded and continuous within any neighborhood of u=o for any fixed $\xi$, since

$$\Phi(\xi+u)-\Phi(\xi)-\frac{u}{1+u^2}\Phi'(u)=u\Phi'(\xi)+\frac{u^2}{2}\Phi''(\xi)-u\Phi(\xi)+o(u^2,$$

$$=\frac{u^2}{2(1+u^2)}\Phi''(\xi)+o(u^2).$$

Therefore we have, by Lemma 2, (17) and (18),

$$\lim_{h\downarrow o}\frac{1}{h}\left[\int\Phi(\xi+u)P_h(du)-\Phi(\xi)\right]=m\Phi'(\xi)+\int_{-\infty}^{\infty}\psi(u,\xi)d\,G(u)$$

$$=m\Phi'(\xi)+\frac{\sigma^2}{2}\Phi''(\xi)+\left(\int_{-\infty}^{-o}+\int_{o}^{\infty}\right)\psi(u,\xi)\,d\,G(u)$$

$$=m\Phi'(\xi)+\frac{\sigma^2}{2}\Phi''(\xi)+\int_{-\infty}^{\infty}\left(\Phi(\xi+u)-\Phi(\xi)-\frac{u}{1+u^2}\Phi'(\xi)\right)n(du),$$

which proves the fundamental lemma.

Proof of II of the above theorem. We can modify the above fundamental lemma without any essential change in the proof as follows:

" Let $\{P_{\Delta_1\Delta_2}\}$ be a system of probability laws, where $o\leq\Delta_1,\Delta_2<c$ and $\Delta_1+\Delta_2>o$, and P be an i.d.l. stated in the lemma. Then

$$(12')\qquad P_{\Delta_1\Delta_2}^{*[\frac{1}{\Delta_1+\Delta_2}]}\longrightarrow P$$

implies

$$(16')\qquad \lim_{\Delta_1+\Delta_2\downarrow o}\frac{1}{\Delta_1+\Delta_2}\left[\Phi(\xi+u)P_{\Delta_1\Delta_2}(du)-\Phi(\xi)\right]$$

$$=m\Phi'(\xi)+\frac{\sigma^2}{2}\Phi''(\xi)+\int_{-\infty}^{\infty}\left(\phi(\xi+u)-\phi(\xi)-\frac{u}{1+u^2}\Phi'(\xi)\right)n(du)."$$

Now put

$$\Phi(\xi) = F(s, \xi; t, E),$$

$$P_{\Delta_1 \Delta_2}(E) = F(s-\Delta_1, \xi; s, E),$$

and

$$P = L(s, \xi).$$

Then (8) i.e. (12') implies (16') i.e. (4). Thus the theorem is completely proved.

Reprinted from
*Memoirs Amer. Math. Soc.* **4**, 1–51 (1951)

# ON A FORMULA CONCERNING STOCHASTIC DIFFERENTIALS

KIYOSI ITÔ

In his previous paper [1][1] the author has stated a formula[2] concering stochastic differentials with the outline of the proof. The aim of this paper is to show this formula in details in a little more general form (Theorem 6).

**1. Definitions.** Throughout this paper we assume that all stochastic processes[3] $\xi(t, \omega)$, $\eta(t, \omega)$, $a(t, \omega)$, $b(t, \omega)$, etc. are measurable in variables $t$ and $\omega$. A system of $r$ one-dimensional Brownian motions independent of each other is called an $r$-dimensional Brownian motion.

Given two system of stochastic processes:

$$\xi=\{\xi_\lambda(t, \omega),\ \lambda\in\Lambda\}, \qquad \eta=\{\eta_\mu(t, \omega),\ \mu\in M\}. \tag{1.1}$$

We say that $\xi$ has the property $\alpha$ with regard to $\eta$ in $u\leqq t\leqq v$, if, for any $t$, the following two systems of random variables are independent of one another:

$$\begin{cases}\varphi_t=\{\xi_\lambda(\tau, \omega),\ \lambda\in\Lambda,\ \eta_\mu(\tau, \omega),\ \mu\in M,\ u\leqq\tau\leqq t\}\\ \psi_t=\{\eta_\mu(\sigma, \omega)-\eta_\mu(t, \omega),\ \mu\in M,\ t\leqq\sigma\leqq v\}.\end{cases} \tag{1.2}$$

Now we shall state an outline[4] of a stochastic integral of the form:

$$\int_s^t \xi(\tau, \omega)d\beta(\tau, \omega), \qquad u\leqq s\leqq t\leqq v, \qquad \omega\in\Omega_1, \tag{1.3}$$

where $\beta(t, \omega)$ is a one-dimensional Brownian motion and $\Omega_1$ is a measurable subset of $\Omega$. We shall set the two conditions on $\xi$;

(C. 1) $\quad$ *$\xi(t, \omega)$ has the property $\alpha$ concerning $\beta(t, \omega)$ in $u\leqq t\leqq v$,*

(C. 2) $\quad$ $\int_u^v \xi(\tau, \omega)^2 d\tau$ *for almost all* $\omega\in\Omega_1$.

---

Received April 16, 1951.

[1] The number in [ ] refers to the Reference at the end of this paper.

[2] Theorem 1.1 in [1].

[3] In the analytical theory of probability any stochastic process is expressed as a function of the time parameter $t$ and the probability parameter $\omega$ which runs over a probability space $\Omega(P)$, $P$ being the probability distribution.

[4] Cf. [2] concerning the details.

*Case 1.* When $\xi$ is uniformly stepwise, that is when there exists a system of time-points:

(1.4) $$u=s_0<s_1<\dots<s_n=v$$

such that

$$\xi(\tau,\omega)=\xi(s_{i-1},\omega),\quad s_{i-1}\leqq\tau<s_i,\quad i=1,2,\dots,n,$$

we define

(1.5) $$\int_s^t \xi(\tau,\omega)d\beta(\tau,\omega)=\sum_{i=k+1}^{l-1}\xi(s_{i-1},\omega)(\beta(s_i,\omega)-\beta(s_{i-1},\omega))$$
$$+\xi(s_{k-1},\omega)(\beta(s_k,\omega)-\beta(s,\omega))+\xi(s_{l-1},\omega)(\beta(t,\omega)-\beta(s_{l-1},\omega))$$
$$(s_{k-1}\leqq s<s_k,\ s_{l-1}\leqq t<s_l).$$

*Case 2.* When

(1.6) $$\int_\Omega\int_u^v \xi(t,\omega)^2 dtP(d\omega)<\infty,$$

there exists a sequence of uniformly stepwise processes $\xi_n(t,\omega)$, $n=1,2,\dots$ whose value at any time-point $t$ is a $B$-measurable function of $\xi(\tau,\omega)$, $u\leqq\tau\leqq t$, such that

(1.7) $$\int_\Omega\int_u^v(\xi_n(t,\omega)-\xi(t,\omega))^2 dtP(dw)<8^{-n}.$$

We define

(1.8) $$\int_s^t \xi(\tau,\omega)d\beta(\tau,\omega)=\lim_n\int_s^t \xi_n(\tau,\omega)d\beta(\tau,\omega).$$

As was proved in our previous paper [2], the sequence:

$$\int_s^t \xi_n(\tau,\omega)d\beta(\tau,\omega),\quad n=1,2,\dots,$$

is uniformly convergent *in* $u\leqq s\leqq t\leqq v$ for almost all $\omega$, and the definition is independent of the special choice of the sequence $\{\xi_n(t,\omega)\}$.

*Case 3.* Now we shall consider the general case. Put

(1.9) $$\phi_n(\lambda)=\begin{cases}1 & (|\lambda|\leqq n)\\ 0 & (|\lambda|>n)\end{cases}$$

and

$$\xi_n(t,\omega)=\phi_n\left(\int_u^t \xi(\tau,\omega)^2 dt\right)\xi(t,\omega),\quad n=1,2,\dots.$$

Then $\xi_n(t,\omega)$ satisfies the conditions (1.6) and (C.1), and so their stochastic integrals are well defined. Since $\xi_n(t,\omega)=\xi(t,\omega)$, $u\leqq t\leqq v$, for a sufficiently large $n$ for almost all $\omega\in\Omega_1$ by (C.2), we define naturally

$$\int_s^t \xi(\tau,\omega)d\beta(\tau,\omega)=\lim_n \int_s^t \xi_n(\tau,\omega)d\beta(\tau,\omega).$$

**2. Preliminary properties of stochastic integrals.** In the following Theorems 1, 3, and 4 we shall assume that $\xi(t,\omega)$ satisfies the conditions (C. 1) and (C. 2).

THEOREM 1. *The stochastic integral (1.3) is continuous in s, t for almost all* $\omega\in\Omega_1$.

THEOREM 2. *If each of* $\xi(t,\omega)$ *and* $\eta(t,\omega)$ *satisfies (C. 1) and (C. 2), and if the system* $\{\xi(t,\omega),\eta(t,\omega)\}$ *has the property* $\alpha$ *with regard to the Brownian motion* $\beta$, *then* $\zeta(t,\omega)\equiv a\xi(t,\omega)+b\eta(t,\omega)$ *(a, b being constants) satisfies (C. 1) and (C. 2) and we obtain*

$$\int_s^t (a\xi(\tau,\omega)+b\eta(\tau,\omega))d\beta(\tau,\omega)=a\int_s^t \xi(\tau,\omega)d\beta(\tau,\omega)+b\int_s^t \eta(\tau,\omega)d\beta(\tau,\omega) \tag{2.1}$$

*for* $u\leqq t\leqq s\leqq v$ *for almost all* $\omega\in\Omega_1$.

THEOREM 3. *We have*

$$\int_{s_1}^{s_2}\xi(\tau,\omega)d\beta(\tau,\omega)+\int_{s_2}^{s_3}\xi(\tau,\omega)d\beta(\tau,\omega)=\int_{s_1}^{s_3}\xi(\tau,\omega)d\beta(\tau.\,\omega) \tag{2.2}$$

*for* $u\leqq s_1\leqq s_2\leqq s_3\leqq v$ *for almost all* $\omega\in\Omega_1$.

THEOREM 4. *If (1.6) is satisfied, then we have*

$$\begin{aligned}\frac{c^2}{2}Pr\Big\{\sup_{u\leqq s\leqq t\leqq v}\Big|\int_s^t\xi(\tau,\omega)d\beta(\tau,\omega)\Big|\geqq 2c\Big\}\\ \leqq c^2 Pr\Big\{\sup_{u\leqq t\leqq v}\Big|\int_u^t\xi(\tau,\omega)d\beta(\tau,\omega)\Big|\geqq c\Big\}\\ \leqq\int_\Omega\Big(\int_u^v\xi(t,\omega)d\beta(t,w)\Big)^2P(d\omega)=\int_\Omega\int_u^v\xi(t,\omega)^2dtP(d\omega).\end{aligned} \tag{2.4}$$

THEOREM 5. *If each of* $\xi_n(t,\omega)$, $n=1, 2, \ldots$, *satisfies (C. 1) and (C. 2) and if the system* $\{\xi_n(t,\omega), n=1, 2, \ldots, \infty\}$ *has the property* $\alpha$ *with regard to* $\beta$ *in* $u\leqq t\leqq v$, *and further if*

$$\int_u^v(\xi_n(t,\omega)-\xi_\infty(t,\omega))^2dt\to 0 \tag{2.5}$$

*for almost all* $\omega\in\Omega_1$, *then*

$$\sup_{u\leqq s<t\leqq v}\Big|\int_s^t\xi_n(\tau,\omega)d\beta(\tau,\omega)-\int_s^t\xi_\infty(\tau,\omega)d\beta(\tau,\omega) \tag{2.6}$$

*tends to 0 in probability over* $\Omega_1$.

Since Theorems 1, 2, 3 and 4 follow at once from the properties of the stochastic integral established in [2], we shall here prove Theorem 5 only.

Since we have

$$\int_u^v \xi_n(t, \omega)^2 dt \to \int_u^v \xi_\infty(t, \omega)^2 dt$$

for almost all $\omega \in \Omega_1$ by the assumption (2.5) and since there exists $M=M(\varepsilon)$ for any $\varepsilon>0$ such that

$$Pr\left\{\omega \in \Omega_1, \int_u^v \xi_\infty(t, \omega)^2 dt < M\right\} > P(\Omega_1) - \varepsilon$$

by the assumption that $\xi_\infty(t, \omega)$ satisfies (C. 2), there exists $N_1 = N_1(\varepsilon)$ for any $\varepsilon>0$

$$Pr\left\{\omega \in \Omega_1, \int_u^v \xi_n(t, \omega)^2 dt < M, N_1 < n \leqq \infty\right\} > P(\Omega_1) - 2\varepsilon. \tag{2.7}$$

Put

$$\xi_n^M(t, \omega) = \phi_M\left(\sup_{n \leqq k \leqq \infty} \int_u^t \xi_k(\tau, \omega)^2 dt\right) \xi_n(t, \omega), \quad n=1, 2, \ldots, \infty,$$

where $\phi_M$ is defined by (1. 9). Then it follows from (2. 7) that

$$Pr\{\omega \in \Omega_1, \xi_n^M(t, \omega) = \xi_n(t, \omega), u \leqq t \leqq v, N_1 < n \leqq \infty\} > P(\Omega_1) - 2\varepsilon. \tag{2.8}$$

Since we have

$$\int_u^v (\xi_n^M(t, \omega) - \xi_\infty^M(t, \omega))^2 dt \leqq \int_u^v \xi_n(t, \omega) - \xi_\infty(t, \omega))^2 dt \to 0,$$

$$\int_u^v (\xi_n^M(t, \omega) - \xi_\infty^M(t, \omega))^2 dt \leqq 2\int_u^v \xi_n^M(t, \omega)^2 dt + 2\int_u^v \xi_\infty^M(t, \omega)^2 dt < 4M,$$

we obtain

$$\int_\Omega \int_u^v (\xi_n^M(t, \omega) - \xi_\infty^M(t, \omega))^2 dt P(d\omega) \to 0.$$

By Theorem 4 there exists $N_2 = N_2(\varepsilon)$ for any $\varepsilon>0$ such that

$$Pr\left\{\sup_{u \leqq s \leqq t \leqq v} \left| \int_s^t \xi_n^M(\tau, \omega) d\beta(\tau, \omega) - \int_s^t \xi_\infty^M(\tau, \omega) d\beta(\tau, \omega) \right| < \varepsilon\right\} > P(\Omega_1) - 3\varepsilon, \tag{2.9}$$

which, combined with (2. 8), proves our theorem.

**3. A formula concerning stochastic differentials.** Let $\beta \equiv (\beta^i(t, \omega), i=1, 2, \ldots, r)$ be an $r$-dimensional Brownian motion, and let the system:

$$\{\xi^i(t, \omega), a^i(t, \omega), b_j^i(t, \omega), i=1, 2, \ldots, n, j=1, 2, \ldots, r\} \tag{3.1}$$

have the property $\alpha$ with regard to $\beta$ in $u \leqq t \leqq v$. When we have

$$\xi^i(s, \omega) - \xi^i(t, \omega) = \int_t^s a^i(\tau, \omega) d\tau + \int_t^s b_j^i(\tau, \omega) d\beta^j(\tau, \omega),^{5)} \quad u \leqq t \leqq s \leqq v, \ 1 \leqq i \leqq n, \tag{3.2}$$

[5] We omit the summation sign $\sum_{j=1}^{r}$ according to the usual rule of tensor calculus.

for almost all $\omega \in \Omega_1$, we write this relation in the differential form as follows:

$$d\xi^i(t, \omega) = a^i(t, \omega)dt + b_j^i(t, \omega)d\beta^j(t, \omega), \quad u \leqq t \leqq v, \quad \omega \in \Omega_1, \quad 1 \leqq i \leqq n. \tag{3.3}$$

THEOREM 6. *Let $\xi^i(t, \omega)$, $i=1, 2, \ldots, n$, satisfy*

$$d\xi^i(t, \omega) = a^i(t, \omega)dt + b_j^i(t, \omega)d\beta^j(t, \omega), \quad i=1, 2, \ldots, n, \tag{3.4}$$

*and G be an open subset of the n-space $R^n$ which contains all the points ($\xi^i(t, \omega)$, $i=1, 2, \ldots, n$) for $u \leqq t \leqq v$, $\omega \in \Omega_1$.*

*Let $f(t, x^1, x^2, \ldots, x^n)$ be a continuous function defined in $u \leqq t \leqq v$, $(x^1, x^2, \ldots, x^n) \in G$, such that*

$$\begin{cases} f_0(t, x^1, \ldots, x^n) = \dfrac{\partial f}{\partial t}(t, x^1, \ldots, x^n) \\ f_i(t, x^1, \ldots, x^n) = \dfrac{\partial f}{\partial x_i}(t, x^1, \ldots, x^n), \quad i=1, 2, \ldots, n, \\ f_{ij}(t, x^1, \ldots, x^n) = \dfrac{\partial^2 f}{\partial x^i \partial x^j}(t, x^1, \ldots, x^n), \quad i, j=1, 2, \ldots, n, \end{cases} \tag{3.5}$$

*are all continous.*

*Then $\eta(t, \omega) \equiv f(t, \xi^1(t, \omega), \ldots, \xi^n(t, \omega))$ satisfies*

$$\begin{aligned} d\eta(t, \omega) = (f_0(t, \xi) + f_i(t, \xi)a^i(t, \omega) + \frac{1}{2}f_{ij}(t, \xi)b_k^i(t, \omega)b_k^j(t, \omega))dt \\ + f_i(t, \xi)b_j^i(t, \omega)d\beta^j(t, \omega), \end{aligned} \tag{3.6}$$

*where $\xi \equiv (\xi^1(t, \omega), \xi^2(t, \omega), \ldots, \xi^n(t, \omega))$.*

LEMMA 1. *For any stochastic process $\xi(t, \omega)$ satisfying*

$$\int_u^v \xi(t, \omega)^2 dt < \infty, \quad \omega \in \Omega_1, \tag{3.7}$$

*there exists a sequence of uniformly stepwise stochastic processes $\xi_n(t, \omega)$, $n=1, 2, \ldots$ whose value at any time-point t is a B-measurable function of $\xi(\tau, \omega)$, $\tau \leqq t$, and such that*

$$\int_u^v |\xi_n(t, \omega) - \xi(t, \omega)|^2 dt \to 0$$

*for almost all $\omega \in \Omega_1$.*

This Lemma follows immediately from Lemma 7.1 in [2].

LEMMA 2. *Let $\xi(t, \omega)$, $\eta(t, \omega)$ be stochastic processes such that the system $\{\xi(t, \omega), \eta(t, \omega)\}$ has the property $\alpha$ with regard to a one-dimensional Brownian motion $\beta(t, \omega)$ in $u \leqq t \leqq v$ and that*

$$\int_u^v \xi(t, \omega)^2 dt < \infty, \quad \int_u^v \eta(t, \omega)^2 dt < \infty$$

*for almost all $\omega \in \Omega_1$. Then we have*

$$(3.8)\quad \int_u^v \xi(t,\omega)d\beta(t,\omega)\int_u^v \eta(s,\omega)d\beta(s,\omega)$$
$$=\int_u^v \xi(t,\omega)\int_u^t \eta(s,\omega)d\beta(s,\omega)d\beta(t,\omega)$$
$$+\int_u^v \eta(s,\omega)\int_u^s \xi(t,\omega)d\beta(t,\omega)d\beta(s,\omega)+\int_u^v \xi(t,\omega)\eta(t,\omega)dt$$

*for almost all.*

*Proof.* Firstly we shall prove (3.8) in the case that both $\xi(t,\omega)$ and $\eta(t,\omega)$ are uniformly stepwise. Then we may assume that

$$(3.9)\quad \xi(t,\omega)=\xi(u_{i-1},\omega),\quad \eta(t,\omega)=\eta(u_{i-1},\omega),\quad u_{i-1}\leqq t\leqq u_i,\ i=1,2,\ldots,n,$$

where

$$(3.10)\quad u=u_0<u_1<\ldots<u_n=v.$$

The left side of (3.8) equals the following: $(u_{ij}=u_i+\frac{j}{N}(u_{i+1}-u_i))$

$$\sum_{ijpq}\int_{u_{i,j-1}}^{u_{ij}}\xi(t,\omega)d\beta(t,\omega)\int_{u_{p,q-1}}^{u_{pq}}\eta(s,\omega)d\beta(s,\omega)$$
$$=\sum_{u_{ij}>u_{pq}}+\sum_{u_{pq}>u_{ij}}+\sum_{p=i,q=j}=\int_u^v\xi(t,\omega)\int_u^{\lambda_n(t)}\eta(s,\omega)d\beta(s,\omega)d\beta(t,\omega)$$
$$+\int_u^v\eta(s,\omega)\int_u^{\lambda_n(s)}\xi(t,\omega)d\beta(t,\omega)d\beta(s,\omega)$$
$$+\sum_i\xi(u_{i-1},\omega)\eta(u_{i-1},\omega)\sum_{j=1}^N(\beta(u_{ij},\omega)-\beta(u_{i,j-1},\omega))^2=I_1+I_2+I_3,$$

where $\lambda_n(t)$ is the maximum $u_{ij}$ which does not exceed $t$.

As $N\to\infty$, $I_1$ tends to the first term of the right side of (3.8) in probability by virtue of Theorem 5, since we have

$$\int_u^v|\xi(t,\omega)\int_u^{\lambda_n(t)}\eta(s,\omega)d\beta(s,\omega)-\xi(t,\omega)\int_u^t\eta(s,\omega)d\beta(s,\omega)|^2dt$$
$$=\int_u^v|\xi(t,\omega)\int_{\lambda_n(t)}^t\eta(s,\omega)d\beta(s,\omega)|^2dt$$
$$=\int_u^v|\xi(t,\omega)\eta(\lambda_n(t),\omega)(\beta(t,\omega)-\beta(\lambda_n(t),\omega))|^2dt\to 0$$

for almost all $\omega$. By the same reason we see that $I_2$ tends to the second term of the right side of (3.8) in probability.

Since $\sum_{j=1}^N(\beta(u_{ij},\omega)-\beta(u_{i,j-1},\omega))^2\to u_i-u_{i-1}$ (in probability), $I_3$ tends to the third term of the right side of (3.8).

Next we shall consider the general case. By Lemma 1 we shall construct $\xi_n(t,\omega)$, $n=1,2,\ldots$, and $\eta_n(t,\omega)$, $n=1,2,\ldots$ for $\xi(t,\omega)$ and $\eta(t,\omega)$ respectively. Since our Lemma 2 holds for uniformly stepwise processes as is proved above, we have

$$(3.11)\quad \int_u^v \xi_n(t,\omega)d\beta(t,\omega)\int_u^v \eta_n(t,\omega)d\beta(s,\omega)$$
$$=\int_u^v \xi_n(t,\omega)\int_u^t \eta_n(s,\omega)d\beta(s,\omega)d\beta(t,\omega)$$
$$+\int_u^v \eta_n(s,\omega)\int_u^s \xi_n(t,\omega)d\beta(t,\omega)d\beta(s,\omega)+\int_u^v \xi_n(t,\omega)\eta_n(t,\omega)dt.$$

Put

$$\zeta_n(t,\omega)=\int_u^t \eta_n(s,\omega)d\beta(s,\omega),\qquad \zeta(t,w)=\int_u^t \eta(s,\omega)d\beta(s,\omega),$$
$$\rho_n(t,\omega)=\int_u^t \xi_n(s,\omega)d\beta(s,\omega),\qquad \rho(t,\omega)=\int_u^t \eta(s,\omega)d\beta(s,\omega).$$

By taking adequate subsequences we see, by Theorem 5, that $\zeta_n(t,\omega)$ and $\rho_n(t,\omega)$ converge uniformly in $t$ to $\zeta(t,\omega)$ and $\rho(t,\omega)$ respectively for almost all $\omega\in\Omega_1$. Therefore we have

$$\int_u^v |\xi_n(t,\omega)\zeta_n(t,\omega)-\xi(t,\omega)\zeta(t,\omega)|^2dt$$
$$\leqq 2\int_u^v |\xi_n(t,\omega)-\xi(t,\omega)|^2|\zeta_n(t,\omega)|^2dt$$
$$+2\int_u^v |\xi(t,\omega)|^2|\zeta_n(t,\omega)-\zeta(t,\omega)|^2dt\to 0$$

for almost all $\omega\in\Omega_1$, from which follows by Theorem 5

$$\int_u^v \xi_n(t,\omega)\zeta_n(t,\omega)d\beta(t,\omega)\to\int_u^v \xi(t,\omega)\zeta(t,\omega)d\beta(t,\omega)\quad \text{(in probability)}.$$

Similaly we have

$$\int_u^v \eta_n(s,\omega)\rho_n(s,\omega)d\beta(s,\omega)\to\int_u^v \eta(s,\omega)\rho(s,\omega)d\beta(s,\omega)\quad \text{(in probability)}.$$

Further we have

$$\left|\int_u^v \xi_n(t,\omega)\eta_n(t,\omega)dt-\int_u^v \xi(t,\omega)\eta(t,\omega)dt\right.$$
$$\leqq\sqrt{\int_u^v \eta_n(t,\omega)^2dt\int_u^v(\xi_n(t,\omega)-\xi(t,\omega))^2dt}$$
$$+\sqrt{\int_u^v \xi(t,\omega)dt\int_u^v(\eta_n(t,\omega)-\eta(t,\omega))^2dt}\to 0.$$

Thus our Lemma 2 is completely proved.

By the same way as above, we obtain the following Lemmas 3 and 4.

LEMMA 3. *Let $\xi(t, \omega)$ and $\eta(t, \omega)$ be stochastic processes such that the system $\{\xi(t, \omega), \eta(t, \omega)\}$ has the property $\alpha$ with regard to the two-dimensional Brownian motion $(\beta(t, \omega), \gamma(t, \omega))$ and that*

$$\int_u^v \xi(t, \omega)^2 dt < \infty, \qquad \int_u^v \eta(t, \omega)^2 dt < \infty$$

*for almost all $\omega \in \Omega_1$.*

*Then we have*

$$\int_u^v \xi(t, \omega) d\beta(t, \omega) \int_u^t \eta(s, \omega) d\gamma(s, \omega) \tag{3.12}$$

$$= \int_u^v \xi(t, \omega) \int_u^t \eta(s, \omega) d\gamma(s, \omega) d\beta(t, \omega)$$

$$+ \int_u^v \eta(s, \omega) \int_u^s \xi(t, \omega) d\beta(t, \omega) d\gamma(s, \omega)$$

*for almost all $\omega \in \Omega_1$.*

LEMMA 4. *Let $a(t, \omega)$ and $b(t, \omega)$ be stochastic processes such that the system $a(t, \omega)$, $b(t, \omega)$ has the property $\alpha$ with regard to a one-dimensional Brownian motion $\beta(t, \omega)$ and that*

$$\int_u^v |a(t, \omega)| dt < \infty, \qquad \int_u^v |b(t, \omega)|^2 dt < \infty$$

*for almost all $\omega \in \Omega_1$. Then we have*

$$\int_u^v a(t, \omega) dt \int_u^v b(s, \omega) d\beta(s, \omega) \tag{3.13}$$

$$= \int_u^v a(t, \omega) \int_u^t b(s, \omega) d\beta(s, \omega) dt + \int_u^v b(s, \omega) \int_u^s a(t, \omega) dt \, d\beta(s, \omega).$$

LEMMA 5. *Let $\xi^i(t, \omega)$, $i = 1, 2, \ldots, n$, be determined as in Theorem 6. Then we have*

$$(\xi^i(s, \omega) - \xi^i(t, \omega))(\xi^j(s, \omega) - \xi^j(t, \omega)) \tag{3.14}$$

$$= \int_t^s \{(\xi^i(\tau, \omega) - \xi^i(t, \omega)) a^j(\tau, \omega) + (\xi^j(\tau, \omega) - \xi^j(t, \omega)) a^i(\tau, \omega) + b_k{}^i(\tau, \omega) b_k{}^j(\tau, \omega)\} d\tau$$

$$+ \int_t^s \{(\xi^i(\tau, \omega) - \xi^i(t, \omega)) b_k{}^j(\tau, \omega) + (\xi^j(\tau, \omega) - \xi^j(t, \omega)) b_k{}^i(\tau, \omega)\} d\beta^k(\tau, \omega)$$

*for almost all $\omega \in \Omega_1$ and for $u \leqq t \leqq s \leqq v$.*

*Proof.* By the assumption we have

$$(\xi^i(s,\omega)-\xi^i(t,\omega))(\xi^j(s,\omega)-\xi^j(t,\omega))$$
$$=\int_t^s a^i(\tau,\omega)d\tau\int_t^s a^j(\sigma,\omega)d\sigma+\int_t^s b_k^i(\tau,\omega)d\beta^k(\tau,\omega)\int_t^s a^j(\sigma,\omega)d\sigma$$
$$+\int_t^s a(\tau,\omega)d\tau\int_t^s b_l^i(\sigma,\omega)d\beta^l(\sigma,\omega)$$
$$+\int_t^s b_k^i(\tau,\omega)d\beta(\tau,\omega)\int_t^s b_b^i(\sigma,\omega)d\beta^l(\sigma,\omega),$$

from which follows (3.14) at once by virtue of Lemmas 2, 3 and 4.

LEMMA 6. *Let $\xi^i(t,\omega)$, $i=1,2,\ldots,n$, be determined as in Theorem 6. For the point-system:*

$$\Delta: t=t_0<t_1<\ldots<t_m=s$$

*we put*

$$S(\Delta,\omega)=\sum_{\mu=1}^m|\xi^i(t_\mu,\omega)-\xi^i(t_{\mu-1},\omega)|\,|\xi^j(t_\mu,\omega)-\xi^j(t_{\mu-1},\omega)|.$$

*Then there exists $M=M(\varepsilon)$ independent of $\Delta$ for any $\varepsilon>0$ such that*

(3.15) $$Pr\{\omega\in\Omega_1,\ S(\Delta,\omega)>M\}<\varepsilon.$$

*Proof.* We may consider the case that

$$\int_\Omega\int_t^s b_k^p(\tau,\omega)dtP(d\omega)<\infty,\qquad p=i,\ j,\qquad k=1,2,\ldots,n,$$

since, if our Lemma is established in this case, we can easily deduce our Lemma in the general case by the definition of stochastic integral. When there is no confusion, we omit the time parameter and the probability parameter $\omega$ in the following.

$$(\xi^i(t_\mu)-\xi^i(t_{\mu-1}))(\xi^j(t_\mu)-\xi^j(t_{\mu-1}))$$
$$=\int_{t_{\mu-1}}^{t_\mu}a^i d\tau\int_{t_{\mu-1}}^{t_\mu}a^j d\sigma+\int_{t_{\mu-1}}^{t_\mu}b_k^i d\beta^k\int_{t_{\mu-1}}^{t_\mu}a^j d\sigma$$
$$+\int_{t_{\mu-1}}^{t_\mu}a^i d\tau\int_{t_{\mu-1}}^{t_\mu}b_k^i d\beta^k+\int_{t_{\mu-1}}^{t_\mu}b_k^i d\beta^k\int_{t_{\mu-1}}^{t_\mu}b_k^j d\beta^k.$$

Since

$$S_1(\Delta,\omega)\equiv\sum_\mu\left|\int_{t_{\mu-1}}^{t_\mu}a^i d\tau\right|\left|\int_{t_{\mu-1}}^{t_\mu}a^j d\sigma\right|\leqq\max_{t\leqq t'\leqq s'\leqq s}\left|\int_{t'}^{s'}a^i d\tau\right|\int_t^s|a^j|\,d\sigma$$

for almost all $\omega\in\Omega_1$ we may find $M_1(\varepsilon)$ independent of $\Delta$ for any $\varepsilon<0$ such that $M>M_1(\varepsilon)$ implies

(3.16) $$Pr\{\omega\in\Omega_1,\ S_1(\Delta,\omega)>M\}<\varepsilon/4.$$

By the same way we may find $M_{2k}(\varepsilon)$ and $M_{3k}(\varepsilon)$ independent of $\Delta$ for any

$\varepsilon<0$ such that $M>M_{2k}(\varepsilon)$ or $M>M_{3k}(\varepsilon)$ implies

(3.17) $$Pr\{\omega\in\Omega_1,\ S_{2k}(\Delta,\ \omega)>M\}<\varepsilon/4r \quad \text{or} \quad Pr\{\omega\in\Omega_2,\ S_{3k}(\Delta,\ \omega)>M\}<\varepsilon/4r$$

respectively, where

$$S_{2k}(\Delta,\ \omega)=\sum_\mu\left|\int_{t_{\mu-1}}^{t_\mu} b_k{}^i d\beta^k\right|\left|\int_{t_{\mu-1}}^{t_\mu} a^j d\sigma\right|,$$

$$S_{3k}(\Delta,\ \omega)=\sum_\mu\left|\int_{t_{\mu-1}}^{t_\mu} a^i d\tau\right|\left|\int_{t_{\mu-1}}^{t_\mu} b_k{}^j d\beta^k\right|.$$

Put

$$S_{4kl}(\Delta,\ \omega)\equiv\sum_\mu\left|\int_{t_{\mu-1}}^{t_\mu} b_k{}^i d\beta^k\right|\left|\int_{t_{\mu-1}}^{t_\mu} b_l{}^j d\beta^l\right|=\sum_\mu A_\mu B_\mu.$$

$$\int S_{4kl}(\Delta,\ \omega)P(d\omega)=\sum_\mu\int_\Omega|A_\mu|^2P(d\omega)+\sum_\mu\int_\Omega|B_\mu|^2P(d\omega)$$

$$\leqq\sum_\mu\int_\Omega\int_{t_{\mu-1}}^{t_\mu}(b_k{}^i)^2d\tau P(d\omega)+\sum_\mu\int_\Omega\int_{t_{\mu-1}}^{t_\mu}(b_l{}^j)^2d\tau P(d\omega)$$

$$\leqq\int_\Omega\int_t^s(b_k{}^i)^2d\tau P(d\omega)+\int_\Omega\int_t^s(b_l{}^j)^2d\sigma P(d\omega).$$

Thus we may find $M_{4kl}(\varepsilon)$ independent of $\Delta$ for any $\varepsilon>0$ such that $M>M_{4kl}(\varepsilon)$ implies

$$Pr\{\omega\in\Omega_1,\ S_{4kl}(\Delta,\ \omega)>M\}<\varepsilon/4r^2.$$

Since

$$S(\Delta,\ \omega)\leqq S_1(\Delta,\ \omega)+\sum_k S_{2k}(\Delta,\ \omega)+\sum_k S_{3k}(\Delta,\ \omega)+\sum_{k,\,l}S_{4kl}(\Delta,\ \omega),$$

we have (3.15) by putting

$$M=M(\varepsilon)=M_1(\varepsilon)+\sum_k M_{2k}(\varepsilon)+\sum_k M_{3k}(\varepsilon)+\sum_{k,\,l}M_{4kl}(\varepsilon).$$

*Proof of* THEOREM 6. By Taylor expansion of $f(t,\ x^1,\ \ldots,\ x^n)$ we have (the probability parameter $\omega$ being omitted)

$$\eta(s,\ \omega)-\eta(t,\ \omega)$$

$$=\sum_{k=1}^m(\eta(t_k^m,\ \omega)-\eta(t_{k-1}^m,\ \omega))\qquad \left(t_k^m=t+\frac{k}{m}(s-t)\right)$$

$$=\sum_{k=1}^m[f(*k)(t_k^m-t_{k-1}^m)+\sum_{i=1}^n f_i(*k)(\xi^i(t_k^m)-\xi^i(t_{k-1}^m))$$

$$+\frac{1}{2}\sum_{ij}f_{ij}(*k)(\xi^i(t_k^m)-\xi^i(t_{k-1}^m))(\xi^j(t_k^m)-\xi^j(t_{k-1}^m))$$

$$+\sum_{ij}\theta_{ijk}^m(\xi^i(t_k^m)-\xi^i(t_{k-1}^m))(\xi^j(t_k^m)-\xi^j(t_{k-1}^m))],$$

where

$$*k=(t_{k-1}^m,\ \xi^1(t_{k-1}^m),\ \ldots,\ \xi^n(t_{k-1}^m)).$$

Since $f_{ij}(t, x^1, \ldots, x^n)$ are continuous and $\xi^i(t, \omega)$, $i=1, 2, \ldots, n$ are all continuous in $t$ for almost all $\omega \in \Omega_1$, $\theta^m_{i,j,k}$ tends to 0 uniformly in $m$ and $k$ as $n \to \infty$ for almost all $\omega \in \Omega_1$. Therefore we have

$$\sum_{kij} \theta^m_{ijk} \, ('') \, ('') \to 0 \text{ (in probability on } \Omega_1)$$

by virtue of Lemma 6.

By Lemma 5 the remainder equals the following expression:

$$\begin{aligned}(3.18)\quad & \int_t^s (f_0(*) + f_i(*) a^i(\tau) + \frac{1}{2} f_{ij}(*) b_k^i(\tau) b_k^j(\tau)) d\tau + \int_t^s f_i(*) b_j^i(\tau) d\beta^j(\tau) \\ & + \frac{1}{2} \int_t^s f_{ij}(*) [(\xi^i(\tau) - \xi^i(\lambda_m(\tau))) a^j(\tau) + (\xi^j(\tau) - \xi^j(\lambda_m(\tau))) a^i(\tau)] d\tau \\ & + \frac{1}{2} \int_t^s f_{ij}(*) [(\xi^i(\tau) - \xi^i(\lambda_m(\tau))) b_k^j(\tau) + (\xi^j(\tau) - \xi^j(\lambda_m(\tau))) b_k^i(\tau)] d\beta^k(\tau),\end{aligned}$$

where $\lambda_m(\tau)$ denotes the maximum $t_k^m$ which dose not exceed $\tau$ and $*$ denotes $(\lambda_m(\tau), \xi^1(\lambda_m(\tau)), \ldots, \xi^n(\lambda_m(\tau)))$.

But $\xi^i(\lambda_n(\tau), \omega) \to \xi^i(\tau, \omega)$ uniformly in $\tau$ for almost all $\omega \in \Omega_1$ as $n \to \infty$. Therefore, by letting $n \to \infty$ in (3.18) we obtain

$$\begin{aligned}\eta(s) - \eta(t) = & \int_t^s (f_0(\tau, \xi) + f_i(\tau, \xi) a^i(\tau) + \frac{1}{2} f_{ij}(\tau, \xi) b_k^i(\tau) b_k^j(\tau)) d\tau \\ & + \int_t^s f_i(\tau, \xi) b_j^i(\tau) d\beta^j(\tau),\end{aligned}$$

which proves our Theorem 5.

## REFERENCES

[1] K. Itô: On stochastic differential equations in a differentiable manifold, this Journal Vol. 1, 1950.

[2] K. Itô: Stochastic differential equations, Memoris of the American Mathematical Society, 4, 1951.

*Mathematical Institute,*
*Nagoya University*

Reprinted from
*Nagoya Math. J.* **3**, 55–65 (1951)

Journal of the Mathematical Society of Japan Vol. 3, No. 1, May, 1951.

# Multiple Wiener Integral

Kiyosi Itô

The notion of *multiple Wiener integral* was introduced first by N. Wiener[1] who termed it *polynomial chaos.* Our definition in the present paper is obtained by a slight modification of Wiener's one, and seems to be more convenient in the point that our integrals of different degrees are orthogonal to each other while Wiener's polynomial chaos has not such a property.

In § 1 we shall define a normal random measure as a generalization of a brownian motion process. In § 2 we shall define multiple Wiener integral and show its fundamental property. In § 3 we shall establish a close relation between our integrals and Hermite polynomials. By making use of this relation we shall give, in § 4, an orthogonal expansion of any $L_2$-functional of the normal random measure, which proves to be coincident with the expansion given by S. Kakutani[2] for the purpose of the spectral resolution of the shift operator in the $L_2$ over the brownian motion process. In § 5 we shall treat the case of a brownian motion process, and in this case we shall show that we can define the multiple Wiener integral by the iteration of stochastic integrals.[3]

## § 1. Normal random measure

A system of real random veriables $\xi_\alpha(\omega)$, $\alpha \in A$, $\omega$ being a probability parameter, is called normal when the joint destribution of $\xi_{\alpha_1}, \ldots, \xi_{\alpha_n}$; $\alpha_1, \ldots, \alpha_n \in A$, is always a multivariate Gaussian distribution (including degenerate cases) with the mean vector $(0, \cdots, 0)$.

By making use of Kolmogoroff's theorem[4] of introducing a probability distribution in $R^A$, we can easily prove the following

1) N. Wiener: The homogeneous chaos, Amer. Journ. Math. Vol. *LV*, No. 4, 1938.

2) S. Kakntani: Determination of the spectrum of the flow of Brownian motion, Proc. Nat. Acad. Sci., U.S.A. **36** (1950), 319–323.

3) K. Itô: Stochastic integral, Proc. Imp. Acad. Tokyo, Vol. *XX*, No. 8, 1944.

4) A. Kolmogoroff: Grundbegriffe der Wahrscheinlichkeitsrechnung, Berlin, 1933. The consistency-condition is well satisfied by virtue of the property of multivariate Gaussian distribution.

**Theorem 1. 1** *If $v_{\alpha\beta}$; $\alpha, \beta \in A$, satisfies the following two conditions:*

*symmetric* : $v_{\alpha\beta}=v_{\beta\alpha}$ ; (1.1)

*positive-definite* : $\sum x_i \bar{x}_j v_{\alpha_i \alpha_j} \geqq 0$ *(for any* $\alpha_1, \cdots, \alpha_n \in A$ *and for any complex numbers* $x_1, x_2, \cdots, x_n$*), then there exists a normal random system* $\xi_\alpha$, $\alpha \in A$, *which satisfies* (1.2)

$$v_{\alpha\beta}=\mathfrak{E}(\xi_\alpha \xi_\beta)=\int \xi_\alpha(\omega)\xi_\beta(\omega)d\omega. \tag{1.3}$$

**Definition.** Let $(T, \boldsymbol{B}, m)$ be a measure space. We denote by $\boldsymbol{B}^*$ the system $\{E\,;E \in \boldsymbol{B},\ m(E)<\infty\}$. A normal system $\beta(E, \omega)$, $E \in \boldsymbol{B}^*$, is called a ***normal random measure*** on $(T, \boldsymbol{B}, m)$, if

$$\mathfrak{E}(\beta(E)\ \beta(E'))=m(E \cap E') \text{ for any } E, E' \in \boldsymbol{B}^*. \tag{1.4}$$

**Remark.** Since we have $m(E\cap E')=m(E'\cap E)$ and $\sum x_i \bar{x}_j m(E_i \cap E_j) = \int |\sum_i x_i C_i(t)|^2\ m(dt) \geqq 0$, $C_i(t)$ being the characteristic function of the set $E_i$, we can see, by Theorem 1.1, the existence of a normal random measure on any measure space $(T, \boldsymbol{B}, m)$.

The following theerem, which can be easily shown, justifies the name of normal random " measure."

**Theorem 1. 2** *Let $\beta(E)$ be a normal random measure on $(T, \boldsymbol{B}, m)$. If $E_1, E_2, \cdots$ are disjoint, then $\beta(E_1), \beta(E_2), \cdots$ are independent. Furthermore if $E=E_1+E_2+\cdots \in \boldsymbol{B}^*$, then $\beta(E)=\sum_n \beta(E_n)$ (in mean convergence).*

**Remark.** Since $\beta(E_1), \beta(E_2), \cdots$, are independent, then the mean convergence of $\sum \beta(E_n)$ implies the almost certain convergence by virtue of Levy's theorem.[5)]

Hereafter we set the following restriction on the measure $m$.

*Continuity.* For any $E \in \boldsymbol{B}^*$ and $\varepsilon>0$ there exists a decomposition of $E$:

$$E=\sum_{i=1}^n E_i \tag{1.5}$$

such that

$$m(E_i)<\varepsilon,\ i=1, 2, \cdots, n. \tag{1.6}$$

5) P. Lévy: Théorie de l'addition des variariables aléatoires, Paris, 1937.

## § 2. Definition of multiple Wiener integral

By $L^2(T^p)$ we denote the totality of square-summable complex-valued functions defined on the product measure space $(T, \boldsymbol{B}, m)^p$. An elementary function[6] $f(t_1,\cdots,t_p)$ is called *special* if $f(t_1,\cdots,t_p)$ vanishes except for the case that $t_1,\cdots,t_p$ are all different. We shall denote by $S_p$ the totality of special elementary functions.

**Theorem 2. 1.** *$S_p$ is a linear manifold dense in $L^2(T^p)$.*

*Proof.* It suffices to show that the characteristic function $c(t_1,\cdots,t_p)$ of any set $E$ of the form:

$$E=E_1\times E_2\times\cdots\times E_p \quad (E_i\in \boldsymbol{B}^*,\ i=1,2,\cdots,p) \tag{2.1}$$

can be approximated (in the $L_2$-norm) by a special elementary function.

For any $\varepsilon>0$ we can determine, by the *continuity condition*, a set-system $\boldsymbol{F}=\{F_1,\cdots,F_n\}\in\boldsymbol{B}^*$ which satisfies

$\boldsymbol{F}$. 1. $F_1, F_2,\cdots,F_n$ are disjoint,

$\boldsymbol{F}$. 2. $m(F_i)<\varepsilon_1\equiv\varepsilon/\binom{p}{2}\cdot(\sum m(E_i))^{p-1},\ \binom{p}{2}=\dfrac{p(p-1)}{2\cdot 1}$,

$\boldsymbol{F}$. 3. each $E_i$ is expressible as the sum of a subsystem of $\boldsymbol{F}$.

Then $c(t_1,\cdots,t_p)$ is expressible in the form:

$$c(t_1,\cdots,t_p)=\sum \varepsilon_{i_1\ldots i_p}c_{i_1}(t_1)\ldots c_{i_1}(t_p) \tag{2.2}$$

where $\varepsilon_{i_1\ldots i_p}=0$ or $1$ and $c_i(t)$ is the characteirstic function of $F_i$, $i=1, 2,\ldots,n$. We devide $\sum$ into two parts: $\sum'$ and $\sum''$: $\sum'$ corresponds to the indices $\{i_1,\ldots,i_p\}$ which are all different, while $\sum''$ corresponds to the others.

We put

$$f(t_1,\ldots,t_p)=\sum{}' \varepsilon_{i_1\ldots i_p}c_{i_1}(t_1)\ldots c_{i_p}(t_p). \tag{2.3}$$

Then $f\in S_p$ and

$$\begin{aligned}\|c-f\|^2&=\int\cdots\int|c(t_1,\ldots,t_p)-f(t_1,\cdots,t_p)|^2\, m(dt_1)\ldots m(dt_p)\\&=\sum{}''\varepsilon_{i_1\ldots i_p}m(F_{i_1})\ldots m(F_{i_p})\end{aligned}$$

6) An elementary function of $(t_1,\ldots,t_p)$ is defined as a linear combination of the characteristic functions of the sets of the form $E_1\times\ldots\times E_p$, $E_i\in\boldsymbol{B}^*$, $i=1\ 2,\ldots,n$.

$$\leqq \binom{p}{2} \sum m(F_i)^2 (\sum m(F_i))^{p-2}$$

$$\leqq \binom{p}{2} \varepsilon_1 (\sum m(F_i))^{p-1} = \binom{p}{2} \varepsilon_1 (\sum m(E_i))^{p-1} < \varepsilon.$$

Now we shall define the multiple wiener integral of $f \in L^2(T^p)$, which we denote by

$$I_p(f) \text{ or } \int \cdots \int f(t_1, \ldots, t_p) d\beta(t_1) \ldots d\beta(t_p)$$

Let $f$ be a special elementary function. Then $f$ can be expressible as follows:

$$f(t_1, \ldots, t_p) = a_{i_1 \ldots i_p} \text{ for } (t_1, \ldots, t_p) \in T_{i_1} \times \ldots \times T_{i_p}, \tag{2.4}$$
$$= 0 \quad \text{elsewhere,}$$

where $T_1, T_2, \ldots, T_n$ are disjoint and $m(T_i) < \infty$, $i = 1, 2, \ldots, n$, and $a_{i_1 \ldots i_p} = 0$ if any two of $i_1, \ldots, i_p$ are equal. We define $I_p(f)$ for such $f$ by

$$I_p(f) = \sum a_{i_1 \ldots i_p} \beta(T_{i_1}) \ldots \beta(T_{i_p}). \tag{2.5}$$

Then we obtain

$$I_p(af + bg) = aI_p(f) + bI_p(g) \tag{I.1}$$

$$I_p(f) = I_p(\tilde{f}), \tag{I.2}$$

where

$\tilde{f}(t_1, \ldots, t_p) = \frac{1}{\underline{|p}} \sum_{(\pi)} f(t_{\pi_1}, \ldots, t_{\pi_p})$, $(\pi) = (\pi_1, \ldots, \pi_p)$ running over all permutations of $(1, 2, \ldots, p)$ $(\underline{|p} = 1 \cdot 2 \cdot \ldots \cdot p)$.

$$(I_p(f), I_p(g)) = \underline{|p} \, (\tilde{f}, \tilde{g}), \tag{I.3}$$

where $(I_p(f), I_p(g)) \equiv \mathfrak{E}(I_p(f) \overline{I_p(g)}) \equiv \int I_p(f) \cdot \overline{I_p(g)} d\omega$, and

$$(\tilde{f}, \tilde{g}) \equiv \int \cdots \int \tilde{f}(t_1, \ldots, t_p) \overline{\tilde{g}(t_1, \ldots, t_p)} m(dt_1) \ldots m(dt_p).$$

$$(I_p(f), I_p(g)) = 0, \text{ if } p \neq q. \tag{I.4}$$

$(I.1)$ is clear. In order to show $(I.2)$ and $(I.3)$ we may assume that $f$ and $g$ are expressible as follows:

$$f(t_1,\dots,t_p)=a_{i_1,\dots,i_p},\ g(t_1,\dots,t_p)=b_{i_1,\dots,i_p}$$

$$\text{for}\quad (t_1,\dots,t_p)\in T_{i_1}\times\cdots\times T_{i_p}$$

and

$$f(t_1,\dots,t_p)=0,\ g(t_1,\dots,t_p)=0\quad\text{elsewhere.}$$

Then we have

$$\begin{aligned}
I_p(f)&=\sum_{i_1<\cdots<i_p}\Big(\sum_{(j)\sim(i)}a_{j_1,\dots,j_p}\Big)\,\beta(T_{i_1})\dots\beta(T_{i_p})^{7)}\\
&=\underline{|p}\sum_{i_1<\cdots<i_p}\Big(\frac{1}{\underline{|p}}\sum_{(j)\sim(i)}a_{j_1,\dots,j_p}\Big)\,\beta(T_{i_1})\dots\beta(T_{i_p})\\
&=\sum_{i_1,\dots,i_p}\Big(\frac{1}{\underline{|p}}\sum_{(j)\sim(i)}a_{j_1,\dots,j_p}\Big)\,\beta(T_{i_1})\dots\beta(T_{i_p})\\
&=I(\tilde f),\ (\underline{|p}=1\cdot 2\cdot\dots\cdot p)
\end{aligned}$$

which proves $(I.2)$.

$$\begin{aligned}
(I_p(f),\ I_p(g))&=\Big(\sum_{i_1<\cdots<i_p}\Big(\sum_{(j)\sim(i)}a_{j_1,\dots,j_p}\Big)\,\beta(T_{i_1})\dots\beta(T_{i_p}),\\
&\qquad\qquad\sum_{i_1<\cdots<i_p}\Big(\sum_{(j)\sim(i)}b_{j_1,\dots,j_p}\Big)\,\beta(T_{i_1})\dots\beta(T_{i_p})\Big)\\
&=\sum_{i_1<\cdots<i_p}\Big(\sum_{(j)\sim(i)}a_{j_1,\dots,j_p}\Big)\cdot\Big(\sum_{(j)\sim(i)}\bar b_{j_1,\dots,j_p}\Big)\,m(T_{i_1})\dots m(T_{i_p})\\
&=\frac{1}{\underline{|p}}\sum_{i_1,\dots,i_p}\Big(\sum_{(j)\sim(i)}a_{j_1,\dots,j_p}\Big)\Big(\sum_{(j)\sim(i)}\bar b_{j_1,\dots,j_p}\Big)\,m(T_{i_1})\dots m(T_{i_p})\\
&=\underline{|p}\sum_{i_1,\dots,i_p}\Big(\frac{1}{\underline{|p}}\sum_{(j)\sim(i)}a_{j_1,\dots,j_p}\Big)\Big(\frac{1}{\underline{|p}}\sum_{(j)\sim(i)}b_{j_1,\dots,j_p}\Big)\,m(T_{i_1})\dots m(T_{i_p})\\
&=\underline{|p}\int\cdots\int\tilde f(t_1,\dots,t_p)\cdot\overline{\tilde g(t_1,\dots,t_p)}\,m(dt_1)\dots m(dt_p)\\
&=\underline{|p}\,(\tilde f,\tilde g).
\end{aligned}$$

7) $(j)\sim(i)$ means that $(j)\equiv(j_1,\dots j_q)$ is a permutation of $(i)\equiv(i_1,\dots,i_p)$.

Thus $(I.3)$ is proved.

By the similar computations we can prove $(I.4)$.

By putting $f=g$ in $(I.3)$, we obtain

$$\|I_p(f)\|^2=\underline{|p}\ \|\tilde{f}\|^2 \leqq \underline{|p}\ \|f\|^2, \tag{I.3'}$$

the last inequality being obtained by virtue of Schwarz' inequality.

Therefore $I_p$ can be considered as a bounded linear operator from $S_p$ into $L_2(\omega)$, and so it can be extended to an operator from the closure of $S_p(=L^2(T^p)$ by Theorem 2.1) into $L_2(\omega)$ which satisfies also $(I.1)$, $(I.2)$, $(I.3)$, $(I.4)$ and $(I.3')$.

For the later use we denote by $L^2_0$ the totality of complex numbers and we define as $I_0(c)=c$. Thus $(I.1)$, $(I.2)$, $(I.3)$, $(I.4)$ and $(I.3')$ are true for $p, q=0, 1, 2, \ldots$.

## § 3. Relation between multiple Wiener integrals and Hermite polynomials.

**Theorem 3.1.** *Let $\varphi_1(t), \varphi_2(t), \ldots, \varphi_n(t)$ be an orthogonal system of real-valued functions in $L^2(T)$ and $H_p(x)$ be the Hermite polynomial of degree $p$. Then we have*

$$\int\cdots\int \varphi_1(t_1)\cdots\varphi_1(t_{p_1})\cdot\varphi_2(t_{p_1+1})\cdots\varphi_2(t_{p_1+p_2})\cdots$$

$$\times\varphi_n(t_{p_1+\cdots+p_{n-1}+1})\cdots\varphi_n(t_{p_1+\cdots+p_n})d\beta(t_1)\cdots d\beta(t_{p_1+\cdots+p_n})$$

$$=\prod_{\nu=1}^{n}\frac{H_{p_\nu}\left(\frac{1}{\sqrt{2}}\int\varphi_\nu(t)d\beta(t)\right)}{\sqrt{2}^{p_\nu}}.$$

For the proof of this theorem we prepare the following

**Theorem 2.2.** I, *If $\varphi(t_1,\ldots,t_p)\in L^2(T^p)$ and $\psi(t)\in L^2(T)$, then*

$$\int\cdots\left[|\varphi(t_1,\ldots,t_p)\psi(t_k)|m(dt_k)\right]^2 m(dt_1)\ldots m(dt_{k-1})m(dt_{k-1})\ldots m(dt_p)$$

$$\leqq\|\psi\|^2\cdot\|\varphi\|^2<\infty.$$

*Therefore*

$$\text{II.}\quad \varphi\underset{(k)}{\times}\psi(t_1\ldots t_{k-1}t_{k+1}\ldots t_p)\equiv\int\varphi(t_1,\ldots,t_p)\psi(t_k)m(dt_k)$$

*is a square-summable function of* $t_1, \cdots, t_{k-1}, t_{k+1}, \cdots, t_p$, *and it holds*

$$\|\varphi \underset{(k)}{\times} \psi\| \leqq \|\varphi\| \cdot \|\psi\|. \tag{3.3}$$

III. *We have*

$$I_{p+1}(\varphi\psi) = I_p(\varphi) \cdot I_1(\psi) - \sum_{k=1}^{p} I_{p-1}(\varphi \underset{(k)}{\times} \psi) \tag{3.4}$$

*Proof.* (3.2) is clear by virtue of Schwarz' inequality and (3.3) is also true by the definition of the norm $\|.\|$ in $L_2$. For the proof of (3.4) we consider firstly the case when $\varphi$ and $\psi$ are special elementary functions. Then we may express $\varphi$ and $\psi$ in the form

$$\varphi(t_1, \ldots, t_p) = a_{i_1 \ldots i_p} \text{ for } (t_1, \ldots, t_p) \in T_{i_1} \times \ldots \times T_{i_p},$$
$$= 0 \quad \text{elsewhere,}$$
$$\psi(t) = b_i \quad \text{for} \quad t_i \in T_i$$
$$= 0 \quad \text{elsewhere,}$$

where $T_1, T_2, \ldots, T_N$ are disjoint and $m(T_i) < \infty$, $i = 1, 2, \ldots, N$ and $a_{i_1 \ldots i_p} = 0$ if any two of $i, \ldots, i_p$ are equal.

Put $S = T_1 + \ldots + T_N$, $A = \max |a_i|$, and $B = \max |b_i|$. Then

$$m(S), A, B < \infty$$

On account of the continuity-condition of $m$ we may assume that

$$m(T_i) < \varepsilon, \; i = 1, 2, \ldots, N,$$

for any asigned $\varepsilon > 0$, by subdividing each $T_i$, if necessary. $S$, $A$ and $B$ remain invariant by this subdivision.

Now we define a special elementary function $\chi_\varepsilon$ by

$$\chi_\varepsilon(t_1, \ldots, t_p, t) = a_{i_1 \ldots i_p} b_i, \text{ if } (t_1, \ldots, t_p, t) \in T_{i_1} \times \ldots \dot{\times} T_{i_p} \times T_i,$$
$$\text{and } i \neq i_1, \ldots, i_p.$$
$$= 0, \text{ if otherwise.}$$

Then we have

$$I_p(\varphi) \cdot I_1(\psi) = \sum a_{i_1 \ldots i_p} \beta(T_{i_1}) \ldots \beta(T_{i_p}) \sum b_i \beta(T_i)$$

$$= \sum_{i \neq i_1, \dots, i_p} a_{i_1 \dots i_p} b_i \beta(T_{i_1}) \dots \beta(T_{i_p}) \beta(T_i)$$

$$+ \sum_{k=1}^{p} \sum a_{i_1 \dots i_p} b_{i_k} \beta(T_{i_1}) \dots \beta(T_{i_{k-1}}) \beta(T_{i_k})^2 \beta(T_{i_{k+1}}) \dots \beta(T_{i_p})$$

$$= I_{p+1}(\chi_\varepsilon) + \sum_{k=1}^{p} \sum a_{i_1 \dots i_p} b_{i_k} \beta(T_{i_1}) \dots \beta(T_{i_{k-1}}) m(T_{i_k}) \beta(T_{i_{k+1}}) \dots \beta(T_{i_p})$$

$$+ \sum_{k=1}^{p} \sum a_{i_1 \dots i_p} b_{i_k} \beta(T_{i_1}) \dots \beta(T_{i_{k-1}}) (\beta(T_{i_k})^2 - m(T_{i_k})) \beta(T_{i_{k+1}}) \dots \beta(T_{i_p})$$

$$= I_{p+1}(\chi_\varepsilon) + \sum_{k=1}^{p} I_{p-1}(\varphi \underset{(k)}{\times} \psi) + \sum R_k$$

$$\| I_{p+1}(\chi_\varepsilon) - I_{p+1}(\varphi\psi) \|^2 = \underline{|p} \, \| \chi_\varepsilon - \varphi \cdot \psi \|^2$$

$$= \sum_{k=1}^{p} \sum a^2_{i_1 \dots i_p} b^2_{i_k} m(T_{i_1}) \dots m(T_{i_{k-1}}) m(T_{i_k})^2 \dots m(T_{i_p})$$

$$\leqq p A^2 B^2 (\sum m(T_i))^{p-1} \cdot (\sum m(T_i))^2$$

$$\leqq \varepsilon p A^2 B^2 (\sum m(T_i))^p = \varepsilon p A^2 B^2 m(S)^p$$

$$\| R_k \|^2 = c \sum a^2_{i_1 \dots i_p} b^2_{i_k} m(T_{i_1}) \dots m(T_{i_{k-1}}) m(T_{i_k})^2 \dots m(T_{i_p})$$

$$\left( c = \frac{1}{2\pi} \int_{-\infty}^{\infty} (x^2 - 1)^2 e^{-x^2/2} dx \right), \quad \leqq \varepsilon \, c \, A^2 B^2 m(S)^p.$$

Thus we obtain, as $\varepsilon \longrightarrow 0$, $I_p(\varphi) \cdot I_1(\psi) = I_{p+1}(\varphi\psi) + \sum_{k=1}^{p} I_{p-1}(\varphi k \psi)$.

Let $\varphi$ and $\psi$ any functions respectively in $L^2(T^p)$ and $L^2(T)$. By virtue of Theorem 2.1 we can find special elementary functions $\varphi_n \in L^2(T^p)$ and $\psi_n \in L^2(T)$ such that $\| \varphi_n - \varphi \| \to 0$, $\| \psi_n - \psi \| \to 0$.

By the above argument we have

$$I_{p+1}(\varphi_n \psi_n) = I_p(\varphi_n) I_1(\psi_n) - \sum_{k=1}^{p} I_{p-1}(\varphi_n \underset{(k)}{\times} \psi_n). \tag{3.5}$$

By making use of (3.2), (3.3) and ($I.3'$) (§ 2) we obtain

$$\| I_{p+1}(\varphi_n \psi_n) - I_{p+1}(\varphi\psi) \|_1 = \| I_{p+1}(\varphi_n \psi_n - \varphi\psi) \|_1 \; (\| . \|_1 \text{ being the } L_1\text{-norm})$$

$$\leqq \| I_{p+1}(\varphi_n \psi_n - \varphi\psi) \| \leqq \sqrt{\underline{|p+1}} \, \| \varphi_n \psi_n - \varphi\psi \|$$

$$\leqq \sqrt{\underline{|p+1}} \, \| \varphi_n (\psi_n - \psi) \| + \sqrt{\underline{|p+1}} \, \| (\varphi_n - \varphi) \psi \|$$

$$=\sqrt{\lfloor p+1}\,\|\varphi_n\|\cdot\|\psi_n-\psi\|+\sqrt{\lfloor p+1}\,\|\varphi_n-\varphi\|\cdot\|\psi\|.$$

$$\|I_p(\varphi_n)\cdot I_1(\psi_n)-I_p(\varphi)\cdot I_1(\psi)\|_1\leqq\|I_p(\varphi_n)I_1(\psi_n-\psi)\|_1$$
$$+\|I_p(\varphi_n-\varphi)I_1(\psi)\|_1$$
$$\leqq\|I_p(\varphi_n)\|\cdot\|I_1(\psi_n-\psi)\|+\|I_p(\varphi_n-\varphi)\|\cdot\|I_1(\psi)\|$$
$$\leqq\sqrt{\lfloor p}\,\|\varphi_n\|\cdot\|\psi_n-\psi\|+\sqrt{\lfloor p}\,\|\varphi_n-\varphi\|\cdot\|\psi\|.$$

$$\|I_{p-1}(\varphi_n\underset{(k)}{\times}\psi)-I_{p-1}(\varphi\underset{(k)}{\times}\psi)\|_1\leqq\|I_{p-1}(\varphi_n\underset{(k)}{\times}\psi_n-\varphi\underset{(k)}{\times}\psi)\|$$
$$\leqq\sqrt{\lfloor p-1}\,\|\varphi_n\underset{(k)}{\times}\psi_n-\varphi\underset{(k)}{\times}\psi\|$$
$$\leqq\sqrt{\lfloor p-1}\,\|(\varphi_n-\varphi)\underset{(k)}{\times}\psi_n\|+\sqrt{\lfloor p-1}\,\|\varphi\underset{(k)}{\times}(\psi_n-\psi)\|$$
$$\leqq\sqrt{\lfloor p-1}\,\|\varphi_n-\varphi\|\cdot\|\psi_n\|+\sqrt{\lfloor p-1}\,\|\varphi\|\cdot\|\psi_n-\psi\|.$$

Thus we see that (3.4) is true in the general case by letting $n$ tend to $\infty$ in (3.5).

*Proof of Theorem* 3.1. We make use of the mathematical induction with regard to $p_1+\dots+p_n$. The theorem is trivially true in case $p_1+\dots+p_n=0$ or 1. It suffices to show that (3.1) is also true for $p_1+\dots+p_n=p+1$ under the assumption that (3.1) is valid for $p_1+\dots+p_n=p-1$, $p$. We may suppose that $p_1\geqq 1$ with no loss of generality.

If we put, in Theorem 3.2,

$$\varphi(t_1,\dots,t_p)=\varphi_1(t_1)\dots\varphi_1(t_{p_1-1})\varphi_2(t_{p_1})\dots\varphi_2(t_{p_1+p_2-1})\dots$$
$$\times\varphi_n(t_{p_1+\dots+p_{n-1}})\dots\varphi_n(t_{p_1+\dots+p_n-1})$$

$$\psi(t)=\varphi_1(t),$$

we obtain, by the assumption of induction

$$\int\dots\int\varphi(t_1,\dots,t_p)\varphi_1(t)d\beta(t_1)\dots d\beta(t_p)d\beta(t)$$
$$=\int\dots\int\varphi(t_1,\dots,t_p)d\beta(t_1)\dots d\beta(t_p)\int\varphi_1(t)d\beta(t)$$
$$-\textstyle\sum_{k=1}^{p}\int\dots\int(\varphi\underset{(k)}{\times}\varphi_1)d\beta(t_1)\dots d\beta(t_{k-1})d\beta(t_{k+1})\dots d\beta(t_p)$$

$$= \prod_{\nu=2}^{n} \frac{H_{p_\nu}\left(\frac{1}{\sqrt{2}}\int \varphi_\nu(t)\,d\beta(t)\right)}{\sqrt{2}^{p_\nu}} \cdot \frac{H_{p_1-1}\left(\frac{1}{\sqrt{2}}\int \varphi_1(t)\,d\beta(t)\right)}{\sqrt{2}^{p_1-1}}$$

$$\cdot \int \varphi_1(t)\,d\beta(t) + (p_1-1)$$

$$\times \prod_{\nu=2}^{n} \frac{H_{p_\nu}\left(\frac{1}{\sqrt{2}}\int \varphi_\nu(t)\,d\beta(t)\right)}{\sqrt{2}^{p_\nu}} \cdot \frac{H_{p_2-2}\left(\frac{1}{\sqrt{2}}\int \varphi_1(t)\,d\beta(t)\right)}{\sqrt{2}^{p_\nu-2}}$$

$$= \prod_{\nu=1}^{n} \frac{H_{p_\nu}\left(\frac{1}{\sqrt{2}}\int \varphi_\nu(t)\,d\beta(t)\right)}{\sqrt{2}^{p_\nu}}$$

in considering

$$\int \varphi(t_1,\ldots,t_p)\varphi_1(t_k)m(dt_k) = \begin{cases} \varphi_1(t_1)\ldots\varphi_1(t_{k-1})\varphi_1(t_{k+1})\ldots\varphi_1(t_{p_1-1})\varphi_2(t_{p_1}) \\ \quad\ldots\varphi_n(t_{p_1+\cdots+p_n-1}) \quad (1 \leqq k \leqq p_1-1) \\ 0 \qquad (k \geqq p_1) \end{cases}$$

(which follows from the orthonormality of $\{\varphi_1,\ldots,\varphi_n\}$)

and

$$H_{p_1}\left(\frac{x}{\sqrt{2}}\right) = \sqrt{2}\,H_{p_1}\left(\frac{x}{\sqrt{2}}\right) - 2(p_1-1)H_{p_1-2}\left(\frac{x}{\sqrt{2}}\right)$$

(which is the recursion formula of Hermite polynomials).

## §4. Orthogonal development of $L_2$-functionals of $\beta$ by multiple Wiener integrals.

A mapping from $R^{B^*}$ into the complex number space $K$ is called *B-measurable* if the inverse image of any Borel set in $K$ is a $B$-measurable set in $R^{B^*}$, which is a set belonging to the least complete additive class that contains all the Borel cylinder sets in $R^{B^*}$. A complex-valued random variable $\xi(\omega)$ is a $B$-measurable function of $\beta$ if it is expressible in the from:

$$\xi(\omega) = f(\beta(E,\omega),\ E \in \boldsymbol{B}^*), \tag{4.1}$$

for any $\omega$, $f$ being a $B$-meassurable mapping from $R^{B^*}$ into $K$.

Furthermore if

$$\|\xi\|^2 = \int |\xi(\omega)|^2 d\omega < \infty, \tag{4.2}$$

then we say that $\xi(\omega)$ is an $L_2$-functional of $\beta$.

**Theorem 4.1** (R. H. Cameron and W. T. Martin)[8] *Let* $\{\varphi_\alpha(t)\}$ *be a complete orthonormal system. Then any* $L_2$*-functional* $\xi(\omega)$ *of* $\beta$ *can be developed as follows*:

$$\xi = \sum_p \sum_{p_1+\cdots+p_n=p} \sum_{\alpha_1,\ldots,\alpha_n} a_{p_1\ldots p_n}^{\alpha_1\ldots\alpha_n} \prod_{\nu=1}^n H_{p_\nu}\left(\frac{1}{\sqrt{2}} \int \varphi_{\alpha_\nu}(t) d\beta(t)\right).$$

Cameron and Martin has shown this theorem in the case when $\beta$ is a normal random measure derived from a brownian motion process, but their proof is available for our general case.

**Theorem 4.2** *Any* $L_2$*-functional* $\xi$ *of* $\beta$ *can be expressible in the form*:

$$\xi = \sum I_p(f_p) = \sum I_p(\tilde{f}_p), \tag{4.3}$$

*where f is given by the following orthogonal development*

$$f_p(t_1,\ldots, t_p) = \sqrt{2} \sum_{p_1+\cdots+p_n=p} \sum_{\alpha_1,\ldots,\alpha_n} a_{p_1\ldots p_n}^{\alpha_1\ldots\alpha_n} \varphi_{\alpha_1}(t_1)\cdots\varphi_{\alpha_1}(t_{p_1})$$

$$\times \varphi_{\alpha_2}(t_{p_1+1})\cdots\varphi_{\alpha_2}(t_{p_1+p_2})\cdots\varphi_{\alpha_n}(t_{p_1+\cdots+p_n+1})\cdots\varphi_{\alpha_n}(t_{p_1+\cdots+p_n}),$$

$\{\varphi_\alpha\}$ *and* $\{a_{p_1\ldots p_n}^{\alpha_1\ldots\alpha_n}\}$ *being the same as those appearing in Theorem* 4.1.

Since $I_p(f_p)$ (or $I_p(\tilde{f}_p)$), $p=0, 1, 2,\ldots$, are orthogonal to each other, (4.3) may be considered as an orthogonal development,

We shall give another method of defining the symmetric functions $\{\tilde{S}\}$ which satisfy $\xi = \sum I_p(\tilde{S}_p)$. Put

$$F(\tilde{h}_p) = \frac{1}{\underline{|p}} (\xi, I_p(\tilde{h}_p)), \tilde{h} \in \tilde{L}^2(T^p),$$

where $\tilde{L}_2(T^p)$ is the totality of symmetric functions in $L^2(T^p)$ which forms a closed linear subspace of $L^2(T^p)$.

---

8) R. H. Cameron and W. T. Martin: The orthogonal development of non-linear functionals in series of Fourier-Hermite functions,

Then $F_p$ is a bounded linear functional on $\tilde{L}^2(T^p)$, since

$$F_p(a\tilde{h}_p+b\tilde{g}_p)=aF_p(\tilde{h}_p)+bF_p(\tilde{g}_p)$$

$$|F_p(\tilde{h}_p)|\leqq\frac{1}{\sqrt{p}}\|\xi\|\cdot\|I_p(\tilde{h}_p)\|=\|\xi\|\cdot\|\tilde{h}_p\|.$$

By Riesz-Fischer's theorem in Hilbert space, we can find $\tilde{s}_p\in\tilde{L}^2(T^p)$ such that

$$F_p(\tilde{h}_p)=(\tilde{s}_p,\ \tilde{h}_p).$$

By (4.3) we have $F_p(\tilde{h}_p)=\frac{1}{\sqrt{p}}(I_p(\tilde{f}_p),\ I_p(\tilde{h}_p))=(\tilde{f}_p,\tilde{h}_p)$.

Thus we have

$$(\tilde{s}_p,\tilde{h}_p)=(\tilde{f}_p,\tilde{h}_p)\ \text{ for } h_p\in\tilde{L}^2(T^p),$$

which proves $\tilde{s}_p=\tilde{f}_p$.

From the above argument follows at once

**Theorem 4.3.** $\xi=\sum I_p(f_p)=\sum I_p(g_p)$ *implies* $\tilde{f}_p=\tilde{g}_p$.

## § 5. The case of a brownian motion process.

Let $\beta(t)$, $a<t<b$, be a brownian motion process.
If we put

$$\beta(E)=\int c_E(t)d\beta(t),$$

where $c_E(t)$ is the characteristic function of the set $E$ and the integral is the so-called Wiener integral. Then $\beta(E)$ is a normal random measure on $T=(a,b)$, the measure $m$ on $T$ being the so-called Lebesgue measure, which clearly fulfills the continuity-condition.

Let $f(t_1,\ldots,t_p)\in L^2(T^p)$. Then we can consider

$$I=\int\cdots\int f(t_1,\ldots,t_p)d\beta(t_1)\ldots d\beta(t_p).$$

**Theorem 5.1.** *The above multiple Wiener integral I is expressible in the form of iterated stochastic intgrals*[9]

9) loc. cit. 2).

$$I=\lfloor p \int_a^b \left(\int_a^{t_p}\left(\cdots\int_a^{t_3}\left(\int_a^{t_2} f(t_1,\dots,t_p)\,d\beta(t_1)\right)d\beta(t_2)\right)\cdots\right)d\beta(t_{p-1})\Big)d\beta(t_p)$$

*Proof.* If $f$ is a special elementary function, this theorem is easily verified by the definitions. In the general case we can show it by approximating $f$ with a special elementary function and making use of the properties of multiple Wiener integrals and stochastic integral.

Any Wiener functional of the brownian motion process is an $L_2$-functional of the normal random measure derived from it, and vice versa. Therefore we see that Theorem 4.2 gives an orthogonal development of Wiener functionals.

I express my hearty thanks to Mr. H. Anzai for his friendly aid and valuable suggestions.

Mathematical Institute, Nagoya University.

MEMORIRS OF THE COLLEGE OF SCIENCE, UNIVERSITY OF KYOTO, SERIES, A
Vol. XXVIII, Mathematics No. 1, 1953.

# Stochastic Differential Equations in a Differentiable Manifold (2)

By

Kiyosi ITÔ

(Received April 20, 1953)

**§ 1.** Let $\pi(t)$ be any Markov process in an $r$-dimensional differentialble manifold $M$ with the transition probability:

$$(1\cdot1)\qquad F(t, p; s, E)=P(\pi(s)\in E/\pi(t)=p).$$

As is well-known, the generating operator $A_t$ of this process is defined as follows:

$$(1\cdot2)\qquad (A_t f)(p)=\lim_{\Delta\to+0}\frac{1}{\Delta}\int_M [f(q)-f(p)]\ F(t, p; t+\Delta, dq).$$

We shall consider here the process whose generating operator $A_t$ is expressible in the form:

$$(1\cdot3)\qquad (A_t f)(x)=a^i(t, x)\frac{\partial f}{\partial x^i}(x)+\frac{1}{2}\ B^{ij}(t, x)\frac{\partial^2 f}{\partial x^i\,\partial x^j}(x),$$

where $x$ is the local coordinate and $f$ is a bounded function of class $C_1$. (1.3) is equiualent to the following (1.3′):

$$(1.3')\quad \begin{cases} \dfrac{1}{\Delta}\displaystyle\int_U (y^i-x^i)F(t, x; t+\Delta, dy)\longrightarrow a^i(t, x), \\ \dfrac{1}{\Delta}\displaystyle\int_U (y^i-x^i)(y^j-x^j)F(t, x; t+\Delta, dy)\longrightarrow B^{ij}(t, x),\ (\Delta\to+0) \\ \dfrac{1}{\Delta}\displaystyle\int_U F(t, x; t+\Delta,\ U^c)\to 0. \end{cases}$$

We can easily see that $(B^{ij})$ is symmetric and positive-definite and that $a^i$ and $B^{ij}$ are transformed in the following way:

$$(1.4')\quad \begin{aligned} \bar{a}^i &= a^k\frac{\partial\bar{x}^i}{\partial x^k}+\frac{1}{2}B^{kl}\frac{\partial^2\bar{x}^i}{\partial x^k\partial x^l} \\ \bar{B}^{ij} &= B^{kl}\frac{\partial\bar{x}^i}{\partial x^k}\ \frac{\partial\bar{x}^j}{\partial x^l} \end{aligned}$$

The purpose of the present paper is to find a continuous Markov process whose generating operator is given by (1.3) when $a^i$ and $B^{ij}$ satisfy some regularity conditions. This problem has been discussed by K. Yosida [4] and S. Itô [6] by means of parabolic differential equations. We shall here make use of stochastic differential equations established in our previous paper [1].

We have shown in [1] that if $B^{ij}$ is written in the form

$$B^{ij}=\sum_{\nu=1}^{r} b_\nu^i\, b_\nu^j \tag{1·5}$$

by a system of $r$ vectors $b_\nu=(b_\nu^1,\cdots, b_\nu^r)$, $\nu=1, 2,\cdots, r$, having some appropriate regularity properties, then such a process is determined as the solution of the following stochastic differential equations:

$$d\xi^i=a^i(t, \xi)dt+b_\nu^i(t, \xi)d\beta^\nu \tag{1·6}$$

where $\beta=(\beta^1(t),\cdots, \beta^r(t))$ is an $r$-dimensional Wiener process.

$(B^{ij})$ being symmetric and positive-definite, it may be always expressible in the form (5) in many different ways, but we cannot always find the continuous vector system $\{b_\nu\}$ even if the tensor $(B^{ij})$ satisfies regularity conditions. Such possibility depends on the topological structure of the space $M$; for example, it is possible in the case of Brownian motions in Lie groups [2], while it is impossible in the case of those on the surface of 3-sphere [3] [5]. At first sight one may consider that this fact raises an essential dfficulty in our method of stochastic differential equations, but as is shown in this paper, we can overcome it by considering the vector system satisfying regularity conditions *locally*, whose existence is easily verified.

§ **2.** THEOREM. *Consider a differential operator*:

$$(A_t f)(x)=a^i(t,x)\frac{\partial f}{\partial x^i}(x)+\frac{1}{2}B^{ij}(t, x)\frac{\partial^2 f}{\partial x^i\, \partial x^j}(x) \tag{2·1}$$

and put

(2.2) $B_{ij}(t, x)=$ *the* $(i, j)$*-component of the inverse matrix of* $(B^{ij}(t, x))$. If $a^i(t, x)$, $B^{ij}(t, x)$ *and* $B_{ij}(t, x)$ *are all bounded*[1] *and*

1) The boundedness of $a^i(t, x)$ etc. is defined as follows.

By a canonical coordinate around p we understand a local coordinate which maps a neighbourhood of $p$ onto the interior of the unit sphere in the $r$-dimensional space $R^r$ and especially transforms $p$ to the centre of the sphere.

$a^i(t, x)$ is called to be bounded in $0\leq t\leq 1$, $x \in M$, if and only if there exist a constant $K$ and a canonical coordinate $(x)$ around any point of $M$ satisfying

$$|a^i(t, x)|<K,\ 0\leq t\leq 1,\ \|x\|<1.$$

*continuous in $t$ for each $x$ and of class $C_1$ in $x$ for each $t$, then there exists a continuous Markov procass $\pi(t)$, $0 \leq t \leq 1$, whose generating operator is given by* (2.1). *The initial distribution i.e. the distribution of $\pi(o)$ is arbitrarily assigned.*

PROOF. 1°. By the assumption of the boundedness of $a^i$, $B^{ij}$ and $B_{ij}$ we shall define a canonical coordinate $(x)$ around any point $p$ of $M$ with the canonical neighbourhood $U(p)$ such that

$$(2.3) \qquad |a^i(t, x)|,\ |B^{ij}(t, x)|,\ |B_{ij}(t, x)|, < K,\ 0 \leq t \leq 1,\ \|x\| < 1, \quad i, j = 1, 2, \cdots, r,$$

$K$ being independent of $p$. Firstly we shall show that there exist a constant $K_1$ $(0 < K_1 < 1)$ independent of $p$ and a vector system $(b^i_\nu(t, x))$ for $0 \leq t \leq 1$ and for $\| x \| < K_1$ which is continuous in $t$ for each $x$ and of class $C_1$ in $x$ for each $t$ and satisfies

$$(2.4) \qquad \sum_{\nu=1}^{n} b^i_\nu(t, x)\ b^j_\nu(t, x) = B^{ij}(t, x).$$

For brevity we shall write as

$$(2\cdot5) \qquad B = B(t, x) = (B^{ij}(t, x)),\ B_0 = B(0, 0).$$

Since $B_0$ is symmetric and positive-definite we can express it as follows:

$$(2.6) \qquad B_0 = UD^2U',\ D = \begin{pmatrix} \lambda_1 & & 0 \\ & \ddots & \\ 0 & & \lambda_r \end{pmatrix},\ U = \text{orthogonal matrix}.$$

Put

$$(2.7) \qquad b_0 = UDU'.$$

Then we have

$$b_0 = b_0',\ B_0 = b_0^2, \text{ and } \|b_0^{-1}\| \leq \sqrt{\|B_0^{-1}\|}.\text{[2)]}$$

Next we put

$$(2.8) \qquad C = b_0^{-1}\ B\ b_0^{-1} - E = b_0^{-1}(B - B_0)b_0^{-1},\ E = \text{unit matrix},$$

which satisfies

$$\|C\| \leq \|B_0\| \cdot \|B - B_0\| \leq K_2\|x\|,$$

---

2) We define $\| A \|$ by

$$\| A \| = \sup\ \{ \| Ax \| ; \| x \| < 1\},$$

for which the following properties are easily verified:

$$\max_{ij} | a^{ij} | \leq \| A \| \leq \sum_{ij} | a^{ij} | \text{ for } A = (a^{ij}),$$

$$\| U \| = 1 \quad \text{for any orthogonal } U,$$

$$\| A\ B \| \leq \| A \| \| B \|,$$

$$\| D_1\ D_2 \| = \| D_1 \| \| D_2 \| \text{ for diagonal } D_1, D_2.$$

$K_2$ being a constant determined only by $K$, and so we obtain

(2.9) $$\|C\| < 1 \text{ for } \|x\| < K_1 = \max(\frac{1}{K_2}, 1).$$

Therefore

(2.10) $$\begin{cases} c = \sum_{n=0}^{\infty} \binom{\frac{1}{2}}{n} C^n \\ c_{(k)} = \sum_{n=0}^{\infty} \binom{\frac{1}{2}}{n} nC \cdot \frac{\partial C}{\partial x^k} \end{cases}$$

is convergent in the above norm for $\|x\| < 1$, and so we easily see that

(2.11) $$\frac{\partial c^{ij}}{\partial x^k} = c^{ij}_{(k)}$$

in considering that $|a^{ij}| \leq \|A\|, i, j = 1, 2, \cdots, r$ for $A = (a^{ij})$. Thus we see that $c^{ij}$ are of class $C_1$ in $x$ and continuous in $t$. We have clearly

$$c = c', \ c^2 = E + C = b_0^{-1} Bb_0^{-1} \ i.e. \ B = b_0 c^2 b_0 = (b_0 c)(b_0 c)'.$$

We put

(2.12) $$b = (b^i_\nu) = b_0 c,$$

which satisfies the above-mentioned conditions.

2°. We shall define the subsets $V$, $W$ and $Q$ of $R^r$ by the following conditions:

$$V: \|x\| < K_1, \ W: \|x\| < \frac{2}{3}K_1, \quad Q: \|x\| < \frac{1}{3}K_1,$$

and denote by $V(p)$, $W(p)$ and $Q(p)$ the parts of the above assigned canonical coordinate neighbourhood $U(p)$ corresponding to $V$, $W$ and $Q$ respectively.

Let $\varphi(x)$ denote the function of $x \in R^r$,

$$\varphi(x) = 1 \ (x \in W), = 0 \ (x \in V^c), = \frac{2K_1 - 3\|x\|}{K_1} (x \in V - W).$$

Then $\varphi(x)$ satisfies the following conditions:

$$0 \leq \varphi(x) \leq 1, \ |\varphi(x) - \varphi(y)| \leq \frac{3}{K_1}\|x - y\|.$$

Since $a^i(t, x)$ and $b^i_\nu(t, x)$ depend on the neighbourhood $U(p)$, we shall denote them with $a^i(t, x; p)$ and $b^i_\nu(t, x; p)$ respectively. We put

$$\tilde{a}^i(t, x: p) = a^i(t, x; p) \ \varphi(x) \ \ (x \in U(p)),\ 0(x \in U(p))$$

$$\tilde{b}^i(t, x; p) = d^i_\nu(i, x; p) \ \varphi(x) \ \ (x \in U(p)),\ 0(x \in U(p))$$

and consider a stochastic differential equation

$$d\xi^i(t) = \tilde{a}^i(t, \xi; p)\ dt + \tilde{b}^i_\nu(t, \xi; p)\ d\beta^\nu(t),\ i=1, 2,\cdots, r,\ t_0 \leq t \leq 1,$$

where $\xi(t_0)$ is any assigned point $q$ in $Q(p)$. As is shown in [1], this equation has a unique solution $\pi(t; U(p); t, q)$ which lies in $V(p)$ for $t_0 \leq t \leq 1$ with probability 1 and in $W(p)$ for $t_0 \leq t \leq t+\Delta$ with probabilty $1-o(\Delta)$, $o(\Delta)$ being independent of $p$ and $q$.

Let $\Phi$ denote the distribution assigned as the probability law of $\pi(0)$. We shall consider an $M$-valued random variable $\pi$ which is subject to $\Phi$ and an $r$-dimensional Wiener process $\beta(t) = (\beta^1(t), \cdots, \beta^r(t))$, $0 \leq t \leq 1$, independent of $\pi$. Since $M$ satisfies the second countability axiom, $M$ is covered by a countable system of $Q(p)$, say $Q(p_1)$, $Q(p_2)$,$\cdots$

Let $Q(p_f)$ be the first of $\{Q(p_n)\}$ that contains $\pi$. We shall define

$$\pi_1(t) = \pi(t;\ U(p_f);\ 0,\ \pi).$$

Next we shall define $\pi_2(t)$ to be equal to $\pi_1(t)$ as far as $\pi_1(t)$ remains in $W(p_f)$ and if $\pi_1(t)$ attains the boundary of $W(p_f)$ at $t=t_1$, then we shall define $\pi_2(t)$ $(t \geq t_1)$ by

$$\pi_2(t) = \pi_2(t;\ U(p_s);\ t_1,\ \pi_1(t_1)),$$

where $Q(p_s)$ is the first of $\{Q(p_n)\}$ that contains $\pi_1(t_1)$. In the same way we shall $\pi_n(t)$, $n=1, 2,\cdots$ recursively. We put

$$\pi(t) = \lim\ \pi_n(t).$$

This limit process exists and satisfies the conditions of our theorem, as is shown by the same idea as our previous paper [1].

## References

1) K. Itô: Stochastic differential equations in a differentiale manifold, Nagoya Math. Journ., **1**, 1950, pp. 35—pp. 47.

2) K. Ito: Brownian motions in a Lie group, Proc. Japan Acad., **26**, 8, 1950, pp. 4—10.

3) K. Yosida: Brownian motion on the surface of 3-sphere, Ann. of Math. Statis., **20**, 2, 1949.

4) K. Yosida: The fundamental solution of the parabolic equation in a Riemannian space, Osaka Math. J. **5**, 1, 1953, pp. 65—74.

5) F. Perrin: Étude mathematique du mouvement brownien de rotation, Ann. Ec. Norm., **45**, 3, 1928.

6) S. Itô: The fundamental solution of the parabolic equation in a differentiable manifold, Osaka Math. J. **5**, 1953, pp. 75—92.

Stochastic Differential Equations in a Differentiable Manifold [illegible]

[illegible]

[illegible]

and consider a stochastic differential equation

[illegible]

where [illegible] is any generalized [illegible] it was shown in [1] that this equation has a unique solution [illegible] which has [illegible] with probability 1 and [illegible] with probability [illegible]

Let [illegible] denote the [illegible] the probability [illegible] We shall [illegible] random variable [illegible] and [illegible] Wiener process [illegible] independent of [illegible] and [illegible] satisfies the second countability axiom [illegible]

Let [illegible] be the first [illegible] that [illegible] We then define

[illegible]

[illegible] the smallest [illegible] and if [illegible] then we shall define

[illegible]

where [illegible] is the [illegible] In the same way we [illegible] We have

[illegible]

This final process [illegible] is shown [illegible] as [illegible] previous paper [1].

## References


[illegible]

MEMOIRS OF THE COLLEGE OF SCIENCE, UNIVERSITY OF KYOTO, SERIES A
Vol. XXVIII, Mathematics No. 3, 1953.

# Stationary random distributions

By

Kiyosi ITÔ

(Received April 15, 1954)

In the same way as the concept of distributions by L. Schwartz [11][1] was introduced as an extended one of functions, we may define stationary random distributions as an extension of stationary random functions viz. stationary processes. Such consideration will enable us to establish a unified theory of stationary processes, Brownian motion processes, processes with stationary increments and other similar ones, as is shown in this paper. We shall first introduce some fundamental notions in § 1. In § 2 we shall define the covariance distribution of stationary random distributions, which corresponds to Khintchine's covariance function [7]. In § 3 and § 4 we shall prove the spectral decomposition theorems of covariance distributions and stationary random distributions respectively. In § 5 we shall discuss the derivatives of stationary distributions. In § 6 we shall show that any stationary distribution is identified with a $k$-th derivative (in the sense of distributions) of a certain continuous process with stationary $k$-th order increments for some $k$.

## § 1. Fundamental Notions and Notations

In this paper we shall restrict ourselves to complex-valued random variables with mean 0 and finite variance. The totality of such variables constitute a *Hilbert space* $\mathfrak{H}$ with the following definition of inner product:

$$(1\cdot1) \qquad (X, Y)=\mathfrak{E}(X\cdot\overline{Y}); \quad \mathfrak{E}: \text{ expectation.}$$

We shall here consider only the strong topology on $\mathfrak{H}$. A continuous random process $X(t)$, $-\infty<t<\infty$, is an $\mathfrak{H}$-valued continuous

1) The numbers in [ ] refer to those of the Bibliography at the end of this paper.

function defined on $R=(-\infty, \infty)$. The set of all continuous processes will be denoted by $\mathfrak{C}(\mathfrak{H})$.

Let $\mathfrak{D}$ denote the space of all complex-valued $C_\infty$-functions defined on $R$ whose carrier is compact. We shall introduce in $\mathfrak{D}$ the topology of Schwartz [11] (III, § 1). An $\mathfrak{H}$-valued continuous linear functional difined on $\mathfrak{D}$ is called a ***random distribution***. We shall here denote with $\mathfrak{D}'(\mathfrak{H})$ the totality of random distributions, while $\mathfrak{D}'$ will denote the set of all complex-valued distributions (continuous linear functionals) defined on $\mathfrak{D}$. $\mathfrak{C}(\mathfrak{H})$ may be considered as a sub-system of $\mathfrak{D}'(\mathfrak{H})$, since we may identify a continuous random process $X(t)$ with the following random distribution $X(\phi)$ determined by it:

$$X(\phi)=\int X(t)\,\phi(t)\,dt\,; \tag{1.2}$$

the integral sign without bounds means the integral from $-\infty$ to $\infty$ in this paper.

The following notations will be often used in the theory of distributions. Let $F \in \mathfrak{D}'$ or $\mathfrak{D}'(\mathfrak{H})$ and $\phi \in \mathfrak{D}$.

| | |
|---|---|
| $\tau_h$ (shift transformation) | : $\tau_h\phi(t)=\phi(t+h)$, $\tau_h F(\phi)=F(\tau_{-h}\phi)$ |
| $D$ (Derivative) | : $D\phi(t)=\phi'(t)$, $DF(\phi)=-F(D\phi)$ |
| $\check{}$ (Inversion) | : $\check{\phi}(t)=\phi(-t)$, $\check{F}(\phi)=F(\check{\phi})$ |
| $-$ (Conjugate) | : $\bar{\phi}(t)=\overline{\phi(t)}$, $\bar{F}(\phi)=\overline{F(\bar{\phi})}$ |
| $\sim$ $(=\check{\bar{}}=\bar{\check{}})$ | : $\tilde{\phi}(t)=\overline{\phi(-t)}$, $\tilde{F}(\phi)=\overline{F(\tilde{\phi})}$ |
| $\mathcal{F}$ (Fourier transform) | : $\mathcal{F}\phi(\lambda)=\int e^{-i2\pi\lambda t}\phi(t)\,dt$ |

The following relations should be noted.

$$(F*\phi)(0)=F(\check{\phi})=\check{F}(\phi),\quad \mathcal{F}(\phi*\psi)=\mathcal{F}\phi\cdot\mathcal{F}\psi,\quad \mathcal{F}\tilde{\phi}=\overline{\mathcal{F}\phi} \tag{1.3}$$

We shall call, following Khintchine, $X \in \mathfrak{C}(\mathfrak{H})$ ***weakly stationary*** or merely ***stationary*** for the brevity if its covariance function is invariant under shift transformations viz. if we have, for any $(t, s)$,

$$(\tau_h X(t),\ \tau_h X(s))=(X(t),\ X(s)), \tag{1.4}$$

and to be ***strictly stationary*** if its probability law is invariant under shift transformations viz. if the joint probability law of

$$(\tau_h X(t_1), \cdots, \tau_h X(t_n)) \tag{1.5}$$

is independent of $h$ for any $n$ and any $(t_1, \cdots, t_n)$. Generalizing these definitions onto random distributions, we shall call $X \in \mathfrak{D}'(\mathfrak{H})$ ***weakly*** *stationary* or merely *stationary* for the brevity if we have, for any $(\phi, \psi)$,

$$(1\cdot 4')\qquad (\tau_h X(\phi),\ \tau_h X(\psi)) = (X(\phi),\ X(\psi))$$

and *strictly stationary* if the joint probability law of

$$(1\cdot 5')\qquad (\tau_h X(\phi_1), \cdots, \tau_h X(\phi_n))$$

is independent of $h$ for any $n$ and $(\phi_1, \cdots, \phi_n)$.

We shall adopt here the following notations:

$\mathfrak{S}$ : the totality of stationary distributions,

$\mathfrak{S}^\circ$ : the totality of stationary processes,

$\bar{\mathfrak{S}}$ : the totality of strictly stationary distributions,

$\bar{\mathfrak{S}}^\circ$ : the totality of strictly stationary processes.

Clearly we have

$$(1\cdot 6)\qquad \mathfrak{S} \supseteq \bar{\mathfrak{S}} \vee \mathfrak{S}^\circ, \qquad \mathfrak{S}^\circ \supseteq \bar{\mathfrak{S}}^\circ.$$

A random distribution $X$ is called a ***complex normal (random) distribution*** if $X(\phi)$, $\phi \in \mathfrak{D}$, constitute a complex normal system [5], and to be a ***real normal (random) distribution*** if $X$ is real viz. $X = \bar{X}$ and $X(\phi)$, $\phi$ running over real functions in $\mathfrak{D}$, constitute a (real) normal system. This is an extention of normal processes or Gaussian processes [3] (II, § 3)[2]. A (complex as well as real) normal distribution will be strictly stationary, if it is weakly stationary. The corresponding fact is well-known regarding stationary processes.

We shall here mention a typical example of real stationary distributions which are not stationary processes. Let $B(t)$ be a (real) Brownian motion process [3] (p. 97). The derivative (in the sense of distribution) of this process $B' \equiv DB$ is a random distribution defined by

$$(1\cdot 7)\qquad B'(\phi) = -B(\phi') = \int \phi(t)\,dB(t) \qquad \text{(Wiener integral) [4].}$$

This is evidently real normal and stationary, since

$$\begin{aligned}(\tau_h B'(\phi),\ \tau_h B'(\psi)) &= (B'(\tau_{-h}\phi),\ B'(\tau_{-h}\psi))\\ &= \left(\int \phi(t-h)\,dB(t),\ \int \psi(t-h)\,dB(t)\right)\\ &= \int \phi(t-h)\,\overline{\psi(t-h)}\,dt = \int \phi(t)\,\overline{\psi(t)}\,dt,\end{aligned}$$

2) This definition of complex normal ones is somewhat different from that of Doob, while both definitions are the same in the real case.

which shows that $B' \in \mathfrak{S}$. The fact that $B' \bar{\in} \mathfrak{S}^\circ$ will be proved in § 2.

## § 2. Covariance Distribution

As is known, Khintchine's covariance function [7] $\rho(t)$ of $X(t)$ is defined by

$$(2\cdot1) \qquad \rho(t) = (X(t), X(0)) = (X(t+s), X(s)).$$

Then we have clearly

$$\begin{aligned}(X(\phi), X(\psi)) &= \left(\int X(t)\,\phi(t)\,dt, \int X(s)\,\psi(s)\,ds\right)\\ &= \iint (X(t), X(s))\,\phi(t)\,\overline{\psi(s)}\,dt\,ds\\ &= \iint \rho(t-s)\,\phi(t)\,\overline{\psi(s)}\,dt\,ds\\ &= \int \rho(\tau) \int \phi(\tau-\sigma)\,\overline{\psi(-\sigma)}\,d\sigma\,d\tau\,.\end{aligned}$$

By considering $\rho$ as a distribution $\in \mathfrak{D}'$, we may write the above identity as

$$(2\cdot2) \qquad (X(\phi), X(\psi)) = \rho(\phi * \check{\bar{\psi}}).$$

This may suggest to us how to define the covariance distribution of our stationary distributions.

THEOREM 2·1. *Let $X(\phi)$ be any stationary distribution. Then there exists one and only one distribution $\rho \in \mathfrak{D}'$ satisfying* (2·2).

DEFINITION. This distribution is called the *covariance distribution* of $X$.

PROOF OF THEOREM. If we put

$$(2\cdot4) \qquad T_\phi(\psi) = (X(\phi), X(\bar{\psi})),$$

then we get a distribution $T_\phi \in \mathfrak{D}'$ for each $\phi \in \mathfrak{D}$. Taking into account the fact that $T_\phi(\psi)$ is continuous in $(\phi, \psi) \in \mathfrak{D} \times \mathfrak{D}$ and $\mathfrak{D}$ is a Montel space [11] (III, § 2), we shall easily see that '$\phi \to T_\phi$' is a continuous linear mapping from $\mathfrak{D}$ into $\mathfrak{D}'$.[3] Furthermore this transformation commutes with the shift transformation;

$$\begin{aligned}(\tau_h T_\phi)(\psi) = T_\phi(\tau_{-h}\psi) &= (X(\phi), X(\overline{\tau_{-h}\psi})) = (X(\phi), X(\tau_{-h}\bar{\psi}))\\ &= (X(\tau_h\phi), X(\bar{\psi})) = T_{\tau_h\phi}(\psi).\end{aligned}$$

Therefore $T_\phi$ is expressible as a convolution of a distribution $T$ and $\phi$:

3) Cf. [11] III, § 3 as to the topology in $\mathfrak{D}'$.

$$T_{\phi}=T*\phi,$$

by a theorem of Schwartz [11] (Vol. 2, VI, § 7, p. 53, Amélioration du théorèm X). Hence it follows that

$$(X(\phi),X(\psi))=T_{\phi}(\bar{\psi})=(T*\phi)(\bar{\psi})=(T*\phi*\tilde{\psi})(0)=\rho(\phi*\tilde{\psi}),$$

where

$$\rho=\check{T}.$$

The uniqueness of $\rho$ follows at once from the fact that the set of all elements of the form $\phi*\psi$, $\phi,\psi\in\mathfrak{D}$, is dense in $\mathfrak{D}$.

THEOREM 2·2. *If $X(\phi)$ is real stationary distribution, then its covariance distribution is real, i.e. $\rho=\bar{\rho}$.*

PROOF. Let $\rho$ be the covariance distribution of $X$. Then that of $\bar{X}$ will become $\bar{\rho}$, since we have

$$(\bar{X}(\phi),\bar{X}(\psi))=(\overline{X(\bar{\phi})},\overline{X(\bar{\psi})})=\overline{(X(\bar{\phi}),X(\bar{\psi}))}=\overline{\rho(\bar{\phi}*\tilde{\bar{\psi}})}$$
$$=\overline{\rho(\bar{\bar{\phi}}*\tilde{\psi})}=\bar{\rho}(\phi*\tilde{\psi}).$$

Thus $X=\bar{X}$ implies $\rho=\bar{\rho}$.

EXAMPLE. The covariance distribution of $B'$ is Dirac's $\delta$-distribution [11] (I, § 1), as

$$(B'(\phi),B'(\psi))=\int\phi(t)\overline{\psi(t)}\,dt=\int\phi(t)\,\tilde{\psi}(-t)\,dt=(\phi*\tilde{\psi})(0)$$
$$=\delta(\phi*\psi).$$

Thus we see that $B'\bar{\in}\mathfrak{S}^{\circ}$, because if $B'\in\mathfrak{S}^{\circ}$ then the covariance distribution would be induced by a continuous function which is Khintchine's covariance function.

## § 3. Spectral Decomposition of Covariance Function

Let $X(t)$ be any stationary continuous random process with covariance function $\rho(t)$. A known theorem of Khintchine [7] shows that $\rho(t)$ is expressible in the form:

$$\rho(t)=\int e^{-i2\pi\lambda t}\,d\mu(\lambda) \tag{3·1}$$

in one and only one way, where $\mu$ is a non-negative measure on $R$ such that $\mu(R)<\infty$. Therefore the distribution $\rho(\phi)$ induced by $\rho(t)$ may be expressed as

$$\rho(\phi)\equiv\int\rho(t)\,\phi(t)\,dt=\int\mathcal{F}\phi(\lambda)\,d\mu(\lambda). \tag{3·2}$$

Let us now consider any stationary distribution with the covariance distribution $\rho$. Then we have

(3·3) $$\rho(\phi * \tilde{\phi}) = (X(\phi), X(\phi)) \geqq 0,$$

which implies that $\rho$ is a positive-definite distribution. By Schwartz' generalization of Bochner's theorem [11] (VII, § 9) we obtain the following theorem.

THEOREM 3·1. *$\rho$ is expressible in the from* (3·2) *in one and only way, where $\mu$ is a non-negative measure satisfying*

(3·4) $$\int \frac{d\mu(\lambda)}{(1+\lambda^2)^k} < +\infty$$

*for some integer $k$.*

DEFINITION. We shall call (3·2) the *spectral decomposition* of $\rho$ and $\mu$ *spectral measure* of $\rho$. When $\mu$ satisfies (3·4), we say that $X \in \mathfrak{S}_k$. We have clearly

(3·5) $$\cdots \mathfrak{S}_{-2} \subseteq \mathfrak{S}_{-1} \subseteq \mathfrak{S}_0 \subseteq \mathfrak{S}_1 \subseteq \mathfrak{S}_2 \subseteq \cdots.$$

Conversely we have

THEOREM 3·2. *Any distribution of the above form is the covariance distribution of a stationary distribution which is complex normal.*

PROOF. Let $\rho$ be a distribution of the above form. Put

$$\Gamma(\phi, \psi) = \rho(\phi * \tilde{\psi}), \ \phi, \psi \in \mathfrak{D}.$$

Then $\Gamma(\phi, \psi)$ is positive-definite in $(\phi, \psi)$, as we have

$$\sum_{i,j=1}^{n} \Gamma(\phi_i, \phi_j) \xi_i \bar{\xi}_j = \rho(\theta * \tilde{\theta}) \geqq 0, \quad \theta = \sum_i \xi_i \phi_i.$$

Therefore we can define a complex normal system $X(\phi)$, $\phi \in \mathfrak{D}$, such that $\mathfrak{E} X(\phi) = 0$ and $\mathfrak{E}(X(\phi) \cdot \overline{X(\psi)}) = \Gamma(\phi, \psi) = \rho(\phi * \tilde{\psi})$ [5]. It remains only to show that $X(\phi)$ is a random distribution. From the identity:

$$\begin{aligned} \|X(c\phi) - cX(\phi)\|^2 &= (X(c\phi), X(c\phi)) - c(X(\phi), X(c\phi)) \\ &\quad - \bar{c}(X(c\phi), X(\phi)) + c\bar{c}(X(\phi), X(\phi)) \\ &= \rho(c\phi * \widetilde{c\phi}) - c\rho(\phi * \widetilde{c\phi}) - \bar{c}\rho(c\phi * \tilde{\phi}) + c\bar{c}\rho(\phi * \tilde{\phi}) \\ &= c\bar{c}\,\rho(\phi * \tilde{\phi}) - c\bar{c}\,\rho(\phi * \tilde{\phi}) - c\bar{c}\,\rho(\phi * \tilde{\phi}) + c\bar{c}\,\rho(\phi * \tilde{\phi}) \\ &= 0, \end{aligned}$$

it follows that $X(c\phi) = cX(\phi)$. By the same way we can see that $X(\phi + \psi) = X(\phi) + X(\psi)$. Therefore $X(\phi)$ is linear in $\phi$. By the identity $\|X(\phi)\|^2=\rho(\phi*\tilde{\phi})$ we obtain the continuity of $X$. Thus our theorem is completely proved.

Next we shall discuss the case of real stationary distributions. By Theorem 2·2 we see that $\rho=\bar{\rho}$ in this case. But we have

$$\bar{\rho}(\phi)=\overline{\rho(\bar{\phi})}=\int\overline{\mathcal{F}\bar{\phi}(\lambda)}\,d\mu(\lambda)=\int\mathcal{F}\phi(-\lambda)\,d\mu(\lambda)$$
$$=\int\mathcal{F}\phi(\lambda)\,d\check{\mu}(\lambda),$$
$$(\check{\mu}(E)=\mu(-E),\quad E=\{t;\ -t\in E\}).$$

By the uniqueness of the spectral measure we shall obtain the following

THEOREM 3·3. *In the case of real stationary distribution the spectral measure $\mu$ is symmetric with respect to* 0, *viz.* $\mu(E)=\mu(-E)$.

Conversely we have

THEOREM 3·4. *Any distribution of the form* (3·2) *with symmetric measure $\mu$ is the covariance function of a certain stationary distribution which is real normal.*

The proof is similar to that of Theorem 3·2; we use the existence theorem of real normal systems instead of that of complex normal ones.

EXAMPLE. $B'$ is a real stationary distribution whose spectral measure is the ordinary Lebesgue measure, because we have

$$\delta(\phi)=\phi(0)=\int\mathcal{F}\phi(\lambda)\,d\lambda\,.$$

## § 4. Spectral Decomposition of Stationary Distribution

We shall first introduce a random measure. Let $\mu$ be a non-negative measure defined for all Borel sets in $R=(-\infty,\infty)$ and $\boldsymbol{B}^*$ denote the system of all Borel sets with finite $\mu$-measure. An $\mathfrak{H}$-valued function $M(E)$ defined for $E\in\boldsymbol{B}^*$ is called a ***random measure*** with respect to $\mu$ if and only if

$$(4\cdot1)\qquad (M(E_1),M(E_2))=\mu(E_1\cap E_2),\ E_1,E_2\in\boldsymbol{B}^*.$$

We get, by the definition,

$$(4\cdot2)\qquad \|M(E)\|^2=\mu(E),$$

$$(4\cdot2')\qquad M(E_1)\perp M(E_2)\quad\text{if}\quad E_1\cap E_2=\phi\,.$$

Making use of the additivity of $\mu$ we have

$$(4\cdot2'')\qquad M(E)=\sum_{n=1}^{\infty}M(E_n)\,,$$

where $E_1, E_2, \cdots$ are disjoint to each other and belong to $B^*$ with their sum $E=\sum_{n=1}^{\infty} E_n$; (4·2″) will justify the term of random measure. We can easily define the integral with respect to the measure [3] (IX, § 2)[4]:

$$(4\cdot 3) \qquad M(f)=\int f(\lambda)\, dM(\lambda)$$

for $f \in L^2(R, \mu)$. $M(f)$ satisfies the following conditions:

$$(4\cdot 4\cdot a) \qquad (M(f_1), M(f_2))=(f_1, f_2) \quad \left(=\int f_1(\lambda)\, \overline{f_2(\lambda)}\, d\mu(\lambda)\right),$$

$$(4\cdot 4\cdot b) \qquad M(c_1 f_1+c_2 f_2)=c_1 M(f_1)+c_2 M(f_2).$$

Let $X(t)$ be any stationary continuous process with the spectral measure $\mu$. Then it is known [3] (XI, § 4) that $X(t)$ may be expressed in one and only one way as

$$(4\cdot 5) \qquad X(t)=\int e^{-i2\pi\lambda t}\, dM(\lambda),$$

where $M$ is a random measure with respect to $\mu$. This fact was proved by many scholars, especially by A. Kolmogorov, H. Cramer [2] and M. Loève (See [9] Th. 27.2, p. 123). Therefore the stationary distribution induced by $X(t)$ may be expressed as

$$(4\cdot 6) \qquad X(\phi)=\int \mathcal{F}\phi(\lambda)\, dM(\lambda)=M(\mathcal{F}\phi).$$

We shall now show that this identity holds for any general stationary distribution, i.e.

THEOREM 4·1. *Let $X$ be any stationary distribution with the spectral measure $\mu$. Then $X(\phi)$ will be expressible in the form* (4·6) *in one and one way, $M$ being a random measure with respect to $\mu$. Conversely, any random distribution of such form is a stationary distribution.*

DEFINITION. We shall call the identity (4·6) the spectral decomposition of $X$ and $M$ the *spectral random measure* of $X$.

PROOF OF THEOREM. We shall first remark that $\mathcal{F}\mathfrak{D}=\{\mathcal{F}\phi;\ \phi \in \mathfrak{D}\}$ is dense in $L^2=L^2(R, \mu)$. Let $\mathfrak{S}$ denote the class of all rapidly decreasing functions [11] (VII, § 3). Then $\mathcal{F}$ will be a homeomorphic mapping from $\mathfrak{S}$ onto itself, where the topology on $\mathfrak{S}$ is what is usually used in the theory of distributions. As $\mathfrak{D}$ is dense in $\mathfrak{S}$, so is $\mathcal{F}\mathfrak{D}$. In making use of the condition (3·5) we shall easily see that $\mathfrak{S}$ is included by $L^2$ and that '$\varphi_n \to \varphi$ in

4) J. L. Doob uses a process with orthogonal increments instead of our random measure, but both concepts are equivalent.

$\mathfrak{S}$' implies '$\varphi_n \to \varphi$ in $L^2$'. Therefore $\mathcal{F}\mathfrak{D}$ is dense in $\mathfrak{S}$ with respect to the norm of $L^2$. As $\mathfrak{S} \supseteq \mathfrak{D}$ and $\mathfrak{D}$ is dense in $L^2$, $\mathfrak{S}$ is dense in $L^2$ and so $\mathcal{F}\mathfrak{D}$ is dense in $L^2$.

The uniqueness of the expression is clear as $\mathcal{F}\mathfrak{D}$ is dense in $L^2$. In order to prove the possiblity of the expression, we shall put

$$T(\psi) = X(\phi) \quad \text{for} \quad \psi = \mathcal{F}\phi .$$

Then $T$ will be a mapping from $\mathcal{F}\mathfrak{D}$ $(\subseteq L^2)$ into $\mathfrak{H}$, which is clearly linear and isometric on account of the identity:

$$\|T(\psi)\|^2 = (X(\phi), X(\phi)) = \rho(\phi * \tilde{\phi}) = \int |\psi(\lambda)|^2 \, d\mu(\lambda) = \|\psi\|^2,$$

since $\mathcal{F}(\phi * \tilde{\phi}) = |F\phi|^2$. $\mathcal{F}\mathfrak{D}$ being dense in $L^2$, we can extend $T(\psi)$ to a linear isometric mapping from $L^2$ into $\mathfrak{H}$. As the characteristic function $\chi_E(\lambda)$ of a set $E \in B^*$ belongs to $L^2$, we may define $M(E)$ as follows:

$$(4\cdot 7) \qquad M(E) = T(\chi_E).$$

Then we have

$$(4\cdot 8) \quad (M(E_1), M(E_2)) = \int \chi_{E_1}(\lambda) \, \overline{\chi_{E_2}(\lambda)} \, d\mu(\lambda) = \mu(E_1 \cap E_2),$$

since $T$ is isometric. In addition to this, we shall have

$$(4\cdot 9) \qquad M(f) = T(f) \quad \text{for} \quad f \in L^2,$$

for this is evidently true for any simple function $f$ in $L^2$ by the definition and we shall easily see that it is also true for any $f \in L^2$, by taking into account the fact that both sides of (4·10) are isometric in $f$ and any $f \in L^2$ is expressed as the $L^2$-limit of a sequence of simple functions. If we put $f = \mathcal{F}\phi$ in (4·9), we obtain (5·6) at once. The last part of the theorem is clear by the definitions.

Making use of this theorem we can prove that the class of stationary processes $\mathfrak{S}^\circ$ coincides with $\mathfrak{S}_0$ in (3·5), i.e.

THEOREM 4·2. $\qquad \mathfrak{S}^\circ = \mathfrak{S}_0 .$

PROOF. As $\mathfrak{S}^\circ \subseteq \mathfrak{S}_0$ is clear by Khintchine's decomposition (3·1), we need only prove its inverse inclusion relation. Let $X$ be any element of $\mathfrak{S}_0$. Then we have

$$(4\cdot 10) \quad X(\phi) = \int \mathcal{F}\phi(\lambda) \, dM(\lambda), \ (M(E_1), M(E_2)) = \mu(E_1 \cap E_2),$$

where

$$(4\cdot 11) \qquad \int d\mu(\lambda) < \infty .$$

Put

$$(4\cdot12) \qquad Y(t)=\int e^{-i2\pi\lambda t}\,dM(\lambda),$$

which may be defined, since the $\lambda$-function $e^{-i2\pi\lambda t}$ belongs to $L^2$ by virtue of (4·11). $Y(t)$ proves to be a stationary continuous random process. Therefore we have, for $\phi \in \mathfrak{D}$,

$$\int Y(t)\,\phi(t)\,dt=\int\phi(t)\int e^{-i2\pi\lambda t}\,dM(\lambda)\,dt=\int\mathcal{F}\phi(\lambda)\,dM(\lambda)=X(\phi),$$

which implies that $X(\phi)$ is induced by a process $Y \in \mathfrak{S}^\circ$.

By the same way as in Theorem 3·3 we shall obtain

THEOREM 4·3. *In case of real stationary distributions the spectral random measure M is hermitian-symmetric i.e.*

$$(4\cdot3) \qquad M(E)=\overline{M(-E)}.$$

## § 5. Derivative of Stationary Random Distribution

Any random distribution has derivatives of any order, which are also random distributions.

THEOREM 5·1. *Let $X$ be a stationary distribution with spectral measure $\mu$ and spectral random measure $M$. Then $X^{(k)}$ $(=D^kX)$ is also a stationary distribution $M$ whose spectral measure $\mu_k$ and spectral random measure $M_k$ are given by*

$$(5\cdot1) \qquad d\mu_k(\lambda)=(2\pi\lambda)^{2k}\,d\mu(\lambda), \qquad dM_k(\lambda)=(i2\pi\lambda)^k\,dM(\lambda).$$

PROOF. We have, by the definition,

$$X^{(k)}(\phi)=(-1)^k X(\phi^{(k)})=(-1)^k\int\mathcal{F}\phi^{(k)}(\lambda)\,dM(\lambda)$$
$$=\int(i2\pi\lambda)^k\,\mathcal{F}\phi(\lambda)\,dM(\lambda),$$

since we have, for $\phi \in \mathfrak{D}$,

$$\mathcal{F}\phi^{(k)}(\lambda)=(-1)^k\,(i2\pi\lambda)^k\,\mathcal{F}\phi(\lambda).$$

Thus $X^{(k)}$ proves to be a stationary distribution satisfying the above conditions.

By this theorem we shall see that '$X \in \mathfrak{S}_n$' implies '$X^{(k)} \in \mathfrak{S}_{n+k}$'. Therefore we shall have, by Theorem 4·2,

THEOREM 5·2.[5] In order that $X^{(k)}$ is a stationary continuous process, it is necessary and sufficient that the spectral measure $\mu$ of $X$ satisfies

5) Cf. [3] XI, § 9 Example 1, p. 535.

(5·2) $$\int \lambda^{2k}\, d\mu(\lambda) < \infty.$$

## § 6. Continuous Process with Stationary $k$-th Order Increments

Let $Y(t)$ be a continuous random process, i.e. $Y \in \mathfrak{C}(\mathfrak{H})$. The *k-th order increment* of $Y$ is defined by

(6·1) $$\Delta^{(k)} Y(t) = \Sigma_{\nu=0}^{k} \binom{k}{\nu} (-1)^{\nu}\, Y(t+\nu\Delta),$$

where $\Delta$ is an arbitrary real number. This is also a continuous random process for any fixed $\Delta$. If $\Delta^{(k)}Y$ is a continuous random process for any fixed $\Delta$, viz.

(6·2) $$(\Delta^{(k)}Y(t+h),\, \Delta^{(k)}Y(s+h)) = (\Delta^{(k)}Y(t),\, \Delta^{(k)}Y(s)),$$

then $Y(t)$ is called a ***process with stationary k-th order increments.*** The totality of such processes will be here denoted with $\mathfrak{J}_k$. The special case $k=1$ has already been discussed as a screw line in Hilbert space by J. v. Neumann and I. J. Schönberg [10] and A. Kolmogorov [8].[6] Now we shall establish a close relation between $\mathfrak{J}_k$ and $\mathfrak{S}_k$, which will give us a generalisation of the results obtained by these scholars.

THEOREM 6·1. *If $Y \in \mathfrak{J}_k$, then $Y^{(k)} \in \mathfrak{S}_k$, and conversely, if $X \in \mathfrak{S}_k$, then there exists $Y \in \mathfrak{J}_k$ such that $X = Y^{(k)}$; Precisely speaking, if*

(6·3) $$X(\phi) = \int \mathcal{F}\phi(\lambda)\, dM(\lambda)$$

*then*

(6·4) $$Y(t) = \sum_{\nu=0}^{k-1} t^{\nu} A_{\nu} + \int \frac{e^{-i2\pi\lambda t} - L_k(\lambda, t)}{(-i2\pi\lambda)^k}\, dM(\lambda),$$

$$(L_k(\lambda, t) = \Sigma_{\nu=0}^{k-1} (-i2\pi\lambda t)^{\nu}/\underline{|\nu}\ \ (|\lambda| \leq 1),\ = 0\ \ (|\lambda| > 1))$$

(6·5) $$\Delta^{(k)} Y(t) = \int e^{-i2\pi\lambda t} \left( \frac{1 - e^{-i2\pi\lambda\Delta}}{-i2\pi\lambda} \right)^k dM(\lambda).$$

PROOF. Suppose that $Y \in \mathfrak{J}_k$. Then we have

$$\begin{aligned} &(\Delta^{(k)}Y(\tau_h\phi),\, \Delta^{(k)}Y(\tau_h\psi)) \\ &= \iint (\Delta^{(k)}Y(t),\, \Delta^{(k)}Y(s))\, \phi(t+h)\, \overline{\psi(s+h)}\, dt\, ds \\ &= \iint (\Delta^{(k)}Y(t+h),\, \Delta^{(k)}Y(s+h))\, \phi(t+h)\, \overline{\psi(s+h)}\, dt\, ds \end{aligned}$$

6) Cf. [3] XI, § 11 and [6].

$$=\iint(\Delta^{(k)}Y(t),\Delta^{(k)}Y(s))\,\phi(t)\,\overline{\psi(s)}\,dt\,ds$$

$$=(\Delta^{(k)}Y(\phi),\Delta^{(k)}Y(\psi)).$$

Hence it follows that

$$(Y^{(k)}(\tau_h\phi),Y^{(k)}(\tau_h\psi))=(Y^{(k)}(\phi),Y^{(k)}(\psi))$$

which implies $Y^{(k)}\in\mathfrak{S}$. By virture of Theorem 4·1, $Y^{(k)}$ is expressible in the following form:

$$(6\cdot 6)\quad Y^{(k)}(\phi)=\int\mathcal{F}\phi(\lambda)\,dM(\lambda),\qquad (M(E),M(E'))=\mu(E\cap E').$$

In order to show that $Y^{(k)}\in\mathfrak{S}_k$, we need only prove that

$$(6\cdot 7)\qquad \int_{|\lambda|\geq 1}\frac{d\mu(\lambda)}{\lambda^{2k}}<\infty.$$

Put

$$(6\cdot 8)\qquad Y_1(\phi)=\int\mathcal{G}_k\phi(\lambda)\,dM(\lambda),$$

where

$$\mathcal{G}_k\phi(\lambda)=\int\frac{e^{-i2\pi\lambda t}-L_k(\lambda,t)}{(-i2\pi\lambda)^k}\phi(t)\,dt\,.$$

$\int\frac{\mu(d\lambda)}{(1+\lambda^2)^l}<\infty$ for some

Since we have

$$\mathcal{G}_k\phi(\lambda)=\frac{\mathcal{F}\phi(\lambda)}{(-i2\pi\lambda)^k},\qquad\text{for}\quad |\lambda|>1,$$

and $\mathcal{F}\phi$ is rapidly decreasing, $\mathcal{G}_k\phi$ belongs to $L^2(R,\mu)$, so that the above integral will be determined. Then we have

$$Y_1^{(k)}(\phi)=(-1)^k\,Y_1(\phi^{(k)})=(-1)^k\int\mathcal{G}_k\phi^{(k)}(\lambda)\,dM(\lambda)$$

$$=\int\mathcal{F}\phi(\lambda)\,dM(\lambda)\qquad\text{(by partial integration)}$$

$$=Y^{(k)}(\phi)\qquad\text{(by (6·6))},$$

which implies

$$(Y-Y_1)^{(k)}(\phi)=0,$$

and so $Y-Y_1$ may be induced by a polynomial of $t$ with coefficients in $\mathfrak{H}$ and the degree less than $k$:

$$(Y-Y_1)(t)=\sum_{\nu=0}^{k-1}t^\nu A_\nu,\qquad A_\nu\in\mathfrak{H},\qquad 0\leq\nu\leq k-1;$$

the proof of this fact is the same as that of the corresponding fact in $\mathfrak{D}'$ [11] (II, § 4). Therefore we get

$$\Delta^{(k)}Y(\phi)=\Delta^{(k)}Y_1(\phi)=Y_1((-\Delta)^{(k)}\phi)$$

$$=\int \mathcal{G}_k(-\Delta)^{(k)}\phi(\lambda)\,dM(\lambda)$$

$$=\int\left(\frac{1-e^{-i2\pi\lambda\Delta}}{-i2\pi\lambda}\right)^k \mathcal{F}\phi(\lambda)\,dM(\lambda)$$

and so

see the end of p. 223

$$(6\cdot 9)\qquad \|\Delta^{(k)}Y(\phi)\|^2=\int\left|\frac{1-e^{-i2\pi\lambda\Delta}}{-i2\pi\lambda}\right|^{2k}|\mathcal{F}\phi(\lambda)|^2\,d\mu(\lambda)$$

$$\geq c_1\int_{|\lambda|\geq 1}|1-e^{-i2\pi\lambda\Delta}|^{2k}\,|\mathcal{F}\phi(\lambda)|^2\,\frac{d\mu(\lambda)}{\lambda^{2k}}\qquad\left(c_1=\left(\frac{1}{2\pi}\right)^{2k}\right).$$

Let $\phi_n(t)$ be a non-negative function $\in\mathfrak{D}$ such that

$$\phi_n(t)=0\quad\left(|t|\geq\frac{1}{n}\right),\qquad \int\phi_n(t)\,dt=1.$$

Then we have

$$|\mathcal{F}\phi_n(\lambda)|\geq\frac{1}{2}\quad\text{for}\quad|\lambda|\leq\frac{n}{16},$$

and in addition

$$\|\Delta^{(k)}Y(\phi_n)\|\leq\left\|\int\Delta^{(k)}Y(t)\,\phi_n(t)\,dt\right\|\leq\int\|\Delta^{(k)}Y(t)\|\,\phi_n(t)\,dt$$

$$=\int_{-1/n}^{1/n}\|\Delta^{(k)}Y(t)\|\,\phi_n(t)\,dt\leq\max_{|t|\leq 1}\|\Delta^{(k)}Y(t)\|\quad(=c_2(\Delta))$$

Thus we obtain, putting $\phi=\phi_n$ in (6·9),

$$c_2(\Delta)^2\geq c_1\int_{|\lambda|\geq 1}|1-e^{-i2\pi\lambda\Delta}|^{2k}\,|\mathcal{F}\phi_n(\lambda)|^2\,\frac{d\mu(\lambda)}{\lambda^{2k}}$$

$$\geq\frac{c_1}{4}\int_{n/16\geq|\lambda|\geq 1}|1-e^{-i2\pi\lambda\Delta}|^{2k}\,\frac{d\mu(\lambda)}{\lambda^{2k}}.$$

Integrating both sides in $0\leq\Delta\leq 1$, we have

$$c_2^2\geq\frac{c_1}{4}\int_{n/16\geq|\lambda|\geq 1}c(\lambda)\,\frac{d\mu(\lambda)}{\lambda^{2k}},\qquad c(\lambda)=\int_0^1|1-e^{-i2\pi\lambda\Delta}|^{2k}\,d\Delta,\ c_2^2=\int_0^1 c_2(\Delta)^2 d\Delta.$$

But it holds that

$$c(\lambda)=\frac{1}{|2\pi\lambda|}\int_0^{|2\pi\lambda|}|1-e^{-i\Delta}|^{2k}\,d\Delta\geq\frac{1}{4\pi}\int_0^{2\pi}|1-e^{-i\Delta}|^{2k}\,d\Delta\quad(=c_3>0)$$

for $|\lambda|\geq 1$, because if we determine a natural number $n$ so that $n\leq|\lambda|\leq n+1$, then we have

$$c(\lambda)\geq\frac{1}{2\pi(n+1)}\int_0^{2\pi n}|1-e^{-i\Delta}|^{2k}\,d\Delta$$

$$=\frac{n}{2\pi(n+1)}\int_0^{2\pi}|1-e^{-i\Delta}|^{2k}\,d\Delta\geq\frac{1}{4\pi}\int_0^{2\pi}|1-e^{-i\Delta}|^{2k}\,d\Delta\,.$$

Consequently we obtain

$$\int_{n/10\geq|\lambda|\geq 1}\frac{d\mu(\lambda)}{\lambda^{2k}}\leq\frac{4c_2^2}{c_1c_3}<\infty\,,$$

which implies (6·7), as we see by letting $n$ tend to $\infty$.

Now we put

$$Y_2(t)=\int\frac{e^{-i2\pi\lambda t}-L_k(\lambda,t)}{(-i2\pi\lambda)^k}\,dM(\lambda)\,;$$

this integral exists and is continuous in $t$ by virtue of (6·7). Then we have

$$Y_2(\phi)=\int\phi(t)\,Y_2(t)\,dt=\int\mathcal{G}_k\phi(\lambda)\,dM(\lambda)=Y_1(\phi).$$

Thus $Y_1(\phi)$ is defined by $Y(t)$ and so we have

$$Y(t)=Y_2(t)+\Sigma_{\nu=0}^{k-1}t^\nu A_\nu\,,$$

which implies (6·4) and accordingly (6·5).

Now we shall prove the second part of the theorem. Suppose that $X(\phi)\in\mathfrak{S}_k$ be expressed as (6·3). Put

$$Y_0(t)=\int\frac{e^{-i2\pi\lambda t}-L_k(\lambda,t)}{(-i2\pi\lambda)^k}\,dM(\lambda)\,;$$

as $X(\phi)\in\mathfrak{S}_k$, we see that this integral exists and is continuous in $t$. Then we have

$$Y_0^{(k)}(\phi)=(-1)^k\,Y_0(\phi^{(k)})=(-1)^k\int\mathcal{G}_k\phi^{(k)}(\lambda)\,dM(\lambda)$$

$$=\int\mathcal{J}\phi(\lambda)\,dM(\lambda)=X(\phi).$$

If $Y^{(k)}(\phi)=X(\phi)$, then $(Y-Y_0)^{(k)}(\phi)=0$ and so

$$Y(t)=Y_0(t)+\sum_{\nu=0}^{k-1}t^\nu A_\nu,\qquad A_\nu\in\mathfrak{H}\,.$$

We have further

$$\Delta^{(k)}Y(t)=\Delta^{(k)}Y_0(t)=\int e^{-i2\pi\lambda t}\left(\frac{e^{-i2\pi\lambda\Delta}-1}{-i2\pi\lambda}\right)^k dM(\lambda)$$

which implies that $\Delta^{(k)}Y(t)$ is a continuous stationary process, viz. $Y\in\mathfrak{F}_k$, since

$$\int \left| \frac{e^{-i2\pi\lambda\Delta}-1}{-i2\pi\lambda} \right|^{2k} d\mu(\lambda) < \infty$$

by virtue of the assumption: $X \in \mathfrak{S}_k$.

By the same way as in Theorem 3·3 we have

THEOREM 6·2. *In case of real processes* $\in \mathfrak{J}_k$ *the above spectral random measure satisfies* $M(E) = \overline{M(-E)}$ *and the above coefficients* $A_\nu$, $\nu = 1, 2, \cdots, k-1$ are real.

Mathematics Department, Kyoto University.

## BIBLIOGRAPHY


1. S. Bochner: Fouriersche Integrale, Leipzig, 1932.
2. H. Cramér: On the theory of stationary random processes, Ann. Math. 41, 215-230 (1940).
3. J. L. Doob: Stochastic processes, New York, 1953.
4. K. Itô: Multiple Wiener integral, Jour. Math. Soc. Japan, **3**, 1, 156-169, (1951).
5. K. Itô: Complex multiple Wiener integral, Jap. Jour. Math. **22**, 63-86, (1952).
6. K. Itô: A screw line in Hilbert space and its application to probability theory, Proc. Imp. Acad. Tokyo **20**, 4, 203-209, (1944).
7. A. Khintchine: Korrelations theorie des stationären stochastischen Prozesse, Math. Ann. 109, 604-615.
8. A. Kolmogorov: Kurven in Hilbertchem Raum die gegenüber eine einparametrigen Gruppe von Bewegungen invariant sind. C. R. (Doklady) Acad. Sci. U.R.S.S. 26, 6-9, (1940).
9. P. Lévy: Processus stochastiques et mouvement brownien, Paris 1948.
10. J. v. Neumann and I. J. Schönberg: Fourier integrals and metric geometry, Trans. Amer. Math. Soc. 50, 226-251 (1941).
11. L. Schwartz: Théorie des distributions 1, 2, Paris, 1950.


We can simplify the part from p.221 line 4 to p.222 line 4. as follows:

"This identity means that

$$\Delta^{(k)} Y(t) = \int e^{-i2\pi\lambda t} \left( \frac{1-e^{-i2\pi\lambda\Delta}}{-i2\pi\lambda} \right)^k dM(\lambda),$$

Therefore

$$\infty > \|\Delta^{(k)} Y(t)\|^2 = \int \left| \frac{1-e^{-i2\pi\Delta}}{-i2\pi\lambda} \right|^k dM(\lambda),$$

which prove (6.7)

# 5. *Complex Multiple Wiener Integral.*

By

Kiyosi Itô.

(Received Dec. 23, 1952)

**Introduction.** It is the aim of our present paper to discuss the properties of multiple Wiener integral on a complex normal random measure and its applications to normal stochastic processes.

We have already treated in our previous paper [3] (abbreviated hereafter as M. W. I.) an integral of the same kind on a real normal random measure, and remarked that it would be more adequate and convenient to consider it in the complex field as is often seen in the theory of Fourier transforms. This is indeed the case, as will be shown in this paper.

Chapter I is mainly concerned with Complex Normal Random Measure on which our integral will be discussed. Roughly speaking, it is an additive set function whose value for any set is a complex-valued random variable subject to an isotropic Gaussian distribution on the complex plane and whose values for disjoint sets are mutually independent. We shall here treat also a more general concept "Complex-normal System" for the later use.

In Chapter II we shall define an integral named as Complex Multiple Wiener Integral:

$$\int\cdots\int f(t_1,\ldots,\ t_p;\ s_1,\ \ldots,\ s_q)dM(t_1)\cdots dM(t_p)d\overline{M(s_1)}\cdots d\overline{M(s_q)},$$

*M* being a normal random measure, and in Chapter III we shall establish fundamental properties of this integral. The idea is the same as in M. W. I. but we shall here make the interpretations clearer and more precise.

In M.W.I. we have shown that real multiple Wiener integrals are closely related to Hermite polynomials. To establish such relations on our present complex case, we shall define Hermite polynomials of complex variables in a natural way; this definition seems new as far as we know.

In Chapter IV we shall make use of the results obtained to generalize and derive systematically various known facts on the ergodicity and spectral structure of temporally homogeneous processes; our method will make clearer the essential points of the problems.

**Contents**

## Notations

$K$ .................... complex number space
$\Omega(B_\Omega, P)$ ........... basic probability space
$\omega$ .................... probability parameter
$\mathcal{E}X=\int_\Omega X(\omega)dP(\omega)$ .... expectation
$e(x)=e^x$ ............ exponential function
$\Re z$ .................. real part of $z$

# Chapter I. Complex Normal Random Measure.

**§1. Complex-normal Random Variables.** Let $Z=Z(\omega)=X(\omega)+iY(\omega)$ be a complex random variable. If the joint distribution of $X$ and $Y$ is a two-dimensional isotropic normal distribution (including degenerate cases), $Z$ will be called a *complex-normal variable.* This condition is expressible by

$$\mathcal{E}e(i(Xu+Yv))=e\left(-\frac{a}{4}(u^2+v^2)\right)\ (u,\ v\in R) \tag{1.1}$$

or equivalently by

$$\mathcal{E}e(i\Re Zw)=e\left(-\frac{a}{4}|w|^2\right)(w\in K) \tag{1.2}$$

for some $a\geqslant 0$. By (1.1) we see that $X$ and $Y$ are independent real random variables and subject to the same distribution $N(0, a/2)$.

THEOREM 1.1. *Let $Z_1, Z_2, \dots, Z_n$ be independent complex normal variables. Then $Z=\sum_\nu c_\nu Z_\nu$, $c_\nu$'s being complex numbers, is also complex-normal.*

PROOF. Assume that

$$\mathcal{E}e(i\Re Z_\nu w)=e\Big(-\frac{a_\nu}{4}|w|^2\Big),\ \nu=1,\ 2,\dots,\ n.$$

By the independence of $\{Z_\nu\}$ we have

$$\mathcal{E}\Pi_\nu e(i\Re(Z_\nu c_\nu w))=\Pi_\nu \mathcal{E}e(i\Re(Z_\nu c_\nu w))=\Pi_\nu e\Big(-\frac{a_\nu}{4}|c_\nu w|^2\Big),$$

that is

$$\mathcal{E}e(i\Re Zw)=e\Big(-\frac{1}{4}\Sigma_\nu a_\nu|c_\nu|^2|w|^2\Big).$$

**Theorem** 1.2. *Suppose that $Z_n\to Z$ and that $Z_n$, $n=1, 2, \dots$, are all complex-normal. Then $Z$ is also complex-normal.*

**Proof.** It follows immediately from

$$\mathcal{E}e(i\Re Zw)=\lim_{n\to\infty}\mathcal{E}e(i\Re Z_n w).$$

**§2, Complex-normal System.** A system of complex random variables $\boldsymbol{Z}=\{Z_\lambda(\omega),\ \lambda\in\Lambda\}$ is called *complex-normal*, if $\Sigma_{i=1}^n c_i Z_{\lambda(i)}$ is complex-normal for any $n$, for any $c_i\in K$ and for any $\lambda(i)\in\Lambda$. By the definition each $Z_\lambda$ is clearly complex-normal. By Theorem 1.1 we have

**Theorem** 2.1. *If $Z_\lambda$, $\lambda\in\Lambda$, are independent, each being complex-normal, then the system $\boldsymbol{Z}=\{Z_\lambda,\ \lambda\in\Lambda\}$ is complex-normal.*

By the definition and Theorem 1.2 we get

**Theorem** 2.2. *Let $\boldsymbol{Z}=\{Z_\lambda,\ \lambda\in\Lambda\}$ be complex-normal and $\boldsymbol{Z}'=\{Z'_\mu,\ \mu\in M\}$ be any system of complex random variables. If each $Z'_\mu$ is of the form $\Sigma_{i=1}^N c_i Z_{\lambda(i)}$ or more generally $\lim_n \Sigma_{i=1}^{N(n)} c_{ni} Z_{\lambda(ni)}$ (limit in probability), then $\boldsymbol{Z}'$ is also complex-normal.*

Let $\boldsymbol{Z}=\{Z_\lambda,\ \lambda\in\Lambda\}$ be any system of complex random variables such that $\mathcal{E}|Z_\lambda|^2<\infty$, $\lambda\in\Lambda$. We put

$$V_{\lambda\mu}=\mathcal{E}(Z_\lambda\overline{Z}_\mu)=(Z_\lambda,\ Z_\mu),\ \lambda,\ \mu\in\Lambda. \tag{2.1}$$

Then $V_{\lambda\mu}$ is positive-definite in $\lambda$, $\mu\in\Lambda$ in the sense that we have

$$\Sigma_{i,j=1}^n V_{\lambda(i)\lambda(j)}\xi_i\overline{\xi}_j\geqq 0 \tag{2.2}$$

for any $n$, for any $\lambda(i)$ and for any $\xi_i\in K$. Conversely, if $V_{\lambda\mu}$ is positive-definite in $\lambda$, $\mu$, then there exists a system $\boldsymbol{Z}=\{Z_\lambda,\ \lambda\in\Lambda\}$ satisfying (2.1). We shall further establish

**Theorem** 2.3 (Existence Theorem). *If $V_{\lambda\mu}$ is positive-definite, then there exists a complex-normal system satisfying (2.1), and the probability distribution of this system is uniquely determined by $V_{\lambda\mu}$.*

**Proof of uniqueness.** We shall first show a preliminary lemma.

LEMMA 2.1. *If $\{Z_1,\ Z_2,\ \dots,\ Z_n\}$ is a complex-normal system, then*

$$\mathcal{E}e(i\Re(\Sigma_{\nu=1}^n c_\nu)Z_\nu)=e\Big(-\frac{1}{4}\Sigma_{\mu\nu}c_\mu\bar{c}_\nu(Z_\mu,\ Z_\nu)\Big). \tag{2.3}$$

**Proof.** $\{Z_\nu\}$ being complex-normal, $\Sigma_{\nu=1}^n c_\nu Z_\nu$ is a complex-normal variable, and so we have

$$\mathcal{E}e(i\Re(\Sigma_\nu c_\nu Z_\nu)w)=e\Big(-\frac{a}{4}|w|^2\Big), \tag{2.4}$$

where

$$a=\|\sum_\nu c_\nu Z_\nu\|^2=\sum_{\mu\nu}c_\mu \bar{c}_\nu(Z_\mu, Z_\nu).$$

Putting $w=1$ in (2.4), we shall get (2.3).

Now, let $\{Z_\lambda=X_\lambda+iY_\lambda\}$ be any complex-normal system for which (2.1) holds. By the above lemma we have

$$\mathcal{E}e(i\sum_\nu(u_\nu X_{\lambda(\nu)}+v_\nu Y_{\lambda(\nu)}))=\mathcal{E}e(i\Re\sum_{\nu=1}^n(u_\nu-iv_\nu)Z_{\lambda(\nu)})$$
$$=e\left(-\frac{1}{4}\sum_{\lambda\nu}(u_\mu-iv_\mu)(u_\nu+iv_\nu)V_{\lambda(\mu)\lambda(\nu)}\right),$$

which implies that $\{V_{\lambda\mu}\}$ determines the joint characteristic function and so the joint distribution of $X_{\lambda(\nu)}$, $Y_{\lambda(\nu)}$, $\nu=1, 2,\dots, n$. Therefore $V_{\lambda\mu}$ determines the probability distribution of the system $\{Z_\lambda\}$.

PROOF OF THE EXISTENCE. We shall begin with a lemma.

LEMMA 2.2. *If $V_{\lambda\mu}$ is positive-definite in $\lambda$, $\mu\in\Lambda$, then there exists a system of (possibly infinite-dimensional) complex vectors*

$$c^\lambda=(c^\lambda{}_\alpha, \ \alpha\in A)$$

*such that*

(2.5) $$\sum_\alpha|c^\lambda{}_\alpha|^2<\infty, \quad \lambda\in\Lambda,$$

(2.6) $$V_{\lambda\mu}=\sum_\alpha c^\lambda{}_\alpha \bar{c}^\mu{}_\alpha, \quad \lambda, \mu\in\Lambda.$$

PROOF. Let $K^\Lambda{}_0$ denote the totality of the complex vector (of finite or infinite dimension) $x=(x_\lambda, \lambda\in\Lambda)\in K^\Lambda$ such that $x_\lambda=0$ except for a finite number of exceptional values of $\lambda$. $K^\Lambda{}_0$ is considered a linear space associated with the inner product:

$$(x, y)=\sum_{\lambda\mu}V_{\lambda\mu}x_\lambda \bar{x}_\mu,$$

which is bilinear, symmetric and non-negative:

$$(\sum_i a_i x_{\lambda_i}, \sum_j b_j y_{\lambda_j})=\sum_{i,j}a_i\bar{b}_j(x_{\lambda_i}, y_{\lambda_j}),$$
$$(x, y)=\overline{(y, x)},$$
$$(x, x)\geqslant 0.$$

Let $N^\Lambda{}_0$ be the set of all elements $x\in K^\Lambda{}_0$ for which $(x, x)=0$. Then $N$ proves to be a linear subspace of $K^\Lambda{}_0$. The factor space $H=K^\Lambda{}_0/N^\Lambda{}_0$ may be considered a linear space with the inner product if we define the operations in the usual way. Then the inner product satisfies the positivity:

$$(x, x)>0 \text{ for } x\neq 0$$

in addition to the above properties.

Consider the completion of $H$: $\mathfrak{H}=\bar{H}$. Then $H$ is imbedded in $\mathfrak{H}$. Let $x^\lambda$ be the element in $K^\Lambda{}_0$ whose $\lambda$-component is equal to 1 and whose other components are all equal to 0. By the natural mapping from $K^\Lambda{}_0$ onto $H\subseteq\mathfrak{H}$, $x^\lambda$ will be mapped to an element of $\mathfrak{H}$, say $\mathfrak{X}^\lambda$. We have clearly

$$(\mathfrak{X}^\lambda, \mathfrak{X}^\mu)=V_{\lambda\mu}.$$

We shall take a complete orthonormal system in $\mathfrak{H}$: $\phi^\alpha$, $\alpha\in A$. $\mathfrak{X}^\lambda$ may be expressible as

$$\mathfrak{X}^\lambda=\sum_{\alpha\in A}c^\lambda{}_\alpha\phi^\alpha,$$

where

$$\sum_\alpha |c^\lambda{}_\alpha|^2 = ||\mathfrak{X}^\lambda||^2 < \infty, \quad V_{\lambda\mu} = (\mathfrak{X}^\lambda, \mathfrak{X}^\alpha) = \sum_\alpha c^\lambda{}_\alpha \bar{c}^\mu{}_\alpha.$$

Thus we obtain a system of vectors $c^\lambda = (c^\lambda{}_\alpha, \alpha \in A)$, $\lambda \in \Lambda$, satisfying (2.5) and (2.6).

Now we shall proceed to the proof of existence. We shall define a normal distribution $\mu$ on $K$ by

$$d\mu(z) = \frac{1}{\pi} e(-x^2 - y^2) dx dy \quad (z = x + iy)$$

and consider the direct product measure $\mu^A$, $A$ being the parameter set mentioned in the above lemma. The existence of $\mu^A$ is assured by the extention theorem of A. Kolmogorov [1] or more directly by a theorem of S. Kakutani [2]. Now we take $K^A(\mu^A)$ as the underlying probability space $\Omega(P)$ and define

$$W_\alpha(\omega) = \text{ the } \alpha\text{-component of } \omega, \ \omega \in \Omega = K.$$

Then $W_\alpha$, $\alpha \in A$, are independent complex random variables, each being complex-normal. Therefore they constitute a complex-normal system as well as an orthogonal system in $L^2(\Omega, P)$.

Making use of the $c_\alpha$'s in the above lemma, we shall define $Z_\lambda$ in $L^2(\Omega, P)$ by

$$Z_\lambda = \sum_\alpha c^\lambda{}_\alpha W_\alpha$$

for each $\lambda$, where the infinite sum is understood in the sense of covergence in $L^2(\Omega, P)$ and so may be considered in the sense of convergence in probability. Therefore we see, by Th. 2.2, that $Z = \{Z_\lambda, \lambda \in \Lambda\}$ is a complete-normal system. $\{W_\alpha\}$ being orthonormal, we get

$$(Z_\lambda, Z_\mu) = \sum_\alpha c^\lambda{}_\alpha \bar{c}^\mu{}_\alpha = V_{\lambda\mu}.$$

THEOREM 2.4. *Given a complex-normal system $\{Z_\lambda, \lambda \in \Lambda\}$. It is necessary and sufficient for $\{Z_\lambda\}$ to be independent that $Z_\lambda$, $\lambda \in \Lambda$, are mutually orthogonal in the Hilbert space $L^2(\Omega, P)$.*

PROOF. The necessity is clear since the independence of $\{Z_\lambda\}$ implies

$$(Z_\lambda, Z_\mu) = \mathcal{E} Z_\lambda \bar{Z}_\mu = \mathcal{E} Z_\lambda \mathcal{E} \bar{Z}_\mu = 0 \quad (\lambda \neq \mu).$$

Conversely assume that $\{Z_\lambda, \lambda \in \Lambda\}$ be mutually orthogonal. Let $\lambda(\nu)$, $1 \leq \nu \leq n$, be any finite subsystem of $\Lambda$. Then we have

$$\begin{aligned}\mathcal{E} e[i\Re(\sum_\nu Z_{\lambda(\nu)} w_\nu)] &= e\left[-\frac{1}{4} ||\sum_\nu Z_{\lambda(\nu)} w_\nu||^2\right] \\ &= e\left[-\frac{1}{4} \sum_\nu ||Z_{\lambda(\nu)}||^2 |w_\nu|^2\right]\end{aligned}$$

by Lemma 2.1. Put $Z_{\lambda(\nu)} = X_{\lambda(\nu)} + iY_{\lambda(\nu)}$ and $w_\nu = u_\nu - iv_\nu$. Then we have

$$\begin{aligned}\mathcal{E} e[i\sum_\nu (X_{\lambda(\nu)} u_\nu + Y_{\lambda(\nu)} v_\nu)] &= e\left[-\frac{1}{4} \sum_\nu ||Z_{\lambda(\nu)}||^2 (u_\nu^2 + v_\nu^2)\right] \\ &= \Pi_\nu e\left(-\frac{1}{4} ||Z_{\lambda(\nu)}||^2 u_\nu^2\right) \Pi_\nu e\left(-\frac{1}{4} ||Z_{\lambda(\nu)}||^2 v_\nu^2\right) \\ &= \Pi_\nu e\left(-\frac{1}{2} ||X_{\lambda(\nu)}||^2 u_\nu^2\right) \Pi_\nu e\left(-\frac{1}{2} ||X_{\lambda(\nu)}||^2 v_\nu^2\right) \\ &= \Pi_\nu \mathcal{E} e(iX_{\lambda(\nu)} u_\nu) \ \Pi_\nu \mathcal{E} e(iY_{\lambda(\nu)} v_\nu),\end{aligned}$$

which implies, by virtue of a theorem of Kac-Steinhaus, the independence of

$X_{\lambda(1)}, \ldots, X_{\lambda(n)}, Y_{\lambda(1)}, \ldots, Y_{\lambda(n)}$ a fortiori that of $Z_{\lambda(1)}, \ldots, Z_{\lambda(n)}$. Thus we see that $X_\lambda$, $\lambda \in \Lambda$, are independent.

§3. **Complex Normal Random Measure**. Let $T(\boldsymbol{B}_T, m)$ be any abstract space, and $\boldsymbol{B}^*_T$ be the totality of the elements in $\boldsymbol{B}_T$ with finite $m$-measure. $m(E \frown F)$ is positive-definite in $E, F \in \boldsymbol{B}^*_T$ since we have

$$\sum_{i,j=1}^{n} \xi_i \bar{\xi}_j m(E_i \frown E_j) = \int_T |\sum_i \xi_i \chi(t; E_i)|^2 dm(t) \geqq 0$$

for any $E_j \in \boldsymbol{B}^*_T$, $\chi(t; E)$ denoting the characteristic function of the set $E_i$. Thus we get, by the existence theorem 2.3.

THEOREM 3.1. *There exists a complex normal system* $\boldsymbol{M} = \{M(E), E \in \boldsymbol{B}^*_T\}$ *such that*

$$(M(E), M(F)) = m(E \frown F).$$

*The probability distribution of* $\boldsymbol{M}$ *is uniquely determined by* $m$.

The above $\boldsymbol{M}$ is defined as a *complex normal random measure* on $T(\boldsymbol{B}_T, m)$; the nomenclature "*random measure*" will be justified by the following theorem.

THEOREM 3.2. *Let* $\boldsymbol{M} = \{M(E), E \in \boldsymbol{B}^*_T\}$ *be a complex normal random measure. If* $E_1, E_2, \ldots$ *are disjoint, then* $M(E_i)$, $i = 1, 2, \ldots$, *are independent, and further if* $E \equiv \sum_i E_i \in \boldsymbol{B}^*_T$, *then*

(3.1) $\quad M(E) = \sum_i M(E_i)$ *(Convergence in* $L^2(\Omega, P)$*)*

PROOF. $\{M(E_1), \ldots, M(E_n)\}$ is complex-normal for any $n$. Since we have

$$(M(E_i), M(E_j)) = m(E_i \frown E_j) = 0, \quad i \neq j,$$

$M(E_1), \ldots, M(E_n)$ are independent by Theorem 2.4. $n$ being arbitrary, $M(E_i)$, $i = 1, 2, \ldots$ are independent.

$$\begin{aligned} ||M(E) - \textstyle\sum_{i=1}^n M(E_i)||^2 &= ||M(E)||^2 + \textstyle\sum_{i=1}^n ||M(E_i)||^2 - 2\textstyle\sum_{i=1}^n (M(E), M(E_i)) \\ &\quad + 2\textstyle\sum_{i<j} (M(E_i), M(E_j)) \\ &= m(E) + \textstyle\sum_{i=1}^n m(E_i) - 2\textstyle\sum_{i=1}^n m(E_i) \\ &= m(E) - \textstyle\sum_{i=1}^n m(E_i) \to 0 \ (n \to \infty), \end{aligned}$$

which proves (3.1).

REMARK. By a theorem of P. Lévy [4] the above convergence in $L^2(\Omega, P)$ (=mean convergence) implies the almost certain convergence on account of the independence of $\{M(E_i)\}$.

The following facts will be proved by simple computations and will become useful in the next §:

(3.2) $\quad \mathcal{E}M(E)^2 = 0, \quad \mathcal{E}|M(E)|^2 = m(E), \quad \mathcal{E}|M(E)|^4 = 2m(E).$

§4. **Continuity**. Let $T(\boldsymbol{B}_T, m)$ be any measure space. A decomposition of a set $E \in \boldsymbol{B}^*_T$ into disjoint parts:

$$E = \sum_{i=1}^n E_i$$

is called an *ε-decomposition*, if $m(E_i) < \varepsilon$, $i = 1, 2, \ldots, n$. If, for any $\varepsilon > 0$ and $E \in \boldsymbol{B}^*_T$, there exists an ε-decomposition of $E$, the measure $m$ is called to be *continuous*.

Let $\boldsymbol{M} = \{M(E), E \in \boldsymbol{B}^*_T\}$ be a complex normal random measure on $T(\boldsymbol{B}_T, m)$. $\boldsymbol{M}$ is said to be *continuous* if the corresponding measure $m$ is continuous in the above sense.

Hereafter we shall often assume the continuity to make use of the following theorem.

THEOREM 4. *Let* $M$ *be any continuous complex normal random measure on* $T(B_T, m)$ *and*

$$\Delta: \quad E=\sum_{i=1}^n E_i$$

*be any* $\varepsilon$*-decomposition of a set* $E\in B^*_T$. *For this decomposition we define*

$$S_\Delta=\sum_{i=1}^n |M(E_i)|^2, \quad S'_\Delta=\sum_{i=1}^n M(E_i)^2.$$

*Then we have, as* $\varepsilon\to 0$,

$$||S_\Delta-m(E)||\to 0, \quad ||S'_\Delta||\to 0$$

PROOF. By virtue of the independence of $\{M(E_i)\}$ and the identities (3.2) we get

$$\begin{aligned} ||S_\Delta-m(E)||^2 &= ||\textstyle\sum_i [M(E_i)-m(E_i)]||^2 \\ &= \textstyle\sum_i ||M(E_i)^2-m(E_i)||^2 \\ &= \textstyle\sum_i [\mathcal{E}|M(E_i)|^4-m(E_i)^2] \\ &= \textstyle\sum_i m(E_i)^2 < \varepsilon \sum_i m(E_i)=\varepsilon m(E), \\ ||S'_\Delta||^2 &= \textstyle\sum_i ||M(E_i)^2||^2=\sum_i 2m(E_i)^2 < \varepsilon 2m(E), \end{aligned}$$

which prove our theorem.

## Chapter II. Definition of Complex Multiple Wiener Integral.

§ 5. $L^2_{pq}$, $S_{pq}$. Let $T(B_T, m)$ be any measure space. Then the product measure spaces $T^p$ and $T^p\times T^q$ may be usually defined and $T^p\times T^q$ is clearly isomorphic to $T^{p+q}$. Let $L^2_{pq}$ be the $L^2$-space over $T^p\times T^q$ which is clearly isomorphic to $L^2(T^{p+q})$. For $f\in L^2_{pq}$ we define $\tilde{f}\in L^2_{pq}$ by the identity:

$$\tilde{f}(t_1, \ldots, t_p; s_1, \ldots, s_q)=\frac{1}{\underline{|p}\,\underline{|q}}\sum_{(i)(j)} f(t_{i_1}, \ldots, t_{i_p}; s_{j_1}, \ldots, s_{j_q}), \tag{5.1}$$

where $(i)=(i_1, \ldots, i_p)$ and $(j)=(j_1, \ldots, j_q)$ run over all permutations of $(1, 2, \ldots, p)$ and $(1, 2, \ldots, q)$ respectively. We have clearly

$$||\tilde{f}||\leqq||f||. \tag{5.2}$$

Next we shall a subsystem $S_{pq}$ of $L^2_{pq}$ which consists of all functions of the following type. Let $E_1, \ldots, E_n$ be any system of disjoint sets $\in B^*_T$ and $a_{i_1\cdots i_p j_1\cdots j_q}$ be a complex-valued function defined for $i_1, \ldots, i_p, j_1, \ldots, j_q= 1, 2, \ldots, n$ such that $a_{i_1\cdots i_p j_1\cdots j_q}=0$ unless $i_1, \ldots, i_p, j_1, \ldots, j_q$ are all different. $f$ is defined by

$$f(t_1, \ldots, t_p; s_1, \ldots, s_q)=a_{i_1\cdots i_p j_1\cdots j_q} \tag{5.3}$$

for $\quad (t_1, \ldots, t_p; s_1, \ldots, s_q)\in E_{i_1}\times\cdots\times E_{i_p}\times E_{j_1}\times\cdots\times E_{j_q}$,

while

$$f(t_1, \ldots, t_p; s_1, \ldots, s_q)=0 \tag{5.4}$$

if some of $t_1, \ldots, t_p, s_1, \ldots, s_q$ lies outsides of $E=\sum_{\nu=1}^n E_\nu$. We shall call the above system $\{E_\nu\}$ the *base* of expression of $f\in S_{pq}$.

From Theorem 2.1 in M.W.I. follows at once

THEOREM 5. *If $m$ is continuous, then $S_{pq}$ is a linear manifold dense in $L^2_{pq}$.*

**§6. Definition of $I_{pq}(f)$ for $f \in S_{pq}$.** Let $f$ be any element of $S_{pq}$ and $f$ be expressed by (5.4) and (5.5). We shall define $I_{pq}(f)$ by

$$I_{pq}(f)=\sum_{(i)(j)} a_{i_1\cdots i_p j_1\cdots j_q} M(E_{i_1})\cdots M(E_{i_p})\overline{M(E_{j_1})}\cdots\overline{M(E_{j_q})}. \tag{6.1}$$

$I_{pq}(f)$ is clearly independent of the expression of $f$ and is well determined by $f$ itself.

We shall here establish the following properties of $I_{pq}$:

$$I_{pq}(f)=I_{pq}(\tilde{f}), \tag{I*.1}$$

$$I_{pq}(\alpha f+\beta g)=\alpha I_{pq}(f)+\beta I_{pq}(g), \tag{I*.2}$$

$$\mathcal{E} I_{pq}(f)=(I_{pq}(f),\ 1)=0, \tag{I*.3}$$

$$(I_{pq}(f),\ I_{pq}(g))=\underline{|p}\,\underline{|q}\,(\tilde{f},\ \tilde{g}), \tag{I*.4}$$

$$||I_{pq}(f)||^2=\underline{|p}\,\underline{|q}\,||\tilde{f}||^2\leqslant \underline{|p}\,\underline{|q}\,||f||^2 \tag{I*.5}$$

$$(I_{pq}(f),\ I_{rs}(g))=0 \quad \text{if } (p,\ q)\neq(r,\ s). \tag{I*.6}$$

Since $M(E_{i_1})\cdots M(E_{i_p})\overline{M(E_{j_1})}\cdots\overline{M(E_{j_q})}$ is symmetric in $(i_1,\ \ldots,\ i_p)$ as well as in $(j_1,\ \ldots,\ j_q)$, (I*.1) is clear. (I*.2) will be easily shown if we express $f$ and $g$ on the same base of expression. Since $M(E_{i_1}),\ \ldots,\ M(E_{i_p}),\ \overline{M(E_{j_1})},\ \ldots,\ \overline{M(E_{j_q})}$ are independent when $i_1,\ \ldots,\ i_p,\ j_1,\ \ldots,\ j_q$ are all different, (I*.3) is evident. By the definition $I_{pq}(f)$ is expressible as

$$I_{pq}(f)=\Sigma'\underline{|p}\,\underline{|q}\,\tilde{a}_{i_1\cdots i_p j_1\cdots j_q}M(E_{i_1})\cdots M(E_{i_p})\overline{M(E_{j_1})}\cdots\overline{M(E_{j_q})}, \tag{6.2}$$

where $\tilde{a}_{i_1\cdots i_p j_1\cdots j_q}$ is the value of $f$ and $\Sigma'$ is the summation over the suffices $(i)$, $(j)$ such that $i_1,\ \ldots,\ i_p,\ j_1,\ \ldots,\ j_q$ are all different and $i_1<\ldots<i_p,\ j_1<\ldots<j_q$. It is clear that the terms in $\Sigma'$ are mutually orthogonal. Expressing $f$ and $g$ on the same base $\{E_\nu\}$ we shall obtain (I*.4) and (I*.6) at once if we remark that

$$\begin{aligned}&||M(E_{i_1})\cdots M(E_{i_p})\overline{M(E_{j_1})}\cdots\overline{M(E_{j_q})}||^2\\&=m(E_{i_1})\cdots m(E_{i_p})m(E_{j_1})\cdots m(E_{j_q})\end{aligned} \tag{6.3}$$

and

$$\begin{aligned}&(M(E_{i_1})\cdots M(E_{i_p})\overline{M(E_{j_1})}\cdots\overline{M(E_{j_q})},\ M(E_{k_1})\cdots M(E_{k_r})\overline{M(E_{l_1})}\cdots\overline{M(E_{l_s})})\\&=0\end{aligned} \tag{6.4}$$

except for the special case that $p=r$, $q=s$ and $i_\pi=k_\pi$, $j_\sigma=l_\sigma$ for all $\pi$, $\sigma$. Putting $f=g$ in (I*.4) we shall get (I*.5).

For the unity of arguments we shall consider also the trivial case $p=q=0$ which was excluded in the above discussions. We put

$$L^2_{00}=K(=\text{the complex number space}),\ I_{00}(c)=c.$$

Then the above properties (I*.1)$\sim$(I*.6) except (I*.3) will hold also in case $p$ or $q$ is equal to 0; (I*.3) will be included in (I*.6) as the special case of $r=s=0$.

**§7 Definition of $I_{pq}(f)$ for $f\in L^2_{pq}$.** In the preceding § we have defined $I_{pq}(f)$ for $f\in S_{pq}$. We shall extend this definition by taking the limit. By virtue of Theorem 5 we may define, for $f\in L^2_{pq}$, a sequence $\{f_n\}$ in $S_{pq}$ such that

$$||f_n-f||\to 0. \tag{7.1}$$

Then we have by (I*.2) and (I*.5), as $m, n\to\infty$,

$$||I_{pq}(f_m)-I_{pq}(f_n)||=||I_{pq}(f_m-f_n)||\leqq||f_m-f_n||$$
$$\leqq||f_m-f||+||f_n-f||\to 0,$$

so that we may define

$$I_{pq}(f)=\lim_n I_{pq}(f_n);$$

this definition is independent of the choice of the sequence $\{f_n\}$.

Now we shall prove

**Theorem 7.**

$$(I.1)\qquad I_{pq}(f)=I_{pq}(\tilde{f})$$

$$(I.2)\qquad I_{pq}(\alpha f+\beta g)=\alpha I_{pq}(f)+\beta I_{pq}(g),$$

$$(I.3)\qquad \mathcal{E}(I_{pq}(f))=(I_{pq}(f),\ 1)=0,\ p+q\neq 0,$$

$$(I.4)\qquad (I_{pq}(f),\ I_{pq}(g))=\underline{|p}\,\underline{|q}\,(\tilde{f},\ \tilde{g}),$$

$$(I.5)\qquad ||I_{pq}(f)||^2=\underline{|p}\,\underline{|q}\,||\tilde{f}||^2\leqq\underline{|p}\,\underline{|q}\,||f||^2$$

$$(I.6)\qquad (I_{pq}(f),\ I_{rs}(g))=0\quad\text{if}\quad (p,\ q)\neq(r,\ s).$$

Proof. These properties were verified in $S_{pq}$ in the preceding §. To prove them in $L^2_{pq}$ we need only remark the continuity of the inner product and the following inequalities:

$$(7.2)\qquad ||\tilde{f}_n-\tilde{f}||\leqq||f_n-f||,$$

$$(7.3)\qquad ||(\alpha f_n+\beta g_n)-(\alpha f+\beta g)||\leqq|\alpha|\,||f_n-f||+|\beta|\,||g_n-g||.$$

We shall call $I_{pq}(f)$ the complex multiple Wiener integral of $f$ and denote it often by the following integral form:

$$(7.4)\qquad \int\cdots\int f(t_1,\ \ldots,\ t_p;\ s_1,\ \cdots,\ s_q)dM(t_1)\cdots dM(t_p)\overline{dM(s_1)}\ldots\overline{dM(s_q)}.$$

**§8. Transformation of Measure.** Let $M=(M(E),\ E\in B^*_T)$ be any complex normal random measure on $T(B_T,\ m)$ and $\lambda(t)$ be any complex-valued measurable function defined on $T(B_T,\ m)$ satisfying

$$(8.1)\qquad |\lambda(t)|\equiv 1.$$

If we define $M_\lambda(E)$ by

$$(8.2)\qquad M_\lambda(E)\equiv\int\lambda(t)\chi_E(t)dM(t),$$

$\chi_E(t)$ being the characteristic function of the set $E\in B^*_T$, then $M_\lambda=(M_\lambda(E),\ E\in B^*_T)$ will be also a complex normal random measure on $T(B_T,\ m)$. Then we obtain immediately

**Theorem 8.**

$$(8.3)\qquad \int\cdots\int f(t_1,\ \ldots,\ t_p;\ s_1,\ \ldots,\ s_q)dM_\lambda(t_1)\cdots dM_\lambda(t_p)\overline{dM_\lambda(s_1)}\cdots\overline{dM_\lambda(s_q)}$$
$$=\int\cdots\int f(t_1,\ \ldots,\ t_p;\ s_1,\ \ldots,\ s_q)\lambda(t_1)\cdots\lambda(t_p)\overline{\lambda(s_1)}\cdots\overline{\lambda(s_q)}$$
$$\times dM(t_1)\cdots dM(t_p)\overline{dM(s_1)}\cdots\overline{dM(s_q)}.$$

## Chapter III. Properties of Complex Multiple Wiener Integral.

**§9. Recurrence Formulae.** For $f \in L^2{}_{pq}$ and $g \in L^2(T)$ we shall define the following four functions:

$$
(9.1)\quad \begin{cases}
f\cdot g(t_1, \ldots, t_{p+1}; s_1, \ldots, s_q) = f(t_1, \ldots, t_p; s_1, \ldots, s_q)g(t_{p+1}), \\
f\bar{\cdot} g(t_1, \ldots, t_p; s_1, \ldots, s_{q+1}) = f(t_1, \ldots, t_p; s_1, \ldots, s_q)g(s_{q+1}), \\
f\circ g(t_1, \ldots, t_{p-1}; s_1, \ldots, s_q) \\
\quad = \sum_{k=1}^{p}\int f(t_1, \ldots, t_{k-1}, t, t_k, \ldots, t_{p-1}; s_1, \ldots, s_q)g(t)dm(t), \\
f\bar{\circ} g(t_1, \ldots t_p, s_1, \ldots, s_{q-1}) \\
\quad = \sum_{k=1}^{q}\int f(t_1, \ldots, t_p; s_1, \ldots, s_{k-1}, s, s_k, \ldots, s_{q-1})g(s)dm(s).
\end{cases}
$$

We shall easily show

$$
(9.2)\quad \begin{cases}
f\cdot g \in L^2{}_{p+1,q}, \quad \|f\cdot g\| = \|f\|\cdot\|g\|, \\
f\bar{\cdot} g \in L^2{}_{p,q+1}, \quad \|f\bar{\cdot} g\| = \|f\|\cdot\|g\|, \\
f\circ g \in L^2{}_{p-1,q}, \quad \|f\circ g\| \leqslant p\|f\|\cdot\|g\|, \\
f\bar{\circ} g \in L^2{}_{p,q-1}, \quad \|f\bar{\circ} g\| \leqslant q\|f\|\cdot\|g\|.
\end{cases}
$$

Any element of $L^2(T)$ may be considered to belong to $L^2{}_{10}$ as well as to $L^2{}_{01}$. Now we shall establish recurrence formulae concerning complex multiple Wiener integral.

THEOREM 9. *For any continuous complex normal random measure we have*

(R.1) $\quad I_{pq}(f)I_{10}(g) = I_{p+1,q}(f\cdot g) + I_{p,q-1}(f\bar{\circ} g), \quad p \geqslant 0, \; q \geqslant 1,$

(R.1′) $\quad I_{p0}(f)I_{10}(g) = I_{p+1,0}(f\cdot g)$

(R.2) $\quad I_{pq}(f)I_{01}(g) = I_{p,q+1}(f\bar{\cdot} g) + I_{p-1,q}(f\circ g), \quad p \geqslant 1, \; q \geqslant 0,$

(R.2′) $\quad I_{0q}(f)I_{01}(g) = I_{0,q+1}(f\bar{\cdot} g)$

PROOF. We shall show only the first formula (R.1) since the others may be proved in the same way. We shall first discuss the following special cases.

*Case* 1. $f(t_1, \ldots, t_p; s_1, \ldots, s_q)$ and $g(t)$ are the characteristic functions of $E_1\times \ldots \times E_p \times F_1 \times \ldots \times F_q$ and $G$ respectively, where $G, E_1, \ldots, E_p, F_1, \ldots, F_q$ are mutually disjoint sets $\in \boldsymbol{B}^*_T$.

In this case both sides of (R.1) are equal to zero.

*Case* 2. $f$ and $g$ are the same as in Case 1 but $E_1, \ldots, E_p, F_1, \ldots, F_q$ are mutually disjoint sets $\in \boldsymbol{B}^*_T$ and $G$ is coincident with some of $E_i$, for example $E_p$.

In this case we get

$$(9.3)\quad f\bar{\circ} g \equiv 0 \quad \text{and so} \quad I_{p,q-1}(f\bar{\circ} g) = 0.$$

On the other hand we have

$$I_{pq}(f)I_{10}(g) = M(E_1)\cdots M(E_{p-1})M(G)^2\overline{M(F_1)}\cdots\overline{M(F_q)}.$$

By the assumption of continuity $G$ may be decomposed as follows:

$$(9.4)\quad G = G_{n1} + G_{n2} + \ldots + G_{n,\nu(n)}, \quad m(G_{ni}) < 1/n, \quad i = 1, 2, \ldots, \nu(n), \quad (n = 1, 2, \ldots).$$

Therefore we have

$$I_{pq}(f)I_{10}(g) = \sum_{i\neq j} M(E_1)\cdots M(E_{p-1})M(G_{ni})M(G_{nj})\overline{M(F_1)}\cdots\overline{M(F_q)}$$

$$+M(E_1)\cdots M(E_{p-1})\overline{M(F_1)}\cdots\overline{M(F_q)}\sum_i M(G_{ni})^2$$
$$=A_n+B_n.$$

Let $h_n(t_1, \ldots, t_{p+1}; s_1, \ldots, s_q)$ be the characteristic function of the set:

$$\sum_{i\neq j} E_1\times\cdots\times E_{p-1}\times G_{ni}\times G_{nj}\times F_1\times\cdots F_q.$$

Then we have

$$\|A_n-I_{p+1,q}(f\cdot g)\|^2=\|h_n-f\cdot g\|^2$$
$$=\sum_{i=1}^{\nu(n)} m(E_1)\cdots m(E_{p-1})m(G_{ni})^2 m(F_1)\cdots m(F_q)$$
$$<\frac{1}{n}m(E_1)\cdots m(E_{p-1})m(G)m(F_1)\cdots m(F_q)$$

and so

$$A_n\to I_{p+1,q}(f\cdot g),$$

while $B_n\to 0$ by Theorem 4. Thus we get

(9.5) $$I_{pq}(f)I_{10}(g)=I_{p+1,q}(f\cdot g),$$

which, combined with (9,3), proves (R.1) in Case 2,

*Case* 3. $f$ and $g$ are the same as in Case 1, but $E_1, \ldots, E_p$, $F_1, \ldots, F_q$ are mutually disjoint sets $\in \boldsymbol{B}^*_T$ and $G$ is coincident with some of $F_j$, for example $F_q$.

By making use of the decompositions in Case 2 we obtain

$$I_{pq}(f)I_{10}(g)=\sum_{i\neq j}M(E_1)\cdots M(E_p)\overline{M(F_1)}\cdots\overline{M(F_{q-1})M(G_{ni})}M(G_{nj})$$
$$+M(E_1)\cdots M(E_p)\overline{M(F_1)}\cdots\overline{M(F_{q-1})}\sum_i|M(G_{ni})|^2,$$

which, as $n\to\infty$, tends to.

$$I_{p+1,q}(f\cdot g)+M(E_1)\cdots M(E_p)\overline{M(F_1)}\cdots\overline{M(F_{q-1})}m(G)$$
$$=I_{p+1,q}(f\cdot g)+I_{pq}(f\circ g)$$

by virtue of Theorem 4.

In considering that both sides of (R.1) are bilinear in $f$, $g$ we shall easily deduce the case $f\in S_{pq}$, $g\in S_{01}$ from the above three cases and further show the most general case $f\in L^2{}_{pq}$, $g\in L^2{}_{10}$ by taking the limit.

**§ 10. The Completeness of Complete Multiple Wiener Integral.** Let $Z=(Z_\lambda(\omega), \lambda\in\Lambda)$ be any system of complex random variables. A complex-valued Baire function of $Z$ is defined in different ways which are equivalent to each other. We shall here adopt the following definition. A complex-random variable $Z(\omega)$ is defined to be a *Baire function* of $\boldsymbol{Z}$, if it belongs to the least class $\mathfrak{B}$ that satifies the following two conditions:

($\mathfrak{B}$.1) When $f$ is a complex-valued Baire function of $n$ complex variables in the usual sense, $f(Z_{\lambda(1)}(\omega), \ldots, Z_{(n)}(\omega))$ belongs to $\mathfrak{B}$.

($\mathfrak{B}$.2) If $\{Z^{(n)}(\omega)\}$ is a sequence in $\mathfrak{B}$ which is convergent for every $\omega$, then the limit belongs to $\mathfrak{B}$.

We shall denote with $L^2(\boldsymbol{Z})$ the totality of the Baire functions of $\boldsymbol{Z}$ belonging to $L^2(\Omega, P)$, We shall easily obtain

LEMMA 10.1. *For any* $Z(\omega)\in L^2(\boldsymbol{Z})$ *and any* $\varepsilon>0$ *there exist* $\lambda(1), \ldots, \lambda(n)\in\Lambda$ *and a Baire function of* $n$ *variables* $f$ *such that*

(10.1) $$\|Z-f(Z_{\lambda(1)}, \ldots, Z_{\lambda(n)})\|<\varepsilon.$$

**Proof.** Let $\varphi_G(x)(0\leqslant x<\infty)$ be defined as

(10.2) $$\varphi_G(x)=1\,(0\leqslant x\leqslant G),\ =0\,(x>G).$$

we shall denote with $\mathfrak{B}$ the totality of the complex random variables $Z$ for which $Z_G=\varphi_G(|Z|)\cdot Z$ may have the property stated in this lemma for every $G>0$. Then $\mathfrak{B}$ will satisfy ($\mathfrak{B}$.1) and ($\mathfrak{B}$.2) and so we see that $\mathfrak{B}\supseteq L^2(\boldsymbol{Z})$, namely that $Z_G\in\mathfrak{B}$ for $Z\in L^2(\boldsymbol{Z})$. But

$$\|Z-Z_G\|\to 0 \quad\text{as}\quad G\to\infty$$

and so we obtain $Z\in\mathfrak{B}$.

Now let $\boldsymbol{M}=(M(E),\ E\in \boldsymbol{B}^*_T)$ be any continuous normal random measure on $T(\boldsymbol{B}_T,\ m)$. Then we obtain the following lemma from the above lemma:

**Lemma 10.2.** *For any $Z\in L^2(\boldsymbol{M})$ and any $\varepsilon>0$ there exists a disjoint system $E_1,\ \dots,\ E_n\in \boldsymbol{B}^*_T$ and a complex-valued Baire function of $n$ complex variables $f$ such that*

(10.3) $$\|Z-f(M(E_1),\ \dots,\ M(E_n))\|<\varepsilon.$$

In making use of the completeness of Hermite polynomials we obtain

**Lemma 10.3.** *If $f(x_1,\ \dots,\ x_m)$ be a complex-valued Baire function of $m$ real variables for which we have*

(10.4) $$\int\cdots\int|f(x_1,\ \dots,\ x_m)|^2 e(-\textstyle\sum x_i^2)dx_1\cdots dx_m<\infty,$$

*then there exists a polynomial of $m$ variables $P(x_1,\ \dots,\ x_m)$ for any $\varepsilon>0$, such that*

(10.5) $$\int\cdots\int|f(x_1,\ \dots,\ x_m)-P(x_1,\ \dots,\ x_m)|^2 e(-\textstyle\sum x_i^2)dx_1\cdots dx_m<\varepsilon.$$

From the above two lemmas follows

**Lemma 10.4.** *For any $Z\in L^2(\boldsymbol{M})$ and any $\varepsilon>0$ there exists a disjoint system $E_1,\ \dots,\ E_n\in B^*_T$ and a polynomial of $2n$ variables $Q$ such that*

(10.6) $$\|Z-Q(M(E_1),\ \dots,\ M(E_n),\ \overline{M(E_1)},\ \dots,\ \overline{M(E_n)})\|<\varepsilon.$$

**Proof.** By Lemma 10.2 we need only to discuss the case in which $Z$ is of the form

$$Z=f(M(E_1),\ \dots,\ M(E_n)),$$

where $f$ is a complex-valued Baire function of $n$ complex variables and $E_1,\ \dots,\ E_n$ are disjoint sets $\in B^*_T$. By denoting the real and imaginary parts of $M(E_i)/m(E_i)^{\frac{1}{2}}$ with $X_i$ and $Y_i$ respectively, we may put

$$Z=g(X_1,\ Y_1,\ \dots,\ X_n,\ Y_n),$$

$g$ being a complex-valued Baire function of $2n$ real variables.

Since $\{M(E)\}$ is a normal random measure, $X_1,\ Y_1,\ \dots,\ X_n,\ Y_n$ are indepedent, each having the probability law $1/\sqrt{2\pi}\,e(-\xi^2)d\xi$. By Lemma 10.3 we can find a polynomial $P(x,\ y,\ \dots,\ x,\ y)$ such that

$$\int\cdots\int|g(x_1,\ y_1,\ \dots,\ x_n,\ y_n)-P(x_1,\ y_1,\ \dots,\ x_n,\ y_n)|^2 \times e(-\textstyle\sum x_i^2-\sum y_i^2)dx_1dy_1\cdots dx_ndy_n<\varepsilon^2,$$

namely that

$$\|Z-P(X_1,\ Y_1,\ \dots,\ X_n,\ Y_n)\|^2<\varepsilon^2.$$

If we define Q by

$$Q(z_1, \ldots, z_n, \bar{z}_1, \ldots, \bar{z}_n) \\ = P((z_1+\bar{z}_1)/2, (z_1-\bar{z}_1)/2i, \ldots, (z_n+\bar{z}_n)/2, (z_n-\bar{z}_n)/2i),$$

then it follows from the above arguments that $Q$ satisfies (10.6).

THEOREM 10. (The Completeness of Multiple Wiener Integral) *The system:*

$$\mathfrak{S}=\{I_{pq}(f_{pq});\ f_{pq}\in L_{pq},\ p,\ q=0,\ 1,\ \ldots\}$$

*is complete in* $L^2(M)$.

PROOF. By the above Lemma 10.4 it is sufficient to show that

$$Z \equiv M(E_1)^{p_1} \ldots M(E_n)^{p_n} \overline{M(E_1)}^{q_1} \cdots \overline{M(E_n)}^{q_n} \tag{10.7}$$

is expressible of the form:

$$Z=\sum_{\rho=0}^{p}\sum_{\sigma=0}^{q} I_{\rho\sigma}(f_{\rho\sigma}),\quad p=\sum p_\nu,\quad q=\sum q_\nu. \tag{10.8}$$

We shall make use of mathematical induction with regard to $p+q$. In case $p+q=0$ our assertion is trivially true. Suppose that it is true for $p+q=r$. If $p+q=r+1$, some of $\{p_i\}$ or $\{q_j\}$ is positive.

*Case* 1. Some of $\{p_i\}>0$. To fix the idea we shall assume that $p_1>0$. By the assumption of induction we have

$$Z' \equiv M(E_1)^{p_1-1} \cdots M(E_n)^{p_n} \overline{M(E_1)}^{q_1} \cdots \overline{M(E_n)}^{q_n} = \sum_{\rho=0}^{p-1}\sum_{\sigma=0}^{q} I_{\rho\sigma}(f_{\rho\sigma}),$$

and so, denoting with $\chi(t)$ the characteristic function of $E_1$ we get, by Theorem 9. (R.1),

$$Z=Z'\cdot I_{10}(\chi(t))=\sum_{\rho=0}^{p}\sum_{\sigma=0}^{q} I_{p+1,\sigma}(f_{\rho\sigma}\cdot\chi)+\sum_{\rho=0}^{p}\sum_{\sigma=1}^{q} I_{\rho,\sigma-1}(f_{\rho\sigma}\bar{\circ}\chi),$$

which proves that $Z$ is also of the above form (10.8).

*Case* 2. some of $\{q_i\}>0$. We can perform the proof in the same way as above by making use of the recurrence formula (R.2) instead of (R.1).

## § 11. Expansion Theorem by Means of Complex Multiple Wiener Integral.

For any system:

$$\mathfrak{F}=\{f_{pq}(\in L_{pq});\ p,\ q=0,\ 1,\ 2,\ \ldots\} \tag{11.1}$$

we shall consider a series

$$\Sigma_{\mathfrak{F}}=\sum_{pq} I_{pq}(f_{pq}). \tag{11.2}$$

Since the terms in the summation are mutually orthogonal by Theorem 7 (I.6) it is necessary and sufficient for $\Sigma_{\mathfrak{F}}$ to be convergent in the norm that

$$\Sigma_{pq}\|I_{pq}(f_{pq})\|^2<\infty, \tag{11.3}$$

which is equivalent to

$$\Sigma_{pq}\,\underline{|p}\,\underline{|q}\,\|f_{pq}\|^2<\infty \tag{11.4}$$

by Theorem 7 (I.5).

If (11.4) is satisfied, $\Sigma_{\mathfrak{F}}$ will clearly belong to $L^2(M)$ since each term of $\Sigma_{\mathfrak{F}}$ does so. We shall now establish the converse of this fact.

THEOREM 11. (Expansion Theorem) *Let* $M$ *be a continuous normal random measure. Every element of* $L^2(M)$ *is expressible of the form* (11.2). *Further we may require that each* $f_{pq}$ *appearing here is symmetric in the sense that*

$$f_{pq}=\tilde{f}_{pq}. \tag{11.5}$$

*Under this restriction the expression* (*11.2*) *i.e. the system* $\mathfrak{F}$ *is uniquely determined.*

Proof. Let $\tilde{L}^2{}_{pq}$ be the totality of the symmetric functions in $L^2{}_{pq}$. Then $\tilde{L}^2{}_{pq}$ is clearly a closed linear manifold $L^2{}_{pq}$ and so a Hilbert space. Let $L^2{}_{pq}(M)$ be the totality of the form $I_{pq}(f_{pq})$, $f_{pq}$ running over $L^2{}_{pq}$ or (by virtue of Theorem 7 (I.1)) equivalently over $\tilde{L}^2{}_{pq}$. $L^2{}_{pq}(M)$ is also a closed linear manifold which is unitary equivalent to $\tilde{L}^2{}_{pq}$ by the correspondence:

$$\tilde{L}^2{}_{pq} \ni \sqrt{\underline{|p}\,\underline{|q}}\,\tilde{f}_{pq} \rightarrow I_{pq}(f_{pq}) \in L^2{}_{pq}(M). \tag{11.6}$$

Let $Z$ be any element of $L^2(M)$ and the projection of $Z$ onto $L^2{}_{pq}(M)$ be $Z_{pq}=I_{pq}(f_{pq})$. Then $\{Z_{pq}\}$ are mutually orthogonal and so we have

$$\sum_{pq}\|Z_{pq}\|^2 \leqslant \|Z\|^2 \quad \text{i.e.} \quad \sum_{pq}\underline{|p}\,\underline{|q}\,\|f_{pq}\|^2 \leqslant \|Z\|^2 < \infty$$

by Bessel's inequality. Therefore we see that

$$Z' = \sum_{pq} Z_{pq}, \quad Z'' = Z - Z'$$

is well defined and that $Z''$ is orthogonal to the system:

$$\mathfrak{S} = \{I_{pq}(f_{pq});\ f_{pq} \in L^2{}_{pq},\ p,\ q=0,\ 1,\ 2,\ \ldots\}.$$

From the completeness of $\mathfrak{S}$ (Theorem 10) follows $Z''=0$ and so

$$Z = \sum_{pq} Z_{pq} = \sum_{pq} I_{pq}(f_{pq}) = \sum_{pq} I_{pq}(\tilde{f}_{pq}).$$

Thus the possibility of the expression was proved.

$$\sum_{pq} I_{pq}(f_{pq}) = \sum_{pq} I_{pq}(g_{pq}),\quad f_{pq},\ g_{pq} \in \tilde{L}^2{}_{pq}.$$

Since $L^2{}_{pq}(M)$, $p$, $q=0, 1, 2, \ldots$ are mutually orthogonal, we see that

$$I_{pq}(f_{pq}) = I_{pq}(g_{pq}),$$

from which we deduce $f_{pq}=g_{pq}$, remembering the unitary equivalence (11.6) between $L^2{}_{pq}(M)$ and $\tilde{L}^2{}_{pq}$.

## § 12. Definition of Hermite Polynomials of Complex Variables.

In M. W.I. we have shown a close relation between real multiple Wiener integral and Hermite polynomials of real variables. We shall here define Hermite polynomials of complex variables to which complex multiple Wiener integrals stand in the similar relation as is shown in the next §.

Definition. Let $z$ be a complex variable and $\bar{z}$ denote its conjugate. For $p$, $q=0, 1, 2, \ldots$ we define

$$H_{pq}(z,\ \bar{z}) = \sum_{n=0}^{p\wedge q}(-1)^n \frac{\underline{|p}\,\underline{|q}}{\underline{|n}\,\underline{|p-n}\,\underline{|q-n}}\, z^{p-n}\bar{z}^{q-n},\quad p\wedge q = \min(p,\ q). \tag{12.1}$$

For the unity of the formulations we shall define trivially

$$H_{pq}(z,\ \bar{z}) \equiv 0 \quad \text{if } p \text{ or } q<0. \tag{12.2}$$

The known identities concerning Hermite polynomials of real variables are extended to our complex case as follows.

Theorem 12.

$$\text{(A)}\qquad e(-t\bar{t}+t\bar{z}+\bar{t}z) = \sum_{p,q=0}^{\infty} \frac{1}{\underline{|p}\,\underline{|q}} H_{pq}(z,\ \bar{z})\bar{t}^p t^q,$$

(B) $$H_{pq}(z, \bar{z})=e(z\bar{z})(-1)^{p+q}\frac{\partial^{p+q}}{\partial\bar{z}^p\partial z^q}e(-z\bar{z}), \ (p, q\geqq 0),$$

(C) $$\begin{cases} H_{p+1,q}(z, \bar{z})-H_{pq}(z, \bar{z})z+qH_{p,q-1}(z, \bar{z})=0 \\ H_{p,q+1}(z, \bar{z})-H_{pq}(z, \bar{z})\bar{z}+pH_{p-1,q}(z, \bar{z})=0, \end{cases}$$

(D) $$\begin{cases} \dfrac{\partial}{\partial z}H_{pq}(z, \bar{z})=pH_{p-1,q}(z, \bar{z}) \\ \dfrac{\partial}{\partial \bar{z}}H_{pq}(z, \bar{z})=pH_{p,q-1}(z, \bar{z}) \end{cases}$$

(E) $$\begin{cases} \dfrac{\partial^2}{\partial z\partial \bar{z}}H_{pq}(z, \bar{z})-\bar{z}\dfrac{\partial}{\partial \bar{z}}H_{pq}(z, \bar{z})+qH_{pq}(z, \bar{z})=0 \\ \dfrac{\partial^2}{\partial \bar{z}\partial z}H_{pq}(z, \bar{z})-z\dfrac{\partial}{\partial z}H_{pq}(z, \bar{z})+pH_{pq}(z, \bar{z})=0. \end{cases}$$

PROOF.

(A) $$e(-t\bar{t}+t\bar{z}+\bar{t}z)=\sum_{n=0}^{\infty}\frac{(-1)^n}{\lfloor n}t^n\bar{t}^n\sum_{q=0}^{\infty}\frac{t^q\bar{z}^q}{\lfloor q}\sum_{p=0}^{\infty}\frac{\bar{t}^pz^p}{\lfloor p}$$
$$=\sum_{p,q=0}^{\infty}\frac{\bar{t}^pt^q}{\lfloor p\lfloor q}\sum_{n=0}^{p\wedge q}(-1)^n\frac{\lfloor p\lfloor q}{\lfloor n\lfloor p-n\lfloor q-n}z^{p-n}\bar{z}^{q-n}.$$

(B) In remarking that
$$e(-t\bar{t}+t\bar{z}+\bar{t}z)=e(-z\bar{z})e\{-(t-z)(\bar{t}-\bar{z})\},$$
we have
$$\frac{\partial}{\partial t}e(-t\bar{t}+tz+\bar{t}z)=e(-z\bar{z})\frac{\partial}{\partial t}e\{-(t-z)(\bar{t}-\bar{z})\}$$
$$=-e(-z\bar{z})\frac{\partial}{\partial \bar{z}}e\{-(t-z)(\bar{t}-\bar{z})\}.$$
By repeating similar computations we get
$$\frac{\partial^{p+q}}{\partial t^p\partial \bar{t}^q}e(-t\bar{t}+t\bar{z}+\bar{t}z)=(-1)^{p+q}e(-z\bar{z})\frac{\partial^{p+q}}{\partial \bar{z}^p\partial z^q}e\{-(t-z)(\bar{t}-\bar{z})\},$$
from which we deduce (B) by putting $t=\bar{t}=0$.

(C) We have
$$\frac{\partial}{\partial \bar{t}}e(-t\bar{t}+t\bar{z}+\bar{t}z)=e(-t\bar{t}+t\bar{z}+\bar{t}z)(z-t).$$
By comparing the coefficients of $\bar{t}^{p-1}t^q$ in both sides we obtain the first identity. The second will be reduced by the same way.

(D) By differentiating both sides of (A) partially with regards to $z$ or $\bar{z}$ we obtain these identities at once.

(E) will be reduced at once from (C) and (D).

## § 13. A Relation Between Complex Multiple Wiener Integral and Hermite Polynomials of Complex Variables.

THEOREM 13.1. Let $M=(M(E),\ E\in E^*_T)$ be any continuous complex normal random measure on $T(B_T,\ m)$ and $f(t)$ be any element of $L^2(T)$ with the norm 1. Then we have

(13.1) $$\int\cdots\int f(t_1)\cdots f(t_p)\overline{f(s_1)}\cdots\overline{f(s_q)}dM(t_1)\cdots dM(t_p)\overline{dM(s_1)}\cdots\overline{dM(s_q)}$$
$$=H_{pq}(Z,\ \bar{Z}),$$

where

$$Z=\int f(t)dM(t). \tag{13.2}$$

PROOF. We shall denote the left side of (13.1) by $K_{pq}$. By the recurrence formulae of § 9 we shall easily verify

$$\begin{cases} K_{pq}Z=K_{p+1,q}+qK_{p,q-1} \ (p\geqq 0,\ q\geqq 1), \\ K_{pq}Z=K_{p,q+1}+pK_{p-1,q} \ (p\geqq 1,\ q\geqq 0); \end{cases} \tag{13.3}$$

if we define $K_{pq}=0$ in case $p$ or $q<0$, the above identities will be also true for all integral values of $p$, $q$. In comparing these identities and (C) in Theorem 12 we shall easily see

$$K_{pq}=H_{pq}(Z, \bar{Z}) \tag{13.4}$$

by means of mathematical induction with respect to $p+q$, since (13.4) is trivially true for $p+q=0$.

By the same idea we shall generalize this theorem as follows.

THEOREM 13.2. Let $M$ be as above and $\{f_1(t), \ldots, f_n(t)\}$ be any orthogonal system in $L^2(T)$. Then we have

$$\int\cdots\int f_{\alpha_1}(t_1)\cdots f_{\alpha_p}(t_p)\overline{f_{\beta_1}(s_1)}\cdots\overline{f_{\beta_q}(s_q)}dM(t_1)\cdots dM(t_p)\overline{dM(s_1)}\cdots\overline{dM(s_q)} \tag{13.4}$$

$$=\Pi_{\nu=1}^{n}H_{p_\nu q_\nu}(Z_\nu, \bar{Z}_\nu),$$

where $p_\nu$ and $q_\nu$ are the number of $\nu$ appearing in $\{\alpha_i\}$ and $\{\beta_j\}$ respectively ($\nu=1, 2, \ldots, n$) and

$$Z_\nu=\int f_\nu(t)dM(t),\ \nu=1, 2, \ldots, n. \tag{13.5}$$

**§ 14. Generalized Cameron-Martin Development.** R. H. Cameron and W. T. Martin have shown the orthogonal development of functionals of a Wiener process by means of Hermite polynomials of real variables. We have extented it to the case of functionals of real normal random measure in M.W.I.. We shall here present the same property over a complex normal random measure by expressing the expansion of Theorem 11 by means of our Hermite polynomials defined in § 2.

Let $T(B_T, m)$ be any measure space and $\{f_\alpha(t), \alpha\in A\}$ be a complete orthonormal system in $L^2(T)$. Then we shall easily see that the system:

$$\begin{cases} f_{(\alpha;\beta)}(t_1, \ldots, t_p; s_1, \ldots, s_q)=f_{\alpha_1}(t_1)\cdots f_{\alpha_p}(t_p)f_{\beta_1}(s_1)\cdots f_{\beta_q}(s_q), \\ (\alpha;\beta)=(\alpha_1, \ldots, \alpha_p; \beta_1, \ldots, \beta_q)\in A^{p+q} \end{cases} \tag{14.1}$$

is a complete orthonormal system in $L^2{}_{pq}$.

It is necessary and sufficient for $\tilde{f}_{(\alpha;\beta)}=\tilde{f}_{(\gamma;\delta)}$ that $(\alpha)\equiv(\alpha_1, \ldots, \alpha_p)$ and $(\beta)\equiv(\beta_1, \ldots, \beta_q)$ are respectively identical with $(\gamma)\equiv(\gamma_1, \ldots, \gamma_p)$ and $(\delta)\equiv(\delta_1 \ldots, \delta_q)$ as the totality including multiplicities. Further we easily see that the totality of different $\tilde{f}_{(\alpha;\beta)}$ is a complete orthonormal system in $\tilde{L}^2{}_{pq}$.

Let $M$ be a continuous normal random measure on $T(B_T, m)$. By the unitary equivalence (11.6) $\sqrt{\underline{p}\,|\underline{q}}\,\tilde{f}_{(\alpha;\beta)}$ corresponds to $I_{pq}(\tilde{f}_{(\alpha;\beta)})=I_{pq}(f_{(\alpha;\beta)})$.

Let $p(\lambda)$ and $q(\lambda)$ be the numbers of $\lambda$ appearing in $(\alpha)$ and $(\beta)$ respectively. Then we have, by Theorem 13.2,

$$I_{pq}(f_{(\alpha;\beta)}) = \prod_{\lambda\in A} H_{p(\lambda)q(\lambda)}(Z_\lambda, \bar{Z}_\lambda);$$

this is a finite product in essential and well defined, since we have $\sum_\lambda p(\lambda)=p$ and $\sum_\lambda q(\lambda)=q$.

Since both completeness and orthonormality are unitary invariants, the system:

$$\prod_{\lambda\in A} H_{p(\lambda)q(\lambda)}(Z_\lambda, \bar{Z}_\lambda), \quad \sum_\lambda p(\lambda)=p, \quad \sum_\lambda q(\lambda)=q, \tag{14.2}$$

constitutes a complete orthonormal system in $L^2_{pq}(\boldsymbol{M})$. $L^2(\boldsymbol{M})$ being a direct sum of $L^2_{pq}(\boldsymbol{M})$, $p$, $q=0, 1, 2, \ldots$, by virtue of Theorem 11, we obtain

THEOREM 14. (Generalized Cameron-Martin Development)

$$\prod_{\lambda\in A} H_{p(\lambda)q(\lambda)}(Z_\lambda, \bar{Z}_\lambda), \quad \sum_\lambda p(\lambda)<\infty, \quad \sum_\lambda q(\lambda)<\infty, \tag{14.3}$$

*constitutes a complete orthonormal system in $L^2(M)$ and so any element of $L^2(M)$ is expressible in the Fourier series by means of (14.3).*

**§ 15. Completeness of Hermite Polynomials of Complex Variables.** As is well-known, Hermite polynomials of real variables constitute a complete orthonormal system in $L^2(R^1, e(-x^2)dx)$. We shall here show that our Hermite polynomials have the same property on the complex plane. This may be established by means of the identities in Theorem 12 in the same way as in the real case, but we shall here prove it by making use of Theorem 13.2.

THEOREM 15. *Let $K$ denote a complex plane and $N$ be a measure defined by*

$$dN(z)=\pi^{-1}e(-x^2-y^2)dxdy, \quad z=x+iy. \tag{15.1}$$

*Then*

$$H_{pq}(z, \bar{z}), \quad p, \quad q=0, 1, 2, \ldots, \tag{15.2}$$

*costitute a complete orthonormal system in $L^2(K, N)$.*

PROOF. Let $T(\boldsymbol{B}_T, m)$ be a measure space associated with a continuous measure $m$, for example

$$T=R^1, \quad m=\text{ordinary Lebesgue measure.}$$

Then there exists a complex normal random measure $\boldsymbol{M}$ on $T(\boldsymbol{B}_T, m)$ by virtue of Theorem 3.1. Let $\{f_\alpha(t),\ \alpha\in A\}$ be a complete orthonormal system in $L^2(T)$. Then we see by § 2, that

$$Z_\alpha=\int f_\alpha(t)dM(t), \quad \alpha\in A, \tag{15.3}$$

constitute a complex normal system and so that $Z_\alpha$, $\alpha\in A$, are independent, each being subject to the probability law $N$, since $(Z_\alpha, Z_\beta)=(f_\alpha, f_\beta)=\delta_{\alpha\beta}$.

We fix an element $f(t)$ in $\{f_\alpha(t)\}$ and denote the corresponding $Z_\alpha$ with $Z$. Let $L^2(Z)$ denote the totality of Baire functions of $Z$ in $L^2(\Omega)$. $L^2(Z)$ is a closed linear manifold in $L^2(\boldsymbol{M})$. Since $Z$ is subject to $N$, we have

$$\|\varphi(Z)\|^2=\iint_K |\varphi(z)|^2\pi^{-1}e(-x^2-y^2)dxdy, \quad z=x+iy,$$

and so we see that

$$L^2(K, N)\ni\varphi(z)\to\varphi(Z)\in L^2(Z) \tag{15.4}$$

determines a unitary equivalence. Therefore in order to prove our theorem, it is sufficient to show that $\{H_{pq}(Z, \bar{Z})\}$ constitute a complete orthogonal system in $L^2(Z)$. It is clear by Theorem 14 that they are mutually orthogonal.

To show the completeness we need only prove that

$$\varphi(Z) \perp H_{pq}(Z, \bar{Z}) \tag{15.5}$$

implies

$$\varphi(Z)=0. \tag{15.6}$$

Since $\varphi(Z) \in L^2(Z) \subseteq L^2(\boldsymbol{M})$ and the system:

$$H_{p_1q_1}(Z_{\alpha_1}, \bar{Z}_{\alpha_1}) \cdots H_{p_nq_n}(Z_{\alpha_n}, \bar{Z}_{\alpha_n}) \tag{15.7}$$

is complete in $L^2(\boldsymbol{M})$, it remains only to show that $\varphi(Z)$ is orthogonal to every element of this system. We may assume that

$$Z_{\alpha_i} \neq Z \quad \text{and} \quad p_{\alpha_i}+q_{\alpha_i}>0 \tag{15.8}$$

for some $i$, for example $i=1$, because, in the contrary case, our assertion will be true by (15.5). By the independence of $\{Z_\alpha\}$ we have

$$\mathcal{E}[H_{p_1q_1}(Z_{\alpha_1}, \bar{Z}_{\alpha_1}) \cdots H_{p_nq_n}(Z_{\alpha_n}, \overline{\bar{Z}_{\alpha_n})\varphi(Z)}]$$

$$= \mathcal{E}H_{p_1q_1}(Z_{\alpha_1}, \bar{Z}_{\alpha_1}) \mathcal{E}[H_{p_2q_2}(Z_{\alpha_2}, \bar{Z}_{\alpha_2}) \cdots H_{p_nq_n}(Z_{\alpha_n}, \overline{\bar{Z}_{\alpha_n})\varphi(Z)}].$$

Since $p_{\alpha_1}+q_{\alpha_1}>0$, $H_{p_1q_1}(\bar{Z}_{\alpha_1}, Z_{\alpha_1}) (\in L^2{}_{p_1q_1}(\boldsymbol{M}))$ is orthogonal to $L^2{}_{00}(\boldsymbol{M})(\equiv K)$, that is $\mathcal{E}H_{p_1q_1}(Z_{\alpha_1}, \bar{Z}_{\alpha_1})=0$. Thus we see that (15.7) is orthogonal to $\varphi(Z)$.

## IV. Spectral Structure and Ergodicity of Normal Screw Line.

§ **16. Normal Screw Line.** A complex-valued stochastic process $\{Z(t)\}$ is called a *normal screw line* if it is a complex normal system and satisfies

$$\mathcal{E}Z(t)=0, \tag{16.1}$$

$$\mathcal{E}[(Z(t+a)-Z(t+b))\overline{(Z(t+c)-Z(t+d))}]=\mathcal{E}[(Z(a)-Z(b))\overline{(Z(c)-Z(d))}]. \tag{16.2}$$

If $Z(t)$ is a normal screw line, then $Z(t)$ is a screw line in Hilbert space $L^2(\Omega)$ in the sense of Kolmogorov [5] [6]. Therefore $Z(t)$ is written in a spectral form:

$$Z(s)-Z(t)=\int_{-\infty}^{\infty}\int_t^s e(i\lambda\tau)d\tau dM(\lambda). \tag{16.3}$$

In considering that $\boldsymbol{M}=\{M(\Lambda)\}$ is a complex normal system, we easily see that $\boldsymbol{M}$ is a complex normal random measure on $R^1(m)$, where $m(\Lambda)=\|M(\Lambda)\|^2$ satisfies

$$\int_{-\infty}^{\infty}\frac{dm(\lambda)}{1+\lambda^2}<\infty, \tag{16.4}$$

$m(\Lambda)$ is called the *spectral measure* of $Z(t)$.

$M(\Lambda)$ is uniquely determined by $Z(t)$ and expressible in the following form:

$$M(\Lambda)=\underset{n\to\infty}{\text{l.i.m.}}\sum_{k=1}^{K(n)} a_{nk}(Z(t_{nk})-Z(s_{nk})),$$

where l.i.m. means the limit in mean i.e. in the norm of $L^2(\Omega)$.

§ **17. Shift Transformation of Normal Screw Line.** Let $\{Z(t)\}$ be a complex normal system. We shall derive a random interval function from $Z(t)$ as follows:

(17.1) $$\boldsymbol{Z}\equiv\{Z(a,\ b)\equiv Z(b)-Z(a),\ -\infty<a<b<\infty\}.$$

Then we shall easily see that

(17.2) $$L^2(\boldsymbol{Z})=L^2(\boldsymbol{M}),$$

where $\boldsymbol{M}$ is the complex normal random measure appearing in the spectral expression (16.3).

We shall now define $T_t\boldsymbol{Z}=\{(T_tZ)(a,\ b),\ -\infty<a<b<\infty\}$ by

$$(T_tZ)(a,\ b)=Z(a+t,\ b+t).$$

$\boldsymbol{Z}$ being a complex normal system, it is so the case with $T_t\boldsymbol{Z}$. Therefore by virtue of Theorem 2.3 it follows from (16.1) and (16.2) that $\boldsymbol{Z}$ and $T_t\boldsymbol{Z}$ are subject to the same probability distribution of $\boldsymbol{Z}$. $T_t$ is called the ***shift transformation*** of $\boldsymbol{Z}$ by definition.

We shall extend this transformation onto $L^2(\boldsymbol{Z})$. Let $L^2_*(\boldsymbol{Z})$ is the totality of $Z$'s in $L^2(\boldsymbol{Z})$ which are expressed in the form:

(17.3) $$Z=f(Z(a_1,\ b_1),\ \ldots,\ Z(a_n,\ b_n)),$$

$f$ being a Baire function of $n$ complex variables; $L^2_*(\boldsymbol{Z})$ is clearly a linear manifold dense in $L^2(\boldsymbol{Z})$. For such $Z$ we shall define

$$T_tZ=f(Z(a_1+t,\ b_1+t),\ \ldots,\ Z(a_n+t,\ b_n+t)).$$

This $T_t$ becomes a mapping from $L^2_*(\boldsymbol{Z})$ onto itself which is isometric, since

(17.4) $$\|T_tZ\|^2=\mathcal{E}|T_tZ|^2=\mathcal{E}|Z|^2=\|Z\|^2$$

on account of the fact that $Z$ and $T_tZ$ have the same distribution. Therefore we shall further extend $T_t$ onto $L^2(\boldsymbol{Z})$ preserving the isometric property (17.4); thus $T_t$ costitute a one-parameter group of unitary operators in $L^2(\boldsymbol{Z})$ i.e. $L^2(\boldsymbol{M})$. Since the above $M(\Lambda)$ belongs to $L^2(\boldsymbol{Z})$, $T_tM(\Lambda)$ and so $T_tM$ are well defined.

**Theorem 17.**

(17.5) $$(T_tM)(\Lambda)=\int_\Lambda e(i\lambda t)dM(\lambda),$$

*or symbolically*

(17.5′) $$T_tdM(\lambda)=e(i\lambda t)dM(\lambda).$$

Proof. Since we have

(17.6) $$Z(a,\ b)=\int_{-\infty}^{\infty}\int_a^b e(i\lambda s)dsdM(\lambda),$$

we get

(17.7) $$\begin{aligned}T_tZ(a,\ b)&=\int_{-\infty}^{\infty}\int_a^b e(i\lambda(s+t))dsdM(\lambda)\\&=\int_{-\infty}^{\infty}\int_a^b e(i\lambda s)ds\cdot e(i\lambda t)dM(\lambda)\\&=\int_{-\infty}^{\infty}\int_a^b e(i\lambda s)dsdM_t(\lambda),\end{aligned}$$

where

(17.8) $$M_t(\Lambda)=\int_\Lambda e(i\lambda t)dM(\lambda).$$

Now suppose that

$$M(\Lambda)=\lim \sum_{k=1}^{K(n)} c_{nk} Z(a_{nk}, b_{nk}).$$

In remembering that $T_t \boldsymbol{Z}$ and $\boldsymbol{Z}$ have the same distribution and that the random measure $\boldsymbol{M}$ is uniquely determined by $\boldsymbol{Z}$, we see that $M_t(\Lambda)$ is also obtained by the same expression from $T_t \boldsymbol{Z}$ as follows:

$$M_t(\Lambda)=\lim \sum_{k=1}^{(K)n} c_{nk} T_t Z(a_{nk}, b_{nk}).$$

By the above two identities we obtain $T_t M(\Lambda)=M_t(\Lambda)$, which was to be proved.

§ **18. Spectral Structure of $T_t$.** We shall use the same notation as in the preceding §. We define a one-parameter group of unitary operators $\{S_t^{(p,q)}\}$ on $\tilde{L}^2{}_{pq}(R^1, m)$ by

(18.1) $$S_t^{(p,q)} f(\lambda_1, \ldots, \lambda_p; \mu_1, \ldots, \mu_q)$$
$$=e(i(\lambda_1+\cdots+\lambda_p-\mu_1-\cdots-\mu_q)t) f(\lambda_1, \ldots, \lambda_p; \mu_1, \ldots, \mu_q).$$

Let $L^2$ denote the direct sum of Hilbert spaces $\tilde{L}^2{}_{pq}(R^1, m)$, $p, q=0, 1, 2, \ldots$, and $S_t$ that of $S_t^{(p,q)}$, $p, q=0, 1, 2, \ldots$. Then we have

THEOREM 18. *If $m$ is continuous except at $\lambda=0$, then $S_t$ is unitary-equivalent with $T_t$.*

PROOF. By the assumption, $m$ and also $\boldsymbol{M}$ are continuous in the sense of § 4. Therefore we may use the results obtained concerning multiple Wiener integral. By § 11 $L^2(\boldsymbol{Z})$ i.e. $L^2(\boldsymbol{M})$ is a direct sum of $L^2{}_{pq}(\boldsymbol{M})$, $p, q=0, 1, 2, \ldots$ Making use of the definition of multiple Wiener integral, Theorem 8 and Theorem 17, we see that

(18.2) $$T_t \int\cdots\int f(\lambda_1, \ldots, \lambda_p; \mu_1, \ldots, \mu_q) dM(\lambda_1)\cdots dM(\lambda_p)\overline{dM(\mu_1)}\cdots\overline{dM(\mu_q)}$$
$$=\iint f(\lambda_1, \ldots, \lambda_p; \mu_1, \ldots, \mu_q) e(i(\lambda_1+\cdots+\lambda_p-\mu_1-\cdots-\mu_q)t)$$
$$\times dM(\lambda_1)\cdots dM(\lambda_p)\overline{dM(\mu_1)}\cdots\overline{dM(\mu_q)},$$

so that $T_t$ makes each $L^2{}_{pq}(\boldsymbol{M})$ invariant. Therefore we need only prove that the transformation $T_t$ restricted on $L^2{}_{pq}(\boldsymbol{M})$ is unitary-equivalent with $S_t^{(p,q)}$. This is evident, since the correspondence:

(18.3) $$L^2{}_{pq} \ni \sqrt{\underline{|p}\ \underline{|q}}\, f \to I_{pq}(f) \in L^2{}_{pq}(\boldsymbol{M})$$

determines a unitary equivalence, as is stated in § 11.

§ **19. Ergodicity of $T_t$.** DEFINITION. If $T_t Z=Z$ implies that $Z$ is a constant, then $T_t$ is defined to be *ergodic.*

THEOREM 19.1. *It is necessary and sufficient for $T_t$ to be ergodic that $m$ is continuous.*

PROOF OF NECESSITY. Suppose that $m$ be discontinuous at $\lambda=\lambda_0$, then $Z=|M(\{\lambda_0\})|$ is invariant by $T_t$, since

$$T_t Z=|T_t M(\{\lambda_0\})|=|e(i\lambda_0 t) M(\{\lambda_0\})|=|M(\{\lambda_0\})|=Z.$$

$Z$ is not constant, since it is subject to the distribution:

$$\frac{2z}{\sqrt{a}} e\left(-\frac{z^2}{a}\right) dz, \quad a=m(\{\lambda_0\})>0.$$

Thus $T_t$ is not ergodic.

PROOF OF SUFFICIENCY. Suppose that $m$ be continuous. Let $Z$ be an invariant element of $T_t$. By Theorem 11, $Z$ and $T_tZ$ are expressed as follows:

$$Z=\sum_{pq} I_{pq}(f_{pq}),\ f_{pq}\in L^2_{pq}(R^1,\ m),$$
$$T_tZ=\sum_{pq} I_{pq}(f_{pq}e(i(\lambda_1+\cdots+\lambda_p-\mu_1-\cdots-\mu_q)t)).$$

By the invariance of $Z$ we obtain

$$f_{pq}=f_{pq}e(i(\lambda_1+\cdots+\lambda_p-\mu_1-\cdots-\mu_q)t))$$

and so, for $(p,\ q)\neq(0,\ 0)$,

$$f_{pq}=0$$

almost everywhere on $R^{p+q}(m^{p+q})$ except on the hyperplane $H_{pq}:\ \lambda_1+\cdots+\lambda_p-\mu_1-\cdots-\mu_q=0$; $m$ being continuous, $H_{pq}$ has the $m^{p+q}$-measure 0. Thus $f_{pq}=0$ almost everywhere on the entire space $R^{p+q}(m^{p+q})$ for $(p,\ q)\neq(0,\ 0)$, which implies that $Z$ is a constant.

DEFINITION. $T_t$ is defined to be *strongly mixing* if we have

$$(T_tY,\ Z)\rightarrow(Y,\ 1)\overline{(Z,\ 1)} \tag{19.1}$$

for any $Y,\ Z\in L^2(M)\equiv L^2(Z)$

THEOREM 19.2. *It is necessary and sufficient for $T_t$ to be strongly mixing that it holds*

$$\int_a^b e(it\lambda)dm(\lambda)\rightarrow 0,\ \textit{as}\ t\rightarrow\infty, \tag{19.2}$$

*for any a,b such that* $-\infty<a<b<\infty$.

PROOF OF NECESSITY. Suppose that (19.2) does not hold for some $a,\ b$. Let $f(\lambda)$ denote the characteristic function of the interval $(a,\ b]$. Put

$$Y=\int f(\lambda)dM(\lambda)=I_{10}(f). \tag{19.3}$$

Then we have

$$T_tY=\int f(\lambda)e(i\lambda t)dM(\lambda), \tag{19.4}$$

and so

$$(T_tY,\ Y)=\int|f(\lambda)|^2e(i\lambda t)dm(\lambda)=\int_a^b e(i\lambda t)dm(\lambda)\nrightarrow 0\equiv(Y,\ 1)\overline{(Y,\ 1)},$$

which implies that $T_t$ is not strongly mixing.

PROOF OF SUFFICIENCY. Suppose that (19.2) hold for any $a,\ b$. Then $m$ is clearly continuous. First we shall show that, for $f(\lambda_1,\ \ldots,\ \lambda_p;\ \mu_1,\ \ldots,\ \mu_q)\in L^1(R^{p+q},\ m^{p+q})$, it holds

$$\int\cdots\int f(\lambda_1,\ \ldots,\ \lambda_p;\ \mu_1,\ \ldots,\ \mu_q)e(i(\textstyle\sum\lambda_\pi-\sum\mu_\rho)t)dm(\lambda_1)\cdots dm(\mu_q)\rightarrow 0, \tag{19.5}$$

as $t\rightarrow\infty$. In case $f$ is the characteristic function of a $(p+q)$-dimensional interval (19.5) follows at once from the assumption (19.2). In the general case we shall approximate $f$ by a linear combination of such functions. Then the integral in (19.4) may be approximated uniformly in $t$. Thus we see that it will also tend to 0 as $t\rightarrow\infty$.

For the proof of sufficiency we shall show that, if $Y$ and $Z$ be any elements in $L^2(M)$, then we get (19.1), as $t\to\infty$.

In case $Y$ and $Z$ are written in the form:

$$Y=\sum_{p,q=0}^{n} I_{pq}(f_{pq}), \qquad Z=\sum_{p,q=0}^{n} I_{pq}(g_{pq}),$$

we have

$$(T_tY,\ Z)=f_{00}\bar{g}_{00}+\sum_{p,q=0}^{n}\int\cdots\int f_{pq}\bar{g}_{pq}e(it(\Sigma\lambda_\pi-\Sigma\mu_\rho))dm(\lambda_1)\ldots dm(\mu_q).$$

By the above remark each term in the summation sign $\Sigma$ will tend to 0 as $t\to\infty$, since $f_{pq}g_{pq}\in L^1(R^{p+q},\ m^{p+q})$ by virtue of the fact that $f_{pq},\ g_{pq}\in L^2{}_{p+q}(R^1,\ m)=L^2(R^{p+q},\ m^{p+q})$. Thus we get

$$(T_tY,\ Z)\to f_{00}g_{00}=(Y,\ 1)\overline{(Z,\ 1)}.$$

Assume that $Y$ and $Z$ be any arbitrary elements of $L^2(M)$. Then there exists $\{Y_n\}$ and $\{Z_n\}$ such that $Y_n$ and $Z_n$ are of the above form and that

$$\|Y_n-Y\|\to 0,\ \|Z_n-Z\|\to 0.$$

Then we have

$$|(T_tY_n,\ Z_n)-(T_tY,\ Z)|\to 0 \text{ (uniformly in } t),$$

and

$$|(Y_n,\ 1)\overline{(Z_n,\ 1)}-(Y,\ 1)\overline{(Z,\ 1)}|\to 0,$$

remembering that $T_t$ is isometric. By the above discussion we see that $(T_tY_n,\ Z_n)\to(Y_n,\ 1)\overline{(Z_n,\ 1)}$ and so that $(T_tY,\ Z)\to(Y,\ 1)\overline{(Z,\ 1)}$.

THEOREM 19.3. *If $m(R^1)<\infty$, then we may replace (19.1) by the following simple condition:*

$$\text{(19.1')}\qquad \int_{-\infty}^{\infty} e(i\lambda t)dm(\lambda)\to 0.$$

PROOF. It is sufficient that (19.1′) are equivalent with (19.2). The former follows from the latter, since

$$\left|\int_{-\infty}^{\infty} e(i\lambda t)dm(\lambda)-\int_{-n}^{n} e(i\lambda t)dm(\lambda)\right| <\int_{|\lambda|\ge n} dm(\lambda)\to 0, \text{ uniformly in } t.$$

Next we shall deduce (19.2) from (19.1′).

If $f(\lambda)$ be of the form:

$$\text{(19.6)}\qquad f(\lambda)=\int_{-\infty}^{\infty} e(i\lambda s)g(s)ds,\ \int_{-\infty}^{\infty}|g(s)|ds<\infty,$$

then we have

$$\int_{-\infty}^{\infty} f(\lambda)e(i\lambda t)dm(\lambda)=\int_{-\infty}^{\infty}\int_{-\infty}^{\infty} e(i\lambda(t+s))g(s)dsdm(\lambda)$$
$$=\int_{-\infty}^{\infty} g(s)\int_{-\infty}^{\infty} e(i\lambda(t+s))dm(\lambda)ds$$

and so, by the assumption (19.1′),

$$\text{(19.7)}\qquad \int_{-\infty}^{\infty} f(\lambda)e(i\lambda t)dm(\lambda)\to 0\ (t\to\infty).$$

If $f(\lambda)$ is a continuous function vanishing outsides of a certain bounded interval, $f(\lambda)$ may be approximated uniformly by a function of the form (19.6), as is well-known in theory of Fourier transforms. Therefore (19.7) will also hold for such $f(\lambda)$. Since $m$ is continuous at any value of $t$ by (19.2′) the characteristic function of any interval will be approximated in the norm of $L^2(R^1, m)$ by a continuous function above mentioned. Thus we see that (19.7) will be also true for such characteristic function $f$, which implies (19. 2)

**§ 20. Application to Complex Wiener Process.** Let $B(t)$ be a complex Wiener process. Then $B(t)$ is a normal screw line. Therefore $B(t)$ is expressible as follows.

$$B(t)-B(s)=\int_{-\infty}^{\infty}\int_{s}^{t} e(i\lambda t)d\tau dM(\lambda),$$

from which follows

$$(B(t)-B(0),\ B(s)-B(0))=\int_{-\infty}^{\infty}\frac{e^{i\lambda t}-1}{i\lambda}\cdot\frac{e^{-i\lambda s}-1}{-i\lambda}dm(\lambda),$$

where $m(\Lambda)=\|M(\Lambda)\|^2$ is the spectral measure of $B(t)$. But we have, as a property of a Wiener process,

$$(B(t)-B(0),\ B(s)-B(0))=\frac{1}{2}(|t|+|s|-|t-s|)=\frac{1}{2\pi}\int_{-\infty}^{\infty}\frac{e^{i\lambda t}-1}{i\lambda}\cdot\frac{e^{-i\lambda s}-1}{i\lambda}d\lambda.$$

Thus we get

(20.1) $m(E)=|E|/2\pi$, $|E|$ = Lebesgue measure of $E$.

By a simple calculation we obtain

$$\int_a^b e(i\lambda t)\,dm(\lambda)=\frac{1}{2\pi}\int_a^b e(i\lambda t)d\lambda\to 0\ \ (t\to\infty),$$

and so, by Theorem 19.2, we see that $T_t$ is strongly mixing; this is a well-known fact.

By making use of Theorem 18 we obtain a theorem of S. Kakutani [7] concerning the spectral structure of $T$.

**§ 21. Application to Complex Normal Stationary Process.** Let $X=\{X(t)\}$ be a complex normal stationary process. Define $Z=\{Z(t)-Z(s)\}$ by

$$Z(t)-Z(s)=\int_s^t X(t)\,d\tau. \tag{21.1}$$

Then $Z(t)$ is a normal screw line. As is well-known, $X(t)$ is expressed in the form :

$$X(t)=\int_{-\infty}^{\infty} e(i\lambda t)dM(\lambda); \tag{21.2}$$

$M(\Lambda)$ is a complex normal random measure on $R^1(m)$, where $m(\Lambda)=\|M(\Lambda)\|^2$ is the spectral measure appearing in Khintchine's canonical form of the autocorrelation function of $X(t)$ and so $m(R^1)<\infty$.

Since we deduce, from (21.2),

$$Z(b)-Z(a)=\int_{-\infty}^{\infty}\int_a^b e(i\lambda t)dtdM(\lambda), \tag{21.3}$$

which is the spectral expression of $Z(t)$ in the sense of § 16. Thus we see that the spectral measure $m'$ of $Z(t)$ is the same as that of $X(t)$ i.e. $m$ and so we have $m'(R_1)<\infty$.

Since $L^2(\boldsymbol{Z})=L^2(\boldsymbol{M})=L^2(\boldsymbol{X})$ and the shift transformation of $\boldsymbol{X}$ has the same effect on $L^2(\boldsymbol{M})$ with that of $\boldsymbol{Z}$, it is sufficient to study about $\boldsymbol{Z}$ in order to investigate $\boldsymbol{X}$. By Theorems 19.1 and 19.3 we shall obtain the following facts.

Let $T_t$ be the group of shift transformations of complex normal stationary process whose spectral measure is denoted by $m$.

(1) *It is necessary and sufficient for $T_t$ to be ergodic that $m$ is continuous.*

(2) *It is necessary and sufficient for $T_t$ to be strongly mixing that the Fourier transform of $m$ vanishes at $\infty$.*

The former has already been obtained by G. Maruyama [8] and by U. Grenander [10] and the latter by the author [9].

## REFERENCES


[1] A. Kolmogorov: Grundbegriffe der Wahrscheinlichkeitsrechnung, Erg. d. Math., 1933.

[2] S. Kakutani: Notes on Infinite Product Measure Spaces I, Proc. Imp. Acad. Tokyo, **19**, 1943.

[3] K. Itô: Multiple Wiener Integral, Jour. of Math. Soc. of Japan, **3**, 1, 1951.

[4] P. Lévy: Théorie de l'addition des variables aléatoires, Paris, 1937.

[5] A. Kolmogorov: Kurven in Hilbertschen Raum, die gegenüber einer einparametrigen Gruppen von Bewegung invariant sind, C.R. (Docklady), **26**, 1, 1940.

[6] K. Itô: A Screw Line in Hilbert Space and its Applicatin to the Probability Theory, Proc. Imp. Acad. Tokyo, **20**, 1944.

[7] S. Kakutani: Determination of the Spectrum of the Flow of Brownian Motion, Proc. Nat. Acad. Sci. U.S.A., **36**, 1950.

[8] G. Maruyama: The Harmonic Analysis of Stationary Stochastic Processes, Memo. Fac. Sci., Kyûsyû Univ., Ser. A, **4**, 1949.

[9] K. Itô: A Kinematical Theory of Turbulence, Proc. Imp. Acad. Tokyo, **20**, 1944.

[10] U. Grenander: Stochastic Processes and Statistical Inference, Arkiv för Math. **1**, 3, 1950.


Mathematical Institute, Kyoto University.

**Reprinted from**
*Japan J. Math.* **22**, 63–86 (1953)

# ISOTROPIC RANDOM CURRENT

KIYOSI ITÔ
PRINCETON UNIVERSITY

## 1. Introduction

The theory of isotropic random vector fields was originated by H. P. Robertson [1] in his theory on isotropic turbulence. He defined the covariance bilinear form of random vector fields which corresponds to Khinchin's covariance function in the theory of stationary stochastic processes. Although in the latter theory the essential point was made clear in connection with the theory of Hilbert space and that of Fourier analysis, we have no corresponding theory on isotropic random vector fields.

Robertson obtained a condition necessary for a bilinear form to be the covariance bilinear form of an isotropic random vector field. Unfortunately his condition is not sufficient; in fact, he took into account only the invariant property of the covariance bilinear form but not its positive definite property. A necessary and sufficient condition was obtained by S. Itô [2]. Although his statement is complicated, he grasped the crucial point. His result corresponds to Khinchin's spectral representation of the covariance function of stationary stochastic processes.

The purpose of this paper is to establish a general theory on homogeneous or isotropic random vector fields, or more generally the homogeneous or isotropic random currents of de Rham [3]. In section 2 we shall give a summary of some known facts on vector analysis for later use. In section 3 we shall define random currents and random measures. The reason we treat random currents rather than random $p$-vector fields or $p$-form fields is that we have no restrictions in applying differential operators $d$ and $\delta$ to random currents. These operators will elucidate the essential point. In section 4 we define homogeneous random currents and give spectral representations. Here we shall explain the relation between homogeneous random currents and random measures. In section 5 we shall show a decomposition of a homogeneous random current into its irrotational part, its solenoidal part and its invariant part. In the next section we shall give a spectral representation of the covariance functional of an isotropic random current. The result here contains S. Itô's formula as a special case. The spectral measure in this representation is decomposed into three parts which correspond to the above three parts in the decomposition of a homogeneous random current. This relation was not known to S. Itô. In [4] we have shown that the Schwartz derivative of the Wiener process is a stationary random distribution which is not itself a process. A similar fact will be seen in section 7 with respect to the gradient of P. Lévy's Brownian motion [5] with a multidimensional parameter.

This work was supported by a research project at Princeton University sponsored by the Office of Ordnance Research, DA-36-034-ORD-1296 RD, and at the University of California, Berkeley, under contract DA-04-200-ORD-355.

## 2. $p$-vector

We shall here summarize some known facts on vector analysis which will be used in this paper. Let $X = R^n$ be the Euclidean $n$-space and $\{e_i\}$ be an orthonormal regular basis, that is,

$$(e_i, e_j) = \delta_{ij}\,; \qquad i, j = 1, 2, \cdots, n\,, \tag{2.1}$$

and the vectors $e_i, \cdots, e_n$, in this order, give a positive orientation. Then any point $x$ of $X$ is expressed uniquely as

$$x = \sum x_i e_i\,. \tag{2.2}$$

The tangent space $T_x$ and its dual space $T_x^*$, that is, the space of differentials at any point $x$ of $X$, are both isomorphic to the space $X$ itself by the following correspondence,

$$\frac{\partial}{\partial x_i} \leftrightarrow dx_i \leftrightarrow e_i\,, \qquad i = 1, 2, \cdots, n\,. \tag{2.3}$$

Therefore, we may identify both the space of $p$-vectors at $x$ and that of $p$-forms with the space $X^{[p]}$ of $p$-vectors with complex coefficients in $X$. $X$ may be considered as a real part of $X^{[1]}$.

Any $p$-vector $\alpha_p \in X^{[p]}$ is expressed uniquely as

$$\alpha_p = \sum_{[i]} \alpha_{i_1 \cdots i_p} e_{i_1} \wedge \cdots \wedge e_{i_p}\,, \tag{2.4}$$

where the coefficients $\alpha_{i_1 \cdots i_p}$ are complex numbers and $[i]$ means that the summation sign $\sum$ refers not to all systems of suffixes, but to those which satisfy $i_1 < i_2 < \cdots < i_p$.

The following notation will often be used in this paper:

$$\delta \begin{pmatrix} i_1 & \cdots & i_s \\ j_1 & \cdots & j_s \end{pmatrix} \tag{2.5}$$

is equal to 1 or $-1$ according to whether $\{j_\nu\}$ is an even or odd permutation of $\{i_\nu\}$, and is equal to 0 in all other cases. We shall state the definitions of exterior product $\alpha_p \wedge \beta_q$, adjoint multivector $\alpha_p^*$, inner product $(\alpha_p, \beta_p)$ and generalized inner product $\alpha_p \vee \beta_q$, which are independent of the choice of the orthonormal regular basis,

$$\alpha_p \wedge \beta_q = \sum_{[i]\,[j]\,[k]} \alpha_{i_1 \cdots i_p} \beta_{j_1 \cdots j_q} \delta \begin{pmatrix} i_1 \cdots i_p & j_1 \cdots j_q \\ k_1 \quad \cdot \quad \cdot & \cdot \quad k_{p+q} \end{pmatrix} e_{k_1} \wedge \cdots \wedge e_{k_{p+q}}\,, \tag{2.6}$$

$$\alpha_p^* = \sum_{[i]\,[j]} \bar{\alpha}_{i_1 \cdots i_p} e_{j_1} \wedge \cdots \wedge e_{j_{n-p}} \delta \begin{pmatrix} 1 \; 2 & \cdot \quad \cdot \quad \cdot & n \\ i_1 \cdots i_p & j_1 \cdots & j_{n-p} \end{pmatrix}, \tag{2.7}$$

$$(\alpha_p, \beta_p) = \sum_{[i]} \alpha_{i_1 \cdots i_p} \bar{\beta}_{i_1 \cdots i_p} = \overline{(\alpha_p \wedge \beta_p^*)^*}\,, \tag{2.8}$$

$$\alpha_p \vee \beta_q = (-1)^{(q-p)\,(n-q)} (\alpha_p \wedge \beta_q^*)^*\,, \qquad q \geqq p\,. \tag{2.9}$$

The product $\vee$ is dual to $\wedge$ in the following sense

$$(\alpha_p \vee \beta_q, \gamma_{q-p}) = (\beta_q, \alpha_p \wedge \gamma_{q-p})\,. \tag{2.10}$$

Now we shall consider a $p$-vector field or equivalently a $p$-form field $\alpha(x)$. Although $\alpha(x)$, in its proper sense, maps each point $x$ of $X$ respectively to a $p$-vector or $p$-form at $x$, it is also considered as a mapping from $X$ to $X^{[p]}$ by the correspondence (2.3). Therefore, $\alpha(x)$ will be expressible as

(2.11) $$\alpha_p(x) = \sum_{[i]} \alpha_{i_1 \cdots i_p}(x)\, e_{i_1} \wedge \cdots \wedge e_{i_p},$$

where the coefficients $\alpha_{i_1 \cdots i_p}(x)$ are complex-valued functions of $x$. The following operations are common in the theory of $p$-forms.

*Differential operators $d$ and $\delta$,*

(2.12) $$d\alpha_p(x) = \sum_{k,\,[i]} \frac{\partial \alpha_p}{\partial x_k}\, e_k \wedge e_{i_1} \wedge \cdots \wedge e_{i_p},$$

(2.13) $$\delta \alpha_p(x) = (-1)^{np+n+1} (d\alpha_p^*)^*.$$

*Inner product,*

(2.14) $$\langle \alpha_p, \beta_p \rangle = \int_{R^n} (\alpha_p, \beta_p)\, dx_1 \cdots dx_n.$$

## 3. Random current and random measure

In this paper we shall treat only complex-valued random variables with mean 0 and finite variance. The totality of such random variables constitutes a Hilbert space which we shall denote by $\boldsymbol{H}$. The inner product of two elements in $\boldsymbol{H}$ is the covariance between them. We shall always refer to the strong topology in $\boldsymbol{H}$.

A random current is an $\boldsymbol{H}$-valued current of de Rham. Let $\mathfrak{D}_p$ be the set of all $C_\infty$ $p$-vector fields with a compact carrier. $\mathfrak{D}_0$ is nothing but the Schwartz $\mathfrak{D}$ space. $\mathfrak{D}_p$ is isomorphic to the ${}_nC_p$th power of $\mathfrak{D}_0$, so that it is a linear topological space. A random current of the $p$th degree is a continuous linear mapping from $\mathfrak{D}_{n-p}$ into $\boldsymbol{H}$. A continuous random $p$-vector field $U_p(x)$ induces a random current of the $p$th degree as follows,

(3.1) $$U_p(\phi_{n-p}) = \int_{R^n} (U_p \wedge \phi_{n-p})^*\, dx_1 \cdots dx_n.$$

A sequence of random currents $\{U_p^{(m)}\}$ is said to converge to $U_p$ if

(3.2) $$U_p^{(m)}(\phi_{n-p}) \to U_p(\phi_{n-p}), \qquad \phi_{n-p} \in \mathfrak{D}_{n-p}.$$

The operations on random currents are defined in the same way as in the case of the currents of de Rham,

(3.3) $$U_p^*(\phi_p) = (-1)^{p(n-p)}\, \overline{U_p(\phi_p^*)},$$

(3.4) $$dU_p(\phi_{n-p-1}) = (-1)^{p+1}\, U_p(d\phi_{n-p-1}),$$

(3.5) $$\delta U = (-1)^{np+n+1} (dU_p^*)^*,$$

(3.6) $$\alpha_q(x) \wedge U_p(\phi_{n-p-q}) = (-1)^{pq}\, U_p(\alpha_q \wedge \phi_{n-p-q}),$$

(3.7) $$\alpha_q(x) \vee U_p(\phi_{n-p+q}) = (-1)^{q(p-q)}\, U_p(\alpha_q \vee \phi_{n-p+q}).$$

For a random current $U_p$ we shall define $(U_p, \alpha_p)$ as a random Schwartz distribution,

(3.8) $$(U_p, \alpha_p)(\phi) = U_p(\phi \cdot \alpha_p^*).$$

A random Schwartz distribution $M(\phi)$ is called a random measure with respect to a measure $m$ if we have, for any pair $\phi, \psi \in \mathfrak{D}$,

$$[M(\phi), M(\psi)]_H = \int_{R^n} \phi(x) \overline{\chi(x)} m(dx), \tag{3.9}$$

where $[M(\phi), M(\psi)]_H$ denotes the inner product in $H$. Putting $M(E) = M(\chi_E)$, where $\chi_E$ is the characteristic function of the set $E$, we get a random set function which is additive in $E$. Further, we have

$$[M(E), M(E')]_H = m(EE'), \qquad M(\phi) = \int_{R^n} \phi(x) M(dx). \tag{3.10}$$

Particularly, if $m$ satisfies

$$\int_{R^n} \frac{m(dx)}{(1+|x|^2)^k} < +\infty, \tag{3.11}$$

that is, $m$ is slowly increasing, then $M(\phi)$ is also said to be slowly increasing. Under this condition, any rapidly decreasing function $\phi$ in Schwartz sense [6] belongs to $L^2(R^n, m)$, so that $M(\phi)$ can be defined.

A random current $M_p(\phi_{n-p})$ is called a random measure (of the $p$th degree) if there exists a complex-valued locally finite measure $m(dx; a_p, b_p)$ for every pair $(a_p, b_p)$ such that we have

$$[(M_p, a_p)(\phi), (M_p, b_p)(\psi)]_H = \int \phi(x) \overline{\psi(x)} m(dx; a_p, b_p). \tag{3.12}$$

If $m(dx; a_p a_p)$ is a slowly increasing measure for every $a_p$, then $M_p$ is also said to be slowly increasing.

## 4. Homogeneous random current: spectral representation

A translation $\tau_h: x \to x + h$ induces a translation $\sigma_h$ of $p$-vector fields in the following usual way,

$$(\sigma_h \cdot \phi_p)(x) = d\tau_h [\phi_p(x+h)]. \tag{4.1}$$

By the identification (2.3), we can easily see that $d\tau_h$ is just the identity mapping. Thus we have

$$(\sigma_h \cdot \phi_p)(x) = \phi_p(x+h). \tag{4.2}$$

Let $U_p$ be a random current. Then we shall define $\sigma_h U_p$ by

$$(\sigma_h U_p)(\phi_{n-p}) = U_p(\sigma_h^{-1} \phi_{n-p}) = U_p(\sigma_{-h} \phi_{n-p}). \tag{4.3}$$

The functional $\rho(\phi_p, \psi_p) = [U_p(\phi_p^*), U_p(\psi_p^*)]_H$ is called the covariance functional of $U_p$. The function defined by

$$\rho(\phi, \psi; a_p, b_p) = [(U_p, a_p)(\phi), (U_p, b_p)(\psi)]_H \tag{4.4}$$

is called the *covariance bilinear form* of $U_p$ and is a generalization of the form Robertson used in his theory of turbulence. If the covariance functional of $\sigma_h U_p$ is independent of $h$, then $U_p$ is said to be *homogeneous*. If this condition is satisfied, the covariance bilinear form of $\sigma_h U_p$ is independent of $h$. The converse is also true.

As in the theory of stationary stochastic processes we have the following theorem of spectral representation.

THEOREM 4.1. *The covariance bilinear form of any homogeneous random current is written as*

$$\rho(\phi, \psi; a_p, b_p) = \int_{R^n} \mathcal{F}\phi(y)\,\overline{\mathcal{F}\psi(y)}\, m(dy; a_p, b_p) \tag{4.5}$$

*where* $m(\Lambda; a_p, b_p)$ *is a positive definite bilinear form in* $(b_p, a_p)$ *and* $m(\Lambda; a_p, a_p)$ *is a slowly increasing nonnegative measure. Conversely the* $\rho$ *defined by* (4.5) *is the covariance bilinear form of a homogeneous current.*

*Remark.* $\mathcal{F}\phi$ is the Fourier transform of $\phi$, that is,

$$(\mathcal{F}\phi)(y) = \int_{R^n} e^{-i2\pi(x,y)}\phi(y)\,dy. \tag{4.6}$$

For a $p$-vector field $\phi_p(y)$, we define

$$(\mathcal{F}\phi_p)(y) = \sum_{[i]} \mathcal{F}\phi_{i_1\cdots i_p}(y)\, e_{i_1}\wedge\cdots\wedge e_{i_p}. \tag{4.7}$$

THEOREM 4.2. *A homogeneous random current* $U_p$ *is the Fourier transform of a slowly increasing random measure* $M_p$, *which is called the spectral measure of* $U_p$, *that is,*

$$U_p(\phi_{n-p}) = M_p = M_p(\mathcal{F}\phi_{n-p}). \tag{4.8}$$

We can prove this theorem easily by remarking that $(U_p, a_p)(\phi)$ is a stationary random distribution [4] with a multidimensional parameter.

## 5. Homogeneous random current: canonical decomposition

In this section we shall discuss a decomposition of a homogeneous random current $U_p$ into an irrotational current, a solenoidal current and an invariant current. This decomposition will be called the canonical decomposition of $U_p$. We shall start from the spectral representation. By using this, we can show the existence of the limit

$$\mathcal{M}U_p(\phi_{n-p}) = \lim_{A\to\infty}\frac{1}{A^n}\int_{-A}^{A}\cdots\int_{-A}^{A}\sigma_h U_p(\phi_{n-p})\,dh_1\cdots dh_n. \tag{5.1}$$

We say that $U_p$ is invariant or unbiased according to whether $\mathcal{M}U_p = U_p$ or 0. An unbiased homogeneous random current is called irrotational or solenoidal according to whether $dU_p = 0$ or $\delta U_p = 0$. We set

$$M_p^0(\Lambda) = M_p(\Lambda \cap \{0\}), \qquad M_p^u(\Lambda) = M_p(\Lambda - \{0\}). \tag{5.2}$$

Then we have $\mathcal{M}U_p = \mathcal{F}M_p^0$.

THEOREM 5.1.

(a) *A homogeneous random current* $U$ *is irrotational if and only if* $M_p^0 = 0$ *and* $y\wedge M_p^u(dy) = 0$.

(b) $U$ *is solenoidal if and only if* $M_p^0 = 0$ *and* $y\vee M_p^u(dy) = 0$.

The essential point of the proof is as follows.

$$\begin{aligned} dU_p(\phi_{n-p-1}) &= (-1)^{p+1}U_p(d\phi_{n-p-1}) = (-1)^{p+1}M_p(\mathcal{F}d\phi_{n-p-1}) \\ &= (-1)^{p+1}M_p(i2\pi y\wedge\mathcal{F}\phi_{n-p-1}). \end{aligned} \tag{5.3}$$

Now we shall introduce two random measures $M_p^i$ and $M_p^s$ by

$$M_p^i(dy) = \frac{y\wedge[y\vee M_p^u(dy)]}{|y|^2}, \qquad M_p^s(dy) = \frac{y\vee[y\wedge M_p^u(dy)]}{|y|^2}. \tag{5.4}$$

THEOREM 5.2. *$U_p^i = \mathcal{A}M_p^i$, $U_p^s = \mathcal{A}M_p^s$ and $U_p^0 = \mathcal{A}M_p^0 = \mathcal{M}U_p$ are respectively irrotational, solenoidal and invariant, and we have*

$$U_p = U_p^i + U_p^s + U_p^0 . \tag{5.5}$$

By using the following identity

$$a_p = a_p' + a_p'' = \frac{y \wedge (y \vee a_p)}{|y|^2} + \frac{y \vee (y \wedge a_p)}{|y|^2} ; \quad (a_p', a_p'') = 0 ; \tag{5.6}$$

we can prove the theorem.

Now we shall define $W_p(\phi_{n-p})$ by the following procedure,

$$G(x, y) = \begin{cases} \dfrac{e^{-2i\pi(x,\, y)} - 1 + 2i\pi\,(x, y)}{|y|^2} & \text{if } |y| < 1 , \\ \dfrac{e^{-2i\pi(x,\, y)}}{|y|^2} & \text{if } |y| \geqq 1 ; \end{cases} \tag{5.7}$$

$$G(\phi_p, y) = \sum_{[i]} [\textstyle\int G(x, y)\, \phi_{i_1 \cdots i_p}(x)\, dx]\, e_{i_1} \wedge \cdots \wedge e_{i_p} ; \tag{5.8}$$

$$W_p(\phi_{n-p}) = \frac{1}{4\pi^2} M_p^u [G(\phi_{n-2}, y)] . \tag{5.9}$$

THEOREM 5.3.

$$d\,\delta W_p = U_p^i , \qquad \delta d W_p = U_p^s . \tag{5.10}$$

*Therefore the canonical decomposition is written as*

$$U_p = d\,\delta W_p + \delta d W_p + U_p^0 . \tag{5.11}$$

## 6. Isotropic random current

To begin with, we shall introduce some preliminary notation. Let $G$ be the whole group of orthogonal transformations (with determinant $\pm 1$) in $X = R^n$. We can carry out the same procedure for $g \in G$ as we did for the translation $\tau_h$ in section 4. Corresponding to (4.1) we have

$$(\sigma_g \phi_p)(x) = d\,\tau_g [\phi_p(g \cdot x)] . \tag{6.1}$$

Although $d\tau_g$ here is not the identity mapping, we can easily see that $d\tau_g = g^{-1}$. Therefore, we have

$$(\sigma_g \cdot \phi_p)(x) = g^{-1} \cdot \phi_p(g \cdot x) , \tag{6.2}$$

where $g^{-1}\phi_p$ is defined by

$$g^{-1}\phi_p = \sum_{[i]} \phi_{i_1 \cdots i_p} g^{-1} e_{i_1} \wedge \cdots \wedge g^{-1} e_{i_p} . \tag{6.3}$$

Generalizing this transformation $\sigma_g$, we define $\sigma_g$ on a current $u_p$,

$$(\sigma_g u_p)(\phi_{n-p}) = u_p(\sigma_{g^{-1}} \phi_{n-p}) = u_p[g \phi_{n-p}(g^{-1}x)] . \tag{6.4}$$

$\sigma_g$ is clearly commutative with the differential operators $d$ and $\delta$.

Now we shall define an isotropic random current. A homogeneous random current is said to be isotropic, if the covariance functional of $\sigma_g U_p$ is independent of $g \in G$. Even if we replace the covariance functional by the covariance bilinear form in the above statement, we shall obtain an equivalent definition.

Let $U_p$ be an isotropic random current. Since $U_p$ is homogeneous, the covariance bilinear form is written as

$$\rho(\phi, \psi; a_p, b_p) = \int \mathfrak{F}\phi \overline{\mathfrak{F}\psi}\, m(dx; a_p, b_p) \tag{6.5}$$

by theorem 4.1. Since $U_p$ is isotropic, we obtain

$$m(g \cdot dx; g \cdot a_p, g \cdot b_p) = m(dx; a_p, b_p). \tag{6.6}$$

By making use of this property we obtain the following theorem.

THEOREM 6.1. *In an isotropic turbulence the* $m(dx; a_p, b_p)$ *in* (6.6) *is expressible as*

$$\begin{aligned} m(dx; a_p, b_p) = {} & (\theta \vee b_p, \theta \vee a_p)\, d\theta\, F_i(dr) \\ & + (\theta \wedge b_p, \theta \wedge a_p)\, d\theta\, F_s(dr) + (b_p, a_p)\, F_0(dx), \end{aligned} \tag{6.7}$$

*where* $r = |x|$, $\theta = x/r$, $x \neq 0$, $d\theta$ *is the surface element of the unit sphere,* $F_1$ *and* $F_2$ *are both slowly increasing nonnegative measures on* $(0, \infty)$, *and* $F_0$ *is a nonnegative measure such that* $F_0(\Lambda) = 0$ *for* $0 \notin \Lambda$.

*Conversely,* $\rho(\phi, \psi; a_p, b_p)$, *if determined by* $m(dx; a_p, b_p)$ *of the form* (6.6), *is the covariance bilinear form of a certain isotropic random current.*

*Remark.* According to whether $p = 0$ or $p = n$, the first or the second term in (6.7) will disappear.

We shall sketch the proof. First we consider the case in which $m(dx; a_p, b_p)$ has a continuous density $f(x; a_p, b_p)$. Here $f(x; a_p, b_p)$ is a positive definite bilinear form in $(b_p, a_p)$ and is invariant in the sense $f(gx; ga_p, gb_p) = f(x; a_p, b_p)$. If we introduce $F(x; \xi_1, \cdots, \xi_p, \eta_1, \cdots, \eta_p) = f(x; \xi_1 \wedge \cdots \wedge \xi_p, \eta_1 \wedge \cdots \wedge \eta_p)$, for real 1-vectors $\xi_i$, $\eta_j$, we obtain a function which is linear in $\xi_i$ and in $\eta_j$ and skew symmetric in each of $\{\xi_i\}$ and $\{\eta_j\}$. Since $F$ is invariant under orthogonal transformations, it is a function of $(\xi_i, \xi_j)$, $(\xi_i, \eta_j)$, $(\eta_i, \eta_j)$, $(x, \xi_i)$, $(x, \eta_j)$ and $(x, x)$. Using these properties we can prove that $f$ can be written in a form similar to (6.7).

We can discuss the case of general measure by approximating it by measures with continuous density in Helly's sense.

The following theorem shows that the decomposition in (6.6) corresponds to the canonical decomposition (5.5).

THEOREM 6.2. *In the case of an isotropic random current* $U_p$, $U_p^i$, $U_p^s$ *and* $U_p^0$ *are all isotropic and orthogonal to each other in* **H**. *The m-measures corresponding to these three parts are respectively the three parts in the decomposition* (6.7).

## 7. The case of $p = 1$

In the case of $p = 1$, the decomposition (6.7) becomes simpler,

$$m(dx; \alpha, \beta) = \sum_{\mu, \nu} \bar{\alpha}_\mu \beta_\nu [\theta_\mu \theta_\nu d\theta\, F_i(dr) + (\delta_{\mu\nu} - \theta_\mu \theta_\nu) F_s(dr) + F_0(dx)], \tag{7.1}$$

where $\alpha = \sum \alpha_\mu e_\mu$ and $\beta = \sum \beta_\mu e_\mu$.

In the case of isotropic 1-vector fields the measures $F_i$ and $F_s$ will be bounded.

(a) *Robertson's isotropic turbulence.* This is an isotropic random 1-vector field in 3-

space. Our decomposition (7.1) is essentially the same as S. Itô's formula (2) but is somewhat simpler.

(b) *The gradient of P. Lévy's Brownian motion with a multidimensional parameter.* Let $B(x)$ be Lévy's Brownian motion with an $n$-dimensional parameter [5]. Then we have

$$[B(x), B(y)]_H = \tfrac{1}{2}(|x| + |y| - |x-y|). \tag{7.2}$$

Therefore, by simple computation we can obtain the following expression of the covariance bilinear form of the gradient $dB$ of $B$,

$$\rho(\phi, \psi, \alpha, \beta) = \int_{R^n} \mathfrak{Q}\phi(x)\,\overline{\mathfrak{Q}\psi(x)}\, m(dx; \alpha, \beta), \tag{7.3}$$

where

$$m(dx; \alpha, \beta) = C_n \sum_{\mu, \nu} \alpha_\mu \beta_\nu \theta_\mu \theta_\nu d\theta\, r^{-(n-1)} dr, \tag{7.4}$$

where $r = |x|$, $\theta = x/r$, and $C_n$ is a positive constant. This is the special case of (7.1) in which

$$F_i(dr) = C_n \cdot r^{-(n-1)} dr, \qquad F_s = F_0 = 0. \tag{7.5}$$

Since $C_n \cdot r^{-(n-1)} dr$ is an unbounded measure, $dB$ is a random proper current, that is, a random current that is not itself a random 1-vector field. Since $F_s = F_0 = 0$ we see that $dB$ is irrotational.

## REFERENCES

[1] H. P. ROBERTSON, "The invariant theory of isotropic turbulence," *Proc. Camb. Phil. Soc.*, Vol. 36 (1940), pp. 209–223.

[2] S. ITÔ, "On the canonical form of turbulence," *Nagoya Math. Jour.*, Vol. 2 (1951), pp. 83–92.

[3] G. DE RHAM and K. KODAIRA, "Harmonic integrals," mimeographed notes of lectures at the Institute for Advanced Study, Princeton, 1940.

[4] K. ITÔ, "Stationary random distributions," *Mem. Coll. Sci., Univ. of Kyoto*, Ser. A, Vol. 28 (1954), pp. 209–223.

[5] P. LÉVY, *Processus Stochastiques et Mouvement Brownien*, Paris, Gauthier-Villars, 1948.

[6] L. SCHWARTZ, *Théorie des Distributions*, Paris, Hermann et Cie, 1950.

Reprinted from
*Proc. Third Berkeley Symp. Math. Stat. Probab.* **II**, 125–132 (1955)

# SPECTRAL TYPE OF THE SHIFT TRANSFORMATION OF DIFFERENTIAL PROCESSES WITH STATIONARY INCREMENTS[1]

BY

KIYOSI ITÔ

1. **Introduction.** Before discussing our problems on stochastic processes, we shall define two kinds of equivalence of groups of measure preserving set transformations following J. v. Neumann and P. R. Halmos [8][2]. By a *measure preserving set transformation* we understand a mapping from the system of measurable sets of one measure space onto that of another measure space modulo null sets which preserves measure and set operations such as countable sum and complement. Given a measure space $\Omega(\boldsymbol{B}, m)$ and a one-parameter group $\{T_\tau\}$ of measure preserving set transformations from $\boldsymbol{B}$ onto itself, let $\boldsymbol{H}$ denote the $L^2$-space over $\Omega$. The above group $\{T_\tau\}$ will induce a group $\{U_\tau\}$ of unitary operators on $\boldsymbol{H}$ such that

$$U_\tau \chi_M = \chi_{M[\tau]} \tag{1.1}$$

where $M$ is any set of finite $m$-measure, $M[\tau] = T_\tau M$, and $\chi_A$ denotes the characteristic function of a set $A$. We consider another measure space $\overline{\Omega}(\overline{\boldsymbol{B}}, \overline{m})$ associated with a group of measure preserving set transformations $\{\overline{T}_\tau\}$ and we define $\overline{\boldsymbol{H}}$ and $\{\overline{U}_\tau\}$ correspondingly. If there exists a measure preserving set transformation $S$ from $\boldsymbol{B}$ onto $\overline{\boldsymbol{B}}$ such that

$$\overline{T}_\tau = S T_\tau S^{-1}, \tag{1.2}$$

then we say that $\{\overline{T}_\tau\}$ and $\{T_\tau\}$ are of the same *spatial type*. We shall also introduce another classification of transformation groups which is rougher than the above. If there exists an isometric linear mapping $V$ from $\boldsymbol{H}$ onto $\overline{\boldsymbol{H}}$ such that

$$\overline{U}_\tau = V U_\tau V^{-1}, \tag{1.3}$$

then we say that $\{T_\tau\}$ and $\{\overline{T}_\tau\}$ are of the same *spectral type*.

Let $X(t, \omega)$, $-\infty < t < \infty$, be a measurable (in two variables $t$ and $\omega$) differential process with stationary increments on a probability space $\Omega(\boldsymbol{B}, P)$. For any finite interval $I = (s, t]$ we shall define the increment $\Delta X(I, \omega)$ as $X(t, \omega) - X(s, \omega)$. By $\boldsymbol{B}_X$ we denote the Borel system of subsets of $\Omega$ generated

---

Received by the editors May 25, 1955.

(1) This work was supported by a research project at Princeton University sponsored by the Office of Ordnance Research, DA-36-034-ORD-1296 RD.

(2) Numbers in brackets refer to the references at the end of this paper.

by all the sets of the form $\{\omega; \Delta X(I, \omega) < c\}$, and we shall consider the measure space $\Omega(\boldsymbol{B}_X, P)$ and the $L^2$-space over it, say $\boldsymbol{H}_X$. Now we shall define a one-parameter group $\{T_\tau\}$ of measure preserving set transformations (*shift transformation*) on $\boldsymbol{B}_X$ by

$$(1.4) \qquad T_\tau\{\omega; \Delta X(I, \omega) < c\} = \{\omega; \Delta X(I + \tau, \omega) < c\},$$

where $I+\tau$ is the interval $\{x+\tau; x \in I\}$. The possibility of this definition follows from the definition of differential processes with stationary increments. From $\{T_\tau\}$ we can derive a group $\{U_\tau\}$ of unitary operators on $\boldsymbol{H}_X$ as above.

The purpose of this paper is to determine the spectral type of this group $\{T_\tau\}$. In §2 we shall summarize some known facts on differential processes as preliminaries. In §3 and §4 we shall introduce a *multiple Wiener integral* which will play an important role in our theory. Our aim will be attained in §5. The fundamental theorem established there generalizes Kakutani's theorem [6] on the spectra of the flow of Brownian motion and implies that the transformation groups induced by different processes are of the same spectral type. But it is still an open question whether they are of the same spatial type or not.

2. **Independent random measure associated with a measurable differential process with stationary increments.** Let $\Xi$ be a *measurable space* on which a class of sets called *measurable*, $\boldsymbol{B}'_\Xi$, is assigned so that (1) the empty set $\emptyset \in \boldsymbol{B}'_\Xi$, (2) $E_1, E_2 \in \boldsymbol{B}'_\Xi$ implies $E_1 \cup E_2$, $E_1 - E_2 \in \boldsymbol{B}'_\Xi$, and (3) any decreasing sequence in $\boldsymbol{B}'_\Xi$ has its limit set in $\boldsymbol{B}'_\Xi$; we do not assume that the whole space $\Xi$ belongs to $\boldsymbol{B}'_\Xi$.

A system of random variables $f(E) = f(E, \omega)$ depending on a set $E \in \boldsymbol{B}'_\Xi$ with a probability parameter $\omega$ is called an *independent random measure* if $f(E_1), f(E_2), \cdots, f(E_n)$ are independent and

$$(2.1.a) \qquad f\left(\bigcup_{i=1}^{n} E_i\right) = \sum_{i=1}^{n} f(E_i)$$

for every finite system $\{E_i\}$ of disjoint sets in $\boldsymbol{B}_\Xi$ and if

$$(2.1.b) \qquad f(E_n) \to 0 \text{ (convergence in probability)}$$

for every decreasing sequence $\{E_n\}$ in $\boldsymbol{B}'_\Xi$ tending to the empty set. We can easily show that the above conditions imply that (2.1.b) holds in the sense of almost everywhere convergence. A *normal random measure* [4] is a special case of an independent random measure.

Let $X(t, \omega)$ be a measurable differential process with stationary increments. By taking Doob's separable modification [1] of this process, we may assume that $X(t, \omega)$ is continuous in $t$ except for discontinuities of the first kind with probability 1. We also assume, as we may, that $X(\tau, \omega)$ is continuous on the right in $\tau$ with probability 1. By a theorem of P. Lévy [7], $\Delta X(I, \omega)$

is subject to an infinitely divisible law whose characteristic function $\phi_I(z)$ is given by

$$\log \phi_I(z) = |I| \left[ i\gamma z + \int_{-\infty}^{\infty} \left( e^{izu} - 1 - \frac{izu}{1+u^2} \right) \frac{1+u^2}{u^2} d\beta(u) \right] \tag{2.2}$$

where $|I|$ means the length of $I$, $\gamma$ is a real constant, and $d\beta$ is a bounded measure on $(-\infty, \infty)$.

We shall consider a plane $\pi$ on which a coordinate system $(t, u)$ is assigned. Let $\boldsymbol{B}_\pi$ denote the class of all Borel subsets of $\pi$. We shall consider two measures $\nu$, $\mu$ on $\pi(\boldsymbol{B}_\pi)$ by

$$\nu(E) = \iint_E \frac{1+u^2}{u^2} dt d\beta(u), \tag{2.3}$$

$$\mu(E) = \iint_E (1+u^2) dt d\beta(u) = \iint_E u^2 d\nu(t, u). \tag{2.4}$$

It is clear that $\mu$ is the product of the one-dimensional measures $dt$ and $d\mu'(u) = (1+u^2)d\beta(u)$. Let $\tilde{\boldsymbol{B}}_\pi$ be the totality of bounded Borel subsets of $\pi$ whose distance from the $t$-axis is positive. We may consider $\pi(\tilde{\boldsymbol{B}}_\pi)$ as a measurable space. If we define $N(E, \omega)$, $E \in \tilde{\boldsymbol{B}}_\pi$, to be the number of points $(t, u) \in E$ for which $X(t, \omega) - X(t-0, \omega) = u$. Then $N(E, \omega)$ is subject to Poisson distribution with mean $\nu(E)$ for $E$ fixed and the system $\{N(E, \omega), E \in \tilde{\boldsymbol{B}}_\pi\}$ is an independent random measure. Further we have the following expression of $X(t, \omega)$:

$$\begin{aligned} \Delta X(I, \omega) &= \gamma |I| + \sigma \Delta B(I, \omega) \\ &\quad + \lim_{n \to \infty} \int_{t \in I} \int_{n^{-1} < |u| < n} \left[ u dN(t, u) - \frac{u}{1+u^2} d\nu(t, u) \right] \end{aligned} \tag{2.5}$$

where $\sigma^2 = \beta(+0) - \beta(-0)$ and $B(t, \omega)$ is a Wiener process which is independent of the system $\{N(E, \omega); E \in \tilde{\boldsymbol{B}}_\pi\}$. We can easily deduce these facts from the results stated in [3].

Next we shall introduce another independent random measure $M(E)$. Let $\boldsymbol{B}_\pi^*$ be the class of all Borel subsets of $\pi$ whose $\mu$-measure is finite. For $E$ in $\boldsymbol{B}_\pi^*$ we define $M(E)$ by

$$M(E) = M(E, \omega) = \int_{E(0)} \sigma \cdot dB(t) + \lim_{n \to \infty} \iint_{(t,u) \in E(n)} [u dN(t, u) - u d\nu(t, u)],$$

where

$$E(n) = \{(t, u) \in E; n^{-1} < |u| < n\}, \qquad n = 1, 2, \cdots,$$
$$E(0) = \{t; (t, 0) \in E\}$$

and the integral based on $dB(t)$ is the Wiener integral [1, IX, 2].

It is to be noted that $N(E)$, $\Delta B(I)$ and $M(E)$ belong to $\boldsymbol{H}_X$ for $E$ and $I$ fixed. Also the expectation of $M(E)$ is 0.

**3. Definition of multiple Wiener integral based on $dM$.** Let $M(E, \omega)$, $E \in \boldsymbol{B}_\pi^*$, be the independent random measure defined in §2. We can easily verify the following properties:

(3.1) $$(M(E_1), M(E_2)) = \mu(E_1 \cap E_2),$$

(3.1′) $$\|M(E)\|^2 = \mu(E),$$

(3.2) $$(M(E), 1) = 0.$$

In this section we shall make use of the following property of $\mu$:

*Continuity*: For every $E \in \boldsymbol{B}_\pi^*$ and every $\epsilon > 0$, there exists a finite subdivision of $E$: $E = E_1 \cup E_2 \cup \cdots \cup E_n$ such that $\mu(E_i) < \epsilon$, $1 \leq i \leq n$.

Let $\pi^p$ be the product measure space $[\pi(\boldsymbol{B}_\pi, \mu)]^p$ and let $L_p^2$ be the $L^2$-space over $\pi^p$. We shall denote points of $\pi$ by $\xi = (t, u)$, $\xi' = (t', u')$, $\xi_i = (t_i, u_i)$, etc. For any $f \in L_p^2$ we shall define the *symmetrized function* $\tilde{f}$ of $f$ to be

(3.3) $$\tilde{f}(\xi_1, \cdots, \xi_p) = \frac{1}{p!} \sum_{(\epsilon)} f(\xi_{\epsilon(1)}, \cdots, \xi_{\epsilon(p)})$$

where $(\epsilon) = (\epsilon(1), \cdots, \epsilon(p))$ runs over all arrangements of $(1, 2, \cdots, p)$. It is clear that $f(\xi_1, \cdots, \xi_p)$ is symmetric in $(\xi_1, \cdots, \xi_p)$ if and only if it coincides with $\tilde{f}$. $\tilde{f}$ is symmetric. We have also

(3.4) $$\|\tilde{f}\| \leq \|f\|, \qquad \| \ \| \text{ being the norm in } L_p^2.$$

Let $C_p$ denote the class of all functions of the form:

(3.5) $$c_p(\xi_1, \xi_2, \cdots, \xi_p) = \chi(\xi_1, E_1)\chi(\xi_2, E_2) \cdots \chi(\xi_p, E_p)$$

where $E_i$, $i = 1, 2, \cdots, p$, are disjoint sets in $\boldsymbol{B}_\pi^*$ and $\chi(\xi, E)$ denotes the characteristic function of the set $E$. Let $S_p$ denote the class of all linear combinations of functions in $C_p$. The continuity of $\mu$ implies that $S_p$ is dense in $L_p^2$ [4, Theorem 2.1].

Now we shall define the multiple Wiener integral of the $p$th degree based on the independent random measure $M$ and denote it by

(3.6) $$I_p(f) = \int \cdots \int f(\xi_1, \cdots, \xi_p) dM(\xi_1) \cdots dM(\xi_p).$$

We define

(3.7) $$I_p(c_p) = M(E_1) \cdots M(E_p)$$

for the $c_p$ of (3.5), then $I_p(s)$ for $s$ in $S_p$ by linearity and finally $I_p(f)$ for $f$ in $L_p^2$ by continuity. For convenience of notation we shall denote by $L_0^2$ the complex number field (one-dimensional Hilbert space) and define $I_0(c)=c$ if $c\in L_0^2$. In exactly the same way as in the case of normal random measure [4] we can show that the definition is possible and that the following theorem holds.

THEOREM 1.

$$I_p(f) = I_p(\tilde{f}), \tag{I.1}$$

$$I_p(af+bg) = aI_p(f)+bI_p(g), \tag{I.2}$$

$$(I_p(f), I_p(g)) = p!\cdot(\tilde{f},\tilde{g}), \tag{I.3}$$

$$\|I_p(f)\| = (p!)^{1/2}\|\tilde{f}\| \leqq (p!)^{1/2}\|f\|, \tag{I.3'}$$

$$(I_p(f), I_q(g)) = 0 \qquad (p\neq q). \tag{I.4}$$

**4. Completeness of multiple Wiener integrals.** Let $\tilde{L}_p^2$ denote the class of all symmetric functions in $L_p^2$. $\tilde{L}_p^2$ is clearly a closed linear manifold of $L_p^2$. By (I.1) in Theorem 1 we see that the image of $\tilde{L}_p^2$ by $I_p$ coincides with that of $L_p^2$, which we shall denote by $H_X^{(p)}$. Since the mapping:

$$V_p: \tilde{L}_p^2 \ni f_p \to (p!)^{-1/2} I_p(\tilde{f}_p) \in H_X \tag{4.1}$$

is isometric by (I.3'), $H_X^{(p)}$ is a closed linear manifold of $H_X$ isomorphic with $\tilde{L}_p^2$ by $V_p$. By (I.4) we see that the $H_X^{(p)}$, $p=0, 1, 2, \cdots$, are orthogonal to each other.

We shall establish the following theorem which implies the completeness of $\{I_p(\tilde{f}_p); \tilde{f}_p\in\tilde{L}_p^2, p=0, 1, 2, \cdots\}$.

THEOREM 2.

$$H_X = \sum_{p\geqq 0}\oplus H_X^{(p)} \qquad (\textstyle\sum\oplus \textit{ means `direct sum'}).$$

**Proof.** 1°. We shall first prove the following lemma.

LEMMA 1. *All the elements of the following form constitute a fundamental set in* $H_X$:

$$N(E_1)^{p_1}\cdots N(E_m)^{p_m}\Delta B(I_1)^{q_1}\cdots\Delta B(I_n)^{q_n}, \tag{4.2}$$

*where* $m, n=1, 2, \cdots, p_i, q_j=0, 1, 2, \cdots$, *and* $\{E_j\}$ *and* $\{I_j\}$ *are each disjoint.*

**Proof**[3]. It is clear that an element of the form (4.2) belongs to $H_X$, since it has a finite norm. By the definition of $H_X$ we can easily see that the totality of the elements of $H_X$ of the following form constitute a fundamental set in $H_X$:

---

[3] The author owes this proof to I. E. Segal who has established a more general fact in his paper [9].

(4.3) $$Y = f(\Delta X(I_1), \cdots, \Delta X(I_n)),$$

where $n = 1, 2, \cdots, \{I_i\}$ are disjoint and $f$ is a bounded continuous function. By the expression of (2.5), $\Delta X(I)$ is the limit of linear combinations of $\{N(E)\}$ and $\{\Delta B(I)\}$. Therefore the elements of the following form also constitute a fundamental set in $\boldsymbol{H}_X$:

(4.4) $$Z = f(N(E_1), \cdots, N(E_m), \quad \Delta B(I_1), \cdots, \Delta B(I_n))$$

where the $E_i$ are pairwise disjoint, the $I_j$ are also pairwise disjoint and $f$ is a bounded continuous function.

Let $\mathfrak{M} = \mathfrak{M}(E_1, \cdots, E_m, I_1, \cdots, I_n)$ be the closed linear manifold spanned by the polynomials in $N(E_1), \cdots, N(E_m), \Delta B(I_1), \cdots, \Delta B(I_n)$. To prove the lemma it is enough to show that $Z$ in (4.4) belongs to $\mathfrak{M}$. We put

$$Z = U + V = g(N(E_1), \cdots, \Delta B(I_n)) + h(N(E_1), \cdots, \Delta B(I_n))$$

where $U \in \mathfrak{M}$, $V \perp \mathfrak{M}$, and both $g$ and $h$ are Baire functions. It is enough to show $V = 0$. To avoid trivial complications we consider the case that $m = n = 1$. It is sufficient to derive $h(N(E), \Delta B(I)) = 0$ for almost all $\omega$ from the following conditions:

(4.5) $$(h(N(E), \Delta B(I)), \quad N(E)^p \Delta B(I)^q) = 0, \qquad p, q = 0, 1, 2, \cdots.$$

We denote by $\sigma_1$ and $\sigma_2$ respectively the distribution of $N(E)$ (Poisson distribution) and that of $\Delta B(I)$ (normal distribution). Then (4.5) can be written as

(4.5′) $$\iint h(x, y) x^p y^q d\sigma_1(x) d\sigma_2(y) = 0.$$

But we have

$$\begin{aligned}
&\iint e^{|tx|+|sy|} \mid h(x, y) \mid d\sigma_1(x) d\sigma_2(y) \\
&\leq \left( \iint \mid h(x, y) \mid^2 d\sigma_1(x) d\sigma_2(y) \right)^{1/2} \left( \iint e^{2|tx|+2|sy|} d\sigma_1(x) d\sigma_2(y) \right)^{1/2} \\
&= \| h(N(E), B(I)) \| \left( \int e^{2|tx|} d\sigma_1(x) \int e^{2|sy|} d\sigma_2(y) \right)^{1/2} \\
&< \infty.
\end{aligned}$$

Therefore (4.5′) will imply

$$\iint h(x, y) e^{i(tx+sy)} d\sigma_1(x) d\sigma_2(y) = 0, \qquad -\infty < t, s < \infty.$$

Thus we have $h(x, y)=0$ for almost all $(x, y)$ with respect to the measure $d\sigma_1(x)d\sigma_2(y)$, that is, $h(N(E), \Delta B(I))=0$ for almost all $\omega$.

2°. LEMMA 2. *The elements of the following form constitute a fundamental set in* $\boldsymbol{H_X}$:

$$Y = N(E_1) \cdots N(E_m)\Delta B(I_1) \cdots \Delta B(I_n), \tag{4.6}$$

*where the* $E_i$ *are pairwise disjoint and the* $I_j$ *are also pairwise disjoint.*

**Proof.** Making use of the fact that the independence of $X$ and $Y$ implies $\|XY\|=\|X\|\cdot\|Y\|$, we can easily see the following fact:

(*) Let $\{X_1, X_2 \cdots\}$ and $\{Y_1, Y_2 \cdots\}$ be two independent sequences of random variables. If $X_n \to X$ and $Y_n \to Y$ in $L^2(\Omega)$, then $X_nY_n \to XY$ in $L^2(\Omega)$.

Let $\boldsymbol{N}$ denote the closed linear manifold in $\boldsymbol{H_X}$ spanned by the elements of the form $N(E_1)N(E_2) \cdots N(E_m)$ with pairwise disjoint $E_i$ and let $\boldsymbol{B}$ be the closed linear manifold spanned by the elements of the form $\Delta B(I_1)\Delta B(I_2) \cdots \Delta B(I_n)$ with pairwise disjoint $I_j$. By the above remark (*) it is enough to show that

$$N(E_1)^{p_1}N(E_2)^{p_2} \cdots N(E_m)^{p_m} \in \boldsymbol{N} \tag{4.7}$$

and

$$\Delta B(I_1)^{q_1}\Delta B(I_2)^{q_2} \cdots \Delta B(I_n)^{q_n} \in \boldsymbol{B} \tag{4.8}$$

whenever the $E_i$ or the $I_j$ are disjoint. (4.8) was proved in [4, Theorem 4.2].

It remains to prove (4.7). Consider a subdivision $\{F_i\}$, $i=1, \cdots, s$, of $\{E_i\}$, $i=1, \cdots, m$, so fine that $\nu(F_i) < \text{Min}\,(\epsilon/\nu(E), 1)$ where $\epsilon>0$ and $E=F_1\cup F_2\cup \cdots \cup F_s$. Then we have the following expression:

$$N = N(E_1)^{p_1}N(E_2)^{p_2} \cdots N(E_m)^{p_m} = \sum N(F_{i(1)})^{j_1} \cdots N(F_{i(r)})^{j_r}$$

with $i(1)<i(2)< \cdots <i(r)$.

Since $N(F_i)$ takes only non-negative integral values, we have

$$N \geqq \sum N(F_{i(1)})N(F_{i(2)}) \cdots N(F_{i(r)}) \equiv N_\epsilon \in \boldsymbol{N},$$

and also

$$\begin{aligned} P(N \neq N_\epsilon) &= P(N(F_i) \geqq 2 \text{ for some } i) \\ &\leqq \sum_{i=1}^{s} P(N(F_i) \geqq 2) \leqq \sum_{i=1}^{s} \nu(F_i)^2 < \epsilon. \end{aligned}$$

Therefore $N_\epsilon \to N$ in probability as $\epsilon \to 0$. Thus we can choose a sequence $\epsilon(n)$ $(\to 0)$ such that $N_{\epsilon(n)} \to N$ almost everywhere in $\omega$. Further we have $0 \leqq N_{\epsilon(n)} \leqq N$ and $N \in L^2(\Omega)$. Therefore we have $\|N_{\epsilon(n)} - N\| \to 0$ and so $N \in \boldsymbol{N}$.

3°. Now we shall deduce Theorem 2 from Lemma 2. It is enough to show that $Y$ in (4.6) is expressible as

$$Y = I_0(f_0) + I_1(f_1) + \cdots + I_{m+n}(f_{m+n}). \tag{4.9}$$

Using the notation in §3, we set

$$f(\xi_1, \cdots, \xi_m, \xi_{m+1}, \cdots, \xi_{m+n})$$
$$= u_1^{-1} \cdots u_m^{-1} \sigma^{-n} \chi_{E_1}(\xi_1) \cdots \chi_{E_n}(\xi_n)\chi_{I_1}(\xi_{m+1}) \cdots \chi_{I_{m+n}}(\xi_{m+n}),$$

where $\chi_I$ denote the characteristic function of the interval $I \times \{0\}$ (in the plane $\pi$) and $\xi_i = (t_i, u_i)$. Then we have

$$I_{m+n}(f) = (N(E_1) - \nu(E_1)) \cdots (N(E_m) - \nu(E_m))\Delta B(I_1) \cdots \Delta B(I_n)$$
$$= N(E_1) \cdots N(E_m)\Delta B(I_1) \cdots \Delta B(I_n) + R,$$

where $R$ is a linear combination of the elements of the form:

$$N(E_{i(1)}) \cdots N(E_{i(p)})\Delta B(I_1) \cdots \Delta B(I_n), \qquad p < m,$$

and $R=0$ in case $m=0$. Therefore we obtain (4.9) by induction on $m$. Thus our theorem is completely proved.

5. **Spectral type of** $\{T_\tau\}$. In the last section we proved that

$$H_X = \sum_{p \geqq 0} \oplus H_X^{(p)} \tag{5.1}$$

and that each $H_X^{(p)}$ is isomorphic to $\tilde{L}_p^2$ by $V_p$. Now we shall investigate the behavior of the group of unitary transformations $U_\tau$ on $H_X$ derived from the shift transformations $T_\tau$ of the process $X(t, \omega)$.

Let $U_\tau^{(p)}$ be the restriction of $U_\tau$ to $H_X^{(p)}$. Then we have

LEMMA 1. *$U_\tau^{(p)}$ is a unitary operator on $H_X^{(p)}$ which is transformed by $V_p$ into the following unitary operator $\tilde{S}_\tau^{(p)}$ on $\tilde{L}_p^2$.*

$$\tilde{S}_\tau^{(p)}\tilde{f}(t_1, u_1, \cdots, t_p, u_p) = \tilde{f}(t_1 - \tau, u_1, \cdots t_p - \tau, u_p), \qquad \tilde{f} \in \tilde{L}_p^2. \tag{5.2}$$

*Corresponding to the decomposition* (5.1), *we have*

$$U_\tau = \sum_{p \geqq 0} \oplus U_\tau^{(p)}. \tag{5.3}$$

**Proof.** By the same expression as in (5.2) we shall define a unitary operator $S_\tau^{(p)}$ on $L_p^2$. It is clear that $S_\tau^{(p)}$ is an extension of $\tilde{S}_\tau^{(p)}$. Then we have

$$I_p(S_\tau^{(p)}f) = U_\tau I_p(f). \tag{5.4}$$

This is clear by the definition if $f$ has the form (3.5), and so it holds also if $f \in L_p^2$, because both sides of (5.4) are bounded and linear in $f$.

Since $U_\tau$ transforms $H_X^{(p)}$ onto $H_X^{(p)}$ by (5.4), the restriction $U_\tau^{(p)}$ of $U_\tau$ to

$H_X^{(p)}$ is a unitary operator on $H_X^{(p)}$. By (5.4) and the definition of $V_p$ and $\tilde{S}_\tau^{(p)}$ we have

$$U_\tau^{(p)} = V_p \tilde{S}_\tau^{(p)} V_p^{-1}.$$

Thus we can easily verify (5.3), since $U_\tau$ is a unitary operator on $H_X$.

Lemma 1 reduces the investigation of $\{U_\tau\}$ to that of $\{\tilde{S}_\tau^{(p)}\}$. $\tilde{S}_\tau^0$ is only the identity operator on $L_0^2$, which is a one-dimensional Hilbert space. We consider $\tilde{S}_\tau^{(p)}$, $p \geqq 1$. We shall introduce two transformation groups, $\{(\tau)\}$ and $\{(\epsilon)\}$, on the $2p$-dimensional space $\pi^p$ as follows:

$$(\tau)(t_1, u_1, t_2, u_2, \cdots, t_p, u_p) = (t_1 - \tau, u_1, t_2 - \tau, u_2, \cdots, t_p - \tau, u_p),$$

$$(\epsilon)(t_1, u_1, t_2, u_2, \cdots, t_p, u_p) = (t_{\epsilon(1)}, u_{\epsilon(1)}, t_{\epsilon(2)}, u_{\epsilon(2)}, \cdots, t_{\epsilon(p)}, u_{\epsilon(p)}),$$

where $\tau$ is a real number and $\epsilon = \{\epsilon(1), \epsilon(2), \cdots, \epsilon(p)\}$ is an arrangement of $\{1, 2, \cdots, p\}$.

To make it easier to see how these transformations act on $\pi^p$, we shall consider the following coordinate transformation:

$$(t_1, u_1, t_2, u_2, \cdots, t_p, u_p)$$
$$\rightarrow (t, s_2, \cdots, s_{p-1}, u_1, \cdots, u_p) = (t, v), \ v \in R^{2p-1}$$

which is defined by

$$t = \frac{1}{p}(t_1 + t_2 + \cdots + t_p), \tag{5.5}$$

$$s_i = \sum_{j=1}^{p} \alpha_{ij} t_j, \qquad i = 1, 2, \cdots, p-1, \tag{5.6}$$

where $(\alpha_{ij};\ 1 \leqq i \leqq p-1,\ 1 \leqq j \leqq p)$ is a certain real matrix satisfying

$$\sum_j \alpha_{ij} = 0, \qquad 1 \leqq i \leqq p-1, \tag{5.7}$$

$$\begin{vmatrix} 1/p & \cdots & 1/p \\ \alpha_{11} & \cdots & \alpha_{1p} \\ \cdot & \cdots & \cdot \\ \alpha_{p-1,1} & \cdots & \alpha_{p-1,p} \end{vmatrix} = 1. \tag{5.8}$$

By (5.8) we have

$$dt_1 d\mu'(u_1) \cdots dt_p d\mu'(u_p) = dt ds_1 \cdots ds_{p-1} d\mu'(u_1) \cdots d\mu'(u_p) = dt d\lambda_p(v)$$

where

$$d\lambda_p(v) = ds_1 \cdots ds_{p-1} d\mu'(u_1) \cdots d\mu'(u_p).$$

Since $t$ is symmetric in $(t_1, \cdots, t_p)$, the transformation $(\epsilon)$ will leave $t$ invariant, and we have

$$(\epsilon)(t, v) = (t, (\bar{\epsilon})v),$$

where $(\bar{\epsilon})$ is defined as follows. Given $t$ and $v=(s_1, \cdots, s_{p-1}, u_1, \cdots, u_p)$, let $(t_1, t_2, \cdots, t_p)$ be the solution of (5.5) and (5.6) and set

$$(\bar{\epsilon})v = (s'_1, \cdots, s'_{p-1}, u_1, \cdots, u_p), \quad s'_i = \sum_j \alpha_{ij} t_{\epsilon(j)}, \qquad i = 1, \cdots, p-1.$$

It follows from (5.7) and (5.8) that $(\bar{\epsilon})v$ is quite independent of $t$, so that we may consider $(\bar{\epsilon})v$ to be a transformation on $v$.

Since the transformation $(\tau)$ will leave $s_i$ and accordingly $v$ invariant by (5.7), we have

$$(\tau)(t, v) = (t + \tau, v).$$

As the transformation $(\epsilon)$ preserves the measure $dt_1 \cdots dt_p d\mu'(u_1) \cdots d\mu'(u_p)$, we have

$$dtd\lambda_p((\bar{\epsilon})v) = dtd\lambda_p(v) \text{ i.e. } d\lambda_p((\bar{\epsilon})v) = d\lambda_p(v).$$

Hereafter we shall consider the functions in $L_p^2$ with respect to these new coordinates $(t, v)$. Then we see that the condition

$$f(t, (\bar{\epsilon})v) = f(t, v)$$

is necessary and sufficient for $f$ to belong to $\tilde{L}_p^2$.

We consider the $L^2$-space over the measure space $(R^{2p-1}, d\lambda_p)$ which will be denoted by $\mathcal{L}_{2p-1}^2$. The totality of the functions in $\mathcal{L}_{2p-1}^2$ invariant under the group $\{(\bar{\epsilon})\}$ constitutes a closed linear manifold in $\mathcal{L}_{2p-1}^2$, say $\tilde{\mathcal{L}}_{2p-1}^2$. Excluding the trivial case that $X(t, \omega) = \gamma t + \alpha$, the measure $d\lambda_p$ is not identically zero for $p = 1, 2, \cdots$, so that $\tilde{\mathcal{L}}_{2p-1}^2$ is at least one-dimensional. In addition to this, the dimension of $\tilde{\mathcal{L}}_{2p-1}^2$ is at most countable. Let $\tilde{\phi}_{pi}(v)$, $i = 1, 2, \cdots$, be a complete orthonormal system in $\tilde{\mathcal{L}}_{2p-1}^2$. Then $f(t, v) \in \tilde{L}_p^2$ is expressed as a sum of orthogonal components:

$$f(t, v) = \sum_n f_n(t)\tilde{\phi}_{pn}(v), \qquad f_n(t) \in L^2(R^1)$$

and we have

$$\|f(t, v)\|^2 = \sum_n \|f_n\|^2$$

$$(\tilde{S}_\tau^{(p)} f)(t, v) = f(t + \tau, v) = \sum_i f_i(t + \tau)\tilde{\phi}_{pi}(v).$$

Thus $\tilde{L}_p^2$ is the direct sum of at most a countable number ($>0$) of subspaces which reduce $\{\tilde{S}_\tau^{(p)}\}$ and on each of which $\tilde{S}_\tau^{(p)}$ acts just as the unitary transformation $f(t) \to f(t+\tau)$ does on $L^2(R^1)$.

Summing up the above arguments we obtain the following

FUNDAMENTAL THEOREM. *Except for the trivial case $X(t, \omega) = \gamma t + \alpha$, $\boldsymbol{H_X}$ is isomorphic to the direct sum of the complex number field $C$ (one-dimensional*

*Hilbert space) and a countable number of Hilbert spaces, each isomorphic to* $L^2(R^1)$, *i.e.*

$$H_X \cong C \oplus L^2(R^1) \oplus L^2(R^1) + \cdots,$$

*in such a way that this isomorphism transforms* $U_\tau$ *into the following operator*:

$$\Phi_\tau:(c, f_1(t), f_2(t), \cdots) \to (c, f_1(t+\tau), f_2(t+\tau), \cdots).$$

In other words, $\{U_\tau\}$ has spectra [2] of multiplicity one over the unitary measure and of uniform multiplicity $\aleph_0$ over the ordinary Lebesgue measure and only these spectra.

## References


1. J. L. Doob, *Stochastic processes*, New York, 1952.
2. P. R. Halmos, *Introduction to Hilbert space and the theory of spectral multiplicity*, New York, 1951.
3. K. Itô, *On stochastic processes*, Jap. J. Math. vol. 18 (1942) pp. 261–301.
4. ———, Multiple Wiener integral, Jour. of Math., Soc. of Japan. vol. 3 (1951) pp. 157–169.
5. ———, *Complex multiple Wiener integral*, Jap. J. of Math. vol. 22 (1952) pp. 63–86.
6. S. Kakutani, *Determination of the spectrum of the flow of Brownian motion*, Proc. Nat. Acad. Sci. U.S.A. vol. 36 (1950) pp. 319–323.
7. P. Lévy, *Théorie de l'addition des variables aléatoires*, Paris, 1937, rev. ed., 1954.
8. J. v. Neumann, *Zur Operatorenmethode der klassischen Mechanik*, Ann. of Math. vol. 33 (1932) pp. 587–648.

   P. R. Halmos and J. v. Neumann, *Operator methods in classical mechanics*, II, Ann. of Math. vol. 43 (1942) pp. 332–350.
9. I. E. Segal, *Tensor algebras over Hilbert spaces*. I, Trans. Amer. Math. Soc. vol. 81 (1956) pp. 106–134.


Princeton University,
Princeton, N. J.

Reprinted from
*Trans. Amer. Math. Soc.* **81**, 253–263 (1956)

Reprinted from ILLINOIS JOURNAL OF MATHEMATICS
Vol. 4, No. 1, March 1960
*Printed in U.S.A.*

# POTENTIALS AND THE RANDOM WALK

BY
KIYOSI ITÔ AND H. P. MCKEAN, JR.[1]

## 1. Introduction

Given an integer $s \geqq 3$, write $e_1, e_2, \cdots, e_s$ for the $s$ coordinate vectors $(1, 0, \cdots, 0), (0, 1, \cdots, 0), \cdots, (0, 0, \cdots, 1)$, spanning the $s$-dimensional lattice of points with integral coordinates, and let $s_n$ denote the position at time $n$ $(= 0, 1, 2, \cdots)$ of a particle performing the standard $s$-dimensional random walk according to the following rule: fixing the first $n - 1$ steps $s_1, s_2, \cdots, s_{n-1}$, the particle starts afresh at $s_{n-1}$, jumping next to one of the $2s$ neighbors $s_n = s_{n-1} \pm e_1, s_{n-1} \pm e_2, \cdots, s_{n-1} \pm e_s$ of $s_{n-1}$, the chance of landing at a particular neighbor being $(2s)^{-1}$.

Given a set $B$ of lattice points, the probability $p_B$ that the random walk hits $B$ at some time $n < +\infty$, as a function of the starting point of the walk, is *excessive* in the sense that $\mathbf{G}p_B \leqq 0$, where $\mathbf{G}$ is Laplace's difference operator:

1.1 $$(\mathbf{G}p)(a) = (2s)^{-1} \sum_{k \leqq s, n=1,2} p(a + (-)^n e_k) - p(a).$$

B. H. Murdoch [1, pp. 13–19] proved that if $p \geqq 0$, and if $\mathbf{G}p = 0$, then $p$ is constant,[2] and, with the help of this result, it follows, as Murdoch himself noted, that $p_B$ is the sum of the *potential* $\mathbf{K}e_B$ and the constant $p_B(\infty)$, where $e_B = -\mathbf{G}p_B$ $(\geqq 0)$, $\mathbf{K}e_B$ is the expectation of $\sum_{n \geqq 0} e_B(s_n)$, as a function of the starting point of the walk, and $p_B(\infty)$ is the (constant) probability $P.(\mathbf{B})$ of the event $\mathbf{B}$ that $s_n \in B$ for an infinite number of integers $n$.

$P.(\mathbf{B})$ is either 0 or 1. When $P.(\mathbf{B}) = 0$, $p_B$ is the greatest *potential* $p \leqq 1$ such that $\mathbf{G}p = 0$ outside $B$, and, on the strength of the example of the Newtonian potential in 3 dimensions, it is natural to think of $e_B$ as the electrostatic distribution of charge on the conductor $B$ and to introduce the total charge (of $e_B$) as the capacity $C(B)$ of $B$.

Given a set $B$, it is an interesting problem to decide whether $P.(\mathbf{B}) = 0$ or 1; the solution is

1.2 $$P.(\mathbf{B}) = 0 \text{ or } 1 \quad \text{according as} \quad \sum_{n \geqq 0} 2^{-n(s-2)} C(B_n) < \text{ or } = +\infty,$$

where $B_n$ is the intersection of $B$ and the spherical shell $2^n \leqq |a| < 2^{n+1}$. Wiener's test for the singular points of the Newtonian electrostatic potential (see Courant and Hilbert [1, p. 286]) served us as a model, and for this reason we call 1.2 Wiener's test also. B. H. Murdoch [1, pp. 45–47] came close to proving 1.2 and used his method to compute $P.(\mathbf{B})$ for sets $B$ similar to those

Received September 25, 1958.

[1] Fulbright grantee 1957–1958.

[2] J. Capoulade [1] also stated this result and S. Verblunsky [1] and R. Duffin [1, pp. 242–245] proved it. Murdoch's results lie much deeper.

figuring in the last example of Section 6. Similar results hold for the Brownian motion and Newtonian potentials in $s\ (\geqq 2)$ dimensions and for stable processes with exponent $< 2\ (< 1)$ and Riesz potentials in $s \geqq 2\ (\geqq 1)$ dimensions; for the identification of hitting probabilities and electrostatic potentials, see J. L. Doob [2] and G. Hunt [1, 2]; the proof of Wiener's test runs along the lines of Section 5. G. Hunt [1, 2, 3] showed the full scope and power of the connection between Markov processes and potentials, and most of the (nonclassical) results of Sections 3 and 4 are from his papers [1, 2] or from B. H. Murdoch. The present paper is based on lectures given at Kyôto and Fukuoka in April, 1958, in which we tried to present Hunt's things in the simplest setting.

We thank J. L. Doob, N. Ikeda, and D. Ray for spotting several misprints.

## 2. The random walk

Write $W$ for the space of paths $w: n = 1, 2, \cdots \to s_n(w)$ with values from the $s$-dimensional lattice, $w_m^-$ for the stopped path $s_n(w_m^-) = s_{n\wedge m}(w)$,[3] $w_m^+$ for the shifted path $s_n(w_m^+) = s_{n+m}(w)$, $\mathbf{C}$ for the class of Borel subsets of $W$, and, for $\mathbf{C}$ measurable $m = m(w) \geqq 0$, write $\mathbf{C}_m$ for the Borel algebra of sets

$$A = (w: w_m^- \in C), \qquad C \in \mathbf{C};$$

let $P_.(C)$ denote the probability of the event $C \in \mathbf{C}$ for the standard random walk as a function of the starting point of the walk; note that $P_.(w_n^+ \in C/\mathbf{C}_n) = P_{s_n}(C)$ for each $n \geqq 0$ and each $C \in \mathbf{C}$; and let $E_.(f) = \int f P_.(dw)$.

$m = m(w) = 0, 1, 2, \cdots$ is a *Markov time* if $(w: m > n) \in \mathbf{C}_n$ for each $n \geqq 0$; for example, given a set $B$ of lattice points, $m = n_B = \min(n: s_n \in B)$ is Markov.

Given a Markov time $m$,

$$P_.(P_.(w_m^+ \in C/\mathbf{C}_m) = P_{s_m}(C)) = 1, \qquad C \in \mathbf{C}; \tag{2.1}$$

in short, $s_n: n \geqq 0$ starts from scratch at time $n = m$ at the place $s_m$.

The proof is simple: $(w: m = n) \in \mathbf{C}_n$ for each $n \geqq 0$, and therefore

$$\begin{aligned} P_.(w_m^+ \in C^+, w_m^- \in C^-) &= \textstyle\sum_{n\geqq 0} P_.(w_n^+ \in C^+, w_n^- \in C^-, m = n) \\ &= \textstyle\sum_{n\geqq 0} E_.(P_{s_n}(C^+), w_n^- \in C^-, m = n) \\ &= E_.(P_{s_m}(C^+), w_m^- \in C^-), \qquad C^-, C^+ \in \mathbf{C}. \end{aligned} \tag{2.2}$$

Given a point $\theta = (\theta_1, \theta_2, \cdots, \theta_s)$ of the $s$-dimensional torus $[-\pi, \pi)^s$,

$$\begin{aligned} E_a(e^{is_n\cdot\theta}) = E_a(E_{s_{n-1}}(e^{is_1\cdot\theta})) &= E_a(e^{is_{n-1}\cdot\theta})f(\theta) = e^{ia\cdot\theta}f(\theta)^n, \\ f(\theta) &= s^{-1}\textstyle\sum_{k\leqq s} \cos\theta_k, \quad n \geqq 0.^{4} \end{aligned} \tag{2.3}$$

[3] $m \wedge n$ means the smaller of $m$ and $n$.

[4] $s_n \cdot \theta$ is the inner product of the vectors $s_n$ and $\theta$.

and inverting 2.3 proves the result of G. Pólya [1, pp. 151–153]:

2.4 $$P_a(s_n = b) = (2\pi)^{-s} \int e^{i(b-a)\cdot\theta} f(\theta)^n \, d\theta, \qquad n \geq 0.$$

Summing 2.4 for $n = 0, 1, 2, \cdots$ proves

2.5 $$\begin{aligned} K(a, b) &= \sum_{n \geq 0} P_a(s_n = b) = (2\pi)^{-s} \int e^{i(b-a)\cdot\theta} \frac{d\theta}{g(\theta)} \\ &\leq (2\pi)^{-s} \int \frac{d\theta}{g(\theta)} < +\infty, \qquad g(\theta) = 1 - f(\theta), \end{aligned}$$

and using the Borel-Cantelli lemma, we infer that

2.6 $$P.(\lim_{n\uparrow+\infty} s_n = \infty) = 1.$$

B. H. Murdoch [1, pp. 23–32] and (for $s = 3$) R. Duffin [1, pp. 238–240, 245–251] estimated $K(a, b)$ for $|a - b| \uparrow +\infty$ up to terms of magnitude $|a - b|^{-s-2}$.

We will want the following simpler result:

2.7 $$\begin{aligned} &\lim_{|b-a|\uparrow+\infty} |b - a|^{s-2} K(a, b) \\ &\quad = \lim_{|b-a|\uparrow+\infty} |b - a|^{s-2} (2\pi)^{-s} \int e^{i(b-a)\cdot\theta} \frac{d\theta}{g(\theta)} = k_1, \\ &\qquad k_1 = s \int_0^{+\infty} (2\pi t)^{-s/2} e^{-1/2t} \, dt.^{5} \end{aligned}$$

Consider, for the proof, the modified Bessel coefficients

2.8 $$\begin{aligned} I_n(t) &= \frac{1}{2\pi} \int_{-\pi}^{\pi} e^{in\theta} e^{t\cos\theta} \, d\theta \\ &= \frac{e^t}{2\pi} \int_{-\pi}^{\pi} e^{in\theta} e^{-2t\sin^2(\theta/2)} \, d\theta, \qquad n = 0, \pm 1, \pm 2, \cdots, \end{aligned}$$

introduce the (positive) Fourier coefficients

2.9 $$\begin{aligned} &s(2\pi)^{-s} \int e^{ic\cdot\theta} e^{-stg(\theta)} \, d\theta \\ &\quad = s(2\pi)^{-1} \int_{-\pi}^{\pi} e^{il_1\theta_1} e^{-2t\sin^2(\theta_1/2)} \, d\theta_1 \cdots (2\pi)^{-1} \int_{-\pi}^{\pi} e^{il_s\theta_s} e^{-2t\sin^2(\theta_s/2)} \, d\theta_s \\ &\quad = se^{-t} I_{l_1}(t) e^{-t} I_{l_2}(t) \cdots e^{-t} I_{l_s}(t), \qquad t \geq 0, \quad c = (l_1, l_2, \cdots, l_s), \end{aligned}$$

and note that

2.10 $$\begin{aligned} (2\pi)^{-s} \int e^{ic\cdot\theta} \frac{d\theta}{g(\theta)} &= \int_0^{+\infty} dt \; s(2\pi)^{-s} \int e^{ic\cdot\theta} e^{-stg(\theta)} \, d\theta \\ &= s \int_0^{+\infty} e^{-st} \, dt \; I_{l_1}(t) I_{l_2}(t) \cdots I_{l_s}(t). \end{aligned}$$

[5] $k_1, k_2, \cdots$ denote positive constants.

Given $\theta \in [-\pi, \pi)^s$, $g(\theta) > k_2|\theta|^2$, so that

2.11 $$e^{-st|c|^2 g(\theta/|c|)} < e^{-stk_2|\theta|^2}, \qquad t \geqq 0, \quad \theta \in [-\pi|c|, \pi|c|)^s,$$

and

$$\Big| |c|^{s-2} \int_{k_3|c|^2}^{+\infty} s(2\pi)^{-s} \int_{[-\pi,\pi)^s} e^{ic\cdot\theta} e^{-stg(\theta)}\, d\theta\, dt - \int_{k_3}^{+\infty} s(2\pi)^{-s} \int_{|\theta|\leqq\pi|c|} e^{i(c/|c|)\cdot\theta} e^{-(t/2)|\theta|^2}\, d\theta\, dt \Big|$$

2.12 $$\leqq s(2\pi)^{-s} \int_{k_3}^{+\infty} dt \Big| \int_{[-\pi|c|,\pi|c|)^s} e^{i(c/|c|)\cdot\theta} e^{-st|c|^2 g(\theta/|c|)}\, d\theta - \int_{|\theta|\leqq\pi|c|} e^{i(c/|c|)\cdot\theta} e^{-(t/2)|\theta|^2}\, d\theta \Big|$$

$$\leqq s(2\pi)^{-s} \int_{k_3}^{+\infty} dt \Big[ \int_{|\theta|\leqq\pi|c|} | e^{-st|c|^2 g(\theta/|c|)} - e^{-(t/2)|\theta|^2} |\, d\theta + \int_{\pi|c|<|\theta|<s^{1/2}\pi|c|} e^{-stk_2|\theta|^2}\, d\theta \Big]$$

$$\to 0, \qquad |c| \uparrow +\infty;$$

this implies

$$\lim_{|c|\uparrow+\infty} |c|^{s-2} \int_{k_3|c|^2}^{+\infty} s(2\pi)^{-s} \int_{[-\pi,\pi)^s} e^{ic\cdot\theta} e^{-stg(\theta)}\, d\theta\, dt$$

2.13 $$= \int_{k_3}^{+\infty} dt\ s(2\pi)^{-s} \int_{R^s} e^{i\theta_1} e^{-(t/2)|\theta|^2}\, d\theta = \int_{k_3}^{+\infty} s(2\pi t)^{-s/2} e^{-1/2t}\, dt$$

$$\uparrow s \int_0^{+\infty} (2\pi t)^{-s/2} e^{-1/2t}\, dt = k_1, \qquad k_3 \downarrow 0;$$

and to complete the proof, it is enough to check that

2.14 $$\lim_{k_3\downarrow 0} \limsup_{|c|\uparrow+\infty} |c|^{s-2} \int_0^{k_3|c|^2} dt\ s(2\pi)^{-s} \int_{[-\pi,\pi)^s} e^{ic\cdot\theta} e^{-stg(\theta)}\, d\theta = 0.$$

But, as is clear from 2.8 and 2.9,

2.15 $$\limsup_{|c|\uparrow+\infty} |c|^{s-2} \int_0^{k_3|c|^2} dt\ s(2\pi)^{-s} \int_{[-\pi,\pi)^s} e^{ic\cdot\theta} e^{-stg(\theta)}\, d\theta \leqq s^{s-1} \limsup_{n\uparrow+\infty} n^{s-2} \int_0^{k_3 s^2 n^2} e^{-t} I_n(t) (e^{-t} I_0(t))^{s-1}\, dt,$$

where $n$ is the greatest of the integers $|l_1|, |l_2|, \cdots, |l_s|$, and since, in view of 2.13,

2.16
$$\begin{aligned}&\lim_{n\uparrow+\infty} n^{s-2}s\int_{k_3 s^2 n^2}^{+\infty} e^{-t}I_n(t)(e^{-t}I_0(t))^{s-1}\,dt\\ &= \lim_{n\uparrow+\infty} n^{s-2}s\int_{k_3 s^2 n^2}^{+\infty}(2\pi)^{-s}\int_{[-\pi,\pi)^s} e^{in\theta_1}e^{-stg(\theta)}\,d\theta\,dt\\ &\uparrow k_1, \qquad\qquad k_3\downarrow 0,\end{aligned}$$

it is enough to show that

2.17
$$k_1 \geqq \limsup_{n\uparrow+\infty} n^{s-2}s\int_0^{+\infty} e^{-t}I_n(t)(e^{-t}I_0(t))^{s-1}\,dt.$$

With the help of

2.18
$$1 \geqq e^{-t}I_0(t) = \frac{1}{2\pi}\int_0^{2\pi} e^{-2t\sin^2(\theta/2)}\,d\theta \sim (2\pi t)^{-1/2}, \qquad t\uparrow+\infty,$$

and of

2.19
$$\int_0^r e^{-t}I_n(t)\,dt = \frac{1}{2\pi}\int_0^{2\pi} e^{in\theta}\frac{1-e^{-2r\sin^2(\theta/2)}}{2\sin^2(\theta/2)}\,d\theta \leqq n^{-s},$$
$$n\uparrow+\infty,\ +\infty > r \geqq 0,$$

our task now simplifies to showing that

2.20
$$k_1 \geqq \limsup_{n\uparrow+\infty} n^{s-2}s\int_0^{+\infty} e^{-t}I_n(t)(2\pi t)^{-(s-1)/2}\,dt,$$

and, consulting A. Erdélyi [2, p. 196 (8) and 1, p. 164 (20)], it is seen that

2.21
$$s\int_0^{+\infty} e^{-t}I_n(t)(2\pi t)^{-(s-1)/2}\,dt \sim k_1 n^{2-s}, \qquad n\uparrow+\infty,$$

which completes the proof.

## 3. Potentials

Write $e$ for nonnegative functions defined for the points of the $s$-dimensional lattice, and consider the Green operator $\mathbf{K}$ defined in

3.1
$$\begin{aligned}(\mathbf{K}e)(a) &= E_a\left(\textstyle\sum_{n\geqq 0} e(s_n)\right)\\ &= \textstyle\sum\left[\sum_{n\geqq 0} P_a(s_n = b)\right]e(b) = \sum K(a,b)e(b).\end{aligned}$$

$-\mathbf{G}$ is inverse to $\mathbf{K}$; for, if $\mathbf{K}e < +\infty$, then

$$\begin{aligned}-\mathbf{GK}e &= E.\left(\textstyle\sum_{n\geqq 0} e(s_n)\right) - E.\,E_{s_1}\left(\textstyle\sum_{n\geqq 0} e(s_n)\right)\\ &= E.\left(\textstyle\sum_{n\geqq 0} e(s_n)\right) - E.\left(\textstyle\sum_{n\geqq 0} e(s_{n+1})\right)\\ &= E.(e(s_0)) = e.\end{aligned}$$

A nonnegative function $p$ is *excessive* if $\mathbf{G}p \leqq 0$; it is a *potential* if $p = \mathbf{K}e$ with $e \geqq 0$; if $p = \mathbf{K}e$ ($e \geqq 0$, $p < +\infty$), then, as the reader will check, $\mathbf{G}p = -e \leqq 0$, so that potentials are excessive.

Given an excessive function $p$,

$$p = E.(\sum_{l \leqq n} (p(s_{l-1}) - p(s_l)) + p(s_n))$$
$$= \sum_{l \leqq n} E.(p(s_{l-1}) - E_{s_{l-1}}(p(s_1))) + E.(p(s_n))$$
$$= E.(\sum_{l \leqq n} -(\mathbf{G}p)(s_{l-1})) + E.(p(s_n)),$$

and, after putting $-\mathbf{G}p = e$ and $p_\infty = \lim_{n \uparrow +\infty} E.(p(s_n))$, it results that

3.2 $$p = \mathbf{K}e + p_\infty .$$

$p_\infty$ is constant; indeed, it is nonnegative, $\mathbf{G}p_\infty = 0$, and, as B. H. Murdoch proved, *such a nonnegative harmonic function is constant.*

J. L. Kelley's proof[6] that a compact convex set is the convex hull of its extreme points served us as a model for the following simple proof of Murdoch's result.

Write $C$ for the class of nonnegative $p$ with $p(0) = 1$ and $\mathbf{G}p = 0$, label the points $c = (l_1, l_2, \cdots, l_s)$ of the $s$-dimensional lattice $c_1$ $(=0)$, $c_2, c_3, \cdots$, and, using the compactness that the estimate

3.3 $$p(l_1, l_2, \cdots, l_s) \leqq (2s)^{|l_1|+|l_2|+\cdots+|l_s|}, \qquad p \in C,$$

provides, put $m_1 = 1$, $C_1 = C$, $m_2 = \max_{p \in C_1} p(c_2)$, $C_2 = C_1 \cap (p: p(c_2) = m_2)$, $m_3 = \max_{p \in C_2} p(c_3)$, $C_3 = C_2 \cap (p: p(c_3) = m_3)$, etc., and select $p_* \in \bigcap_{n \geqq 1} C_n$. $p_*(\pm e_k)$ is then $> 0$, $p_*(\pm e_k)^{-1} p_*(\cdot \pm e_k) \in C$ for each $k \leqq s$, and

3.4 $$p_* = \sum_{k \leqq s, n=1,2} (2s)^{-1} p_*((-)^n e_k) \frac{p_*(\cdot + (-)^n e_k)}{p_*((-)^n e_k)}.$$

Since $\sum_{k \leqq s, n=1,2} (2s)^{-1} p_*((-)^n e_k) = p_*(0) = 1$, the definition of $p_*$ now implies that $p_*(\pm e_k)^{-1} p_*(\cdot \pm e_k) \in \bigcap_{n \geqq 1} C_n$ for each $k \leqq s$; in short, $p_*(\pm e_k)^{-1} p_*(\cdot \pm e_k) = p_*$ for each $k \leqq s$, and we infer that

3.5 $$p_*(l_1, l_2, \cdots, l_s) = p_*(e_1)^{l_1} p_*(e_2)^{l_2} \cdots p_*(e_s)^{l_s}.$$

But

3.6 $$1 = p_*(0) = E_0(p_*(s_1)) = (1/s) \sum_{k \leqq s} \tfrac{1}{2}(p_*(e_k)^{-1} + p_*(e_k)),$$

proving that $p_*(e_k) = 1$ for each $k \leqq s$; therefore $p_* \equiv 1$; since $c_2$ was chosen at pleasure, each $p \in C$ is $\leqq 1$; and since, for $p \in C$, $\mathbf{G}p \equiv 0$, $p \equiv 1$ is the sole member of $C$.

Keeping this result in mind, it is clear from 3.2 that, with the notation $p(\infty) = \liminf_{c \to \infty} p(c)$,

3.7 $$p(\infty) \geqq p_\infty = \lim_{n \uparrow +\infty} E.(p(s_n)) \geqq E.(\liminf_{n \uparrow +\infty} p(s_n)) \geqq p(\infty),$$

and 3.2 goes over into

3.8 $$p = \mathbf{K}e + p(\infty), \qquad p(\infty) = \lim_{n \uparrow +\infty} E.(p(s_n)).$$

[6] See P. T. Bateman [1, pp. 14–15].

Given a set $B$ of lattice points, let $n_B$ denote the hitting time $\min(n: s_n \in B)$ and $p_B$ the hitting probability $P.(n_B < +\infty)$.

$p_B$ is excessive; in fact,

$$
\begin{aligned}
e_B &= -\mathbf{G}p_B = p_B - E.(p_B(s_1)) \\
&= P.(n_B < +\infty) - P.(n_B(w_1^+) < +\infty) \qquad (3.9)\\
&= P.(n_B < +\infty, n_B(w_1^+) = +\infty) \geqq 0,
\end{aligned}
$$

and it follows from 2.6 and 3.9 that $e_B = 0$ off the points of[7] $\partial B$ neighboring the (connected) part of the complement of $B$ reaching out to $\infty$.

Writing $\mathbf{B}$ for the event, $\bigcap_{n \geq 1} (w: n_B(w_n^+) < +\infty)$ that $s_n \in B$ for an infinite number of times $n$, it results from

$$
3.10 \quad p_B(\infty) = \lim_{n\uparrow+\infty} E.(p_B(s_n)) = \lim_{n\uparrow+\infty} P.(n_B(w_n^+) < +\infty) = P.(\mathbf{B})
$$

and from

$$
\begin{aligned}
P.(\mathbf{B}) &= \lim_{n\uparrow+\infty} P.(n_B < +\infty, n_B(w_{n_B+n}^+) < +\infty) \\
&= \lim_{n\uparrow+\infty} E.(n_B < +\infty, P_{s_{n_B}}(n_B(w_n^+) < +\infty)) \qquad (3.11)\\
&= p_B P.(\mathbf{B})
\end{aligned}
$$

that $p_B(\infty) = P.(\mathbf{B})$ is 0 or 1.[8]

Using these results, it is not difficult to prove that, for excessive $p = \mathbf{K}e + p(\infty)$,

$$
3.12 \qquad P.(\lim_{n\uparrow+\infty} p(s_n) = p(\infty)) = 1;^{9}
$$

indeed, $p \geqq p(\infty)$, and if $\alpha > p(\infty)$, if $A$ is the set where $p \geqq \alpha$, and if $p_A(\infty) = P.(\mathbf{A}) = 1$, then

$$
3.13 \qquad (w: l \leqq n \wedge n_A) \in \mathbf{C}_l, \qquad l \geqq 1,
$$

and

$$
\begin{aligned}
p &= E.\left[\textstyle\sum_{l \leqq n\wedge n_A} (p(s_{l-1}) - p(s_l)) + p(s_{n\wedge n_A})\right] \\
&= E.\left[\textstyle\sum_{l \leqq n\wedge n_A} e(s_{l-1})\right] + E.(p(s_{n_A}), n_A \leqq n) \qquad (3.14)\\
&\quad + E.(p(s_n), n < n_A) \\
&\geqq \alpha P.(n_A \leqq n) \uparrow \alpha p_A = \alpha, \qquad n \uparrow +\infty,
\end{aligned}
$$

violating $\alpha > p(\infty)$, and we infer that $P.(\mathbf{A}) = 0$ for each $\alpha > p(\infty)$, completing the proof.

We give the proof of the general *maximum principle* of which 3.14 is a special case.

---

[7] $\partial B$ is the set of points of $B$ not all of whose neighbors belong to $B$.

[8] $P.(\mathbf{B}) = 0$ or 1 is a special case of the 0-or-1 law of Hewitt and Savage [1, pp. 493–494].

[9] 3.12 is a special case of the result of J. L. Doob [1, pp. 324–326] that a nonnegative lower semimartingale converges.

Given excessive $p_1 = \mathbf{K}e_1 + p_1(\infty)$, $p_2 = \mathbf{K}e_2 + p_2(\infty)$, if $p_2 \geqq p_1$ on the support $B$ of $e_1$, and if $p_2(\infty) \geqq p_1(\infty)$ in case $P.(\mathbf{B}) = 0$, then $p_2 \geqq p_1$ on the whole of the $s$-dimensional lattice; in fact, $e_1(s_n) = 0$ for $n < n_B$, and by using the fact that, for excessive $p$, $p \geqq p(\infty)$ and $\lim_{n\uparrow+\infty} E.(p(s_n)) = p(\infty)$, it develops that, for $a \notin B$,

$$
\begin{aligned}
p_2(a) &= E_a\big[\textstyle\sum_{l\leqq n\wedge n_B}(p_2(s_{l-1}) - p_2(s_l)) + p_2(s_{n\wedge n_B})\big] \\
&= E_a\big[\textstyle\sum_{l\leqq n\wedge n_B} e_2(s_{l-1})\big] + E_a(p_2(s_{n_B}), n \geqq n_B) \\
&\qquad\qquad + E_a(p_2(s_n), n < n_B) \\
&\rightarrow E_a\big[\textstyle\sum_{l\leqq n_B} e_2(s_{l-1})\big] + E_a(p_2(s_{n_B}), n_B < +\infty) \\
&\qquad\qquad + p_2(\infty)P_a(n_B = +\infty) \qquad (n\uparrow+\infty) \\
&\geqq E_a\big[\textstyle\sum_{l\leqq n_B} e_1(s_{l-1})\big] + E_a(p_1(s_{n_B}), n_B < +\infty) \\
&\qquad\qquad + p_1(\infty)P_a(n_B = +\infty) \\
&= p_1(a).
\end{aligned} \tag{3.15}
$$

We learn from 3.15 that $p_B$ is the greatest excessive $p = \mathbf{K}e + p(\infty)$ with $e = 0$ off $B$, $p \leqq 1$ on $B$, and $p(\infty) \leqq P.(\mathbf{B})$.

Also (and this will be useful for us in Section 4), $p_B = 1$ on $B$, so that, if, for two *potentials* $p_1$ and $p_2$, $p_2 \geqq p_1$ on $B$ and $e_1, e_2 = 0$ off $B$, then, writing $ep$ for $\sum_{b\in B} e(b)p(b)$,

3.16 $$e_2(B) = e_2\,p_B = e_B\,p_2 \geqq e_B\,p_1 = e_1\,p_B = e_1(B);$$

in short, $e_2(B) \geqq e_1(B)$, a fact due to Gauss [1, pp. 37–39] for the case of Newtonian potentials.

## 4. Capacities

Given a set $B$ of lattice points for which $P.(\mathbf{B}) = 0$, its *capacity* $C(B)$ is the total charge

4.1 $$C(B) = e_B(B) = \textstyle\sum_{a\in\partial B} P_a(n_B(w_1^+) = +\infty)$$

of the *electrostatic distribution* $e_B$.

When $|B|$ (= the number of points of $B$) $= +\infty$, $C(B) = +\infty$; for, if $C(B) < +\infty$, then (use 2.7) $p_B$ converges to 0 at $\infty$, and, since $p_B = 1$ on $B$, $|B| < +\infty$.

We shall therefore confine our attention to the capacities of finite sets $B$.

The following rules are helpful for computing $C(B)$:

4.2 $$C(B) = C(\partial B),$$

4.3 $$C(B_1) = C(B_2), \qquad\qquad B_1 \equiv B_2,$$

4.4 $$C(B) = \max e(B):\ e \geqq 0,\ \ e = 0 \text{ off } B,\ \ p = \mathbf{K}e \leqq 1,$$

4.5 $$C(B_1 \cap B_2) + C(B_1 \cup B_2) \leqq C(B_1) + C(B_2).$$

4.2 is clear. $B_1 \equiv B_2$ means that $B_1$ is congruent to $B_2$ (with respect to orthogonal transformations with integral entries). 4.3 is then clear. $p = \sum_{b \in B} K(a, b,)e(b) \leqq 1$ implies $p_B - p \geqq 0$, and using 3.16 to compute the (nonnegative) total charge $C(B) - e(B)$ of $e_B - e$ proves 4.4. 4.5 gets a similar proof: the inclusion

$$(w: n_{B_1 \cup B_2} < +\infty, n_{B_2} = +\infty) \subset (w: n_{B_1} < +\infty, n_{B_1 \cap B_2} = +\infty)$$

implies $p_{B_1 \cup B_2} - p_{B_2} \leqq p_{B_1} - p_{B_1 \cap B_2}$; now compute the (nonnegative) total charge of $e_{B_1} + e_{B_2} - e_{B_1 \cup B_2} - e_{B_1 \cap B_2}$.

Given $B, B_1, B_2, \cdots, B_n$, let us write $\mathfrak{m}$ for subsets of $1, 2, \cdots, n$, $|\mathfrak{m}| = l$ for the number of points in $\mathfrak{m}$, and $B_\mathfrak{m} = \bigcup_{i \in \mathfrak{m}} B_i$; it is clear from[10]

4.6 $$P.(\bigcap_{l \leqq n} C) = -\sum_{l \leqq n} (-)^l \sum_{|\mathfrak{m}|=l} P.(C_\mathfrak{m})$$

that

4.7 $$\begin{aligned} 0 &\leqq P.(n_B = +\infty, n_{B \cup B_l} < +\infty, l \leqq n) \\ &= -P.(n_B < +\infty) + P.(n_{B \cup B_l} < +\infty, l \leqq n) \\ &= -P.(n_B < +\infty) - \sum_{l \leqq n} (-)^l \sum_{|\mathfrak{m}|=l} P.(n_{B \cup B_\mathfrak{m}} < +\infty) \\ &= p_B - \sum_{l \leqq n} (-)^l \sum_{|\mathfrak{m}|=l} p_{B \cup B_\mathfrak{m}}, \end{aligned}$$

and by using 3.16 to compute the (nonnegative) total charge of $-e_B - \sum_{l \leqq n} (-)^l \sum_{|\mathfrak{m}|=l} e_{B \cup B_\mathfrak{m}}$, it results that

4.8 $$C(B) + \sum_{l \leqq n} (-)^l \sum_{|\mathfrak{m}|=l} C(B \cup B_\mathfrak{m}) \leqq 0.$$

G. Choquet [1, pp. 147–153] proved the counterpart of 4.8 for Newtonian potentials. 4.7 imitates G. Hunt [1, p. 53]. 4.5 is a special case of 4.8 ($n = 2, B = B_1 \cap B_2$).

The following technique for estimating $C(B)$ is useful for Section 6. Given $B$, if $A$ is the sum of $n$ $(= |B|)$ solid cubes $[0, 1]^s$ centered at the points of $B$, and if $\hat{C}(A)$ is the Newtonian capacity:

4.9 $$\hat{C}(A) = \max \hat{e}(A): \hat{e} \geqq 0, \hat{e} = 0 \text{ off } A, \hat{p}(\xi) = \int_A |\xi - \eta|^{2-s} \hat{e}(d\eta) \leqq 1,$$

then

4.10 $$k_4 \hat{C}(A) \leqq C(B) \leqq k_5 \hat{C}(A),$$

with $k_4$, $k_5$ depending on the dimension number $s$, but not on $B$.

To prove the *over*estimate, choose $k_5$ such that, for $\xi$ in the cube centered at $a$, the integral $\int |\xi - \eta|^{2-s} d\eta$ extended over the cube centered at $b$ is $\leqq k_5 K(a, b)$, let $\hat{e}(d\eta) = e_B(b) d\eta$ on the cube centered at $b$, and estimate $\hat{p}(\xi) = k_5^{-1} \int_A |\xi - \eta|^{2-s} e(d\eta)$ in terms of $p_B$; the result is $\hat{p} \leqq 1$, and we conclude that

4.11 $$C(B) = e_B(B) = \hat{e}(A) \leqq k_5 \hat{C}(A).$$

[10] 4.6 is dual to the classical inclusion and exclusion formula.

To prove the *under*estimate, choose $k_4$ such that, for $\xi$ in the cube centered at $a$ and $\eta$ in the cube centered at $b$, $k_4 K(a, b) \leqq |\xi - \eta|^{2-s}$, let $e(b)$ be the charge that the Newtonian electrostatic distribution $\hat{e}$ places on the cube centered at $b$, and estimate $p = k_4 \sum_{b \in B} K(a, b)e(b)$ in terms of

$$\hat{p} = \int |\xi - \eta|^{2-s} \hat{e}(d\eta);$$

the result is $p \leqq 1$, and we conclude that

4.12 $$C(B) \geqq k_4 e(B) = k_4 \hat{e}(A) = k_4 \hat{C}(A).$$

Given compact $A \subset R^s$,

4.13 $$\hat{C}(\alpha A) = \alpha^{s-2} \hat{C}(A), \qquad \alpha > 0,$$

where $\alpha A$ is the set of points $\alpha x$ with $x \in A$.

We will use 4.13 for getting *under*estimates of $C(B)$; for example, if $B_n$ is the disc

$$(l_1, l_2, \cdots, l_s): \ (l_1^2 + l_2^2)^{1/2} \leqq n, \ \ l_3 = l_4 = \cdots = l_s = 0,$$

then $C(B_n) \geqq k_6 n^{s-2}$ for $n \uparrow +\infty$.

## 5. Wiener's test

Given a set $Q$ of lattice points, clustering to $\infty$, let $Q_l$ denote the intersection of $Q$ and the spherical shell $2^l \leqq |a| < 2^{l+1}$, and let us prove Wiener's test:

5.1 $P.(\mathbf{Q}) = 1$ or $0$ according as $\sum_{l \geq 1} 2^{-l(s-2)} C(Q_l) =$ or $< +\infty$.

When

5.2 $$\sum_{l \geq 1} 2^{-l(s-2)} C(Q_l) < +\infty,$$

5.3 $$\begin{aligned} p_{Q_l}(a) &= \sum_{b \in Q_l} K(a, b) e_{Q_l}(b) \leqq \sum_{b \in Q_l} k_7 |a - b|^{2-s} e_{Q_l}(b) \\ &\leqq 2k_7 2^{-l(s-2)} C(Q_l), \qquad l \uparrow +\infty, \end{aligned}$$

is the general term of a convergent sum, and an application of the first Borel-Cantelli lemma implies

5.4 $$P.(n_{Q_l} = +\infty, l \uparrow +\infty) = 1;$$

2.6 implies $P.(\bigcup_{l \leqq n} \mathbf{Q}_l) = 0$ for each $n \geqq 1$; and we infer that

5.5 $$P.(\mathbf{Q}) = 0.$$

When

5.6 $$\sum_{l \geq 1} 2^{-l(s-2)} C(Q_l) = +\infty,$$

$\sum_{l \geq 1} 2^{-(4l+k)(s-2)} C(Q_{4l+k}) = +\infty$ for $k = 0, 1, 2,$ or $3$, and if we suppose, as we can, that $\sum_{l \geq 1} 2^{-(4l+1)(s-2)} C(Q_{4l+1}) = +\infty$, it is clear that, if $m_l$ is the crossing *time*[11] min $(n:2^l \leqq |s_n| < 2^l + 1)$, and if, for the moment, $s_l$ stands for the crossing *place* $s_{m_l}$, then for $l \uparrow +\infty$,

$$\begin{aligned} P_{s_{4l}}(n_{Q_{4l+1}} < m_{4l+4}) &\geqq p_{Q_{4l+1}}(s_{4l}) - E_{s_{4l}}[p_{Q_{4l+1}}(s_{4l+4})] \\ &= \textstyle\sum_{b \in Q_{4l+1}} (K(s_{4l}, b) - E_{s_{4l}}[K(s_{4l+4}, b)])e_{Q_{4l+1}}(b) \\ &\geqq k_8[\tfrac{2}{3}(2^{4l+2} - 2^{4l})^{2-s} - \tfrac{3}{2}(2^{4l+4} - 2^{4l+2})^{2-s}]C(Q_{4l+1}) \\ &\geqq k_9\, 2^{-(4l+1)(s-2)}C(Q_{4l+1}) = t_l, \end{aligned} \tag{5.7}$$

which implies

$$\begin{aligned} P_{s_{4l}}(n_Q = +\infty) &\leqq E_{s_{4l}}[n_{Q_{4l+1}} \geqq m_{4l+4}, P_{s_{4l+4}}(n_Q = +\infty)] \\ &\leqq E_{s_{4l}}[n_{Q_{4l+1}} \geqq m_{4l+4}, E_{s_{4l+4}}(n_{Q_{4l+5}} \geqq m_{4l+8}, P_{s_{4l+8}}(n_Q = +\infty))], \quad \text{etc.} \\ &\leqq (1 - t_l)(1 - t_{l+1})(1 - t_{l+2}) \cdots \\ &= 0, \end{aligned} \tag{5.8}$$

and we conclude that

$$P.(n_Q(w^+_{m_{4l}}) < +\infty) = E.(P_{s_{4l}}(n_Q < +\infty)) = 1, \qquad l \uparrow +\infty. \tag{5.9}$$

which completes the proof of 5.1.

## 6. Thorns

Given nonnegative $i(1) \leqq i(2) \leqq \cdots$, let $Q$ denote the *thorn*

$$(l_1, l_2, \cdots, l_s): \quad (l_1^2 + l_2^2 + \cdots + l_{s-1}^2)^{1/2} \leqq i(l_s), \quad l_s \geqq 1.$$

We use Wiener's test to prove that for $s \geqq 4$ dimensions

$$P.(\mathbf{Q}) = 1 \text{ or } 0 \quad \text{according as} \quad \textstyle\sum_{n \geqq 1}(2^{-n}i(2^n))^{s-3} = \text{or} < +\infty; \tag{6.1}$$[12]

as an example, if $l \geqq 1$, and if $i(n) = n(\lg n \lg_2 n \cdots (\lg_k n)^\alpha)^{-1/s-3}$,[13] then $P.(\mathbf{Q}) = 1\ (0)$ for $\alpha \leqq 1\ (>1)$.

When $\limsup_{n\uparrow+\infty} n^{-1}i(n) > \alpha > 0$, $2^{-n}i(2^n) > \alpha/2$ for an infinite number of integers $n$; for such $n$, $Q_n$ contains the set $Q_n^-$ of lattice points of a sphere of diameter $\geqq \min(1, \alpha)2^n$; $C(Q_n) \geqq C(Q_n^-)$; $C(Q_n)$ is then $\geqq k_{10}\, 2^{n(s-2)}$; and the upshot is

$$+\infty = \textstyle\sum_{n\geqq 1} (2^{-n}i(2^n))^{s-3} = \sum_{n \geqq 1} 2^{-n(s-2)}C(Q_n), \tag{6.2}$$

which checks with 6.1.

---

[11] $m_l < +\infty$ for paths crossing from $|a| \leqq 2^l$ to $|a| > 2^l + 1$; for if

$$(l_1^2 + l_2^2 + \cdots + l_s^2)^{1/2} \leqq 2^l,$$

then $(l_1 \pm 1)^2 + l_2^2 + \cdots + l_s^2 = l_1^2 + l_2^2 + \cdots + l_s^2 \pm 2l_1 + 1 \leqq 2^{2l} + 2\cdot 2^l + 1 = (2^l + 1)^2$.

[12] 6.1 is to be compared with Lebesgue's thorn: see Courant and Hilbert [1, pp. 272–274].

[13] $\lg_1 = \lg$ and $\lg_n = \lg(\lg_{n-1})$ for $n \geqq 2$.

When

6.3 $$\lim_{n\uparrow+\infty} n^{-1}i(n) = 0,$$

$Q_n$ is long and thin, a sphere is not a good approximation, and we consider instead the ellipsoids $Q_n^- \subset Q$ and $Q_n^+ \supset Q$:

$$Q_n^- = (l_1, l_2, \cdots, l_s): \quad \frac{l_1^2 + l_2^2 + \cdots + l_{s-1}^2}{i(2^n)^2} + \frac{(l_s - 3\cdot 2^{-n-1})^2}{2^{2(n-1)}} \leqq 1,$$

$$Q_n^+ = (l_1, l_2, \cdots, l_s): \quad \frac{l_1^2 + l_2^2 + \cdots + l_{s-1}^2}{2\cdot i(2^{n+1})^2} + \frac{(l_s - 3\cdot 2^{n-1})^2}{2\cdot 2^{2(n-1)}} \leqq 1$$

and compare the capacities $C(Q_n^-)$, $C(Q_n^+)$ to the Newtonian capacity of the solid ellipsoid

$$E = (x_1, x_2, \cdots, x_s): \quad \frac{x_1^2}{e_1^2} + \frac{x_2^2}{e_2^2} + \cdots + \frac{x_s^2}{e_s^2} \leqq 1, \qquad e_1, e_2, \cdots, e_s > 0.$$

The Newtonian capacity of $E$ is known; up to a factor depending on the dimension number, it is the reciprocal of the elliptic integral

6.4 $$\mathbf{e} = \int_0^{+\infty} \frac{dt}{\sqrt{(e_1^2 + t)(e_2^2 + t)\cdots(e_s^2 + t)}}.$$

The reader will find a neat proof of 6.4 for $s = 3$ in G. Chrystal [1, p. 30].

When $\alpha = e_1/e_s = e_2/e_s = \cdots = e_{s-1}/e_s$,

6.5 $$\begin{aligned} \mathbf{e} &= \int_0^{+\infty} (\alpha^2 e_s^2 + t)^{-(s-1)/2}(e_s^2 + t)^{-1/2}\,dt \\ &= e_s^{2-s}\int_0^{+\infty} (\alpha^2 + t)^{-(s-1)/2}(1 + t)^{-1/2}\,dt \\ &\sim e_s^{2-s}\int_{\alpha^2}^{1} t^{-(s-1)/2}\,dt \\ &\sim e_s^{2-s}\tfrac{1}{2}(s-3)\alpha^{-(s-3)}, \qquad \alpha \downarrow 0; \end{aligned}$$

by using 4.10, it is plain that

6.6 $$\begin{aligned} &k_{11}\, 2^{+n(s-2)}(2^{-n}i(2^n))^{s-3} < C(Q_n^-) \\ &\leqq C(Q) \leqq C(Q_n^+) < k_{12}\, 2^{+n(s-2)}(2^{-(n+1)}i(2^{n+1}))^{s-3}, \qquad n\uparrow+\infty; \end{aligned}$$

and an application of Wiener's test completes the proof of 6.1.

When $s = 3$, $P.(\mathbf{Q}) = 1$ even for the thinnest thorn $Q = \bigcup_{n\geqq 1}(0, 0, n)$; $C(Q_n)$ is then $> k_{13}\, 2^n$, and Wiener's sum is $+\infty$.

We consider, instead, the set $Q = \bigcup_{n\geqq 1}(0, 0, i(n))$ with integral $i(1) < i(2) < \cdots$ and prove that if

6.7 $$i(n) - i(n-1) \geqq \lg i(n-1), \qquad n\uparrow+\infty,$$

then

6.8 $P.(\mathbf{Q}) = 1$ or $0$ according as $\sum_{n\geqq 1} i(n)^{-1} =$ or $< +\infty$;

as an example, if $i(n) = [n \lg n \lg_2 n \cdots (\lg_k n)^{\alpha}]$,[14] then $P.(\mathbf{Q}) = 1$ (0) for $\alpha \leqq 1$ ($>1$).

Granting

6.9 $$k_{14} \mid Q_n \mid \leqq C(Q_n) \leqq k_{15} \mid Q_n \mid, \qquad n \uparrow +\infty,$$

it is clear that

6.10 $$k_{14} \sum_{2^n \leqq i(l) < 2^{n+1}} i(l)^{-1} \leqq k_{14}\, 2^{-n} \mid Q_n \mid \leqq 2^{-n} C(Q_n) \leqq k_{15}\, 2^{-n} \mid Q_n \mid \leqq 2k_{15} \sum_{2^n \leqq i(l) < 2^{n+1}} i(l)^{-1},$$

and an application of Wiener's test proves 6.8. $C(Q_n) \leqq k_{15} \mid Q_n \mid$ with $k_{15} = K(0, 0)^{-1}$ is immediate from 4.5, and to complete the proof, it is enough to use 6.7 and 2.7 to check the estimate

6.11 $$\sum_{b \in Q_n} K(a, b) \leqq k_{16} \sum_{1 \leqq l \leqq |Q_n|} (ln \lg 2)^{-1} < k_{17}\, n^{-1} \lg \mid Q_n \mid \leqq k_{17} \lg 2, \qquad a \in Q_n, \quad n \uparrow +\infty,$$

and to infer, from 4.4, that

6.12 $$C(Q_n) \geqq k_{14} \mid Q_n \mid, \qquad k_{14} = (k_{17} \lg 2)^{-1}, \qquad n \uparrow +\infty.$$

*Problem.* When $i(n): n \geqq 1$ is the set of prime numbers, is $P.(\mathbf{Q}) = 1$? Were Gauss's law $n/\lg n$ for the number of primes $\leqq n$ *exact*, we could assert that $+\infty > k_{18} \geqq \sum_{b \in Q_n} K(a, b)$ for $a \in Q_n$ and conclude, as in 6.12, that $C(Q_n) \geqq k_{18}^{-1} \mid Q_n \mid \geqq k_{19}\, n^{-1} 2^n$ and that $\sum_{n \geqq 1} 2^{-n} C(Q_n) = +\infty$.

## References


P. Bateman et al.

1. *Seminar on convex sets*, The Institute for Advanced Study, Princeton, 1949–1950.

J. Capoulade

1. *Sur quelques propriétés des fonctions harmoniques et des fonctions préharmoniques*, Mathematica (Cluj), vol. 6 (1932), pp. 146–151.

G. Choquet

1. *Theory of capacities*, Ann. Inst. Fourier, Grenoble, vol. 5 (1953–1954), pp. 131–295.

G. Chrystal

1. *Electricity*, Encyclopaedia Britannica, 9th ed. (1879), vol. 8, pp. 3–104.

R. Courant and D. Hilbert

1. *Methoden der mathematischen Physik, Bd.* 2, Berlin, Springer, 1937.

J. L. Doob

1. *Stochastic processes*, New York, Wiley, 1953.
2. *Semimartingales and subharmonic functions*, Trans. Amer. Math. Soc., vol. 77 (1954), pp. 86–121.

R. J. Duffin

1. *Discrete potential theory*, Duke Math. J., vol. 20 (1953), pp. 233–251.


[14] $[\gamma]$ is the greatest integer $\leqq \gamma$.

A. ERDÉLYI ET AL. (Bateman Manuscript Project)
1. *Higher transcendental functions, Vol. I*, New York, McGraw-Hill, 1953.
2. *Tables of integral transforms, Vol. I*, New York, McGraw-Hill, 1954.

K. F. GAUSS
1. *Allgemeine Lehrsätze in Beziehung auf die im verkehrten Verhältnisse des Quadrats der Entfernung wirkenden Anziehungs- und Abstossungs-Kräfte*, Ostwald's Klassiker der exakten Wissenschaften, Nr. 2, Leipzig, 1889.

E. HEWITT AND L. J. SAVAGE
1. *Symmetric measures on Cartesian products*, Trans. Amer. Math. Soc., vol. 80 (1955), pp. 470–501.

G. A. HUNT
1. *Markoff processes and potentials I*, Illinois J. Math., vol. 1 (1957), pp. 44–93.
2. *Markoff processes and potentials II*, Illinois J. Math., vol. 1 (1957), pp. 316–369.
3. *Markoff processes and potentials III*, Illinois J. Math., vol. 2 (1958), pp. 151–213.

B. H. MURDOCH
1. *Preharmonic functions*, Thesis, Princeton, 1952.

G. PÓLYA
1. *Über eine Aufgabe der Wahrscheinlichkeitsrechnung betreffend die Irrfahrt in Strassennetz*, Math. Ann., vol. 84 (1921), pp. 149–160.

S. VERBLUNSKY
1. *Sur les fonctions préharmoniques* (*Deuxième note*), Bull. Sci. Math. (2), vol. 74 (1950), pp. 153–160.

KYÔTO UNIVERSITY
KYÔTO, JAPAN

# WIENER INTEGRAL AND FEYNMAN INTEGRAL

KIYOSI ITÔ
KYOTO UNIVERSITY

## 1. Introduction

Consider, for example, a classical mechanical system with Lagrangian

$$L(x, \dot{x}) = \frac{\dot{x}^2}{2} - U(x). \tag{1.1}$$

The wave function of the quantum mechanical system corresponding to this classical one changes with time $t$ according to the Schrödinger equation

$$\frac{\hbar}{i}\frac{\partial \psi}{\partial t} = \frac{\hbar^2}{2}\frac{\partial^2 \psi}{\partial x^2} - U\psi, \qquad \psi(0+, x) = \varphi(x). \tag{1.2}$$

Feynman [3] expressed this wave function $\psi(t, x)$ in the following integral form, which we shall here call the *Feynman integral*

$$\psi(t, x) = \frac{1}{N}\int_{\Gamma_x} \exp\left\{\frac{i}{\hbar}\int_0^t \left[\frac{\dot{x}_\tau^2}{2} - U(x_\tau)\right] d\tau\right\} \varphi(x_\tau) \prod_\tau dx_\tau, \tag{1.3}$$

where $\Gamma_x$ is the space of paths $X = (x_\tau, 0 < \tau \leqq t)$ with $x_0 = x$, $\prod_\tau dx_\tau$ is a uniform measure on $R^{(0,t]}$, and $N$ is a normalization factor. It should be noted that the integral $\int_0^t [\dot{x}_\tau^2/2 - U(x_\tau)]\, d\tau$ is the classical action integral along the path $X$. (This idea goes back to Dirac [1].) It is easy to see that (1.3) solves (1.2) unless we require mathematical rigor. It is our purpose to define the generalized measure $\prod_\tau dx_\tau/N$, that is, the integral $\int_{\Gamma_x} F(X) \prod_\tau dx_\tau/N$, rigorously and to prove that (1.3) solves (1.2) in case $U(x) \equiv 0$ (case of no force) or $U(x) \equiv x$ (case of constant force). See theorem 5.2 and theorem 5.3 below. We hope this fact will be proved for a general $U(x)$ with some appropriate regularity conditions.

Our definition is also applicable to the *Wiener integral*; namely, using it, we shall prove that the solution of the heat equation

$$\frac{\partial u}{\partial t} = \frac{1}{2}\frac{\partial^2 u}{\partial x^2} - Uu, \qquad u(0+, x) = f(x), \tag{1.4}$$

is given by

$$u(t, x) = \frac{1}{N}\int_{\Gamma_x} \exp\left\{-\int_0^t \left[\frac{\dot{x}^2}{2} + U(x_\tau)\right] d\tau\right\} f(x_t) \prod_\tau dx_\tau \tag{1.5}$$

for any bounded continuous function $U(x)$. See theorem 4.3. This should be called the Feynman version of Kac's theorem that

$$u(t, x) = \int_{\Gamma_x} \exp\left[-\int_0^t U(x_\tau)\, d\tau\right] f(x_t) W_x(dX) \tag{1.5'}$$

solves (1.4). In this paper the paths, that is, the points in $\Gamma_x$, are denoted with capital letters $X, Y, \cdots$ and their values at time $\tau$ are denoted with the corresponding small letters with the suffix $\tau$ such as $x_\tau, y_\tau, \cdots$. Now that Kac's theorem is well known to probabilists, no one bothers with its Feynman version. However, it is interesting that Kac had the Feynman version (1.5) in mind and formulated it as (1.5') to make it rigorous [5].

Gelfand and Yaglom [4] proposed a method of defining the Feynman integral. They replaced $\hbar$ with $\hbar - i\sigma$ in (1.3) to reduce the Feynman integral to the Wiener integral and defined the Feynman integral as a limit of the Wiener integral by letting $\sigma \downarrow 0$. Our method is different from theirs in the point that we define $\prod_\tau dx_\tau/N$ directly and treat both the Feynman integral and the Wiener integral on the same level.

## 2. The mathematical meaning of $\prod_\tau \mathbf{dx}_\tau/\mathbf{N}$

What Feynman had in mind for $\prod_\tau dx_\tau$ must be a uniform measure on $R^{(0,t]}$. Rigorously speaking, this measure does not exist. Therefore, we should define it as an ideal limit of a sequence of measures on $R^{(0,t]}$. In order to be able to compute the integral (1.3) or (1.5), the approximating measures should be concentrated on the set $L_x$ of all $X = (x_\tau, 0 < \tau \leqq t) \in R^{(0,t]}$ satisfying

(L.1) $x_\tau$ is absolutely continuous in $\tau$,

(L.2) $\dot{x}_\tau \equiv dx_\tau/d\tau \in L^2(0, t]$,

(L.3) $\lim_{\tau\downarrow 0} x_\tau = x$.

We shall now construct a sequence of probability measures $\{P_n^{(x)}\}$ on $L_x$ whose ideal limit is the uniform distribution on $R^{(0,t]}$. Let $\rho(\tau, \sigma)$, with $\sigma, \tau \in (0, t]$, be strictly positive definite and continuous; for example, $\rho(\tau, \sigma) = \exp(-|\tau - \sigma|)$. Let $\xi_\tau(\omega)$, $\omega \in \Omega(\mathbf{B}, P)$, be a Gaussian process with

$$E(\xi_\tau) = 0, \qquad E(\xi_\tau \xi_\sigma) = \rho(\tau, \sigma). \tag{2.1}$$

It is well known that such a Gaussian process exists. Since the continuity of $\rho(\tau, \sigma)$ implies the continuity of $\xi_\tau$ in the mean, there exists a measurable version [2] of $\xi_\tau$. Denote that version with the same symbol $\xi_\tau$.

Noting that

$$E\left(\int_0^t \xi_\tau^2\, d\tau\right) = \int_0^t \rho(\tau, \tau)\, d\tau < +\infty, \tag{2.2}$$

we can see that

$$P\left\{\int_0^t \xi_\tau^2\, d\tau < +\infty\right\} = 1. \tag{2.3}$$

Put

$$x_\tau^{(n)} = x + n \int_0^\tau \xi_\theta \, d\theta, \qquad 0 < \tau \leqq t. \tag{2.4}$$

Then $x_\tau^{(n)}$, for $0 < \tau \leqq t$, is also a Gaussian process with

$$\begin{aligned} E[x_\tau^{(n)}] &= x, \\ E\{[x_\sigma^{(n)} - x][x_\sigma^{(n)} - x]\} &= n^2 \int_0^\tau \int_0^\sigma \rho(\theta_1, \theta_2) \, d\theta_1 \, d\theta_2. \end{aligned} \tag{2.5}$$

Denote by $P_n^{(x)}$ the probability distribution of the sample function $X^{(n)}$ of the process $x_\tau^{(n)}$, with $0 < \tau \leqq t$. Then $P_n^{(x)}$ is concentrated on $L_x$ and any finite dimensional marginal distribution of $P_n^{(x)}$, say over coordinates $\tau_1, \tau_2, \cdots, \tau_m$, is Gaussian with the density

$$\frac{b^{1/2}}{(2\pi n^2)^{m/2}} \exp\left[\frac{-1}{2n^2} \sum_{i,j=1}^{m} b_{ij}(x_i - x)(x_j - x)\right], \tag{2.6}$$

where the matrix $(b_{ij})$ is the inverse of the matrix $(v_{ij})$ with

$$v_{ij} = \int_0^{\tau_i} \int_0^{\tau_j} \rho(\theta_1, \theta_2) \, d\theta_1 \, d\theta_2, \qquad i, j = 1, 2, \cdots, m \tag{2.7}$$

and $b$ is the determinant of $(b_{ij})$. The existence of $(b_{ij})$, that is, the nonvanishing of the determinant of $(v_{ij})$ results from the assumption that $\rho(\tau, \sigma)$ is strictly positive definite.

Since the Gaussian distribution (2.6) tends to a uniform distribution on the $m$-space in the sense that, for any almost periodic function $f(x_1, x_2, \cdots, x_m)$,

(2.8)

$$\int \cdots \int f(x_1, \cdots, x_m) \frac{b^{1/2}}{(2\pi n^2)^{m/2}} \exp\left[\frac{-1}{2n^2} \sum_{i,j=1}^{m} b_{ij}(x_i - x)(x_j - x)\right] dx_1 \cdots dx_m$$

tends to the Bohr mean $\mathfrak{M}(f)$ of $f$ as $n \to \infty$, it is reasonable to say that $P_n^{(x)}$, for $n = 1, 2, \cdots$ approximates the uniform distribution on $R^{(0,t]}$ and that $\prod_\tau dx_\tau$ is an ideal limit of this sequence.

$N$ must be also an ideal limit of a sequence of numbers $\{N_n\}$ such that $P_n^{(x)}/N_n$ tends to $\prod_\tau dx_\tau/N$ in some sense.

Keeping these heuristic considerations in mind, we shall give a mathematical meaning to $\prod_\tau dx_\tau/N$, that is, to the linear functional $I(F) = \int F(X) \prod_\tau dx_\tau/N$. There are many ways of defining this functional in accordance with the choice of the sequence $\{N_n\}$. We shall express $I(F)$ as $I(F, N_n)$ referring to the sequence $\{N_n\}$.

DEFINITION.

$$I(F, N_n) = \lim_{n\to\infty} \frac{1}{N_n} \int_{L_x} F(X) P_n^{(x)}(dX). \tag{2.9}$$

The domain $\mathfrak{D}(N_n)$ of this functional $I(F, N_n)$ is the set of all $F$ for which the limit in (2.9) exists and is finite.

Fixing $\{N_n\}$, we shall write $\mathfrak{D}(N_n)$ as $\mathfrak{D}$ and $I(F, N_n)$ as

$$\frac{1}{N} \int_{\Gamma_x} F(X) \prod_\tau dx_\tau. \tag{2.10}$$

We shall mention three interesting cases.

(i) *Uniform integral.* $N_n = 1$, with $n = 1, 2, \cdots$. If $F(X)$ is of the form $f(x_{\tau_1}, x_{\tau_2}, \cdots, x_{\tau_m})$ with an almost periodic function $f(x_1, x_2, \cdots, x_m)$, then $F \in \mathfrak{D}$ and

$$\frac{1}{N} \int_{\Gamma_x} F(X) \prod_\tau dx_\tau = \mathfrak{M}(f), \tag{2.11}$$

where $\mathfrak{M}(f)$ is the Bohr mean of $f$.

(ii) *Wiener integral.* $N_n = 1/\prod_\nu (1 + n^2\lambda_\nu)^{1/2}$, $n = 1, 2, \cdots$, where $\lambda_\nu$ will be defined in the next section. We shall discuss the Wiener integral in section 4.

(iii) *Feynman integral.* $N_n = 1/\prod_\nu (1 + n^2\lambda_\nu/\hbar i)^{1/2}$, with $n = 1, 2, \cdots$, with the same $\lambda_\nu$ as in (ii). This will be discussed in section 5.

## 3. Orthogonalization method

In the following sections we shall be faced with the integrals of the type

$$I = \int_{L_x} G(X) P_n^{(x)}(dX). \tag{3.1}$$

Recalling that $P_n^{(x)}$ was defined as the probability distribution of the sample path $X^{(n)}$ of the process

$$x_\tau^{(n)}(\omega) = x + n \int_0^\tau \xi_\theta(\omega)\, d\theta \tag{3.2}$$

introduced in section 2, the integral $I$ can be expressed as the mean value of $G[X^{(n)}(\omega)]$ on $\Omega(\mathbf{B}, P)$

$$I = \int_\Omega G[X^{(n)}(\omega)] P(d\omega) \equiv E\{G[X^{(n)}]\}. \tag{3.3}$$

To compute this, we shall use the usual *orthogonalization method.* The idea is as follows. Let $T$ denote the operator from $L^2(0, t]$ into itself,

$$(T\eta)(\tau) = \int_0^t \rho(\tau, \sigma)\eta(\sigma)\, d\sigma. \tag{3.4}$$

Then $T$ is a strictly positive-definite compact operator. Therefore $T$ has positive eigenvalues $\{\lambda_\nu\}$ whose eigenfunctions $\{\eta_\nu\}$ constitute a complete orthonormal system in $L^2(0, t]$.

Now put

$$a_\nu(\omega) = (\xi, \eta_\nu) = \int_0^t \xi_\tau(\omega)\eta_\nu(\tau)\, d\tau; \tag{3.5}$$

this inner product is well defined, thanks to (2.3). Then $\{a_\nu\}$ is a Gaussian system, since $\xi_\tau$ is a Gaussian process. Equation (2.1) implies that

$$E(a_\nu a_\mu) = \int_0^t \int_0^t \rho(\tau, \sigma)\eta_\nu(\tau)\eta_\mu(\sigma)\, d\tau\, d\sigma. \tag{3.6}$$

Therefore $a_\nu$, with $\nu = 1, 2, \cdots$, are independent and each $a_\nu$ is Gaussian with mean 0 and variance $\lambda_\nu$. Since we have

$$\sum_\nu \lambda_\nu = \sum_\nu E(a_\nu^2) = E(\sum_\nu a_\nu^2) = E\left(\int_0^t \xi_\tau^2\, d\tau\right) = \int_0^t \rho(\tau, \tau)\, d\tau, \tag{3.7}$$

the continuity of $\rho(\tau, \sigma)$ implies

$$\sum_\nu \lambda_\nu < +\infty; \tag{3.8}$$

this fact will be useful in the following sections.

Noting that

$$\xi_\tau(\omega) = \sum_\nu a_\nu(\omega)\eta_\nu(\tau) \tag{3.9}$$

and that

$$x_\tau^{(n)}(\omega) = x + n \int_0^\tau \xi_\theta(\omega)\, d\theta. \tag{3.10}$$

we can express $I$ in the form

$$I = E\{H(a_1, a_2, \cdots)\} \tag{3.11}$$

with some $H$. Using the independence and the normality of $\{a_\nu\}$, we can compute (3.11) more easily than the original form (3.1).

## 4. Wiener integral

We shall now discuss the integral (2.10) for

$$N_n = \frac{1}{\prod_\nu (1 + n^2\lambda_\nu)^{1/2}}, \qquad n = 1, 2, \cdots, \tag{4.1}$$

where $\lambda_\nu$, with $\nu = 1, 2, \cdots$, are the eigenvalues introduced in section 3.

We can verify easily the convergence of the infinite sums and infinite products appearing in this section by appealing to (3.8).

LEMMA 4.1.

$$Q_n^{(x)}(dX) \equiv \frac{1}{N_n} \exp\left(-\int_0^t \frac{\dot{x}_\tau^2}{2}\, d\tau\right) P_n^{(x)}(dX) \tag{4.2}$$

*is a probability distribution on* $L_x$.

PROOF. Using the orthogonalization method, we get

$$\begin{aligned} I_n &= \int_{L_x} \exp\left(-\int_0^t \frac{\dot{x}^2}{2}\, d\tau\right) P_n^{(x)}(dX) \\ &= E\left[\exp\left(-n^2 \int_0^t \frac{\xi_\tau^2}{2}\, d\tau\right)\right] \\ &= E\left[\exp\left(-n^2 \sum_\nu \frac{a_\nu^2}{2}\right)\right]. \end{aligned} \tag{4.3}$$

Noting that the $a_\nu$, for $\nu = 1, 2, \cdots$, are independent, we have

$$I_n = \prod_\nu E\left[\exp\left(-\frac{n^2a_\nu^2}{2}\right)\right] \tag{4.4}$$
$$= \prod_\nu \int_{-\infty}^{+\infty} \frac{1}{(2\pi\lambda_\nu)^{1/2}} \exp\left(\frac{-\alpha^2}{2\lambda_\nu} - \frac{n^2\alpha^2}{2}\right) d\alpha$$
$$= \prod_\nu \frac{1}{(1 + n^2\lambda_\nu)^{1/2}}$$
$$= N_n,$$

which proves our lemma.

LEMMA 4.2. *For any $g \in L^2(0, t]$, we have*

$$\int_{L_x} \exp\left[i \int_0^t \dot{x}_\tau g(\tau)\, d\tau\right] Q_n^{(x)}(dX) = \exp\left[-\sum_\nu \frac{n^2\lambda_\nu g_\nu^2}{2(n^2\lambda_\nu + 1)}\right], \tag{4.5}$$

*where $g_\nu = (g, \eta_\nu) \equiv \int_0^t g(\tau)\eta_\nu(\tau)\, d\tau$, for $\nu = 1, 2, \cdots$.*

PROOF. By the same idea as in lemma 1, we have

$$I_n = \int_{L_x} \exp\left[i \int_0^t \dot{x}_\tau g(\tau)\, d\tau\right] Q_n^{(x)}(dX) \tag{4.6}$$
$$= \frac{1}{N_n} \int_{L_x} \exp\left[i \int_0^t \dot{x}_\tau g(\tau)\, d\tau - \int_0^t \frac{\dot{x}_\tau^2}{2}\, d\tau\right] P_n^{(x)}(dX)$$
$$= \frac{1}{N_n} E\left[\exp\left(in \sum_\nu g_\lambda a_\nu - n^2 \sum_\nu \frac{a_\nu^2}{2}\right)\right]$$
$$= \frac{1}{N_n} \prod_\nu E\left[\exp\left(ing_\nu a_\nu - n^2 \frac{a_\nu^2}{2}\right)\right]$$
$$= \frac{1}{N_n} \prod_\nu \int \frac{1}{(2\pi\lambda_\nu)^{1/2}} \exp\left(\frac{-\alpha^2}{2\lambda_\nu} + ing_\nu\alpha - \frac{n^2\alpha^2}{2}\right) d\alpha$$
$$= \frac{1}{N_n} \prod_\nu \frac{1}{(1 + n^2\lambda_\nu)^{1/2}} \exp\left[\frac{-n^2\lambda_\nu g_\nu^2}{2(n^2\lambda_\nu + 1)}\right],$$

which proves (4.5) by virtue of (4.1).

As an immediate result from lemma 4.2, we obtain the following lemma, noting that $\sum_\nu g_\nu^2 = \int_0^t g(\tau)^2\, d\tau$ and that $\int_0^t g(\tau)\, dB(\tau)$ is Gaussian distributed with mean 0 and variance $\int_0^t g(\tau)^2\, d\tau$ for the Brownian motion $B(\tau)$.

LEMMA 4.3. *For any $g \in L^2(0, t]$,*

$$\exp\left[i \int_0^t \dot{x}_\tau g(\tau)\, d\tau - \int_0^t \frac{\dot{x}_\tau^2}{2}\, d\tau\right] \in \mathfrak{D}, \tag{4.7}$$

and

$$\frac{1}{N}\int_{L_x}\exp\left[i\int_0^t \dot{x}_\tau g(\tau)\,d\tau-\int_0^t\frac{\dot{x}_\tau^2}{2}\,d\tau\right]\prod_\tau dx_\tau \\ =\int_{L_x}\exp\left[i\int_0^t g(\tau)\,dx_\tau\right]W_x(dX), \tag{4.8}$$

where $W_x$ is the probability measure for the Brownian motion process starting at $x$, namely the Wiener measure.

THEOREM 4.1.

$$\frac{1}{N}\exp\left(-\int_0^t\frac{\dot{x}_\tau^2}{2}\,d\tau\right)\prod_\tau dx_\tau=W_x(dX); \tag{4.9}$$

*rigorously speaking, we have, for any continuous bounded tame function* $F(X)$,

$$F(X)\exp\left(-\int_0^t\frac{\dot{x}_\tau^2}{\tau}\,d\tau\right)\in\mathfrak{D} \tag{4.10}$$

$$\frac{1}{N}\int_{\Gamma_x}F(X)\exp\left(-\int_0^t\frac{\dot{x}_\tau^2}{2}\,d\tau\right)d\prod_\tau x_\tau=\int_{\Gamma_x}F(X)W_x(dX). \tag{4.11}$$

A tame function is a function defined on an infinite dimensional space which depends only on a finite number of coordinates.

PROOF. $F(X)$ can be expressed as

$$F(X)=f(x_{\tau_1},x_{\tau_2},\cdots,x_{\tau_m}),\quad 0<\tau_1<\cdots<\tau_m\leqq t, \tag{4.12}$$

with a continuous bounded function $f$ of $m$ real variables. To obtain theorem 4.1, it is enough to prove that

$$\lim_{n\to\infty}\int_{L_x}f(x_{\tau_1},x_{\tau_2},\cdots,x_{\tau_m})Q_n^{(x)}(dX)=\int_{\Gamma_x}f(x_{\tau_1},x_{\tau_2},\cdots,x_{\tau_m})W_x(dX). \tag{4.13}$$

Let $\tilde{Q}_n^{(n)}$ and $\tilde{W}_x$ denote the marginal distributions of $Q_n^{(x)}$ and $W_x$ over the coordinates $\tau_1,\tau_2,\cdots,\tau_n$ respectively. Then

$$\begin{aligned} I_n &= \int\cdots\int\exp\left[i(z_1\alpha_1+\cdots+z_m\alpha_m)\right]\tilde{Q}_n^{(x)}(d\alpha_1\cdots d\alpha_m) \\ &= \int_{L_x}\exp\left[i(z_1x_{\tau_1}+\cdots+z_mx_{\tau_m})\right]Q_n^{(x)}(dX) \\ &= \exp\left[i(z_1+\cdots+z_m)x\right]\int_{L_x}\exp\left[i\int g(\tau)\dot{x}_\tau\,d\tau\right]Q_n^{(x)}(dX), \end{aligned} \tag{4.14}$$

where $g(\tau)=\sum_{j=1}^m z_j\varphi_j(\tau)$ and $\varphi_j(\tau)$ is the indicator function of the set $(0,\tau_j)$. Using lemma 4.3, we get

$$\begin{aligned} I_n &\to \int_{\Gamma_x}\exp\left[i(z_1+\cdots+z_m)x+i\int g(\tau)\,dx_\tau\right]W_x(dX) \\ &= \int_{\Gamma_x}\exp\left[i(z_1x_{\tau_1}+\cdots+z_mx_{\tau_m})\right]W_x(dX) \\ &= \int\cdots\int\exp\left[i(z_1\alpha_1+\cdots+z_m\alpha_m)\right]\tilde{W}_x(d\alpha_1\,d\alpha_2\cdots d\alpha_n). \end{aligned} \tag{4.15}$$

Therefore $\tilde{Q}_n^{(x)}\to\tilde{W}_x$ in the weak sense as $n\to\infty$, which implies (4.13).

THEOREM 4.2. *If $f(x): R^1 \to C$ is continuous and bounded, then*

$$\exp\left(-\int_0^t \frac{\dot{x}_\tau^2}{2}\, d\tau\right) f(x_t) \in \mathfrak{D} \tag{4.16}$$

*and*

$$u(t, x) \equiv \frac{1}{N} \int_{\Gamma_x} \exp\left(-\int_0^t \frac{\dot{x}_\tau^2}{2}\, d\tau\right) f(x_t) \prod_\tau dx_\tau \tag{4.17}$$

*solves*

$$\frac{\partial u}{\partial t} = \frac{1}{2} \frac{\partial^2 u}{\partial x^2}, \qquad u(0+, x) = f(x). \tag{4.18}$$

PROOF. Using the previous theorem, we have

$$u(t, x) = \int_{\Gamma_x} f(x_t) W_x(dX) = \int f(y) \frac{1}{(2\pi t)^{1/2}} \exp\left[-\frac{(x-y)^2}{2t}\right] dy \tag{4.19}$$

and this solves (4.18).

THEOREM 4.3. (*Feynman's version of Kac's theorem.*) *If $f(x)$ and $U(x)$ are continuous and bounded, then*

$$\exp\left\{-\int_0^t \left[\frac{\dot{x}_\tau^2}{2} + U(x_\tau)\right] d\tau\right\} f(x_t) \in \mathfrak{D} \tag{4.20}$$

*and*

$$u(t, x) = \frac{1}{N} \int_{\Gamma_x} \exp\left\{-\int_0^t \left[\frac{\dot{x}_\tau^2}{2} + U(x_\tau)\right] d\tau\right\} f(x_t) \prod_\tau dx_\tau \tag{4.21}$$

*solves*

$$\frac{\partial u}{\partial t} = \frac{1}{2} \frac{\partial^2 u}{\partial x^2} - U(x) u, \qquad u(0+, x) = f(x). \tag{4.22}$$

PROOF. It is enough, by virtue of Kac's theorem, to prove that

$$\lim_{n\to\infty} \frac{1}{N_n} \int_{L_x} \exp\left\{-\int_0^t \left[\frac{\dot{x}_\tau^2}{2} + U(x_\tau)\right] d\tau\right\} f(x_t) P_n(dX) = \int_{\Gamma_x} \exp\left[-\int_0^t U(x_\tau)\, d\tau\right] f(x_t) W_x(dX). \tag{4.23}$$

Denoting the integrals in (4.23) by $I_n$ and $I$ respectively, we obtain

$$\begin{aligned} I_n &= \int_{L_x} \exp\left[-\int_0^t U(x_\tau)\, d\tau\right] f(x_t) Q_n^{(x)}(dX) \\ &= I_{nm} + R_{nm}, \end{aligned} \tag{4.24}$$

where

$$I_{nm} = \sum_{\nu=1}^m \frac{(-1)^\nu}{\nu!} \int_0^t \cdots \int_0^t \int_{L_x} U(x_{\tau_1}) \cdots U(x_{\tau_\nu}) Q_n^{(x)}(dX)\, d_{\tau_1} \cdots d\tau_\nu, \tag{4.25}$$

$$|R_{nm}| \le \frac{t^{m+1}}{(m+1)!} ||U||_\infty^{m+1} ||f||_\infty \exp(||U||_\infty t), \tag{4.26}$$

where $||\ \ ||_\infty$ means the uniform norm. Therefore $I_{nm}$ tends to $I_n$ uniformly in $n$ as $m \to \infty$. Using theorem 4.1, we have

$$(4.27)\qquad I_{nm} \to I_{\infty m} \equiv \sum_{\nu=1}^{\infty} \frac{(-1)^\nu}{\nu!} \int_0^t \cdots \int_0^t \int_{L_x} U(x_{\tau_1}) \cdots U(x_{\tau_\nu}) f(x_t) W_x(dX)\, d\tau_1 \cdots d\tau_\nu$$

and it is easy to see that $I_{\infty m} \to I$ as $n \to \infty$. Taking the uniform convergence of $\lim_{m\to\infty} I_{nm} = I_n$ into account, we have

$$(4.28)\qquad \lim_{n\to\infty} I_n = \lim_{n\to\infty} \lim_{m\to\infty} I_{nm} = \lim_{m\to\infty} \lim_{n\to\infty} I_{nm} = \lim_{m\to\infty} I_{\infty m} = I,$$

which completes our proof.

## 5. Feynman integral

In this section we shall discuss (2.10) for

$$(5.1)\qquad N_n = \frac{1}{\prod_\nu \left(1 + \frac{n^2\lambda_\nu}{\hbar i}\right)^{1/2}}, \qquad n = 1, 2, \ldots.$$

As in section 4, we can easily verify the convergence of the infinite sums and infinite products, by appealing to (3.8).

LEMMA 5.1. *If* $\mathrm{Re}(b) > 0$ *and* $c$ *is real, then*

$$(5.2)\qquad \int_{-\infty}^{\infty} \exp(-b\alpha^2 + ic)\, d\alpha = \left(\frac{\pi}{b}\right)^{1/2} \exp\left(-\frac{c^2}{4b}\right).$$

PROOF. This is true if $b > 0$. By analytic continuation, we can verify (5.2) for $\mathrm{Re}(b) > 0$.

LEMMA 5.2. *If* $g(\tau) \in L^2(0, t]$, *then*

$$(5.3)\qquad \frac{1}{N_n} \int_{L_x} \exp\left[\frac{i}{\hbar} \int_0^t \frac{\dot{x}_\tau^2}{2}\, d\tau + i \int_0^t x_\tau g(\tau)\, d\tau\right] P_n^{(n)}(dX) = \exp\left[-\sum_\nu \frac{n^2\lambda_\nu \hbar i g_\nu}{2(n^2\lambda_\nu + \hbar i)}\right],$$

*where*

$$(5.4)\qquad g_\nu = (g, \eta_\nu) = \int_0^t g(\tau)\eta_\nu(\tau)\, d\tau.$$

PROOF. We shall use the orthogonalization method introduced in section 3.

$$(5.5)\qquad I_n = \frac{1}{N_n} \int_{L_x} \exp\left[\frac{i}{\hbar} \int_0^t \frac{\dot{x}_\tau^2}{2}\, d\tau + i \int_0^t \dot{x}_\tau g(\tau)\, d\tau\right] P_n^{(x)}(dX)$$
$$= \frac{1}{N_n} E\left\{\exp\left[\frac{in^2}{2\hbar} \sum_\nu (a_\nu^2 + ing_\nu a_\nu)\right]\right\}$$
$$= \frac{1}{N_n} \prod_\nu E\left[\exp\left(\frac{in^2}{2\hbar} a_\nu^2 + ing_\nu a_\nu\right)\right]$$
$$= \frac{1}{N_n} \prod_\nu \int_{-\infty}^{\infty} \frac{1}{(2\pi\lambda_\nu)^{1/2}} \exp\left(-\frac{\alpha^2}{2\lambda_\nu} + \frac{in^2\alpha^2}{2\hbar} + ing_\nu a_\nu\right) d\alpha.$$

Using lemma 5.1 to evaluate this integral, we have

$$I_n = \exp\left[-\sum_\nu \frac{n^2\lambda_\nu \hbar i g_\nu^2}{2(n^2\lambda_\nu + \hbar i)}\right]. \tag{5.6}$$

Noting that $\sum g_\nu^2 = \int_0^t g(\tau)^2\, d\tau$, we obtain, as an immediate result from lemma 5.2,

THEOREM 5.1. *If* $g(\tau) \in L^2(0, t]$, *then*

$$\exp\left[\frac{i}{\hbar}\int_0^t \frac{\dot{x}_\tau^2}{2}\, d\tau + i\int_0^t \dot{x}_\tau g(\tau)\, d\tau\right] \in \mathfrak{D} \tag{5.7}$$

*and*

$$\frac{1}{N}\int_{\Gamma_x} \exp\left[\frac{i}{\hbar}\int_0^t \frac{\dot{x}_\tau^2}{2}\, d\tau + i\int_0^t \dot{x}_\tau g(\tau)\, d\tau\right]\prod_\tau dx_\tau = \exp\left[-\hbar i\int_0^t \frac{g^2(\tau)}{2}\, d\tau\right]. \tag{5.8}$$

THEOREM 5.2. *If the Fourier transform of* $\varphi(x)$ *is a continuous function with compact support, then*

$$\exp\left(\frac{i}{\hbar}\int_0^t \frac{\dot{x}_\tau^2}{2}\, d\tau\right)\varphi(x_\tau) \in \mathfrak{D}, \tag{5.9}$$

*and*

$$\psi(t, x) = \frac{1}{N}\int_{\Gamma_x} \exp\left(\frac{i}{\hbar}\int_0^t \frac{\dot{x}_\tau^2}{2}\, d\tau\right)\varphi(x_t)\prod_\tau dx_\tau \tag{5.10}$$

*solves*

$$\frac{\hbar}{i}\frac{\partial\psi}{\partial t} = \frac{\hbar^2}{2}\frac{\partial^2\psi}{\partial x^2}, \qquad \psi(0, x) = \varphi(x). \tag{5.11}$$

PROOF. It is enough to prove that

$$I_n = \frac{1}{N_n}\int_{L_x} \exp\left(\frac{i}{\hbar}\int_0^t \frac{\dot{x}_\tau^2}{2}\, d\tau\right)\varphi(x_t) P_n^{(x)}(dX) \tag{5.12}$$

tends to

$$\int \frac{1}{(2\pi\hbar i t)^{1/2}} \exp\left[\frac{-(x-y)^2}{2\hbar i t}\right]\varphi(y)\, dy. \tag{5.13}$$

Denoting the Fourier transform of $\varphi(x)$ by $\hat{\varphi}(\hat{x})$ or $(\mathfrak{F}\varphi)(x)$ as

$$\hat{\varphi}(\hat{x}) = (\mathfrak{F}\varphi)(\hat{x}) = \int_{-\infty}^{\infty} \exp(2\pi i \hat{x} x)\varphi(x)\, dx, \tag{5.14}$$

we have

$$\begin{aligned} I_n &= \frac{1}{N_n}\int_{L_x} \exp\left(\frac{i}{\hbar}\int_0^t \frac{\dot{x}_\tau^2}{2}\, d\tau\right)\int_{-\infty}^{\infty} \exp(-2\pi i \hat{x} x_t)\hat{\varphi}(x)\, d\hat{x}\, P_n^{(x)}(dX) \\ &= \frac{1}{N_n}\int_{-\infty}^{\infty} \hat{\varphi}(\hat{x})\exp(-2\pi i \hat{x} x)\, d\hat{x}\int_{L_x} \exp\left(\frac{i}{\hbar}\int_0^t \frac{\dot{x}_\tau^2}{2}\, d\tau \right. \\ &\qquad \left. - 2\pi i \hat{x}\int_0^t \dot{x}_\tau\, d\tau\right) P_n^{(x)}(dX). \end{aligned} \tag{5.15}$$

Putting $g(t) = -2\pi\hat{x}\dot{x}_\tau$ in lemma 5.2 to compute this integral over $L_x$, we have

$$(5.16)\qquad I_n = \int_{-\infty}^{\infty} \varphi(\hat{x}) \exp\left[-2\pi i\hat{x}x - \sum_\nu \frac{n^2\hbar i\lambda_\nu g_\nu^2}{2(n^2\lambda_\nu + \hbar i)}\right] dx.$$

Recalling the assumption that $\hat{\varphi}(\hat{x})$ has a compact support, and noting that

$$(5.17)\qquad \sum_\nu g_\nu^2 = \int_0^t g^2(\tau)\, d\tau = 4\pi^2\hat{x}^2 t,$$

we have

$$(5.18)\qquad \lim_{n\to\infty} I_n = \int_{-\infty}^{\infty} \hat{\varphi}(\hat{x}) \exp(-2\pi i\hat{x}x - 2\pi^2\hbar it\hat{x}^2)\, d\hat{x}.$$

Since the Fourier transform of

$$(5.19)\qquad N(x, \hbar it) = \frac{1}{(2\pi\hbar it)^{1/2}} \exp\left(\frac{-x^2}{2\hbar it}\right)$$

in the Schwartz distribution sense [6] is $\exp(-2\pi^2\hbar it\hat{x}^2)$, we obtain

$$(5.20)\qquad \begin{aligned}\lim_{n\to\infty} I_n &= \mathfrak{F}^{-1}\{(\mathfrak{F}\varphi)\mathfrak{F}[N(\cdot, \hbar it)]\}\\ &= \varphi(x) * N(x, \hbar it),\end{aligned}$$

namely

$$(5.21)\qquad \psi(t, x) = N(x, \hbar it) * \varphi(x),$$

which completes our proof.

THEOREM 5.3. *If the Fourier transform of $\varphi(x)$ has compact support, then we have*

$$(5.22)\qquad \exp\left[\frac{i}{\hbar}\int_0^t \left(\frac{\dot{x}_\tau^2}{2} - x_\tau\right) d\tau\right] \varphi(x_t) \in \mathfrak{D}$$

*and*

$$(5.23)\qquad \psi(t, x) = \frac{1}{N}\int_{\Gamma_x} \exp\left[\frac{i}{\hbar}\int_0^t \left(\frac{\dot{x}_\tau^2}{2} - x_\tau\right) d\tau\right] \varphi(x_t) \prod_\tau dx_\tau$$

*solves*

$$(5.24)\qquad \frac{\hbar}{i}\frac{\partial\psi}{\partial t} = \frac{\hbar^2}{2}\frac{\partial^2\psi}{\partial x^2} - x\psi, \qquad \psi(0+, x) = \varphi(x).$$

PROOF. Defining $\hat{\varphi}(\hat{x})$ as in (5.14), we obtain

$$(5.25)\qquad \begin{aligned} I_n &= \frac{1}{N_n}\int_{L_x} \exp\left[\frac{i}{\hbar}\int_0^t \left(\frac{\dot{x}_\tau^2}{2} - x_\tau\right) d\tau\right] \varphi(x_t) P_n^{(x)}(dX)\\ &= \frac{1}{N_n}\int \hat{\varphi}(\hat{x})\, d\hat{x} \int_{L_x} \exp\left(\frac{i}{\hbar}\int_0^t \frac{\dot{x}_\tau^2}{2}\, d\tau - \frac{i}{\hbar}\int_0^t x_\tau\, d\tau - 2\pi i\hat{x}x_t\right) P_n^{(x)}(dX)\\ &= \frac{1}{N_n}\int \hat{\varphi}(\hat{x}) \exp\left[-2\pi i\left(\hat{x} + \frac{t}{2\pi\hbar}\right)x\right] d\hat{x} \int_{L_x} \exp\left[\frac{i}{\hbar}\int_0^t \frac{\dot{x}_\tau^2}{2}\, d\tau \right.\\ &\qquad\qquad \left. + i\int_0^t g(\tau)\dot{x}_\tau\, d\tau\right] P_n^{(x)}(dX),\end{aligned}$$

where $g(\tau) = -(t-\tau)/\hbar - 2\pi\hat{x}$. Using lemma 5.2 to evaluate this integral over $L_x$ and recalling that $\hat{\varphi}$ has compact support to take the limit as $n \to \infty$, we have

$$\begin{aligned}(5.26)\quad \psi(t, x) &= \int \hat{\varphi}(\hat{x}) \exp\left[-2i\left(\hat{x} + \frac{t}{2\pi\hbar}\right)x - \frac{\hbar i}{2}\int_0^t \left(-\frac{t-\tau}{\hbar} - 2\pi\hat{x}\right)^2 d\tau\right] d\hat{x} \\ &= \int \hat{\varphi}\left(\hat{x} - \frac{t}{2\pi\hbar}\right) \exp\left[-2\pi i\hat{x}x - \frac{\hbar i}{2}\int_0^t \left(\frac{\tau}{\hbar} - 2\pi\hat{x}\right)^2 d\tau\right] d\hat{x} \\ &= \int \hat{\varphi}\left(\hat{x} - \frac{t}{2\pi\hbar}\right) \exp\left\{-2\pi i\hat{x}x - \frac{\hbar^2 i}{6}\left[\left(-2\pi\hat{x} + \frac{t}{\hbar}\right)^3 - (-2\pi\hat{x})^3\right]\right\} d\hat{x}.\end{aligned}$$

Thus we get

$$\begin{aligned}(5.27)\quad \hat{\psi}(t, \hat{x}) &\equiv \mathfrak{F}[\psi(t, \cdot)] \\ &= \hat{\varphi}\left(\hat{x} - \frac{t}{2\pi\hbar}\right) \exp\left\{-\frac{\hbar^2 i}{6}\left[\left(-2\pi\hat{x} + \frac{t}{\hbar}\right)^3 - (-2\pi\hat{x})^3\right]\right\}.\end{aligned}$$

Simple computation shows that $\hat{\psi}(t, \hat{x})$ satisfies

$$(5.28)\quad \frac{\hbar}{i}\frac{\partial\hat{\psi}}{\partial t} = \frac{\hbar^2}{2}(-2\pi i\hat{x})^2\hat{\psi} - \frac{1}{2\pi i}\frac{\partial\hat{\psi}}{\partial\hat{x}}, \qquad \hat{\varphi}(0+, \hat{x}) = \hat{\varphi}(\hat{x}).$$

This implies (5.13).

I would like to thank Professor G. Baxter for his valuable suggestions and assistance in writing the final version of this paper.

## REFERENCES

[1] P. A. M. Dirac, *The Principles of Quantum Mechanics*, Oxford, Clarendon, 1935 (2nd ed.).
[2] J. L. Doob, *Stochastic Processes*, New York, Wiley, 1953.
[3] R. P. Feynman, "Space-time approach to non-relativistic quantum mechanics," *Rev. Mod. Phys.*, Vol. 20 (1948), pp. 368–387.
[4] I. M. Gelfand and A. M. Yaglom, "Integration in function spaces and its application to quantum physics," *Uspehi. Mat. Nauk*, Vol. 11 (1956), pp. 77–114. (In Russian.)
[5] M. Kac, "On the distribution of certain Wiener integrals," *Trans. Amer. Math. Soc.*, Vol. 65 (1949), pp. 1–13.
[6] L. Schwartz, *Théorie des Distributions*, 2 vols., Paris, Hermann, 1950, 1951.

Reprinted from
*Proc. Fourth Berkeley Symp. Math. Stat. Probab.* **II**, 228–238 (1960)

# CONSTRUCTION OF DIFFUSIONS

Kiyosi ITO
Professeur à l'Université de Kyoto (Japon)

## 1 - INTRODUCTION -

The generator $\mathcal{G}$ of one-dimensional classical diffusions is given by a second order differential operator :

$$\mathcal{G}u(\xi) = \frac{a(\xi)^2}{2}\frac{d^2u}{d\xi^2} + b(\xi)\frac{du}{d\xi}, \tag{1}$$

as one finds in the systematic discussions by A. N. Kolmogorov [1]. W. Feller [2] extended the concept of classical diffusions, introducing a topologically invariant definition of general diffusions and determined their generator $\mathcal{G}$ in the form :

$$\mathcal{G}u(\xi) = D_m D_s u(\xi). \tag{2}$$

In order to construct Kolmogorov's diffusions, we can use the method of stochastic differential equations [3]. In fact, solving the stochastic differential equation :

$$dx_t = a(x_t)d\beta_t + b(x_t)dt, \tag{3}$$

where $\beta_t$ is the standard Brownian motion, we can construct the paths of the diffusions with the generator (1).

However, this method does not apply to Feller's diffusions. To construct the paths of Feller's diffusions the stochastic time substitution is a powerful tool. This was discussed by K. Ito and H. P. McKean, Jr. [4] using Lévy-Trotter's local time of Brownian motions [5]. The time substitution is availble to general diffusions, as V. A. Volkonski [6], H. P. McKean, JR. and H. Tanaka [7] and R. K. Blumenthal, R. K. Getoor and H. P. McKean, Jr. [8] discussed.

A third method is to construct Feller's diffusion as a projective limit of processes of simpler type. F. Knight [10] constructed the Brownian motion as a projective limit of random walks whose time and space scales get smaller in a certain way. Using the same idea we shall construct Feller's diffusions as a projective limit of semi-Markov processes with polygonal paths. This construction is the aim of our paper and will be discussed in Section 4. We shall show the background of our method in Section 2 and prepare in Section 3 some properties of the solutions of $\alpha u - D_m D_s u = 0$ which will be useful in Section 4.

## 2 - THE POLYGONAL SEMI-MARKOV PROCESSES DERIVED FROM FELLER'S DIFFUSION -

Consider a strong Markov process $\mathfrak{M}$ with the state space $[0,1]$ and with continuous paths. We shall use the following notations :

- $w$ : continuous path
- $x_t(w)$ : the value of $w$ at $t$
- $W$ : the space of all continuous paths
- $\mathbf{B}$ : the Borel algebra generated by all cylindrical subsets of $W$
- $P_a$ : the probability law of the path of $\mathfrak{M}$ starting at $a$
- $\sigma_a$ : the first passage time for $a$.

23

We shall assume :

$$P_a(\sigma_0 < \infty) > 0, \quad P_a(\sigma_1 < \infty) > 0 \qquad a \in (0,1) \tag{1}$$

$$E_a(\tau_{01}) < \infty \qquad \tau_{01} = \min(\sigma_0, \sigma_1) \tag{2}$$

and :

$$P_0(x_t = 0) = 1, \qquad P_1(x_t = 1) = 1, \tag{3}$$

and call $\mathfrak{M}$ Feller's diffusion in this paper, though Feller discussed other types of diffusions.

Introducing the scales and speed measure dm as

$$s(a) = P_a(\sigma_1 < \sigma_0) \tag{4}$$

$$dm(a) = -d\, D_s E_a(\tau_{01}), \tag{5}$$

we can express the generator $\mathfrak{G}$ of $\mathfrak{M}$ as

$$\mathfrak{G}u(a) = D_m D_s u(a) \qquad a \in (0,1) \tag{6}$$

$$\mathfrak{G}u(0) = \mathfrak{G}u(1) = 0. \tag{7}$$

Given Feller's diffusion $\mathfrak{M}$ mentioned above and given a set of division points of $[0,1]$

$$\delta : 0 = a_0 < a_1 < \dots < a_N = 1, \tag{8}$$

we shall define a semi-Markov process with polygonal paths. Take any path w of $\mathfrak{M}$ starting at a point in $\delta$, and introduce

$$T_1(w) = \min(\sigma_{a_{k-1}}, \sigma_{a_{k+1}}) \qquad \text{if } x_0(w) = a_k \qquad 0 < k < N \tag{9}$$

$$= \infty \qquad \text{if } x_0(w) = a_0 \quad \text{or} \quad a_N$$

$$T_2(w) = T_1(w^+_{T_1}) \qquad \text{if } T_1(w) < \infty \tag{10}$$

$$T_3(w) = T_1(w^+_{T_1+T_2}) \qquad \text{if } T_1(w) + T_2(w) < \infty \tag{11}$$

and so on, where $w^+_s$ is the *shifted path* defined as

$$x_t(w^+_s) = x_{t+s}(w).$$

Connecting $(0, x_0(w))$, $(T_1(w), x_{T_1}(w))$, $(T_1 + T_2, x_{T_1+T_2}(w)), \dots$, we shall get a polygonal path which will be denoted with $\pi_\delta w$. Here we shall define

$$x_t(\pi_\delta w) = x_T(w) \qquad \text{if } T = T_1 + T_2 + \dots + T_n < t < \infty = T_{n+1} \tag{12}$$

$\mathfrak{M}_\delta \equiv [x_t(\pi_\delta w),\ w \in W(P_a),\ a \in \delta]$ gives a semi-Markov polygonal process. It is *semi-Markov* in the sense that it starts afresh not at every Markov time but at every Markov time of the form $T_1$, $T_1 + T_2$, $T_1 + T_2 + T_3, \dots$

It is clear by the definition that

$$|x_t(\pi_\delta w) - x_t(w)| < \|\delta\| = \max_i |S(a_i) - S(a_{i-1})| \tag{13}$$

and that the probability law governing $\mathfrak{M}_\delta$ is determined completely by :

$$u_{k\alpha} = E_{a_k}(e^{-\alpha T_1},\ x(T_1) = a_{k+1}) \tag{14}$$

$$v_{k\alpha} = E_{a_k}(e^{-\alpha T_1},\ x(T_1) = a_{k-1}) \qquad (k = 1, 2, \dots N-1) \tag{15}$$

But $u_{k\alpha}$ is the value at $a_k$ of the solution of

$$\alpha u - D_m D_s u = 0, \quad u(a_{k-1}) = 0, \quad u(a_{k+1}) = 1, \tag{16}$$

while $v_{k\alpha}$ is the value at $a_k$ of the solution of

$$\alpha v - D_m D_s v = 0, \quad v(a_{k-1}) = 1, \quad v(a_{k+1}) = 0. \tag{17}$$

Such observation is the background of our construction which will be carried out in Section 4.

## 3 - THE EQUATION $\alpha u - D_m D_s u = 0$ -

Consider the equation :

$$\alpha u(\xi) - D_m D_s u(\xi) = 0 \qquad b < \xi < c, \ \alpha \geq 0, \tag{1}$$

and its solutions $u_\alpha = u_\alpha(\xi)$ and $v_\alpha = v_\alpha(\xi)$ with the following boundary conditions

$$u_\alpha(b) = 0, \quad u_\alpha(c) = 1 \tag{2}$$

$$v_\alpha(b) = 1, \quad v_\alpha(c) = 0. \tag{3}$$

$u_\alpha$ is an increasing solution of (1), while $v_\alpha$ is a decreasing solution of (1), and $u_\alpha$ and $v_\alpha$ constitute a fundamental system of solutions of (1).

Define the Green function $G^\circ_\alpha(\xi, \eta)$ as

$$G^\circ_\alpha(\xi,\eta) = G^\circ_\alpha(\eta, \xi) = u_\alpha(\xi) v_\alpha(\eta) / [u_\alpha, v_\alpha] \qquad b \leq \xi \leq \eta \leq c, \tag{4}$$

where $[u_\alpha, v_\alpha]$ is the Wronskian (constant)

$$[u_\alpha, v_\alpha] = v_\alpha D_s u_\alpha - u_\alpha D_s v_\alpha, \tag{5}$$

and the Green operator $G^\circ_\alpha$ as

$$(G^\circ_\alpha f)(\xi) = \int_b^c G^\circ_\alpha(\xi, \eta) f(\eta)\, dm(\eta). \tag{6}$$

As McKean proved, we have

$$\partial^n u_\alpha / \partial \alpha^n = (-)^n\, n!\, (G^\circ_\alpha)^n u_\alpha \qquad (\alpha \geq 0) \tag{7}$$

and so $u_\alpha$ can be expressed as

$$u_\alpha(\xi) = \int_0^\infty e^{-\alpha t} \mu_\xi(dt), \tag{8}$$

where $\mu_\xi(dt)$ is a bounded measure on $[0, \infty)$. Similarly we have

$$\partial^n v_\alpha / \partial \alpha^n = (-)^n\, n!\, (G^\circ_\alpha)^n v_\alpha \tag{9}$$

$$v_\alpha(\xi) = \int_0^\infty e^{-\alpha t} \nu_\xi(dt) \tag{10}$$

Setting

$$\lambda_\alpha(\xi) = u_\alpha(\xi) + v_\alpha(\xi) \tag{11}$$

$$\vartheta_\xi(dt) = \mu_\xi(dt) + \nu_\xi(dt), \tag{12}$$

we have

$$\alpha \lambda_\alpha - D_m D_s \lambda_\alpha = 0 \qquad \text{in } (b,c) \tag{13}$$

$$\lambda_\alpha(b) = \lambda_\alpha(c) = 1 \tag{14}$$

$$\lambda_\alpha(\xi) = \int_0^\infty e^{-\alpha t} \vartheta_\xi(dt). \tag{15}$$

Since $\lambda_0(\xi) = 1$, $\vartheta_\xi$ is a probability measure on $[0,\infty)$.

Let $G(\xi,\eta)$ denote $G_0^\circ(\xi,\eta)$. Then

$$G(\xi,\eta) = G(\eta,\xi) \equiv (s(c) - s(\eta))(s(\xi) - s(b))/(s(c) - s(b)) \qquad (b \le \xi \le \eta \le c) \tag{16}$$

Introducing the integral operator

$$Gf(\xi) = \int_b^c g(\xi,\eta)f(\eta)dm(\eta), \tag{17}$$

we have

$$\int_0^\infty t^n \, \vartheta_\xi(dt) = (-)^n \partial^n \lambda_\alpha / \partial \alpha^n \Big|_{\alpha=0} = n!\,(G^n 1)\,(\xi), \tag{18}$$

so that we have

$$\int_0^\infty t^n \, \vartheta_\xi(dt) \le n!\, \gamma^{n-1}\,(G1)\,(\xi), \qquad \gamma = (s(c)-s(b))m(b,c). \tag{19}$$

It follows from this inequality that

$$\begin{aligned}\int_0^\infty e^{\alpha t}\,(dt) &= 1 + \alpha\,(G1)\,(\xi) + \alpha^2(G^2 1)\,(\xi) + \dots \\ &\le 1 + \alpha\,(G1)\,(\xi)\,[1 + \alpha\gamma + (\alpha\gamma)^2 + \dots]\end{aligned} \tag{20}$$

and

$$\begin{aligned}\int_0^\infty e^{-t}\,\vartheta_\xi(dt) &= 1 - (G1)\,(\xi) + (G^2 1)\,(\xi) - \dots \\ &\ge 1 - (G1)\,(\xi)\,[1 + \gamma + \gamma^2 + \dots],\end{aligned} \tag{21}$$

so that

$$\int_0^\infty e^{\alpha t}\,\vartheta_\xi(dt) - 1 \le 3\alpha\,(1 - \int_0^\infty e^{-t}\,\vartheta_\xi(dt)) \ \text{ if } \gamma,\ \alpha\gamma < 1/3. \tag{22}$$

Using this estimate and noticing

$$(e^{\alpha\epsilon} - 1)\int_\epsilon^\infty \vartheta\,(dt) \le \int_0^\infty e^{\alpha t}\,\vartheta_\xi\,(dt) - 1 \tag{23}$$

and

$$\lim_{\alpha\uparrow\infty} \alpha/(e^{\alpha\epsilon} - 1) = 0, \tag{24}$$

we can immediately prove the following

LEMMA 1 - For any $\varepsilon$, $\eta > 0$, we can determine $\zeta$ depending only on $\varepsilon$ and $\eta$ and independent of b, c, and $\xi$ such that

$$(s(c) - s(b))\,m(b,c) < \zeta \tag{25}$$

implies

$$\int_\epsilon^\infty \vartheta_\xi(dt) < \eta\,(1 - \int_0^\infty e^{-t}\,\vartheta_\xi(dt)). \tag{26}$$

## 4 - CONSTRUCTION OF DIFFUSIONS -

(i) *Construction of* $\mathfrak{M}_\delta$ *and its properties.*

Consider a set $\delta$ of division points of $[0,1]$

$$\delta = (0 = a_0 < a_1 < \dots < a_n = 1), \tag{1}$$

the equation :

$$\alpha u(\xi) - D_m D_s u(\xi) = 0 \qquad \xi \in (a_{k-1}, a_{k+1}), \tag{2}$$

its solutions $u^k(\xi)$ and $v^k(\xi)$ and the measures $\mu_\xi^k$, $\nu_\xi^k$ and $\vartheta_\xi^k$ introduced in Section 3. Let $\mu_k$, $\nu_k$ and $\vartheta_k$ denote respectively $\mu_\xi^k$, $\nu_\xi^k$ and $\vartheta_\xi^k$ evaluated at $a_k$.

Starting at $a_k$, $k = 1,2,\dots,N-1$, we shall construct a polygonal path $w_\delta$ by connecting $(0, a_k)$, $(T_1, Y_1)$, $(T_1 + T_2, Y_2), \dots$, where $(T_1, Y_1, T_2, Y_2, \dots)$ is governed by the following probability law :

$$P^\delta_{a_k}(T_1 \in dt,\ Y_1 = a_{k\pm1}) = \mu_k(dt) \quad \text{or} \quad \nu_k(dt) \tag{3}$$

$$P^\delta_{a_k}(T_{n+1} \in dt,\ Y_{n+1} = a_{j_n \pm 1} / T_1 = t_1, \dots\ T_n = t_n,\ Y_1 = a_{j_1}, \dots,\ Y_n = a_{j_n}) = \mu_{j_n}(dt) \quad \text{or} \quad \nu_{j_n}(dt) \tag{4}$$

If $Y_n = 0$ or $1$, then we define $T_{n+1} = \infty$ and $Y_{n+1} = Y_n$ by convention.

Such a measure $P^\delta_{a_k}$ on the space $W_\delta$ of all such paths $W_\delta$ can be easily constructed by means of Kolmogorov's extension theorem. We shall use $x_t(w_\delta)$ or $x(t, w_\delta)$ to indicate the value of $w_\delta$ at t, and the process thus obtained is denoted by $\mathfrak{M}_\delta = (W_\delta, P^\delta_a, a \in \delta)$. $\mathfrak{M}_\delta$ is semi-Markov in the sense explained in Section 2, as is clear by the definition.

LEMMA 2 -

$$P^\delta_{a_k}(x(T_1 + \dots + T_n) = 0 \quad \text{or } 1 \text{ for some } n) = 1. \tag{5}$$

Proof. It is clear by the construction that the above probability $p_k$ satisfies

$$p_k = u_o^k(a_k)\, p_{k+1} + v_o^k(a_k)\, p_{k-1} \qquad k = 1,2,\dots, N-1. \tag{6}$$

Since $u_o^k(a_k), v_o^k(a_k) > 0$ and their sum is 1, (6) implies that $p_k$ is increasing or decreasing. Therefore $p_k \geq p_o = 1$ or $p_k \geq p_1 = 1$, so that $p_k \equiv 1$.

Now we shall introduce

$$\mathbf{m} = \mathbf{m}(\varepsilon, w_\delta) = \min(n : T_n > \varepsilon). \tag{7}$$

Then we get

LEMMA 3 - Let $\zeta$ denote the $\zeta$ determined in Lemma 1. If

$$\|\delta\| = \max_k |s(a_k) - s(a_{k-1})| < \zeta / 2m(0,1), \tag{8}$$

then :

$$E^\delta_{a_k}(e^{-(T_1 + \dots + T_{\mathbf{m}-1})}) < \eta. \tag{9}$$

Proof. Using Lemma 1, (8) implies

$$P^\delta_{a_k}(T_1 > \varepsilon) \leq \eta(1 - E^\delta_{a_k}(e^{-T_1})). \tag{10}$$

Noticing the semi-Markov property of $\mathfrak{M}$, we have

$$\begin{aligned}
&E^\delta_{a_k}(e^{-(T_1 + \dots + T_{\mathbf{m}-1})}) \\
&= P^\delta_{a_k}(T_1 > \varepsilon) + E^\delta_{a_k}(e^{-T_1}, T_1 \leq \varepsilon, T_2 > \varepsilon) + E^\delta_{a_k}(e^{-(T_1+T_2)}, T_1, T_2 \leqslant \varepsilon, T_3 > \varepsilon) + \dots \\
&\leq P^\delta_{a_k}(T_1 > \varepsilon) + E^\delta_{a_k}(e^{-T_1}, P^\delta_{x(T_1)}(T_1 > \varepsilon)) + E^\delta_{a_k}(e^{-(T_1+T_2)} P^\delta_{x(T_1+T_2)}(T_1 > \varepsilon)) + \dots \\
&\leq \eta\,[(1 - E^\delta_{a_k}(e^{-T_1})) + E^\delta_{a_k}(e^{-T_1}(1 - E^\delta_{x(T_1)}(e^{-T_1}))) + E^\delta_{a_k}(e^{-(T_1+T_2)}(1 - E^\delta_{x(T_1+T_2)}(e^{-T_1}))) \dots] \\
&= \eta\,[(1 - E^\delta_{a_k}(e^{-T_1})) + (E^\delta_{a_k}(e^{-T_1}) - E^\delta_{a_k}(e^{-T_1-T_2})) + (E^\delta_{a_k}(e^{-T_1-T_2})) - E^\delta_{a_k}(e^{-T_1-T_2-T_3})) + \dots] = \eta.
\end{aligned}$$

We shall introduce a random variable T(t) as

$$T(t) = T_1 + T_2 + \dots + T_n \quad \text{if} \quad T_1 + \dots + T_n \leq t < T_1 + \dots + T_{n+1} \tag{12}$$

and prove

LEMMA 4 - If (8) is satisfied, then

$$P^{\delta}_{a_k}(T(t) - t > \varepsilon) \leq \eta\ e^{t} \tag{13}$$

Proof. It is clear by the definition and Lemma 3 that

$$P^{\delta}_{a_k}(T(t) - t > \varepsilon)$$

$$\leq P^{\delta}_{a_k}(T_1 + T_2 + \ldots + T_{m-1} \leq t)$$

$$\leq E^{\delta}_{a_k}(e^{-(T_1+T_2+\ldots+T_{m-1})})\, e^{t} \leq \eta\ e^{t}$$

Let us take three points $b, a, c \in \delta$ such that $b \leq a \leq c$. Let $\tau$ be the smaller of $\sigma_b$ and $\sigma_c$. $\tau$ is the minimum of $T_1 + T_2 + \ldots + T_n$ such that $x(T_1 + \ldots + T_n) = b$ or $c$. It follows from Lemma 2 that :

$$P^{\delta}_{a}(T_1 + \ldots + T_n) < \infty) = 1,$$

and we have

LEMMA 5 -

$$E^{\delta}_{a}(e^{-\alpha\tau}, x(\tau) = b) = u_\alpha(a) \tag{14.a}$$

$$E^{\delta}_{a}(e^{-\alpha\tau}, x(\tau) = c) = v_\alpha(a) \tag{14.b}$$

$$E^{\delta}_{a}(e^{-\alpha\tau}) = \lambda_\alpha(a), \tag{14.c}$$

where $u_\alpha$, $v_\alpha$, and $\lambda_\alpha$ are defined in Section 3.

Proof. Suppose that $b = a_q$, $a = a_k$, and $c = a_r$, $q \leq k \leq r$.

Then we see by the definition that $\tilde{u}_k \equiv E^{\delta}_{a_k}(e^{-\alpha\tau},\ x(\tau) = b)$ satisfies

$$\tilde{u}_k = u^k(a_k)\tilde{u}_{k+1} + v^k(a_k)\tilde{u}_{k-1}, \quad \tilde{u}_q = 0, \quad \tilde{u}_r = 1$$

and it follows from the property of the solutions of linear homogeneous second order differential equations that $\hat{u}_k = u_\alpha(a_k)$ satisfies the same difference equation and the same boundary conditions. The uniqueness of the solution of such difference equation with fixed boundary conditions implies $\tilde{u}_k = \hat{u}_k$ which shows (14.a). Similarly for (14.b) and (14.c).

Setting $\alpha = 0$ in (14.a) we have

LEMMA 6 - $P^{\delta}_{a}(\sigma_b < \sigma_c) = (s(a) - s(b))/(s(c) - s(b))$.

Differentiating both sides of (14.c) in $\alpha$ and setting $\alpha = 0$, we have

LEMMA 7 -

$$E^{\delta}_{a}(\tau_{bc}) = \int_b^c G_{bc}(a,\eta)dm(\eta) \qquad \tau_{bc} = \min(\tau_b, \tau_c)$$

where

$$G_{bc}(\xi,\eta) = G_{bc}(\eta,\xi) = (s(c) - s(\eta))\,(s(\xi) - s(b))/(s(c) - s(b)) \quad (b \leq \xi \leq \eta \leq c \tag{15}$$

Noticing the semi-Markov property of $\mathfrak{M}_\delta$ we have

LEMMA 8 - If $g^{\delta}(a) = E^{\delta}_{a}\left(\int_0^\infty e^{-\alpha t} f(x_t)dt\right)$, then

$$g^{\delta}(a) = E_a\left(\int_0^\tau e^{-\alpha t} f(x_t)\,dt\right) + u_\alpha(a)g(c) + v_\alpha(a)g(b)$$

where $b < a < c$, $b, a, c \in \delta$ and $u_\alpha(a)$ and $v_\alpha(a)$ were defined in Section 3.

(ii) *Projection* $\pi_{\delta\Delta}$ *from* $\mathfrak{M}_\Delta$ *onto* $\mathfrak{M}_\delta$ *in case* $\Delta \supset \delta$.

Let $\delta$ and $\Delta$ be two sets of division points of $[0,1]$ such that $\Delta \supset \delta$, and $\mathfrak{M}_\delta = (W_\delta, P^\delta_a, a \in \delta)$ and $\mathfrak{M}_\Delta = (W_\Delta, P^\Delta_b, b \in \Delta)$ be the corresponding semi-Markov processes defined above.

Define a projection $\pi_{\delta\Delta}$ which carries $w_\Delta (\in W_\Delta)$ starting at $a \in \delta$ to $\pi_{\delta\Delta}\ w_\Delta (\in W_\delta)$ starting at a just as we defined $\pi_\delta$ in Section 2.

We shall now prove.

LEMMA 9 - $P^\delta_a = P^\Delta_a \pi^{-1}_{\delta\Delta}$. $\qquad a \in \delta$

Proof. It is enought to prove that $\tilde{\mathfrak{M}}_\delta = (W_\delta, P^\Delta_a \pi^{-1}_{\Delta\delta}, a \in \delta)$ is the same as $\mathfrak{M}_\delta$. Since $\tilde{\mathfrak{M}}_\delta$ is a semi-Markov process by the definition, it is enough to show that

$$E^\delta_{a_k}(e^{-\alpha\tau_k},\ x(\tau_k) = a_{k+1}) = E^\Delta_{a_k}(e^{-\alpha\tau_k},\ x(\tau_k) = a_{k+1}) \tag{16.a}$$

$$E^\delta_{a_k}(e^{-\alpha\tau_k},\ x(\tau_k) = a_{k-1}) = E^\delta_{a_k}(e^{-\alpha\tau_k},\ x(\tau_k) = a_{k-1}) \tag{16.b}$$

for $\tau_k = \min(\sigma_{a_{k+1}}, \sigma_{a_{k-1}})$. But (16.a) is clear, because both sides are the solution evaluated at $a_k$ of the equation :

$$\alpha u - D_m D_s u = 0 \qquad \text{in } (a_{k-1}, a_{k+1})$$

$$u(a_{k-1}) = 0, \qquad u(a_{k+1}) = 1$$

by Lemma 5. Similarly for (16.b).

(iii) *The projective limit of* $\mathfrak{M}_\delta$ *for* $\|\delta\| \longrightarrow 0$.

We shall define Feller's diffusion $\mathfrak{M} = (W, P_a, a \in [0,1])$ as the projective limit of $\mathfrak{M}_\delta$ for $\|\delta\| \longrightarrow 0$. For each $a \in [0,1]$, we shall define $P_a$ as follows.

Consider the class $C_a$ of all sets of division points containing a and $P_a^\delta$ (defined in (i)) for $\delta \in C_a$. As we proved in (i), we have

$$P^\delta_a = P^\Delta_a \pi^{-1}_{\delta\Delta} \qquad \delta, \Delta \in C_a, \quad \delta \subset \Delta. \tag{17}$$

Applying Bochner's theorem [11], a generalized version of Kolmogorov's extension theorem, we can construct a probability measure space $\Omega(P)$ on which a system of stochastic processes $y^\delta_t(\omega)$, $t \geq 0$ depending on $\delta \in C_a$, is defined such that each $y^\delta_t(\omega)$ is the version of $x_t(w_\delta)$, $w_\delta \in W\ P^\delta_a)$, and that, if $a \in \delta \subset \Delta$, then

$$\pi_{\delta\Delta}\ y^\Delta_\cdot(\omega) = y^\delta_\cdot(\omega), \tag{18}$$

so that

$$|s(y^\Delta_t(\omega)) - s(y^\delta_t(\omega))| < 2\varepsilon \quad \text{if } \|\delta\|, \|\Delta\| < \varepsilon . \tag{19}$$

Since $s(\xi)$ is continuous in $\xi \in [0,1]$ and one to one, $y^\delta_t(\omega)$ converges uniformly in $(t,\omega)$, as $\|\delta\| \longrightarrow 0$. Let $y_t(\omega)$ denote the limit. Since $y^\delta_t(\omega)$ is continuous in t, its uniform limit $y_t(\omega)$ is also continuous in t.

It is easily seen that

$$\pi_\delta\ y_\cdot(\omega) = y_\cdot(\omega), \tag{20}$$

where $\pi_\delta$ is the mapping defined in Section 2.

Let $P_a$ be the probability law the stochastic process $y_t(\omega)$, $t \geq 0$, $\omega \in \Omega(P)$, yields on the space W of continuous paths. It is clear that

$$P^\delta_a = P_a \pi^{-1}_\delta \qquad a \in \delta . \tag{21}$$

Now it remains to prove that $\mathfrak{M} = (W, P_a, a \in [0,1])$ is Feller's process we wished to construct.

For the proof we shall start with

29

LEMMA 10 -

$$g(a) = E_a\Big(\int_0^\infty e^{-\alpha t} f(x_t)dt\Big) \tag{22}$$

is continuous if f is continuous.

Proof. It follows from Lemma 7 and Lemma 8,

$$\begin{aligned} &|g^\delta(a) - u_\alpha(a)g^\delta(c) - v_\alpha(a)g^\delta(b)| \\ &\le E_a\Big(\int_0^\tau e^{-\alpha t} f(x_t)\, dt\Big) \\ &\le \|f\|\ E_a^\delta(\tau) \qquad\qquad \|f\| = \max_\xi |f(\xi)| \\ &= \|f\| \int_b^c G_{(b,c)}(a,\xi)\, dm(\xi) \end{aligned} \tag{23}$$

But as far as $\delta \ni \xi$,

$$\begin{aligned} g^\delta(\xi) &= E_\xi^\delta\Big(\int_0^\infty e^{-\alpha t} f(x_t)dt\Big) \\ &= E_\xi\Big(\int_0^\infty e^{-\alpha t} f(x_t(\pi_\delta w))\, dt\Big) \\ &\longrightarrow E_\xi\Big(\int_0^\infty e^{-\alpha t} f(x_t(w))\, dt\Big) = g(a) \qquad \text{as } \|\delta\| \to 0. \end{aligned}$$

Letting $\|\delta\| \longrightarrow 0$ in (23) under the condition that $\delta \ni a,b,c$, we have

$$\begin{aligned} &|g(a) - u_\alpha(a)g(c) - v_\alpha(a)g(b)| \\ &\le \|f\| \int_b^c G_{bc}(a,\xi)\, dm(\xi) \\ &\le \|f\|\, (s(c) - s(b))\, m(b,c) \\ &\le \|f\|\, m(0,1)\, (s(c) - s(b)), \end{aligned}$$

from which we can see the continuity of g, using the fact that $u_\alpha(\xi)$ and $v_\alpha(\xi)$ are continuous in $\xi \in [b,c]$ and that $s(\xi)$ is continuous in $\xi \in [0,1]$.

Now we shall prove the Markov property of $\mathfrak{M}$.

Let $f(\xi)$ and $F(\xi_1, \xi_2, \ldots, \xi_n)$ be continuous. We shall prove that

$$E_a(f(x_{t_n+s})\ F(x_{t_1}, x_{t_2}, \ldots, x_{t_n})) = E_a[E_{x(t_n)}(f(x_s))\ F(x_{t_1}, x_{t_2}, \ldots, x_{t_n})] \qquad (0 \le t_1 < t_2 < \ldots < t_n + s) \tag{24}$$

Since both sides are continuous in s, it is enough to prove that

$$\int_0^\infty e^{-\alpha s}\ E_a(f(x_{t_n+s})\ F(x_{t_1}, x_{t_2}, \ldots, x_{t_n}))\, ds = E_a(g(x_{t_n})\ F(x_{t_1}, x_{t_2}, \ldots x_{t_n})), \tag{25}$$

where g is a continuous function introduced in Lemma 10.

Let $T_\delta(t)$ be the minimum of u such that $x(u) \in \delta$ and that $u \ge t$. Then it is clear that :

$$T_\delta(t, w) = T_\delta(t, \pi_\delta w). \tag{26}$$

Using Lemma 4, we have :

$$\lim_{\substack{\delta \ni a \\ \|\delta\| \to 0}} P_a^\delta(T_\delta(t) - t > \varepsilon) = 0 \qquad (\varepsilon > 0)$$

and so

$$\lim_{\substack{\delta \ni a \\ \|\delta\| \to 0}} P_a(T_\delta(t, \pi_\delta w) - t > \varepsilon) = 0.$$

30

Using (26), we get

$$\lim_{\substack{\delta \ni a \\ \|\delta\| \to 0}} P_a(T_\delta(t, w) - t > \varepsilon) = 0.$$

Therefore we can take a sequence $\delta_m (\ni a)$, $m = 1, 2, \dots$, such that

$$P_a(\lim_{m \to \infty} T_{\delta_m}(t, w) = t) = 1,$$

for each t, so that

$$P_a(\lim_{m \to \infty} T_{\delta_m}(t_i, w) = t_i, \ i = 1, 2, \dots, n) = 1.$$

Write $T_m$, $P_a^m$ and $\pi_m$ respectively for $T_{\delta_m}$, $P_a^{\delta_m}$ and $\pi_{\delta_m}$ and notice that

$$|s(x_t(\pi_\delta w)) - s(x_t(w))| < 2\|\delta\|. \tag{27}$$

Since $f(\xi)$ and $g(\xi)$ are uniformly continuous in $\xi$ and so in $s(\xi)$ and $F(\xi_1, \xi_2, \dots, \xi_n)$ is uniformly continuous in $(\xi_1, \dots, \xi_n)$ and so in $(s(\xi_1), \dots, s(\xi_n))$, we get :

$$\int_0^\infty e^{-\alpha s} E_a(f(x_{t_n+s}) F(x_{t_1}, \dots, x_{t_n})) ds$$

$$= \lim_{m \to \infty} \int_0^\infty e^{-\alpha s} E_a[f(x_{T_m(t_n)+s}) F(x_{T_m(t_1)}, \dots, x_{T_m(t_n)})] ds$$

$$= \lim_{m \to \infty} \int_0^\infty e^{-\alpha s} E_a[f(x_{T_m(t_n, \pi_m w)+s}(\pi_m w)) F(x_{T_m(t_1, \pi_m w)}(\pi_m w), \dots, x_{T_m(t_n, \pi_m w)}(\pi_m w))] ds$$

$$= \lim_{m \to \infty} \int_0^\infty e^{-\alpha s} E_a^m[f(x_{T_m(t_n)+s}) F(x_{T_m(t_1)}, \dots, x_{T_m(t_n)})] ds$$

$$= \lim_{m \to \infty} \int_0^\infty e^{-\alpha s} E_a^m[E_{x(T_m(t_n))}^m (f(x_s)) F(x_{T_m(t_1)}, \dots, x_{T_m(t_n)})] ds$$

$$= \lim_{m \to \infty} \int_0^\infty e^{-\alpha s} E_a[E_{x(T_m(t_n))} (f(x_s)) F(x_{T_m(t_1)}, \dots, x_{T_m(t_n)})] ds$$

$$= E_a(g(x_{t_n}) F(x_{t_1}, \dots, x_{t_n})).$$

We used the semi-Markov property of $\mathfrak{M}_{\delta_n}$, (26), (27) and the continuity of g and F in the last four steps.

Thus we have proved that $\mathfrak{M}$ is Markov, and therefore $\mathfrak{M}$ is also strong Markov by virtue of Lemma 10.

To identify our process with Feller's diffusion, it is enough to observe :

$$P_a(\sigma_1 < \sigma_0) = P_a(\sigma_1(\pi_\delta w) < \sigma_0(\pi_\delta w)) \qquad (a \in \delta)$$

$$= P_a^\delta(\sigma_1 < \sigma_0)$$

$$E_a(\tau_{01}) = E_a(\tau_{01}(\pi_\delta w)) = E_a^\delta(\tau_{01}) \qquad (\tau_{01} = \min(\sigma_0, \sigma_1)),$$

so that

$$P_a(\sigma_1 < \sigma_0) = s(a)$$

$$E_a(\tau_{01}) = \int_0^1 G_{01}(a, \xi) dm(\xi)$$

and so $-dD_s E_a(\tau_{01}) = dm(a)$.

Thus we have proved

31

THEOREM - $\mathfrak{M}$, defined above, is Feller's diffusion with the generator $D_m D_s$ and with sticking boundaries.

## LITERATURES

[1] KOLMOGOROV, A. N. - Uber die analytischen Methoden in der Wahrsheinlichkeitsrechnung. Math. Ann. *104*, 415-458 (1931).

[2] FELLER, W. - On second order differential operators. Ann. Math. *31* 90-105 (1955).

[3] ITO, K. - Stochastic differential equations. Mem. Amer. Math. Soc. *4* (1951).

[4] ITO, K. and McKEAN, H. P. Jr - Diffusion.

[5] TROTTER, H. - A property of Brownian motion paths. Ill. Jour. Math. *2*, 425-433 (1958).

[6] VOLKONSKI, V. A. - Random substitution of time in strong Markov processes. Th. Prob. Appl. *3* 332-350 (1958).

[7] McKEAN, H. P. Jr. and TANAKA, H - Additive functionals of Brownian path. Mem. Univ. Kyoto, A. Math. *33*, 479-506 (1961).

[8] BLUMENTHAL, R. M., GETOOR, R. K. and McKEAN, H. P., Jr. - Markov processes with identical hitting distributions.

[9] McKEAN, H. P., Jr. - Elementary solutions for certain parabolic differential equations. Trans. Amer. Math. Soc. *82*, 519-548 (1956).

[10] KNIGHT, F. - On the random walk and Brownian motion, to appear in Trans. Amer. Math. Soc.

[11] BOCHNER, S. - Harmonic analysis and the theory of probability, Berkeley, Univ. Cal. Press, 1955.

## DISCUSSION

M. NEVEU - Que sait-on sur la représentation explicite du générateur infinitésimal d'un processus de diffusion pluri-dimensionnel ?

M. ITO - mentionne la formule générale de représentation de Dynkin et cite les travaux concernant les processus de diffusion ayant les mêmes probabilités d'absorption que le mouvement brownien.



Reprinted from
*Ann. Fac. Sci. Univ. Clermont.* **II,** 23–32 (1962)

# THE BROWNIAN MOTION AND TENSOR FIELDS ON RIEMANNIAN MANIFOLD

By KIYOSI ITÔ

## 1. Introduction

The Brownian motion $\xi(t)$ on an $r$-dimensional Riemannian manifold $M^r$ is defined as a diffusion process with the generator

$$g = \tfrac{1}{2} g^{ij} \nabla_i \nabla_j, \tag{1.1}$$

where $\nabla_i$ denotes covariant differentiation. The paths of the Brownian motion can be constructed by means of the stochastic differential equation [1,2,3]. We shall denote with $P_a$ the probability law of the Brownian motion starting at a point $a \in M^r$ and the corresponding expectation with $E_a$.

Let $f$ be a scalar field on $M^r$. It is clear by the definition that

$$u(t,a) = E_a[f(\xi(t))] \tag{1.2}$$

satisfies a heat equation.

$$\frac{\partial}{\partial t} u = \tfrac{1}{2} g^{ij} \nabla_i \nabla_j u \tag{1.3 a}$$

with the initial condition

$$u(0+,a) = f(a). \tag{1.3}$$

Our problem is to establish a similar fact in case $f$ is a vector field $f_k$ or more generally a tensor field $f_{k_1 k_2 \ldots k_p}$. We cannot replace $f$ in (1.2) with $f_{k_1 k_2 \ldots k_p}$, because $f_{k_1 k_2 \ldots k_p}(\xi(t))$ is a tensor attached to the point $\xi(t)$ which varies with $t$, while $u(t,a)$ should be attached to $a = \xi(0)$. Therefore we shall shift $f_{k_1 \ldots k_p}(\xi(t))$ back to $a = \xi(0)$ along the path $\xi$ by parallel displacement. Denoting with $T_{\xi,t}$ the parallel displacement along $\xi$ from $a = \xi(0)$ to $\xi(t)$, we can prove that

$$u_{k_1 \ldots k_p}(t, a) = E_a[T_{\xi,t}^{-1} f_{k_1 \ldots k_p}(\xi(t))] \tag{1.4}$$

satisfies a differential equation of the same form as (1.3):

$$\frac{\partial}{\partial t} u_{k_1 \ldots k_p} = \tfrac{1}{2} g^{ij} \nabla_i \nabla_j u_{k_1 \ldots k_p}. \tag{1.5}$$

In order to define the parallel displacement $T_{\xi,t}$ used above, we cannot use Levi-Civita's definition in its usual form because of the non-differentiability of the Brownian paths. We shall define $T_{\xi,t}$ as follows. Let

$$\Delta : 0 = t_0 < t_1 < \ldots < t_n = t \tag{1.6}$$

be a division of the time interval $[0,t]$, make a polygonal curve $\xi_\Delta(t)$ which consists of the geodesic curves connecting $\xi(t_{i-1})$ with $\xi(t_i)$, $i=1,2,\ldots,n$, and denote with $T_{\xi,t}$ the Levi-Civita parallel displacement along $\xi_\Delta$. Then we can prove that $T_{\xi_\Delta,t}$ tends to a limit in probability as $|\Delta| = \max_i(t_i - t_{i-1})$ tends to 0. This limit is defined to be $T_{\xi,t}$.

Noticing that the operator $\Delta = d\delta + \delta d$ introduced in the Hodge theory of harmonic tensor fields by Kodaira and also by Bidal and de Rham is expressed in the form [5]

$$\Delta = -g^{ij}\nabla_i\nabla_j + \text{linear transformation} \tag{1.7}$$

and modifying the transformation $T_{\xi,t}$ slightly in the above discussion, we can also get a solution of

$$\frac{\partial}{\partial t}u_{k_1\ldots k_p} = -\tfrac{1}{2}\Delta u_{k_1\ldots k_p} \tag{1.8 a}$$

with the initial condition

$$u_{k_1\ldots k_p}(0+,a) = f_{k_1\ldots k_p}(a) \tag{1.8 b}$$

As was proved by Milgram and Rosenbloom [4], $u_{k_1\ldots k_p}(+\infty, a)$ is the harmonic tensor field with the same periods as $f_{k_1\ldots k_p}(a)$ in case $M^r$ is compact. We shall discuss this in a separate paper.

## 2. The paths of the Brownian motion on a Riemannian manifold

The Brownian motion on a Riemannian manifold $M^r$ is a diffusion (a strict Markov process with continuous paths) with the generator

$$\begin{aligned} g &= \tfrac{1}{2}g^{ij}\nabla_i\nabla_j \\ &= \tfrac{1}{2}g^{ij}\frac{\partial^2}{\partial x^i \partial x^j} - \tfrac{1}{2}g^{ij}\Gamma^k_{ij}\frac{\partial}{\partial x^k}. \end{aligned} \tag{2.1}$$

Introducing $\sigma^i_k$ and $m^k$ by

$$\sum_k \sigma^i_k \sigma^j_k = g^{ij}, \tag{2.2 a}$$

$$m^k = -\tfrac{1}{2}g^{ij}\Gamma^k_{ij} \tag{2.2 b}$$

and solving a stochactic differential equation

$$d\xi^i(t) = \sigma^i_k(\xi(t))\,d\beta^k(t) + m^i(\xi(t))\,dt, \tag{2.3 a}$$

$$\xi^i(0) = a^i \tag{2.3 b}$$

($\beta$ is the $n$-dimensional Wiener process),

we can construct the paths of the Brownian notion starting at $a \in M^r$. See [1, 2 and 3] for the details.

## 3. The parallel displacement along the Brownian paths

Since almost all paths are non-differentiable, we shall define the parallel displacement along the Brownian paths as follows.

Let $(\xi^k(t),\ k=1,2,\dots,r)$ be the Brownian notion on $M^r$ starting at $a=(a^k)$ and $\alpha_{k_1 k_2 \dots k_p}$ be a tensor attached to $a$. Let

$$\Delta: 0=t_0<t_1<t_2<\dots<t_n=t \tag{3.1}$$

be a division of the time interval $[0,t]$, make a polygonal curve $\xi_\Delta(s),\ 0\leqslant s\leqslant t$, which consists of the geodesic curves connecting $\xi(t_{i-1})$ with $\xi(t_i)$, $i=1, 2,\dots,n$, and denote with $\alpha^\Delta_{k_1\dots k_p}(t)$ be the tensor obtained from $\alpha_{k_1\dots k_p}$ by the Levi-Civita parallel displacement along $\xi_\Delta$. Then we have

THEOREM 1. *As* $|\Delta|=\max_i(t_i-t_{i-1})$ *tends to* 0, $\alpha^\Delta_{i_1\dots i_p}(t)$ *tends in probability for each fixed* $t$ *to the solution* $\alpha_{i_1\dots i_p}(t)$ *of the stochastic differential equation*

$$\begin{aligned} d\alpha_{k_1\dots k_p}(t) &= \sum_\nu \Gamma^k_{ik_\nu}(\xi(t))\,\alpha_{k_1\dots k_{\nu-1}kk_{\nu+1}\dots k_p}(t)\,d\xi^i(t) \\ &\quad + \tfrac12 \sum_\nu g^{mi}(\xi(t))\left[\frac{\partial \Gamma^k_{ik_\nu}}{\partial x^m}(\xi(t) + \Gamma^l_{ik_\nu}(\xi(t))\Gamma^k_{ml}(\xi(t))\right] \\ &\quad\quad \times \alpha_{k_1\dots k_{\nu-1}kk_{\nu+1}\dots k_p}(\xi(t))\,dt \\ &\quad + \tfrac12 \sum_{\mu\neq\nu} g^{mi}(\xi(t))\,\Gamma^k_{mk_\mu}(\xi(t))\,\Gamma^l_{ik_\nu}(\xi(t)) \\ &\quad\quad \times \alpha_{k_1\dots k_{\mu-1}kk_{\mu+1}\dots k_{\nu-1}lk_{\nu+1}\dots k_p}(\xi(t))\,dt, \end{aligned} \tag{3.2 a}$$

with the initial condition

$$\alpha_{k_1\dots k_p}(0)=\alpha_{k_1\dots k_p}. \tag{3.2 b}$$

See [1,2] for the meaning of the this equation.

*Definition.* $\alpha_{k_1\dots k_p}(t)$ is called the parallel displacement of $\alpha_{k_1\dots k_p}$ along $\xi$.

We shall sketch the proof of this theorem.

Noticing the symbolic estimation

$$\Delta\xi^i(t)\,\Delta\xi^j(t) \sim g^{ij}(\xi(t))\,\Delta t \tag{3.3}$$

(see [2] for the exact meaning), we can ignore $o((\Delta\xi^i(t))^2)$ as well as $o(\Delta t)$. Therefore the geodesic curve $C$ connecting $a=\xi(t)$ with $b=\xi(t+\Delta t)$ is approximately

$$x^i(t)=a^i+(b^i-a^i)t+\frac{t(1-t)}{2}\Gamma^i_{jk}(a)(b^j-a^j)(b^k-a^k). \tag{3.4}$$

Therefore the tensor $\beta_{k_1\dots k_p}$ at $b$ obtained from $\alpha_{k_1\dots k_p}$ at $a$ by the Levi-Civita parallel displacement along $C$ is approximately given by

$$\beta_{k_1\ldots k_p}\sim\alpha_{k_1\ldots k_p}+\sum_{\nu}\Gamma^k_{ik_\nu}(a)\alpha_{k_1\ldots k_{\nu-1}kk_{\nu+1}\ldots k_p}(b^i-a^i)$$
$$+\frac{1}{2}\frac{\partial\Gamma^k_{ik_\nu}}{\partial x^m}(a)+\Gamma^l_{ik_\nu}(a)\,\Gamma^k_{ml}(a)$$
$$\times\alpha_{k_1\ldots k_{\nu-1}kk_{\nu+1}\ldots k_p}(a)\,(b^m-a^m)\,(b^i-a^i)$$
$$+\tfrac{1}{2}\sum_{\mu\neq\nu}\Gamma^l_{ik_\nu}(a)\,\Gamma^l_{mk_\mu}(a)\,\alpha_{i_1\ldots i_{\mu-1}ki_{\mu+1}\ldots i_{\nu-1}li_{\nu+1}\ldots i_p}(a)$$
$$\times(b^m-a^m)\,(b^i-a^i). \tag{3.5}$$

using (3.3) we shall get the stochastic differential equation (3.2 a) for $\alpha_{i_1\ldots i_p}(t)$.

## 4. Heat equation

Using the notations as in section 3, consider the mapping

$$T_{\xi,t}:\alpha_{k_1\ldots k_p}\to\alpha_{k_1\ldots k_p}(t) \tag{4.1}$$

and the tensor

$$u_{k_1\ldots k_p}(t,a)=E_a(T^{-1}_{\xi,t}f_{k_1\ldots k_p}(\xi(t))) \tag{4.2}$$

Then we have

THEOREM 2. $u_{k_1\ldots k_p}(t,a)$ satisfies the heat equation

$$\frac{\partial}{\partial t}u_{k_1\ldots k_p}=\tfrac{1}{2}g^{ij}\nabla_i\nabla_j u_{k_1\ldots k_p}. \tag{4.3}$$

The Markov property of the Brownian motion implies that

$$\frac{\partial}{\partial t}u_{k_1\ldots k_p}=Au_{k_1\ldots k_p}, \tag{4.4}$$

where

$$Af_{k_1\ldots k_p}=\lim_{t\to 0}\frac{f_{k_1\ldots k_p}(t,a)-f_{k_1\ldots k_p}(a)}{t}. \tag{4.5}$$

To identify $A$ with $\frac{1}{2}g^{ij}\nabla_i\nabla_j$, it is enough to compute $E(T^{-1}_{\xi,t}\,f_{i_1\ldots i_p}(\xi(t))$ neglecting $o((d\xi^i)^2)$ as well as $o(dt)$. Though the computation is complicated, the technique is essentially the same as in [1].

### REFERENCES

[1]. Itô, K., Stochastic differential equations in a differentiable manifold. *Nagoya Math. J.*, 1 (1950), 35–47.

[2]. —— On a formula concerning stochastic differentials. *Nagoya Math. J.*, 3 (1951), 55–65.

[3]. —— Stochastic differential equations in a differentiable manifold. *Mem. Coll. Sci. Univ. Kyoto, Ser. A. Math.*, 28, (1953), 81–85.

[4]. Milgram, A. N. & Rosenbloom, P. C., Harmonic forms and heat conduction. I. Closed Riemannian manifolds. *Proc. Nat. Acad. Sci. U.S.A.*, 37 (1951), 180–184.

[5]. De Rham, G., Variétés différentiables. *Actualités Sci. Ind.*, 1222 (1955), Paris.

Reprinted from
*Proc. Int. Congr. Mathemati.* **536–539** (1962)

Reprinted from ILLINOIS JOURNAL OF MATHEMATICS
Vol. 7, No. 2, June 1963
*Printed in U.S.A.*

# BROWNIAN MOTIONS ON A HALF LINE

**Dedicated to W. Feller**

BY

K. ITÔ AND H. P. MCKEAN, JR.[1]

## Contents



*Numbering.* 1 means formula 1 of the present section; 2.1 means formula 1 of Section 2, etc.; the numbering of the diagrams is similar.

## 1. The classical Brownian motions

Consider the space of all (continuous) *sample paths* $w:[0, +\infty) \to R^1$

Received December 4, 1961.

[1] Fulbright grantee 1957–58 during which time the major part of this material was obtained; the support of the Office of Naval Research, U.S. Govt. during the summer of 1961 is gratefully acknowledged also.

181

with coordinates $\mathfrak{x}(t, w) = \mathfrak{x}(t)$ $(t \geqq 0)$, the field $\mathsf{A}$ of events

1. $$B = w^{-1}_{t_1 t_2 \cdots t_n}(A) = (w: (\mathfrak{x}(t_1), \mathfrak{x}(t_2), \cdots, \mathfrak{x}(t_n)) \in A)$$
$$0 < t_1 < t_2 < \cdots < t_n, \quad A \in \mathsf{B}(R^n), \quad n \geqq 1,$$ [2]

and the Gauss kernel

2. $$g(t, a, b) = e^{-(b-a)^2/2t}/(2\pi t)^{1/2}, \quad (t, a, b) \in (0, +\infty) \times R^2.$$

Because of

3a. $$g(t, a, b) > 0,$$

3b. $$\int g(t, a, b)\, db = 1,$$

3c. $$g(t, a, b) = \int g(t - s, a, c) g(s, c, b)\, dc \qquad (t > s),$$

the function

4. $$P_a(B) = \int_A g(t_1, a, b_1) g(t_2 - t_1, b_1, b_2) \cdots g(t_n - t_{n-1}, b_{n-1}, b_n) \cdot db_1\, db_2 \cdots db_n$$

of $B = w^{-1}_{t_1 t_2 \cdots t_n}(A) \in \mathsf{A}$ is well-defined, nonnegative, additive, and of total mass $+1$ for each $a \in R^1$, and, as N. Wiener [1] discovered, the estimate

5. $$\int_{|a-b|>\varepsilon} g(t, a, b)\, db < \text{constant} \times \varepsilon^{-1} t^{1/2} e^{-\varepsilon^2/2t}, \qquad t \downarrow 0,$$

permits us to extend it to a nonnegative Borel measure $P_a(B)$ of total mass $+1$ on the Borel extension $\mathsf{B}$ of $\mathsf{A}$ (see P. Lévy [3] for an alternative proof).

Granting this, it is apparent that $P_a(\mathfrak{x}(0) \in db)$ is the unit mass at $b = a$. $P_a(B)$ is now interpreted as *the chance of the event $B$ for paths starting at the point $a$* and the sample path $w: t \to \mathfrak{x}(t)$ with these probabilities imposed is called *standard Brownian motion starting at $a$.*

Given $t \geqq 0$, if $B \in \mathsf{B}$ and if $w_t^+$ denotes the shifted path $w_t^+: s \to \mathfrak{x}(t + s, w)$, then 4 implies

6. $$P_a(w_t^+ \in B \mid \mathfrak{x}(s): s \leqq t) = P_b(B), \qquad b = \mathfrak{x}(t),$$

i.e., the *law* of the *future* $\mathfrak{x}(s): s > t$ *conditional on the past* $\mathfrak{x}(s): s \leqq t$ depends upon the *present* $b = \mathfrak{x}(s)$ alone (in short, *the Brownian traveller starts afresh at each constant time $t \geqq 0$*).

Because the Gauss kernel $g(t, a, b)$ is the fundamental solution of the heat flow problem

7. $$\frac{\partial u}{\partial t} = \frac{1}{2} \frac{\partial^2 u}{\partial a^2}, \qquad (t, a) \in (0, +\infty) \times R^1,$$

[2] $\mathsf{B}(R^n)$ is the usual topological Borel field of the $n$-dimensional euclidean space $R^n$.

the operator $\mathfrak{G} = D^2/2$ acting on[3] $C^2(R^1)$ is said to *generate* the standard Brownian motion, and it is natural to seek other differential operators $\mathfrak{G}^\bullet$ giving rise via the fundamental solution of $\partial u/\partial t = \mathfrak{G}^\bullet u$ and the rule 4 to similar (stochastic) motions.

Consider, for example, the operator[4]

8. $$\mathfrak{G}^+ = \mathfrak{G} \mid C^2[0, +\infty) \cap (u : u^+(0) = 0):$$

the fundamental solution of $\partial u/\partial t = \mathfrak{G}^+ u$ is

9. $$g^+(t, a, b) = e^{-(b-a)^2/2t}/(2\pi t)^{1/2} + e^{-(b+a)^2/2t}/(2\pi t)^{1/2}, \qquad t > 0 \leqq a, b,$$

which satisfies 3a, 3b, and 3c, and the corresponding (*reflecting Brownian*) *motion* is identical in law to

10. $$\mathfrak{x}^+ = |\mathfrak{x}|,$$

where $\mathfrak{x}$ is a standard Brownian motion.

Consider next the operator

11. $$\mathfrak{G}^- = \mathfrak{G} \mid C^2[0, +\infty) \cap (u : u(0) = 0):$$

the fundamental solution of $\partial u/\partial t = \mathfrak{G}^- u$ is

12. $$g^-(t, a, b) = e^{-(b-a)^2/2t}/(2\pi t)^{1/2} - e^{-(b+a)^2/2t}/(2\pi t)^{1/2}, \qquad t > 0 \leqq a, b,$$

which satisfies 3 with

3b(bis). $$\int g^-(t, a, b)\, db < 1$$

in place of 3b, and the corresponding (*absorbing Brownian*) *motion* is identical in law to

13. $$\begin{aligned} \mathfrak{x}^-(t) &= \mathfrak{x}^+(t) \quad \text{if} \quad t < \mathfrak{m}_0, \\ &= \infty \qquad \text{if} \quad t \geqq \mathfrak{m}_0, \end{aligned}$$

where $\mathfrak{x}^+$ is the reflecting Brownian motion described above, $\mathfrak{m}_0$ is its passage time $\mathfrak{m}_0 = \min(t : \mathfrak{x}^+(t) = 0)$, and $\infty$ is an extra state adjoined to $R^1$.

Given $0 < \gamma < +\infty$, the operator

14. $$\mathfrak{G}^\gamma = \mathfrak{G} \mid C^2[0, +\infty) \cap (u : \gamma u(0) = u^+(0))$$

is also possible: the fundamental solution of $\partial u/\partial t = \mathfrak{G}^\gamma u$ is

15a. $$\begin{aligned} g^\gamma(t, a, b) &= g^\gamma(t, b, a) \\ &= g^-(t, a, b) + \int_0^t \frac{a}{(2\pi s^3)^{1/2}} e^{-a^2/2s} g^\gamma(t - s, 0, b)\, ds, \quad t > 0 \leqq a, b, \end{aligned}$$

[3] $C^d(R^1)$ is the space of bounded continuous functions $f : R^1 \to R^1$ with $d$ bounded continuous derivatives.

[4] $C^2[0, +\infty)$ is the space of functions $u \in C[0, +\infty)$ with $D^2u \in C(0, +\infty)$ and $(D^2u)(0) \equiv (D^2u)(0+)$ existing. $u^+(0) = \lim_{\varepsilon \downarrow 0} \varepsilon^{-1}[u(\varepsilon) - u(0)]$.

15b. $g^{\gamma}(t, 0, 0) = 2 \int_0^{+\infty} e^{-\gamma c} \frac{c}{(2\pi t^3)^{1/2}} e^{-c^2/2t}\, dc, \qquad t > 0,$

which satisfies 3 with 3b(bis) in place of 3b, and the corresponding (*elastic Brownian*) *motion* is identical in law to

16a. $$\mathfrak{x}^{\gamma}(t) = \mathfrak{x}^{+}(t) \quad \text{if} \quad t < \mathfrak{m}_{\infty}, \qquad = \infty \quad \text{if} \quad t \geqq \mathfrak{m}_{\infty},$$

16b. $$\mathfrak{m}_{\infty} = \mathfrak{t}^{-1}(\mathfrak{e}/\gamma),$$

where $\mathfrak{e}$ is an exponential holding time independent of the reflecting Brownian motion $\mathfrak{x}^{+}$ with law $P.(\mathfrak{e} > t) = e^{-t}$ and $\mathfrak{t}^{-1}$ is the inverse function of the *reflecting Brownian local time*:

17. $$\mathfrak{t}^{+}(t) = \lim_{\varepsilon \downarrow 0} (2\varepsilon)^{-1} \text{ measure } (s : \mathfrak{x}^{+}(s) < \varepsilon, s \leqq t)$$

(see Sections 3, 4, 14 for additional information about local times).

## 2. Feller's Brownian motions

W. Feller [1] discovered that the classical Brownian generators $\mathfrak{G}^{\pm}$ and $\mathfrak{G}^{\gamma}$ $(0 < \gamma < +\infty)$ of Section 1 are the simplest members of a wide class of restrictions $\mathfrak{G}^{\bullet}$ of $\mathfrak{G} \mid C^2[0, +\infty)$ which generate what could be called Brownian motions on $[0, +\infty)$. Feller found that the domain $D(\mathfrak{G}^{\bullet}) \subset C^2[0, +\infty)$ of such a generator could be described in terms of three nonnegative numbers $p_1, p_2, p_3$, and a nonnegative mass distribution $p_4(dl)$ $(l > 0)$ subject to[5]

1. $$p_1 + p_2 + p_3 + \int_{0+} (l \wedge 1) p_4(dl) = 1$$

as follows:

2. $$D(\mathfrak{G}^{\bullet}) = C^2[0, +\infty) \cap \left( u : p_1 u(0) - p_2 u^{+}(0) + p_3(\mathfrak{G}u)(0) = \int_{0+} [u(l) - u(0)] p_4(dl) \right).$$

M. Kac [1] cited the problem of describing the sample paths of the elastic Brownian motion $(p_3 = p_4 = 0 < p_1 p_2)$, and it was W. Feller's (private) suggestion that these should be the reflecting Brownian sample paths, killed at the instant some increasing function $\mathfrak{t}^{+}(\mathfrak{Z}^{+} \cap [0, t])$ of the visiting set $\mathfrak{Z}^{+} \equiv (t : \mathfrak{x}^{+}(t) = 0)$ hits a certain level, that was the starting point of this paper.

P. Lévy's profound studies [3] had clarified the fine structure of the standard and reflecting Brownian motions and their local times, the papers of E. B. Dynkin [1] and G. Hunt [1] on Markov times provided an indispensable

[5] $a \wedge b$ is the smaller of $a$ and $b$. $\int_{0+}$ means $\int_{0<l<+\infty}$.

tool, H. Trotter [1] proved a deep result about local times, and W. Feller [2] had presented a (partial) description of the sample paths of the Brownian motion associated with $\mathfrak{G}^{\bullet}$ in the special case $p_4(0, +\infty) < +\infty$ (the case $p_4(0, +\infty) = +\infty$ was not discovered in Feller's original proof of 2, but this error was corrected by W. Feller [3] and A. D. Ventsell [1]).

It was left to use these ideas (and some new ones) to build up the sample paths of Feller's Brownian motions from the reflecting Brownian motion and its local time and (independent) exponential holding times and differential processes; that is the aim of the present paper.

## 3. Outline

Brownian motions on $[0, +\infty)$ are defined from a probabilistic point of view in Section 5, and a special case is disposed of in Section 6. Green operators

$$G_\alpha^{\bullet}: f \rightarrow E_{\cdot}^{\bullet}\left(\int_0 e^{-\alpha t} f(\mathfrak{x}^{\bullet})\, dt\right)$$

and the generator $\mathfrak{G}^{\bullet}$ $(= \alpha - G_\alpha^{\bullet\,-1})$ are introduced in Section 7 and computed in Section 8 using a method of E. B. Dynkin [1]. $\mathfrak{G}^{\bullet}$ turns out to be the restriction of $\mathfrak{G} \mid C^2[0, +\infty)$ to a domain $D(\mathfrak{G}^{\bullet})$ as described in 2.2; it is the *simplest complete invariant* of the motion, i.e., the associated sample paths can be built up from

(a) *a reflecting Brownian motion* $\mathfrak{x}^+$,
(b) *a differential process* $\mathfrak{p}$ *with increasing sample paths based on* $p_2$ *and* $p_4$,
(c) *a stochastic clock* $\mathfrak{f}^{-1}$ *based on* $\mathfrak{x}^+$, $\mathfrak{p}$, *and* $p_3$,
(d) *a killing time based on* $\mathfrak{x}^+$, $\mathfrak{p}$, $\mathfrak{f}^{-1}$, *and* $p_1$

(see Sections 9–15)

Consider, for the sake of conversation, the case:

1. $$p_4(0, +\infty) = +\infty \quad \text{if} \quad p_2 = 0,$$

introduce the reflecting Brownian motion $\mathfrak{x}^+$ as described in Section 1 $(u^+(0) = 0)$, and let $\mathfrak{t}^+$ be P. Lévy's *mesure du voisinage* (*local time*)

2. $$\mathfrak{t}^+(t) = \lim_{\varepsilon \downarrow 0} (2\varepsilon)^{-1} \text{ measure } (s : \mathfrak{x}^+(s) < \varepsilon, s \leqq t)$$

as described in Section 4.

Given $p_1 = p_3 = 0$, if $\mathfrak{p}(dt \times dl)$ is a Poisson measure as described in Section 11 with mean $dt \times p_4(dl)$ indepedent of $\mathfrak{x}^+$, if $\mathfrak{p}$ is the (increasing) differential process

3. $$\mathfrak{p}(t) = p_2 t + \int_{0+} l\mathfrak{p}([0, t] \times dl), \qquad t \geqq 0,$$

and if $\mathfrak{p}^{-1}$ is its inverse function, then the desired motion is identical in law to[6]

4. $$\mathfrak{x}^{\bullet} = \mathfrak{p}\mathfrak{p}^{-1}\mathfrak{t}^{+} - \mathfrak{t}^{+} + \mathfrak{x}^{+},$$

which could be described as *a reflecting Brownian motion jumping out from $l = 0$ like the germ of the differential process $\mathfrak{p}$ run with the clock $\mathfrak{p}^{-1}\mathfrak{t}^{+}$* (see Section 12 for pictures).

$\mathfrak{p}^{-1}\mathfrak{t}^{+}$ can be interpreted as a *local time* for the new sample path $\mathfrak{x}^{\bullet}$ (see Section 14), and, with its help, the description of the sample paths can be completed as follows: in case $p_1 = 0$, the desired motion is identical in law to

5a. $$\mathfrak{x}^{\bullet}(\mathfrak{f}^{-1}), \qquad \mathfrak{x}^{\bullet} = \mathfrak{p}\mathfrak{p}^{-1}\mathfrak{t}^{+} - \mathfrak{t}^{+} + \mathfrak{x}^{+},$$

where the stochastic clock $\mathfrak{f}^{-1}$ is the inverse function of

5b. $$\mathfrak{f} = t + p_3\mathfrak{p}^{-1}(\mathfrak{t}^{+}(t)),$$

while, in case $p_1 > 0$, it is identical in law to $\mathfrak{x}^{\bullet}(\mathfrak{f}^{-1})$ *killed* (i.e., sent off to an extra state $\infty$) at a time $\mathfrak{m}_{\infty}^{\bullet}$ ($< +\infty$) with conditional distribution

6. $$P.(\mathfrak{m}_{\infty}^{\bullet} > t \mid \mathfrak{x}^{\bullet}(\mathfrak{f}^{-1})) = e^{-p_1\mathfrak{p}^{-1}\mathfrak{t}^{+}\mathfrak{f}^{-1}}.$$

Here are two simple cases to be treated in Section 10.

Given $p_1 = p_4 = 0 < p_2\, p_3$ (i.e., $u^{+}(0) = (p_3/p_2)(\mathfrak{G}^{\bullet}u)(0)$), the desired motion is identical in law to

7a. $$\mathfrak{x}^{\bullet} = \mathfrak{x}^{+}(\mathfrak{f}^{-1}),$$

7b. $$\mathfrak{f} = t + (p_3/p_2)\mathfrak{t}^{+}.$$

*$\mathfrak{f}^{-1}$ counts standard time while $\mathfrak{x}^{\bullet}(t) > 0$ but runs slow on the barrier, and hence, compared to the reflecting Brownian motion, $\mathfrak{x}^{\bullet}$ lingers at $l = 0$ a little longer than it should*; as a matter of fact,

8. $$\text{measure}\,(s: \mathfrak{x}^{\bullet}(s) = 0,\ s \leqq t) = p_3\,\mathfrak{t}^{+}(\mathfrak{f}^{-1}(t)) > 0$$

if $t > \min\,(s: \mathfrak{x}^{\bullet}(s) = 0)$.

Given $p_3 = p_4 < p_1\, p_2$ (i.e., $(p_1/p_2)\, u(0) = u^{+}(0)$), the desired (elastic Brownian) motion is identical in law to a reflecting Brownian motion, killed at time $\mathfrak{m}_{\infty}^{\bullet}$ with conditional distribution

9. $$P.(\mathfrak{m}_{\infty}^{\bullet} > t \mid \mathfrak{x}^{+}) = e^{-(p_1/p_2)\mathfrak{t}^{+}(t)},$$

i.e., *killed on the barrier $l = 0$ at a rate $(p_1/p_2)\mathfrak{t}^{+}(dt){:}dt$ proportional to the local time.*

Brownian motions with similar barriers at both ends of $[-1, +1]$ or with a two-sided barrier on the line or the unit circle are studied in Sections 16 and 17, Section 18 treats a wider class of Brownian motions on $[0, +\infty)$, substantiating a conjecture of N. Ikeda, Section 19 describes the sample paths in case a diffusion operator $\mathfrak{G}u = u^{+}(dl)/e(dl)$ is used in place of

[6] $\mathfrak{p}\mathfrak{p}^{-1}\mathfrak{t}^{+}$ means $\mathfrak{p}(\mathfrak{p}^{-1}(\mathfrak{t}^{+}))$.

the reflecting Brownian generator $\mathfrak{G}^+$, and Section 20 indicates how to adapt the method to birth and death processes.

## 4. Standard Brownian motion: stopping times and local times

Before coming to Brownian motions on a half line, it is convenient to collect in one place some facts about the standard Brownian motion on the line (see K. Itô and H. P. McKean, Jr. [1] for the proofs and additional information).

Consider a standard Brownian motion with sample paths $w:t \to \mathfrak{x}(t)$, universal field $\mathsf{B}$, and probabilities $P_a(B)$ as described in Section 1, define[7] $\mathsf{B}_t = \mathsf{B}[\mathfrak{x}(s):s \leqq t]$, and, if $\mathfrak{m} = \mathfrak{m}(w)$ is a stopping time, i.e., if

1a. $$0 \leqq \mathfrak{m} \leqq +\infty,$$

1b. $$(\mathfrak{m} < t) \in \mathsf{B}_t, \qquad t \geqq 0,\text{[8]}$$

then introduce the associated field

2. $$\mathsf{B}_{\mathfrak{m}+} = \mathsf{B} \cap (B:(\mathfrak{m} < t) \cap B \in \mathsf{B}_t, t \geqq 0).$$

$\mathsf{B}_{\mathfrak{m}+} = \bigcap_{s>t} \mathsf{B}_s$ in case $\mathfrak{m} \equiv t$; in general, $(\mathfrak{m} < t) \in \mathsf{B}_{\mathfrak{m}+}$ $(t \geqq 0)$, and, with the aid of

3a. $$\mathsf{B}_{\mathfrak{a}+} \subset \mathsf{B}_{\mathfrak{b}+}, \qquad \mathfrak{a} \leqq \mathfrak{b},$$

3b. $$\mathsf{B}_{\mathfrak{a}+} = \bigcap_{\varepsilon>0} \mathsf{B}_{\mathfrak{b}+}, \qquad \mathfrak{b} = \mathfrak{a} + \varepsilon,$$

it is not hard to see that $\mathsf{B}_{\mathfrak{m}+}$ measures the *past* $x(t):t \leqq \mathfrak{m}+$, i.e.,

4. $$\mathsf{B}_{\mathfrak{m}+} \supset \bigcap_{\varepsilon>0} \mathsf{B}[\mathfrak{x}(t \wedge (\mathfrak{m} + \varepsilon)):t \geqq 0].$$

E. B. Dynkin [1] and G. Hunt [1] discovered that *the Brownian traveller starts afresh at a stopping time*; this means that for each stopping time $\mathfrak{m}$, each $a \in R^1$, and each $B \in \mathsf{B}$,

5. $$P_a(w_{\mathfrak{m}}^+ \in B \mid \mathsf{B}_{\mathfrak{m}+}) = P_b(B), \qquad b = \mathfrak{x}(\mathfrak{m})$$

where $w_{\mathfrak{m}}^+$ denotes the shifted path[9] $w_{\mathfrak{m}}^+:t \to \mathfrak{x}(t + \mathfrak{m})$, $\mathfrak{x}(+\infty) \equiv \infty$, and $P_\infty(\mathfrak{x}(t) \equiv \infty, t \geqq 0) = 1$. Because $\mathfrak{m} \equiv t$ is a stopping time, 5 includes the *simple Markovian evolution* noted in 1.6; an alternative statement is that *conditional on* $\mathfrak{m} < +\infty$ *and on the present state* $b = \mathfrak{x}(\mathfrak{m})$, *the future* $\mathfrak{x}(t + \mathfrak{m}):t \geqq 0$ *is a standard Brownian motion, independent of* $\mathfrak{m}$ *and of the past* $\mathfrak{x}(t):t \leqq \mathfrak{m}+$.

Given $l > 0$, the *passage time* $\mathfrak{m}_l = \min(t:\mathfrak{x}(t) = l)$ is a stopping time, and the motion $[\mathfrak{m}_l:l \geqq 0, P_0]$ is a differential process, homogeneous in the parameter $l$; it is, in fact, *the one-sided stable process* with exponent $\frac{1}{2}$, rate

[7] $\mathsf{B}[\mathfrak{q}(t):a \leqq t < b]$ means the smallest Borel subfield of $\mathsf{B}$ measuring the motion indicated inside the brackets.

[8] $(\mathfrak{m} < t)$ is short for $(w:\mathfrak{m} < t)$.

[9] $\infty$ is an extra state $\notin R^1$.

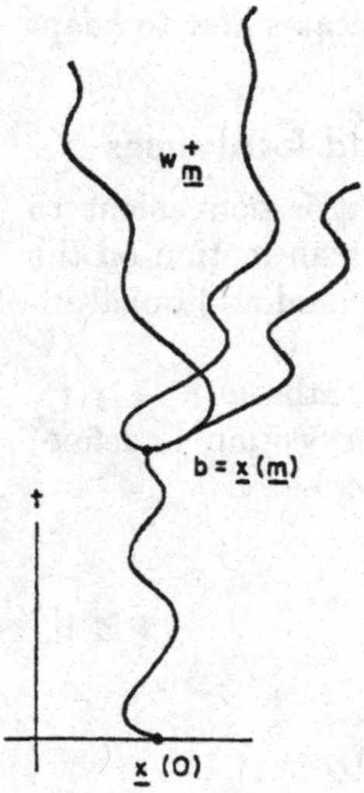


Diagram 1

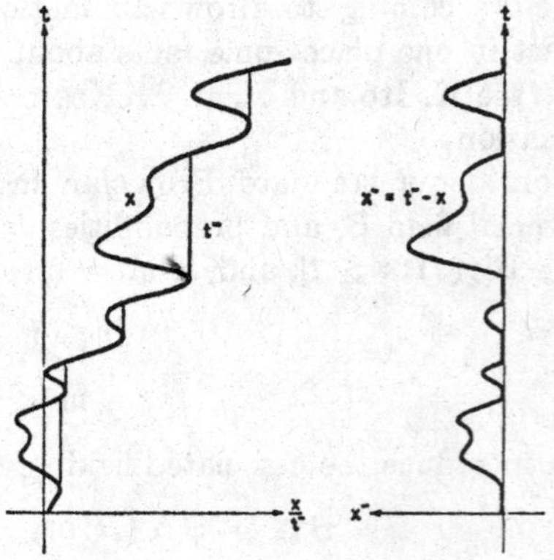


Diagram 2

$\sqrt{2}$, and law

$$6. \qquad P_0(\mathfrak{m}_l \in dt) = \frac{l}{(2\pi t^3)^{1/2}} e^{-l^2/2t}\, dt$$

as P. Lévy [2] discovered.

$\mathfrak{m}.$, itself, is a sum of positive jumps (see Section 11 for information on this point), and its inverse function $\mathfrak{t}^-(t) = \max_{s \leq t} \mathfrak{x}(s)$ is continuous and flat outside a (Cantor-like) set of times of Hausdorff-Besicovitch dimension number $\frac{1}{2}$; the joint law

$$7. \qquad P_0[\mathfrak{x}(t) \in da, \mathfrak{t}^-(t) \in db] = 2\frac{2b - a}{(2\pi t^3)^{1/2}} e^{-(2b-a)^2/2t}\, da\, db, \qquad b \geqq 0,\ a \leqq b$$

is cited for future use.

Consider, next, the reflecting Brownian motion $\mathfrak{x}^+ = |\mathfrak{x}|$.

Given a reflecting Brownian stopping time $\mathfrak{m}$, i.e., a time $0 \leqq \mathfrak{m} \leqq +\infty$ with $(\mathfrak{m} < t) \in \mathsf{B}[\mathfrak{x}^+(s): s \leqq t]$ $(t \geqq 0)$, $\mathfrak{m}$ is likewise a standard Brownian stopping time, and it follows that, conditional on $\mathfrak{m} < +\infty$ and $b = \mathfrak{x}^+(\mathfrak{m})$, the shifted path $\mathfrak{x}^+(t + \mathfrak{m}): t \geqq 0$ is a reflecting Brownian motion, independent of $\mathfrak{m}$ and of the past $\mathfrak{x}^+(t): t \leqq \mathfrak{m}$; *in brief, the reflecting Brownian motion starts afresh at its stopping times.*

P. Lévy [3] observed that if $\mathfrak{x}$ is a standard Brownian motion starting at 0, then $\mathfrak{x}^- = \mathfrak{t}^- - \mathfrak{x}$ $(\mathfrak{t}^- = \max_{s \leq t} \mathfrak{x}(s))$ is identical in law to the reflecting Brownian motion $\mathfrak{x}^+$ starting at 0. Diagram 2 is a mere caricature of the path, the actual visiting set $(t:\mathfrak{x} = 0)$ being a closed Cantor-like set of Lebesgue measure 0.

P. Lévy also indicated a proof of

8. $$P_0[\lim_{\varepsilon\downarrow 0}(2\varepsilon)^{-1} \text{ measure } (s:\mathfrak{x}^-(s) < \varepsilon, s \leq t) = \mathfrak{t}^-(t), t \geq 0] = 1,$$

which implies that $\mathfrak{t}^-$ is a function of $\mathfrak{x}^-$ alone, and deduced the existence of *the reflecting Brownian local time* (*mesure du voisinage*):

9. $$\mathfrak{t}^+(t) = \lim_{\varepsilon\downarrow 0}(2\varepsilon)^{-1} \text{ measure } (s:\mathfrak{x}^+(s) < \varepsilon, s \leq t)$$

(see H. Trotter [1] for a complete proof). $\mathfrak{t}^+$ grows on the visiting set $\mathfrak{Z}^+ = (t:\mathfrak{x}^+(t) = 0)$; it is identical in law to $\mathfrak{t}^-$, and its inverse function $\mathfrak{t}^{-1}$ is identical in law to the standard Brownian passage times; especially, the joint law

10. $$P_0[\mathfrak{x}^+(t) \in da, \mathfrak{t}^+(t) \in db] = 2\,\frac{b+a}{(2\pi t^3)^{1/2}}\,e^{-(b+a)^2/2t}\,da\,db, \qquad a,b \geq 0,$$

is deduced from the joint law of $\mathfrak{x}$ and $\mathfrak{t}^-$ above.

Skorokhod [1] has made the point that if $\mathfrak{x}$ is a standard Brownian motion, if $0 \leq \mathfrak{x}^\bullet$ is continuous, if $0 \leq \mathfrak{t}^\bullet$ is continuous, increasing, and flat outside $\mathfrak{Z}^\bullet = (t:\mathfrak{x}^\bullet = 0)$, and if $\mathfrak{x}^\bullet = \mathfrak{t}^\bullet - \mathfrak{x}$, then $\mathfrak{x}^\bullet = \mathfrak{x}^-$ and $\mathfrak{t}^\bullet = \mathfrak{t}^-$.

## 5. Brownian motions on $[0, +\infty)$

Given *probabilities* $P_a^\bullet(B)$ $(a \in [0, +\infty) \cup \infty)$ defined on the natural *universal field* $\mathsf{B}^\bullet$ of the *path space* comprising all *sample paths*

1a. $$w^\bullet: t \to \mathfrak{x}^\bullet(t) \equiv \mathfrak{x}^\bullet(t+) \in [0, +\infty) \cup \infty,$$

1b. $$\mathfrak{x}^\bullet(t) \equiv \infty, \qquad t \geq \mathfrak{m}_\infty^\bullet \equiv \inf\,(t:\mathfrak{x}^\bullet = \infty)$$

and subject to

2a. $P_a^\bullet(B)$ *is a Borel function of* $a$,

2b. $P_a^\bullet[\mathfrak{x}^\bullet(0) \in db]$ *is the unit mass at* $b = a$ $(a \neq 0)$,

let us speak of the associated motion as

(a) *simple Markov* if it starts afresh at constant times:

3a. $$P_\cdot^\bullet(w_s^{\bullet+} \in B \mid \mathsf{B}_s^\bullet) = P_a^\bullet(B), \qquad s \geq 0, B \in \mathsf{B}^\bullet, a = \mathfrak{x}^\bullet(s),$$

where $w_s^{\bullet+}$ is the shifted path $t \to \mathfrak{x}^\bullet(t+s)$ and $\mathsf{B}_s^\bullet$ is the field of $\mathfrak{x}^\bullet(t): t \leq s$,

(b) *strict Markov* if it starts afresh at its stopping times:

3b. $$P_\cdot^\bullet(w_{\mathfrak{m}^\bullet}^{\bullet+} \in B \mid \mathsf{B}_{\mathfrak{m}^\bullet+}^\bullet) = P_a^\bullet(B), \qquad B \in \mathsf{B}^\bullet, a = \mathfrak{x}^\bullet(\mathfrak{m}^\bullet),$$

for each stopping time

4a. $$0 \leq \mathfrak{m}^\bullet \leq +\infty,$$

4b. $$(\mathfrak{m}^\bullet < t) \in \mathsf{B}_t^\bullet \quad (t \geq 0),$$

where $\mathfrak{x}^\bullet(+\infty) \equiv \infty$ and $\mathbf{B}^\bullet_{\mathfrak{m}^\bullet+}$ is the field of events

5a. $$B \in \mathbf{B}^\bullet,$$

5b. $$B \cap (\mathfrak{m}^\bullet < t) \in \mathbf{B}^\bullet_t \quad (t \geqq 0),$$

(c) *a Brownian motion* if, in addition to (b), the stopped path

6a. $$\mathfrak{x}^\bullet(t) : t < \mathfrak{m}_{0+} = \lim_{\varepsilon \downarrow 0} \inf (t : \mathfrak{x}^\bullet < \varepsilon), \qquad \mathfrak{x}^\bullet(0) = l > 0,$$

is identical in law to the stopped standard Brownian motion

6b $$\mathfrak{x}(t) : t < \mathfrak{m}_0 = \min (t : \mathfrak{x} = 0), \qquad \mathfrak{x}(0) = l.$$

$E^\bullet_\cdot$ denotes the integral (expectation) based upon $P^\bullet_\cdot$, and $E^\bullet_\cdot(e, B) = E^\bullet_\cdot(B, e)$ denotes the integral of $e = e(w^\bullet)$ extended over $B$; the subscript . as in 3a and 3b stands for an unspecified point of $[0, +\infty) \cup \infty$ with the understanding that if several dots appear in a *single* formula, then it is the *same* point that is meant each time.

## 6. Special case: $p_+(0) < 1$

Given a Brownian motion as described above and a sample path $\mathfrak{x}^\bullet$ starting at $\mathfrak{x}^\bullet(0) = l > 0$, the *crossing time*

1. $$\mathfrak{m}^\bullet = \mathfrak{m}^\bullet_\varepsilon = \inf (t : \mathfrak{x}^\bullet(t) < \varepsilon), \qquad 0 < \varepsilon < l,$$

is a stopping time, $P^\bullet_l[\mathfrak{x}^\bullet(\mathfrak{m}^\bullet_\varepsilon) = \varepsilon] = 1$, $\mathfrak{m}^\bullet_{0+} = \lim_{\delta \downarrow 0} \mathfrak{m}^\bullet_\delta = \mathfrak{m}^\bullet_\varepsilon + \mathfrak{m}^\bullet_{0+}(w^{\bullet +}_{\mathfrak{m}^\bullet})$, and, since the stopped path $\mathfrak{x}^\bullet(t) : t < \mathfrak{m}^\bullet_{0+}$ is standard Brownian,

2. $$\begin{aligned} E^\bullet_l[e^{-\alpha \mathfrak{m}^\bullet_{0+}}, \mathfrak{x}^\bullet(\mathfrak{m}^\bullet_{0+}) \in B] &= E^\bullet_l(e^{-\alpha \mathfrak{m}^\bullet_\varepsilon} E^\bullet_l[\exp(-\alpha \mathfrak{m}^\bullet_{0+}(w^{\bullet +}_{\mathfrak{m}^\bullet})), \mathfrak{x}^\bullet(\mathfrak{m}^\bullet_{0+}(w^{\bullet +}_{\mathfrak{m}^\bullet}), w^{\bullet +}_{\mathfrak{m}^\bullet}) \in B \mid \mathbf{B}^\bullet_{\mathfrak{m}^\bullet+}]) \\ &= E^\bullet_l(e^{-\alpha \mathfrak{m}^\bullet_\varepsilon}) E^\bullet_\varepsilon[e^{-\alpha \mathfrak{m}^\bullet_{0+}}, \mathfrak{x}^\bullet(\mathfrak{m}^\bullet_{0+}) \in B] \\ &\to E^\bullet_l(e^{-\alpha \mathfrak{m}^\bullet_{0+}}) P^\bullet_\varepsilon[\mathfrak{x}^\bullet(\mathfrak{m}^\bullet_{0+}) \in B] \qquad (\varepsilon \downarrow 0) \\ &= e^{-(2\alpha)^{1/2} l} P^\bullet_l[\mathfrak{x}^\bullet(\mathfrak{m}^\bullet_{0+}) \in B], \end{aligned}$$ [10]

i.e., $\mathfrak{x}^\bullet(\mathfrak{m}^\bullet_{0+})$ *is independent of* $\mathfrak{m}^\bullet_{0+}$, *and its law* $p_+(B) = P^\bullet_l[\mathfrak{x}^\bullet(\mathfrak{m}^\bullet_{0+}) \in B]$ *does not depend on* $l > 0$.

Consider the law $p(dl) \equiv P^\bullet_0[\mathfrak{x}^\bullet(0) \in dl]$, and, in case $p(0) = 1$, let $\mathfrak{e}$ be the *exit time* $\inf(t : \mathfrak{x}^\bullet(t) \neq 0)$.

Because

3a. $$p_+(0) = P^\bullet_l[\mathfrak{x}^\bullet(\mathfrak{m}^\bullet_{0+}) = 0, \mathfrak{x}^\bullet(0, w^{\bullet +}_{\mathfrak{m}^\bullet 0+}) = 0] = p_+(0) p(0), \qquad l > 0,$$

and

3b. $$p(0) = P^\bullet_0[\mathfrak{x}^\bullet(0) = 0, \mathfrak{x}^\bullet(0, w^{\bullet +}_0) = 0] = p(0)^2,$$

---

[10] $\mathfrak{m}^\bullet_{0+}$ is identical in law to the standard Brownian passage time $\mathfrak{m}_0 = \min(t : \mathfrak{x}(t) = 0)$, and hence $E^\bullet_l(\exp(-\alpha \mathfrak{m}^\bullet_{0+})) = \exp(-(2\alpha)^{1/2} l)$ (see 4.6).

the possibilities are

4a. $$p(0) = p_+(0) = 0,$$

4b. $$p(0) = 1 > p_+(0),$$

4c. $$p(0) = p_+(0) = 1.$$

4a is the simplest case. Diagram 1 shows the motion $[\mathfrak{x}^\bullet, P_0^\bullet]$: the jumps $l_1, l_2$, etc. are independent with common law $p_+(dl)$, the initial position $l_0$ is independent of $l_1, l_2$, etc. with law $p(dl)$, and the excursions leading back to $l = 0+$ are standard Brownian.

4b is more interesting. $\mathfrak{e}$ *is an exponential holding time independent of* $\mathfrak{x}^\bullet(\mathfrak{e})$ *with law* $e^{-t/p_3}$ ($0 \leqq p_3 \leqq +\infty$); indeed, if $s \geqq 0$, then $(\mathfrak{e} > s) \in \mathsf{B}^\bullet_{s+} = \bigcap_{t>s} \mathsf{B}^\bullet_t$, whence

5. $$P_0^\bullet(\mathfrak{e} > t + s) = P_0^\bullet(\mathfrak{e} > s, \mathfrak{e}(w_s^{\bullet +}) > t) = P_0^\bullet(\mathfrak{e} > s)P_0^\bullet(\mathfrak{e} > t)$$

and

6. $$\begin{aligned} P_0^\bullet[\mathfrak{e} > s, \mathfrak{x}^\bullet(\mathfrak{e}) \in dl] &= P_0^\bullet[\mathfrak{e} > s, \mathfrak{x}^\bullet(\mathfrak{e}(w_s^{\bullet +}) + s) \in dl] \\ &= P_0^\bullet(\mathfrak{e} > s)P_0^\bullet[\mathfrak{x}^\bullet(\mathfrak{e}) \in dl], \end{aligned}$$

completing the proof.

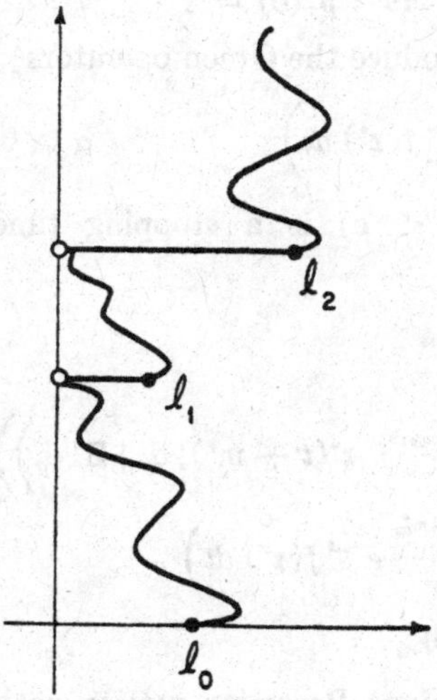


DIAGRAM 1

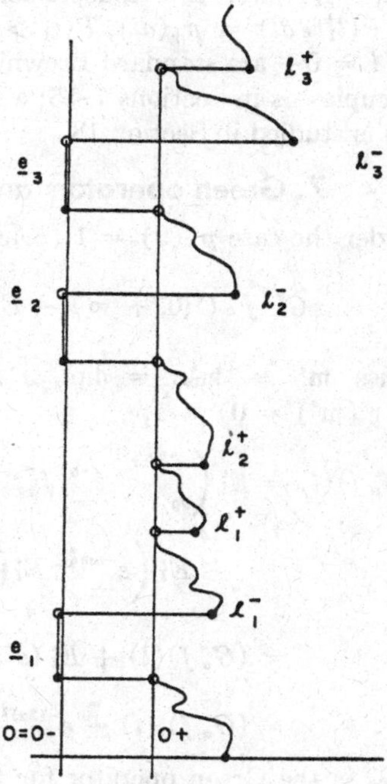


DIAGRAM 2

*$p_3$ has to be positive*; in the opposite case,

$$P_0^\bullet(\mathfrak{e} = 0) = p(0) = P_0^\bullet(\lim_{\varepsilon\downarrow 0} \mathfrak{m}_\varepsilon^\bullet = 0) = 1,$$

where now $\mathfrak{m}_\varepsilon^\bullet$ is the sum of the crossing time $\mathfrak{m}^\bullet = \inf(t:\mathfrak{x}^\bullet(t) > \varepsilon)$ and $\mathfrak{m}_{0+}^\bullet(w_{\mathfrak{m}^\bullet}^{\bullet +})$, and hence

7. $$1 = p(0) = P_0^\bullet(\lim_{\varepsilon\downarrow 0} \mathfrak{x}^\bullet(\mathfrak{m}_\varepsilon^\bullet) = 0)$$
$$= \lim_{\delta\downarrow 0} \lim_{\varepsilon\downarrow 0} P_0^\bullet(\mathfrak{x}^\bullet(\mathfrak{m}_\varepsilon^\bullet) < \delta)$$
$$= \lim_{\delta\downarrow 0} p_+[0, \delta)$$
$$= p_+(0),$$

contradicting $p_+(0) < 1$.

$p_-(dl) \equiv P_0^\bullet[\mathfrak{x}^\bullet(\mathfrak{e}) \in dl, \mathfrak{e} < +\infty]$ *attributes no mass to* $l = 0$ as is clear from

8a. $$P_0^\bullet(\mathfrak{e} > 0) = \lim_{t\downarrow 0} e^{-t/p_3} = 1$$

and

8b. $$p_-(0) = P_0^\bullet[\mathfrak{x}^\bullet(\mathfrak{e}) = 0, \mathfrak{e} < +\infty, \mathfrak{e}(w_{\mathfrak{e}}^{\bullet +}) = 0] \leq P_0^\bullet(\mathfrak{e} = 0).$$

Diagram 2 is now evident; the jumps $l_1^-$, $l_2^-$, etc., $l_1^+$, $l_2^+$, etc., and the holding times $\mathfrak{e}_1$, $\mathfrak{e}_2$, etc. are independent with common laws $P(l_1^- \in dl) = p_-(dl)$, $P(l_1^+ \in dl) = p_+(dl)$, $P(\mathfrak{e}_1 > t) = e^{-t/p_3}$, and the excursions leading back to $l = 0+$ are standard Brownian.

4c occupies us in Sections 7–15; a further class of ramified simple Markov motions is studied in Section 18.

## 7. Green operators and generators: $p_+(0) = 1$

Consider the case $p_+(0) = 1$ (6.4c), and introduce the Green operators

1. $$G_\alpha^\bullet: f \in C[0, +\infty) \to E_\cdot^\bullet\left(\int_0^{\mathfrak{m}_\infty^\bullet} e^{-\alpha t} f(\mathfrak{x}^\bullet)\, dt\right), \qquad \alpha > 0.$$

Because $\mathfrak{m}^\bullet \equiv \mathfrak{m}_{0+}^\bullet = \lim_{\varepsilon\downarrow 0} \inf(t:\mathfrak{x}^\bullet(t) < \varepsilon)$ is a stopping time and $P_\cdot^\bullet(\mathfrak{x}^\bullet(\mathfrak{m}^\bullet) = 0) \equiv 1$,

2. $$(G_\alpha^\bullet f)(l) = E_l^\bullet\left(\int_0^{\mathfrak{m}_{0+}^\bullet} e^{-\alpha t} f(\mathfrak{x}^\bullet)\, dt\right)$$
$$+ E_l^\bullet\left(e^{-\alpha \mathfrak{m}_{0+}^\bullet} E_l^\bullet\left(\int_0^{\mathfrak{m}_\infty^\bullet(w_{\mathfrak{m}^\bullet}^{\bullet +})} e^{-\alpha t} f[\mathfrak{x}^\bullet(t + \mathfrak{m}^\bullet)]\, dt \mid \mathbf{B}_{\mathfrak{m}^\bullet +}^\bullet\right)\right)$$
$$= (G_\alpha^- f)(l) + E_l^\bullet(e^{-\alpha \mathfrak{m}_{0+}^\bullet}) E_0\left(\int_0^{\mathfrak{m}_\infty^\bullet} e^{-\alpha t} f(\mathfrak{x}^\bullet)\, dt\right)$$
$$= (G_\alpha^- f)(l) + e^{-(2\alpha)^{1/2} l} (G_\alpha^\bullet f)(0),$$

where $G_\alpha^-$ is the Green operator for the (absorbing) Brownian motion with instant killing at $l = 0$:

3. $$(G^-_\alpha f)(a) = E_a\left(\int_0^{\mathfrak{m}_0} e^{-\alpha t} f(\mathfrak{x})\,dt\right) = \int_0^{+\infty} \frac{e^{-(2\alpha)^{1/2}|b-a|} - e^{-(2\alpha)^{1/2}|b+a|}}{(2\alpha)^{1/2}} f\,db, \quad a \geqq 0;$$

especially, $G^\bullet_\alpha$ maps $C[0, +\infty)$ *into* $C^2[0, +\infty)$.

Given $\alpha, \beta > 0$ and $f \in C[0, +\infty)$,

4. $$\begin{aligned}(\alpha - \beta)\, G^\bullet_\alpha G^\bullet_\beta f &= (\alpha - \beta)\, E^\bullet_\cdot\left(\int_0^{\mathfrak{m}^\bullet_\infty} e^{-\alpha t} (G^\bullet_\beta f)(\mathfrak{x}^\bullet)\,dt\right) \\ &= (\alpha - \beta)\, E^\bullet_\cdot\left(\int_0^{\mathfrak{m}^\bullet_\infty} e^{-\alpha t}\,dt\, E^\bullet_{\mathfrak{x}^\bullet(t)}\left(\int_0^{\mathfrak{m}^\bullet_\infty} e^{-\beta s} f(\mathfrak{x}^\bullet)\,ds\right)\right) \\ &= (\alpha - \beta)\, E^\bullet_\cdot\left(\int_0^{\mathfrak{m}^\bullet_\infty} e^{-(\alpha-\beta)t}\,dt \int_t^{\mathfrak{m}^\bullet_\infty} e^{-\beta s} f(\mathfrak{x}^\bullet)\,ds\right) \\ &= E^\bullet_\cdot\left(\int_0^{\mathfrak{m}^\bullet_\infty} e^{-\beta s} f(\mathfrak{x}^\bullet)\,ds\,(\alpha - \beta)\int_0^s e^{-(\alpha-\beta)t}\,dt\right) \\ &= G^\bullet_\beta f - G^\bullet_\alpha f,\end{aligned}$$

i.e.,

5. $$G^\bullet_\alpha - G^\bullet_\beta + (\alpha - \beta) G^\bullet_\alpha G^\bullet_\beta = 0, \qquad \alpha, \beta > 0,$$

proving that the *range* $G^\bullet_\alpha\, C[0, +\infty) \equiv D(\mathfrak{G}^\bullet)$ and the *null-space* $G^{\bullet\,-1}_\alpha(0)$ are both independent of $\alpha > 0$; in fact, $G^{-1}_\beta(0) = \bigcap_{\alpha>0} G^{\bullet\,-1}_\alpha(0) = 0$ because if $f$ belongs to it, then

6. $$0 = \lim_{\alpha\uparrow+\infty} \alpha (G^\bullet_\alpha f)(l) = \lim_{\alpha\uparrow+\infty} E^\bullet_l\left(\alpha\int_0^{\mathfrak{m}^\bullet_\infty} e^{-\alpha t} f(\mathfrak{x}^\bullet)\,dt\right) = f(l), \quad l \geqq 0,$$

thanks to $P^\bullet_l(\mathfrak{x}^\bullet(0+) = l) \equiv 1 \quad (l \geqq 0)$.

$G^\bullet_\alpha$ is now seen to be invertible, and another application of 5 implies that

7. $$\mathfrak{G}^\bullet \equiv \alpha - G^{\bullet\,-1}_\alpha : D(\mathfrak{G}^\bullet) \to C[0, +\infty)$$

is likewise independent of $\alpha > 0$.

$\mathfrak{G}^\bullet$ *is the generator cited in the section title; it is a contraction of* $\mathfrak{G} = D^2/2$ *acting on* $C^2[0, +\infty)$ because

8a. $$D(\mathfrak{G}^\bullet) = G^\bullet_1\, C[0, +\infty) \subset C^2[0, +\infty)$$

and

8b. $$(\alpha - \mathfrak{G}) G^-_\alpha = 1, \qquad \alpha > 0.$$

Given two Brownian motions with the *same generator*, their Green operators and hence their transition probabilities and laws in function space are the same, i.e., $\mathfrak{G}^\bullet$ *is a complete invariant of the Brownian motion.*

## 8. Generator and Green operators computed: $p_+(0) = 1$

$D(\mathfrak{G}^\bullet)$ can be described in terms of three nonnegative numbers $p_1$, $p_2$, $p_3$ and a nonnegative mass distribution $p_4(dl)$ $(l > 0)$ subject to

1a. $$p_1 + p_2 + p_3 + \int_{0+} (l \wedge 1)\, p_4(dl) = 1$$

and

1b. $$p_4(0, +\infty) = +\infty \quad \text{in case} \quad p_2 = p_3 = 0,$$

namely, $D(\mathfrak{G}^\bullet)$ is *the class of functions* $u \in C^2[0, +\infty)$ *subject to*[11]

2a. $$p_1\, u(0) + p_3(\mathfrak{G}u)(0) = p_2\, u^+(0) + \int_{0+} [u(l) - u(0)]\, p_4(dl),$$

as will now be proved.

1b is automatic from the rest because if $p_2 = p_3 = 0$ and $p_4(0, +\infty) < +\infty$, then an application of 2a to $u = \alpha G^\bullet_\alpha f \in D(\mathfrak{G}^\bullet)$ implies, on letting $\alpha \uparrow +\infty$, that

$$[p_1 + p_4(0, +\infty)]\, f(0) = \int_{0+} f p_4(dl) \qquad \text{for each } f \in C[0, +\infty),$$

which is absurd in view of 1a. Besides, it is enough to prove that

2b. $$D(\mathfrak{G}^\bullet) \subset C^2[0, +\infty) \cap \Big( u : p_1\, u(0) + p_3(\mathfrak{G}u)(0) = p_2 u^+(0) + \int_{0+} [u(l) - u(0)]\, p_4(dl) \Big)$$

or some choice of $p_1$, $p_2$, $p_3$, $p_4$ subject to 1a, because, if $u$ is a member of the second line, then so is the bounded solution $u^\bullet = G^\bullet_1(1 - \mathfrak{G})u - u$ of $\mathfrak{G}u^\bullet = u^\bullet$, and, expressing $u^\bullet$ as $c_1 e^{2^{1/2}l} + c_2 e^{-2^{1/2}l}$, it is found that $c_1 = c_2 = u^\bullet \equiv 0$, i.e., $u = G^\bullet_1(1 - \mathfrak{G})u \in D(\mathfrak{G}^\bullet)$.

Consider, for the proof of 2b, the *exit time*

3. $$\mathfrak{e} = \inf(t : \mathfrak{x}^\bullet(t) \neq 0)$$

and its law

4. $$P^\bullet_0(\mathfrak{e} > t) = e^{-t/k} \qquad (0 \leqq k \leqq +\infty),$$

and bear in mind that $\mathfrak{x}^\bullet(\mathfrak{e})$ is independent of $\mathfrak{e}$:

5. $$P^\bullet_0[\mathfrak{e} > t,\, \mathfrak{x}^\bullet(\mathfrak{e}) \in dl] = e^{-t/k}\, p(dl).$$

If $k = +\infty$ $(\mathfrak{e} \equiv +\infty)$, then $(\mathfrak{G}^\bullet u)(0) = 0$ for each $u \in D(\mathfrak{G}^\bullet)$, and 2b holds with $p_1 = p_2 = p_4 = 0$ and $p_3 = 1$.

[11] $\mathfrak{G} = D^2/2$.

If $0 < k < +\infty$, then

6. $$p(0) = P_0^\bullet[\mathfrak{x}^\bullet(\mathfrak{e}) = 0, \mathfrak{e}(w_\mathfrak{e}^{\bullet +}) = 0] \leq P_0^\bullet(\mathfrak{e} = 0) = 0,$$

and choosing $u = G_\alpha^\bullet f \in D(\mathfrak{G}^\bullet)$, it appears that

7a. $$u(0) = f(0)E_0^\bullet\left(\int_0^\mathfrak{e} e^{-\alpha t}\,dt\right) + E_0^\bullet[e^{-\alpha \mathfrak{e}}u(\mathfrak{x}^\bullet(\mathfrak{e})), \mathfrak{e} < \mathfrak{m}_\infty^\bullet]$$
$$= \frac{\alpha u(0) - (\mathfrak{G}^\bullet u)(0)}{\alpha + k} + \frac{k}{\alpha + k}\int_{0+} up(dl),$$ [12]

or, what is the same,

7b. $$u(0) + k^{-1}(\mathfrak{G}^\bullet u)(0) = \int_{0+} up(dl),$$

i.e., 2b holds with $p_1:p_2:p_3:p_4 = 1:0:k^{-1}:p$.

But, if $k = 0$ ($\mathfrak{e} \equiv 0$), the proof is less simple; the method used below is due to E. B. Dynkin [1].

$(\mathfrak{G}^\bullet u)(0) < -1$ for some $u \in D(\mathfrak{G}^\bullet)$ (if not, then $(\mathfrak{G}^\bullet u)(0) \equiv 0$, $f(0) = (1 - \mathfrak{G}^\bullet)G_1^\bullet f(0) = (G_1^\bullet f)(0)$ for each $f \in C[0, +\infty)$, and $P_0^\bullet(\mathfrak{e} = +\infty) = 1$), so, choosing $\varepsilon > 0$ so small that $(\mathfrak{G}^\bullet u)(l) < -1$ $(l \leq \varepsilon)$ and introducing the *crossing time* $\mathfrak{m}_\varepsilon^\bullet = \inf(t: \mathfrak{x}^\bullet(t) > \varepsilon)$, it is clear from

8. $$u(0) = E_0^\bullet\left(\int_0^{\mathfrak{m}_\infty^\bullet} e^{-\alpha t}f(\mathfrak{x}^\bullet)\,dt\right), \qquad f = (\alpha - \mathfrak{G}^\bullet)u,$$
$$= E_0^\bullet\left(\int_0^{\mathfrak{m}_\varepsilon^\bullet \wedge \mathfrak{m}_\infty^\bullet} e^{-\alpha t}(\alpha - \mathfrak{G}^\bullet)u(\mathfrak{x}^\bullet)\,dt\right)$$
$$+ E_0^\bullet[e^{-\alpha \mathfrak{m}_\varepsilon^\bullet}u(\mathfrak{x}^\bullet(\mathfrak{m}_\varepsilon^\bullet)), \mathfrak{m}_\varepsilon^\bullet < \mathfrak{m}_\infty^\bullet]$$

that

9. $$E_0^\bullet(\mathfrak{m}_\varepsilon^\bullet \wedge \mathfrak{m}_\infty^\bullet) \leq \lim_{\alpha \downarrow 0} E_0^\bullet\left(\int_0^{\mathfrak{m}_\varepsilon^\bullet \wedge \mathfrak{m}_\infty^\bullet} e^{-\alpha t}(\alpha - \mathfrak{G}^\bullet)u(\mathfrak{x}^\bullet)\,dt\right) < +\infty.$$

$(\mathfrak{G}^\bullet u)(0) < -1$ has no special advantage for the derivation of 8 which holds for each $u \in D(\mathfrak{G}^\bullet)$ and $\varepsilon > 0$; thus, keeping $\varepsilon > 0$ so small that $E_0^\bullet(\mathfrak{m}_\varepsilon^\bullet \wedge \mathfrak{m}_\infty^\bullet) < +\infty$ and letting $\alpha \downarrow 0$ in 8 implies

10. $$u(0) = -E_0^\bullet\left(\int_0^{\mathfrak{m}_\varepsilon^\bullet \wedge \mathfrak{m}_\infty^\bullet} (\mathfrak{G}^\bullet u)(\mathfrak{x}^\bullet)\,dt\right) + E_0^\bullet[u(\mathfrak{x}^\bullet(\mathfrak{m}_\varepsilon^\bullet)), \mathfrak{m}_\varepsilon^\bullet < +\infty],$$
$$u \in D(\mathfrak{G}^\bullet),$$

and letting $\varepsilon \downarrow 0$ in 10 establishes E. B. Dynkin's *formula for the generator*:

11a. $$(\mathfrak{G}^\bullet u)(0) = \lim_{\varepsilon \downarrow 0} \int_{[\varepsilon, +\infty) \cup \infty} [u(l) - u(0)]p_\varepsilon(dl), \quad u \in D(\mathfrak{G}^\bullet), u(\infty) \equiv 0,$$

[12] $(\alpha - \mathfrak{G}^\bullet)G_\alpha^\bullet = 1.$

11b. $$p_\varepsilon(dl) = E_0^\bullet(\mathfrak{m}_\varepsilon^\bullet \wedge \mathfrak{m}_\infty^\bullet)^{-1} P_0^\bullet[\mathfrak{x}^\bullet(\mathfrak{m}_\varepsilon^\bullet \wedge \mathfrak{m}_\infty^\bullet) \in dl],$$

or, what is better for the present purpose,

12a. $$\lim_{\varepsilon \downarrow 0} \left[ \frac{p_\varepsilon(\infty)}{D} u(0) + \frac{(\mathfrak{G}^\bullet u)(0)}{D} - \int_{[\varepsilon,+\infty)} u^\bullet(l)(l \wedge 1) \frac{p_\varepsilon(dl)}{D} \right] = 0,$$

12b. $$D = p_\varepsilon(\infty) + 1 + \int_{0+} (l \wedge 1) p_\varepsilon(dl),$$

12c. $$u^\bullet(l) = \frac{u(l) - u(0)}{l \wedge 1} \quad \text{if} \quad l > 0,$$
$$= u^+(0) \quad \text{if} \quad l = 0.$$

Because $D(\mathfrak{G}^\bullet) \subset C^2[0, +\infty)$, $u^\bullet \in C[0, +\infty)$ and selecting $\varepsilon = \varepsilon_1 > \varepsilon_2 >$ etc. $\downarrow 0$ so as to have

13a. $$\lim_{\varepsilon \downarrow 0} p_\varepsilon(\infty)/D = p_1,$$

13b. $$\lim_{\varepsilon \downarrow 0} 1/D = p_3,$$

13c. $$\lim_{\varepsilon \downarrow 0} (l \wedge 1) p_3(dl)/D = p_*(dl)$$ [13]

existing, it is clear from 12 that

14a. $$p_1 u(0) + p_3(\mathfrak{G}^\bullet u)(0) = p_2 u^+(0) + \int_{(0,+\infty]} [u(l) - u(0)] p_4(dl),$$

14b. $$p_2 = p_*(0), \qquad p_4(dl) = p_*(dl)/(l \wedge 1) \quad (l > 0),$$

14c. $$p_1 + p_2 + p_3 + \int_{(0,+\infty]} (l \wedge 1) p_4(dl) = 1$$

for each $u \in D(\mathfrak{G}^\bullet)$ having a limit $u(+\infty)$ at $l = +\infty$.

But $p_4(+\infty) = 0$ because, if $f = e^{-n/l}$, then $u = G_1^\bullet f \in D(\mathfrak{G}^\bullet)$, $u(+\infty) = 1$, and at the same time $u(0)$, $u^+(0)$, $\mathfrak{G}^\bullet u(0)$, and $\int_{l<+\infty} [u(l) - u(0)] p_4(dl)$ are all small for large $n$, and this permits us to derive 14a anew for *each* $u \in D(\mathfrak{G}^\bullet)$, completing the proof of 2b.

Given $u \in D(\mathfrak{G}^\bullet)$ and inserting 7.2 into 2a, a little algebra justifies

15. $$(G_\alpha^\bullet f)(0) = \frac{p_2 2 \int_{0+} e^{-(2\alpha)^{1/2} l} f(l)\, dl + p_3 f(0) + \int_{0+} (G_\alpha^- f)(l) p_4(dl)}{p_1 + (2\alpha)^{1/2} p_2 + \alpha p_3 + \int_{0+} [1 - e^{-(2\alpha)^{1/2} l}] p_4(dl)},$$

which finishes the computation of the Green operators.

## 9. Special case: $p_2 = 0 < p_3$ and $p_4 < +\infty$

Consider the special case

1a. $$p_2 = 0 < p_3,$$

[13] $\int_0 f(l \wedge 1) D^{-1} p_\varepsilon(dl)$ converges as $\varepsilon \downarrow 0$ to $\int f p_*(dl)$ extended over $[0, +\infty]$ for each $f \in C[0, +\infty]$.

1b. $$p_4 = p_4(0, +\infty) < +\infty,$$

and introduce a motion $\mathfrak{x}^\bullet$ based on a reflecting Brownian motion with sample paths $t \to \mathfrak{x}^+(t)$ and probabilities $P_a(B)$ $(a \geqq 0)$ as follows.

Given a sample path $\mathfrak{x}^+$ starting at a point of $[0, +\infty)$, let $\mathfrak{x}^\bullet = \mathfrak{x}^+$ up to the passage time $\mathfrak{m}_0 = \min(t: \mathfrak{x}^+(t) = 0)$; then make $\mathfrak{x}^\bullet$ wait at 0 for an exponential holding time $\mathfrak{e}_1$ with conditional law

1. $$P.(\mathfrak{e}_1 > t \mid \mathfrak{x}^+) = e^{-((p_1+p_4)/p_3)t};$$

at the end of that time let it jump to a point $l_1 \in (0, +\infty) \cup \infty$ with conditional law

2. $$\begin{aligned} P.(l_1 \in dl \mid \mathfrak{e}_1, \mathfrak{x}^+) &= p_4(dl)/(p_1 + p_4) \quad \text{if} \quad l > 0, \\ &= p_1/(p_1 + p_4) \qquad \text{if} \quad l = 0, \end{aligned}$$

and, if $+\infty > l_1 > 0$, let it start afresh, while, if $l_1 = \infty$, let $\mathfrak{x}^\bullet = \infty$ at all later times (see Diagram 1).

Because $\mathfrak{x}^\bullet$ starts afresh at the passage time $\mathfrak{m}_0$,

3. $$\begin{aligned} (G_\alpha^\bullet f)(l) &= E_l\left(\int_0^{\mathfrak{m}_\infty^\bullet} e^{-\alpha t} f(\mathfrak{x}^\bullet)\, dt\right) \qquad (\mathfrak{m}_\infty^\bullet = \min(t: \mathfrak{x}^\bullet(t) = \infty)) \\ &= E_l\left(\int_0^{\mathfrak{m}_0} e^{-\alpha t} f(\mathfrak{x}^+)\, dt\right) + E_l(e^{-\alpha \mathfrak{m}_0}) E_0\left(\int_0^{\mathfrak{m}_\infty^\bullet} e^{-\alpha t} f(\mathfrak{x}^\bullet)\, dt\right) \\ &= (G_\alpha^- f)(l) + e^{-(2\alpha)^{1/2} l} (G_\alpha^\bullet f)(0) \end{aligned}$$

as in 7.2, whence

4. $$\begin{aligned} (G_\alpha^\bullet f)(0) &= f(0) E_0\left(\int_0^{\mathfrak{e}_1} e^{-\alpha t}\, dt\right) + E_0(e^{-\alpha \mathfrak{e}_1}) E_0[(G_\alpha^\bullet f)(l_1), \mathfrak{e}_1 < \mathfrak{m}_\infty^\bullet] \\ &= \frac{p_3 f(0)}{p_1 + \alpha p_3 + p_4} \\ &\quad + \frac{1}{p_1 + \alpha p_3 + p_4}\left[\int_{0+} (G_\alpha^- f)(l) p_4(dl) + \int_{0+} e^{-(2\alpha)^{1/2} l} p_4(dl) (G_\alpha^\bullet f)(0)\right], \end{aligned}$$

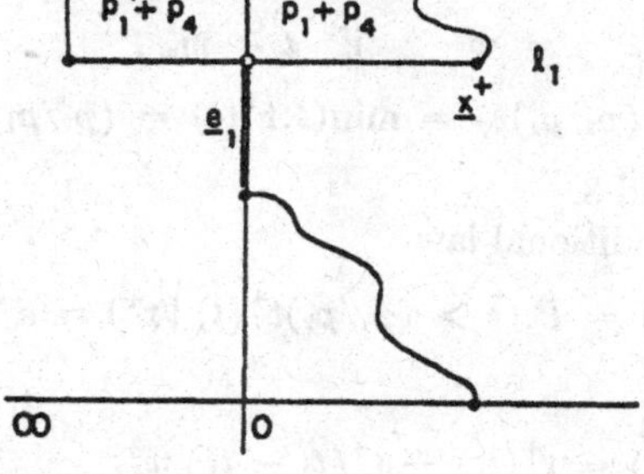


DIAGRAM 1

and, solving for $(G^\bullet_\alpha f)(0)$, one finds

5. $$(G^\bullet_\alpha f)(0) = \frac{p_3 f(0) + \int_{0+} (G^-_\alpha f)(l) p_4(dl)}{p_1 + \alpha p_3 + \int_{0+} [1 - e^{-(2\alpha)^{1/2} l}] p_4(dl)}.$$

Granting that the dot motion starts afresh at constant times (the reader will fill this gap), a comparison of 5 and 8.15 permits its identification as the Brownian motion associated with the operator $\mathfrak{G}^\bullet$ with domain

6. $$D(\mathfrak{G}^\bullet) = C^2[0, +\infty) \cap \left(u: p_1 u(0) + p_3(\mathfrak{G}u)(0) = \int_{0+} [u(l) - u(0)] p_4(dl)\right);$$

the proof that $\mathfrak{x}^\bullet$ is a Brownian motion can be based on the fact, used several times below, that if a motion is *simple* Markov and if its Green operators map $C[0, +\infty)$ into itself, then it is also *strict* Markov (see, for example K. Itô and H. P. McKean, Jr. [1]).

## 10. Special case: $p_2 > 0 = p_4$

Given a reflecting Brownian motion with sample paths $t \to \mathfrak{x}^+(t)$, probabilities $P_a(B)$, and local time

1. $$\mathfrak{t}^+(t) = \lim_{\varepsilon \downarrow 0} (2\varepsilon)^{-1} \text{ measure } (s: \mathfrak{x}^+(s) < \varepsilon, s \leq t),$$

it is possible to build up all the Brownian motions attached to the generators

2. $$\mathfrak{G}^\bullet = \mathfrak{G} \mid C^2[0, +\infty) \cap (u: p_1 u(0) - p_2 u^+(0) + p_3(\mathfrak{G}u)(0) = 0), \quad p_2 > 0$$

with the aid of an extra exponential holding time $\mathfrak{e}$ with conditional law

3. $$P.(\mathfrak{e} > t \mid \mathfrak{x}^+) = e^{-t}.$$

Beginning with the elastic Brownian case ($p_1 > 0 = p_3$), the desired motion is

4a. $$\mathfrak{x}^\bullet(t) = \mathfrak{x}^+(t) \quad \text{if} \quad t < \mathfrak{m}^\bullet_\infty, \qquad = \infty \quad \text{if} \quad t \geq \mathfrak{m}^\bullet_\infty,$$

4b. $$\mathfrak{m}^\bullet_\infty = \mathfrak{t}^{-1}((p_2/p_1)\mathfrak{e}) = \min(t: \mathfrak{t}^+(t) = (p_2/p_1)\mathfrak{e})$$

as stated in Sections 1 and 3.

With the aid of the conditional law

5. $$P.(\mathfrak{m}^\bullet_\infty > t \mid \mathfrak{x}^+) = P.(\mathfrak{e} > (p_1/p_2)\mathfrak{t}^+(t) \mid \mathfrak{x}^+) = e^{-(p_1/p_2)\mathfrak{t}^+(t)}$$

and the *addition rule*

6. $$\mathfrak{t}^+(t_2) = \mathfrak{t}^+(t_1) + \mathfrak{t}^+(t_2 - t_1, w^+_{t_1}), \qquad t_2 \geq t_1,$$

it is clear that, if $db \subset [0, +\infty)$ and if $\mathfrak{m}_\infty^\bullet > t_1 \leqq t_2$, then

7a. $$P.[\mathfrak{x}^\bullet(t_2) \in db \mid \mathfrak{x}^+(s): s \leqq t_1, \mathfrak{m}_\infty^\bullet \wedge t_1, \mathfrak{m}_\infty^\bullet > t_1]$$
$$= \frac{P.[\mathfrak{x}^+(t_2) \in db, \mathfrak{m}_\infty^\bullet > t_2 \mid \mathfrak{x}^+(s): s \leqq t_1]}{P.(\mathfrak{m}_\infty^\bullet > t_1)}$$
$$= E.[\mathfrak{x}^+(t_2) \in db, e^{-(p_1/p_2)\mathfrak{t}^+(t_2)} \mid \mathfrak{x}^+(s): s \leqq t_1] e^{+(p_1/p_2)\mathfrak{t}^+(t_1)}$$
$$= E.[\mathfrak{x}^+(t_2) \in db, e^{-(p_1/p_2)\mathfrak{t}^+(t_2-t_1, w_{t_1}^+)} \mid \mathfrak{x}^+(s): s \leqq t_1]$$
$$= E_a[\mathfrak{x}^+(t_2 - t_1) \in db, e^{-(p_1/p_2)\mathfrak{t}^+(t_2-t_1)}], \qquad a = \mathfrak{x}^+(t_1),$$
$$= P_a[\mathfrak{x}^\bullet(t_2 - t_1) \in db], \qquad a = \mathfrak{x}^\bullet(t_1),$$

while, if $\mathfrak{m}_\infty^\bullet \leqq t_1$, then $\mathfrak{x}^\bullet(t_1) = \infty$, and

7b. $$P.[\mathfrak{x}^\bullet(t_2) \in db \mid \mathfrak{x}^+(s): s \leqq t_1, \mathfrak{m}_\infty^\bullet \wedge t_1, \mathfrak{m}_\infty^\bullet \leqq t_1]$$
$$= 0 = P_\infty[\mathfrak{x}^\bullet(t_2 - t_1) \in db].$$[14]

Since $\mathfrak{x}^\bullet(s): s \leqq t_1$ is a Borel function of $\mathfrak{x}^+(s): s \leqq t_1$, $\mathfrak{m}_\infty^\bullet \wedge t_1$, and of the indicator of $(\mathfrak{m}_\infty^\bullet < t_1)$, it follows that

8. $$P.[\mathfrak{x}^\bullet(t_2) \in db \mid \mathfrak{x}^\bullet(s): s \leqq t_1] = P_a[\mathfrak{x}^\bullet(t_2 - t_1) \in db], \qquad a = \mathfrak{x}^\bullet(t_1),$$

*establishing the simple Markovian nature of the dot motion.*

Consider for the next step, its Green operators

$$G_\alpha^\bullet f = E.\left(\int_0^{\mathfrak{m}_\infty^\bullet} e^{-\alpha t} f(\mathfrak{x}^\bullet)\, dt\right),$$

and use the conditional law of $\mathfrak{m}_\infty^\bullet$ to check

9. $$G_\alpha^\bullet f = E.\left(\int_0^{+\infty} e^{-(p_1/p_2)\mathfrak{t}^+(s)} \frac{p_1}{p_2} \mathfrak{t}^+(ds) \int_0^s e^{-\alpha t} f(\mathfrak{x}^+)\, dt\right)$$
$$= E.\left(\int_0^{+\infty} e^{-\alpha t} f(\mathfrak{x}^+)\, dt \int_t^{+\infty} e^{-(p_1/p_2)\mathfrak{t}^+} \mathfrak{t}^+(ds)\right)$$
$$= E.\left(\int_0^{+\infty} e^{-\alpha t} e^{-(p_1/p_2)\mathfrak{t}^+} f(\mathfrak{x}^+)\, dt\right).$$

Because $\mathfrak{m}_0 = \min(t: \mathfrak{x}^+(t) = 0)$ is a stopping time and $\mathfrak{t}^+(t) = 0$ $(t \leqq \mathfrak{m}_0)$,

10. $$(G_\alpha^\bullet f)(l) = E_l\left(\int_0^{\mathfrak{m}_0} e^{-\alpha t} f(\mathfrak{x}^+)\, dt\right)$$
$$+ E_l\left(e^{-\alpha \mathfrak{m}_0} \int_0^{+\infty} e^{-\alpha t} \exp\{-(p_1/p_2)\mathfrak{t}^+(t, w_{\mathfrak{m}_0}^+)\} f[\mathfrak{x}^+(t + \mathfrak{m}_0)]\, dt\right)$$
$$= (G_\alpha^- f)(l) + E_l(e^{-\alpha \mathfrak{m}_0}) E_0\left(\int_0^{+\infty} e^{-\alpha t} e^{-(p_1/p_2)\mathfrak{t}^+} f(\mathfrak{x}^+)\, dt\right)$$
$$= (G_\alpha^- f)(l) + e^{-(2\alpha)^{1/2} l} (G_\alpha^\bullet f)(0), \qquad l \geqq 0,$$

[14] $P_\infty[\mathfrak{x}^\bullet \equiv \infty] = 1$ as usual.

and now the identification of the dot motion as the elastic Brownian motion will be complete as soon as it is verified that

11. $$(G^{\bullet}_{\alpha} f)(0) = \frac{p_2 \, 2 \int_0^{+\infty} e^{-(2\alpha)^{1/2} l} f(l) \, dl}{p_1 + (2\alpha)^{1/2} p_2};$$

in fact, this will prove that the dot motion is simple Markov with the correct (elastic Brownian) Green operators, and the proof can be completed as at the end of Section 9.

But 11 is trivial; in fact, using the joint law 4.10,

12. $$(G^{\bullet}_{\alpha} f)(0) = E_0\left(\int_0^{+\infty} e^{-\alpha t} e^{-(p_1/p_2)t^+} f(\mathfrak{x}^+) \, dt\right)$$

$$= \int_0^{+\infty} e^{-\alpha t} \, dt \int_0^{+\infty} db \int_0^{+\infty} da \, 2 \frac{b+a}{(2\pi t^3)^{1/2}} e^{-(b+a)^2/2t} e^{-(p_1/p_2)b} f(a)$$

$$= 2 \int_0^{+\infty} db \int_0^{+\infty} da \, e^{-(2\alpha)^{1/2}(b+a)} e^{-(p_1/p_2)b} f(a)$$

$$= \frac{p_2 \, 2 \int_{0+} e^{-(2\alpha)^{1/2} l} f(l) \, dl}{p_1 + (2\alpha)^{1/2} p_2}$$

as stated.

Consider next, the case $p_3 > 0 = p_1$, and let us prove the desired motion to be[15]

13. $$\mathfrak{x}^{\bullet} = \mathfrak{x}^+(\mathfrak{f}^{-1}), \qquad \mathfrak{f} = t + (p_3/p_2)\mathfrak{t}^+.$$

Beginning, as before, with the proof that the dot motion is *simple Markov*, if $t_2 \geqq t_1$ and if $\mathfrak{m} = \mathfrak{f}^{-1}(t_1)$, then

(a) $(\mathfrak{m} < t) = (t_1 < \mathfrak{f}(t)) \in \mathsf{B}[\mathfrak{x}^+(s): s \leqq t]$, i.e., $\mathfrak{m}$ *is a stopping time*;

(b) $\mathfrak{f}(\mathfrak{m} + s)) = \mathfrak{f}(\mathfrak{m}) + \mathfrak{f}(s, w^+_{\mathfrak{m}}) = t_1 + (t_2 - t_1)$ *if* $s = \mathfrak{f}^{-1}(t_2 - t_1, w^+_{\mathfrak{m}})$ *and so* $\mathfrak{f}^{-1}(t_2) = \mathfrak{m} + s = \mathfrak{m} + \mathfrak{f}^{-1}(t_2 - t_1, w^+_{\mathfrak{m}})$;

(c) $\mathfrak{x}^{\bullet}(t_2) = \mathfrak{x}^+[\mathfrak{f}^{-1}(t_2 - t_1, w^+_{\mathfrak{m}}) + \mathfrak{m}]$;

(d) $\mathfrak{x}^{\bullet}(s): s \leqq t_1$ *is a Borel function of the stopped path* $t \to \mathfrak{x}^+(t \wedge \mathfrak{m})$ *and of* $\mathfrak{f}^{-1}(s): s \leqq t_1$;

(e) $\mathfrak{f}^{-1}(s)$ *is the solution* $r$ *of* $f(r) = s$ $(\leqq t_1 = f(\mathfrak{m}))$ *and, as such, it is likewise a Borel function of the stopped path*;

and now, using the strict Markovian nature of $\mathfrak{x}^+$, the law of $\mathfrak{x}^{\bullet}(t_2)$ conditional on $\mathsf{B}_{\mathfrak{m}+} \supset \mathsf{B}[\mathfrak{x}^{\bullet}(s): s \leqq t_1]$ is found to be

14a. $$P_{\cdot}(\mathfrak{x}^+[\mathfrak{f}^{-1}(t_2 - t_1, w^+_{\mathfrak{m}}) + \mathfrak{m}] \in db \mid \mathsf{B}_{\mathfrak{m}+})$$

$$= P_a(\mathfrak{x}^+[\mathfrak{f}^{-1}(t_2 - t_1)] \in db), \qquad a = \mathfrak{x}^+(\mathfrak{m}),$$

$$= P_a(\mathfrak{x}^{\bullet}(t_2 - t_1) \in db), \qquad a = \mathfrak{x}^{\bullet}(t_1),$$

[15] $\mathfrak{f}^{-1}$ is the inverse function of $\mathfrak{f}$.

whence, taking the expectation of both sides conditional on $\mathsf{B}[\mathfrak{x}^\bullet(s):s \leqq t_1]$,

14b. $$P_\cdot(\mathfrak{x}^\bullet(t_2) \in db \mid \mathfrak{x}^\bullet(s):s \leqq t_1) = P_a(\mathfrak{x}^\bullet(t_2 - t_1) \in db), \qquad a = \mathfrak{x}^\bullet(t_1),$$

i.e., $\mathfrak{x}^\bullet = \mathfrak{x}^+(\mathfrak{f}^{-1})$ starts afresh at time $t_1$, as was to be proved.

Coming to the Green operators

$$G_\alpha^\bullet f = E_\cdot\left(\int_0^{+\infty} e^{-\alpha t} f(\mathfrak{x}^\bullet)\,dt\right),$$

since $\mathfrak{m}_0 = \min(t:\mathfrak{x}^+(t) = 0)$ is a stopping time and $\mathfrak{f}^{-1} \equiv t \ (t \leqq \mathfrak{m}_0)$,

15. $$\begin{aligned}(G_\alpha^\bullet f)(l) &= E_l\left(\int_0^{\mathfrak{m}_0} e^{-\alpha t} f(\mathfrak{x}^+)\,dt\right)\\ &\quad + E_l\left(e^{-\alpha \mathfrak{m}_0}\int_0^{+\infty} e^{-\alpha t} f[\mathfrak{x}^+(\mathfrak{f}^{-1}(t, w_{\mathfrak{m}_0}^+) + \mathfrak{m}_0)]\,dt\right)\\ &= (G_\alpha^- f)(l) + e^{-(2\alpha)^{1/2} l}(G_\alpha^\bullet f)(0)\end{aligned}$$

as in the elastic Brownian case, and to complete the identification of $\mathfrak{x}^\bullet$ it is sufficient to check that

16. $$\begin{aligned}(G_\alpha^\bullet f)(0) &= E_0\left(\int_0^{+\infty} e^{-\alpha t} f[\mathfrak{x}^+(\mathfrak{f}^{-1})]\,dt\right)\\ &= E_0\left(\int_0^{+\infty} e^{-\alpha t} f(\mathfrak{x}^+)\mathfrak{f}(dt)\right)\\ &= E_0\left(\int_0^{+\infty} e^{-\alpha[t+(p_3/p_2)\mathfrak{t}^+]} f(\mathfrak{x}^+)\,dt\right)\\ &\quad + f(0)E_0\left(\int_0^{+\infty} e^{-\alpha[t+(p_3/p_2)\mathfrak{t}^+]}\frac{p_3}{p_2}\,\mathfrak{t}^+(dt)\right)^{16}\\ &= \frac{p_2\, 2\int_{0+} e^{-(2\alpha)^{1/2} l} f(l)\,dl}{(2\alpha)^{1/2}p_2 + \alpha p_3}\\ &\quad + \frac{f(0)}{\alpha}\left[1 - E_0\left(\int_0^{+\infty} e^{-\alpha[t+(p_3/p_2)\mathfrak{t}^+]}\,dt\right)\right]^{17,18}\\ &= \frac{p_2\, 2\int_{0+} e^{-(2\alpha)^{1/2} l} f(l)\,dl + p_3 f(0)}{(2\alpha)^{1/2}p_2 + \alpha p_3}\end{aligned}$$

as it should be.

Consider now the case $0 < p_1\, p_2\, p_3$; this time the motion is

---

[16] $\mathfrak{t}^+(dt) = 0$ off $\mathfrak{Z}^+ = (t:\mathfrak{x}^+(t) = 0)$.

[17] Use 12 with $\alpha p_3$ in place of $p_1$.

[18] Do a partial integration under the expectation sign.

17a. $$\mathfrak{x}^\bullet(t) = \mathfrak{x}^+(\mathfrak{f}^{-1}) \quad \text{if} \quad t < \mathfrak{m}^\bullet_\infty,$$
$$= \infty \quad \text{if} \quad t \geqq \mathfrak{m}^\bullet_\infty,$$

17b. $$\mathfrak{m}^\bullet_\infty = \mathfrak{f}[\mathfrak{t}^{-1}((p_2/p_1)\mathfrak{e})] = [\mathfrak{t}^+(\mathfrak{f}^{-1})]^{-1}((p_2/p_1)\mathfrak{e}),$$

as will still be proved.

$\mathfrak{x}^+(\mathfrak{f}^{-1})$ is a Brownian motion, its *local time*

18. $$\begin{aligned} \mathfrak{t}^\bullet(t) &= \text{measure } (s : \mathfrak{x}^+(\mathfrak{f}^{-1}) = 0, s \leqq t) \\ &= \text{measure } (s : \mathfrak{f}^{-1}(s) \in \mathfrak{Z}^+, s \leqq t) \\ &= \text{measure } \mathfrak{f}(\mathfrak{Z}^+) \cap [0, t] \\ &= \int_{\mathfrak{Z}^+ \cap [0, \mathfrak{f}^{-1}(t)]} \mathfrak{f}(ds) \\ &= (p_3/p_2)\mathfrak{t}^+[\mathfrak{f}^{-1}(t)] \end{aligned}$$ [19]

satisfies the addition rule 6, and, substituting them in place of $\mathfrak{x}^+$ and $\mathfrak{t}^+$ in the derivation of the simple Markovian nature of the elastic Brownian motion, it is found that the present motion is likewise simple Markov.

$G^\bullet_\alpha f = G^-_\alpha f + e^{-(2\alpha)^{1/2} l}(G^\bullet_\alpha f)(0)$ is derived as before, that the dot motion is Brownian follows, and now, using the evaluation 12 with $p_1 + \alpha\, p_3$ in place of $p_1$ in conjunction with the conditional law

19. $$P_\cdot(\mathfrak{m}^\bullet_\infty > t \mid \mathfrak{x}^+(\mathfrak{f}^{-1})) = P_\cdot(\mathfrak{e} > (p_1/p_2)\mathfrak{t}^+(\mathfrak{f}^{-1}) \mid \mathfrak{x}^+(\mathfrak{f}^{-1}))$$
$$= e^{-(p_1/p_2)\mathfrak{t}^+(\mathfrak{f}^{-1})} = e^{-(p_1/p_3)\mathfrak{t}^\bullet(t)},$$

it develops that

20. $$\begin{aligned} (G^\bullet_\alpha f)(0) &= E_0\left(\int_0^{\mathfrak{m}^\bullet_\infty} e^{-\alpha t} f[\mathfrak{x}^+(\mathfrak{f}^{-1})]\, dt\right) \\ &= E_0\left(\int_0^{+\infty} e^{-\alpha t} e^{-(p_1/p_2)\mathfrak{t}^+(\mathfrak{f}^{-1})} f[\mathfrak{x}^+(\mathfrak{f}^{-1})]\, dt\right) \\ &= E_0\left(\int_0^{+\infty} e^{-\alpha t} e^{-((p_1+\alpha p_3)/p_2)\mathfrak{t}^+} f(\mathfrak{x}^+)\, dt\right) \\ &\qquad + f(0) E_0\left(\int_0^{+\infty} e^{-\alpha t} e^{-((p_1+\alpha p_3)/p_2)\mathfrak{t}^+} \frac{p_3}{p_2}\, \mathfrak{t}^+(dt)\right) \\ &= \frac{p_2\, 2 \int_{0+} e^{-(2\alpha)^{1/2} l} f(l)\, dl + p_3 f(0)}{p_1 + (2\alpha)^{1/2} p_2 + \alpha p_3}, \end{aligned}$$

completing the proof.

A second description of the present motion is available: *it is the elastic*

[19] measure $(\mathfrak{Z}^+) = 0$. $\mathfrak{t}^+(dt) = 0$ outside $\mathfrak{Z}^+$.

*Brownian motion* $\mathfrak{x}^\bullet$ *described in 4 run with the new stochastic clock* $\mathfrak{f}^{-1}$ *which is the inverse function of*

21a. $\mathfrak{f} = t + (p_3/p_2) \times$ *the elastic Brownian local time* $\mathfrak{t}^\bullet$,

21b. $\mathfrak{t}^\bullet(t) = \lim_{\varepsilon\downarrow 0} (2\varepsilon)^{-1}$ measure $(s: \mathfrak{x}^\bullet(s) < \varepsilon, s \leqq t)$
$= \mathfrak{t}^+(t \wedge \mathfrak{m}_\infty^\bullet), \qquad \mathfrak{m}_\infty^\bullet = \min(t: \mathfrak{x}^\bullet = \infty).$

## 11. Increasing differential processes

Before describing the sample paths in the case $p_4 = p_4(0, +\infty) = +\infty$, it will be helpful to list some properties of differential processes with increasing sample paths.

Given a stochastic process with universal field B, probabilities $P$, and sample paths $t \to \mathfrak{p}(t)$:

1a. $$\mathfrak{p}(0) = 0,$$

1b. $$\mathfrak{p}(s) \leqq \mathfrak{p}(t), \qquad s \leqq t,$$

1c. $$\mathfrak{p}(t+) = \mathfrak{p}(t) < +\infty, \qquad t \geqq 0,$$

which is *differential* in the sense that *the shifted path* $\mathfrak{p}_+(t) \equiv \mathfrak{p}(t+s) - \mathfrak{p}(s)$ *is independent of its past* $\mathfrak{p}(t): t \leqq s$ *and identical in law to* $\mathfrak{p}$, P. Lévy [1][20] proved that

2a. $$E(e^{-\alpha\mathfrak{p}(t)}) = \exp\left\{-t\left[p_2\,\alpha + \int_{0+} (1 - e^{-\alpha l})p(dl)\right]\right\}, \qquad \alpha > 0,$$

2b. $$p_2 \geqq 0, \quad p(dl) \geqq 0, \quad \int_{0+} (l \wedge 1)p(dl) < +\infty$$

and expressed $\mathfrak{p}$ as

3. $$\mathfrak{p}(t) = p_2\,t + \int_{0+} l\mathfrak{p}([0, t] \times dl), \qquad t \geqq 0,$$

in which $\mathfrak{p}(dt \times dl) =$ *the number of jumps of* $\mathfrak{p}$ *of magnitude* $\in dl$ *occurring in time* $dt$ is differential in the pair $(t, l) \in [0, +\infty) \times (0, +\infty)$ and Poisson distributed with mean $dt\, p(dl)$, i.e., if $Q_1$, $Q_2$, etc. are disjoint figures of $[0, +\infty) \times (0, +\infty)$, then $\mathfrak{p}(Q_1)$, $\mathfrak{p}(Q_2)$, etc., are independent, and

4. $$P(\mathfrak{p}(Q) = n) = (|Q|^n/n!)\, e^{-|Q|}, \qquad n \geqq 0,\ |Q| = \int_Q dt\, p(dl);$$

in short, $\mathfrak{p}(t)$ is *the (direct) integral* $\int_{0+} l\mathfrak{p}([0, t] \times dl)$ *of the differential Poisson processes* $\mathfrak{p}([0, t] \times dl)$ *with rates* $p(dl)$ *plus a linear part* $p_2 t$.

Given nonnegative $p_2$ and $p(dl)$ with $\int_{0+} (l \wedge 1)p(dl) < +\infty$ as in 2b, it is possible to make a Poisson measure $\mathfrak{p}(dt \times dl)$ with mean $dt\, p(dl)$ as de-

---

[20] See also K. Itô [1].

scribed above; the associated $\mathfrak{p}(t) = p_2 t + \int_{0+} l\mathfrak{p}([0, t] \times dl)$ is a differential process having 2a as its Lévy formula.

G. Hunt [1] discovered that if $\mathfrak{m}$ is a *stopping time*, i.e., if

5. $$(\mathfrak{m} < t) \in \mathbf{B}[\mathfrak{p}(s): s \leqq t] \times \mathbf{B}^{\bullet}, \qquad t \geqq 0,$$

for some field $\mathbf{B}^{\bullet}$ independent of $\mathfrak{p}$, then $\mathfrak{p}$ *starts afresh at time* $t = \mathfrak{m}$, *i.e., the shifted path* $\mathfrak{p}_+(t) \equiv \mathfrak{p}(t + \mathfrak{m}) - \mathfrak{p}(\mathfrak{m})$ *is independent of the past* $\mathfrak{p}(t): t \leqq \mathfrak{m}$ *and identical in law to* $\mathfrak{p}$ *itself.*

Given $a \geqq 0$, if $P_a$ is the law that $P$ induces on the space of sample paths $\mathfrak{q} \equiv \mathfrak{p} + a$, then

6. $$P.(\mathfrak{q}(t_2) \in db \mid \mathfrak{q}(s): s \leqq t_1) = P_a(\mathfrak{q}(t_2 - t_1) \in db), \quad t_2 \geqq t_1, \ a = \mathfrak{q}(t_1),$$

the associated Green operators $f \to E(\int_0^{+\infty} e^{-\alpha t} f(\mathfrak{q})dt)$ map $C[0, +\infty)$ into itself, and the associated generator $\mathfrak{Q}$ is

7. $$(\mathfrak{Q}f)(a) = p_2 f^+(a) + \int_{0+} [f(b + a) - f(a)]p(db), \qquad f \in C^1[0, +\infty).$$

Given $t \geqq 0$, $\mathfrak{p}([0, t] \times [\varepsilon, +\infty))$ is Poisson distributed and differential in $\varepsilon$ with mean $tp[\varepsilon, +\infty)$; as such, it is identical in law to a standard Poisson process $\mathfrak{q}$ with unit jumps and unit rate run with the clock $tp[\varepsilon, +\infty)$, and, using the strong law of large numbers, it follows that

8. $$\lim_{\varepsilon \downarrow 0} \frac{\mathfrak{p}([0, t] \times [\varepsilon, +\infty))}{p[\varepsilon, +\infty)} = \lim_{\varepsilon \downarrow 0} \frac{\mathfrak{q}(tp[\varepsilon, +\infty))}{p[\varepsilon, +\infty)} = t,$$

which will be helpful to us in Section 14.

Consider the special case $p(0, +\infty) < +\infty$ pictured in Diagram 1: the exponential holding times $\mathfrak{e}_1$, $\mathfrak{e}_2$, etc. between jumps are independent with common law $P(\mathfrak{e}_1 > t) = e^{-p(0,+\infty)t}$, the jumps $l_1$, $l_2$, etc. are likewise independent with common law $P(l_1 \in dl) = p(0, +\infty)^{-1}p(dl)$, and the slope of the slanting lines is $1/p_2$.

Consider, as a second example, the standard Brownian passage times $\mathfrak{m}_a =$

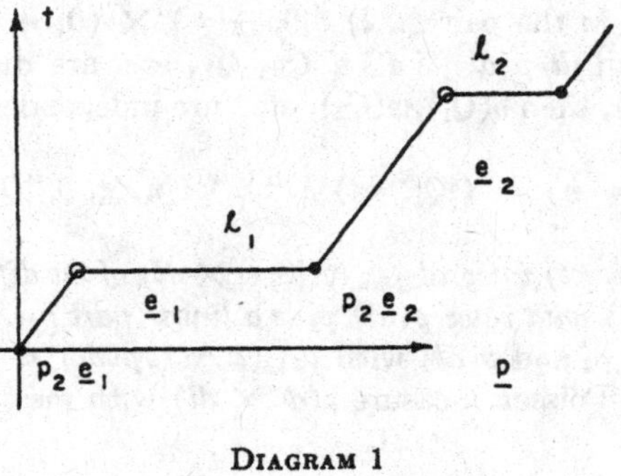


DIAGRAM 1

$\min(t: \mathfrak{x} = a)$ $(a \geq 0)$ under the law $P = P_0$. Because the Brownian traveller starts afresh at its passage times, the shifted path $\mathfrak{m}_{b+a} - \mathfrak{m}_a = \mathfrak{m}_{b+a}(w^+_{\mathfrak{m}_a})$ is independent of $\mathfrak{m}_b : b \leq a$ and identical in law to $\mathfrak{m}.$, i.e., $\mathfrak{m}.$ is differential (it is the *one-sided stable process with exponent* $\frac{1}{2}$ *and rate* $\sqrt{2}$ as noted in Section 4);

9a. $$p_2 = 0,$$

9b. $$p(dl) = dl/(2\pi l^3)^{1/2}$$

can be read off

10. $$E_0(e^{-\alpha \mathfrak{m}_a}) = e^{-(2\alpha)^{1/2}a} = \exp\left\{-a \int_{0+} (1 - e^{-\alpha l}) \frac{dl}{(2\pi l^3)^{1/2}}\right\}.$$

$\mathfrak{m}_a$ is left-continuous, so in the direct integral $[0, a)$ must be used in place of $[0, a]$:

$$\mathfrak{m}_a = \int_{0+} l\mathfrak{p}([0, a) \times dl).$$

## 12. Sample paths: $p_1 = p_3 = 0 < p_4$ $(p_2 > 0/p_4 = +\infty)$

Given a reflecting Brownian motion with local time $\mathfrak{t}^+$, a nonnegative number $p_2$, and a nonnegative mass distribution $p_4(dl)$ $(l > 0)$ with $p_4 = p_4(0, +\infty) = +\infty$ in case $p_2 = 0$, introduce the Poisson measure $\mathfrak{p}(dt \times dl)$ with mean $dt\, p_4(dl)$, make up the associated differential process

1. $$\mathfrak{p}(t) = p_2 t + \int_{0+} l\mathfrak{p}([0, t] \times dl),$$

and consider the sample path[21]

2a. $$\mathfrak{x}^\bullet(t) = \mathfrak{p}\mathfrak{p}^{-1}\mathfrak{t}^+(t) - \mathfrak{t}^+(t) + \mathfrak{x}^+(t), \qquad t \geq 0,$$

2b. $$\mathfrak{p}^{-1}(l) = \inf(t: \mathfrak{p}(t) > l)$$

and its alternative description

3. $$\mathfrak{x}^\bullet(t) = \mathfrak{p}\mathfrak{p}^{-1}\mathfrak{t}^-(t) + \mathfrak{x}^-(t), \qquad t \geq 0,$$

in terms of the *standard Brownian motion* $\mathfrak{x}^- = -\mathfrak{t}^+ + \mathfrak{x}^+$ and its *minimum function* $\mathfrak{t}^-(t) = \mathfrak{t}^+(t) = -(\min_{s \leq t} \mathfrak{x}^-(s) \wedge 0)$; it is to be proved that $\mathfrak{x}^\bullet$ *is the Brownian motion associated with*

4. $$p_2 u^+(0) + \int_{0+} [u(l) - u(0)]p_4(dl) = 0,$$

but before doing that let us look at some pictures of the sample path.

Consider the case $p_4 < +\infty$: the jumps $l_1, l_2$, etc. of $\mathfrak{p}$ are finite in number per unit time and can be labelled in their correct temporal order. $\mathfrak{p}$ and

[21] $\mathfrak{p}\mathfrak{p}^{-1}\mathfrak{t}^+(t)$ is short for $\mathfrak{p}(\mathfrak{p}^{-1}(\mathfrak{t}^+(t)))$.

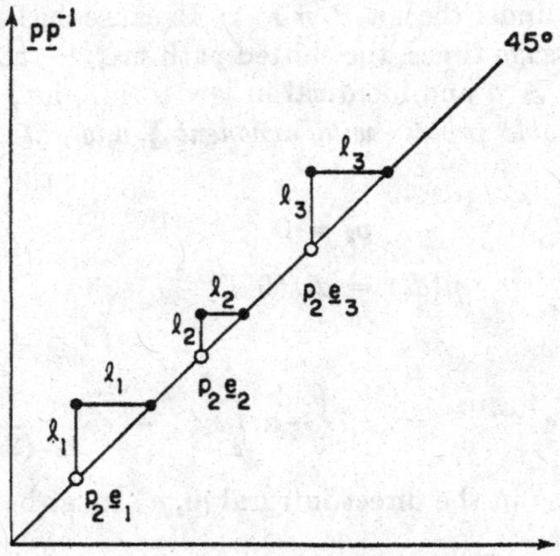


DIAGRAM 1

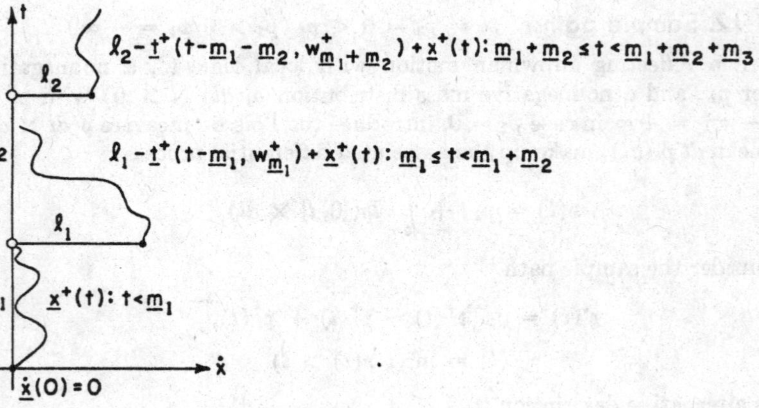


$$\begin{aligned} \mathfrak{m}_1 &= \mathfrak{t}^{-1}(p_2\, \mathfrak{e}_1) \\ \mathfrak{m}_2 &= \mathfrak{t}^{-1}(p_2\, \mathfrak{e}_2 + l_1\,,\, w^+_{\mathfrak{m}_1}) \\ \mathfrak{m}_3 &= \mathfrak{t}^{-1}(p_2\, \mathfrak{e}_3 + l_2\,,\, w^+_{\mathfrak{m}_1+\mathfrak{m}_2}) \\ &\text{etc.} \end{aligned}$$

DIAGRAM 2

$\mathfrak{p}^{-1}$ are seen in Diagram 11.1, $\mathfrak{p}\mathfrak{p}^{-1}$ in Diagram 1 of the present section, and the $\mathfrak{x}^{\bullet} = \mathfrak{p}\mathfrak{p}^{-1}\mathfrak{t}^{+} - \mathfrak{t}^{+} + \mathfrak{x}^{+}$ path in Diagram 2, in which $\mathfrak{t}^{-1}$ is left-continuous as usual and $\mathfrak{e}_1$, $\mathfrak{e}_2$, etc. are the exponential holding times between jumps of $\mathfrak{p}$.

Coming to the case $p_4 = +\infty$, $\mathfrak{p}(t)$ experiences an infinite number of jumps during each time interval $[t_1, t_2)$ $(t_1 < t_2)$, but

$$\mathfrak{p}([t_1, t_2) \times [\varepsilon, +\infty)) < +\infty \qquad (t_2 < +\infty, \varepsilon > 0),$$

and so it is legitimate to label the jumps as follows:

(a) arrange in separate rows the jumps occurring in $(0, 1]$, $(1, 2]$, etc.;

(b) in each row, arrange the jumps in order of magnitude beginning with the largest one;

(c) if several jumps of the same magnitude occur in a single row, arrange them in correct temporal order;

(d) number the rows as indicated below:

$$l_1 \geqq l_3 \geqq l_6 \geqq l_{10},$$
$$l_2 \geqq l_5 \geqq l_9,$$
$$l_4 \geqq l_8,$$
$$l_7 \quad \text{etc.}$$

Diagram 2 gives an approximate idea of the sample path in the case $p_2 = 0$. Diagram 3 ($p_2 = 0$, $\mathfrak{x}(0) = 0$) is based on the alternative description 3: the standard Brownian path $\mathfrak{x}^-$ has been slanted off to the left for the purposes of the picture, and the rule is to translate the excursions of $\mathfrak{x}^-$ between the endpoints of the flat stretches of $\mathfrak{p}^{-1}$ until the left legs of the hatched curvilinear triangles abut on the time axis and then to fill up the gaps with $\mathfrak{x}^\bullet = 0$. The picture is not so simple in case $p_2 > 0$: then $\mathfrak{Q} = (l : \mathfrak{p}\mathfrak{p}^{-1}(l) = l)$ has positive measure, and, on $\mathfrak{Q}^- = (t : \mathfrak{t}^-(t) \in \mathfrak{Q})$, $\mathfrak{x}^\bullet = \mathfrak{p}\mathfrak{p}^{-1}\mathfrak{t}^- - \mathfrak{x}^-$ reduces to the reflecting Brownian motion $\mathfrak{t}^- - \mathfrak{x}^- = \mathfrak{x}^+$.

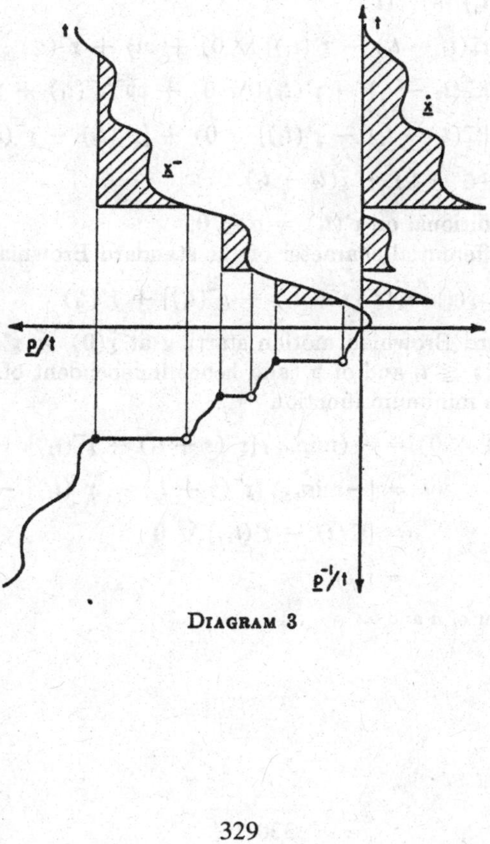


DIAGRAM 3

## 13. Simple Markovian character: $p_1 = p_3 = 0\ (p_2 > 0/p_4 = +\infty)$

Consider the sample path

1. $$\mathfrak{x}^\bullet = \mathfrak{p}\mathfrak{p}^{-1}\mathfrak{t}^+ - \mathfrak{t}^+ + \mathfrak{x}^+ = \mathfrak{p}\mathfrak{p}^{-1}\mathfrak{t}^- + \mathfrak{x}^-$$

described in Section 12.

Given $t_2 \geqq t_1 \geqq 0$, if $\mathfrak{m} = \mathfrak{p}^{-1}\mathfrak{t}^-(t_1)$, if $\mathfrak{p}_+(t) = \mathfrak{p}(t + \mathfrak{m}) - \mathfrak{p}(\mathfrak{m})$, and if $\mathfrak{t}_+^-(t) = -\min_{s \leqq t}[\mathfrak{x}^-(s + t_1) - \mathfrak{x}^-(t_1)]$, then, as the reader will check,

2. $$\begin{aligned}\mathfrak{p}^{-1}\mathfrak{t}^-(t_2) - \mathfrak{p}^{-1}\mathfrak{t}^-(t_1) &= \inf(s:\mathfrak{p}(s) > \mathfrak{t}^-(t_2)) - \mathfrak{p}^{-1}\mathfrak{t}^-(t_1)) \\ &= \inf(s:\mathfrak{p}(s + \mathfrak{m}) > \mathfrak{t}^-(t_2)) \\ &= \inf(s:\mathfrak{p}_+(s) + \mathfrak{p}(\mathfrak{m}) > [\mathfrak{t}_+^-(t_2 - t_1) - \mathfrak{x}^-(t_1)] \vee \mathfrak{t}^-(t_1)) \quad [22] \\ &= \inf(s:\mathfrak{p}_+(s) > [\mathfrak{t}_+^-(t_2 - t_1) - \mathfrak{x}^\bullet(t_1)] \vee [\mathfrak{t}^-(t_1) - \mathfrak{p}(\mathfrak{m})]) \\ &= \inf(s:\mathfrak{p}_+(s) > [\mathfrak{t}_+^-(t_2 - t_1) - \mathfrak{x}^\bullet(t_1)] \vee 0),\end{aligned}$$

where the last step is justified as follows: $\mathfrak{a} = \mathfrak{t}^-(t_1) - \mathfrak{p}(\mathfrak{m}) \leqq 0$ since *either* $p_2 > 0$ *or* $p_4(0, +\infty) = +\infty$, $\mathfrak{p}^{-1}(0) = 0$, and it follows that *either* $\mathfrak{b} = \mathfrak{t}_+^-(t_2 - t_1) - \mathfrak{x}^\bullet(t_1) < 0$ and $\inf(s:\mathfrak{p}_+(s) > \mathfrak{a} \vee \mathfrak{b}) = \inf(s:\mathfrak{p}_+ > 0) = 0$ *or* $\mathfrak{b} \geqq 0$ and $\mathfrak{a} \vee \mathfrak{b} = \mathfrak{b}$.

Coming to the sample path, itself, an application of 2 implies

3. $$\begin{aligned}\mathfrak{x}^\bullet(t_2) &= \mathfrak{p}\mathfrak{p}^{-1}\mathfrak{t}^-(t_2) + \mathfrak{x}^-(t_2) \\ &= \mathfrak{p}(\mathfrak{p}_+^{-1}([\mathfrak{t}_+^-(t_2 - t_1) - \mathfrak{x}^\bullet(t_1)] \vee 0) + \mathfrak{m}) + \mathfrak{x}^-(t_2) \\ &= \mathfrak{p}_+\mathfrak{p}_+^{-1}([\mathfrak{t}_+^-(t_2 - t_1) - \mathfrak{x}^\bullet(t_1)] \vee 0) + \mathfrak{p}\mathfrak{p}^{-1}\mathfrak{t}^-(t_1) + \mathfrak{x}^-(t_2) \\ &= \mathfrak{p}_+\mathfrak{p}_+^{-1}([\mathfrak{t}_+^-(t_2 - t_1) - \mathfrak{x}^\bullet(t_1)] \vee 0) + [\mathfrak{x}^-(t_2) - \mathfrak{x}^-(t_1)] + \mathfrak{x}^\bullet(t_1) \\ &= \mathfrak{p}_+\mathfrak{p}_+^{-1}\mathring{\mathfrak{t}}(t_2 - t_1) + \mathring{\mathfrak{x}}(t_2 - t_1).\end{aligned}$$

Consider this conditional on $\mathfrak{x}^\bullet(t_1) = a \geqq 0$.

Because of the differential character of the standard Brownian motion $\mathfrak{x}^-$,

4a. $$t \to \mathring{\mathfrak{x}}(t) = [\mathfrak{x}^-(t + t_1) - \mathfrak{x}^-(t_1)] + \mathfrak{x}^\bullet(t_1)$$

is likewise a standard Brownian motion starting at $\mathring{\mathfrak{x}}(0) = \mathfrak{x}^\bullet(t_1) = a$, independent of $\mathfrak{x}^-(s):s \leqq t_1$ and of $\mathfrak{p}$ (and hence independent of $\mathfrak{x}^\bullet(s):s \leqq t$, and of $\mathfrak{p}_+$ also) with minimum function

4b. $$\begin{aligned}-(\min_{s \leqq t} \mathring{\mathfrak{x}}(s) \wedge 0) &= -(\min_{s \leqq t} [\mathfrak{x}^-(s + t_1) - \mathfrak{x}^-(t_1)] + \mathfrak{x}^\bullet(t_1) \wedge 0) \\ &= [-\min_{s \leqq t} [\mathfrak{x}^-(s + t_1) - \mathfrak{x}^-(t_1)] - \mathfrak{x}^\bullet(t_1)] \vee 0 \\ &= [\mathfrak{t}_+^-(t) - \mathfrak{x}^\bullet(t_1)] \vee 0 \\ &= \mathring{\mathfrak{t}}(t).\end{aligned}$$

[22] $a \vee b$ is the larger of $a$ and $b$.

Given $t \geqq 0$, the indicator of the event

5. $$(\mathfrak{m} > t) = (\mathfrak{p}^{-1}\mathfrak{t}^{-}(t_1) > t) = (\mathfrak{t}^{-}(t_1) > \mathfrak{p}(t))$$

is a Borel function of $\mathfrak{p}(s):s \leqq t$ and $\mathfrak{x}^{-}(s):s \leqq t_1$, and, since $\mathfrak{x}^{-}$ and $\mathfrak{p}$ are independent, $\mathfrak{m}$ is a *stopping time* for $\mathfrak{p}$, i.e., $\mathfrak{p}_{+}$ is identical in law to $\mathfrak{p}$ and independent of $\mathfrak{x}^{-}$ and of $\mathfrak{p}(s):s \leqq \mathfrak{m}$ and hence independent of $\mathfrak{x}^{\bullet}(s):s \leqq t_1$ and of $\mathring{\mathfrak{x}}$.

But now it is clear that, conditional on $\mathfrak{x}^{\bullet}(t_1) = a$, $\mathfrak{x}^{\bullet}(t_2)$ is independent of the past $\mathfrak{x}^{\bullet}(s):s \leqq t_1$ with law

6. $$P_a[\mathfrak{x}^{\bullet}(t) \in db], \qquad a = \mathfrak{x}^{\bullet}(t_1), \quad t = t_2 - t_1$$

as was to be proved.

## 14. Local times: $p_1 = p_3 = 0 \ (p_2 > 0/p_4 = +\infty)$

Because the reflecting Brownian local time $\mathfrak{t}^{+}$ was central to the construction of the Brownian motions in the case $p_4 = 0$ treated in Section 10, one expects that a similar local time $\mathfrak{t}^{\bullet}$ based upon the path $\mathfrak{x}^{\bullet} = \mathfrak{p}\mathfrak{p}^{-1}\mathfrak{t}^{+} - \mathfrak{t}^{+} + \mathfrak{x}^{+}$ should figure in the general case; the purpose of this section is to prove its existence.

Given $p_2 > 0$, the contention is that the *local time*

1a. $$\mathfrak{t}^{\bullet}(t) = \lim_{\varepsilon \downarrow 0}(2\varepsilon p_2)^{-1} \text{ measure } (s:\mathfrak{x}^{\bullet}(s) < \varepsilon, s \leqq t), \qquad t \geqq 0$$

exists and can be expressed as

1b. $$\begin{aligned} \mathfrak{t}^{\bullet}(t) &= p_2^{-1}\mathfrak{t}^{+}(\mathfrak{Q}^{+} \cap [0, t]) \\ &= p_2^{-1} \,|\, \mathfrak{Q} \cap [0, \mathfrak{t}^{+}(t)]| \\ &= \mathfrak{p}^{-1}\mathfrak{t}^{+}(t), \end{aligned}$$

in which

2a. $$\mathfrak{Q} = (t:\mathfrak{p}\mathfrak{p}^{-1}(t) = t),$$

2b. $$\mathfrak{Q}^{+} = (t:\mathfrak{t}^{+}(t) \in \mathfrak{Q}).$$

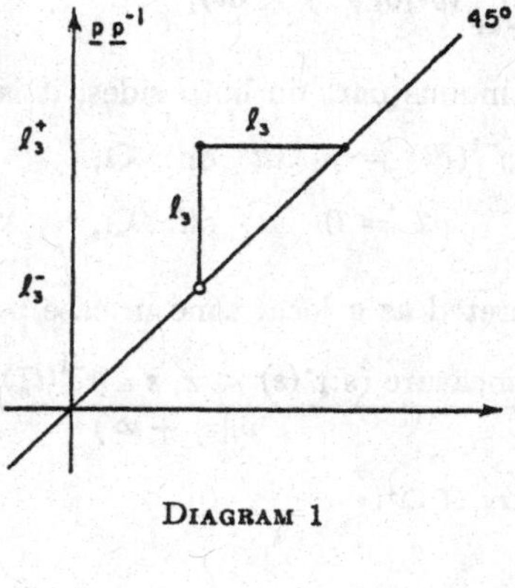


DIAGRAM 1

Consider, for the proof, the intervals $[l_1^-, l_1^+)$, $[l_2^-, l_2^+)$, etc. of the complement of $\mathfrak{Q}$, and note that the complement of $\mathfrak{Q}^+$ is the union of the intervals $[\mathfrak{t}^{-1}(l_1^-), \mathfrak{t}^{-1}(l_1^+))$, $[\mathfrak{t}^{-1}(l_2^-), \mathfrak{t}^{-1}(l_2^+))$, etc., whence[23] $\partial\mathfrak{Q}^+$ is *countable*.

Because $\mathfrak{t}^+$ is continuous and $\mathfrak{x}^\bullet = \mathfrak{x}^+$ on $\mathfrak{Q}^+$,

$$3.\quad \lim_{\varepsilon\downarrow 0}(2\varepsilon)^{-1}\text{ measure }(s:\mathfrak{x}^\bullet(s) < \varepsilon, s \in \mathfrak{Q}^+ \cap [0, t]) = \lim_{\varepsilon\downarrow 0}(2\varepsilon)^{-1}\text{ measure }(s:\mathfrak{x}^+(s) < \varepsilon, s \in \mathfrak{Q}^+ \cap [0, t]) = \mathfrak{t}^+(\mathfrak{Q}^+ \cap [0, t]).$$

Consider, next,

$$4.\quad \begin{aligned}\mathfrak{Q}_\varepsilon^- &= (t: t \notin \mathfrak{Q}^+, \mathfrak{x}^\bullet < \varepsilon)\\ &= \cup_{n\ge 1}[\mathfrak{t}^{-1}(l_n^-), \mathfrak{t}^{-1}(l_n^+)) \cap (t: l_n^+ - \mathfrak{t}^+ < \varepsilon)\\ &= \cup_{n\ge 1}[\mathfrak{t}^{-1}(l_n^-), \mathfrak{t}^{-1}(l_n^+)) \cap [\mathfrak{t}^{-1}(l_n^+ - \varepsilon), +\infty)\\ &= \cup_{n\ge 1}[\mathfrak{t}^{-1}(l_n^- \vee (l_n^+ - \varepsilon)), \mathfrak{t}^{-1}(l_n^+)).\end{aligned}$$

Because $\mathfrak{t}^{-1}$ is left-continuous, $\cap_{\varepsilon>0}\mathfrak{Q}_\varepsilon^- = \emptyset$, and, seeing as $\partial\mathfrak{Q}_\varepsilon^-$ is countable and $\mathfrak{t}^+$ is continuous, it develops, much as in 3, that

$$5.\quad \begin{aligned}&\overline{\lim}_{\varepsilon\downarrow 0}(2\varepsilon)^{-1}\text{ measure }(s:\mathfrak{x}^\bullet(s) < \varepsilon, s \in [0, t] - \mathfrak{Q}^+)\\ &\le \overline{\lim}_{\varepsilon\downarrow 0}(2\varepsilon)^{-1}\text{ measure }(s:\mathfrak{x}^+(s) < \varepsilon, s \in \mathfrak{Q}_\delta^- \cap [0, t])\\ &= \mathfrak{t}^+(\mathfrak{Q}_\delta^- \cap [0, t])\\ &\downarrow 0 \quad (\delta \downarrow 0),\end{aligned}$$

which justifies the definition 1a and the first line of 1b; the second line of 1b is immediate from the definition of $\mathfrak{Q}^+$, and, as to the third line,

$$6.\quad \begin{aligned}\mathfrak{p}\mathfrak{p}^{-1} &= t, && t \in \mathfrak{Q},\\ &= l_n^+, && t \in [l_n^-, l_n^+) \quad (n \ge 1),\\ &= p_2\mathfrak{p}^{-1} + \int_{0+} l\mathfrak{p}([0, \mathfrak{p}^{-1}] \times dl),\end{aligned}$$

and, picking out the continuous part on both sides, it is clear that

$$7.\quad \begin{aligned}\mathfrak{p}^{-1}(dt) &= p_2^{-1}\,dt \quad \text{on } \mathfrak{Q},\\ &= 0 \quad \text{off } \mathfrak{Q},\end{aligned}$$

completing the proof.

$\mathfrak{p}^{-1}\mathfrak{t}^+$ can still be interpreted as a local time in case $p_2 = 0$ $(p_4 = +\infty)$:

$$8.\quad \mathfrak{p}^{-1}\mathfrak{t}^+(t) = \lim_{\varepsilon\downarrow 0}\frac{\sum_{l_n>\varepsilon}\text{ measure }(s:\mathfrak{x}^\bullet(s) < \varepsilon, s \in [\mathfrak{t}^{-1}(l_n^-), \mathfrak{t}^{-1}(l_n^+) \cap [0, t])}{\varepsilon^2 p_4[\varepsilon, +\infty)}.$$

[23] $\partial\mathfrak{Q}^+$ denotes the boundary of $\mathfrak{Q}^+$.

Consider, for the proof, the scaled *visiting times*:

9. $$\mathfrak{d}_n = \varepsilon^{-2} \text{ measure } (s:\mathfrak{x}^\bullet(s) < \varepsilon,\ s \in [\mathfrak{t}^{-1}(l_n^-), \mathfrak{t}^{-1}(l_n^+))).$$

Conditional on $\mathfrak{p}$ (i.e., conditional on $l_1^\pm$, $l_2^\pm$, etc.), the visiting times $\mathfrak{d}_n$ are independent because $\mathfrak{x}^+$ starts from scratch at the place $\mathfrak{x}^+(\mathfrak{m}) = 0$ at time $\mathfrak{m} = \mathfrak{t}^{-1}(l_n^-)$ $(n \geqq 1)$; in addition, if $l_n > \varepsilon$, then $\mathfrak{d}_n$ is identical in law to measure $(s:\mathfrak{x}(s) > 0,\ s < \mathfrak{m}_1)$, where $\mathfrak{x}$ is a standard Brownian motion starting at 0 and $\mathfrak{m}_1$ is its passage time to 1, as will now be verified.

Given $\sigma > 0$, the *scaling*

10. $$\mathfrak{x}(t) \to \sigma\mathfrak{x}(t/\sigma^2)$$

preserves the Wiener measure for standard Brownian paths starting at 0 and sends

11a. $$\mathfrak{x}^+(t) \to \sigma\mathfrak{x}^+(t/\sigma^2),$$

11b. $$\mathfrak{t}^+(t) \to \sigma\mathfrak{t}^+(t/\sigma^2),$$

11c. $$\mathfrak{t}^{-1}(t) \to \sigma^2\mathfrak{t}^{-1}(t/\sigma),$$

12a. $$\mathfrak{x}^-(t) \to \sigma\mathfrak{x}^-(t/\sigma^2),$$

12b. $$\mathfrak{t}^-(t) \to \sigma\mathfrak{t}^-(t/\sigma^2),$$

12c. $$\mathfrak{m}_l \to \sigma^2\mathfrak{m}_{l/\sigma},$$

where $\mathfrak{x}^-$ is the standard Brownian motion $\mathfrak{t}^+ - \mathfrak{x}^+$, $\mathfrak{t}^- = \mathfrak{t}^+ = \max_{s \leq t} \mathfrak{x}^-(s)$, and $\mathfrak{m}_l = \min(t:\mathfrak{x}^- = l)$, and, using $\mathfrak{a} \equiv \mathfrak{b}$ to indicate that $\mathfrak{a}$ and $\mathfrak{b}$ are identical in law, it follows from the rules 11 and 12 that in case $l_n > \varepsilon$,

13. $$\begin{aligned}
\mathfrak{d}_n &\equiv \varepsilon^{-2} \text{ measure } (s:l_n - \mathfrak{t}^+(s) + \mathfrak{x}^+(s) < \varepsilon,\ s < \mathfrak{t}^{-1}(l_n)), \quad \mathfrak{x}^+(0) = 0,\\
&\equiv \varepsilon^{-2} \text{ measure } (s:1 - \mathfrak{t}^+(s/\sigma^2) + \mathfrak{x}^+(s/\sigma^2) < \varepsilon/\sigma,\ s/\sigma^2 < \mathfrak{t}^{-1}(1)),\\
&\qquad\qquad \sigma = l_n,\\
&= \varepsilon^{-2} l_n^2 \text{ measure } (s:1 - \mathfrak{t}^+(s) + \mathfrak{x}^+(s) < \varepsilon/l_n,\ s < \mathfrak{t}^{-1}(1))\\
&= \varepsilon^{-2} l_n^2 \text{ measure } (s:\mathfrak{x}^-(s) > 1 - \varepsilon/l_n,\ s < \mathfrak{m}_1)\\
&= \varepsilon^{-2} l_n^2 \text{ measure } (s:\mathfrak{x}^-(s) > 1 - \varepsilon/l_n,\ \mathfrak{m}_{1-\varepsilon/l_n} \leqq s < \mathfrak{m}_1)\\
&\equiv \varepsilon^{-2} l_n^2 \text{ measure } (s:\mathfrak{x}(s) > 0,\ s < \mathfrak{m}_{\varepsilon/l_n}), \qquad \mathfrak{x}(0) = 0,\\
&\equiv \text{measure } (s:\mathfrak{x}(s) > 0,\ s < \mathfrak{m}_1),
\end{aligned}$$

where the scaling 10 was used in step 2 ($\sigma = l_n$) and in step 7 ($\sigma = \varepsilon/l_n$).

Coming back to 8, the strong law of large numbers combined with the rule

14. $$\lim_{\varepsilon \downarrow 0} p_4[\varepsilon, +\infty)^{-1}\mathfrak{p}([0, t] \times [\varepsilon, +\infty)) = t \qquad (t \geqq 0)$$

(see Section 11) and the simple evaluation

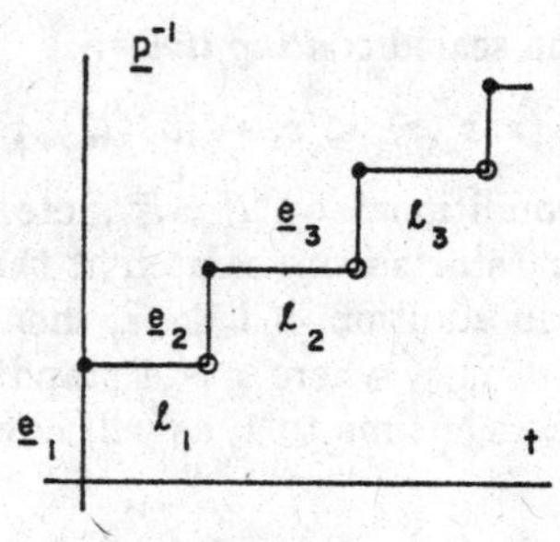


DIAGRAM 2

15. $$E_0(\mathfrak{d}_1) = \int_0^{+\infty} dt\, P_1[\mathfrak{x}(t) < 1,\ \mathfrak{m}_0 > t] = \int_0^{+\infty} dt \int_0^1 \frac{e^{-(a-1)^2/2t} - e^{-(a+1)^2/2t}}{(2\pi t)^{1/2}}\, da = \int_0^1 2a\, da = 1,$$

justifies

16. $$\lim_{\varepsilon \downarrow 0} \frac{\sum_{l_n > \varepsilon} \text{measure } (s : \mathfrak{x}^{\bullet}(s) < \varepsilon,\ s \in [\mathfrak{t}^{-1}(l_n^-),\ \mathfrak{t}^{-1}(l_n^+)) \cap [0, t])}{\varepsilon^2 p_4[\varepsilon, +\infty)} = \lim_{\varepsilon \downarrow 0} \sum_{\substack{l_n > \varepsilon \\ \mathfrak{t}^{-1}(l_n^+) \leqq t}} \mathfrak{d}_n / p_4[\varepsilon, +\infty)$$
$$= E_0(\mathfrak{d}_1) \lim_{\varepsilon \downarrow 0} \frac{\#(l_n : l_n > \varepsilon, \mathfrak{t}^{-1}(l_n^+) \leqq t)}{p_4[\varepsilon, +\infty)}\ {}^{24}$$
$$= \lim_{\varepsilon \downarrow 0} \frac{\#(l_n : l_n > \varepsilon, l_n^+ < \mathfrak{t}^+(t))}{p_4[\varepsilon, +\infty)} = \lim_{\varepsilon \downarrow 0} \frac{\mathfrak{p}([0, \mathfrak{p}^{-1}\mathfrak{t}^+(t)] \times [\varepsilon, +\infty))}{p_4[\varepsilon, +\infty)} = \mathfrak{p}^{-1}\mathfrak{t}^+(t),$$

where the use of $l_n^+ < \mathfrak{t}^+(t)$ in place of $\mathfrak{t}^{-1}(l_n^+) \leqq t$ in step 3 is justified because both describe the same class of jumps plus or minus a single jump and $p_4[\varepsilon, +\infty) \uparrow +\infty$ as $\varepsilon \downarrow 0$; a picture helps to see that $l_n^+ < \mathfrak{t}^+$ and $\mathfrak{p}^{-1}(l_n^+) < \mathfrak{p}^{-1}(\mathfrak{t}^+)$ are identical as needed in step 4.

$\mathfrak{p}^{-1}\mathfrak{t}^+$ *cannot be computed from the sample path if* $p_2 = 0$ *and* $p_4 < +\infty$, as is clear from Diagram 2 in which

17. $$\mathfrak{p}^{-1}\mathfrak{t}^+(t) = \mathfrak{e}_1 + \cdots + \mathfrak{e}_n, \quad \mathfrak{t}^{-1}(l_1 + \cdots + l_{n-1}) \leqq t < \mathfrak{t}^{-1}(l_1 + \cdots + l_n),$$

and $\mathfrak{x}^{\bullet}$ is independent of the holding times $\mathfrak{e}_1$, $\mathfrak{e}_2$, etc. But it still has some features of a local time: it is the sum of $n$ independent holding times $\mathfrak{e}$ with

[24] $\#(l_n$: etc.) denotes the number of jumps $l_n$ with the properties described inside.

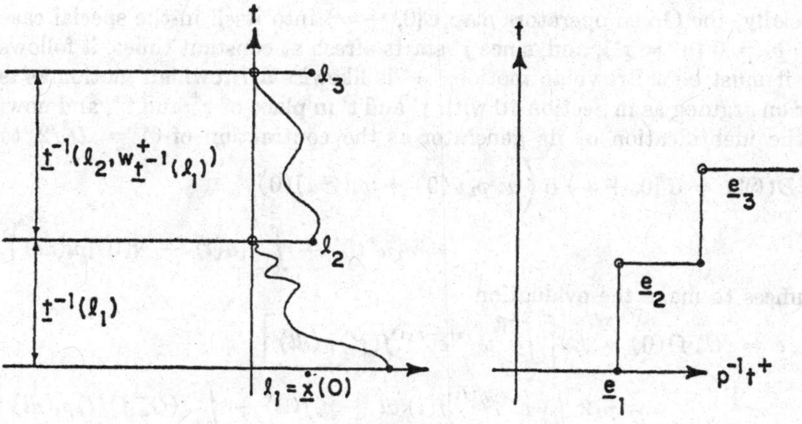


DIAGRAM 3

common conditional law $P.(\mathfrak{e}_1 > t \mid \mathfrak{x}^{\bullet}) = e^{-p_4 t}$, where $n$ is the number of times that the sample path approaches 0 before time $t$ (see Diagram 3).

## 15. Sample paths and Green operators: $p_1 u(0) + p_3(\mathfrak{G}u)(0) = p_2 u^+(0) + \int_{0+} [u(l) - u(0)]p_4(dl)$ $(p_2 > 0/p_4 = +\infty)$

Consider the motion $\mathfrak{x}^{\bullet} = \mathfrak{p}\mathfrak{p}^{-1}\mathfrak{t}^+ - \mathfrak{t}^+ + \mathfrak{x}^+$ and its local time $\mathfrak{t}^{\bullet} = \mathfrak{p}^{-1}\mathfrak{t}^+$, and let us use them to build up the sample paths in the general case $(p_2 > 0/p_4 = +\infty)$ imitating the prescription of Section 10:

1a. $$\mathfrak{y}^{\bullet}(t) = \mathfrak{x}^{\bullet}[\mathfrak{f}^{-1}(t)] \quad \text{if} \quad t < \mathfrak{m}_\infty^{\bullet},$$
$$= \infty \quad \text{if} \quad t \geqq \mathfrak{m}_\infty^{\bullet},$$

1b. $$\mathfrak{f}(t) = t + p_3 \mathfrak{t}^{\bullet}(t),$$

1c. $$P.(\mathfrak{m}_\infty^{\bullet} > t \mid \mathfrak{x}^{\bullet}) = e^{-p_1 \mathfrak{t}^{\bullet}[\mathfrak{f}^{-1}(t)]}.$$

Given $l \geqq 0$,

2. $$(G_\alpha^{\bullet} f)(l) = E_l\left(\int_0^{\mathfrak{m}_\infty^{\bullet}} e^{-\alpha t} f(\mathfrak{y}^{\bullet})\, dt\right)$$
$$= E_l\left(\int_0^{+\infty} e^{-\alpha t} e^{-p_1 \mathfrak{t}^{\bullet}(\mathfrak{f}^{-1})} f[\mathfrak{x}^{\bullet}(\mathfrak{f}^{-1})]\, dt\right)$$
$$= E_l\left[\int_0^{+\infty} e^{-\alpha \mathfrak{f}} e^{-p_1 \mathfrak{t}^{\bullet}} f(\mathfrak{x}^{\bullet}) \mathfrak{f}(dt)\right]$$
$$= E_l\left[\int_0^{\mathfrak{m}_0} e^{-\alpha t} f(\mathfrak{x}^+)\, dt\right]$$
$$+ E_l(e^{-\alpha \mathfrak{m}_0}) E_0\left[\int_0^{+\infty} e^{-\alpha t} e^{-p_1 \mathfrak{t}^{\bullet}} f(\mathfrak{x}^{\bullet}) \mathfrak{f}\,(dt)\right] \quad [25]$$
$$= (G_\alpha^- f)(l) + e^{-(2\alpha)^{1/2} l} (G_\alpha^{\bullet} f)(0),$$

[25] $\mathfrak{x}^{\bullet} = \mathfrak{x}^+$ and $\mathfrak{t}^{\bullet} = 0$ up to time $\mathfrak{m}_0 = \min(t: \mathfrak{x}^+ = 0)$, and $\mathfrak{x}^{\bullet}$ starts afresh at that moment.

especially, the Green operators map $C[0, +\infty)$ into itself in the special case $p_1 = p_3 = 0$ ($\mathfrak{y}^\bullet = \mathfrak{x}^\bullet$), and, since $\mathfrak{x}^\bullet$ starts afresh at constant times, it follows that it must be a Brownian motion. $\mathfrak{y}^\bullet$ is likewise a Brownian motion as is clear on arguing as in Section 10 with $\mathfrak{x}^\bullet$ and $\mathfrak{t}^\bullet$ in place of $\mathfrak{x}^+$ and $\mathfrak{t}^+$, and now, for the identification of its generator as the contraction of $\mathfrak{G} = D^2/2$ to

3. $$D(\mathfrak{G}^\bullet) = C^2[0, +\infty) \cap \Big(u\colon p_1 u(0) + p_3(\mathfrak{G}u)(0) = p_2 u^+(0) + \int_{0+} [u(l) - u(0)]p_4(dl)\Big),$$

it suffices to make the evaluation

4. $$e = (G^\bullet_\alpha f)(0) = E_0\Big[\int_0^{+\infty} e^{-\alpha \mathfrak{f}} e^{-p_1 \mathfrak{t}^\bullet} f(\mathfrak{x}^\bullet)\mathfrak{f}(dt)\Big] = \frac{p_2 2 \int_{0+} e^{-(2\alpha)^{1/2}l} f(l)\,dl + p_3 f(0) + \int_{0+} (G^-_\alpha f)(l) p_4(dl)}{p_1 + (2\alpha)^{1/2}p_2 + \alpha p_3 + \int_{0+} [1 - e^{-(2\alpha)^{1/2}l}]p_4(dl)}.$$

$e$ is decomposed into simpler integrals in several steps (see the explanation below):

5. $$\begin{aligned} e &= E_0\Big[\int_0^{+\infty} e^{-\alpha t} e^{-(p_1+\alpha p_3)\mathfrak{t}^\bullet} f(\mathfrak{x}^\bullet)\,dt\Big] + p_3 f(0) E_0\Big[\int_0^{+\infty} e^{-\alpha t} e^{-(p_1+\alpha p_3)\mathfrak{t}^\bullet}\mathfrak{t}^\bullet(dt)\Big] \\ &= \sum_{n\geq 1} E_0\Big[\int_{[\mathfrak{t}^{-1}(l_n^-), \mathfrak{t}^{-1}(l_n^+))} e^{-\alpha t}\, e^{-(p_1+\alpha p_3)\mathfrak{p}^{-1}\mathfrak{t}^+} f(l_n^+ - \mathfrak{t}^+ + \mathfrak{x}^+)\,dt\Big] \\ &\quad + E_0\Big[\int_0^{+\infty} e^{-\alpha t} e^{-(p_1+\alpha p_3)\mathfrak{p}^{-1}\mathfrak{t}^+} f(\mathfrak{x}^+)\,dt\Big] \\ &\quad - \sum_{n\geq 1} E_0\Big[\int_{[\mathfrak{t}^{-1}(l_n^-), \mathfrak{t}^{-1}(l_n^+))} e^{-\alpha t} e^{-(p_1+\alpha p_3)\mathfrak{p}^{-1}\mathfrak{t}^+} f(\mathfrak{x}^+)\,dt\Big] \\ &\quad + p_3 f(0) E_0\Big[\int_0^{+\infty} e^{-\alpha t} e^{-(p_1+\alpha p_3)\mathfrak{p}^{-1}\mathfrak{t}^+} \mathfrak{p}^{-1}\mathfrak{t}^+(dt)\Big] \\ &= \sum_{n\geq 1} E_0\Big[e^{-\alpha \mathfrak{t}^{-1}(l_n^-)}\, e^{-(p_1+\alpha p_3)\mathfrak{p}^{-1}(l_n^+)} \cdot E_0\Big(\int_0^{\mathfrak{t}^{-1}(l_n)} e^{-\alpha t} f[l_n - \mathfrak{t}^+ + \mathfrak{x}^+]\,dt \,\Big|\, l_n\Big)\Big] \\ &\quad + E_0\Big[\int_0^{+\infty} e^{-\alpha t} e^{-(p_1+\alpha p_3)\mathfrak{p}^{-1}\mathfrak{t}^+} f(\mathfrak{x}^+)\,dt\Big] \\ &\quad - \sum_{n\geq 1} E_0\Big[e^{-\alpha \mathfrak{t}^{-1}(l_n^-)}\, e^{-(p_1+\alpha p_3)\mathfrak{p}^{-1}(l_n^+)} E_0\Big(\int_0^{\mathfrak{t}^{-1}(l_n)} e^{-\alpha t} f(\mathfrak{x}^+)\,dt \,\Big|\, l_n\Big)\Big] \\ &\quad + \frac{p_3 f(0)}{p_1 + \alpha p_3}\Big[1 - \alpha E_0\Big(\int_0^{+\infty} e^{-\alpha t} e^{-(p_1+\alpha p_3)\mathfrak{p}^{-1}\mathfrak{t}^+}\,dt\Big)\Big]^{26} \\ &= e_1 + e_2 - e_3 + e_4, \end{aligned}$$

[26] $p_3/(p_1 + \alpha p_3) = 0$ if $p_3 = 0$.

where $\mathfrak{t}^{\bullet}(dt) = 0$ *outside* $\mathfrak{Z}^{\bullet} \equiv (t : \mathfrak{x}^{\bullet} = 0)$ was used in step 1, $[0, +\infty)$ was split into $\mathfrak{O}^+ + \cup_{n\geq 1} [\mathfrak{t}^{-1}(l_n^-), \mathfrak{t}^{-1}(l_n^+))$ in step 2, and $\mathfrak{p}\mathfrak{p}^{-1}\mathfrak{t}^+$ was evaluated as $\mathfrak{t}^+$ or $l_n^+$ according as $t \in \mathfrak{O}^+$ or $\mathfrak{t}^{-1}(l_n^-) \leqq t < \mathfrak{t}^{-1}(l_n^+)$, and, in step 3, it was noted that, conditional on $\mathfrak{p}$, the standard Brownian traveller starts afresh at time $\mathfrak{m} = \mathfrak{t}^{-1}(l_n^-)$ at the place $l = 0$; the addition rule

$$\mathfrak{t}^{-1}(l_n^+) = \mathfrak{t}^{-1}(l_n^-) + \mathfrak{t}^{-1}(l_n, w^+_{\mathfrak{t}^{-1}(l_n^-)})$$

was also used in step 3, and a partial (time) integration was performed under the expectation sign in $e_4$.

To compute $e_1$, substitute the standard Brownian motion $\mathfrak{x}^- = \mathfrak{t}^+ - \mathfrak{x}^+$ and its passage times $\mathfrak{m}_l = \mathfrak{t}^{-1}(l)$ into the conditional expectation and integrate them out, next integrate out $\mathfrak{t}^{-1}(l_n^-)$ conditional on $\mathfrak{p}$, express the integral in terms of the Poisson measure $\mathfrak{p}(dt \times dl)$, and use the differential character of the latter to integrate it out also:

$$6.\quad e_1 = \sum_{n\geq 1} E_0\left[e^{-\alpha\mathfrak{t}^{-1}(l_n^-)}e^{-(p_1+\alpha p_3)\mathfrak{p}^{-1}(l_n^+)}E_0\left(\int_0^{\mathfrak{m}_{l_n}} e^{-\alpha t}f(l_n - \mathfrak{x}^-)\,dt \mid l_n\right)\right]$$

$$= \sum_{n\geq 1} E_0[e^{-\alpha\mathfrak{t}^{-1}(l_n^-)}e^{-(p_1+\alpha p_3)\mathfrak{p}^{-1}(l_n^+)}(G_\alpha^- f)(l_n)]$$

$$= \sum_{n\geq 1} E_0[e^{-(2\alpha)^{1/2}l_n^-}e^{-(p_1+\alpha p_3)\mathfrak{p}^{-1}(l_n^+)}(G_\alpha^- f)(l_n)]$$

$$= E_0\left[\int_{[0,+\infty)\times(0,+\infty)} \mathfrak{p}(dt\times dl)e^{-(2\alpha)^{1/2}\mathfrak{p}(t-)}e^{-(p_1+\alpha p_3)t}(G_\alpha^- f)(l)\right]$$

$$= \lim_{\varepsilon\downarrow 0} E_0\left[\int_{[\varepsilon,+\infty)\times(0,+\infty)} \mathfrak{p}(dt\times dl)e^{-(2\alpha)^{1/2}\mathfrak{p}(t-\varepsilon)}e^{-(p_1+\alpha p_3)t}(G_\alpha^- f)(l)\right]$$

$$= \int_{[0,+\infty)\times(0,+\infty)} dt\, p_4(dl)\exp\left\{-t\left[p_2(2\alpha)^{1/2} + \int_{0+}(1 - e^{-(2\alpha)^{1/2}l})p_4(dl)\right]\right\} \cdot e^{-(p_1+\alpha p_3)t}(G_\alpha^- f)(l)$$

$$= \frac{\int_{0+}(G_\alpha^- f)(l)p_4(dl)}{p_1 + (2\alpha)^{1/2}p_2 + \alpha p_3 + \int_{0+}(1 - e^{-(2\alpha)^{1/2}l})p_4(dl)}.$$

To compute $e_2$, use the joint law $\dfrac{2(b+a)}{(2\pi t^3)^{1/2}}e^{-(b+a)^2/2t}\,da\,db$ of $\mathfrak{x}^+$ and $\mathfrak{t}^+$:

$$7.\quad e_2 = E_0\left[\int_0^{+\infty} e^{-\alpha t}\,dt\int_0^{+\infty} db\int_0^{+\infty} da\; 2\frac{b+a}{(2\pi t^3)^{1/2}}e^{-(b+a)^2/2t} \cdot e^{-(p_1+\alpha p_3)\mathfrak{p}^{-1}(b)}f(a)\right]$$

$$= E_0\left[\int_0^{+\infty} e^{-(2\alpha)^{1/2}b} e^{-(p_1+\alpha p_3)\mathfrak{p}^{-1}(b)}\, db\right] 2\int_{0+} e^{-(2\alpha)^{1/2}a} f(a)\, da$$

$$= E_0\left[\int_0^{+\infty} e^{-(2\alpha)^{1/2}\mathfrak{p}} e^{-(p_1+\alpha p_3)t}\, \mathfrak{p}(dt)\right] 2\int_{0+} e^{-(2\alpha)^{1/2}l} f(l)\, dl$$

$$= \frac{p_1+\alpha p_3}{(2\alpha)^{1/2}} E_0\left[\int_0^{+\infty} e^{-(p_1+\alpha p_3)t}(1-e^{-(2\alpha)^{1/2}\mathfrak{p}})\, dt\right] 2\int_{0+} e^{-(2\alpha)^{1/2}l} f(l)\, dl$$

$$= \frac{(2\alpha)^{1/2}p_2 + \int_{0+}(1-e^{-(2\alpha)^{1/2}l})p_4(dl)}{p_1+(2\alpha)^{1/2}p_2+\alpha p_3+\int_{0+}(1-e^{-(2\alpha)^{1/2}l})p_4(dl)} \frac{2}{(2\alpha)^{1/2}}\int_{0+} e^{-(2\alpha)^{1/2}l} f(l)\, dl.$$

To compute $e_3$, use the same manipulations as for $e_1$ together with the lemma[27]

8. $$E_0\left(\int_0^{t^{-1}(l)} e^{-\alpha t} f(\mathfrak{x}^+)\, dt\right) = (G_\alpha^+ f)(0)[1-e^{-(2\alpha)^{1/2}l}],$$

obtaining

9. $$e_3 = \sum_{n\geq 1} E_0[e^{-(2\alpha)^{1/2}l_n^-} e^{-(p_1+\alpha p_3)\mathfrak{p}^{-1}(l_n^+)}(1-e^{-(2\alpha)^{1/2}l_n})](G_\alpha^+ f)(0)$$

$$= E_0\left[\int_{[0,+\infty)\times(0,+\infty)} \mathfrak{p}(dt\times dl)e^{-(2\alpha)^{1/2}\mathfrak{p}(t-)} e^{-(p_1+\alpha p_3)t}(1-e^{-(2\alpha)^{1/2}l})\right] \cdot (G_\alpha^+ f)(0)$$

$$= \frac{\int_{0+}(1-e^{-(2\alpha)^{1/2}l})p_4(dl)}{p_1+(2\alpha)^{1/2}p_2+\alpha p_3+\int_{0+}(1-e^{-(2\alpha)^{1/2}l})p_4(dl)} \frac{2}{(2\alpha)^{1/2}}\int_{0+} e^{-(2\alpha)^{1/2}l} f(l)\, dl.$$

To compute $e_4$, use 6 with $f = 1$:

10. $$e_4 = \frac{p_3 f(0)}{p_1+\alpha p_3}\left[1 - \frac{(2\alpha)^{1/2}p_2 + \int_{0+}(1-e^{-(2\alpha)^{1/2}l})p_4(dl)}{p_1+(2\alpha)^{1/2}p_2+\alpha p_3+\int_{0+}(1-e^{-(2\alpha)^{1/2}l})p_4(dl)}\right]$$

$$= \frac{p_3 f(0)}{p_1+(2\alpha)^{1/2}p_2+\alpha p_3+\int_{0+}(1-e^{-(2\alpha)^{1/2}l})p_4(dl)}.$$

Combining 5, 6, 7, 9, and 10 verifies 4, and that finishes the proof.

## 16. Bounded interval: [− 1, + 1]

A Brownian motion on $[-1, +1]$ is defined as in Section 5 except that

1. $$|\mathfrak{x}^\bullet| \leq 1, \qquad t < \mathfrak{m}_\infty^\bullet,$$

[27] $G_\alpha^+$ is the reflecting Brownian Green operator.

and the stopped path

2a. $$\mathfrak{x}^\bullet(t) : t < \mathfrak{e}^\bullet = \lim_{\varepsilon \downarrow 0} \inf(t : |\mathfrak{x}^\bullet| > 1 - \varepsilon), \quad -1 < \mathfrak{x}^\bullet(0) = l < +1$$

is now identical in law to the stopped standard Brownian path

2b. $$\mathfrak{x}(t) : t < \mathfrak{e} = \min(t : |\mathfrak{x}| = 1), \qquad \mathfrak{x}(0) = l.$$

Except in the case $P^\bullet_\cdot[|\mathfrak{x}^\bullet(\mathfrak{e}^\bullet)| = 1] < 1$ which can be treated as in Section 6, $C[-1, +1]$ is mapped into itself under the Green operators, $\mathfrak{G}^\bullet$ can be defined as before, and $D(\mathfrak{G}^\bullet)$ can be described in terms of six nonnegative numbers $p_{\pm 1}$, $p_{\pm 2}$, $p_{\pm 3}$ and two nonnegative mass distributions $p_{\pm 4}(dl)$ subject to

3a. $$p_{-1} + p_{-2} + p_{-3} + \int_{-1}^{+1} (1 + l) p_{-4}(dl) = 1, \qquad p_{-4}(-1) = 0,$$

3b. $$p_{+1} + p_{+2} + p_{+3} + \int_{-1}^{+1} (1 - l) p_{+4}(dl) = 1, \qquad p_{+4}(+1) = 0,$$

4a. $$p_{-4}(-1, +1] = +\infty \quad \text{in case} \quad p_{-2} = p_{-3} = 0,$$

4b. $$p_{+4}[-1, +1) = +\infty \quad \text{in case} \quad p_{+2} = p_{+3} = 0$$

as follows. $D(\mathfrak{G}^\bullet)$ is the class of functions $u \in C^2[-1, +1]$ subject to

5a. $$p_{-1} u(-1) - p_{-2} u^+(-1) + p_{-3}(\mathfrak{G}u)(-1) = \int_{-1}^{+1} [u(l) - u(-1)]\, p_{-4}(dl),$$

5b. $$p_{+1} u(+1) + p_{+2} u^-(+1) + p_{+3}(\mathfrak{G}u)(+1) = \int_{-1}^{+1} [u(l) - u(+1)] p_{+4}(dl).$$ [28]

$\mathfrak{G}^\bullet$ is the contraction of $\mathfrak{G} = D^2/2$ to $D(\mathfrak{G}^\bullet)$,

6. $$(G^\bullet_\alpha f)(l) = (G^-_\alpha f)(l) + e_-(l)(G^\bullet_\alpha f)(-1) + e_+(l)(G^\bullet_\alpha f)(+1), \qquad |l| \leqq 1,$$

in which

7a. $$(G^-_\alpha f)(a) = E_a\left(\int_0^{\mathfrak{e}} e^{-\alpha t} f(\mathfrak{x})\, dt\right) = 2 \int_{-1}^{+1} G(a, b) f(b)\, db,$$ [29]

7b. $$G(a, b) = G(b, a) = \frac{\sinh(2\alpha)^{1/2}(1 + a) \sinh(2\alpha)^{1/2}(1 - b)}{(2\alpha)^{1/2}}, \quad a \leqq b,$$

[28] $u^-(+1) = \lim_{\varepsilon \downarrow 0} \varepsilon^{-1}[u(1) - u(1 - \varepsilon)]$.

[29] $P_\cdot$, $E_\cdot$, $\mathfrak{x}$, $\mathfrak{m}$ are the standard Brownian probabilities, expectations, sample paths, and passage times.

is the Green operator for the Brownian motion with instant killing at $\pm 1$ and

8a. $$e_-(l) = \frac{\sinh (2\alpha)^{1/2}(1-l)}{\sinh 2(2\alpha)^{1/2}} = E_l(e^{-\alpha \mathfrak{m}_{-1}}, \mathfrak{m}_{-1} < \mathfrak{m}_{+1}),$$

8b. $$e_+(l) = \frac{\sinh (2\alpha)^{1/2}(1+l)}{\sinh 2(2\alpha)^{1/2}} = E_l(e^{-\alpha \mathfrak{m}_{+1}}, \mathfrak{m}_{+1} < \mathfrak{m}_{-1}),$$

and, substituting 6 into 5 and solving for $(G^\bullet_\alpha f)(\pm 1)$, one obtains

9. $$\begin{bmatrix} (G^\bullet_\alpha f)(-1) \\ (G^\bullet_\alpha f)(+1) \end{bmatrix} = \begin{bmatrix} e_{11} & e_{12} \\ e_{21} & e_{22} \end{bmatrix}^{-1} \cdot \begin{bmatrix} p_{-2}(G^-_\alpha f)^+(-1) + p_{-3}f(-1) + \int_{-1}^{+1} (G^-_\alpha f)(l)p_{-4}(dl) \\ -p_{+2}(G^-_\alpha f)^-(+1) + p_{+3}f(+1) + \int_{-1}^{+1} (G^-_\alpha f)(l)p_{+4}(dl) \end{bmatrix},$$

where the exponent $-1$ indicates the inverse of $\begin{bmatrix} e_{11} & e_{12} \\ e_{21} & e_{22} \end{bmatrix}$, and

10a. $$e_{11} = p_{-1} - e_-^+(-1)p_{-2} + \alpha p_{-3} + \int_{-1}^{+1} (1 - e_-)p_{-4}(dl),$$

10b. $$e_{12} = -p_{-2}e_+^+(-1) - \int_{-1}^{+1} e_+ p_{-4}(dl),$$

10c. $$e_{21} = p_{+2}e_-^-(+1) - \int_{-1}^{+1} e_- p_{+4}(dl),$$

10d. $$e_{22} = p_{+1} + e_+^-(+1)p_{+2} + \alpha p_{+3} + \int_{-1}^{+1} (1 - e_+)p_{+4}(dl),$$

all of which is due to W. Feller [1], [3]; the proofs can be carried out as in Section 8.

Coming to the sample paths, let us confine our attention to the case $p_{-4}(-1, +1] = p_{+4}[-1, +1) = +\infty$, leaving the opposite case to the reader.

Given a standard Brownian motion with sample paths $t \to \mathfrak{x}(t)$ and probabilities $P_a(B)$, if $f$ is the map: $R^1 \to [-1, +1]$ defined by folding the line at $\pm 1, \pm 3, \pm 5$, etc. as in Diagram 1, then $\mathfrak{x}^+ = f(\mathfrak{x})$ is the (reflecting) Brownian motion on $[-1, +1]$ associated with 5 in the special case $p_{\pm 1} = p_{\pm 3} = p_{\pm 4} = 0$ $(u^+(-1) = u^-(+1) = 0)$; the dot sample path will be made up using $\mathfrak{x}^+$ and its local times

11a. $$\mathfrak{t}^-(t) = \lim_{\varepsilon \downarrow 0}(2\varepsilon)^{-1} \text{ measure } (s: \mathfrak{x}^+(s) < -1 + \varepsilon, s \leqq t),$$

11b. $$\mathfrak{t}^+(t) = \lim_{\varepsilon \downarrow 0}(2\varepsilon)^{-1} \text{ measure } (s: \mathfrak{x}^+(s) > 1 - \varepsilon, s \leqq t),$$

a pair of independent Poisson measures $\mathfrak{p}_\pm(dt \times dl)$ with means $dt\ p_{\pm 4}(dl)$,

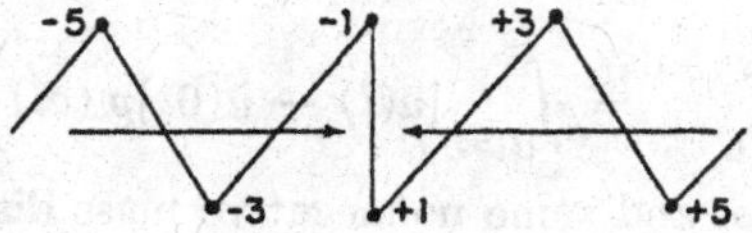


DIAGRAM 1

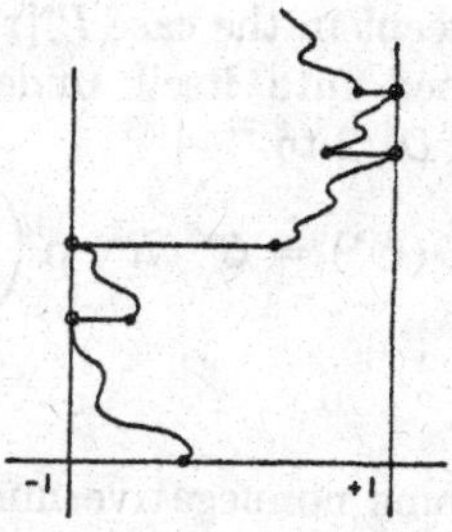


DIAGRAM 2

and the associated differential processes

12a. $$\mathfrak{p}_-(t) = p_{-2}t + \int_{l>-1} (1 + l)\mathfrak{p}_-([0, t] \times dl),$$

12b. $$\mathfrak{p}_+(t) = p_{+2}t + \int_{l<+1} (1 - l)\mathfrak{p}_+([0, t] \times dl).$$

Diagram 2 depicts the sample paths associated with 5 if $p_{\pm 1} = p_{\pm 3} = 0$: $\mathfrak{x}^\bullet$ and $\mathfrak{x}^+$ agree up to time $\mathfrak{m}_{\pm 1} = \min(t:|\mathfrak{x}^+| = 1)$; if $\mathfrak{m}_{-1} < \mathfrak{m}_{+1}$, as in the picture, then $\mathfrak{x}^\bullet$ changes over into $\mathfrak{p}_-\mathfrak{p}_-^{-1}\mathfrak{t}^- - \mathfrak{t}^- + \mathfrak{x}^+$ until it hits $+1$, at which instant it changes over into $-\mathfrak{p}_+\mathfrak{p}_+^{-1}\mathfrak{t}^+ + \mathfrak{t}^+ + \mathfrak{x}^+$ until it hits $-1$ for the second time, etc.

If $p_{-1} = p_{+1} = 0$ and $p_{-3} + p_{+3} > 0$, then the desired motion is as in Diagram 2 but run with the new clock $\mathfrak{f}^{-1}$ which is the inverse function of

13a. $$\mathfrak{f} = t + p_{-3}\,\mathfrak{t}^{\bullet -} + p_{+3}\,\mathfrak{t}^{\bullet +},$$

13b. $$\mathfrak{t}^{\bullet\pm} = \mathfrak{p}_\pm^{-1}\mathfrak{t}^{\pm},$$

while, if $p_{-1} + p_{+1} > 0$, then one has just to kill the above motion $\mathfrak{x}^\bullet(\mathfrak{f}^{-1})$ at time $\mathfrak{m}_\infty^\bullet$ with conditional law

14. $$P_\bullet(\mathfrak{m}_\infty^\bullet > t \mid \mathfrak{x}^\bullet) = e^{-[p_{-1}\mathfrak{t}^{\bullet-}(\mathfrak{f}^{-1}) + p_{+1}\mathfrak{t}^{\bullet+}(\mathfrak{f}^{-1})]};$$

the proofs are left to the industrious reader.

## 17. Two-sided barriers

A Brownian motion on $R^1$ with a *two-sided barrier* at $l = 0$ is defined as in Section 5 except that

1. $$\mathfrak{x}^\bullet \in R^1, \qquad t < \mathfrak{m}_\infty^\bullet,$$

and the stopped path

2a. $$\mathfrak{x}^\bullet(t): t < \mathfrak{e}^\bullet = \lim_{\varepsilon \downarrow 0} \inf(t: |\mathfrak{x}^\bullet| < \varepsilon), \qquad \mathfrak{x}^\bullet(0) = l \in R^1 - 0$$

is identical in law to the stopped standard Brownian motion

2b. $$\mathfrak{x}(t): t < \mathfrak{e} = \min(t: \mathfrak{x} = 0), \qquad \mathfrak{x}(0) = l.$$

Except in the case $P_\cdot^\bullet[\mathfrak{x}^\bullet(\mathfrak{e}^\bullet) = 0] < 1$, which is ignored as before, $C(R^1)$ is mapped into itself under the Green operators, $\mathfrak{G}^\bullet$ is the contraction of $\mathfrak{G} = D^2/2$ to [30]

$$3.\quad D(\mathfrak{G}^\bullet) = C^{\bullet 2}(R^1) \cap \Big(u: p_1 u(0) + p_{-2} u^-(0) - p_{+2} u^+(0) + p_3(\mathfrak{G}u)(0\pm) = \int_{|l|>0} [u(l) - u(0)] p_4(dl)\Big)$$

for some nonnegative numbers $p_1$, $p_{\pm 2}$, $p_3$ and some nonnegative mass distribution $p_4(dl)$ subject to

$$4a.\quad p_1 + p_{-2} + p_{+2} + p_3 + \int (|l| \wedge 1) p_4(dl) = 1, \qquad p_4(0) = 0,$$

$$4b.\quad p_4(R^1) = +\infty \quad \text{in case} \quad p_{\pm 2} = p_3 = 0,$$

and the Green operators are

$$5.\quad (G_\alpha^\bullet f)(l) = (G_\alpha^- f)(l) + e^{-(2\alpha)^{1/2}|l|} (G_\alpha^\bullet f)(0),$$

where

$$6.\quad (G_\alpha^- f)(a) = \int_{ab>0} \frac{e^{-(2\alpha)^{1/2}|b-a|} - e^{-(2\alpha)^{1/2}|b+a|}}{(2\alpha)^{1/2}} f(b)\, db$$

is the Green operator for the Brownian motion with instant killing at $l = 0$ and

$$7a.\quad (G_\alpha^\bullet f)(0) = \frac{-p_{-2}(G_\alpha^- f)^-(0) + p_{+2}(G_\alpha^- f)^+(0) + p_3 f(0) + \int_{|l|>0} (G_\alpha^- f)(l) p_4(dl)}{p_1 + (2\alpha)^{1/2}(p_{-2} + p_{+2}) + \alpha p_3 + \int_{|l|>0} (1 - e^{-(2\alpha)^{1/2}|l|}) p_4(dl)},$$

$$7b.\quad \pm(G_\alpha^- f)^\pm(0) = 2 \int_{\pm l>0} e^{-(2\alpha)^{1/2}|l|} f(l)\, dl.$$

Coming to the sample paths, P. Lévy [3] proved that if $t \to \mathfrak{x}(t)$ is a standard Brownian path starting at 0 and if $\mathfrak{Z}_1$, $\mathfrak{Z}_2$, etc. are the (open) intervals of the complement of $\mathfrak{Z} = (t: \mathfrak{x} = 0)$, then the ***signs*** $e_1$, $e_2$, etc. of the ***excursions*** $\mathfrak{x}(t): t \in \mathfrak{Z}_1$, etc., are independent Bernouilli trials with common law $P_0(e_1 = \pm 1) = \frac{1}{2}$ (standard coin-tossing game), independent of $\mathfrak{Z}$ and of the (unsigned) ***scaled excursions***

$$8.\quad \mathfrak{x}_1(t) = |\mathfrak{Z}_1|^{-1/2} |\mathfrak{x}(t|\mathfrak{Z}_1| + \inf \mathfrak{Z}_1)|, \qquad 0 \leqq t \leqq 1,$$

etc.

which are independent, identical in law, and likewise independent of $\mathfrak{Z}$ (see Diagram 1).

Given $p_{-2} + p_{+2} > 0$, it is not difficult to see that if $e_1$, $e_2$, etc. is now a ***skew coin-tossing game*** independent of the scaled excursions and of $\mathfrak{Z}$ (i.e.,

[30] $C^{\bullet 2}(R^1) = C^2(-\infty, 0] \cap C^2[0, +\infty) \cap (u: u''(0-) = u''(0+))$.

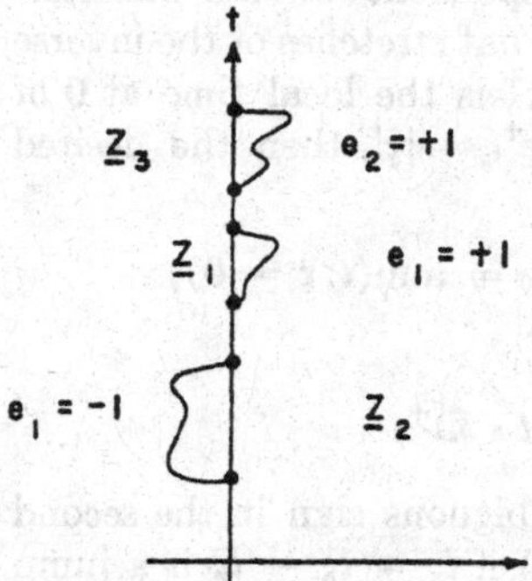


DIAGRAM 1

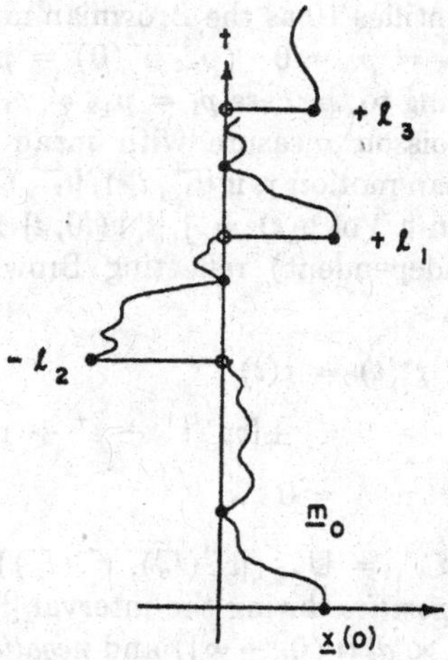


DIAGRAM 2

independent of $|\mathfrak{x}|$) with law

9. $$P_0(e_1 = -1) : P_0(e_1 = +1) = p_{-2} : p_{+2} ;$$

then the *skew Brownian motion*

10. $$\begin{aligned} \mathfrak{x}^{\bullet}(t) &= e_n\,|\mathfrak{x}(t)| \quad \text{if} \quad t \in \mathfrak{Z}_n, \quad n \geqq 1, \\ &= 0 \quad \text{if} \quad t \in \mathfrak{Z}, \end{aligned}$$

starts afresh at each *constant* time $t \geqq 0$; in addition, its Green operators decompose as in 5, and evaluating $(G^{\bullet}_{\alpha} f)(0)$ as [31]

11. $$\begin{aligned} (G^{\bullet}_{\alpha} f)(0) &= \sum_{n \geqq 1} E_0\left(\int_{\mathfrak{Z}_n} e^{-\alpha t} f(e_n\,|\mathfrak{x}|)\, dt\right) \\ &= \sum_{n \geqq 1}\left(\frac{p_{-2}}{p_{-2} + p_{+2}} E_0\left[\int_{\mathfrak{Z}_n} e^{-\alpha t} f(-|\mathfrak{x}|)\, dt\right]\right. \\ &\qquad\qquad \left. + \frac{p_{+2}}{p_{-2} + p_{+2}} E_0\left[\int_{\mathfrak{Z}_n} e^{-\alpha t} f(+|\mathfrak{x}|)\, dt\right]\right) \\ &= \frac{p_{-2}}{p_{-2} + p_{+2}} E_0\left[\int_0^{+\infty} e^{-\alpha t} f(-|\mathfrak{x}|)\, dt\right] \\ &\qquad\qquad + \frac{p_{+2}}{p_{-2} + p_{+2}} E_0\left[\int_0^{+\infty} e^{-\alpha t} f(+|\mathfrak{x}|)\, dt\right] \\ &= \frac{2p_{-2} \int^{0-} e^{-(2\alpha)^{1/2} l} f(l)\, dl + 2p_{+2} \int_{0+} e^{-(2\alpha)^{1/2} l} f(l)\, dl}{(2\alpha)^{1/2}(p_{-2} + p_{+2})} \\ &= \frac{-p_{-2}(G_{\alpha} f)^{-}(0) + p_{+2}(G_{\alpha} f)^{+}(0)}{(2\alpha)^{1/2}(p_{-2} + p_{+2})}, \end{aligned}$$

[31] $|\mathfrak{Z}| = 0$.

one identifies 10 as the Brownian motion associated with 3 in the special case $p_1 = p_3 = p_4 = 0 \quad (p_{-2}u^-(0) = p_{+2}u^+(0))$.

Coming to the case $p_1 = p_{\pm 2} = p_3 = 0 \quad (p_4(R^1 - 0) = +\infty)$, if $\mathfrak{p}(dt \times dl)$ is a Poisson measure with mean $dt\, p_4(dl)$ independent of the standard Brownian motion $\mathfrak{x}$, if $[l_1^-, l_1^+)$, $[l_2^-, l_2^+)$, etc. are the flat stretches of the inverse function $\mathfrak{p}^{-1}$ of $\mathfrak{p}(t) = \int |l|\, \mathfrak{p}([0, t] \times dl)$, and if $\mathfrak{t}^+$ is the local time at 0 of the (independent) reflecting Brownian motion $\mathfrak{x}^+ = |\mathfrak{x}|$, then the desired motion is

$$\begin{aligned}
12. \qquad \mathfrak{x}^\bullet(t) &= \mathfrak{x}(t) && \text{if } \ t < \mathfrak{m}_0 = \min(t : \mathfrak{x} = 0), \\
&= \pm[\mathfrak{p}\mathfrak{p}^{-1}\mathfrak{t}^+ - \mathfrak{t}^+ + \mathfrak{x}^+] && \text{if } \ t \in \mathfrak{Q}^+, \\
&= 0 && \text{if } \ \mathfrak{m}_0 \leqq t \notin \mathfrak{Q}^+,
\end{aligned}$$

where $\mathfrak{Q}^+ = \bigcup_{n \geq 1} [\mathfrak{t}^{-1}(l_n^-), \mathfrak{t}^{-1}(l_n^-))$, and the ambiguous sign in the second line is *positive* during the interval $[\mathfrak{t}^{-1}(l_n^-), \mathfrak{t}^{-1}(l_n^+))$ if $l_n = l_n^+ - l_n^-$ is a jump of $\mathfrak{p}(dt \times dl \cap (0, +\infty])$ and *negative* otherwise (see Diagram 2).

Granting that 12 is simple Markov (the proof is left to the reader), it is enough for its identification to evaluate[32]

$$\begin{aligned}
13. \quad (G_\alpha^\bullet f)(0) &= \sum_{n \geq 1} E_0\left(\int_{\mathfrak{t}^{-1}(l_n^-)}^{\mathfrak{t}^{-1}(l_n^+)} e^{-\alpha t} f[\pm(l_n^+ - \mathfrak{t}^+ + \mathfrak{x}^+)]\, dt\right) \\
&= \sum_{n \geq 1} E_0[e^{-(2\alpha)^{1/2} l_n^-} (G_\alpha^- f)(\pm l_n)] \\
&= E_0\left[\int_0^{+\infty} \int_{R^1 - 0} \mathfrak{p}(dt \times dl) e^{-(2\alpha)^{1/2}\mathfrak{p}(t-)} (G_\alpha^- f)(l)\right] \\
&= \int_{|l|>0} (G_\alpha^- f)(l) p_4(dl) \Big/ \int_{|l|>0} (1 - e^{-(2\alpha)^{1/2} l}) p_4(dl)
\end{aligned}$$

with the aid of the tricks developed in Section 15.

Coming to the case $p_1 = p_3 = 0$, it suffices to combine the special cases $p_1 = p_3 = p_4 = 0$ and $p_1 = p_{\pm 2} = p_3 = 0$ as follows.

Given $\mathfrak{p}(dt \times dl)$, $\mathfrak{x}$, and $\mathfrak{t}^+$ as above, if $\mathfrak{x}_2^\bullet$ is the skew Brownian motion based upon $p_{\pm 2}$ and $\mathfrak{x}$, if $\mathfrak{x}_4^\bullet$ is the motion of 12 based upon $\mathfrak{p}^\bullet(t) = \int |l|\, \mathfrak{p}([0, t] \times dl)$ and $\mathfrak{x}$, if $[l_1^-, l_1^+)$, $[l_2^-, l_2^+)$, etc. are the flat stretches of the inverse function of $\mathfrak{p} = p_2 t + \mathfrak{p}^\bullet \quad (p_2 = p_{-2} + p_{+2})$, and if $\mathfrak{Q}^+ = \bigcup_{n \geq 1} [\mathfrak{t}^{-1}(l_n^-), t^{-1}(l_n^+))$, then the desired motion is

$$\begin{aligned}
14. \qquad \mathfrak{x}^\bullet(t) &= \mathfrak{x}(t) && \text{if } \ t < \mathfrak{m}_0, \\
&= \mathfrak{x}_4^\bullet(t^\bullet) && \text{if } \ t \in \mathfrak{Q}^+, \ t^\bullet = |\mathfrak{Q}^+ \cap [0, t)|, \\
&= \mathfrak{x}_2^\bullet(t) && \text{if } \ t \in [\mathfrak{m}_0, +\infty) - \mathfrak{Q}^+;
\end{aligned}$$

the reader will check that this sample path starts afresh at each *constant*

[32] $|[0, +\infty) - \mathfrak{Q}^+| = 0$ because $\mathfrak{p}(t)$ has no linear part ($p_2 t$).

time $t \geqq 0$ and will complete its identification with the aid of

15. $$(G^{\bullet}_{\alpha} f)(0) = E_0\left[\int_0^{+\infty} e^{-\alpha t} f(\mathfrak{x}^{\bullet})\,dt\right]$$
$$= \sum_{n\geqq 1} E_0\left[\int_{\mathfrak{t}^{-1}(l_n^-)}^{\mathfrak{t}^{-1}(l_n^+)} e^{-\alpha t} f[\mathfrak{x}_4^{\bullet}(t)]\,dt\right]$$
$$+ E_0\left[\int_0^{+\infty} e^{-\alpha t} f(\mathfrak{x}_2^{\bullet})\,dt\right] - \sum_{n\geqq 1} E_0\left[\int_{\mathfrak{t}^{-1}(l_n^-)}^{\mathfrak{t}^{-1}(l_n^+)} e^{-\alpha t} f(\mathfrak{x}_2^{\bullet})\,dt\right]$$
$$= \sum_{n\geqq 1} E_0[e^{-\alpha \mathfrak{t}^{-1}(l_n^-)} (G_\alpha^- f)(\pm l_n)]$$
$$+ E_0\left[\int_0^{+\infty} e^{-\alpha t} f(\mathfrak{x}_2^{\bullet})\,dt\right]\left(1 - \sum_{n\geqq 1} E_0[e^{-\alpha \mathfrak{t}^{-1}(l_n^-)} - e^{-\alpha \mathfrak{t}^{-1}(l_n^+)}]\right)$$
$$= E_0\left[\int_0^{+\infty}\int_{R^1 - 0} \mathfrak{p}(dt \times dl) e^{-(2\alpha)^{1/2}\mathfrak{p}(t-)} (G_\alpha^- f)(l)\right]$$
$$+ E_0\left[\int_0^{+\infty} e^{-\alpha t} f(\mathfrak{x}_2^{\bullet})\,dt\right]$$
$$\times \left(1 - E_0\left[\int_0^{+\infty}\int_{R^1 - 0} \mathfrak{p}(dt \times dl) e^{-(2\alpha)^{1/2}\mathfrak{p}(t-)} e^{-(2\alpha)^{1/2}|l|}\right]\right)$$
$$= \frac{-p_{-2}(G_\alpha^- f)^-(0) + p_{+2}(G_\alpha^- f)^+(0) + \int (G_\alpha^- f)(l) p_4(dl)}{(2\alpha)^{1/2} p_2 + \int (1 - e^{-(2\alpha)^{1/2}|l|}) p_4(dl)}.$$

If $p_3 > 0 = p_1$, it is clear that the desired motion is the sample path $\mathfrak{x}^{\bullet}$ of 14 run with the stochastic clock $\mathfrak{f}^{-1}$ inverse to $\mathfrak{f} = t + p_3\,\mathfrak{p}^{-1}\mathfrak{t}^+$ (see Section 14 for the interpretation of $\mathfrak{p}^{-1}\mathfrak{t}^+$ as a local time), while, if $p_1 > 0$ also, the motion $\mathfrak{x}^{\bullet}(\mathfrak{f}^{-1})$ has to be annihilated at time $\mathfrak{m}^{\bullet}_\infty$ with conditional law

16. $$P.(\mathfrak{m}^{\bullet}_\infty > t \mid \mathfrak{x}^{\bullet}(\mathfrak{f}^{-1})) = e^{-p_1 \mathfrak{p}^{-1}\mathfrak{t}^+ \mathfrak{f}^{-1}(t)}.$$

The reader is invited to furnish the proofs.

Brownian motions with the same kind of two-sided barrier can be defined on the unit circle $S^1 = [0, 1)$ as W. Feller [1], [3] pointed out.

Given a standard Brownian motion on $R^1$, its projection onto[33] $S^1 = R^1/Z^1$ is the so-called *standard circular Brownian motion*; its generator is the contraction of $\mathfrak{G} = D^2/2$ to $C^2(S^1)$.

Consider now the general circular Brownian motion with a two-sided barrier at $l = 0$ (i.e., the obvious circular analogue of a Brownian motion with two-sided barrier on $R^1$), and, as before, single out the case

17. $$P^{\bullet}_{\cdot}[\mathfrak{x}^{\bullet}(\mathfrak{e}^{\bullet}) = 0] = 1, \qquad \mathfrak{e}^{\bullet} = \lim_{\varepsilon \downarrow 0} \inf(t: |\mathfrak{x}^{\bullet}| < \varepsilon).$$

[33] $Z^1$ is the integers.

$\mathfrak{G}^{\bullet}$ is the contraction of $\mathfrak{G} = D_2/2$ to[34]

$$18.\quad D(\mathfrak{G}^{\bullet}) = C^{\bullet 2}(S^1) \cap \Big(u : p_1 u(0) + p_{-2} u^-(0) - p_{+2} u^+(0) + p_3(\mathfrak{G}u)(0\pm) = \int [u(l) - u(0)] p_4(dl)\Big)$$

for some nonnegative numbers $p_1$, $p_{\pm 2}$, $p_3$ and some nonnegative mass distribution $p_4(dl)$ subject to

$$19a.\quad p_1 + p_{-2} + p_{+2} + p_3 + \int_0^1 l(1-l) p_4(dl) = 1, \quad p_4(0) = p_4(1) = 0,$$

$$19b.\quad p_4(S^1) = +\infty \quad \text{in case} \quad p_{\pm 2} = p_3 = 0,$$

and an application of 18 to

$$20a.\quad (G^{\bullet}_{\alpha} f)(l) = (G^-_{\alpha} f)(l) + \frac{\sinh (2\alpha)^{1/2} l + \sinh (2\alpha)^{1/2}(1-l)}{\sinh (2\alpha)^{1/2}} (G^{\bullet}_{\alpha} f)(0), \qquad 0 \leqq l < 1,$$

$$20b.\quad (G^-_{\alpha} f)(a) = 2 \int_0^1 G(a, b) f(b)\, db, \qquad 0 \leqq a < 1,$$

$$20c.\quad G(a, b) = G(b, a) = \frac{\sinh (2\alpha)^{1/2} a \sinh (2\alpha)^{1/2}(1-b)}{(2\alpha)^{1/2} \sinh (2\alpha)^{1/2}}, \qquad 0 \leqq a \leqq b < 1,$$

establishes the formula

$$21.\quad (G^{\bullet}_{\alpha} f)(0) = \Big[2p_{-2} \int_0^1 \frac{\sinh (2\alpha)^{1/2}(1-l)}{\sinh (2\alpha)^{1/2}} f(l)\, dl + 2p_{+2} \int_0^1 \frac{\sinh (2\alpha)^{1/2} l}{\sinh (2\alpha)^{1/2}} f(l)\, dl + p_3 f(0) + \int_0^1 (G^-_{\alpha} f)(l) p_4(dl)\Big] \Big/ \Big[p_1 + (2\alpha)^{1/2} \frac{\cosh (2\alpha)^{1/2} - 1}{\sinh (2\alpha)^{1/2}} (p_{-2} + p_{+2}) + \alpha p_3 + \int_0^1 \Big(1 - \frac{\sinh (2\alpha)^{1/2} l + \sinh (2\alpha)^{1/2}(1-l)}{\sinh (2\alpha)^{1/2}}\Big) p_4(dl)\Big].$$

Given a standard circular Brownian motion $\mathfrak{x}$ with local time

$$22.\quad \mathfrak{t}(t) = \lim_{\varepsilon \downarrow 0} (2\varepsilon)^{-1} \text{ measure } (s : |\mathfrak{x}(s)| < \varepsilon,\ s \leqq t)$$

and a (circular) differential process $\mathfrak{p}$ based on $p_{\pm 2}$ and $p_4$, it is possible to build up the circular Brownian sample paths as in the linear case, but a second method suggests itself: *the method of images.*

Consider for this purpose a Brownian motion on $R^1$ with two-sided barriers at the integers having as its generator the contraction of $\mathfrak{G} = D^2/2$ to the class of functions $u \in C(R^1) \cap C^2(R^1 - Z^1)$ such that

[34] $C^{\bullet 2}(S^1) = C(S^1) \cap C^2(S^1 - 0) \cap (u : u''(0-) = u''(0+))$.

23a. $$(\mathfrak{G}u)(n-) = (\mathfrak{G}u)(n+),$$
23b. $$p_1 u(n) + p_{-2}u^-(n) - p_{+2}u^+(n) + p_3(\mathfrak{G}u)(n\pm) = \int_0^1 [u(l+n) - u(n)]p_4(dl)$$

at each integer $n = 0, \pm 1, \pm 2$, etc. (the reader is invited to build up the sample paths for himself). Because the barriers are periodic, the projection of this motion onto $S^1 = R^1/Z^1$ is (simple) Markov, and its identification as the desired circular Brownian motion is immediate.

## 18. Simple Brownian motions

Given a *simple Brownian motion* on $[0, +\infty)$, described as in Section 5 except that *it need not start afresh at nonconstant stopping times*,

1. $$(G_\alpha^\bullet f)(l) = (G_\alpha^- f)(l) + e^{-(2\alpha)^{1/2}l}(G_\alpha^\bullet f)(0+), \qquad l > 0,$$

as will now be proved with a view to the classification of all such Brownian motions.

Given $\alpha > 0$, a nonnegative Borel function $f$, and $t_2 \geqq t_1 \geqq 0$,

2. $$\begin{aligned} E_\cdot^\bullet[e^{-\alpha t_2}(G_\alpha^\bullet f)(\mathfrak{x}^\bullet(t_2)) \mid \mathsf{B}_{t_1}^\bullet] &= e^{-\alpha t_2}E_l^\bullet[(G_\alpha^\bullet f)(\mathfrak{x}^\bullet(t))], \qquad l = \mathfrak{x}^\bullet(t_1), t = t_2 - t_1, \\ &= e^{-\alpha t_2}\int_0^{+\infty} e^{-\alpha s}\, ds\, E_l^\bullet(E_{\mathfrak{x}^\bullet(t)}^\bullet[f(\mathfrak{x}^\bullet(s))]) \\ &= e^{-\alpha t_2}\int_0^{+\infty} e^{-\alpha s}\, ds\, E_l^\bullet[f(\mathfrak{x}^\bullet(t+s))] \\ &= e^{-\alpha t_1}\int_t^{+\infty} e^{-\alpha s}\, ds\, E_l^\bullet[f(\mathfrak{x}^\bullet(s))] \\ &\leqq e^{-\alpha t_1}(G_\alpha^\bullet f)(\mathfrak{x}^\bullet(t_1)), \end{aligned}$$

i.e., $e^{-\alpha t}(G_\alpha^\bullet f)(\mathfrak{x}^\bullet)$ is a (nonnegative) *supermartingale*; as such, it possesses one-sided limits as[35] $t = k2^{-n} \downarrow s$ $(s \geqq 0)$, and it follows that if $l > \varepsilon > 0$ and if $\mathfrak{m}^\bullet$ is the crossing time $\inf(t: \mathfrak{x}^\bullet < \varepsilon)$, then

3. $$\begin{aligned} (G_\alpha^\bullet f)(l) = E_l^\bullet\left[\int_0^{\mathfrak{m}^\bullet} e^{-\alpha t}f(\mathfrak{x}^\bullet)\, dt\right] &+ \lim_{n\uparrow+\infty} \sum_{k\geq 0} E_l^\bullet\Big[(k-1)2^{-n} \leqq \mathfrak{m}^\bullet < k2^{-n}, \\ &\qquad e^{-\alpha k2^{-n}}\int_0^{\mathfrak{m}_\infty^\bullet(w_{k2^{-n}}^+)} e^{-\alpha t}f(\mathfrak{x}^\bullet(t+k2^{-n}))\, dt\Big] \\ = E_l^\bullet\left[\int_0^{\mathfrak{m}^\bullet} e^{-\alpha t}f(\mathfrak{x}^\bullet)\, dt\right] &+ \lim_{n\uparrow+\infty} \sum_{k\geq 0} E_l^\bullet[(k-1)2^{-n} \leqq \mathfrak{m}^\bullet < k2^{-n}, \\ &\qquad e^{-\alpha k2^{-n}}(G_\alpha^\bullet f)(\mathfrak{x}^\bullet(k2^{-n}))] \end{aligned}$$

---

[35] See J. L. Doob [1].

$$= E_l\left[\int_0^{\mathfrak{m}} e^{-\alpha t} f(\mathfrak{x})\,dt\right] + E_l[e^{-\alpha\mathfrak{m}} \lim_{k2^{-n}\downarrow\mathfrak{m}} (G^\bullet_\alpha f)(\mathfrak{x}(k2^{-n}))],$$

where $\mathfrak{x}$ is a standard Brownian motion, $E.$ its expectation, and $\mathfrak{m}$ its passage time $\min(t: \mathfrak{x} = \varepsilon)$.

But, in the standard Brownian case, $\lim_{k2^{-n}\downarrow\mathfrak{m}}(G^\bullet_\alpha f)(\mathfrak{x}(k2^{-n}))$ is measurable over $\mathsf{B}_{\mathfrak{m}+}$ and also independent of $\mathsf{B}_{\mathfrak{m}+}$ (i.e., it is measurable over $\mathsf{B}[\mathfrak{x}(t+\mathfrak{m}): t \geqq 0]$ which is independent of $\mathsf{B}_{\mathfrak{m}+}$ conditional on the *constant* $\mathfrak{x}(\mathfrak{m}) = \varepsilon$); as such, it is constant, and inserting this information back into 3 and letting $\varepsilon \downarrow 0$ establishes

4. $$(G^\bullet_\alpha f)(l) = (G^-_\alpha f)(l) + e^{-(2\alpha)^{1/2} l} \times \text{constant},$$

which implies the existence of $(G^\bullet_\alpha f)(0+)$ and leads at once to 1.

Given a bounded function $f$ on $[0, +\infty)$, continuous apart from a possible jump at $l = 0$, define a new function $\hat{f}$ on $(-1) \cup [0, +\infty)$ as

5. $$\begin{aligned} \hat{f}(l) &= f(0) \quad &\text{if } l = -1, \\ &= f(0+) \quad &\text{if } l = 0, \\ &= f(l) \quad &\text{if } l > 0, \end{aligned}$$

and introduce the new Green operators

6. $$\hat{G}_\alpha \hat{f} = (G^\bullet_\alpha f)^\wedge$$

mapping $C((-1) \cup [0, +\infty))$ into itself.

$\hat{G}_\alpha$ is the Green operator of a *strict* Markov motion on $(-1) \cup [0, +\infty)$ with sample paths $t \to \hat{\mathfrak{x}}(t) = \hat{\mathfrak{x}}(t+) \in (-1) \cup [0, +\infty) \cup \infty$, and $\mathfrak{x}^\bullet$ is identical in law to the projection of $\hat{\mathfrak{x}}$ under the identification $-1 \to 0$, as the reader can check for himself or deduce from the general embedding of D. Ray [1].

One now computes the domain $D(\hat{\mathfrak{G}})$ of the generator $\hat{\mathfrak{G}}$ of this *covering motion* and finds that it is the class of functions

$$u \in C((-1) \cap [0, +\infty)) \cup C^2[0, +\infty)$$

subject to

7a. $$-p_{+2}\, u^+(0) + p_{+3}(\mathfrak{G}u)(0) = \int_{(-1)\cup(0,+\infty)\cup\infty} [u(l) - u(0)]\, p_{+4}(dl),$$

$$p_{+4}(0) = 0 \leqq p_{+2},\, p_{+3},\, p_{+4}(dl)$$

$$p_{+2} + p_{+3} + p_{+4}(-1) + \int_{0+} (l \wedge 1)\, p_{+4}(dl) + p_{+4}(\infty) = 1,$$

7b. $$p_{-3}(\hat{\mathfrak{G}}u)(-1) = \int_{[0,+\infty)\cup\infty} [u(l) - u(-1)]\, p_{-4}(dl),$$

$$p_{-4}(-1) = 0 \leqq p_{-3},\, p_{-4}(dl),$$

$$p_{-3} + p_{-4}[0, +\infty) + p_{-4}(\infty) = 1,$$

where $u(\infty) \equiv 0$.

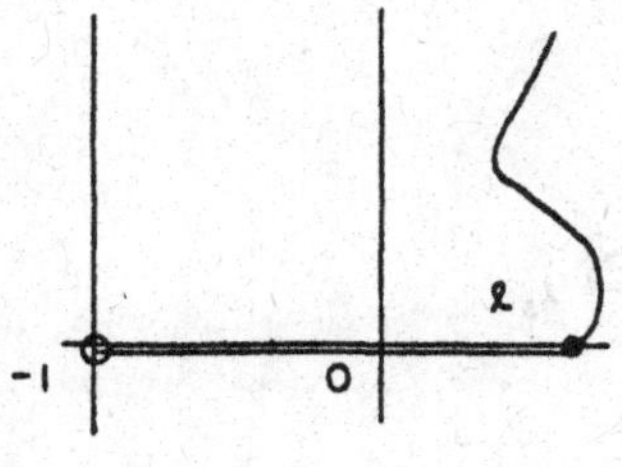


DIAGRAM 1

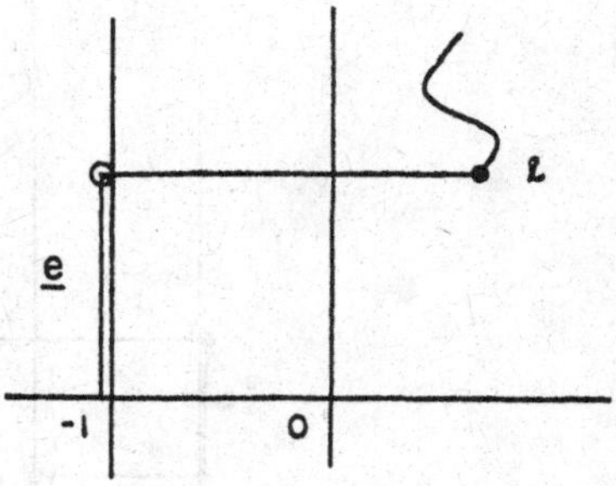


DIAGRAM 2

If $p_{-3} = 0$, the motion starting at $-1$ begins with a jump $l \in [0, +\infty) \cup \infty$ with law $p_{-4}(dl)$ as in Diagram 1, $u(-1) = \int_{[0,+\infty)} u(l)p_{-4}(dl)$, and 7a goes over into

8a. $$p_1^\bullet u(0) - p_2^\bullet u^+(0) + p_3^\bullet(\mathfrak{G}u)(0) = \int_{0+} [u(l) - u(0)]p_4^\bullet(dl),$$

8b. $$p_1^\bullet = p_{+4}(\infty) + p_{+4}(-1)p_{-4}(\infty),$$
$$p_2^\bullet = p_{+2}, \qquad p_3^\bullet = p_{+3},$$
$$p_4^\bullet(dl) = p_{+4}(dl) + p_{+4}(-1)p_{-4}(dl), \qquad l > 0,$$

i.e., the covering motion *does not land at* $-1$ which is a superfluous state, and $\hat{\mathfrak{x}} = \mathfrak{x}^\bullet$ is a *strict Brownian motion on* $[0, +\infty)$ as in Sections 5–16.

If $p_{-3} > 0$, then $\hat{\mathfrak{G}}$ is the contraction of $\mathfrak{G} = D^2/2$ to $D(\hat{\mathfrak{G}})$ with the added specification

9. $$(\hat{\mathfrak{G}}u)(-1) = \int_{[0,+\infty)} [u(l) - u(-1)] \frac{p_{-4}(dl)}{p_{-3}}, \qquad u(\infty) \equiv 0,$$

at $-1$, and the particle starting at $-1$ waits there for an exponential holding time $\mathfrak{e}$ with law $e^{-p_{-4}t/p_{-3}}$ ($p_{-4} = p_{-4}([0, +\infty) \cup \infty)$), and then jumps to $l \in [0, +\infty) \cup \infty$ with law $p_{-4}(dl)/p_{-4}$ as in Diagram 2.

If, in addition to $p_{-3} > 0$, one has $p_2 = 0$ and $p_4(0, +\infty) < +\infty$, then the motion starting at 0 is of the same kind, and it is clear that the projection of this motion down to $[0, +\infty)$ $(-1 \to 0)$ cannot even be *simple* Markov unless $p_{-3} = p_{+3}$ and $p_{-4}(dl) = p_{+4}(dl)$ $(l \neq 0)$ up to a common multiplicative constant, in which case the projection is the Brownian motion associated with

9a. $$p_1 u(0) + p_{+3}(\mathfrak{G}u)(0) = \int_{0+} [u(l) - u(0)]p_{+4}(dl),$$

9b. $$p_1 = p_{+4}(\infty)$$

studied in Section 9.

If $p_{-3} > 0$ and either $p_{+2} > 0$ or $p_{+4}(0, +\infty) = +\infty$, the particle starting at $-1$ waits for an exponential holding time $\mathfrak{e}_1$ and then jumps as in Diagram

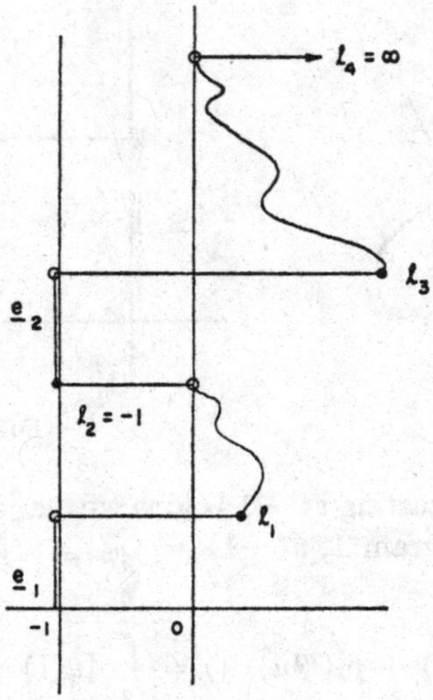


DIAGRAM 3

3 to $l_1 \in [0, +\infty) \cup \infty$ and starts afresh; if $0 \leqq l_1 < +\infty$, the particle performs the Brownian motion on $[0, +\infty)$ associated with

10a. $p_1 u(0) - p_{+2} u^+(0) + p_{+3}(\mathfrak{G}u)(0) = \int_{0+} [u(l) - u(0)]p_{+4}(dl),$

10b. $$p_1 = p_{+4}(-1 \cup \infty)$$

up to the killing time of that motion, at which instant it jumps to $l_2 = \infty$ or $-1$ with probabilities $p_{+4}(\infty) : p_{+4}(-1)$, and, if $l_2 = -1$, it starts afresh as in Diagram 3, while if $l_2 = \infty$, then the motion rests at that place at all later times.

Now the projection $\mathfrak{x}^\bullet$ of this motion onto $[0, +\infty)$ $(-1 \rightarrow 0)$ is simple Markov if the Brownian motion attached to 10 does not spend positive (Lebesgue) time at $l = 0$; otherwise the knowledge that $\mathfrak{x}^\bullet(s) = 0$ is not sufficient to discriminate between the two possible coverings, and the law of $\mathfrak{x}^\bullet(t) : t \geqq s$ is moot. But if $e$ is the indicator of $l = 0$, and if

11. $$\mathfrak{x}^\bullet(\mathfrak{f}^{-1}) \quad (t < \mathfrak{m}_\infty^\bullet), \qquad \infty \quad (t \geqq \mathfrak{m}_\infty^\bullet)$$

12a. $$\mathfrak{f} = t + p_3 \mathfrak{p}^{-1}\mathfrak{t}^+,$$

12b. $$\mathfrak{x}^\bullet = \mathfrak{p}\mathfrak{p}^{-1}\mathfrak{t}^+ - \mathfrak{t}^+ + \mathfrak{x}^+$$

is the motion attached to 10, then, in the notation of Section 14,

13. $$\text{measure}\,(s:\mathfrak{x}^\bullet(\mathfrak{f}^{-1}) = 0, s \leq t) = \int_0^t e[\mathfrak{x}^\bullet(\mathfrak{f}^{-1})]\,ds$$

$$= \int_0^{\mathfrak{f}^{-1}(t)} e(\mathfrak{x}^\bullet)\,\mathfrak{f}\,(ds)$$

$$= p_{+3}\int_{\mathfrak{Z}^+\cap\mathfrak{Q}^+\cap[0,\mathfrak{f}^{-1}(t))} \mathfrak{p}^{-1}\mathfrak{t}^+\,(dt)$$

$$= p_{+3}\mathfrak{p}^{-1}\mathfrak{t}^+[\mathfrak{Q}^+\cap[0,\mathfrak{f}^{-1}(t))], \qquad t \leq \mathfrak{m}_\infty^\bullet,$$

and this cannot be positive unless $p_{+3} > 0$ and $0 < \mathfrak{t}^+(\mathfrak{Q}^+) = |\mathfrak{Q}|$, i.e., unless $p_{+2} > 0$ also; in short, *the projection is simple Markov unless* $p_{+2}\,p_{+3} > 0$, and now the classification is complete.

N. Ikeda had conjectured part of our classification (private communication); the case of a two-sided barrier on $R^1$ is similar except that three covering points lie over 0.

## 19. Feller's differential operators

Given a nonnegative mass distribution $e$ on the open half line $(0, +\infty)$ with $0 < e(a, b]$ $(a < b)$, let $D(\mathfrak{G})$ be the class of functions $u \in C[0, +\infty)$ such that

1. $$u^+(b) - u^+(a) = \int_{(a,b]} f\,de, \qquad a < b,$$

for some $f \in C[0, +\infty)$, and introduce the differential operator $\mathfrak{G}u = f$:

2. $$(\mathfrak{G}u)(a) = \lim_{b\downarrow a}\frac{u^+(b) - u^+(a)}{e(a, b]}.$$

W. Feller [3] proved that if $e(0, 1] < +\infty$, and if $p_1, p_2, p_3, p_4(dl)$ are nonnegative with $p_4(0) = 0$ and $p_1 + p_2 + p_3 + \int_{0+}(l \wedge 1)p_4(dl) = 1$, then the contraction $\mathfrak{G}^\bullet$ of $\mathfrak{G}$ to

3. $$D(\mathfrak{G}^\bullet) = D(\mathfrak{G}) \cap \Big(u: p_1u(0) - p_2u^+(0) + p_3(\mathfrak{G}u)(0) = \int_{0+}[u(l) - u(0)]p_4(dl)\Big)$$

is the generator of a strict Markov motion (diffusion) on $[0, +\infty)$.

Given a reflecting Brownian motion $\mathfrak{x}^+$ on $[0, +\infty)$, the *local time*

4. $$\mathfrak{t}^+(t, l) = (\text{measure}\,(s:\mathfrak{x}^+(s) \in dl, s < t))/2\,dl$$

is continuous in the pair $(t, l) \in [0, +\infty)^2$ (see H. Trotter [1]), and the motion associated with $\mathfrak{G}^\bullet$ in the special case $p_1 = p_3 = p_4 = 0$ $(u^+(0) = 0$ is identical in law to $\mathfrak{x}^\bullet = \mathfrak{x}^+(\mathfrak{f}^{-1})$ where $\mathfrak{f} = \int_{0+}\mathfrak{t}^+(t, l)e(dl)$ (see V. A. Volkonskii [1] and K. Itô and H. P. McKean, Jr. [1]).

Because $\mathfrak{t}^+(dt, l) = 0$ outside $\mathfrak{Z} = (t : \mathfrak{x}^+ = l)$,

5. $$\int_0^t f(\mathfrak{x}^\bullet)\, ds = \int_0^{\mathfrak{f}^{-1}(t)} f(\mathfrak{x}^+) \int_{0+} \mathfrak{t}^+(ds, l) e(dl)$$
$$= \int_{0+} \left( \int_0^{\mathfrak{f}^{-1}(t)} \mathfrak{t}^+(ds, l) \right) f(l) e(dl)$$
$$= \int_{0+} \mathfrak{t}^+[\mathfrak{f}^{-1}(t), l] f(l) e(dl);$$

hence the *local time*

6. $$\mathfrak{t}^\bullet(t) = \lim_{\varepsilon \downarrow 0} e(0, \varepsilon]^{-1} \text{ measure } (s : \mathfrak{x}^\bullet(s) < \varepsilon,\ s \leqq t)$$
$$= \mathfrak{t}^+(\mathfrak{f}^{-1}, 0)$$

exists, and now it is clear that the discussion of the Brownian case can be adapted with little change.

## 20. Birth and death processes

Quite a general birth and death process on the nonnegative integers can be changed via a scale substitution into a motion on a discrete series $Q : 0 = l_0 < l_1 < l_2 < \cdots < 1$ having as its generator

1. $$\mathfrak{G}^\bullet u = (u^+ - u^-)/e,$$

2a. $$u^+(l_n) = u^-(l_{n+1}) = (l_{n+1} - l_n)^{-1}[u(l_{n+1}) - u(l_n)],$$

2b. $$e = e(l_n) > 0,$$

2c. $$e(l_0) + e(l_1) + \cdots < +\infty,$$

subject to

3a. $$u^+(0) = 0,$$

3b. $$p_1 u(1) + p_3(\mathfrak{G}^\bullet u)(1) = -p_2 u^-(1) + \int_Q [u(l) - u(1)] p_4(dl),$$
$$p_1 + p_2 + p_3 + \int_Q (1 - l) p_4(dl) = 1$$

(see W. Feller [4]). In the special case $p_1 = p_3 = p_4 = 0$ the corresponding motion is just the reflecting Brownian motion on [0, 1] run with the inverse function of $\mathfrak{f} = \int_Q \mathfrak{t}^+(t, l) e(dl)$, $\mathfrak{t}^+$ being the reflecting Brownian local time. Once this motion has been obtained, the general path can be built up using local times and differential processes as before.

### References

J. L. Doob

1. *Stochastic processes*, New York, Wiley, 1953.

E. B. Dynkin

1. *Infinitesimal operators of Markov processes*, Teor. Veroyatnost. i Primenen., vol. 1 (1956), pp. 38–60 (in Russian with English summary).

W. FELLER

1. *The parabolic differential equations and the associated semi-groups of transformations*, Ann. of Math. (2), vol. 55 (1952), pp. 468–519.
2. *Diffusion processes in one dimension*, Trans. Amer. Math. Soc., vol. 77 (1954), pp. 1–31.
3. *Generalized second order differential operators and their lateral conditions*, Illinois J. Math., vol. 1 (1957), pp. 459–504.
4. *The birth and death processes as diffusion processes*, J. Math. Pures Appl. (9), vol. 38 (1959), pp. 301–345.

G. A. HUNT

1. *Some theorems concerning Brownian motion*, Trans. Amer. Math. Soc., vol. 81 (1956), pp. 294–319.

K. ITÔ

1. *On stochastic processes (I) (Infinitely divisible laws of probability)*, Jap. J. Math., vol. 18 (1942), pp. 261–301.

K. ITÔ AND H. P. MCKEAN, JR.

1. *Diffusion*, to appear.

M. KAC

1. *On some connections between probability theory and differential and integral equations*, Proceedings of the Second Berkeley Symposium on Mathematical Statistics and Probability, 1950, pp. 189–215, University of California Press, 1951.

P. LÉVY

1. *Sur les intégrales dont les éléments sont des variables aléatoires indépendantes*, Ann. Scuola Norm. Sup. Pisa (2), vol. 3 (1934), pp. 337–366.
2. *Sur certains processus stochastiques homogènes*, Compositio Math., vol. 7 (1940), pp. 283–339.
3. *Processus stochastiques et mouvement brownien; suivi d'une note de M. Loève*, Paris, Gauthier-Villars, 1948.

D. RAY

1. *Resolvents, transition functions, and strongly Markovian processes*, Ann. of Math. (2), vol. 70 (1959), pp. 43–72.

A. V. SKOROKHOD

1. *Stochastic equations for diffusion processes in a bounded region I and II*, Teor. Veroyatnost. i Primenen., vol. 6 (1961), pp. 287–298 and vol. 7 (1962), pp. 5–25 (in Russian with English summary).

H. TROTTER

1. *A property of Brownian motion paths*, Illinois J. Math., vol. 2 (1958), pp. 425–433.

A. D. VENTSELL

1. *Semi-groups of operators corresponding to a generalized differential operator of second order*, Dokl. Akad. Nauk SSSR (N.S.), vol. 111 (1956), pp. 269–272 (in Russian).

V. A. VOLKONSKII

1. *Random substitution of time in strong Markov processes*, Teor. Veroyatnost. i Primenen., vol. 3 (1958), pp. 332–350 (in Russian with English summary).

N. WIENER

1. *Differential-space*, J. Math. Phys., vol. 2 (1923), pp. 131–174.

KYÔTO UNIVERSITY
KYÔTO, JAPAN
MASSACHUSETTS INSTITUTE OF TECHNOLOGY
CAMBRIDGE, MASSACHUSETTS

J. Math. Kyoto Univ.
3-2 (1964) 207-216.

# The expected number of zeros of continuous stationary Gaussian processes*

Dedicated to Professor Y. Akizuki for his sixtieth birthday

By

Kiyosi ITO

(Reccived Jan. 16, 1964)

## 1. Introduction

Let $x(t)=x(t, \omega)$, $\omega \in \Omega(\mathcal{B}, P)$ be a stationary Gaussian process with continuous sample paths. Then the mean $a=E[x(t)]$ is independent of $t$ and the covariance function $r(t)=E[(x(s+t)-a)(x(s)-a)]$ is an even function of $t$, independent of $s$, expressible in the form

$$r(t) = \int_{-\infty}^{\infty} e^{i\lambda t} dF(\lambda) \tag{1}$$

with a bounded measure $dF$ symmetric with respect to 0.

Let $N=N(\omega)$ be the number of zeros of the sample path of $x(t)$ in $0 \leq t \leq T$ and $N_c=N_c(\omega)$ the number of crossings of the level $c$ by the sample path of $x(t, \omega)$.

The purpose of this paper is to prove

**Theorem.**

$$E(N) = E(N_c) = \frac{T}{\pi}\sqrt{-\frac{r''(0)}{r(0)}} \exp\left(-\frac{a^2}{2r(0)}\right) \tag{2}$$

*where $r''(0)$ is the second Schwarz derivative*, i.e.,

* This work was supported in part by the National Science Foundation, Grant 16319, and in part by the Office of Naval Research, Contract Nonr 225 (28) at Stanford University.

$$(3)\qquad r''(0)=\lim_{h\downarrow 0}\frac{r(h)-2r(0)+r(-h)}{h^2}\qquad \left(=-\int_{-\infty}^{\infty}\lambda^2 dF(\lambda)\right).$$

$r''(0)$ is finite or $-\infty$ according as $\int_{-\infty}^{\infty}\lambda^2 dF(\lambda)<\infty$ or $=\infty$; in the latter case (2) shows that $E(N)=E(N_c)=\infty$.

Given a constant $a$ and a function $r(t)$ in the form (1), there exists essentially a unique separable stationary process with the mean 0 and the covariance fouction $r(t)$. If Hunt's condition [2]:

$$(4)\qquad \int_{-\infty}^{\infty}[\log(1+|\lambda|)]^{1+\alpha}dF(\lambda)<\infty \qquad \text{for some}\quad \alpha>0,$$

is fulfilled, then almost all sample functions of $x(t,\omega)$ are continuous and so (2) holds.

Let us give a historical account of the formula (2).

In 1944 S. O. Rice [1] (see pp. 271-3) proved (2) in case $F(\lambda)$ increases only with a finite number of jumps, i.e.,

$$(5)\qquad r(t)=\sum_{n=1}^{m}e^{i\lambda_n t}\sigma_n\,.$$

Noticing that $(x(t), x'(t))$ is Gauss distributed with

$$(6)\qquad \begin{cases} E(x(t))=a\,, & E(x'(t))=0\\ E(x(t)^2)=r(0)\,, & E(x'(t)^2)=-r''(0)\\ \qquad E(x(t)x'(t))=r'(0)=0 & \end{cases}$$

we have

$$(7)\qquad E(N)=\int_0^T E(N(dt))\qquad (N(dt)=\text{the number of zeros of } x(t) \text{ in } (t,t+dt))$$

$$\sim\int_0^T P(N(dt)=1)$$

$$\sim\int_0^T\int_0^\infty P(x'(t)\in d\eta,\ -\eta dt<x(t)<0)$$

$$+\int_0^T\int_{-\infty}^0 P(x'(t)\in d\eta,\ 0<x(t)<\eta dt)$$

$$=2\int_0^T\int_0^\infty\frac{1}{2\pi\sqrt{-r''(0)r(0)}}\exp\left\{-\frac{(0-a)^2}{2r(0)}-\frac{\eta^2}{2(-r''(0))}\right\}\eta\,d\eta\,dt$$

$$=\frac{T}{\pi}\sqrt{-\frac{r''(0)}{r(0)}}\exp\left\{-\frac{a^2}{2r(0)}\right\}.$$

This is the outline of Rice's proof.

In 1957 U. Grenander and M. Rosenblatt [3] (see p. 271–3) gave a neater proof to the formula in the same special cace using the following formula on the number $N$ of zeros of a function $f(t)$ due to M. Kac [4]:

$$(8) \qquad N \sim \frac{1}{2\varepsilon}\int_0^T |f'(t)| e_{(-\varepsilon,\varepsilon)}(f(t))\,dt$$

where $e_{(-\varepsilon,\varepsilon)}$ is the indicator function of the interval $(-\varepsilon, \varepsilon)$.

There were some technical gaps in both the proof of Rice and that of Grenander-Rosenblatt in connection with the estimation and interchangeability of the limit and the expectation.

In 1960 V. A. Ivanov [5] gave the first rigorous proof to the formula in cace

$$(9) \qquad \int_{-\infty}^{\infty} \lambda^4 dF(\lambda) < \infty$$

carrying out the argument of Grenander-Rosenblatt carefully.

On 1961 E. V. Bulinskaya [6] proved it in a more general case

$$(10) \qquad \int_{-\infty}^{\infty} \log(1+|\lambda|)^{1+\alpha}\lambda^2 dF(\lambda) < \infty \qquad \text{for some} \quad \alpha > 0\,.$$

She used a lower estimate of the number of zeros of $f(t)$:

$$(11) \qquad N \sim \lim_{p\to\infty} \sum_{q=1}^{2^p} c\left(f\left(\frac{(q-1)T}{2^p}\right), f\left(\frac{qT}{2^p}\right)\right)$$

$$(c(\xi, \eta) = 1 \text{ or } 0 \text{ according as } \xi\eta < 0 \text{ or } \geq 0)$$

in addition to the estimate (8) which is actually an upper one.

Both Ivanov and Bulinskaya imposed the conditions ((9), (10)) to let almost all sample functions be continuously differential, but a slight modification of their methods will enable us to prove the formula in complete generality, as we can see in our paper.

We wish to express our thanks to Professor J. Chover who gave us a lot of information on this subject.

## 2. The upper and lower bounds of the number of zeros

Let $n$ denote the number of zeros of $f(t)$ in $0 < t < T$ and $n_c$ the number of crossings of the level $c$ by $f(t)$. The numbers $n$ and $n_c$ may be finite but $n \geq n_c$ is clear in any case.

* and vanishes only on a t-set which contains no interval

The following lemmas will be used later.

**Lemma 1.** *If $f(t)$ is continuous,* then*

$$(1)\quad n \geq n_c \geq \lim_{p\to\infty} \sum_{q=1}^{2^p} c\left(f\left(\frac{(q-1)T}{2^p}\right), f\left(\frac{qT}{2^p}\right)\right) \quad \textit{(monotone limit)}$$

*where $c(\xi, \eta)$ is the indicator function of the two-dimensional region: $\xi\eta < 0$.*

**Lemma 2.** *If $f(t)$ is absolutely continuous and vanishes only on a $t$-set which contains no interval, then*

$$(2)\quad n_c \leq n \leq \varliminf_{\varepsilon\downarrow 0} \frac{1}{2\varepsilon}\int_0^T |f'(t)| e_{(-\varepsilon,\varepsilon)}(f(t))dt\,,$$

*where $e_{(-\varepsilon,\varepsilon)}(\xi)$ is the indicator function of the interval $(-\varepsilon, \varepsilon)$.*

*Proof.* Lemma 1 is evident by the intermediate value theorem for continuous functions.

To prove Lemma 2, it is enough to show that

$$(2)\quad \varliminf_{\varepsilon\downarrow 0} \frac{1}{2\varepsilon}\int_0^T |f'(t)| e_{(-\varepsilon,\varepsilon)}(f(t))dt \geq m$$

for any finite number $m \leq n$. If $m \leq n$, we can take a sequence of zeros $t_1 < t_2 < \cdots < t_m$ of $f(t)$ in $0 < t < T$. Let $u_i = u_i(\varepsilon)$ be the greatest number $u < t_i$ with $|f(u)| = \varepsilon$ ($u_i = -\infty$ if there exists no such $u$) and $v_i = v_i(\varepsilon)$ be the least number $v \nless t_i$ with $|f(v)| = \varepsilon$ ($v_i = \infty$ if there exists no such $v$). Then we have some $\varepsilon_0 > 0$ such that, for any positive $\varepsilon < \varepsilon_0$,

$$(3)\quad 0 < u_1 < t_1 < v_1 < u_2 < t_2 < v_2 < \cdots < u_m < t_m < v_m < T;$$

if otherwise, $f(t)$ would vanish on a subinterval of $(0, T)$ in contradiction with our assumption. It follows from (3) that

$$\begin{aligned}\frac{1}{2\varepsilon}&\int_0^T |f'(t)| e_{(-\varepsilon,\varepsilon)}(f(t))dt \\ &\geq \frac{1}{2\varepsilon}\sum_{i=1}^m \left[\int_{u_i}^{t_i} |f'(t)|dt + \int_{t_i}^{v_i} |f'(t)|dt\right] \\ &\geq \frac{1}{2\varepsilon}\sum_{i=1}^m \left[|f(t_i) - f(u_i)| + |f(v_i) - f(t_i)|\right] = m\,,\end{aligned}$$

which completes the proof.

**3. Proof of Theorem in case $\int_{-\infty}^{\infty} \lambda^2 dF(\lambda) < \infty$.**

We shall consider a stationary Gaussian process described in the introduction and assume that

$$(1) \qquad \int_{-\infty}^{\infty} \lambda^2 dF(\lambda) < \infty .$$

We shall prove several preliminary propositions.

*Proposition* 1. *$x(t)$ is differentiable at every time point $t$ in the sense of limit in the mean, i.e., limit with respect to the norm in $L^2(\Omega, \mathfrak{B}, P)$.*

By our assumption we have

$$\left\| \frac{x(t+h)-x(t)}{h} - \frac{x(t+k)-x(t)}{k} \right\|^2$$

$$= \int_{-\infty}^{\infty} \left| \frac{e^{i\lambda h}-1}{h} - \frac{e^{i\lambda k}-1}{k} \right|^2 dF(\lambda) \to 0 \qquad (h, k \to 0)$$

since the integrand tends to 0 and is bounded by $4\lambda^2$.

*Proposition* 2. *$x(t)$, $x'(t)$, $t \in [-\infty, \infty)$ form a Gaussian system with*

$$(2) \quad \begin{cases} E(x(t)) = n, \qquad E(x'(t)) = 0 \\ E((x(t)-a)(x(s)-a)) = r(t-s) = \int e^{i\lambda(t-s)} dF(\lambda) \\ E(x'(t)(x(s)-a)) = r'(t-s) = \int e^{i\lambda(t-s)} i\lambda dF(\lambda) \\ E(x'(t)x'(s)) = -r''(t-s) = \int e^{i\lambda(t-s)} \lambda^2 dF(\lambda) . \end{cases}$$

This is clear, because $x(t)$, $t \in (-\infty, \infty)$ is a Gaussian system and $x'(t)$ belongs to the closed linear subspace of $L^2(\Omega, \mathfrak{B}, P)$ spanned by $x(t)$, $t \in (-\infty, \infty)$.

*Proposition* 3. *$x'(t)$ has a measurable version, namely we can construct a function $y(t, \omega)$ measurable in $(t, \omega)$ such that $P(x'(t) = y(t))$ $=1$ for every $t$.*

Write $y_h(t)$ for $h^{-1}(x(t+h)-x(t))$. $y_h(t, \omega)$ is measurable in $(t, \omega)$

because it is measurable in $\omega$ for each value of $t$ and continuous in $t$ for almost every value of $\omega$. It follows from Proposition 2 that

$$\|y_h(t)-x'(t)\|^2 = \int_{-\infty}^{\infty} \left|\frac{e^{i\lambda h}-1}{h}-i\lambda\right|^2 dF(\lambda) \leq \frac{h^2}{4}\int_{-\infty}^{\infty}\lambda^2 dF(\lambda).$$

Writing $y_n(t)$ for $y_{2^{-n}}(t)$, we have, for every $t$,

$$\|y_n(t)-x'(t)\|^2 \leq 2^{-2n-2}\int_{-\infty}^{\infty}\lambda^2 dF(\lambda)$$

and so

$$P(y_n(t)\to x'(t) \quad \text{as} \quad n\to\infty)=1,$$

i.e.

$$P(\overline{\lim_{n\to\infty}}\, y_n(t)=x'(t))=1.$$

Thus $y(t)\equiv\overline{\lim_{n\to\infty}}\, y_n(t)$ is what we wanted to construct.

From now on we shall denote the measurable version $y(t)$ of $x'(t)$ by $x'(t)$ itself.

*Proposition* 4. *For almost all* $\omega$, $x(t)$ *is absolutely continuous in* $t$ *and* $dx(t)/dt=x'(t)$ *for almost all* $t$.

Noticing that $x'(t)$ and $x(t)$ are measurable in $(t,\omega)$, we have, by Proposition 2 and simple computations,

$$\left\|\int_s^t x'(\theta)d\theta-(x(t)-x(s))\right\|^2=0$$

and so

$$P\left(\int_s^t x'(\theta)d\theta = x(t)-x(s)\right)=1$$

for every pair $(s,t)$ fixed. Therefore

$$P\left(\int_s^t x'(\theta)d\theta = x(t)-x(s) \quad \text{for rational} \quad t,s\right)=1.$$

Since $\int_s^t x'(\theta)d\theta$ and $x(t)-x(s)$ are both continuous in $t, s$, we have

$$P\left(\int_s^t x'(\theta)d\theta = x(t)-x(s) \quad \text{for real} \quad t,s\right)=1$$

which proves our proposition.

*Proposition* 5. *For almost all* $\omega$, $x(t)=0$ *only on a t-set of Lebesgue measure* 0 *in* $0<t<T$.

Let $e_0(\xi)$ be the indicator of the single point 0. Then $e_0(x(t,\omega))$ is measurable in $(t,\omega)$, $\int_0^T e_0(x(t,\omega))\,dt$ is the Lebesgue measures of the $t$-set: $x(t,\omega)=0$ and

$$E\left(\int_0^T e_0(x(t,\omega))\,dt\right) = \int_0^T E(e_0(x(t,\omega)))\,dt = \int_0^T P(x(t)=0)\,dt = 0\,,$$

because $x(t,\omega)$ is Gauss distributed.

By Proposition 4 and 5 we can apply Lemma 2 in Section 2 to the sample functions of our process.

$$\begin{aligned} E(N_c) &\leq E(N) \\ &\leq E\left(\lim_{\varepsilon\downarrow 0}\frac{1}{2\varepsilon}\int_0^T |x'(t)|\,e_{(-\varepsilon,\varepsilon)}(x(t))\,dt\right) \\ &\leq \lim_{\varepsilon\downarrow 0}\frac{1}{2\varepsilon}\int_0^T E[\,|x'(t)|\,e_{(-\varepsilon,\varepsilon)}(x(t))]\,dt \\ &\qquad\qquad \text{(by Lebesgue-Fatou's theorem)} \\ &= \lim_{\varepsilon\downarrow 0}\frac{T}{2\varepsilon}E[\,|x'(0)|\,e_{(-\varepsilon,\varepsilon)}(x(0))]\,. \end{aligned}$$

By Proposition 2, $(x(0), x'(0))$ is Gauss distributed with

$$\begin{aligned} E(x(0)) &= a\,, & E(x'(0)) &= 0 \\ E((x(0)-a)^2) &= r(0)\,, & E(x'(0)^2) &= -r''(0) \end{aligned}$$

$$E(x'(0)(x(0)-a)) = r'(0) = \int_{-\infty}^{\infty} i\lambda\, dF(\lambda) = 0\,,$$

and so we have

$$\begin{aligned} &E[\,|x'(0)|\,e_{(-\varepsilon,\varepsilon)}(x(0))] \\ &= \frac{1}{2\pi\sqrt{-r''(0)r(0)}}\int\!\!\int_{R^2}\exp\left[-\frac{(\xi-a)^2}{2r(0)}-\frac{\eta^2}{-2r''(0)}\right]|\eta|\,e_{(-\varepsilon,\varepsilon)}(\xi)\,d\xi\, d\eta \\ &= \frac{1}{\pi}\left(-\frac{r''(0)}{r(0)}\right)^{1/2}\int_{-\varepsilon}^{\varepsilon}\exp\left[-\frac{(\xi-a)^2}{2r(0)}\right]d\xi\,. \end{aligned}$$

Thus we have

$$(3) \qquad E(N_c) \leq E(N) \leq \frac{T}{\pi}\sqrt{-\frac{r''(0)}{r(0)}}\exp\left[-\frac{a^2}{2r(0)}\right].$$

By Lemma 1 in Section 2, we have

$$(4)\qquad E(N)\geq E(N_c)\geq E\left[\lim_{p\to\infty}\sum_{q=1}^{2^p} c\left(x\left(\frac{(q-1)T}{2^p}\right),\ x\left(\frac{qT}{2^p}\right)\right)\right]$$

$$=\lim_{p\to\infty} E\left[\sum_{q=1}^{2^p} c\left(x\left(\frac{(q-1)T}{2^p}\right),\ x\left(\frac{qT}{2^p}\right)\right)\right]$$

$$=\lim_{p\to\infty} T\cdot\frac{2^p}{T}E[c(x(0),\ x(2^{-p}T))].$$

If we can prove

$$(5)\qquad \varliminf_{t\downarrow 0}\frac{1}{t}E[c(x(0),\ x(t))]\geq\frac{1}{\pi}\sqrt{-\frac{r''(0)}{r(0)}}\exp\left[-\frac{a^2}{2r(0)}\right]$$

then it follows from (4) that

$$(6)\qquad E(N)\geq E(N_c)\geq\frac{T}{\pi}\sqrt{-\frac{r''(0)}{r(0)}}\exp\left[-\frac{a^2}{2r(0)}\right]$$

which, combined with (3), completes the proof of our theorem in case $\int_{-\infty}^{\infty}\lambda^2 dF(\lambda)<\infty$.

By Proposition 2, $(x(0),\ x(t))$ is Gauss distributed with

$$E(x(0))=E(x(t))=a$$

$$E((x(0)-a)^2)=E((x(t)-a)^2)=r(0)\qquad(=\alpha)$$

$$E((x(0)-a)(x(t)-a))=r(t)\qquad(=\beta)$$

and so

$$\frac{1}{t}E[c(x(0),\ x(t))]$$

$$=\frac{1}{t}\iint_{\xi\eta<0}\frac{1}{2\pi\sqrt{\alpha^2-\beta^2}}\exp\left[-\frac{\alpha((\xi-a)^2+(\eta-a)^2)-2\beta(\xi-a)(\eta-a)}{2(\alpha^2-\beta^2)}\right]d\xi\, d\eta$$

$$=e^{-a^2/\alpha+\beta}\frac{1}{t}\iint_{\xi\eta<0}\frac{1}{2\pi\sqrt{\alpha^2-\beta^2}}\exp\left[-\frac{\alpha(\xi^2+\eta^2)-2\beta\xi\eta}{2(\alpha^2-\beta^2)}+\frac{(\xi+\eta)a}{\alpha+\beta}\right]d\xi\, d\eta$$

$$=e^{-a^2/\gamma}\frac{1}{t}\iint_{|v/u|<\sqrt{\delta/\gamma}}\frac{1}{2\pi}\exp\left[-\frac{u^2+v^2}{2}+\frac{\sqrt{2}\,va}{\sqrt{\gamma}}\right]du\, dv$$

$$\left(\gamma=\alpha+\beta,\ \delta=\alpha-\beta,\ u=\frac{\xi-\eta}{\sqrt{2\delta}},\ v=\frac{\xi+\eta}{\sqrt{2\gamma}}\right)$$

$$=e^{-a^2/\gamma}\frac{1}{2\pi}\int_0^{\infty}e^{-\rho^2/2}\rho\, d\rho\,\frac{1}{t}\iint_{|\tan\theta|<-\sqrt{\delta/\gamma}}\exp\left[\frac{\sqrt{2}\,\rho\sin\theta\, a}{\sqrt{\gamma}}\right]d\theta\,.$$

Since $|\sin\theta|\leq|\tan\theta|$, we have

$$\frac{1}{t}E[c(x(0),x(t))]$$
$$\geq \exp(-a^2/\gamma)\frac{1}{2\pi}\int_0^\infty e^{-\rho^2/2}\rho\,d\rho\,\exp\left(-\frac{\sqrt{2\delta}\,\rho a}{\gamma}\right)\frac{4}{t}\arctan\sqrt{\frac{\delta}{\gamma}}\,.$$

As $t\to 0$, we have

$$\gamma\to 2r(0)\,,\qquad \delta\to 0\,,$$

$$\frac{1}{t}\arctan\sqrt{\frac{\delta}{\gamma}}\sim\frac{1}{t}\sqrt{\frac{\delta}{\gamma}}=\sqrt{\frac{r(0)-r(t)}{(r(0)+r(t))t^2}}$$
$$=\sqrt{\frac{2r(0)-r(t)-r(-t)}{t^2}}\sqrt{\frac{1}{2(r(0)+r(t))}}$$
$$(\text{notice } r(-t)=r(t))$$
$$\to\frac{1}{2}\sqrt{\frac{-r''(0)}{r(0)}}\,,$$

and so

$$\lim_{t\downarrow 0}\frac{1}{t}E[c(x(0),x(t))]\geq\exp\left(-\frac{a^2}{2r(0)}\right)\frac{1}{\pi}\sqrt{\frac{-r''(0)}{r(0)}}\int_0^\infty\exp(-\rho^2/2)\rho\,d\rho$$
$$=\frac{1}{\pi}\sqrt{\frac{-r''(0)}{r(0)}}\exp\left(-\frac{a^2}{2r(0)}\right).$$

**4. Proof of Theorem in case $\int_{-\infty}^{\infty}\lambda^2 dF(\lambda)=\infty$**

In this case

$$\lim_{h\downarrow 0}\frac{r(h)-2r(0)+r(-h)}{h^2}=-\int_{-\infty}^{\infty}\lambda^2 dF(\lambda)=-\infty\,.$$

Therefore it is enough to prove

$$E(N)=E(N_c)=\infty\,.$$

By virtue of the continuity of the sample paths, we can apply Lemma 1 to get

$$N\geq N_c\geq\lim_{p\to\infty}\sum_{q=1}^{2^p}c\left(x\left(\frac{(q-1)T}{2^p}\right),\,x\left(\frac{qT}{2^p}\right)\right)$$

and so, using the same arguments as in the last part of Section 3, we get

$$E(N)\geq E(N_c)\geq E\left[\lim_{p\to\infty}\sum_{q=1}^{2^p} c\left(x\left(\frac{(q-1)T}{2^p}\right), x\left(\frac{qT}{2^p}\right)\right)\right]$$

$$= \lim_{p\to\infty} T\cdot\frac{2^p}{T}E[c(x(0), x(2^{-p}T))]$$

$$\geq \frac{2T}{\pi}\lim_{t\downarrow 0}\sqrt{\frac{2r(0)-r(t)-r(-t)}{t^2}}\sqrt{\frac{1}{2(r(0)+r(t))}} = \infty .$$

Department of Mathematics, Kyoto University
and
Department of Mathematics, Stanford University

## REFERENCES

[1] S. O. Rice : Mathematical analysis of random noise. Bell System Tech J. **24** (1945), 46–156.

[2] G. Hunt : Random Fourier transforms. Trans. Amer. Math. Soc. **71** (1956), 38–69.

[3] U. Grenander and M. Rosenblatt : Statistical analysis of stationary time series. New York, 1957.

[4] M. Kac : On the average number of real roots of a random algebraic equation. Bull. Amer. Math. Soc. **49** (1943), 314–319.

[5] V. A. Ivanov : On the average number of crossings of a level by sample functions of a stochastic processes. Th. Prob. Appl. **5** (1960), 319–323 (English translation).

[6] E. V. Bulinskaya : On the mean number of crossings of a level by a stationary Gaussian process. Th. Prob. Appl. **6** (1961), 435–438 (English translation).

J. Math. Kyoto Univ.
4-1 (1964) 1-75

# On stationary solutions of a stochastic differential equation*

Dedicated to Professor A. Kobori on his sixtieth birthday

By

Kiyosi ITÔ and Makiko NISIO

(Received July, 1964)

CONTENTS



## 1. Introduction

To begin with, let us introduce some preliminary notions.

Given a stochastic process $X(t)$, $-\infty < t < \infty$, $\mathcal{B}_{uv}(X)$ denotes the least Borel algebra for which $X(t)$ is measurable for every $t \in [u, v]$.

* This work was supported in part by NSF 16319, NONR 225(28), AFOSR 49(638)-1339 and the National Institutes of Health Grant 10452-02 at Stanford University; by NONR 562(29) at Brown University.

The Wiener process (Brownian motion) is denoted by $B(t)$, $-\infty<t<\infty$, and we normalize it as $B(0)\equiv 0$. In case we consider sevel Wiener processes at the same time, we shall write them as $B_1(t)$, $B_2(t)$, etc. $\mathcal{B}_{uv}(dB)$ denotes the least Borel algebra for which $B(s)-B(t)$ is measurable for every $(t, s)$ with $u\leq t<s\leq v$.

The least Borel algebra that contains $\mathcal{B}_1, \mathcal{B}_2, \cdots$ is denoted by $\mathcal{B}_1\vee\mathcal{B}_2\vee\cdots$ or $\bigvee_k \mathcal{B}_k$.

Given a stochastic process $X(t)$, $-\infty<t<\infty$, and any fixed $s$, we shall call the stochastic process

(1.1) $$Y(\theta)=X(s+\theta), \qquad \theta\leq 0,$$

the *past process* of $X$ at time $s$ and denote it by $\pi_s X$.

$C_-$ denotes the space of all continuous functions defined on the negative half-line $(-\infty, 0]$. $C_-$ is a metric space with the metric

(1.2) $$\rho_-(f, g)=\sum_{n=1}^{\infty}2^{-n}\frac{||f-g||_n}{1+||f-g||_n}$$

where $||h||_n=\max_{-n\leq t\leq 0}|h(t)|$.

Let $a(f)$ and $b(f)$ be continuous functionals defined on $C_-$. A stochastic process $X(t)$ is called a solution of a stochastic differential equation:

(1.3) $$dX(t)=a(\pi_t X)dt+b(\pi_t X)dB(t) \qquad (-\infty<t<\infty),$$

if

(1.4) $\mathcal{B}_{-\infty t}(X)\vee\mathcal{B}_{-\infty t}(dB)$ is independent of $\mathcal{B}_{t\infty}(dB)$ for every $t\in(-\infty, \infty)$,

and if

(1.5) $$X(t)-X(s)=\int_s^t a(\pi_\theta X)d\theta+\int_s^t b(\pi_\theta X)dB(\theta) \quad \text{for} \quad -\infty<s<t<\infty.$$

A solution $X(t)$ of (1.3) is called a stationary solution if $X(t)$ and $B(t)$ are *strictly stationarily correlated*, i.e., if the probability law of the system

(1.6) $(X, dB)\equiv(X(t), -\infty<t<\infty, B(v)-B(u), -\infty<u<v<\infty)$

is invariant under the time shift. A stationary solution is clearly a strictly stationary process. We are here mainly concerned with stationary solutions.

Consider a stochastic integral equation

$$X(t) = X(0)+\int_0^t a(\pi_s X)ds+\int_0^t b(\pi_s X)dB(s) \tag{1.7}$$

with the *past condition*:

$$X(t) = X_-(t)\,, \quad t\leq 0 \tag{1.8}$$

where $X_-(t)$, $t\leq 0$ is a given process. Since a solution of this equation satisfies (1. 3) on $t\geq 0$, it is called a *one-sided solution.*

Now we shall describe the outline of our paper.

In Sections 2, 3 and 4 we prove some facts on stochastic integrals and on the topology of stochastic processes which will be useful in the subsequent sections.

In Section 5 we shall prove the existence of a one-sided solution under a condition which requires that the coefficients $a(f)$ and $b(f)$ grows with $f$ in at most linear order. This condition prevents the solution from *blowing up* in finite time.

Our method is almost the same as Skorokhod's [18] and uses the Prohorov-Skorokhod theory of the totally bounded sets of stochastic processes [13] [17]. Though we are discussing the case in which the coefficients depend not only on the present value of $X(t)$ but also on its past behavior, this generalization does not create any serious difficulty.

In Sections 6 and 7 we shall find out some conditions for the existence of a stationary solution in terms of the one-sided solution discussed in Section 5. Roughly speaking, if the one-sided solution has a bounded moment, then there exists a stationary solution.

For the proof we shift the probability law $P$ of the joint process of the one-sided solution $Y$ and $dB$ by $-s$ to define $P_s$, then take the time average of $P_s$ to define $Q_s=s^{-1}\int_0^s P_u du$ and finally take a convergent subsequence of $Q_t$ to get the probability law governing the joint process of the stationary solution $X$ and $dB$. We can prove the existence of such convergent subsequence observing that $\{P_s, s>0\}$ and $\{Q_s, s>0\}$ are both totally bounded in the Prohorov metric.

The *shifting and averaging method* is well-known as a tool to find an invariant measure, but our method used here seems to be new in that we applied it to the probability measure on the function space to get a stationary solution.

In Section 8 we shall give some direct conditions for the existence of a stationary solution in terms of the coefficients $a(f)$ and $b(f)$. Our conditions require that $a(f)$ is of the form $-a_0(f)f(0)+a_1(f)$ where $a_0(f)$ is dominant in a certain sense. However, these sufficient conditions are far from necessary.

It seems to be an interesting problem to find out a nice weak sufficient condition for the existence of a stationary solution, even though it is not necessary.

The condition (1.4) is the minimum requirement to make the stochastic integral in (1.5) meaningful but it does not always imply

$$(1.4') \qquad \mathcal{B}_{-\infty t}(X) \subset \mathcal{B}_{-\infty t}(dB) \qquad -\infty < t < \infty\,;$$

even a weaker condition:

$$(1.4'') \qquad \mathcal{B}_{-\infty t}(X) \subset \mathcal{B}_{-\infty s}(X) \vee B_{st}(dB) \qquad -\infty < s < t < \infty$$

does not follow from (1.4). Therefore it occurs usually that in order to solve (1.3) for a given Wiener process $B(t)$, we must enlarge the given probability measure space on which $B(t)$ is defined.

The relationship between $\mathcal{B}_{-\infty t}(X)$ and $B_{-\infty t}(dB)$ is discussed in Section 9, and the non-linear backward representation will be also mentioned in this connection.

In Section 10 we shall treat the case that the coefficients satisfy the Lipschitz condition. Even in this case we should impose almost the same condition as in the general cases mentioned above to prove the existence of a stationary solution. However, the solution obtained satisfies (1.4″), or sometimes even (1.4′).

Sections 11 and 12 are devoted to the case that $a(f)$ and $b(f)$ are linear in $f$.

In Section 13 we shall give a complete picture of the possible stationary solutions of a stochastic differential equation of the Markov type (i.e., the coefficients $a(f)$ and $b(f)$ depending only on the present value $f(0)$) under the Lipschitz condition.

In the last three sections we shall mention some pathological examples in connection with the relationship between $\mathcal{B}_{-\infty t}(X)$ and $\mathcal{B}_{-\infty t}(dB)$.

We put the time parameter $t$ always in the bracket as $X(t)$ and the suffix $s$ used in $X_s$ or $X_s(t)$ is to be considered as a parameter to distinguish stochastic processes from each other. Therefore $X_s$ means the process itself, while $X_s(t)$ means the value of the process at time $t$.

*Acknowledgement*. We thank Professor S. Karlin at Stanford University and Professor M. Rosenblatt at Brown University for their kind support and valuable suggestions during our work on this paper.

## 2. Inequalities Concerning Stochastic Integrals

Let $U$ be a random variable, $Y(t)$ and $Z(t)$ stochastic processes and $B(t)$ a Wiener process, where the time parameter $t$ moves on a bounded interval $[u, v]$.

Suppose that

$$\mathcal{B}(U) \vee \mathcal{B}_{ut}(Y, Z, dB) \tag{2.1}$$

is independent of $\mathcal{B}_{tv}(dB)$ for every $t \in [u, v]$ and

$$\int_u^v E[|Y|(s)]ds + \int_u^v E[Z(s)^2]ds < \infty \tag{2.2}$$

Now define $X(t)$ by

$$X(t) = U + \int_u^t Y(s)ds + \int_u^t Z(s)dB(s), \qquad u \leq t \leq v. \tag{2.3}$$

$X(t)$ is well defined and we have $\mathcal{B}_{ut}(X) \subset \mathcal{B}(U) \vee \mathcal{B}_{ut}(Y, Z, dB)$. Then we have

LEMMA 2.1.

(2.4)

$$E(X(v)^4) \leq E(X(u)^4) + 4\int_u^v E[|X(s)^3 Y(s)|](ds + 6\int_u^v E(X(s)^2 Z(s)^2)ds$$

(Notice that both sides may be infinite.)

PROOF: Consider a random time

(2.5) $$\sigma_A = \sup \{t \in [u, v] : \sup_{u \leq s \leq t} |X(s)| \leq A\};$$

if there exists no such $t$, then put $\sigma_A = u$. Since almost all paths of $X(t)$ are continuous, we have

(2.6) $$(\sigma_A > t) \in \mathcal{B}_{ut}(X)$$

and

(2.7) $$P(\sigma_A = v \text{ for sufficiently big } A) = 1.$$

We shall define $U_A$, $Y_A(t)$, $Z_A(t)$ and $X_A(t)$ as follows

$$\begin{aligned} &U_A = U \ (|U| \leq A), \quad = A \ (U > A), \quad = -A \ (U < -A) \\ &Y_A(t) = Y(t), \qquad Z_A(t) = Z(t) \qquad (t < \sigma_A) \\ &Y_A(t) = Z_A(t) = 0 \qquad (t \geq \sigma_A) \\ &X_A(t) = U_A + \int_u^t Y_A(s)ds + \int_u^t Z_A(s)dB(s). \end{aligned}$$

The stochastic integral $\int Z_A(s)dB(s)$ is well defined by virtue of (2.6) and $X_A(t) = X(t)$ or $\pm A$ according as $t < \sigma_A$ or $t \geq \sigma_A$. Therefore it holds with probability 1 that

(2.8) $$|X_A(t)| \leq A, \quad \lim_{A \uparrow \infty} X_A(t) = X(t) \qquad \text{for every } t.$$

Using a formula on stochastic differentials [4], we get

(2.9) $$\begin{aligned} &X_A(v)^4 \\ &= U_A^4 + \int_u^v [4X_A(s)^3 Y_A(s) + 6X_A(s)^2 Z_A(s)^2] ds + \int_u^v 4X_A(s)^3 Z_A(s) dB(s) \end{aligned}$$

By virtue of $|X_A(t)| \leq A$ and the assumption (2.2), we can take the expectation on both sides of (2.9) to get

(2.10) $$\begin{aligned} &E(X_A(v)^4) \\ &= E(U_A^4) + E\left(\int_u^v 4X_A(s)^3 Y_A(s) ds\right) + E\left(\int_u^v 6X_A(s)^2 Z_A(s)^2 ds\right) \\ &= E(U_A^4) + E\left(\int_u^{\sigma_A} 4X(s)^3 Y(s) ds\right) + E\left(\int_u^{\sigma_A} 6X(s)^2 Z(s)^2 ds\right) \\ &\leq E(U^4) + E\left(\int_u^v 4|X(s)^3 Y(s)| ds\right) + E\left(\int_u^v 6X(s)^2 Z(s)^2 ds\right) \\ &= E(X(u)^4) + 4\int_u^v E(|X(s)^3 Y(s)|) ds + 6\int_u^v E(X(s)^2 Z(s)^2) ds \end{aligned}$$

Recalling $X(t)=\lim_{A\to\infty} X_A(t)$, we have

$$E(X(v)^4)\leq \varliminf_{A\to\infty} E(X_A(v)^4)$$
$$\leq E(X(u)^4)+4\int_u^v E(|X(s)^3Y(s)|)ds+6\int_u^v E(X(s)^2Z(s)^2)ds$$

LEMMA 2.2. *If* $\int_u^v E(|X(s)^3Y(s)|)ds<\infty$, *then*

$$E(X(v)^4) \tag{2.11}$$
$$\leq E(X(u)^4)+4\int_u^v E(X(s)^3Y(s))ds+6\int_u^v E(X(s)^2Z(s)^2)ds.$$

PROOF: In (2.10) we get

$$E(X_A(v)^4)$$
$$\leq E(U^4)+4E\left(\int_u^{\sigma_A} X(s)^3Y(s)ds\right)+6\int_u^v E(X(s)^2Z(s)^2)ds$$

But

$$X(s)^3Y(s)\geq -|X(s)^3Y(s)|$$
$$\int_u^v E(|X(s)^3Y(s)|)ds<\infty$$

and $\sigma_A\uparrow v$ as $A\uparrow\infty$. Therefore

$$\lim_{A\uparrow\infty} E\left(\int_u^{\sigma_A} X(s)^3Y(s)ds\right)=E\left(\int_u^v X(s)^3Y(s)ds\right)$$

The rest of the proof is the same as in Lemma 2.1.

Using $4X^3Y\leq 3X^4+Y^4$ and $2X^2Z^2\leq X^4+Z^4$ we can also derive from Lemma 2.1

LEMMA 2.3.

$$E(X(v)^4)\leq E(X(u)^4)+6\int_u^v E(X(s)^4+Y(s)^4+Z(s)^4)ds.$$

Using the Schwartz inequality, $4XYZ^2\leq X^4+Y^4+2Z^4$ and $2X^2Y^2\leq X^4+Y^4$, we can derive from Lemma 2.1

LEMMA 2.4.

$$E(X(v)^4)\leq E(X(u)^4)+8\int_u^v \sqrt{E(X(s)^4)}\sqrt{E(X(s)^4+Y(s)^4+Z(s)^4)}ds.$$

## 3. Totally Bounded Sets of Stochastic Processes

Let $S$ be a separable complete metric space with the metric $\rho$ and $\mathfrak{B}_\rho(S)$ the topological Borel algebra of subsets of $S$, i.e., the least Eorel algeara containing all open subsets of $S$.

A mapping $X(\omega)$ from a probability measure space $\Omega(\mathcal{B}, P)$ into $S$ is called an $S$-valued random variable if it is measurable in the sense that $\{\omega : X(\omega) \in B\} \in \mathcal{B}$ for every $B \in \mathcal{B}_\rho(S)$. The probability law $\mu_X$ of $X$ is defined as the probability measure

$$\mu_X(B) = P\{\omega : X(\omega) \in B\} .$$

Given two probability measures $\mu_1, \mu_2$ on $S(\mathcal{B}_\rho(S))$, the Prohorov distance $L(\mu_1, \mu_2)$ is defined as follows. Let $\varepsilon_{12}$ be the infimum of $\varepsilon$ such that, for every closed subset $F$ of $S$

$$\mu_1(F) < \mu_2(U_\varepsilon(F)) + \varepsilon$$

where $U_\varepsilon(F)$ is the $\varepsilon$-neighborhood of $F$. Define $\varepsilon_{21}$ by switching $\mu_1$ and $\mu_2$ in $\varepsilon_{12}$ and set

$$L(\mu_1, \mu_2) = \max(\varepsilon_{12}, \varepsilon_{21}) . \tag{3.1}$$

The law-metric between two $S$-valued random variables $X_1, X_2$ (whether or not they are defined on the same probability measure space) is defined as the Prohorov distance between their probability laws and is denoted by $L(X_1, X_2)$. Notice that $L(X_1, X_2) = 0$ means that $X_1$ and $X_2$ has the same probability law.

Let $\mathfrak{X}(S)$ be the system of all $S$-valued random variables. We can define $L$-convergence, $L$-Cauchy sequence, etc. on $\mathfrak{X}(S)$.

If $X_n$, $n=1, 2, \cdots$ and $X$ are all defined on the same probability measure space and if $P(\rho(X_n, X) \to 0) = 1$, then $X_n$ is clearly an $L$-Cauchy sequence. The converse is also true in the following sense:

THEOREM OF SKOROKHOD [17]. *If $X_n$, $n=1, 2, \cdots$ (whether or not they are definned on the same probability measure space) is an $L$-Cauchy sequence, then we can construct a sequence $Y_n$, $n=1, 2, \cdots$ and $Y$ on the same probability measure space such that*

$$L(Y_n, X_n) = 0 \quad \textit{and} \quad P(\rho(Y_n, Y) \to 0) = 1 .$$

This can be also stated as follows.

If $\mu_n$ is a Cauchy sequence of probability measures on $S(\mathcal{B}_\rho(S))$ in the Prohorov metric, then we can construct a sequence $Y_n$,

$n=1, 2, \cdots$ and $Y$ on the same probability measure space such that

$$\mu_{Y_n} = \mu_n$$

and

$$P(\rho(Y_n, Y) \to 0) = 1.$$

A subsystem $\mathfrak{N} = \{X_\alpha, \alpha \in A\}$ of $\mathfrak{X}(S)$ is called totally $L$-bounded if every infinite sequence $X_{\alpha_n}$, $n=1, 2, \cdots$ taken from $\mathfrak{N}$ has an $L$-Cauchy subsequence.

THEOREM OF PROHOROV [13]. *In order for* $\mathfrak{D} = \{X_\alpha, \alpha \in A\}$ *to be totally L-bounded in* $\mathfrak{X}(S)$, *it is necessary and sufficient that for every* $\varepsilon > 0$, *there exists a compact subset* $K_\varepsilon$ *of S (independent of* $\alpha$*) with*

$$P(X_\alpha \in K_\varepsilon) > 1 - \varepsilon \qquad \text{for every } \alpha \in A. \tag{3.2}$$

Let $S_i(\rho_i)$, $i=1, 2, \cdots, n$, be complete separable metric spaces. Then the direct $S = S_1 \times S_2 \times \cdots \times S_n$ is also a complete separable metric space with the metric

$$\rho(x, y) = \sum_{i=1}^{n} \rho_i(x_i, y_i), \qquad x = (x_1, \cdots, x_n), \;\; y = (y_1, \cdots, y_n) \in S$$

It is clear that $X = (X_1, X_2, \cdots, X_n) \in \mathfrak{X}(S)$ if and only if $X_i \in \mathfrak{X}(S_i)$, $i=1, 2, \cdots, n$.

Let $\mathfrak{N} = \{X_\alpha = (X_{\alpha,1}, X_{\alpha,2}, \cdots, X_{\alpha,n}), \alpha \in A\}$ be a subsystem of $\mathfrak{X}(S)$. Using Prohorov's theorem and recalling the fact that the direct product of compact sets and the projection of a compact set are also compact, we can easily see

LEMMA 3.1. $\mathfrak{N}$ *is totally bounded if and only if* $\mathfrak{N}_i = \{X_{\alpha,i}; \alpha \in A\}$ $(\subset \mathfrak{X}(S_i))$ *is totally L-bounded for every* $i=1, 2, \cdots, n$.

Hereafter we are concerned with the metric spaces $C_- = C(-\infty, 0]$, $C_+ = C[0, \infty)$ and $C = (-\infty, \infty)$. The metric $\rho_-$ on $C_-$ was defined in Section 1, namely

$$\rho_-(f, g) = \sum_{n=1}^{\infty} 2^{-n} \frac{\|f-g\|_n}{1+\|f-g\|_n} \tag{3.3}$$

$$\|h\|_n = \max_{-n \le t \le 0} |h(t)|$$

Similarly for $\rho_+$ and $\rho$ with

$$||h||_n = \max_{0 \leq t \leq n} |h(t)| \quad \text{or} \quad \max_{-n \leq t \leq n} |h(t)|.$$

Using Kolmogorov's idea, Prohorov [13] obtained a useful criterion for $\mathfrak{N} \subset \mathfrak{X}(C)$ to be totally $L$-bounded. We shall state it in a rather restricted form which will be useful in Sections 4, 5 and 6.

LEMMA 3.2. *$\mathfrak{N} \subset \mathfrak{X}(C)$ is totally bounded if there exist $c > 0$ and $c_n > 0$, $n = 1, 2, \cdots$ such that, for every $X \equiv (X(t), t \in (-\infty, \infty)) \in \mathfrak{N}$, we have*

$$E[X(0)^4] \leq c \tag{3.4}$$

*and*

$$E[[|X(t) - X(s)|^4] < c_n |t-s|^{3/2} \quad \textit{for } |t|, |s| \leq n. \tag{3.5}$$

Similarly for $C_-$ and $C_+$.

The idea of the proof is as follows. Using Chebyshev's inequality, Borel-Cantelli's lemma and the usual technique of diadic division, we can prove that, for every $\varepsilon > 0$, there exist $\gamma(\varepsilon)$ and $\gamma_n(\varepsilon)$, $n = 1, 2, \cdots$ (independent of $X \in \mathfrak{N}$) with

$$P\left[ |X(0)| \leq \gamma(\varepsilon), \sup_{-n \leq t < s \leq n} \frac{|X(t) - X(s)|}{|t-s|^{1/16}} \leq \gamma_n(\varepsilon) \right] > 1 - \varepsilon \tag{3.6}$$

for every $X \in \mathfrak{N}$.

Setting

$$K_\varepsilon = \left\{ f \in C; |f(0)| \leq \gamma(\varepsilon), \sup_{-n \leq t < s \leq n} \frac{|f(t) - f(s)|}{|t-s|^{1/16}} \leq \gamma_n(\varepsilon) \right\}$$

we can write (3.6) as $P(X \in K_\varepsilon) > 1 - \varepsilon$. On the other hand Ascoli-Arzela's theorem shows that $K_\varepsilon$ is a compact subset of $C(\rho)$. To complete the proof, it is enough to apply Prohorov's theorem stated above.

Using the same technique we can prove the following lemma which will be useful in Section 7.

LEMMA 3.3. *$\mathfrak{N} \subset \mathfrak{X}(C)$ is totally bounded, if there exist $c > 0$, $c_n > 0$, $n = 1, 2, \cdots$ and $A_n \in \mathfrak{B}(C)$, $n = 1, 2, \cdots$ such that, for every $X = (X(t), t \in (-\infty, \infty)) \in \mathfrak{N}$, we have*

$$E[X(0)^2] \leq c \tag{3.7}$$

(3.8) $E[|X(t)-X(s)|^4,\ X\in A_n]\leq c_n|t-s|^2$ *for* $|t|, |s|\leq n$,

*and*

(3.9) $\sum_n (1-P(X\in A_n))$ *is convergent uniformly in* $X\in\mathfrak{N}$.

## 4. The Approximate Sum of a Stochastic Integral

In this section $a(f)$ and $b(f)$ denote continuous functions of $f\in C_-(\rho_-)$, $A$ is a parameter set and $[u, v]$ is a bounded interval. For each $\alpha\in A$ we have a stochastic process $X_\alpha=X_\alpha(t)$, $-\infty<t<\infty$, with continuous paths and a Wiener process $B_\alpha$ such that

(4.1) $\mathcal{B}_{-\infty t}(X_\alpha)\vee\mathcal{B}_{ut}(d\mathcal{B}_\alpha)$ is independent of $\mathcal{B}_{tv}(dB_\alpha)$, $u\leq t\leq v$.

Then $a(\pi_t X_\alpha)$ and $b(\pi_t X_\alpha)$ are continuous in $t$ with probability 1 and so

$$I_\alpha=\int_u^v a(\pi_t X_\alpha)dt+\int_u^v b(\pi_t X_\alpha)dB_\alpha(t) \tag{4.2}$$

is well defined.

Let $L_\alpha(\Delta)$ be the approximate sum of $I_\alpha$ for $\Delta\equiv\{u=u_0<u_1<\cdots u_n=v\}$:

$$I_\alpha(\Delta)=\sum_i a(\pi_{u_{i-1}}X_\alpha)(u_i-u_{i-1})+\sum_i b(\pi_{u_{i-1}}X_\alpha)(B_\alpha(u_i)-B_\alpha(u_{i-1})). \tag{4.3}$$

It is clear that $I_\alpha(\Delta)\to I_\alpha$ in probability for each $\alpha$ as $\|\Delta\|\equiv(u_i-u_{i-1})\to 0$, i.e., there exists $\delta=\delta(\varepsilon,\alpha)$ such that $\|\Delta\|<\delta(\varepsilon,\alpha)$ implies $P(|I_\alpha(\Delta)-I_\alpha|>\varepsilon)<\varepsilon$.

LEMMA 4.1. *If $\{X_\alpha, \alpha\in A\}$ is totally L-bounded, then we can take $\delta=\delta(\varepsilon)$ independent of $\alpha$ such that*

(4.4) $\|\Delta\|<\delta(\varepsilon)$ *implies* $P(|I_\alpha(\Delta)-I_\alpha|>\varepsilon)<\varepsilon$

*for every $\alpha\in A$.*

To prove this we shall first prove

LEMMA 4.2. *If $a(f)$ is bontinuous in $f\in C$, then $a(\pi_t f)$ is continuous in two variables $(t, f)$.*

PROOF: Assume that $t_n\to t_0$, $f_n\to f_0$. Set $l=\sup(|t_0|, |t_1|, \cdots)$. Given any $m$ fixed, we have

$$\|\pi_{t_n}f_n-\pi_{t_0}f_0\|_m$$
$$\leq\|\pi_{t_n}f_n-\pi_n f_0\|_m+\|\pi_{t_n}f_0-\pi_{t_0}f_0\|_m$$
$$\leq\|f_n-f_0\|_{m+l}+\sup_{-m\leq t\leq 0}|f_0(t+t_n)-f_0(t+t_0)|\to 0 \qquad (n\to\infty).$$

Therefore $\rho_-(\pi_{t_n}f_n, \pi_{t_0}f_0)\to 0$ and so $a(\pi_{t_n}f_n)\to a(\pi_{t_0}f_0)$.

Now we shall prove Lemma 4.1. Since $\{X_\alpha\}$ is totally $L$-bounded, there exists a compact subset $K(\varepsilon)$ of $C$ such that

(4.5) $$P(X_\alpha\in K)>1-\frac{\varepsilon}{2} \quad \text{for every} \quad \alpha\in A$$

Since $a(\pi_t f)$ and $b(\pi_t f)$ are continuous in $(t, f)$ by Lemma 4.2, they are uniformly continuous in $u\leq t\leq v$ and $f\in K$, namely they exists $\delta=\delta(\eta)>0$ for $\eta>0$ such that

$$t-s|<\delta, \quad u\leq t,\ s\leq v, \qquad \rho(f, g)<\delta, \quad f, g\in K$$

implies

$$|a(\pi_t f)-a(\pi_s g)|<\eta, \quad |b(\pi_t f)-b(\pi_s g)|<\eta,$$

in particular

$$|t-s|<\delta, \qquad u\leq t,\ s\leq v, \quad f\in K$$

implies

$$|a(\pi_t f)-a(\pi_s f)|<\eta, \quad |b(\pi_t f)-b(\pi_s f)|<\eta.$$

Using a step function $\varphi_\Delta(t)$:

$$\varphi_\Delta(t)=u_{i-1} \qquad \text{on} \quad u_{i-1}\leq t<u_i,$$

$I_\alpha(\Delta)$ can be written in the integral form

$$I_\alpha(\Delta)=\int_u^v a(\pi_{\varphi_\Delta(t)}X_\alpha)dt+\int_u^v b(\pi_{\varphi_\Delta(t)}X_\alpha)dB_\alpha(t)$$

and so

$$I_\alpha-I_\alpha(\Delta)=\int_u^v(a(\pi_t(X_\alpha)-a(\pi_{\varphi_\Delta(t)}X_\alpha))dt+\int_u^v(b(\pi_t X_\alpha)-b(\pi_{\varphi_\Delta(t)}X_\alpha))dB_\alpha(t)$$

Writing $[\varphi]_\eta$ for $\varphi$ truncated by $\eta$, i.e.,

$$[\varphi]_\eta=\varphi \quad (|\varphi|\leq\eta), \quad =0 \quad (|\varphi|>\eta)$$

and setting

$$J_\alpha(\Delta)=\int_u^v[a(\pi_t X_\alpha)-a(\pi_{\varphi_\Delta(t)}X_\alpha)]_\eta dt+\int_u^v[b(\pi_t X_\alpha)-b(\pi_{\varphi_\Delta(t)}X_\alpha)]_\eta dB_\alpha(t),$$

we can see that $||\Delta||<\delta(\eta)$ implies

$$P(J_\alpha(\Delta)\neq I_\alpha - I_\alpha(\Delta)) \leq P(X_\alpha \notin K) < \frac{\varepsilon}{2}$$

because $|\varphi_{\Delta(t)}-t|<||\Delta||<\delta(\eta)$. On the other hand, $||\Delta||<\delta(\eta)$ implies

$$E[|J_\alpha(\Delta)|^2] \leq 2\left[\int_u^v \eta\, dt\right]^2 + 2\int_u^v \eta^2\, dt \leq \eta^2 \cdot c$$
$$(c = 2(v-u)^2 + 2(v-u))$$

and so

$$P(|J_\alpha(\Delta)|>\varepsilon) \leq \frac{\eta^2 c}{\varepsilon^2} = \frac{\varepsilon}{2} \qquad \text{if} \quad \eta = \sqrt{\varepsilon^3/2c},$$

Thus $||\Delta||<\delta(\sqrt{\varepsilon^3/2})$ implies that for every $\alpha \in A$,

$$P(|I_\alpha - I_\alpha(\Delta)|>\varepsilon) \leq P(J_\alpha(\Delta) \neq I_\alpha - I_\alpha(\Delta)) + P(|J_\alpha(\Delta)|>\varepsilon) < \varepsilon.$$

## 5. One-sided Solutions

In this section we shall solve

$$(5.1) \qquad dX(t) = a(\pi_t X)dt + b(\pi_t X)dB(t) \qquad (t \geq 0)$$

with the *past condition*:

$$(5.2) \qquad X(t) = X_-(t) \qquad (t \leq 0),$$

where $X_-(t)$, $t \leq 0$ is a given stochastic process with continuous paths and $B(t)$, $-\infty<t<\infty$, is a Wiener process with $B(0)=0$. To do this, it is enough to solve the stochastic integral equation

$$(5.1') \quad X(t) = X(0) + \int_0^t a(\pi_s X)ds + \int_0^t b(\pi_s X)dB(s) \qquad t \geq 0)$$

with (5.2) and

$$(5.3) \qquad \mathcal{B}_{-\infty t}(X) \vee \mathcal{B}_{0t}(dB) \text{ is independent of } \mathcal{B}_{t\infty}(dB) \text{ for every } t \geq 0$$

Let $\Omega(\mathcal{B}, P)$ be the probability measure space on which $X_-(t)$, $t \leq 0$ and $B(t)$, $-\infty<t<\infty$ are defined. We can assume that

$$\mathcal{B}_{-\infty 0}(X_-) \vee \mathcal{B}(dB) = \mathcal{B},$$

since this does not make any essential restriction.

In order to solve our equation, it might be necessary to enlarge $\Omega(\mathcal{B}, P)$, as we have explained in Section 1.

Let us impose the following assumptions.

(A. 1) $a(f)$ and $b(f)$ are both continuous on $C_-(\rho_-)$.

(A. 2) There exist a positive number $M_1$ and a bounded measure $dK_1$ on $(-\infty, 0]$ such that

$$|a(f)|+|b(f)|\leq M_1+\int_{-\infty}^0 |f(t)|dK_1(t).$$

(A. 3) $E(X_-(t)^4)\leq c \qquad t\leq 0$

with a constant $c$.

(A. 4) $\mathcal{B}_{-\infty 0}(X_-)$ is indepenpent of $\mathcal{B}_{0\infty}(B)$.

The assumption (A. 4) is automatically necessary by (5. 2) and (5. 3).

THEOREM 1. (*Existence of a one-sided solution*). *Under* (A. 1), (A. 2), (A. 3) *and* (A. 4), *we can find a solution* $X(t)$ *of the stochastic integral equation* (5. 1′) *with* (5. 2), (5. 3) *and*

(5. 4) $$E[X(t)^4]\leq \gamma e^{\gamma t} \qquad (t\geq 0),$$

*where* $\gamma$ *is a positive constant.*

*Remark*: As we can see in the proof given below, (A. 2) can be replaced by a weaker condition:

(A. 2′) There exist a positive $M$ and a bounded measure $dK$ on $(-\infty, 0]$ such that $|a(f)|^4+|b(f)|^4\leq M+\int_{-\infty}^0 f(t)|^4 dK(t)$.

We shall divide the proof into two steps.

(i) *Approximate polygonal solutions.* Take $h>0$ and define an approximate solution $X_h(t)$ by Cauchy's polygonal method.

(5. 5) $$\begin{aligned} X_h(t) &= X_-(t) \qquad t\leq 0 \\ &= X_h(nh)+a(\pi_{nh}X_h((t-nh)+b(\pi_{nh}X_h)(B(t)-B(nh)) \\ &\qquad nh\leq t\leq (n+1)h;\quad n=0,1,2,\cdots \end{aligned}$$

Take the step function $\varphi_h(t)$:

(5. 6) $$\varphi_h(t) = nh \qquad nh\leq t<(n+1)h;\quad n=0,1,2,\cdots.$$

Then $X_h(t)$ satisfies

(5.7) $$X_h(t) = X_h(0) + \int_0^t a(\pi_{\varphi_h(s)} X_h)ds + \int_0^t b(\pi_{\varphi_h(s)} X_h)dB(s) \qquad (t \geq 0)$$

LEMMA 5.1.

(5.8) $$c_h(t) \equiv \sup_{s \geq t} E[X_h(t)^4] < \gamma e^{\gamma t}, \qquad t \geq 0$$

*with a constant* $\gamma = \gamma(M, dK, c)$ *which does not depend on h.*

PROOF: We shall first show that

(5.9) $$c_h(t) < \infty \qquad -\infty < t < \infty .$$

Since $c_h(t)$ is increasing, it is enough to show (5.9) for $t = nh$, $n = 0, 1, 2, \cdots$, by induction. It is clearly by (A.3) that $c_h(0) \leq c < \infty$. If $c_n(nh) < \infty$, then $c_n((n+1)h) < \infty$, because we have, for $nh \leq t \leq (n+1)h$,

$$\begin{aligned} E(X_h(t)^4) &\leq 3^3 [E(X_h(nh)^4 + E(a(\pi_{nh} X_h)^4) h^4 + 3E(b(\pi_{nh} X_h)^4) h^2] \\ &\leq 3^3 [c_h(nh) + (M + \|K\| c_n(nh))(h^4 + 3h^2)] < \infty \end{aligned}$$

by virtue of (5.5) and (A.2′), where $\|K\| = \int_{-\infty}^0 dK(t)$.

Applying Lemma 2.3 to (5.7), we have, for $t \geq 0$,

$$\begin{aligned} E(X_h(t)^4) &\leq E(X_h(0)^4) + 6\int_0^t E[X_h(s)^4 + a(\pi_{\varphi_n(s)} X_h)^4 + b(\pi_{\varphi_h(s)} X_h)^4] ds \\ &\leq c + 6\int_0^t [c_h(s)(1 + \|K\|) + M] ds , \end{aligned}$$

and so

$$c_h(t) \leq c + 6\int_0^t [c_h(s)(1 + \|K\|) + M] ds .$$

Since $c_h(t) < \infty$ for all $t \geq 0$, we have

$$c_h(t) \leq c_h(t) + M(1 + \|K\|)^{-1} \leq (c + M(1 + \|K\|)^{-1} e^{6(1+\|K\|)t}$$

by the method of iteration. Setting

$$\gamma = \max(c + M(1 + \|K\|)^{-1}, 6(1 + \|K\|)),$$

we get (5.8).

LEMMA 5.2.

(5.10) $$E((X_h(t) - X_h(s))^4) \leq \gamma_n |t - s|^{3/2}$$
$$0 \leq s < t \leq n, \quad n = 1, 2, \cdots ,$$

*where $\gamma_n$ is a constant which does not depend on $h$ but only on $n$, $M$, $\|K\|$ and $c$.*

PROOF: It follows from (5.7)

$$X_h(t)-X_h(s) = \int_s^t a(\pi_{\varphi_h(u)}X_h)du+\int_s^t b(\pi_{\varphi_h(u)}X_h)dB(u).$$

Applying Lemma 2.4 to this, we have, for $0\leq s<t\leq n$,

$$\begin{aligned}
&E[(X_h(t)-X_h(s))^4]\\
&\leq 8\int_s^t \sqrt{E[(X_h(u)-X_h(s))^4]}\\
&\qquad\qquad \sqrt{E[(X_h(u)-X_h(s))^4+a(\pi_{\varphi_h(u)}X_h)^4+b(\pi_{\varphi_h(u)}X_h)^4]}\,du\\
&\leq 8\int_s^t \sqrt{E[(X_h(u)-X_h(s))^4]}\sqrt{M+(\|K\|+16)c_h(u)}\,du\\
&\leq 8\sqrt{M+(\|K\|+16)\gamma e^{\gamma n}}\int_s^t \sqrt{E[(X_h(u)-X_h(s))^4]}\,du
\end{aligned}$$

by Lemma 5.1.

Using Lemma 5.1 again, we have

$$E[(X_h(t)-X_h(s))^4]\leq 2^4 E[X_h(t)^4+X_n(s)^4]\leq 2^5\gamma e^{\gamma n} \qquad (0\leq s<t<n)$$

Putting this in the above integral, we get

$$E[(X_h(t)-X_h(s))^4]\leq \gamma_n'\cdot(t-s) \qquad (0\leq s<t\leq n)$$

with a constant $\gamma_n'=\gamma_n'(M, dK, c)$ independent of $h$ and putting this in the integral again, we get

$$E[(X_h(t)-X_h(s))^4]\leq \gamma_n(t-s)^{3/2}$$

with another constant $\gamma_n=\gamma_n(m, dK, c)$ independent of $h$.

Applying Lemma 3.2 to the class of stochastic processes $\{X_{+,h}\equiv(X_h(t), t>0); h>0\}$ $(\subset\mathcal{X}(C_+))$ in view of Lemma 5.1 and Lemma 5.2, we can see that $\{X_{+,h}; h>0\}$ is totally $L$-bounded. It is evident that $\{X_{-,h}\equiv X_- : h>0\}$ $(\subset\mathcal{X}(C_-))$ is totally $L$-bounded. Since $X_h=(X_{-,h}, X_{+,h})$, we have

LEMMA 5.3. *$\{X_h; h>0\}$ is totally $L$-bounded.*

(ii) *Finding a solution.*

It is evident that $\{B_h\equiv B; h>0\}$ and $\{X_{-,h}\equiv X_-; h>0\}$ are totally $L$-bounded. Since $\{X_h: h>0\}$ is totally $L$-bounded by Lemma 5.3, $\{(X_h, B, X_-): h>0\}$ is also totally $L$-bounded by Lemma 3.1, so that we can find an $L$-Cauchy sequence $(X_{h(n)}, B, X_-)$, $n=1, 2, \cdots$ with $h(n)\downarrow 0$. By Skorokhod's theorem (see Section 3), we can construct $(Y_n, B_n, Y_{-,n})$, $n=1, 2, \cdots, \infty$ on a certain probability measure space such that

$$(5.11) \qquad L((Y_n, B_n, Y_{-,n}), (X_{h(n)}, B, X_-)) = 0 \qquad n=1, 2, \cdots$$

$$(5.12) \qquad P((Y_n, B_n, Y_{-,n}) \to (Y_\infty, B_\infty, Y_{-,\infty})) = 1$$

where the convergence is to be understood in the sense of the metric in $C\times C\times C_-$.

Since $L((B_n, Y_{-,n}), (B, X_-))=0$ by (5.11) and $P((B_n, Y_{-,n})\to (B_\infty, Y_{-,\infty}))=1$ by (5.12), we get

$$(5.13) \qquad L((B_\infty, Y_{-,\infty}), (B, X_-)) = 0 .$$

If we can prove that

(5.14) $\mathcal{B}_{-\infty t}(Y_\infty)\vee\mathcal{B}_{0t}(dB_\infty)$ is independent of $\mathcal{B}_{t\infty}(dB_\infty)$,

(5.15) with probability 1,

$$Y_\infty(t) = Y_{-,\infty}(t), \qquad t\leq 0,$$

(5.16) with proaability 1,

$$Y_\infty(t) = Y_\infty(0)+\int_0^t a(\pi_s Y_\infty)ds+\int_0^t b(\pi_s Y_\infty)dB_\infty(s), \qquad t\leq 0,$$

and

$$(5.17) \qquad E(Y_\infty(t)^4)\leq \gamma e^{\gamma t}, \qquad t\geq 0,$$

then we can conclude that $X(t)\equiv Y_\infty(t)$ is a solution of (5.1) which is to be constructed, by identifying $(B_\infty, Y_{-,\infty})$ with $(B, X_-)$ as is justified by (5.13).

$\mathcal{B}_{-\infty,t}(X_h)\vee\mathcal{B}_{0t}(dB)$ is independent of $\mathcal{B}_{t\infty}(dB)$ by the definition of $X_h$ and $\mathcal{B}_{-\infty t}(Y_n)\vee\mathcal{B}_{0s}(dB_n)$ is therefore independent of $\mathcal{B}_{t\infty}(dB_n)$ by virtue of (5.11) and so (5.14) holds by (5.12).

Since $Y_\infty(t)$ and $Y_{-,\infty}(t)$ are continuous with probability 1, it is enough for the proof of (4.15) to prove

$$P(Y_\infty(t) = Y_{-,\infty}(t)) = 1 \qquad \text{for each } t\leq 0.$$

But this is immediate from (5. 11), (5. 12) and the definition of $X_h$.

By the same reason as above, it is enough for the proof of (5. 16) to prove for each $t$ that

(5. 18) with probability 1,

$$Y_\infty(t) = Y_\infty(0)+\int_0^t a(\pi_s Y_\infty)ds+\int_0^t b(\pi_s Y_\infty)dB_\infty(s)\,.$$

Set

(5. 19) $$I_n = \int_0^t a(\pi_s Y_n)ds+\int_0^t b(\pi_s Y_n)dB_n(s)$$

(5. 20)

$$\begin{aligned}I_n(h) &= \int_0^t a(\pi_{\varphi_h}(s)Y_n)ds+\int_0^t b(\pi_{\varphi_h(s)}Y_n)dB_n(s)\\ &= \sum_{k=0}^{m-1} a(\pi_{kh}Y_n)\cdot h+a(\pi_{mh}Y_n)\cdot(t-mh) \qquad (mh\leq t<(m+1)h)\\ &\quad+\sum_{k=0}^{m-1} b(\pi_{kh}Y_n)[B((k+1)h)-B(kh)]+b(\pi_{mh}Y_n)[B(t)-B(mh)]\\ &\qquad\qquad (n=1,2,\cdots,\infty)\end{aligned}$$

Since $P(\rho(Y_n, Y_\infty)\to 0)=1$ by (5. 12), $\{Y_n, n=1, 2, \cdots, \infty\}$ is totally $L$-bounded and we can use Lemma 4. 1 to get $\delta(\varepsilon)$ such that $|h|<\delta(\varepsilon)$ implies

(5. 21) $$P(|I_n(h)-I_n|>\varepsilon)<\varepsilon, \qquad n=1, 2, \cdots, \infty\,.$$

Since $Y_n(t)=Y_n(0)+I_n(h(n))$ by (5. 7), (5. 11) and (5. 20), we have

$$\begin{aligned}&P(|Y_\infty(t)-Y_\infty(0)-I_\infty|>6\varepsilon)\\ &=P(|Y_\infty(t)-Y_n(t))-(Y_\infty(0)-Y_n(0))-(I_\infty-I_n(h(n))|>6\varepsilon)\\ &\leq P(|Y_\infty(t)-Y_n(t)|>\varepsilon)+P(|Y_\infty(0)-Y_n(0)|>\varepsilon)\\ &\qquad+P(|I_\infty-I_n(h(n))|>4\varepsilon)\end{aligned}$$

and so by (5. 12)

$$P(|Y_\infty(t)-Y_\infty(0)-I_\infty|>6\varepsilon)\leq \varliminf_{n\to\infty} P(|I_\infty-I_n(h(n))|>4\varepsilon)\,.$$

But it holds by (5. 21) that

$$\begin{aligned}&P(|I_\infty-I_n(h(n))|>4\varepsilon)\\ &\quad\leq P(|I_\infty-I_\infty(h)|>\varepsilon)+P(|I_\infty(h)-I_n(h)|>\varepsilon)\\ &\qquad\quad+P(|I_n(h)-I_n|>\varepsilon)+P(|I_n-I_n(h(n))|>\varepsilon)\\ &\quad\leq 3\varepsilon+P(|I_\infty(h)-I_n(h)|>\varepsilon)\end{aligned}$$

for $h=\delta(\varepsilon)/2<\delta(\varepsilon)$ and $n\geq n_0(\varepsilon)$, where $n_0(\varepsilon)$ is the maximum number $n$ for which $h(n)<\delta(\varepsilon)$.

By (5.20), (5.12) and the continuity of $a(\pi_s f)$ and $b(\pi_s f)$ in $f$ (see Lemma 4.2), $I_n(h)\to I_\infty(h)$ with probability 1 as $n\to\infty$ by Lemma 4.1. Therefore

$$\varlimsup_{n\to\infty} P(|I_\infty - I_n(h(n))|>4\varepsilon)\leq 3\varepsilon$$

and so

$$P(|Y_\infty(t)-Y_\infty(0)-I_\infty|>6\varepsilon)\leq 3\varepsilon\,.$$

Since $\varepsilon$ is arbitrary, this inequality implies (5.18).

Using Lemma 5.1, we have (5.17) as follows

$$E(Y_\infty(t)^4)\leq \varliminf_{n\to\infty}(Y_n(t)^4) = \varliminf_{n\to\infty} E(X_{h(n)}(s)^4)\leq \gamma e^{\gamma s}\,.$$

## 6. Stationary Solutions (1)

Consider a stochastic differential equation:

$$dX(t) = a(\pi_t X)\,dt + b(\pi_t X)\,dB(t) \tag{6.1}$$

under the assumptions (A.1) and (A.2) (or (A.2′)) imposed in Section 5. Since $X_-(t)\equiv 0$ satisfies (A.3) and (A.4) in that section, Theorem 1 shows that there exists a one-sided solution of (6.1) with the past condition:

$$X(t)\equiv 0 \qquad (t\leq 0) \tag{6.2}$$

The solution constructed in this theorem satisfies

$$E[X(t)^4]\leq \gamma e^{\gamma t} \qquad (t>0) \tag{6.3}$$

with a constant $\gamma\in(0,\infty)$.

No we shall prove

THEOREM 2. *If the one-sided solution $X(t)$ mentioned above is bounded in the 4-th moment, i.e.,*

$$E[X(t)^4]\leq \alpha \qquad (t\geq 0) \tag{S}$$

*with a donstant $\alpha\in(0,\infty)$, then there exists a stationary solution of* (6.1). (*See Section* 1 *for the definition of a stationary solution.*)

PROOF: Write $X_s(t)$ and $B_s(t)$ for $X(s+t)$ and $B(s+t)-B(s)$, respectively. Then we have

$$dX_s(t) = a(\pi_t X_s)dt + b(\pi_t X_s)dB_s(t) \qquad t \geq -s \tag{6.3}$$

$$X_s(t) \equiv 0 \qquad t \leq -s \tag{6.4}$$

and $B_s$ is clearly a Wiener process with $B_s(0)=0$.

Let $P_s$ denote the probability law of $(X_s, B_s)$ which is a probability measure on $C^2$ $(=C\times C)$ and $\theta_s$ the shift operator on $C^2$, i.e.,

$$(\tilde{f}, \tilde{g}) = \theta_s(f, g)$$

if and only if $\tilde{f}(t)=f(s+t)$, $\tilde{g}(t)=g(s+t)-g(s)$. Noticing that $(X_s, B_s)=\theta_s(X, B)\equiv\theta_s(X_0, B_0)$, we have

$$P_s(E) = P_0(\theta_{-s}E). \tag{6.5}$$

Since $\pi_s f$ is continuous in $(s, f)\in(-\infty, \infty)\times C$, it is easy to see that $P_s(E)$ is Borel-measurable in $s$ for every $E\in\mathcal{B}(C^2)$. Therefore

$$Q_T(E) = \frac{1}{T}\int_0^T P_s(E)ds \tag{6.6}$$

is also a probability measure on $(C^2, \mathcal{B}(C^2))$.

$B_s$ is clearly totally $L$-bounded by Lemma 3.2, because $E(B_s(0)^4)=0$ and

$$E((B_s(v)-B_s(u))^4) = 3(v-u)^2 \leq 3\sqrt{2n}\,|v-u|^{3/2}$$

for $|u|, |v|\leq n$.

To prove that $\{X_s, s>0\}$ is also totally $L$-bounded, we shall verify

$$E(X_s(t)^4) \leq \alpha \tag{6.7}$$

$$E((X_s(v)-X_s(u))^4) \leq \gamma|v-u|^{3/2} \tag{6.8}$$

with some constant $\gamma$. (6.7) is clear by (S). Since $X_s(t)=X(s+t)$, it is enough to prove that

$$E((X(v)-X(u))^4) \leq \gamma|v-u|^{3/2}. \tag{6.9}$$

Noticing $X(t)\equiv0$ for $t\leq0$, it is also enough to prove (6.9) for

$v>u\geq 0$. Since $X$ satisfies (6. 1), we have

$$X(v)-X(u)=\int_u^v a(\pi_t X)\,dt+\int_u^v b(\pi_t X)\,dB(t)\,.$$

Using Lemma 2. 4, we get

$$\begin{aligned}&E((X(v)-X(u))^4)\\ &\leq 8\int_u^v \sqrt{E((X(t)-X(u))^4)}\sqrt{E((X(t)-X(u))^4+a(\pi_t X)^4+b(\pi_t X)^4)}\,dt\\ &\leq 8\sqrt{(16+\|K\|)\alpha+M}\int_u^v \sqrt{E((X(t)-X(u))^4)}\,dt\,.\end{aligned}$$

But

$$E((X(t)-X(u))^4)\leq 16\alpha$$

by (S).

Replacing the integrand in the above integral with $\sqrt{16\alpha}$, we get

$$E((X(v)-X(u))^4)\leq \gamma_1(v-u)\qquad (\gamma_1=\text{constant})$$

and replacing the integrand with $\sqrt{\gamma_1(t-u)}$, we have

$$E((X(v)-X(u))^4)\leq \gamma(v-u)^{3/2}\qquad (\gamma=\text{constant})\,.$$

Thus we have proved that both $\{X_s,\ s\geq 0\}$ and $\{B_s,\ s\geq 0\}$ are totally $L$-bounded. Hence it follows by Lemma 3. 1 that $\{(X_s, B_s),\ s\geq 0\}$ is also totally $L$-bounded. Therefore we can find a compact subset $K=K(\varepsilon)$ of $C^2$ independent of $s$ for every $\varepsilon>0$ that $P_s(K)>1-\varepsilon$, so that $Q_T(K)>1-\varepsilon$. Since $K$ depends only on $\varepsilon$, $\{R_T,\ T>0\}$ is also totally bounded. Therefore there exists $T_n\uparrow\infty$ such that $Q_{T_n}$ converges to a certain probability measure $Q$ on $C^2$ in the Prohorov distance.

Let $(\tilde{X}, \tilde{B})$ be a $C^2$-valued random variable whose probability measure is $Q$. Now we shall prove that

(6. 10) $\tilde{B}(t)$ is a Wiener process with $\tilde{B}(0)=0$,

(6. 11) $\tilde{X}(t)$ is strictly stationarily correlated to $d\tilde{B}$,

(6. 12) $\mathcal{B}_{-\infty t}(\tilde{X}, d\tilde{B})$ is independent of $\mathcal{B}_{t\infty}(d\tilde{B})$,

and

(6. 13) $d\tilde{X}(t)=a(\pi_t\tilde{X})\,dt+b(\pi_t\tilde{X})\,d\tilde{B}(t)\qquad (-\infty<t<\infty)$.

Once we prove this, we can see that $\tilde{X}(t)$ is a stationary solution which is to be sought, by identifying $\tilde{B}$ and $B$.

Taking any bounded continuous function $\varphi$ on the $m$-space and observing

$$\begin{aligned} &E[\varphi(\tilde{B}(t_1), \cdots, \tilde{B}(t_m))] \\ &= \lim_{n\to\infty} \frac{1}{T_n}\int_0^{T_n} E[\varphi(B_s(t_1), \cdots, B_s(t_m))]ds \\ &= E[\varphi(B(t_1), \cdots, B(t_m))], \end{aligned}$$

we can see that $\tilde{B}(t)$ has the same probability law as $B(t)$, so that $\tilde{B}(t)$ is a Wiener process with $\tilde{B}(0)=0$.

Observing also, for any bounded continuous $\varphi$ and $\psi$ and for any fixed $t$,

$$\begin{aligned} &E[\varphi(\tilde{X}(t_1+t), \cdots, \tilde{X}(t_m+t))\psi(\tilde{B}(v_1+t)-\tilde{B}(u_1+t), \cdots, \tilde{B}(u_k+t)-\tilde{B}(u_k+t))] \\ &= \lim_{n\to\infty} \frac{1}{T_n}\int_0^{T_n} E[\varphi(X_s(t_1+t), \cdots)\psi(B_s(v_1+t)-B_s(u_1+t), \cdots)]ds \\ &= \lim_{n\to\infty} \frac{1}{T_n}\int_0^{T_n} E[\varphi(X_{s+t}(t_1), \cdots)\psi(B_{s+t}(v_1)-B_{s+t}(u_1), \cdots)]ds \\ &= \lim_{n\to\infty} \frac{1}{T_n}\int_t^{T_n+t} E[\varphi(X_s(t_1), \cdots)\psi(B_s(v_1)-B_s(u_1), \cdots)]ds \\ &= \lim_{n\to\infty} \frac{1}{T_n}\int_0^{T_n} E[\varphi(X_s(t_1), \cdots)\psi(B_s(v_1)-B_s(u_1), \cdots)]ds \\ &= E[\varphi(\tilde{X}(t_1), \cdots)\psi(\tilde{B}(v_1)-\tilde{B}(u_1), \cdots)], \end{aligned}$$

we can see that $\tilde{X}$ and $d\tilde{B}$ are strictly stationarily correlated.

In order to prove (6.12) and (6.13), it is enough to show that, for every $S$,

(6.12′) $\mathcal{B}_{-\infty t}(\tilde{X}) \vee \mathcal{B}_{-St}(\tilde{X}, d\tilde{B})$ is independent of $\mathcal{B}_{t\infty}(d\tilde{B})$,

$$-S \le t < \infty,$$

and

(6.13′) $$\tilde{X}(t) = \tilde{X}(-S) + \int_{-S}^t a(\pi_u \tilde{X})du + \int_{-S}^t b(\pi_u \tilde{X})d\tilde{B}(u), \quad -S \le t < \infty$$

Use $P_s$ and $T_n$ determined above to define the probability measures on $C^2$:

(6.14) $$Q_{S,T_n}(E)=\frac{1}{T_n-S}\int_S^{T_n}P_s(E)ds\,,\qquad n\geq n_0(S)$$

where $n_0(S)$ is the minimum number $n$ for which $T_n > S$.

For any bounded continuous function $f$ on $C^2$, we have

$$\begin{aligned}&\lim_{n\to\infty}\int_{C^2}f(\omega)Q_{S,T_n}(d\omega)\\&=\lim_{n\to\infty}\frac{1}{T_n-S}\int_S^{T_n}\int_{C^2}f(\omega)P_s(d\omega)ds\\&=\lim_{n\to\infty}\frac{1}{T_n}\int_0^{T_n}\int_{C^2}f(\omega)P_s(d\omega)ds\\&=\lim_{n\to\infty}\int_{C^2}f(\omega)Q_{T_n}(d\omega)\\&=\int_{C^2}f(\omega)Q(d\omega)\,.\end{aligned}$$

$Q_{S,T_n}\to Q$ in the weak* convergence and so in the Prohorov distance. Using Skorokhod's theorem, we can construct $(Y_n, \beta_n)$, $n=n_0(S)$, $n_0(S)+1, \cdots, \infty$ on a certain probability measure space such that (6.15) the probability law of $(Y_n, \beta_n)$ in $Q_{S,T_n}$ for $n=n_0(S)$, $n_0(S)+1, \cdots$ and that of $(Y_\infty, \beta_\infty)$ is $Q$, and

(6.16) $$(Y_n, \beta_n)\to(Y_\infty, \beta_\infty)\quad\text{in the metric in } C^2$$

with probability 1.

$(Y_\infty, \beta_\infty)$ and $(\tilde{X}, \tilde{B})$ have the same probability law $Q$ and therefore $\beta_\infty$ is a Wiener process with $\beta_\infty(0)=0$.

$\beta_n$ in also a Wiener process with $\beta_n(0)=0$ for $n\geq n_0(S)$, because we have for any bounded continuous function $\varphi(\xi_1, \cdots, \xi_m)$,

$$\begin{aligned}&E[\varphi(\beta_n(t_1), \cdots, \beta_n(t_m))]\\&=\frac{1}{T_n-S}\int_S^{T_n}E[\varphi(B_s(t_1), \cdots, B_s(t_m))]ds\\&=E[\varphi(B(t_1), \cdots, B(t_m))]\end{aligned}$$

Now we shall prove

(6.12″) $\mathcal{B}_{-\infty t}(Y_n)\vee\mathcal{B}_{-St}(Y_n, d\beta_n)$ is independent of $\mathcal{B}_{t\infty}(d\beta_n)$, $\quad -S\leq t<\infty$

and

$$(6.13'') \quad Y_n(t) = Y_n(-S) + \int_{-S}^{t} a(\pi_u Y_n) du + \int_{-S}^{t} b(\pi_u Y_n) d\beta_n(u), \qquad -S \le t < \infty$$

with probability 1.

Noticing that $\mathcal{B}_{-\infty t}(X_s) \vee \mathcal{B}_{-St}(X_s, dB_s)$ is independent of $\mathcal{B}_{t\infty}(dB_s)$ for $-S \le t < \infty$ as far as $s \ge S$, and observing, for any bounded continuous $\varphi(\xi_1, \cdots, \xi_p, \eta_1, \cdots, \eta_q)$ and $\psi(\zeta_1, \cdots, \zeta_r)$ and for $s_1, \cdots, s_p \le t$, $-S \le t_1, \cdots, t_q \le t$ and $t \le u_1, \cdots, u_r$,

$$\begin{aligned} &E\{\varphi(Y_n(s_1), \cdots, Y_n(s_p), \beta_n(t_1, t), \cdots, \beta_n(t_q, t))\psi(\beta_n(t, u_1), \cdots, \beta_n(t, u_r))\} \\ &= \frac{1}{T_n - S}\int_S^{T_n} E[\varphi(X_s(s_1), \cdots, B_s(t_1, t), \cdots)\psi(B_s(t, u_1), \cdots)] ds \\ &= \frac{1}{T_n - S}\int_S^{T_n} E[\varphi(X_s(s_1), \cdots, B_s(t_1, t), \cdots)] E[\psi(B_s(t, u_1), \cdots)] ds \\ &= \frac{1}{T_n - S}\int_S^{T_n} E[\varphi(X_s(s_1), \cdots, B_s(t_1, t), \cdots)] ds \cdot E[\psi(\beta_n(t, u_1), \cdots)] \\ &= E[\varphi(Y_n(s_1), \cdots, \beta_n(t_1, t), \cdots)] E(\psi(\beta_n(t, u_1), \cdots)], \end{aligned}$$

we can see that (6.12″) is true where $\beta_n(u, v)$ and $B_s(u, u)$ stand for $\beta_n(v) - \beta_n(u)$ and $B_s(v) - B_s(u)$ respectively; in the computation above, we used the fact that both $B_s$ and $\beta_n$ are Wiener processes and so have the same probability law.

It is clear by the definition of $(X_s, B_s)$ that

$$(6.17) \quad X_s(t) = X_s(-S) + \int_{-S}^{t} a(\pi_u X_s) du + \int_{-S}^{t} b(\pi_u X_s) dB_s(u)$$

with probability 1, as far as $s < S$.

Noticing that both sides of (6.17) are $\mathcal{B}(X_s, B_s)$-measurable by the definition, there exist Borel measurable functions $\varphi$ and $\psi$ defined on $C^2$ such that, with probability 1, $\varphi(X_s, B_s)$ and $\psi(X_s, B_s)$ are equal to the left and right sides respectively for each $t > -S$, as far as $s > S$. Therefore we have

$$E(|\varphi(Y_n, \beta_n) - \psi(Y_n, \beta_n)|) = \frac{1}{T_n - S}\int_S^{T_n} E(|\varphi(X_s, B_s) - \psi(X_s, B_s)|) ds = 0$$

by (6.17). Thus $\varphi(Y_n, \beta_n) = \psi(Y_n, \beta_n)$ with probability 1 and so (6.13″) holds with probability 1 for each $t \in (-S, \infty)$. But both sides of (6.13″) are continuous in $t$ with probability 1, we can see that the exceptional set of probability 0 can be taken indently of $t$. Thus (6.13″) is proved.

Noticing (6.16), we can derive (6.12′) and (6.13′) from (6.12″) and (6.13″) respectively by the argument we used at the end of Section 5.

## 7. Stationary Solutions (2)

We shall find another condition for the existence of a stationary solution of the stochastic differential equation (6.1).

(A.1) and (A.2) are assumed here as in the previous section. We can replace (A.2) by a weaker condition:

$$\text{(A.2″)} \qquad |a(t)|^2+|b(f)|^2 \leq M+\int_{-\infty}^{0}|f(t)|^2 dK(t)$$

with a positive constant $M$ and a bounded measure $dK$ on $(-\infty, 0]$. Since (A.2″) is stronger than (A.2′) (with different $M$ and $dK$) in Section 5, we can construct a one-sided solution $X(t)$ of (6.1) with the past condition: $X(t)\equiv 0$, $t\leq 0$ under the assumption (A.1) and (A.2) (or (A.2″)), by virtue of Theorem 1 and the remark immediately following it.

Now we shall prove

THEOREM 3. *If the one-sided solution $X(t)$ mentioned above is bounded in the second moment, i.e.,*

$$(\tilde{S}) \qquad E(X(t)^2)\leq \alpha \qquad (t\geq 0)$$

*with a constant $\alpha$, then there exists a stationary solution, provided $dK$ has a compact support.*

PROOF: Take $\sigma>0$ such that the support of $dK$ is contained in $[-\sigma, 0]$. The proof goes in the same way as in the proof of Theorem 2 except in proving the fact that $\{X_s\equiv\theta_s X;\ s>0\}$ $(\theta_s X(t)\equiv X(s+t))$ is totally $L$-bounded. To prove this, we shall use Lemma 3.3.

We shall start with two lemmas.

LEMMA 7.1. *There exists $G=G(\varepsilon, l)$ such that*

$$(7.1) \qquad P(\sup_{s\leq t\leq s+l}|X(t)|>G)\leq\varepsilon \quad \textit{for every } s>0.$$

PROOF: Introducing two processes

$$\tilde{a}(t) = a(\pi_t X) \quad (t \geq 0), \qquad = 0 \quad (t < 0)$$

and

$$\tilde{b}(t) = b(\pi_t X) \quad (t \geq 0), \qquad = 0 \quad (t < 0),$$

we obtain

$$X(t) = X(s) + \int_s^t \tilde{a}(u)du + \int_s^t \tilde{b}(u)dB(u) \qquad (-\infty < t < \infty)$$

Since

$$\begin{aligned} S \equiv & \sup_{s \leq t \leq s+l} |X(t)| \\ & \leq |X(s)| + \int_s^{s+l} |\tilde{a}(u)|du + \sup_{s \leq t \leq s+l} \left| \int_s^t \tilde{b}(u)dB(u) \right| \\ & (= U + V + W), \end{aligned}$$

we have

$$P(S > G) \leq P\left(U > \frac{G}{3}\right) + P\left(V > \frac{G}{3}\right) + P\left(W > \frac{G}{3}\right).$$

Using (A. 2′), ($\tilde{S}$) and the property of stochastic integrals, we have

$$P\left(U > \frac{G}{3}\right) \leq \frac{3E(U)}{G} \leq \frac{3\sqrt{E(X(s)^2)}}{G} \leq \frac{3\sqrt{\alpha}}{G}$$

$$\begin{aligned} P\left(V > \frac{G}{3}\right) & \leq \frac{3E(V)}{G} \leq \frac{3}{G} \int_s^{s+l} \sqrt{E(\tilde{a}(u))^2}du \\ & \leq \frac{3l}{G} \sqrt{M + \|K\| \alpha} \end{aligned}$$

and

$$P\left(W > \frac{G}{3}\right) \leq \left(\frac{3}{G}\right)^2 \int_s^{s+l} E(\tilde{b}(u)^2)du \leq \frac{9l}{G^2}(M + \|K\| \alpha).$$

Therefore we get

$$P(s > G) \leq \frac{\delta_1}{G} + \frac{\delta_2 l}{G} + \frac{\delta_3 l}{G^2} \qquad (\delta_i : \text{constant})$$

and so we can find $G = G(\varepsilon, l)$ for which (7. 1) holds.

LEMMA 7. 2. *There exists $\mu = \mu(G, l)$ such that for every $s$ and for every $u, v \in [s, s+l]$, we have*

(7. 2) $$E[(X(v) - X(u))^4, \sup_{s-\gamma \leq t \leq s+l} |X(t)| \leq G] \leq \mu |u - u|^2.$$

PROOF: We shall use $\tilde{a}(t)$ and $\tilde{b}(t)$ in the same meaning as above. We shall define $X_G(t)$ by truncating $X(t)$ by $G$, i.e.,

$$X_G(t) = X(t) \quad \text{or} \quad 0$$

according as $|X(t)| \leq G$ or not. Similarly we shall define $\tilde{a}_G(t)$ and $\tilde{b}_G(t)$ respectively by truncating $\tilde{a}(t)$ and $\tilde{b}(t)$ by $\alpha_G = \sqrt{M + \|K\| G}$. Now we shall consider the stochastic integral:

$$L_G(u, v) = \int_u^v \tilde{a}_G(t)\,dt + \int_u^v \tilde{b}_G(t)\,dB(t),$$

and observe

$$E[I_G(u, v)^4] \leq 8E\left[\left(\int_u^v \tilde{a}_G(u)\,dt\right)^4\right] + 8E\left[\left(\int_u^v \tilde{a}_G(t)\,dB(t)\right)^4\right]$$

$$E\left[\left(\int_u^v \tilde{a}_G(u)\,dt\right)^2\right] \leq \alpha_G^4 (v-u)^4$$

and

$$E\left[\left(\int_u^v \tilde{b}_G(t)\,dB(t)\right)^4\right] \leq 6\int_u^v E\left[\left(\int_u^t \tilde{b}_G(s)\,dB(s)\right)^2 \tilde{b}_G(t)^2\right] dt$$

(use Lemma 2. 1)

$$\leq 6\alpha_G^2 \int_u^v E\left[\left(\int_u^t \tilde{b}_G(s)\,dB(s)\right)^2\right] dt$$

$$\leq 6\alpha_G^2 \int_u^v \int_u^t E(\tilde{b}_G(s)^2)\,ds\,dt = 3\alpha_G^4 (v-u)^2.$$

Thus we get

$$E(I_G(u, v)^4) \leq 8\alpha_G^4 (l^2+3)(v-u)^2 \qquad (s \leq u < v \leq s+l)$$

If $\sup_{s-\sigma \leq t \leq s+l} |X(t)| \leq G$, then we have $X_G = X(t)$, $\tilde{a}_G(t) = \tilde{a}(t)$, $\tilde{b}_G(t) = \tilde{b}(t)$ and so $I_G(u, v) = X(v) - X(u)$, as far as $t, u, v \in [s, s+l]$, because the support $dK$ is contained in $[-\sigma, 0]$. Thus we get

$$E[(X(v)-X(u))^4, \sup_{s-\sigma \leq t \leq s+l} |X(t)| \leq G]$$
$$\leq E[I_G(u, v)^4] \leq 8\alpha_G^4 (l^2+3)(v-u)^2,$$

which proves Lemma 7. 2.

Now let us prove that $\{X_s, s>0\}$ is totally $L$-bounded. Using Lemma 7. 1, we have, for $G_n = G(2^{-n}, 2n+\sigma)$,

(7. 3) $\quad P(\sup_{-n-\sigma \leq t \leq n} |X_s(t)| \leq G_u) = P(\sup_{s-n-\sigma \leq t \leq s+n} |X(t)| < G_u) < 1-2^{-n}.$

Using Lemma 7.2, we have, for $u, v \in [s-n, s+n]$

$$E[(X(v)-X(u))^4 : \sup_{s-n-\sigma \leq t \leq s+n} |X(t)| \leq G_n] \leq \gamma_n(a-u)^2$$
$$(\gamma_n = \mu(G_n, 2n+\sigma)),$$

and so, for $u, v \in [-n, n]$

(7.4) $$E[(X_s(v)-X_s(u))^4, \sup_{-n-\sigma \leq t \leq n} |X_s(t) \leq G_n] \leq \gamma_n(v-u)^2 .$$

Writing $A_n$ for the set $\{f \in C: \sup_{-n-\sigma \leq t \leq n} |f)| \leq G_n\}$, we have

$$E[(X_s(v)-X_s(u))^4, X_s \in A_n] \leq \gamma_n(v-u)^2 \qquad (u, v \in [-n, n])$$

by (7.4) and

$$P(X_s \in A_n) > 1-2^{-n} \qquad (s>0)$$

by (7.3). It is clear by $(\tilde{S})$ that $E(X_s(t)^2) \leq \alpha$ for every $t \in (-\infty, \infty)$ and $s>0$. Thus we have proved that $\{X_s, s>0\}$ satisfies the assumptions in Lemma 3.3, so that it is totally $L$-bounded.

## 8. Stationary Solutions (3).

In the previous section 6 and 7 we presented some sufficient conditions for the existence of a stationary solution of the stochastic differential equation:

$$dX(t) = a(\pi_t X)dt + b(\pi_t X)dB(t),$$

using a one-sided solution of this equation with the past conditions: $X(t) \equiv 0$, $t \leq 0$; the existence of a one-sided solution was discussed in Section 5.

In the present section we shall use the results obtained in Sections 6 and 7 to give sufficient conditions for the existence of a stationary solution in terms of the coefficients $a$ and $b$.

We shall begin with three lemmas.

LEMMA 8.1. *If $r(t)$ and $\xi(t)$ are continuous on $[0, \infty)$, and if*

(8.1) $$r(t)-r(s) \leq -\beta\int_s^t r(u)du + \int_s^t \xi(u)du \qquad (0 \leq s < t < \infty),$$

*where $\beta$ is a positive constant, then*

(8.2) $$r(t) \leq r(0) + \int_0^t e^{-\beta(t-u)}\xi(u)du .$$

PROOF: Write $r_0(t)$ for the integral appearing in (8.2) and set $r_1(t)=r(t)-r_0(t)-r(0)$. $r_0(t)$ satisfies

$$r_0'(t) = -\beta r_0(t)+\xi(t)$$

and so

$$r_0(t)-r_0(s) = -\beta\int_s^t r_0(u)du+\int_s^t \xi(u)du\,.$$

It follows from this and (8.1) that $r_1(t)$ is continuous and

$$r_1(t)-r_1(s)\leq -\beta\int_s^t r_1(u)du \quad \text{and} \quad r_1(0)=0\,. \tag{8.3}$$

To prove (8.2), it is enough to prove that $r_1(t)\leq 0$. If $r_1(t)>0$ for some $t$, then $r_1(0)=0$ implies that there exists an interval $[s_1, t_1]\subset[0, \infty)$ with $r_1(t_1)>r_1(s_1)$ and $r_1(t)>0$ on $[s_1, t_1]$, which contradict (8.3).

LEMMA 8.2. *Suppose that $\rho(t)$ is continuous and satisfies*

$$\rho(t)\leq \alpha+\beta\int_0^t e^{-\gamma(t-s)}\rho(s)ds \qquad (0\leq t<\infty), \tag{8.4}$$

*weere $\alpha$, $\beta$ and $\gamma$ are all positive constants. If*

$$\gamma>\beta, \tag{8.5}$$

*then $\rho(t)$ is bounded; in fact,*

$$\rho(t)\leq\frac{\alpha\gamma}{\gamma-\beta} \tag{8.6}$$

PROOF: It is easy to verify that

$$\rho_0(t)\equiv\frac{\alpha}{\gamma-\beta}(\gamma-\beta e^{-(\gamma-\beta)t}) \tag{8.7}$$

satisfies

$$\rho_0(t) = \alpha+\beta\int_0^t e^{-\gamma(t-s)}\rho_0(s)ds \qquad (0\leq t<\infty). \tag{8.8}$$

Therefore $\rho_1(t)\equiv\rho(t)-\rho_0(t)$ satisfies

$$\rho_1(t)\leq\beta\int_0^t e^{-\gamma(t-s)}\rho_1(s)ds \qquad (0\leq t<\infty).$$

Hence it follows by the method of iteration that

$$\rho_1(t) \leq \frac{\beta^n t^n}{n!} \max_{0 \leq s \leq t} |\rho_1(s)| \to 0 \qquad (n \to \infty).$$

Thus we get $\rho(t) \leq \rho_0(t) \leq \alpha\gamma/(\gamma - \beta)$.

LEMMA 8.3. *Suppose that $r(t)$ is continuous and non-negative on $[0, \infty)$ and satisfies*

$$r(t) - r(s) \leq \int_s^t (-\alpha\gamma(u) + \beta\sqrt{r(u)} + \gamma)\, du \qquad (0 \leq s < t < \infty), \tag{8.9}$$

*where $\alpha$, $\beta$ and $\gamma$ are constants and*

$$\alpha > 0, \qquad \gamma > 0. \tag{8.10}$$

*Then $r(t)$ is bounded.*

PROOF: It is clear that $Q(\theta) \equiv -\alpha\theta^2 + \beta\theta + \gamma < 0$ on a certain half line $[\theta_0, \infty)$. We shall prove that $r(t) \leq \max(r(0), \theta_0^2)$. If $r(t) > \max(r(0), \theta_0^2)$ for some $t \geq 0$, then there exists an interval $[r_1, t_1] \subset [0, \infty]$ such that "$r(t_1) > r(s_1)$" and "$\sqrt{r(t)} > \theta_0$, i.e., $Q(\sqrt{r(t)}) > 0$ on $[s_1, t_1]$". It follows from the second condition and (8.9) that

$$r(t_1) - r(s_1) \leq \int_{s_1}^{t_1} Q(\sqrt{r(u)})\, du \leq 0,$$

which contradicts the first condition $r(t_1) > r(s_1)$.

Now we shall use Theorem 2 to derive

THEOREM 4. *Suppose that $a(f)$ is of the form*

$$a(f) = -a_0(f) f(0) + a_1(f) \tag{8.11}$$

*with $a_0(f)$ and $a_1(f)$ continuous in $f \in C_-$ and that $b(f)$ is continuous iu $f \in C_-$. Furthermore we assume the existence of positive constants $m$, $M$, $M_1$ and $M_2$ and bounded measures $dK_1$ and $dK_2$ on $(-\infty, 0]$ for which*

$$m \leq a_0(f) \leq M, \tag{8.12}$$

$$a_1(f)^4 \leq M_1 + \int_{-\infty}^0 f(t)^4\, dK_1(t), \tag{8.13}$$

*and*

$$b(f)^4 \leq M_2 + \int_{-\infty}^0 f(t)^4\, dK_2(t). \tag{8.14}$$

*Then there exists a stationary solution of*

$$dX(t) = a(\pi_t X)dt + b(\pi_t X)dB(t),$$

*provided*

(8.15) $$m > \|K_1\|^{1/4} + \frac{3}{2}\|K_2\|^{1/2}.$$

*Remark.* (8.13) can be replaced by

$$|a_1(f)| \leq M_1 + \int_{-\infty}^{0} |f(t)| dK_1(t)$$

or

$$|a_1(f)|^2 \leq M_1 + \int_{-\infty}^{0} |f(t)|^2 dK_1(t),$$

because each of these conditions implies (8.13) with *different* $M_1$ and $dK_1$. Similarly for (8.14).

PROOF: It follows from (8.11), (8.12) and (8.13) that

(8.16) $$\begin{aligned} a(f)^4 &\leq 8a_0(f)^4 f(0)^4 + 8a_1(f)^4 \\ &\leq 8M^4 f(0)^4 + 8M_1 + 8\int_{-\infty}^{0} f(t)^4 dK_1(t) \\ &= \tilde{M} + \int_{-\infty}^{0} f(t)^4 d\tilde{K}(t), \end{aligned}$$

where $\tilde{M} = 8M_1$ and $d\tilde{K} = 8dK_1 + 8M^4\delta_0$ ($\delta_0$ = the $\delta$-measure concentrated at 0). By virtue of (8.16) and (8.14) we can use Theorem 1 (and the remark immediately following it) to find a one-sided solution with the past condition: $X(t) \equiv 0$, $t \leq 0$ such that $E(X(t)^4) \leq \gamma e^{\gamma t}$ $(-\infty < t < \infty)$ for some $\gamma > 0$. To prove the existence of a stationary solution, it is enough to verify (*S*) in Theorem 2 for this one-sided solution.

Write $r(t)$ for $E(X(t)^4)$ and set

$$\rho(t) = \sup_{s \leq t} r(s).$$

It is clear that $r(t) \leq \rho(t) \leq \gamma e^{\gamma t}$. Now we shall prove that $\rho(t)$ is bounded.

By an assumption we have

(8.17) $$E[a(\pi_t X)^4] \leq \tilde{M} + \|\tilde{K}\|\rho(t),$$

(8.18) $$E[a_1(\pi_t x)^4] \leq M_1 + \|K_1\| \rho(t),$$

and

(8.19) $$E[b(\pi_t X)^4] \leq M_2 + \|K_2\| \rho(t).$$

Applying Lemma 2.1 to

$$X(t) - X(s) = \int_s^t a(\pi_u X) du + \int_s^t b(\pi_u X) dB(u) \qquad (0 \leq s \leq t),$$

we get

$$E[(X(t)-X(s))^4]$$
$$\leq 4\int_s^t E[|(X(u)-X(t))^3 a(\pi_u X)|] du + 6\int_s^t E[(X(u)-X(t))^2 b(\pi_u X)^2] du.$$

Using

$$4|(X-Y)^3 Z| \leq 3(X-Y)^4 + Z^4 \leq 24X^4 + 24Y^4 + Z^4,$$
$$2(X-X)^2 Z^2 \leq (X-Y)^4 + Z^4 \leq 8X^4 + 8Y^4 + Z^4,$$

(8.17) and (8.19), we get

$$E[(X(t)-X(s))^4] \leq \int_s^t (\gamma_1 + \gamma_2 e^{\gamma u}) du$$

with some constants $\gamma_1$ and $\gamma_2$, and so

$$|r(t)^{1/4} - r(s)^{1/4}|^4 \leq E((X(t)-X(s))^4) \leq \int_s^t (\gamma_1 + \gamma_2 e^{\gamma u}) du.$$

This shows that $r(t)$ is continuous in $t \geq 0$.

By similar arguments we can easily see that $E([|X(u)^3 a(\pi_u X)|]$, $E[|X(u)^4 a_0(\pi_u X)|]$, $E[|X(u)^3 a_1(\pi_u X)|]$ and $E[X(u)^2 b(\pi_u X)^2]$ are all bounded by a function of the form $\gamma_1 + \gamma_2 e^{\gamma t}$. Therefore we can apply Lemma 2.2 to

$$X(t) = X(s) + \int_s^t a(\pi_u X) du + \int_s^t b(\pi_u X) dB(u)$$

to get

$$E(X(t)^4) - E(X(s)^4)$$
$$\leq -4\int_s^t E[a_0(\pi_u X) X(u)^4] du + 4\int_s^t E[X(u)^3 a_1(\pi_u X)] du$$
$$+ 6\int_s^t E[X(u)^2 b(\pi_u X)^2] du$$

$$\leq -4m\int_s^t E(X(u)^4)du + \int_s^t E[3c\cdot X(u)^4 + \frac{1}{c^3}a_1(\pi_u X)^4]du$$
$$+3\int_s^t E[d\cdot X(u)^4 + \frac{1}{d}b(\pi_u X)^4]du,$$

where $c$ and $d$ are positive constants and will be determined later; we used $4X^3Y \leq 3X^4 + Y^4$ and $2X^2Y^2 \leq X^4 + Y^4$ in the above observation.

Using the results obtained above, we have

$$r(t)-r(s)$$
$$\leq -(4m-3c-3d)\int_s^t r(u)du + \int_s^t \Big[k + \Big(\frac{\|K_1\|}{c^3} + \frac{3\|K_2\|}{d}\Big)\rho(u)\Big]du,$$

where $k = M_1/c^3 + 3M_2/d$ (=constant).

If we assume that

(8.20) $$4m-3c-3d > 0,$$

then we can apply Lemma 8.1 to get

(8.21)
$$r(t) \leq \int_0^t \exp[-(4m-3c-3d)(t-u)]\cdot\Big[k + \Big(\frac{\|K_1\|}{c^3} + \frac{3\|K_2\|}{d}\Big)\rho(u)\Big]du.$$

Noticing that if $\xi(t)$ is increasing, then

$$\int_0^t e^{-\beta(t-s)}\xi(s)ds \equiv \frac{1-e^{-\beta t}}{\beta}\xi(0) + \int_0^t \frac{1-e^{-\beta(t-s)}}{\beta}d\xi(s)$$

is also increasing in $t$, we can see that the right side of (8.21) is also increasing, so that the left side of (8.21) can be replaced by $r(s)$ for every $s \leq t$ and so we get, with some constant $\tilde{k}$,

(8.22)
$$\rho(t) \leq \tilde{k} + \Big(\frac{\|K_1\|}{c^3} + \frac{3\|K_2\|}{d}\Big)\int_0^t \exp[-(4m-3c-3d)(t-u)]\rho(u)du.$$

Applying Lemma 8.2 to this, we can see that $\rho(t)$ is bounded, provided

(8.23) $$m > \frac{1}{4}\Big(3c + \frac{\|K_1\|}{c^3}\Big) + \frac{3}{4}\Big(d + \frac{\|K_2\|}{d}\Big);$$

notice that (8.23) includes (8.20).

Setting $c=\|K_1\|^{1/4}$ and $d=\|K_2\|^{1/2}$, we can see that $\rho(t)$ is bounded under the condition (8.15); in fact (8.15) is the best condition among all conditions of the form (8.23).

As an application of Theorem 3 we shall prove

THEOREM 5. *Suppose that $a(f)$ and $b(f)$ are written in the form:*

$$a(f)=a_0(f)f(0)+a_1(f), \quad b(f)=b_0(f)f(0)+b_1(f) \tag{8.24}$$

*with $a_0(f)$, $a_1(f)$, $b_0(f)$ and $b_1(f)$ continuous and bouhded in $f\in C_-$. Then it is sufficient for the existence of a stationary solution of the stochastic differential equation:*

$$dX(t)=a(\pi_t X)dt+b(\pi_t X)dB(t)$$

*that there exists a constant $m>0$ such that*

$$2a_0(f)+b_0(f)^2\leq -m \tag{8.25}$$

*for every $f\in C_-$.*

PROOF: By our assumption we have a constant $M$ which bounds $|a_0(f)|$, $|a_1(f)|$, $|b_0(f)|$ and $|b_1(f)|$ from above. Therefore

$$|a(f)|+|b(f)|\leq 2M+2M|f(0)|, \tag{8.26}$$

and so the conditions (A.1) and (A.2) in Theorem 3 are satisfied. To prove our theorem, it is enough to prove that a one-sided solution $X(t)$ with $X(t)\equiv 0$, $t\leq 0$, satisfies the condition that $E(X(t)^2)$ is bounded in $t\geq 0$; we can assume that $X(t)$ satisfies $E(X(t)^4)\leq \gamma e^{\gamma t}$ for some constant $\gamma>0$.

Using a formula on stochastic integral [4], we have

$$X(t)^2-X(s)^2=2\int_s^t X(u)a(\pi_u X)du+2\int_s^t X(u)b(\pi_u X)dB(u) + \int_s^t b(\pi_u X)^2 du. \tag{8.27}$$

Taking the expectations of both sides of (8.27) in view of

$$\begin{aligned} &E(|X(u)a(\pi_u X)|)+E(|X(u)b(\pi_u X)|)\\ &\leq 2ME(X(u)^2)+2ME(|X(u)|)\leq ME(X(u)^4+1)+\frac{1}{2}ME(X(u)^4+3)\\ &\leq \frac{5}{2}M+\frac{3}{2}ME(X(u)^4)\leq \frac{5}{2}M+\frac{3}{2}\gamma e^{\gamma u}, \end{aligned}$$

we get, for $0 \leq s \leq t < \infty$,

$$(8.28)\quad E(X(t)^2)-E(X(s)^2)$$

$$= 2\int_s^t E(X(u)a(\pi_u X))du + \int_s^t b(\pi_u X)^2 du$$

$$= \int_s^t E[(2a_0(\pi_u X)+b_0(\pi_u X)^2)X(u)^2]du$$

$$+2\int_s^t E[(a_1(\pi_u X)+b_1(\pi_u X)b_0(\pi_u X))X(u)]du$$

$$+\int_s^t E[b_1(\pi_u X)^2]du$$

$$\leq \int_s^t [-mE(X(u)^2)+2(M+M^2)E(|X(u)|)+M^2]du$$

$$\leq \int_s^t [-mE(X(u)^2)+2(M+M^2)\sqrt{E(X(u)^2)}+M^2]du\,.$$

The continuity of $E(X(t)^2)$ follows from

$$|E(X(t)^2)^{1/2}-E(X(s)^2)^{1/2}|^2 \leq E[(X(t)-X(s))^2]$$

$$= 2\int_s^t E[(X(u)-X(s))a(\pi_u X)]du + \int_s^t E[b(\pi_u X)^2]du$$

which can be verified in the same way as above.

Therefore we can apply Lemma 8.3 to $r(t) \equiv E(X(t)^2)$ to see that $E(X(t)^2)$ is bounded.

As a second application of Theorem 3 we can prove the following theorem which is similar to Theorem 4.

THEOREM 6. *Under the same assumptions as in Theorem* 4 *except*

$$(8.13')\qquad a_1(f)^2 \leq M_1 + \int_{-\infty}^0 f(t)^2 dK_1(t)$$

$$(8.14')\qquad b(f)^2 \leq M_2 + \int_{-\infty}^0 f(t)^2 dK_2(t)$$

*instead of* (8.13) *and* (8.14), *there exists a stationary solution of the same stochastic differential equation, provided*

$$(8.15')\qquad m > \|K_1\|^{1/2} + \frac{1}{2}\|K_2\|$$

*and*

(C) *the supports of* $dK_1$ *and* $dK_2$ *are bounded.*

PROOF: Since the integrability of the integrals appearing here will be proved in the same way as in Theorem 4, we shall carry out only the formal computations.

Using Theorem 3 in view of (*C*), we can see that it is enough to show the boundedness of $E(X(t)^2)$.

By virtue of a formula on stochastic integrals [4] we get

$$X(t)^2 - X(s)^2 = 2\int_s^t X(u)a(\pi_u X)dt + 2\int_s^t X(u)b(\pi_u X)dB(u) + \int_s^t b(\pi_u X)^2 du$$

and so

$$\begin{aligned} E(X(t)^2 - E(X(s)^2) &= \int_s^t E[2X(u)a(\pi_u X) + b(\pi_u X)^2]du \\ &\leq -2m\int_s^t E[X(u)^2]du + \int_s^t E[2X(u)a_1(\pi_u X) + b(\pi_u X)^2]du \\ &\leq -2m\int_s^t E[X(u)^2]du + \int_s^t E[cX(u)^2 + \frac{1}{c}a_1(\pi_u X)^2 + b(\pi_u X)^2]du \\ &\qquad (c = ||K_1||^{1/2}). \end{aligned}$$

Setting $r(t) = E(X(t)^2)$ and $\rho(t) = \max_{0\leq s\leq t} r(s)$, we get

$$\begin{aligned} &r(t) - r(s) \\ &\quad \leq -2m\int_s^t r(u)du + \int_s^t \left[\left(c + \frac{1}{c}||K_1|| + ||K_2||\right)c(u) + \gamma\right]du \\ &\qquad\qquad \left(\gamma = \frac{1}{c}M_1 + M_2\right) \\ &\quad = -2m\int_s^t r(u)du + \int_s^t [(2||K_1||^{1/2} + ||K_2||)\rho(u) + \gamma]du\,. \end{aligned}$$

Hence it follows by Lemma 8.1 that

$$\begin{aligned} r(t) &\leq \int_0^t e^{-2m(t-s)}[(2||K_1||^{1/2} + ||K_2||)\rho(s) + \gamma]ds \\ &\leq \frac{\gamma}{2m} + (2||K_1||^{1/2} + ||K_2||)\int_0^t e^{-2m(t-s)}\rho(s)ds\,. \end{aligned}$$

Now use Lemma 8.2 to conclude that $r(t)$ is bounded if $m > ||K_1||^{1/2} + ||K_2||/2$.

The following theorem is an immediate result from Theorem 4 and Theorem 6.

THEOREM 7. *Under the same assumptions as in Theorem 4 except*

$$(8.13'') \qquad |a_1(f)| \leq M_1 + \int_{-\infty}^{0} |f(t)|\, dK_1(t)$$

$$(8.14'') \qquad |b(f)| \leq M_2 + \int_{-\infty}^{0} |f(t)|\, dK_2(t)$$

*instead of* (8.13) *and* (8.14), *there exists a stationary solution of the same stochastic differential equation, if*

$$(8.15'') \qquad m > \|K_1\| + \frac{3}{2}\|K_2\|^2 .$$

*In case the supports of $dK_1$ and $dK_2$ are bounded,* (8.15″) *can be replaced by a weaker condition*:

$$(8.15''') \qquad m > \|K_1\| + \frac{1}{2}\|K_2\|^2 .$$

PROOF: By virtue $(X+Y)^2 < \left(1+\frac{1}{\varepsilon}\right)X^2 + (1+\varepsilon)Y^2$ ($\varepsilon > 0$) and the Schwartz inequality, (8.13″), and (8.14″) imply

$$|a_1(f)|^2 \leq \left(1+\frac{1}{\varepsilon}\right)M_1^2 + (1+\varepsilon)\|K_1\| \int_{-\infty}^{0} |f(t)|^2 dK_1(t),$$

$$|b(f)|^2 \leq \left(1+\frac{1}{\varepsilon}\right)M_2^2 + (1+\varepsilon)\|K_2\| \int_{-\infty}^{0} |f(t)|^2 dK_2(t),$$

$$|a_1(f)|^4 \leq \left(1+\frac{1}{\varepsilon}\right)^3 M_1^4 + (1+\varepsilon)^3\|K_1\|^3 \int_{-\infty}^{0} |f(t)|^4 dK_1(t)$$

and

$$|b(f)|^4 \leq \left(1+\frac{1}{\varepsilon}\right)^3 M_2^4 + (1+\varepsilon)^3\|K_2\|^3 \int_{-\infty}^{0} |f(t)|^4 dK_2(t).$$

If (8.15″) holds, then we have

$$m > (1+\varepsilon)^{3/4}\|K_1\| + \frac{3}{2}(1+\varepsilon)^{3/2}\|K_2\|^2$$

i.e.,

$$m > ((1+\varepsilon)^3\|K_1\|^4)^{1/4} + \frac{3}{2}[(1+\varepsilon)^3\|K_2\|^4]^{1/2}$$

by taking small $\varepsilon > 0$. Now we can apply Theorem 4 to prove first part of our theorem.

The second part will be proved similarly by Theorem 6.

## 9. Borel Algebras Related to the Stationary Solution

Let $X(t)$ be a stationary solution of the stochastic differential equation:

$$dX(t) = a(\pi_t X)dt + b(\pi_t X)dB(t). \tag{9.1}$$

We shall prove some facts concerning the Borel algebras related to the solution $X$ and the Wiener process $B$.

The following three relations seem to be interesting:

$$\mathcal{B}_{-\infty t}(X) \subset \mathcal{B}_{-\infty t}(dB) \qquad (-\infty < t < \infty), \tag{9.2}$$

$$\bigwedge_t \mathcal{B}_{-\infty t}(X) = \mathfrak{N}, \tag{9.3}$$

$$\mathcal{B}_{-\infty t}(X) \subset \mathcal{B}_{-\infty s}(X) \vee \mathcal{B}_{st}(dB) \qquad (-\infty < s < t < \infty), \tag{9.4}$$

where $\mathfrak{N}$ is a *triuial algebra* which contains only sets of probability 0 or 1.

It is evident that (9.2) implies (9.3).

In the general cases discussed in the previous sections we neither proved nor disproved these relations. In the next three sections we shall discuss some special cases in which (9.2) is true. In Section 13 we shall discuss the diffusion defined by a stochastic differential equation with coefficients satisfying the Lipschitz condition for which (9.4) holds but (9.3) does not always hold. In Section 14 we shall mention an example for which (9.4) does not hold. In Section 16 we shall give a two-dimensional stochastic differential equation for which both (9.3) and (9.4) are true but (9.2) is false.

If $X$ is a stationary solution of (9.1), then $X$ is strictly stationarily correlated to $dB$. In the following Theorems 8 and 9 we shall only assume this property.

The following theorem will be used in Section 16 and a similar fact for the discrete time parameter case was discussed by M. Rosenblatt [14] [15] [16].

THEOREM 8. *Assume that $X$ is strictly stationarily correlated to $dB$.*

(i) *If* (9.2) *holds, then*

(9.5) $$E[|h(X(t))-E[h(X(t))|\mathcal{B}_{0t}(dB)]|]\to 0 \qquad (t\to\infty)$$

*for every bounded Borel measurable* $h(\xi)$ *defined on* $R^1$.

(ii) *If* (9.5) *holds for one bounded monotone function* $h$ (*for example* $h(\xi)\equiv\arctan\xi$), *then* (9.2) *holds.*

PROOF: It is enough to observe that Lévy-Doob's theorem shows

(9.6) $$E[h(X(0))|\mathcal{B}_{-\infty 0}(dB)] = \lim_{t\to\infty} E[h(X(0)|\mathcal{B}_{-t0}(dB)]$$

with probability 1 and that the strictly stationary correlation between $X$ and $dB$ implies

(9.7) $$\begin{aligned}&E[|h(X(t))-E[h(X(t))|\mathcal{B}_{0t}(dB)]|]\\ &\quad= E[|h(0))-E[h(X(0))|\mathcal{B}_{-t0}(dB)]|].\end{aligned}$$

As an immediate result from the definitions we get

THEOREM 9. *Assume that* $X$ *is strictly stationarily correlated to* $dB$. *If* (9.2) *holds, then there exists a Borel measurable functional* $F$ *on* $C_-$ $(\mathcal{B}(C_-))$ *such that*

(9.8) $$X(t) = F(\pi_t B)$$

*where*

(9.9) $$\pi_t B(s) = B(s+t)-B(t), \qquad s\leq 0.$$

*Furthermore, if* $E[X(0)^2]<\infty$, *then* $X(t)$ *can be expanded in multiple Wiener integrals* (see K. Itô [5] [6]):

(9.8′) $$X(t) = \sum_{n\geq 0}\int\cdots\int_{t_1\leq\cdots\leq t_n\leq n} f(t-t_1, t-t_2, \cdots, t-t_n)dB(t_1)\cdots dB(t_n).$$

We shall call (9.8) a *backward representation* of $X(t)$ and (9.8′) a *backward representation by multiple Wiener integrals* (see N. Wiener [19]).

The representation (9.8) or (9.8′) is called *properly canonical* (see M. Nisio [12]) if it holds that

(9.10) $$\mathcal{B}_{-\infty t}(X) = \mathcal{B}_{-\infty t}(dB) \qquad (-\infty<t<\infty)$$

By a theorem of Doob on stochastic integrals we can easily prove

THEOREM 10. *If $X(t)$ is a stationary solution of* (9.1) *satisfying* (9.2) *and if $b(f)>0$ (or at least $P(b(\pi_0 X)>0)=1$), then $X(t)$ has a properly canonical backward representation. Furthermore if $E(X(0)^2)<\infty$, then $X(t)$ has a properly canonical backward representa tion by multiple Wiener integrals.*

PROOF: It is enough to prove that

$$B(t)-B(s)=\int_s^t b(\pi_u X)^{-1}dX(u)$$

following Doob (see page 448 in his book [1]).

## 10. Lipschitz Conditions

In this section we shall discuss the stochastic differential equation:

$$dX(t)=a(\pi_t X)dt+b(\pi_t X)dB(t) \tag{10.1}$$

in case $a(f)$ and $b(f)$ satisfy the Lipschitz condition.

Let $C^-(dK_1 dK_2)$ denote the subspace of $C_-$:

$$\left\{f\in C_-:\ \int_{-\infty}^0 |f(t)dK_1(t)+\int_{-\infty}^0 |f(t)|dK_2(t)<\infty\right\}$$

and $\mathfrak{B}_{\rho_-}(C_-(dK_1 dK_2))$ the Borel algebra of sets of the form $C_-(dK_1 dK_2)\wedge B$, $B\in\mathfrak{B}_{\rho_-}(C_-)$. Notice that $C_-(dK_1 dK_2)=C_-$ in case the supports of $dK_1$ and $dK_2$ are bounded.

We shall first discuss one-sided solutions.

THEOREM 11. *Suppose that $a(f)$ and $b(f)$ are measurable in $f\in C_-(dK_1 dK_2)$ with respect to $\mathfrak{B}_{\rho_-}(C_-(dK_1 dK_2))$ and satisfy the Lipschitz condition:*

$$|a(f)-a(g)|^2\le\int_{-\infty}^0 |f(t)-g(t)|^2 dK_1(t) \tag{L.1}$$

*and*

$$|b(f)-b(g)|^2\le\int_{-\infty}^0 |f(t)-g(t)|^2 dK_2(t) \tag{L.2}$$

*with bounded measures $dK_1$ and $dK_2$ on $(-\infty, 0]$. Then there exists one and only one solution of* (10.1) *with the past condition:*

(10.2) $$X(t) = X_-(t) \qquad t \leq 0,$$

*where* $X_-(t)$, $t \leq 0$, *is a given stochastic process independent of* $\mathcal{B}_{0\infty}(dB)$ *and is bounded in the second order moment, i.e.,*

(10.3) $$E[X_-(t)^2] \leq \alpha \qquad (\alpha = constant)$$

*Furthermore this solution satisfies*

(10.4) $$\mathcal{B}_{-\infty t}(X) \subset \mathcal{B}_{-\infty 0}(X_-) \vee \mathcal{B}_{0t}(dB)$$

*Remark* 1. (L. 1) can be replaced by

(L. 1′) $$|a(f)-a(g)| \leq \int_{-\infty}^{0} |f(t)-g(t)| d\tilde{K}_1(t),$$

because (L. 1′) implies (L. 1) with $dK_1 = \|\tilde{K}_1\| dK_1$. Similarly for (L. 2).

*Remark* 2. If $dK_1$ and $dK_2$ are concentrated at 0, then $a(\pi_s X)$ and $b(\pi_t X)$ can be written as $\alpha(X(t))$ with $\beta(X(t))$ with $\alpha(\xi)$ and $\beta(\xi)$ satisfying the usual Lipschitz condition. This case was treated by K. Itô [8] in connection with the construction of diffusion processes attached to the infinitesimal generator:

$$\mathcal{G} = a(\xi)\frac{d}{d\xi} + \frac{1}{2} b(\xi)^2 \frac{d^2}{d\xi^2}.$$

Since the method of the proof for this special case will work in our present case with no essential change, we shall give an outline of the proof of the existence only.

*Proof of the existence part of Theorem* 11. It is enough to solve the the stochastic integral equation

(10.5) $$X(t) = X(0) + \int_0^t a(\pi_s X) ds + \int_0^t b(\pi_s X) dB(s)$$

with (10.2). To do this, we shall use the successive approximation.

Let us define a sequence of approximate solutions $X_n(t)$, $n=0, 1, 2, \cdots$ as follows:

(10.6) $$X_0(t) = X_-(t) \qquad (t \leq 0)$$
$$= X_-(0) \qquad (t \geq 0)$$

(10.7) $$X_n(t) = X_-(t) \qquad (t \leq 0)$$
$$= X_-(0) + \int_0^t a(\pi_s X_{n-1}) ds + \int_0^t b(\pi_s X_{n-1}) dB(s) \qquad (t \geq 0)$$

The existence of the integral in (10.7) and the following inequality will be proved by induction:

$$(10.8)\quad E[|X_n(t)-X_{n-1}(t)|^2]\leq \frac{\gamma_1\gamma_2^n(t+1)^n t^n}{n!} \qquad (t>0,\ n=1,2,\cdots),$$

where $\gamma_1$ and $\gamma_2$ are some constants determined by $\alpha$, $\|K_1\|$, $\|K_2\|$, $|a(0)|$ and $|b(0)|$ (0 in $a(0)$ and $b(0)$ stands for the function identically equal to 0 on $(-\infty, 0]$).

Using the property of stochastic integrals, we get

$$\begin{aligned}
&P(\sup_{0\leq s\leq t}|X_{n+1}(t)-X_n(s)|>2\varepsilon)\\
&\leq P\Big(\int_0^t|a(\pi_sX_n)-a(\pi_sX_{n-1})|ds>\varepsilon\Big)\\
&\quad+P\Big(\sup_{0\leq s\leq t}\Big|\int_0^t(b(\pi_sX_n)-b(\pi_sX_{n-1}))dB(s)\Big|>\varepsilon\Big)\\
&\leq \varepsilon^{-2}t\int_0^t E[(a(\pi_sX_n)-a(\pi_sX_{n-1}))^2]ds\\
&\qquad+\varepsilon^{-2}\int_0^t E[(b(\pi_sX_n)-(b(\pi_sX_{n-1}))^2]ds\\
&\leq \frac{\varepsilon^{-2}\gamma_3\gamma_2^n(t+1)^{n+1}t^{n+1}}{(n+1)!}
\end{aligned}$$

with some constant $\gamma_3>0$.

Setting $\varepsilon=n^{-2}$ and $t=\log n$ in this inequality, we get

$$\begin{aligned}
(10.9)\quad &P(\sup_{0\leq s\leq \log n}|X_{n+1}(s)-X_n(s)|\geq 2n^{-2})\\
&\leq \frac{\gamma_4\gamma_2^{n+1}(\log n)^{n+1}n^2(1+\log n)^{n+1}}{(n-1)!} \qquad (\gamma_4=\text{constant}>0)
\end{aligned}$$

It is clear by Stirling's formula on $n!$ that the right side of (10.9) is the $n$-th term of a convergent series. Hence it follows by Borel-Cantelli's lemma that

$$P(\sum_n \sup_{r\leq s\leq \log n}|X_{n+1}(s)-X_n(s)|<\infty)=1,$$

which shows that, with probability 1,

$$X_n(s)\equiv X_0(s)+\sum_{k=1}^n(X_k(s)-X_{k-1}(s)), \qquad n=1,2,\cdots$$

is convergent uniformly on every bounded subinterval of $[0,\infty)$.

Therefore the limit process $X(t) \equiv \lim_{n\to\infty} X_n(t)$ exists and is continuous in $t$ with probability 1.

(10. 8) implies that

$$\sum_n [E[(X_n(t)-X_{n-1}(t))^2]]^{1/2} < \infty,$$

so that

$$E((X(t)-X_n(t))^2 \to 0 \qquad (n\to\infty). \tag{10.10}$$

Letting $n$ tend to $\infty$ in (10.7) in view of (10.8), (10.9) and (10.10), we can conclude that $X(t)$ satisfies our stochastic integral equation with (10.2).

Since $\mathcal{B}_{-\infty t}(X_n) \subset \mathcal{B}_{-\infty t}(X_-) \vee \mathcal{B}_{t\infty}(dB)$ by the definition of $X_n$, (10.4) is obvious.

Now we shall discuss stationary solutions.

THEOREM 12. *Suppose that $a(f)$ is of the form:*

$$a(f) = -cf(0)+a_1(f) \qquad (c = \textit{positive constant}) \tag{10.11}$$

*and that $a_1(f)$ and $b(f)$ are measurable in $f\in C_-(dK_1 dK_2)$ with respect to $\mathcal{B}_{\rho\_}(C_-(dK_1 dK_2))$. If following Lipschitz conditions are satisfied for $a_1$ and $b$:*

$$|a_1(f)-a_1(g)|^2 \leq \int_{-\infty}^0 |f(t)-g(t)|^2 dK_1(t) \tag{L. 1'}$$

*and*

$$|b(f)-b(g)|^2 \leq \int_{-\infty}^0 |f(t)-g(t)|^2 dK_2(t) \tag{L. 2}$$

*and if*

$$c > ||K_1||^{1/2} + \frac{1}{4}||K_2|| + \frac{1}{4}(||K_2||^2 + 8||K_1||^{1/2}||K_2||)^{1/2}, \tag{10.12}$$

*then there exists one and only one stationary solution of the stochastic differential equation* (10. 1) *with*

$$E(X(0)^2) < \infty. \tag{10.13}$$

*This solution satisfies $\mathcal{B}_{-\infty t}(X) \subset \mathcal{B}_{-\infty t}(dB)$ and has a backward representation by multiple Wiener integrals on $dB$ (see Section* 9 *for the definition of representations).*

PROOF:

(i) *Existence.* Consider the stochastic integral equation:

$$X(t)=\int_{-\infty}^{t}e^{-c(t-s)}a_1(\pi_sX)ds+\int_{-\infty}^{t}e^{-c(t-s)}b(\pi_sX)dB(s). \tag{10.14}$$

Any solution of this equation satisfies (10. 1), because it follows from (10. 14) that

$$X(t)=e^{-c(t-s)}X(s)+e^{-ct}\int_s^t e^{cu}a_1(\pi X)du+e^{-ct}\int_s^t e^{cu}b(\pi_uX)dB(u) \qquad (s<t),$$

so that a formula on stochastic differentials [4] shows that

$$\begin{aligned}dX(t)&=-ce^{-c(t-s)}X(s)dt-ce^{-ct}dt\int_s^t e^{cu}a_1(\pi_uX)du+e^{-ct}e^{ct}a_1(\pi_tX)dt\\&\quad-ce^{-ct}dt\int_s^t e^{cu}b(\pi_uX)dB(u)+e^{-ct}e^{ct}b(\pi_tX)dB(t)\\&=-cX(t)dt+a_1(\pi_tX)dt+b(\pi_tX)dB(t)\\&=a(\pi_tX)dt+b(\pi_tX)dB(t).\end{aligned}$$

Therefore if we can find a stationary process $X(t)$ which statisfies (10. 14), then it will be a solution of (10. 1).

In order to solve (10. 14) we shall use the method of successive approximation. Define $X_n(t)$, $n=0,1,2,\cdots$ as

$$X_0(t)\equiv 0, \tag{10.15}$$

$$X_n(t)=\int_{-\infty}^{t}e^{-c(t-s)}a_1(\pi_sX_{n-1})ds+\int_{-\infty}^{t}e^{-c(t-s)}b(\pi_sX_{n-1})dB(s). \tag{10.10}$$

Let $\theta_\tau$ be the shift operator on the space of all random variables measurable with respect to $\mathcal{B}(dB)$ that is determined by

$$\theta_\tau B(s,t)=B(s+\tau,t+\tau) \qquad (B(s,t)=B(t)-B(s)) \tag{10.17}$$

The precise definition of $\theta_\tau$ was given in Section 9.

We shall easily verify the existence of the stochastic integral in (10. 16) and the following properties of $X_n(t)$ by induction:

$$\mathcal{B}_{-\infty t}(X_n)\subset\mathcal{B}_{-\infty t}(dB) \tag{10.18}$$

$$\theta_\tau X(t)=X_n(t+\tau) \tag{10.19}$$

and

(10.20) $$E[(X_n(t)-X_{n-1}(t))^2]<\infty$$

Setting $\rho_n=E[(X_n(t)-X_{n-1}(t))^2]$, noticing that $\rho_n$ is finite by (10.21) and independent of $t$ by (10.21) and using $(X+Y)^2\leq(1+\varepsilon)X^2+(1+\varepsilon^{-1})Y^2$ and the Schwarz inequality, we get

$$\begin{aligned}\rho_{n+1}&=E[(X_{n+1}(t)-X_n(t))^2]\\&\leq E\Big\{(1+\varepsilon)\Big[\int_{-\infty}^t e^{-c(t-s)}(a(\pi_sX_n)-a(\pi_sX_{n-1}))ds\Big]^2\Big\}\\&\quad+E\Big\{(1+\varepsilon^{-1})\Big[\int_{-\infty}^t e^{-c(t-s)}(b(\pi_sX_n)-b(\pi_sX_{n-1}))dB(s)\Big]^2\Big\}\\&\leq[(1+\varepsilon)(c^2)^{-1}\|K_1\|+(1+\varepsilon^{-1})(2c)^{-1}\|K_2\|]\rho_n\qquad(\varepsilon>0)\end{aligned}$$

Setting $\varepsilon=(c\|K_2\|/2\|K_1\|)^{1/2}$ to get the best estimate:

$$\rho_{n+1}\leq\Big(\frac{1}{c}\|K_1\|^{1/2}+\frac{1}{\sqrt{2c}}\|K_2\|\Big)^2\rho_n,$$

we obtain

(10.21) $$\rho_{n+1}\leq A^{2n}\rho_1\qquad\Big(A=\frac{1}{c}\|K_1\|^{1/2}+\frac{1}{\sqrt{2c}}\|K_2\|^{1/2}\Big)$$

It is easy to see that "$A<1$" is equivalent to (10.12), so that $\rho_n$ tends to 0 exponentially fast under the assumption (10.22).

Noticing

$$\begin{aligned}&\sup_{|t|\leq T}|X_{n+1}(t)-X_n(t)|\\&\leq e^{2cT}\int_{-\infty}^T e^{-c(T-s)}|a_1(\pi_sX_n)-a_1(\pi_s(X_{n-1})|ds\\&\qquad+e^{2cT}\sup_{-\infty\leq t\leq T}\Big|\int_{-\infty}^t e^{-c(T-s)}(b(\pi_sX_n)-b(\pi_sX_{n-1}))dB(s)\Big|,\end{aligned}$$

we can easily see

$$P(\sup_{|s|\leq\log n}|X_n(t)-X_{n-1}(t)|>n^{2c}A^{n/2})\leq\gamma A^{-n}\rho_n\leq\gamma A^n\rho_1$$
$$(\gamma=\text{constant});$$

we should use here the fact that

$$P\Big[\sup_{s<t\leq T}\Big|\int_S^t Y(s)dB(s)\Big|>\varepsilon\Big]\leq\varepsilon^{-2}\int_S^T E(Y(t)^2)dt$$

is true even if $S=-\infty$, which can be easily seen as a limit of the usual case that $S>-\infty$.

Since $\sum_n \gamma A^n \rho_1 < \infty$ and $\sum_n n^{2c} A^{n/2} < \infty$, we can use Eorel-Cantelli's lemma to see that with probability 1, $X_n(t)$ converges uniformly on every bounded interval and that the limit process $X(t)$ is a solution of (10. 14).

It follows at once from (10. 18), (10. 19) and (10. 21) that $\mathcal{B}_{-\infty t}(X) \subset \mathcal{B}_{-\infty t}(dB)$, $\theta_\tau X(t) = X(t+\tau)$ and $E(X(0)^2) < \infty$, so that $X(t)$ has a backward non-linear representation by multiple Wiener integrals.

*Uniqueness.* Let $X(t)$ be the stationary solution obtained above and $Y(t)$ any stationary solution. Using a formula on stochastic integrals [2], we can verify

$$X(t) = e^{-c(t-s)}X(s) + \int_s^t e^{-c(t-u)} a_1 \pi_u(X)\,du + \int_s^t e^{-c(t-u)} b(\pi_u X)\,dB(u)$$

and the same equation for $Y(t)$, from which we get

(10. 22)

$$\begin{aligned} Y(t)-X(t) = e^{-c(t-s)}(Y(s)-X(s)) &+ \int_s^t e^{-c(t-s)}[a_1(\pi_u Y)-a_1(\pi_u X)]\,du \\ &+ \int_s^t e^{-c(t-s)}[b(\pi_u Y)-b(\pi_u X)]\,dB(u); \end{aligned}$$

notice that the integrals are meaningful, because $\mathcal{B}_{-\infty t}(X, Y, dB)$ $(\subset \mathcal{B}_{-\infty t}(Y, dB))$ is independent of $\mathcal{B}_{t\infty}(dB)$,

$$\begin{aligned} E[|a_1(\pi_u Y)-a_1(\pi_u X)|^2] &\leq \int_{-\infty}^0 E[|Y(u+v)-X(u+v)|^2]\,dK_1(u) \\ &\leq 2\|K_1\|[E(Y(0)^2)+E(X(0)^2)] \end{aligned}$$

and similarly

$$E[|b(\pi_u Y)-b(\pi_u X)|^2] \leq 2\|K_2\|[E(Y(0)^2)+E(X(0)^2)].$$

Using (10.22) and the same method as in the proof of (10.21), we can see that $r(t) = E[(Y(t)-X(t))^2]$ and $\rho(t) \equiv \sup_{s \leq t} r(s)$ $(\leq 2[E(Y(0)^2)+E(X(0)^2)] < \infty)$ satisfy

$$r(t) \leq e^{-(t-t)}\rho(t) + (1+\varepsilon)\frac{1}{c^2}\|K_1\|\rho(t) + (1+\varepsilon^{-1})\frac{1}{2c}\|K_2\|\rho(t)$$
$$(t > s,\ \varepsilon > 0).$$

Now letting $s$ tend to $-\infty$ in this inequality and taking the same $\varepsilon$ as before, we get

$$r(t) \leq A^2\rho(t) \qquad \left(A \equiv \frac{1}{c}\|K_1\|^{1/2} + \frac{1}{\sqrt{2c}}\|K_2\|^{1/2} < 1\right)$$

and so

$$r(s) \leq A^2\rho(s) \leq A^2\rho(t) \quad \text{for} \quad s \leq t,$$

i.e.,

$$\rho(t) \leq A^2\rho(t).$$

Since $A<1$, we get $\rho(t)=0$, which shows $P(Y(t)=X(t))=1$ for each $t$. Hence it follows that

$$P(Y(t) = X(t) \text{ for every } t) = 1,$$

because $Y(t)$ and $X(t)$ are continuous, in $t$ with probability 1. This completes the proof of Theorem 12.

The following theorem which will be useful later is an immediate consequence of Theorem 12 by virtue of the Schwartz inequality.

THEOREM 13. *In Theorem* 9 *we can replace* (L. 1′), (L. 2) *and* (10. 14) *with the following conditions*

(L*. 1′) $$|a_1(f)-a_1(g)| \leq \int_{-\infty}^{0} |f(t)-g(t)|dK_1(t)$$

(L*. 2) $$|b(f)-b(g)| \leq \int_{-+}^{0} |f(t)-g(t)|dK_2(t)$$

*and*

(10. 14*) $$c > \|K_1\| + \frac{1}{4}\|K_2\|^2 + \frac{1}{4}[\|K_2\|^4 + 8\|K_1\|\,\|K_2\|^2]^{1/2}.$$

## 11. Linear Coefficients (1).

In this section we shall solve the stochastic differential equation:

(11. 1) $$dX(t) = a(\pi_t X)dt + b(\pi_t X)dB(t)$$

in case $a(f)$ and $b(f)$ are *linear*, i.e.,

(11. 2) $$a(f) = M_1 + \int_{-\infty}^{0} f(t)dK_1(t), \qquad b(f) = M_2 + \int_{-\infty}^{0} f(t)dK_2(t),$$

where $dK_1$ and $dK_2$ are bounded signed measures on $(-\infty, 0]$. $a(f)$ is defined for every Borel measurable function $f$ integrable on $(-\infty, 0]$ with respect to the total variation measure $|dK_1|$ of $dK_1$. Similarly for $b(f)$.

It is obvious that

$$
\begin{aligned}
|a(f)| &\leq |M_1| + \int_{-\infty}^{0} |f(t)|\,|dK_1(t)|\,, \\
|b(f)| &\leq |M_1| + \int_{-\infty}^{0} |f(t)|\,|dK_2(t)|\,,
\end{aligned} \tag{11.3}
$$

and

$$
\begin{aligned}
|a(f)-a(g)| &\leq \int_{-\infty}^{0} |f(t)-g(t)|\,|dK_1(t)|\,, \\
|b(f)-b(g)| &\leq \int_{-\infty}^{0} |f(t)-g(t)|\,|dK_2(t)|\,.
\end{aligned} \tag{11.4}
$$

Since the Lipschitz conditions for $a(f)$ and $b(f)$ are satisfied by (11.4), all the results obtained in the previous section hold here.

If the supports of $dK_1$ and $dK_2$ are bounded, then $a(f)$ and $b(f)$ are defined for every $f \in C_-$ and the condition (A.2) (or (A.2′)) is satisfied, so that the results in Section 5, 6, 7 and 8 hold here.

The following theorem for the stationary solutions holds only in our case of linear coefficients.

THEOREM 14. *If $x(t)$ is any stationary solution of the stochastic differential equation* (11.1) *with $a(f)$ and $b(f)$ linear in $f$ and if $\gamma_0 = E(X(0)^2) < \infty$, then $Y(t) = E[X(t)\,|\,\mathfrak{B}(dB)]$ has a version with continuous paths which is a stationary solution of* (11.1) *and has a backward representation by multiple Wiener integrals.*

PROOF: We shall use the following facts on the conditional expectation:

(C.1) if $E(X^2) < \infty$, then $E[E(X\,|\,\mathfrak{C})^2] \leq E(X^2)$,

(C.2) if $\mathfrak{C} \supset \mathfrak{D}$, if $\theta$ is a measure preserving set transformation (up to measure 0) of $\mathfrak{C}$ onto itself ($\theta$ denoting also the transformation of the space of $\mathfrak{C}$-measurable function onto itself induced by the set transformation $\theta$) and if $X$ is $\mathfrak{C}$-measurable and integrable, then

$$E(\theta X\,|\,\theta\mathfrak{D}) = \theta E(X\,|\,\mathfrak{D})\,,$$

(C. 3) if $X$ is $\mathcal{C}$–measurable and integrable, if $\mathcal{C} \vee \mathfrak{D}$ is independent of $\mathcal{E}$, then $E(X|\mathfrak{D} \vee \mathcal{E}) = E(X|\mathfrak{D})$.

(C. 1) and (C. 2) follow at once from the definitions. To prove (C. 3), let $Y$ be bounded and $\mathfrak{D}$-measurable and $Z$ bounded and $\mathcal{E}$-measurable. Then

$$E[E(X|\mathfrak{D} \vee \mathcal{E})YZ] = E(XYZ) = E(XY)E(Z) = E[E(X|\mathfrak{D})Y]E(Z) = E[E(X|\mathfrak{D})YZ]$$

which proves (C. 3).

Since $E[X(t)^2] = E[X(0)^2] = \gamma_0 < \infty$, we have

$$E[a(\pi_t X)^2] \leq \gamma_1, \qquad E[b(\pi_t X)^2] \leq \gamma_1 \tag{11.5}$$

by (11. 3); $\gamma_1, \gamma_2, \cdots$ stand for positive constants in this proof.

Using (11. 5) and

$$X(t) - X(s) = \int_s^t a(\pi_u X)du + \int_s^t b(\pi_u X)dB(u) \qquad (-\infty < s < t < \infty), \tag{11.6}$$

we get

$$E[|X(t) - X(s)|^2] \leq \gamma_2 |t-s| \qquad (|t-s| < 1) \tag{11.7}$$

with some constant $\gamma_2$ and so

$$\begin{aligned} &E[|a(\pi_t X) - a(\pi_s X)|^2] \leq \gamma_3 |t-s|; \\ &E[|b(\pi_c X) - b(\pi_c X)|^2] \leq \gamma_3 |t-s| \qquad (|t-s| < 1), \end{aligned} \tag{11.8}$$

by virtue of (11. 2).

Since $E[X(t)^2] = E[(X(0)^2] = \gamma_0 < \infty$, $Y(t) \equiv E[X(t)|\mathcal{B}(dB)]$ is well defined and so

$$E[Y(t)^2] \leq E[X(t)^2] = \gamma_0 \tag{11.9}$$

$$E[|Y(t) - Y(s)|^2] \leq \gamma_2 |t-s| \tag{11.7'}$$

by (C. 1) and (11. 7), so that $Y(t)$ has a measurable version which we shall denote with $Y(t)$ again. Because of

$$E\left\{\left[\int_{-\infty}^0 |\pi_t Y(s)|\, |dK_i(s)|\right]^2\right\} \leq \|K_i\| \int_{-\infty}^0 E[Y(t+s)^2]\, |dK_i(s)| \leq \|K_i\|^2 \gamma_0 < \infty,$$

$\pi_t Y(s)$ is integrable with respect to $|dK_i(s)|$ for every $t$ and $i=1,2$. Therefore $a(\pi_t Y)$ and $b(\pi_t Y)$ are well defined and we have

$$(11.10)\quad a(\pi_t Y) = E[a(\pi_t X)|\mathcal{B}(dB)], \quad b(\pi_t Y) = E[b(\pi_t X)|\mathcal{B}(dB)]$$

by (11.2).

Using (C.1) and (11.10), we can derive

$$(11.5')\qquad E[a(\pi_t Y)^2] \leq \gamma_1, \quad E[b(\pi_t Y)^2] \leq \gamma_1$$

$$(11.8')\qquad \begin{array}{l} E[|a(\pi_t Y)-a(\pi_s Y)|^2] \leq \gamma_3|t-s| \\ E[|b(\pi_t Y)-b(\pi_s Y)|^2] \leq \gamma_3|t-s| \end{array} \qquad (|t-s|<1)$$

from (11.5) and (11.8). This completes the proof of Theorem 14.

Setting $X=X(t)$, $\mathcal{C}=\mathcal{B}_{-\infty t}(X)$, $\mathcal{D}=\mathcal{B}_{-\infty t}(dB)$ and $\mathcal{E}=\mathcal{B}_{t\infty}(dB)$ in (C.3), we have

$$(11.11)\qquad Y(t) = E[X(t)|\mathcal{B}_{-\infty t}(dB)],$$

so that $Y(t)$ is measurable with respect to $\mathcal{B}_{-\infty t}(dB)$ and $\mathcal{B}_{-\infty t}(Y, dB)$ is therefore independent of $\mathcal{B}_{t\infty}(dB)$. Thus we can see that

$$(11.12)\qquad I(Y) \equiv \int_s^t a(\pi_u Y)du + \int_s^t b(\pi_u Y)dB(u)$$

is well-defined.

Let $I(X)$ denote the right side of (11.6), $I_\Delta(X)$ be the approximate sum for $I(X)$ with respect to the division $\Delta \equiv (s=u_0<u_1<\cdots<u_u=t)$ and $I_\Delta(Y)$ that for $I(Y)$. Then (11.8) and (11.8′) imply that, as $||\Delta|| \equiv \max(u_i-u_{-1}) \to 0$,

$$(11.13)\qquad E[|I_\Delta(X)-I(X)|^2] \to 0,$$

$$(11.13')\qquad E[|I_\Delta(Y)-I(Y)|^2] \to 0.$$

Observing the forms of $I_\Delta(X)$ and $I_\Delta(Y)$, we get

$$(11.14)\qquad I_\Delta(Y) = E[I_\Delta(X)|\mathcal{B}(dB)]$$

and so

$$(11.15)\qquad \begin{array}{l} E\{[I_\Delta(Y)-E(I(X)|\mathcal{B}(dB))]^2\} \\ \qquad \leq E[(I_\Delta(X)-I(X))^2] \to 0 \end{array}$$

by (C.1) and (11.13), as $||\Delta|| \to 0$. Hence it follows by (11.13′) and (11.6) that

$$I(Y) = E[I(X)|\mathcal{B}(dB)] = E[X(t)-X(s)|\mathcal{B}(dB)] = Y(t)-Y(s).$$

Since $X$ and $dB$ are strictly stationarily correlated, the shift operator $\theta_\tau$ on $\mathcal{B}(X, dB)$ is measure preserving and $\theta_\tau(t)=X(t+\tau)$ and $\theta_\tau\mathcal{B}(dB)=\mathcal{B}(dB)$. Hence it follows by (C. 2) that

$$Y(t+\tau) = E(X(t+\tau)|\mathcal{B}(dB)) = E(\theta_\tau X(t)|\theta_\tau\mathcal{B}(dB)) = \theta_\tau Y(t),$$

so that $Y(t)$ has a baceward representation by multiple Wiener integrals by virtue of $\mathcal{B}_{-\infty t}(Y)\subset\mathcal{B}_{-\infty t}(dB)$ and $E(Y(0)^2)<\infty$.

Now we shall combine Theorem 14 with Theorem 8 and Theorem 13 to prove

THEOREM 15. *Suppose that $a(f)$ and $b(f)$ are of the form* (11. 2) *and set*

(11. 16) $$-c = \text{the jump of } dK_1 \text{ at } 0$$

(11. 16′) $$c_1 = \int_{-\infty}^{0-} |dK_1(t)|, \quad c_2 = \int_{-\infty}^{0} |dK_2(t)|.$$

*Then there exists a stationary solution of* (11. 1) *with a backward representation by multiple Wiener integrals integrals in each of the cases*:

(i) $c>c_1+\frac{1}{2}c_2^2$ *and the supports of $dK_1$ and $dK_2$ are bounded,*

(ii) $c>c_1+\frac{1}{4}c_2^2+\frac{1}{4}(c_2^4+8c_1c_2^2)^{1/2}$.

*In the case* (ii) *this solution is the only one solution of* (11. 1) *with* $\sup_t E(X(t)^2)<\infty$.

PROOF: The assertion in the case (i) follows at once from Theorem 8 and Theorem 14; notice that $a(f)$ and $b(f)$ are continuous with respect to the $\rho_-$-topology, since $dK_1$ and $dK_2$ have bounded supports. The assertion in the case (ii) and the second part of our theorem follow from Theorem 13 and Theorem 14 (see also Remarks 3 and 4 at the end of Section 10).

## 12. Linear Coefficients (2).

In this section we shall discuss the case that $a(f)$ is linear and $b(f)\equiv 1$. Then our equation turns out to be

(12.1) $$dX(t) = \left[\mu + \int_{-\infty}^{0} X(t+s)dK(s)\right]dt + dB(t).$$

Before stating our results we shall examine this problem heuristically.

First of all it should be noted that once we get a solution with a backward moving average representation, then it should be canonical, i.e., $\mathcal{B}_t(X) = \mathcal{B}_t(dB)$ by virtue of Theorem 10.

Let $X(t)$ be any stationary solution of (12.1) with $E(X(0)^2) < \infty$. If we assume

(12.2) $$k \equiv \int_{-\infty}^{0} dK(s) \neq 0,$$

then the stationary process

(12.3) $$Y(t) \equiv X(t) + \frac{\mu}{k}$$

satisfies

(12.1′) $$dY(t) = \left(\int_{-\infty}^{0} Y(t+s)dK(s)\right)dt + dB(t).$$

Using the properties of stationary random distributions [7] we can write (12.1) as

(12.1″) $$DY(\varphi) = \int_{-\infty}^{\infty} \varphi(t) \int_{-\infty}^{0} Y(t+s)dK(s)dt + DB(\varphi),$$

where $D$ is the Schwartz derivative and $\varphi$ moves over the space of all $C^\infty$-functions of compact support.

Express the stationary process $Y(t)$ and the stationary random distribution $DB$ in the Fourier transforms:

(12.4) $$Y(t) = c + \int_{-\infty}^{\infty} e^{i\lambda t} dM(\lambda)$$

and

(12.5) $$DB(\varphi) = \int_{-\infty}^{\infty} \mathfrak{F}\varphi(\lambda) dU(\lambda)$$

where $dM(\lambda)$ is an orthogonal random measure with

(12.6) $$\int_{-\infty}^{\infty} E[|dM(\lambda)|^2] < \infty,$$

$dU(\lambda)$ is a complex Gaussian orthogonal random measure with

$$E(|dU)\lambda)|^2 = \frac{1}{2\pi}d\lambda \tag{12.7}$$
$$U(\Lambda) = \overline{U(-\Lambda)}$$

and $\mathfrak{F}\varphi(\lambda)$ is the Fourier transform of $\varphi$:

$$\mathfrak{F}\varphi(\lambda) = \int_{-\infty}^{\infty} e^{i\lambda t}\varphi(\lambda)d\lambda\,. \tag{12.8}$$

Then it is easy to see that

$$DY(\varphi) = \int_{-\infty}^{\infty} \mathfrak{F}\varphi(\lambda)i\lambda dM(\lambda)$$

and

$$\int_{-\infty}^{\infty}\varphi(t)\int_{-\infty}^{0} Y(t+s)dK(s)dt \tag{12.9}$$
$$= \int_{-\infty}^{\infty}\mathfrak{F}\varphi(\lambda)\mathfrak{F}K(\lambda)dM(\lambda)+c\cdot k\int_{-\infty}^{\infty}\varphi(t)dt\,,$$

where $\mathfrak{F}K(\lambda)$ is the Fourier transform of the signed measure $dK$:

$$\mathfrak{F}K(\lambda) = \int_{-\infty}^{0} e^{i\lambda t}dK(t)\,; \tag{12.10}$$

notice that $\mathfrak{F}K(\lambda)$ is bounded:

$$|\mathfrak{F}K(\lambda)| \leq \int_{-\infty}^{0} |dK(t)| \equiv k_0 < \infty\,. \tag{12.11}$$

Putting (12. 8) and (12. 1″), we get

$$\int\mathfrak{F}\varphi(\lambda)i\lambda dM(\lambda) = \int\mathfrak{F}\varphi(\lambda)\mathfrak{F}K(\lambda)dM(\lambda)+ck\int\varphi(t)dt+\int\mathfrak{F}\varphi(\lambda)dU(\lambda)\,,$$

i.e.

$$ck\int\varphi(t)dt = 0 \qquad \text{and so} \quad c = 0\,,$$
$$i\lambda dM(\lambda) = \mathfrak{F}K(\lambda)dM(\lambda)+dU(\lambda)$$

and so

$$(i\lambda-\mathfrak{F}K(\lambda))dM(\lambda) = dU(\lambda)\,.$$

If

$$i\lambda-\mathfrak{F}K(\lambda) \neq 0\,, \tag{12.12}$$

then we get

$$dM(\lambda) = (i\lambda-\mathfrak{F}K(\lambda))^{-1}dU(\lambda)$$

and it is conceivable that our solution $X(t)$ is expressible in

$$X(t) \equiv -\frac{M}{k} + \int_{-\infty}^{\infty} e^{i\lambda t}(i\lambda - \mathfrak{F}K(\lambda))^{-1} dU(\lambda) \tag{12.13}$$

In order for $X(t)$ to be a stationary process, we need the assumption:

$$\int_{-\infty}^{\infty} |i\lambda - \mathfrak{F}k(\lambda)|^{-2} d\lambda < \infty . \tag{12.14}$$

The following Lemma will give reasonable sufficient conditions for (12.14).

LEMMA 12.1. *If*

$$k = \int_{-\infty}^{0} dK(t) < 0 \tag{K.1}$$

*and*

$$k_1 = \int_{-\infty}^{0} |t|\,|dK(t)| < 1,^{1)} \tag{K.2}$$

*then* $H(\nu) = i\nu - \int_{-\infty}^{0} e^{i\nu t} dK(t)$ $(\mathrm{Im}\,\nu \leq 0)$ *has the following properties*:

(H.1) $H(\nu)$ *is continuous in* $\mathrm{Im}\,\nu \leq 0$ *and analytic in* $\mathrm{Im}\,\nu < 0$

(H.2) $H(\nu) \neq 0$ *in* $\mathrm{Im}\,\nu \leq 0$ *and* $\int_{-\infty}^{\infty} |H(\lambda + i\nu)|^{-2} d\lambda$ *is bounded in* $\mu \leq 0$

(H.3) *The inverse Fourier transform of* $H(\lambda)$ *vanishes on* $t \leq 0$.

PROOF: (H.1) is clear. To prove (H.2), denote the real and imaginary parts of $H(\lambda + i\mu)$ by $R(\lambda, \mu)$ and $I(\lambda, \mu)$ respectively. Then

$$R(\lambda, \mu) = -\mu - \int_{-\infty}^{0} \cos \lambda t\, e^{-\mu t} dK(t)$$

and

$$I(\lambda, \mu) = \lambda - \int_{-\infty}^{0} \sin \lambda t\, e^{-\mu t} dK(t) .$$

It holds for $\mu \leq 0$ that

$$\frac{\partial R}{\partial \mu} = -1 + \int_{-\infty}^{0} t \cos \lambda t\, e^{-\mu t} dK(t) \leq -1 + k_1 < 0$$

1) These conditions, (K.1) and (K.2), were sought by our conversation with J. McGregor.

and

$$\frac{\partial R}{\partial \lambda} = \left| \int_{-\infty}^{0} t \sin \lambda t e^{-\mu t} dK(t) \right| \leq k_1 ,$$

so that

$$|H(\lambda+i\mu)| \geq R(\lambda,\mu) \geq R(\lambda, 0) \geq R(0, 0)-k_1|\lambda| = -k-k_1(\lambda)$$

On the other hand

$$|H(\lambda+i\mu)| \geq |I(\lambda, \mu)| \geq |\lambda| (1-k_1) \quad (\text{by } |\sin \lambda t| \leq |\lambda| |t|)$$

Taking a combination, we get

$$|H(\lambda+i\mu)| \geq \frac{1-k_1}{2}(-k-k_1|\lambda|)+\frac{1+k_1}{2}|\lambda| (1-k_1) = \frac{1-k_1}{2}(-k+|\lambda|)$$

This proves (H. 2). (H. 3) follows at once from (H. 1) and (H. 2) by Paley-Wiener's theorem.

Using this Lemma we shall prove

THEOREM 16. *Under the conditions* (K. 1) *and* (K. 2), *there exists one and only one stationary solution of* (12. 1) *with* $E(X(0)^2) < \infty$.

*The solution has a canonical linear backward representation* [10]

$$X(t) = -\frac{\mu}{k}+\int_{-\infty}^{t} g(s-t)dB(s), \tag{12. 15}$$

*where* $g(t)$ *is a real valued square summable function vanishing on* $t \geq 0$ *and is given by*

$$g(t) = \frac{1}{2\pi} \operatorname*{l.i.m.}_{A\to\infty} \int_{-\Lambda}^{\Lambda} e^{-i\lambda t} H(\lambda)^{-1} d\lambda \quad (H(\lambda) = i\lambda - \mathfrak{F}K(\lambda)) \tag{12. 16}$$

PROOF: The proof of the first part of the theorem is included in the above explanation. The only point we should check is the equivalence of (12. 1″) with (12. 1′), but we can prove it by noticing that (12. 1′) means

$$Y(t)-Y(s) = \int_s^t du \int_{-\infty}^{0} Y(u+v)dK(v)+B(t)-B(s)$$

and (12. 1″) means

$$\int_{-\infty}^{\infty} Y(u)\varphi'(u)du = \int_{-\infty}^{\infty} \varphi(u) \int_{-\infty}^{0} Y(u+v)dK(v)du+\int_{-\infty}^{\infty} B(u)\varphi'(u)du$$

which can be expressed as the limit of the Riemann sum over $(k/n,\ k=\cdots -2, -1, 0, 1, 2, \cdots)$ as $n\to\infty$.

We can also prove the second part by the usual argument except the fact that $g(t)$ is real and vanishes on $t\geq 0$. $g(t)$ is real because

$$H(-\lambda) = -i\lambda - \int e^{-i\tau t} dK(t) = \overline{H(\lambda)}$$

and $g(t)$ vanishes on $t\geq 0$, as we can see from (H.3) in Lemma 2.1.

We should mention one word on the continuity of almost all paths of the solution $X(t)$ obtained above. For any given pair $s<t$, we have

$$(12.17)\quad X(t)-X(s) = \int_s^t \left(M+\int_{-\infty}^0 X(u+v)dK(v)\right) dU + B(t)-B(s)$$

with probability 1, but the usual argument will show that $X(t)$ has a version with continuous paths for which, with probability 1, (12.17) holds simultaneously for every pair $s<t$.

## 13. Diffusions

In this section we consider the special case that $a(f)$ and $b(f)$ depend only on the value $f(0)$, i.e.

$$(13.1)\qquad a(f) = \alpha(f(0)), \qquad b(f) = \beta(f(0))$$

Therefore our stochastic differential equation will be of the form:

$$(13.2)\qquad dX(t) = \alpha(X(t))\,dt + \beta(X(t))dB(t).$$

We shall assume the Lipschitz condition

$$(13.3)\qquad |\alpha(\xi)-\alpha(\eta)| + |\beta(\xi)-\beta(\eta)| \leq K|\xi-\eta|.$$

The one-sided solution of this equation was treated by K. Itô [8] in connection with the theory of diffusions. Now we shall discuss its stationary solutions.

Before entering our problem, let us examine the nature of the one-sided solutions, using Feller's theory of general one-dimensional diffusions.

Let $P_\xi$ be the probability law governing the solution $X(t)$,

$t \geq 0$ with the initial condition $X(0) \equiv \xi$, where $\xi$ is a real constant. Then the system $(P_\xi, \xi \in R^1)$ defines a *diffusion* (=strict Markov process with continuous paths) on $R^1$, the domain $\mathfrak{D}(\mathcal{G})$ of its generator $\mathcal{G}$ contains the space $\mathcal{C}_0^2$ of all twice continuously differentiable functions of compact support and

$$\mathcal{G}u(\xi) = \frac{1}{2}\beta(\xi)^2 \frac{d^2u}{d\xi^2} + \alpha(\xi)\frac{du}{d\xi} \qquad u \in \mathcal{C}. \tag{13.4}$$

(see K. Itô [8], K. Itô and H. P. McKean, Jr. [9] and E. B. Dynkin [2]).

Our diffusion does not exhibit all different behaviors of Feller's general diffusion; for example, no shunt of the reflecting boundary type can occur, as we can see in

THEOREM 17. *Denote the path of our diffusion $(P_\xi, \xi \in R^1)$ with $x_t$. Then we have the following cases.*

(i) *If $\beta(\xi) \neq 0$, then $\xi$ is a non-singular point, i.e.*

$$P_\xi(x_t > \xi \text{ and } x_s < \xi \text{ for some } t, s > 0) = 1.$$

(ii) *If $\beta(\xi) = 0$ and $(\alpha(\xi) > 0$, then $\xi$ is a strict right shunt, i.e.*

$$P_\xi(x_t > \xi \text{ for every } t > 0) = 1.$$

(iii) *If $\beta(\xi) = 0$ and $\alpha(\xi) < 0$, then $\xi$ is a strict left shunt, i.e.*

$$P_\xi(x_t < \xi \text{ for every } t > 0) = 1.$$

(iv) *If $\beta(\xi) = \alpha(\xi) = 0$, then $\xi$ is a trap, i.e.*

$$P_\xi(x_t = \xi \text{ for every } t \geq 0) = 1.$$

PROOF: As we mentioned above, $P_\xi$ is the probability law of of the sample path of the unique solution of the stochastic integral equation:

$$X(t) = \xi + \int_0^t \alpha(X(s))ds + \int_0^t \beta(X(s))dB(s). \tag{13.5}$$

In this proof $c_1, c_2, \cdots$ stand for positive constants and $t$ and $s$ move only on [0, 1], unless stated otherwise. For example, $F(t) \leq c_1 G(t)$ means that there exists a positive constant $c_1$ such that $F(t) \leq c_1 G(t)$ for every $t \in [0, 1]$.

*Case (i).* $\beta(\xi)\neq 0$.

Consider a Gaussian process $Y(t)$:

$$(13.6)\quad Y(t)\equiv\xi+\int_0^t\alpha(\xi)ds+\int_0^t\beta(\xi)dB(s)=\xi+\alpha(\xi)t+\beta(\xi)B(t).$$

Then it is clear that

$$(13.7)\quad \begin{aligned} &P(Y(t)>\xi+\alpha(\xi)t+|\beta(\xi)|t^{1/2}) \\ &=P(Y(t)<\xi+\alpha(\xi)t-|\beta(\xi)|t^{1/2}) \\ &=\frac{1}{\sqrt{2\pi}}\int_1^\infty\exp(-y^2/2)dy\equiv c_1>0. \end{aligned}$$

If follows from (13.5) that

$$(13.8)\quad E[(X(t)-\xi)^2]\leq c_2t$$

as we saw in Section 10, and so

$$(13.9)\quad E[(X(t)-Y(t))^2]\leq c_3t^2$$

by virtue of the Lipschitz condition. Hence it follows that

$$(13.10)\quad P(|X(t)-Y(t)|>t^{3/4})\leq c_3t^{1/2}.$$

Using (13.7) and (13.10), we get

$$P(X(t)>\xi+\alpha(\xi)t+|\beta(\xi)|t^{1/2}-t^{3/4})>c_1-c_3t^{1/2}$$

and

$$P(X(t)<\xi+\alpha(\xi)t-|\beta(\xi)|t^{1/2}+t^{3/4})>c_1-c_3t^{1/2}.$$

By taking $t>0$ so small that

$$t<1,\quad |\beta(\xi)|t^{1/2}>|\alpha(\xi)|t+t^{3/4}\ \text{ and }\ c_1-c_3t^{1/2}>0,$$

we get $P(X(t)>\xi)>0$ and $P(X(t)<\xi)>0$ for some $t$, i.e., $P_\xi(x_t>\xi)>0$ and $P_\xi(x_t<\xi)>0$ for some $t$. Therefore $\xi$ is a non-singular point and

$$P_\xi(x_t>\xi\ \text{ and }\ x_s<\xi\ \text{ for some }\ t,s>0)=1,$$

as is well known in the theory of diffusions (see E. B. Dynkin [2], and K. Itô and H. P. McKean, Jr. [9]).

*Case (ii).* $\beta(\xi)=0$, $\alpha(\xi)>0$.

Consider the same $Y(t)$ as in (13.6). Since $\beta(\xi)=0$, $Y(t)$ is a fixed linear function:

$$(13.6') \qquad Y(t) = \xi + \alpha(\xi)t$$

(13.8) can be sharpened in this case by virtue of $\beta(\xi)=0$. As we saw in Section 10, we have

$$E[(X(t)-\xi)^2] \leq c_2 t$$

as before and

$$E[\alpha(X(t))^2] \leq c_3$$

$$E[\beta(X(t))^2] \equiv E[(\beta(X(t))-\beta(\xi))^2] \leq K^2 E[(X(t)-\xi)^2] \leq c_4 t\,.$$

Hence it follows that

$$(13.8') \qquad E[(X(t)-\xi)^2]$$

$$\leq 2t\int_0^t E[\alpha(X(s))^2]\,ds + 2\int_0^t E[\beta(X(s))^2]\,ds$$

$$\leq 2c_3 t^2 + c_4 t^2 = c_5 t^2\,.$$

Writing $X(t)$ in the form:

$$(13.11) \quad X(t) = Y(t) + (X(t) - Y(t))$$

$$= \xi + \alpha(\xi)t + \int_0^t (\alpha(X(s)) - \alpha(\xi))\,ds + \int_0^t (\beta(X(s)) - \beta(\xi))\,dB(s)$$

$$= \xi + \alpha(\xi)t + I_1(t) + I_2(t)\,,$$

we shall examine the behavior of $X(t)$ near $t=0$.

Since $\alpha(X(s)) - \alpha(\xi)$ ($\varepsilon \downarrow 0$) with probability 1, we have

$$(13.12) \qquad |I_1(t)| < \frac{1}{3}\alpha(\xi)t \qquad 0 \leq t \leq \tau_1$$

where $\tau_1$ is a certain positive random variable.

Now we shall estimate $I_2(t)$ by means of (13.8'). Writing $\bar{I}_2(t)$ for $\sup_{0\leq s\leq t} |I_2(s)|$, we get

$$P(\bar{I}_2(t) > \varepsilon) \leq \varepsilon^{-2}\int_0^t E[(\beta(X(s)) - \beta(\xi))^2]\,ds$$

$$\leq \varepsilon^{-2} K^2 \int_0^t E[(X(s)-\xi)^2]\,ds$$

$$\leq \varepsilon^{-2} K^2 c_5 \frac{t^3}{3} \; c_6 t^3 \varepsilon^{-2}$$

and so

$$P(\bar{I}_2(t) > t^{4/3}) \leq c_6 t^{1/2}$$

Hence it follows that

$$\sum_n P(\bar{I}_2(2^{-3n}) > 2^{-4n}) \leq \sum c_6 2^{-n} < \infty ,$$

so that we can apply the Borel-Cantelli lemma to get a positive random variable $\nu$ (which is finite with probability 1) such that

$$\bar{I}_2(2^{-3n}) \leq 2^{-4n} \quad \text{for} \quad n \leq \nu .$$

Then we have

(13. 13) $$\bar{I}_2(t) \leq 2^4 t^{4/3} \quad \text{for} \quad 0 \leq t \leq 2^{-3\nu} ;$$

in fact, taking $m = m(t) \leq \nu$ with

$$2^{-3(m+1)} \leq t \leq 2^{-3m} ,$$

we can get

$$I_2(t) \leq \bar{I}_2(2^{-3m}) \leq 2^{-4m} \leq 2^4 2^{-4(m+1)} \leq 2^4 t^{4/3} .$$

Setting $\tau_2 = \min(3^{-3} 2^{-12} \alpha(\xi)^3, 2^{-3\nu})$, we have

(13. 14) $$|I_2(t)| \leq \bar{I}_2(f) \leq 2^4 t^{1/3} t \leq 2^4 \tau_2^{1/3} t \leq \frac{1}{3} \alpha(\xi) t$$

for $0 \leq t \leq \tau_2$.

Putting (13. 12) and (13. 14) in (13. 11) we get

$$P(X(t) > \xi + \frac{1}{3}\alpha(\xi)t \quad \text{for} \quad 0 \leq t \leq \min(\tau_1, \tau_2) = 1 ,$$

a fortiori

(13. 15) $$P_\xi(x_t > \xi \quad \text{for sufficiently small} \quad t > 0) = 1 .$$

By the property of the one-dimensional diffusion (see [9] or [2] for example), (13. 15) implies the conclusion of (ii).

*Case (iii).* $\beta(\xi) = 0$, $\alpha(\xi) < 0$. The same as (ii).

*Case (iv).* $\beta(\xi) = \alpha(\xi) = 0$. Since $X(t) \equiv \xi$ satisfies (13. 5), $P_\xi$ is concentrated on the single path identically equal to $\xi$.

Thus Theorem 17 is completely proved.

Now consider the set of all non-singular points. It is a finite or countable sum of disjoint open intervals $I_n = (i_n, j_n)$, $n = 1, 2, \cdots$.

Each $I=I_n$ is the maximal non-singular interval on which Feller's scale $ds$ and speed measure $dm$ are given as

$$ds(\xi) = e^{-\gamma(\xi)}d\xi\,, \qquad dm(\xi) = \beta(\xi)^{-2}e^{\gamma(\xi)}d\xi \tag{13.16}$$

with

$$\gamma(\xi) = \int_\zeta^\xi \alpha(\eta)\beta(\eta)^{-2}\,d\eta \tag{13.17}$$

where $\zeta$ is any point assigned arbitrarily in $I$.

In the discussion of the stationary solution of our stochastic differential equation (13.2), a non-singular interval $I=(i,j)$ *of positive recurrent type* is important. It is characterized in terms of the scale $ds$ and the speed measure $dm$ as

$$\int_i^j ds = \infty\,, \quad \int_\zeta^j ds = \infty\,, \quad \int_i^j dm < \infty\,. \tag{13.18}$$

In case both $i$ and $j$ are finite, $(i, j)$ is of such type if $\alpha(i)>0$ and $\alpha(j)<0$, but not so if either $\alpha(i)<0$ or $\alpha(j)>0$.

From now on we shall denote all non-singular intervals of positive recurrent type by $J_1, J_2, \cdots$ and the scale and the speed measure on each $J_n$ by $ds_n$ and $dm_n$ respectively.

Let $P(t, \xi, E)$ be the transition probability measure of our diffusion $(P_\xi, \xi\varepsilon(-\infty, \infty))$. A probability measure $\mu$ on the real line called an *invariant measure,* if

$$\mu(E) = \int_{-\infty}^{\infty} P(t, \xi, E)\,d\mu(\xi)\,. \tag{13.19}$$

Hereafter we denote the set of all traps with $\theta$. Take any trap $\theta$. Then the $\delta$-distribution concentrated at $\theta$ is a trivial invariant measure and $X(t)\equiv\theta$ is a trivial stationary solution of our equation (13.2).

Take any non-singular interval $J=J_n$ of positive recurrent type. Then the probability measure $d\nu=d\nu_n$ proportional to its speed measure $dm=dm_n$ is the only one invariant measure concentrated on this interval.

The following lemma 13.1 which will be useful here follows easily from the eigen-differential expansion of $P(t, \xi, E)$ due to H. P. McKean, Jr. [11].

LEMMA 13.1. *Let $J$ be a non-singular interval of positive recurrent type and $d\nu$ its invariant measure.*

(i) *Given any $\varepsilon > 0$ and any compact subinterval $J'$ of $J$, there exists a compact subinterval $J'' = J''(\varepsilon, J')$ such that*

$$P(t, \xi, J'') > 1 - \varepsilon$$

*for every $t \geq 0$ and every $\xi \in J'$.*

(ii) *Given any $\varepsilon > 0$, any bounded continuous function $h$ defined on $J$ and any compact subinterval $J'$, there exists $t_0 = t_0(\varepsilon, h, J') > 0$ such that*

$$\left| \int_J h(\eta) P(t, \xi, d\eta) - \int_J h(\eta) d\nu(\eta) \right| < \varepsilon$$

*for every $t > t_0$ and ever $\xi \in J'$.*

LEMMA 13.2. *Let $J$ be a non-singular interval of positive recurrent type, $J'$ a compact subinterval of $J$ and $Y_\xi(t)$ the solution of*

$$Y(t) = \xi + \int_0^t \alpha(Y(s)) + \int_0^t \beta(Y(s)) dB(s) . \tag{13.20}$$

Then

$$E(|\arctan Y_{\xi_2}(t) - \arctan Y_{\xi_1}(t)|) \to 0 \tag{13.21}$$

*uniformly on $(\xi_1, \xi_2) \in J' \times J'$ as $t \to \infty$.*

PROOF: Since $Y_\xi(t)$ can be approximated by successive approximation, $Y_\xi(t)$ is $\mathcal{B}_{0,t}(dB)$-measurable. Take $\xi_1 < \xi_2$ in $J$. Let $\sigma$ be $\sup \{t : Y_{\xi_1}(s) < Y_{\xi_2}(s) \text{ for } s \leq t\}$. Using the strong Markov property of the Wiener process $B(t)$ and definition of stochastic integral, we can see that with probability 1, *either* $\sigma = \infty$ or

$$Y(t+\sigma) = Y(\sigma) + \int_0^t \alpha(Y(s+\sigma)) ds + \int_0^t \beta(Y(s+\sigma)) dB_\sigma(s) \tag{13.22}$$

for $Y = Y_{\xi_1}, Y_{\xi_2}$, where $B_\sigma(s) = B(s+\sigma) - B(\sigma)$; we can prove by routine argument that $B_\sigma(s)$, $s \geq 0$ is a Wiener process and that $[\bigvee_{0 \leq s \leq t} \mathcal{B}(Y_\xi(s+\sigma))] \vee \mathcal{B}_{0t}(dB_\sigma)$ is independent of $\mathcal{B}_{t\infty}(dB_\sigma)$, so that the stochastic integral in (13.22) is meaningful.

By the uniqueness of the solution of the stochastic integral equation and the fact that $Y_{\xi_2}(\sigma)=Y_{\xi_1}(\sigma)$ if $\sigma<\infty$, we get from (13.22)

(13.23) $P(\text{either } \sigma=\infty \text{ or } Y_{\xi_2}(t)=Y_{\xi_1}(t) \text{ for } t\geq\sigma)=1$;

see the remark at the end of this section.

Since $Y_{\xi_2}(t)>Y_{\xi_1}(t)$ for $t<\sigma$, we get

(13.24) $$P(Y_{\xi_2}(t)\geq Y_{\xi_1}(t) \text{ for every } t\geq 0)=1$$

Since $h(x)\equiv\arctan x$ is continuous, bounded and increasing, it follows from (13.24) that

$$E[|h(Y_{\xi_2}(t))-h(Y_{\xi_1}(t))|]=|E[h(Y_{\xi_2}(t))]-E[h(Y_{\xi_1}(t))]|$$

for $\xi_1\leq\xi_2$. Noticing the symmetry of $\xi_1$ and $\xi_2$ in this identity, we can see that it is true for $(\xi_1,\xi_2)\in J\times J$, a fortiori for $(\xi_1,\xi_2)\in J'\times J'$. To complete the proof, it is enough to observe

$$\begin{aligned}|E[h(Y_{\xi_2}(t))]-E[h(Y_{\xi_1}(t))]| &= \left|\int_J h(\xi)P(t,\xi_2,d\xi)-\int_J h(\xi)P(t,\xi_1,d\xi)\right| \\ &\leq \sum_{n=1}^{2}\left|\int_J h(\xi)P(t,\xi_i,d\xi)-\int_J h(\xi)d\nu(\xi)\right|\end{aligned}$$

and to use Lemma 13.1 (ii).

Using the equivalence in law of $(B(t), t\geq 0)$ and $(B(t-T)-B(-T))$ and the fact that the one-sided solution of our stochastic differential equation is determined by successive approximation, we can derive the following lemma immediately from Lemma 13.2.

LEMMA 13.3. *Let $J$ be a non-singular interval of positive recurrent type, $J'$ a compact subinterval of $J$ and $X_{T,\xi}(t)$ the solution of*

(13.25) $$X(t)=\xi+\int_{-T}^{t}\alpha(X(s))ds+\int_{-T}^{t}\beta(X(s))dB(s).$$

*Then*

(13.26) $$E[|\arctan X_{T,\xi_2}(t)-\arctan X_{T,\xi_1}(t)|]\to 0$$

*uniformly in $(\xi_1,\xi_2)\in J'\times J'$ as $t\to\infty$.*

Now we shall discuss the stationary solution on a single non-singular interval of positive recurrent type.

THEOREM 18. *If $J$ be a non-singular interval of positive re-*

*current type, then the stochastic differential equation* (13.2) *has one and only one solution whose sample path is confined in $J$ with probability* 1.

*This solution satisfies*

$$(13.27)\qquad \mathcal{B}_{-\infty t}(X) = \mathcal{B}_{-\infty t}(dB) \qquad -\infty < t < \infty .$$

*Furthermore if the invariant measure on $J$ has finite second order moment, then this solution has a properly canonical backward representation by multiple Wiener integrals.*

PROOF:

(i) *Existence.* We shall consider the process $X_{T,\xi}(t)$, $t \geq -T$, introduced in Lemma 13.2. Take a point $\zeta \in J$ and fix it. The limit of $X_{T,\zeta}(t)$ as $T \to \infty$ is supposed to be a solution.

First of all we shall prove that $X_{T,\zeta}(t)$ converges to a certain random variable $X(t)$ in probability for each $t$.

Take $S$ and $T$ with $-S < -T < t$. Then

$$(13.28)\quad X_{S,\zeta}(t) = X_{S,\zeta}(-T) + \int_{-T}^{t} \alpha(X_{S,\zeta}(s))ds + \int_{-T}^{t} \beta(X_{S,\eta}(s))dB(s).$$

Since $X_{S,\zeta}(-T)$ is independent of $\mathcal{B}_{-T\infty}(dB)$, we can obtain $X_{S,\zeta}(t)$ by replacing $\xi$ in $X_{T,\xi}$ with $X_{S,\zeta}(-T)$ by virtue of the uniqueness of the one-sided solution and the Borel-measurability of $X_{T,\xi}(t, \omega)$ in $(\xi, \omega)$ on the product measurable space $R^1(\mathcal{B}(R^1)) \times \Omega(\mathcal{B}(dB))$ (see K. Itô [8]). Noticing that $X_{T,\zeta}(t)$ is $\mathcal{B}_{-T\infty}(dB)$-measurable, we get

$$\begin{aligned} &E[|\arctan X_{S,\zeta}(t) - \arctan X_{X,\zeta}(t)|] \\ &\quad = \int_J P(X_{S,\zeta}(-T) \in d\xi) E[|\arctan X_{T,\xi}(t) - \arctan X_{T,\zeta}(t)|] \\ &\quad = \int_J P(S-T, \zeta, d\xi) E[|\arctan X_{T,\xi}(t) - \arctan X_{T,\zeta}(t)|]. \end{aligned}$$

Using Lemma 13.1 (i) and Lemma 13.3, we can easily see that this tends to 0 as $S > T \to \infty$. Therefore the limit in probability of $\arctan X_{T,\zeta}(t)$ as $T \to \infty$ exists, so that $X_{T,\zeta}(t)$ tends to a certain random variable $\tilde{X}(t)$ in probability as $T \to \infty$. It is clear that $\tilde{X}(t)$ is $\mathcal{B}_{-\infty t}(dB)$-measurable, so that (13.27) holds for $X = \tilde{X}$.

To prove that $\tilde{X}(t)$ has a version with continuous paths which is a stationary solution of (13.2), we shall express $\tilde{X}(t)$ by the limit of another sequence of one-sided solution.

Take a random variable $A$ which is governed by the invariant distribution $\nu$ on $J$, by extending the basic probability measure space if necessary. Let $X_{T,A}(t)$, $t\geq -T$, denote the solution of

$$X(t) = A+\int_{-T}^{t}\alpha(X(s))ds+\int_{-T}^{t}\beta(X(s))dB(s). \tag{13.28}$$

It is easy to see that if $S>T$, then

$$(X_{S,A}(t),\ t\geq -T,\quad B(v)-B(u),\quad v>u\geq -T)$$

is equivalent in law with

$$(X_{T,A}(t),\ t\geq -T,\quad B(v)-B(u),\quad v>u\geq -T)$$

because $A$ is $\nu$-distributed. This is a nice property of $X_{T,A}$ which $X_{T,\zeta}(t)$ did not enjoy; on the other hand $X_{T,\zeta}(t)$ was $\mathcal{B}_{-T,t}(dB)$-measurable, while $X_{T,A}$ is not so.

Now we can prove that for each $t$, $X_{T,A}(t)$ also converges to $\tilde{X}(t)$ in probability as $T\to\infty$, by observing, as above,

$$\begin{aligned}
&E[|\arctan X_{T,A}(t)-\arctan X_{T,\zeta}(t)|]\\
&=\int_J P(A\in d\xi)E[|\arctan X_{T,\xi}(t)-\arctan X_{T,\zeta}(t)|]\\
&=\int_J \nu(d\xi)E[|\arctan X_{T,\xi}(t)-\arctan X_{T,\zeta}(t)|]\\
&\to 0 \qquad\qquad (T\to\infty),
\end{aligned}$$

by virtue of Lemma 13.3.

Take any time points $t_1<t_2<\cdots<t_n$. Then the joint distribution of

$$(\tilde{X}(t_1),\ \cdots,\ \tilde{X}(t_n),\ B(t_2)-B(t_1),\ \cdots,\ B(t_n)-B(t_{n-1}))$$

is clearly the weak* limit of that of

$$(X_{T,A}(t_1),\ \cdots,\ X_{T,A}(t_n),\ B(t_2)-B(t_1),\ \cdots,\ B(t_n)-B(t_{n-1}))$$

as $T\to\infty$. But the latter one is independent of $T$ as far as $-T<t_1$ because of the nice property of $X_{T,A}$ mentioned above.

This means that any finite dimensional joint distribution of the system

$$(\tilde{X}(t), B(v)-B(u), -\infty<t<\infty, -\infty<u<v<\infty)$$

is the same as the corresponding joint distribution of

$$(X_{T,A}(t), B(v)-B(u), -\infty<t<\infty, -\infty<u<v<\infty)$$

for big enough $T$ which depends on the time points referred to in the joint distribution.

Since $X_{T,A}(t)$, $t\geq -T$ is continuous in $t$ with probability 1, it is strictly stationarily correlated to $dB(t)$, $t\geq -T$, and it satisfies (13.2) on $t\geq -T$, we can use routine argument to prove that $\tilde{X}(t)$ has a version $X(t)$ with continuous paths and that it is a stationary solution of (13.2). It is clear that $X(t)$ has also the property $\mathcal{B}_{-\infty t}(X)\subset\mathcal{B}_{-\infty t}(dB)$ as a version of $\tilde{X}(t)$.

Hence it follows by Theorem 10 in Section 9 that $\mathcal{B}_{-\infty t}(X) =\mathcal{B}_{-\infty t}(dB)$.

If $\nu$ has finite second order moment, then $E(X(t)^2)<\infty$, so that $X(t)$ has a backward representation by multiple Wiener integral which is properly canonical by $\mathcal{B}_{-\infty t}(X)=\mathcal{B}_{-\infty t}(dB)$.

(ii) *Uniqueness.* Take any stationary solution $Y(t)$. Then it is clear that $Y(t)$ is $\nu$-distributed. Since it satisfies

$$Y(t) = (-T)+\int_{-T}^{t}\alpha(Y(s))ds+\int_{-T}^{t}\beta(Y(s))dB(s) \tag{13.29}$$

and $Y(-T)$ is independent of $\mathcal{B}_{-T\infty}(dB)$, we get

$$\begin{aligned}&E[|\arctan Y(t)-\arctan X_{T,\zeta}(t)|]\\&\quad=\int_J \nu(d\xi)E[|\arctan X_{T,\xi}(t)-\arctan X_{T,\zeta}(t)|]\end{aligned}$$

as above. This proves that $P(Y(t)=X(t))=1$ for each $t$ and so $P(Y(t)=X=(t)$ for every $t)=1$ by virtue of the continuity of the paths.

Take any invariant probability measure $\mu$. It is easy to see that $\mu$ can be uniquely decomposed as a convex comaination of the *extremal invariant measures* introduced above:

(13.30) $$\mu(\cdot)=\int_{\Theta}\delta_{\theta}(\cdot)dk(\theta)+\sum_{n}k_n\nu_n(\cdot),$$

where $dk\geq 0$, $k_n\geq 0$ and $\int_{\Theta}dk(\theta)+\sum k_n=1$.

For the convenience of notations, we shall take a point $\zeta_n$ from each $J_n$ as a representative, and write (13.30) as

(13.30′) $$\mu(\cdot)=\int_{\Theta\vee\Pi}\nu_{\zeta}(\cdot)dk(\zeta),$$

where $\Pi=\{\zeta_1,\zeta_2,\cdots\}$, $k$ on $\Theta$ is as in (13.30), $k(\{\xi_n\})=k_n$, and $\nu_{\zeta}=\delta_{\zeta}$ or $\nu_n$ according as $\zeta\in\Theta$ or $\zeta=\zeta_n$. The expression (13.30) is called the *canonical decomposition* of $\mu$ and $dk$ is called the *coefficient measure.*

Let $\Phi$ be the mapping: $\Theta\vee(\bigvee_n J_n)\to\Theta\vee\Pi$ defined as

(13.31) $$\Phi(\zeta)=\zeta\quad(\zeta\in\Theta),\quad=\zeta_n\quad(\zeta\in J_n).$$

Let $X_{\zeta}(t)$, $\zeta\in\Theta\vee\Pi$ denote the *extremal stationary solution* we introduced above, i.e.,

(13.32) $$X_{\zeta}(t)\equiv\zeta\qquad(\zeta\in\Theta)$$
$$=\text{the stationary determined for } J_n \text{ in Theorem 18}\qquad(\zeta=\zeta_n\in\Pi).$$

Now we shall prove the following theorem which gives a complete description of stationary solutions.

THEOREM 19.

(i) *Let* $X(t)$ *be any stationary of* (13.2) *and set* $Z=\Phi(X(0))$. *Then*

(*a*) $P(Z=\Phi(X(t))$ *for every* $t)=1$

(*b*) $\mathcal{B}(Z)=\bigwedge_t\mathcal{B}_{-\infty,t}(X)$ (=*the remote past Borel algebra*)

(*c*) $Z$ *is independent of* $\mathcal{B}(dB)$.

(*d*) $\mu$ *has a canonical decomposition with the coefficient measure* $k(\cdot)=P(Z\in\cdot)$:

$$\mu(E)=P(X(t)\in E)=\int\nu_{\zeta}(E)(PZ\in d\zeta)$$

(*e*) $P(X(t)=X_Z(t)$ *for every* $t)=1$

(ii) *Consider any invariant probability measure* $\mu$ *and decompose it as* (13.30′). *Take a random variable* $Z$ *which is independent of* $\mathcal{B}(dB)$ *and* $k$*-distributed by extending the basic probability measure space if necessary. Then* $X(t)\equiv X_Z(t)$ *is a stationary solution of* (13.3) *with* $P(X(t)\in E)=\mu(E)$.

*Furthermore if* $\mu$ *has finite second order moment, then this solution has a backward representation by multiple Wiener integrals whose integrands contain* $Z$ *as a parameter. If and only if* $\mu$ *is extremal,* $X$ *has a properly canonical backward representation by multiple Wiener integrals.*

PROOF *of (i).*

(a) is clear by the definitions.

(b) follows at once from (a) by virtue of Theorem 18.

(c) follows at once from (b), because $\mathcal{B}_{-\infty t}(X)$ is independent of $\mathcal{B}_{t\infty}(dB)$.

(d) is clear by the definition of $\Phi$.

(e) Consider the solution $X_{T,\xi}(t)$ of

$$X_{T,\xi}(t) = \xi+\int_{-T}^{t}\alpha(X_{T,\xi}(s))ds+\int_{-T}^{t}\beta(X_{T,\xi}(s))dB(s)$$

If $\xi\in\Theta$, then $X_{T,\xi}(t)\equiv\xi\equiv X_\xi(t)$. If $\xi\in J_n$, then $X_{T,\xi}\to X_\xi(t)$ in probability as we proved in Theorem 18.

Since $P(Z\in\Theta\vee(\bigvee_n J_n))=1$ and $Z$ is independent of $\mathcal{B}(dB)$, $X_{T,Z}(t)\to X_Z(t)$ in probability as far as $t\geq -T$.

It is also clear that $X_{T,X(-T)}(t)=X(t)$ $(t\geq -T)$. Observing

$$\begin{aligned}
&E[|\arctan X(t(-\arctan X_{T,Z}(t)|]\\
&\quad=\sum_n E[|\arctan X_{T,X(-T)}-\arctan X_{T,Z}(t)|,\ X(-T)\in J_n]\\
&\quad=\sum_n E[|\arctan X_{T,X(-T)}-\arctan X_{T,\zeta_n}(t)|,\ X(-T)\in J_n]\\
&\quad=\sum_n \int_{J_n} P(X(-T)\in d\xi)E[|\arctan X_{T,\zeta}-\arctan X_{T,\zeta_n}|]\\
&\quad=\sum_n \int_{J_n} \mu(d\xi)E[|\arctan X_{T,\xi}-\arctan X_{T,\zeta_n}|]\\
&\quad=\sum_n k_n\int_{J_n} \nu_n(d\xi)E[|\arctan X_{T,\xi}-\arctan X_{T,\zeta_n}|]\\
&\quad\to 0,
\end{aligned}$$

we can see that $X(t)=X_Z(t)$.

PROOF *of (ii)*.

Using the process $X_{T,\xi}(t)$ as above, we have

$$X_Z(t) = \lim_{T\to\infty} \text{ in probability } X_{T,Z}.$$

Consider a random variable $A$ which is $\mu$-distributed and independent of $\mathcal{B}(dB)$. By the argument used above, we get

$$E[|\arctan X_{T,A}(t) - \arctan X_{T,Z}(t)|] \to 0$$

as $t \to \infty$, and so

$$X_Z(t) = \lim_{A\to\infty} \text{ in probability } X_{T,Z}(t).$$

The rest of the proof is the same as in the proof of the existence part of Theorem 18; notice that $X_Z(t)$ is already continuous in $t$ with probability 1 by the definition.

The statement for the backward representation is now easy to see.

*Remark*. Let us give a detailed proof of (13.23). Since $Y_\xi(t)$ is a function of $\xi, t, B$, we shall write it as $f(\xi, t, B)$. Because of the uniqueness of the solution of (13.20), it is clear that, with probability 1,

(13.33) $$f(\xi, s+t, B) = f(f(\xi, s, B), t, B_s)$$

where $B_s(t) = B(t+s) - B(s)$.

Since $f(\xi, t, B)$ is $B_{0t}(B)$-measurable, the event $(\sigma < t)$, is also $\mathcal{B}_{0t}(B)$-measurable. Approximating $\sigma$ with discrete-valued random variables $\sigma_n = [n\cdot\sigma]/n$ as we usually do in deriving the strict Markov property from the Markov property in the Feller process (see [9] or [2]), we can easily see that for any fixed $\xi$, with probability 1,

(13.34) $$f(\xi, \sigma+t, B) = f(f(\xi, \sigma, B), t, B_\sigma) \quad (t \geq 0)$$

by interpreting both sides as $\infty$ for convention when $\sigma = \infty$. Therefore with probability 1,

(13.35) $$\begin{aligned} f(\xi_1, \sigma+t, B) &= f(f(\xi_1, \sigma, B), t, B_\sigma) \\ &= f(f(\xi_2, \sigma, B), t, B_\sigma) \\ &= f(\xi_2, \sigma+t, B), \end{aligned}$$

which proves (13.23).

## 14. A Modified Girsanov Example

In the general cases treated in Sections 6, 7 and 8 we neither proved nor disproved

$$\mathcal{B}_{-\infty t}(X) \subset \mathcal{B}_{-\infty s}(X) \wedge \mathcal{B}_{st}(dB) \qquad (s < t) \tag{14.1}$$

for the stationary solutions obtained there, while (14. 1) was true for the special cases discussed in Sections 10, 11, 12 and 13.

Let us consider the equation:

$$dX(t) = h(X(t))dB(t) \tag{14.2}$$
$$\left(h(\xi) = \frac{|\xi|^\alpha}{1+|\xi|^\alpha} + |\xi|, \quad \alpha = \text{positive constant} < \frac{1}{2}\right)$$

and call it a *modified Girsanov equation.* We put an extra term $|\xi|$ in $h(\xi)$ in the Girsanov equation [3] in order to be able to get a stationary solution without any essential change of $h(\xi)$ near $\xi=0$.

First of all we shall consider two solutions of the stochastic integral equation:

$$X(t) = \int_0^t h(X(s))dB(s). \tag{14.3}$$

One is the trivial one $X^0(t) \equiv 0$. The other one is a Girsanov solution $X^1(t)$ which corresponds to the diffusion with the scale $ds = d\xi$ and the speed measure $2h(\xi)^{-2}d\xi$; Girsanov also constructed a solution which corresponds to the diffusion with the speed measure $2h(\xi)^{-2}d\xi + c\cdot\delta_0$ for every $c > 0$, but we will not use this here.

Now we shall combine these two solutions by Ikeda's method (see Ikeda's example [9]).

Take two random variables $\tau^0, \tau^1$ which are exponentially distributed with mean $\lambda^0$ and $\lambda^1$ respectively.

Take a countable number of copies $(X_n^i, B_n^i, \tau_n^i)$, $n=1, 2, \cdots$ of $(X^i, B^i, \tau^i)$ for $i=0, 1$. We shall also assume that $\mathcal{B}(X_n^i, B_n^i, \tau_n^i)$, $n=1, 2, \cdots$, $i=0, 1$, are independent.

Let $T(t)$ be the local time at 0 for the process $X_n^1$. It is clear that $X_n^1(T^{-1}(t_n^1)) = 0$. In view of this fact, we shall define $(X_+, B)$ as follows:

$$
(14.4)\quad
\begin{aligned}
&X_+(t) = X_1^0(t)\,, && B(t) = B_1^0(t)\,,\\
& && (0 \leq t \leq \tau_1^0)\,;\\
&X_+(t+\tau_1^0) = X_1^1(t)\,, && B(t+\tau_1^0) - B(\tau_1^0) = B_1^1(t)\,,\\
& && (0 \leq t \leq T^{-1}(\tau_1^1) \equiv \tilde{\tau}_1^1)\,;\\
&X_+(t+\tau_1^0+\tilde{\tau}_1^1) = X_2^0(t)\,, && B(t+\tau_1^0+\tilde{\tau}_1^1) - B(\tau_1^0+\tilde{\tau}_1^1) = B_2^0(t)\,,\\
& && (0 \leq t \leq \tau_2^0)\,;\\
&X_+(t+\tau_1^0+\tilde{\tau}_1^1+\tau_2^0) = X_2^1(t)\,, &&\\
& && B(t+\tau_1^0+\tilde{\tau}_1^1+\tau_2^0) - B(\tau_1^0+\tilde{\tau}_1^1+\tau_2^0) = B_2^1(t)\,,\\
& && (0 \leq t \leq T^{-1}(\tau_2^1) \equiv \tilde{\tau}_2^1)
\end{aligned}
$$

and so on. Then $B(t)$ is a Wiener process and (14.3) holds for the pair $(X_+, B)$.

Because of the property of the non-singular diffusion, the set of zero points of $X_n^1(t)$ is non-dense for every $n=1, 2, \cdots$. Therefore $\tau_1^0, \tau_2^0, \cdots$ are determined as the lengths of the successive zero intervals of $X_+(t)$, so that $\mathcal{B}(X_+) \supset \mathcal{B}(\tau_1^0, \tau_2^0, \cdots)$. On the other hand the construction (14) shows that $\mathcal{B}(\tau_1^0, \tau_2^0, \cdots)$ is independent of $\mathcal{B}(dB)$.

Noticing that the probability measures $P(X_+(t) \in \cdot)$, $t \geq 0$, form a totally bounded set, we can apply the shifting and averaging method used in Secrion 6 to our pair $(X_+, B)$ in order to get a stationary solution $X$ of (14.2).

After the shift the lengths $\tau_1^0, \tau_2^0, \cdots$ of zero intervals $\subset [0, \infty)$ of the solution are independently and exponentially distributed with the mean $\lambda^0$ and independent of the Wiener process $B$, so that this fact remains to be true even after the averaging. This means that we have, for the stationary solution $X$ obtained iu such a way,

$$\mathcal{B}_{0\infty}(X) - \mathcal{B}_{-\infty 0}(X) \supset \mathcal{B}(\tau_1^0, \tau_2^0, \cdots)$$

and

$$\mathcal{B}(\tau_1^0, \tau_2^0, \cdots) \text{ is independent of } \mathcal{B}_{0\infty}(dB)\,,$$

so that

$$\mathcal{B}_{0\infty}(X) \not\subset \mathcal{B}_{-\infty 0}(X) \vee \mathcal{B}_{0\infty}(dB)\,,$$

This implies that (14.1) does not hold for any pair $s<t$, because $X$ and $dB$ are stationarily correlated.

## 15. A Deterministic Example

Since we did not require $b(f)\neq 0$ in our stochastic differential equation:

$$dX(t) = a(\pi_t X)dt + b(\pi_t X)dB(t), \tag{15.1}$$

even the deterministic equation (differential-difference equation):

$$dX(t) = a(\pi_t X)dt, \tag{15.2}$$

i.e.

$$\frac{dX(t)}{dt} = a(\pi_t X)$$

lies in our frame.

Take an example

$$\frac{dX(t)}{dt} = -X\left(t-\frac{\pi}{2}\right). \tag{15.3}$$

The one sided solution is uniquely determined; in fact, if $X(t)$ is given for $t\leq 0$, then $X(t)$ is determined for $0\leq t\leq \pi/2$ by integration and then for $\pi/2\leq t\leq \pi$ and so on. Besides the trivial stationary solution $X(t)\equiv 0$, we have another stationary solution

$$X(t) = \sin(\omega+t+\alpha) \qquad (\alpha = \text{constant}) \tag{15.4}$$

where $\omega$ is a probability parameter moving on $\Omega=[0, \pi/2]$ associated with the uniform distribution. This stationary process is ergodic.

## 16. A Two-dimensional Example

Consider a two-dimensional stochastic differential equation:

$$\begin{aligned} dX &= \frac{X}{2}dt - YdB_2 \\ dY &= \frac{Y}{2}dt + XdB_2 \end{aligned} \qquad (R\equiv\sqrt{X^2+Y^2}\leq 1) \tag{16.1}$$

$$\begin{aligned} dX &= \left(\frac{X}{\sqrt{R}}-\frac{X}{2}\right)dt+\left(X-\frac{X}{R}\right)dB_1 - YdB_2 \\ dY &= \left(\frac{Y}{\sqrt{R}}-\frac{Y}{2}\right)dt+\left(Y-\frac{Y}{R}\right)dB_1 + XdB_2 \end{aligned} \qquad (R\geq 1) \tag{16.2}$$

We can easily see that the coefficients satisfy the Lipschitz condition in the whole plane, so that the one-sided solution can be determined uniquely by successive approximation.

In order to fined out a stationary solution, we shall write the equations in the polar coordinates $(R, \Theta)$. (16. 1) and (16. 2) are transformed into

(16. 3) $dR = R\cdot dt \quad (R\leq 1), \qquad dR = \sqrt{R}\,dt + (R-1)dB_1 \quad (R\geq 1)$

(16. 4) $d\Theta = dB_2 \pmod{2\pi}$

Observing the form of these equations, we can easily see that

(16. 5) $\Theta(t)$ corresponds to a Brownian motion on the unit circle,

(16. 6) $R(t)$ corresponds to a one-dimensional diffusion,

and

(16. 7) $R(t)$ and $\Theta(t)$ are independent processes.

Using Theorem 17 in Section 13, we can easily see that for the diffusion $R(t)$, 0 is a trap, every point in $(0, 1]$ is a strict right shunt and $(1, \infty)$ is a non-singular interval of positive recurrent type.

If $(X(t), Y(t))$ is a stationary solution of (16. 1) and (16. 2), then the corresponding $R(t)$ and $\Theta(t)$ should be stationary solutions of (16. 3) and (16. 4) respectively. Therefore either $R(t)\equiv 0$ or $R(t)$ is a unique stationary solution with the invariant measure $d\mu(r)$ proportional to the speed measure of the non-singular interval $(1, \infty)$.

$R(t)\equiv 0$ corresponds to the trivial stationary solution $(X(t), Y(t))\equiv(0, 0)$.

The second stationary solution $R(t)$ corresponds to a stationary solution $(X(t), Y(t))\equiv(R(t), \Theta(t))$ with the invariant measure $d\mu(r)\cdot d\theta/2\pi$ concentrated in the outside of the unit circle.

We shall verify for this stationary solution

(16. 8) $\bigwedge_t \mathcal{B}_{-\infty t}(X, Y) = \mathfrak{N} \qquad (=\text{trivial algebra})$

(16. 9) $\mathcal{B}_{-\infty t}(X, Y) \subset \mathcal{B}_{-\infty s}(X, Y) \vee \mathcal{B}_{st}(dB_1, dB_2) \qquad (-\infty < s < t < \infty)$

and

(16. 10) $\mathcal{B}_{-\infty t}(X, Y) \not\subset \mathcal{B}_{-\infty t}(dB_1, dB_2) \qquad (-\infty < t < \infty)$

The transition probability of $R(t)$ tends to its invariant distribution as $t \to \infty$, and a similar fact holds for $\Theta(t)$. Since $R(t)$ and $\Theta(t)$ are independent, it is also true for the joint process $(R(t), \Theta(t)) = (X(t), Y(t))$. Using this, we can verify (16.8) easily.

(16.9) follows at once from the fact that the one-sided solution can be uniquely determined by successive approximation.

To prove (16.10), it is enough to prove it for $t=0$ by virtue of the stationarity. If $\mathcal{B}_{-\infty 0}(X, Y) \subset \mathcal{B}_{-\infty 0}(dB_1, dB_2)$ holds, then we have

$$E\left[\left|\frac{X(t)}{R(t)} - E\left(\frac{X(t)}{R(t)}\middle| \mathcal{B}_{0t}(dB_1 dB_2)\right)\right|\right] \to 0 \qquad (R = \sqrt{X^2+Y^2})$$

as $t \to \infty$ by Theorem 8 in Section 9; notice here that $R(t) \geq 1$ and that $|X(t)/R(t)| \leq 1$

$$E\left(\frac{X(t)}{R(t)}\middle| \mathcal{B}_{0t}(dB_1, dB_2)\right) = E[\cos(\Theta(0)+B_2(t)-B_2(0)) \,|\, \mathcal{B}_{0t}(dB_1, dB_2))]$$
$$= \frac{1}{2\pi}\int_0^{2\pi} \cos(\theta + B_2(t) - B_2(0))\, d\theta = 0\,;$$

recall that $\Theta(0)$ is independent of $\mathcal{B}_{0\infty}(dB_1, dB_2)$. Therefore we get

$$E\left[\left|\frac{X(t)}{R(t)}\right|\right] \to 0 \qquad (t \to \infty),$$

and similarly

$$E\left[\left|\frac{Y(t)}{R(t)}\right|\right] \to 0,$$

and so $E[(|X(t)|+|Y(t)|)/R(t)] \to 0$ in contradiction with $|X|+|Y| \geq R$. Thus (16.10) is proved.

Department of Mathematics, Stanford University
and
Division of Applied Mathematics, Brown University

**BIBLIOGRAPHY**

[1] J. L. Doob: Stochastic processes, New York 1953.
[2] E. B. Dynkin: Markov processes, 2nd Vol. Moscow (1962), English translation to be published by Springer, Berlin.
[3] I. V. Girsanov: An example of non-uniqueness of the solution of the stochastic equation of K. Itô. Th. Prob. Appl. 7 (1962), 325–331.

[4] K. Itô: On a formula concerning stochastic differentials, Nagoya Math. Jour. 3 (1951), 55-65.

[5] K. Itô: Multiple Wiener integral, Jour. Math. Soc. Japan 3 (1951), 157-169.

[6] K. Itô: Complex multiple Wiener integral. Jap. Jour. Math. 22 (1952), 63-86.

[7] K. Itô: Stationary random distributions, Memoirs Fac. Science, Kyoto Univ. Ser. A. Math. 28 (1953), 209-223.

[8] K. Itô: On stochastic differential equations, Mem. Amer. Math. Soc. 4 (1954).

[9] K. Itô and H. P. McKean, Jr.: Diffusion processes and their sample paths, to be published by Springer, Berlin.

[10] K. Karhunen: Über die Struktur stationären zufälliger Funktionen. Arkiv för mat. 1 (1950), 114-160.

[11] H. P. McKean, Jr.: Elementary solutions for certain parabolic partial differential equations, Trans. Amer. Math. Soc. 82 (1956), 519-548.

[12] M. Nisio: Remark on the canonical representation of strictly stationary processes, Jour. Math. Kyoto Univ. 1 (1961), 129-146.

[13] Yu V. Prohorov: Convergence of random processes and limit theorems in probability theory, Th. Prob. Appl. 1 (1956), 157-214 (English translation).

[14] M. Rosenblatt: Stationary processes as shifts of functions of independent random variables, Jour. Math. Mech. 8 (1959), 665-682.

[15] M. Rosenblatt: Stationary Markov chains and independent random variables. Jour. Math. Mech. 9 (1960), 945-950.

[16] M. Rosenblatt: The representation of a class of two state stationary processes in terms of independent random variables, Jour. Math. Mech. 12 (1963), 721-730.

[17] A. V. Skorokhod: Limit theorems for stochastic processes, Th. Prob. Appl. 1 (1956), 261-290 (English translation).

[18] A. V. Skorokhod: On the existence and uniqueness of solutions of stochastic differential equations (Russian), Sibirsk. Mat. Ž. 2 (1961), 129-137.

[19] N. Wiener: Non-linear problems in random theory, Cambridge, Mass. 1958.

Ann. Inst. Fourier, Grenoble
15, 1 (1965), 13-30.

# TRANSFORMATION OF MARKOV PROCESSES BY MULTIPLICATIVE FUNCTIONALS

by Kiyosi ITÔ and Shinzo WATANABE

## 1. Introduction.

Let $X_t = X_t(S, P_a)$ be a Markov process with the state space S and the probability law $P_a$ of the path starting at $a$ and let $\alpha_t$ be a multiplicative functional (m.f.) of $X_t$. A Markov process $X_t^* = X_t^*(S, P_a^*)$ is called the *α-subprocess* of $X_t$ if

$$(1.1) \qquad P_a^*[X_t^* \in E] = \int_{X_t \in E} \alpha_t \, dP_a \qquad (E \subset S).$$

An important example of the α-subprocess is the Doob $h$-process which is the subprocess of the Brownian motion with respect to the m.f.

$$(1.2) \qquad \alpha_t = \frac{h(X_t)}{h(X_0)},$$

where $h$ is a positive harmonic function. The $h$-process plays an important role in the potential theory. Another important example is a diffusion process with the generator

$$(1.3) \qquad u = \tfrac{1}{2}\Delta u - ku \qquad (k \geqslant 0)$$

This is the subprocess of the Brownian motion with respect to the m.f.

$$(1.4) \qquad \alpha_t = \exp\left[\int_0^t - k(X_s)\, ds\right].$$

The transformation to get this subprocess from the Brownian motion is *killing* with the killing rate $k$.

The general α-subprocess was discussed by E. B. Dynkin [1] and independently by H. Kunita and T. Watanabe [4] under the natural assumption

$$E_a(\alpha_t) \leqslant 1. \tag{1.5}$$

In this note we shall give another general method of the transformation which seems to be more probabilistic than theirs.

Let us now describe the outline of our construction of the $\alpha$-subprocess.

If there exists an increasing sequence of Markov times $T_n$ whose limit is no less than the least zero point $T_\alpha$ of the given m.f. $\alpha$ such that $\alpha_{t \wedge T_n}$ is a martingale, we call $\alpha_t$ *regular*. We shall define the *subregularity* and the *superregularity* of $\alpha_t$ by replacing martingale with submartingale and supermartingale respectively in this definition; the superregularity of $\alpha_t$ is equivalent to the condition (1.5). We can prove that the *factorization theorem* that any superregular m.f. is expressed as the product of a regular m.f. $\alpha_t^{(0)}$ and a decreasing m.f. $\alpha_t^{(1)}$. $\alpha_t^{(0)}$ is called the *regular factor* of $\alpha_t$ and $\alpha_t^{(1)}$ the *decreasing factor*.

In order to construct the $\alpha$-subprocess, we shall first distort the probability law $P_a$ of the original process as

$$dP_a^{(0)} = \lim_{n\to\infty} \alpha_{T_n}^{(0)} \, dP_a|F_{T_n}, \tag{1.6}$$

$$T_n \uparrow T_\alpha, \; P_a|F_n = \text{restriction of } P_a \text{ to } F_{T_n}$$

to get a Markov process $X_t^{(0)}(P_a^{(0)}, S)$ which is *semi-conservative* in the sense that the life time $T_\Delta$ can be approximated strictly from below by a sequence of Markov times almost surely and *next* kill $X_t^{(0)}$ by the rate $d\alpha_t^{(1)}/\alpha_t^{(1)}$ to get the $\alpha$-subprocess $X_t^*$; the precise meaning of (1.6) will be given in Section 4.

In Section 2 we shall prove a factorization theorem for positive supermartingales. In Section 3 we shall use this to get a similar theorem for superregular m.f.'s which will be useful in the construction of the $\alpha$-subprocess in Section 4. Interesting examples are given in Section 5. Our idea can apply to the transformation by a subregular m.f. in which case we should introduce a creation instead of killing as is seen in Section 6.

We would like to express our hearty thanks to Professors M. Brelot, G. Choquet and J. Deny who organized the Colloquium of Potential Theory at University of Paris where the idea of our construction was presented by one of us, Itô, and also to Professors S. Karlin and

K. L. Chung at Stanford University for their friendly aid during our preparation of this note.

## 2. Factorization theorem for positive supermartingales.

Let $\Omega(\mathbf{F}, P)$ be the basic probability measure space where $\mathbf{F}$ is complete with respect to P and suppose that we are given an increasing and right-continuous family of $\sigma$-algebras $\mathbf{F}_t \subset \mathbf{F}$, $0 \leqslant t < \infty$, each containing all null sets. A non-negative random variable T is called a *Markov time* if $\{T < t\} \in \mathbf{F}_t$ for every $t$. Given a Markov time T we shall define $\mathbf{F}_T$ as the system of all sets $A \in \mathbf{F}$ for which $A \cap \{T < t\} \in \mathbf{F}_t$ for every $t \geqslant 0$. $\mathbf{F}_T$ is clearly a $\sigma$-algebra complete with respect to P and it is easy to see by the right-continuity of $\mathbf{F}_t$ that $\mathbf{F}_T = \mathbf{F}_t$ for $T \equiv t$. $\mathbf{F}_T$ is clearly *strongly right-continuous* in the sense that $\bigcap_n \mathbf{F}_{T_n} = \mathbf{F}_T$ for $T_n \downarrow T$. However, we shall here assume that it is also *strongly left-continuous* i.e. that $\bigvee_n \mathbf{F}_{T_n} = \mathbf{F}_T$ for $T_n \uparrow T$.

To avoid constant repetition of qualifying phrases, we assume that T, $T_1$, $T_2$, etc., denote Markov times and that $A_t$, $X_t$, etc., are stochastic processes measurable $(\mathbf{F}_t)$ at each time point $t$.

By a theorem due to P. Meyer [6] we have a decomposition of $\Omega$ for a given T

$$(2.1) \qquad \Omega = \Omega_T \vee (\Omega - \Omega_T) \qquad \Omega_T \in \mathbf{F}_T$$

such that

$$(2.2) \qquad \Omega_T \underset{\text{a.s.}}{\subset} \{\forall_n \quad T_n < T\}$$

for some $T_n \uparrow T$ and that

$$(2.3) \qquad \Omega - \Omega_T \underset{\text{a.s.}}{\subset} \{\exists n \quad T_n = T\}$$

for every $T_n \uparrow T$, where « a.s. » means the inclusion modulo null sets.

Following P. Meyer we shall call a right-continuous process $A_t$, $t \geqslant 0$, *natural* if

$$(2.4) \qquad \{\Delta A_T > 0, T < \infty\} \underset{\text{a.s.}}{\subset} \Omega_T$$

for every T, where $\Delta A_T = A_T - A_{T-}$.

Since every supermartingale has a right-continuous version, we assume that the supermartingale in consideration is always right continuous.

Given a positive (= nonnegative) supermartingale, we set

$$\text{(2.5)} \qquad T_X = \inf(t : X_t = 0) \qquad \text{if } X_t = 0 \text{ for some } t$$
$$= \infty \qquad \text{if otherwise}$$

and denote with $\Omega_X$ the $\Omega_T$ in (2.1) for $T = T_X$. It is clear that $X_t = 0$ for $t \geqslant T_X$.

DEFINITION. — $X_t$, $0 \leqslant t < \infty$, is called a *local martingale* if there exists a sequence $T_n \uparrow \infty$ with $P(T_n < \infty) = 1$ such that $X_{T_n \wedge t}$, $0 \leqslant t < \infty$, is a martingale.

*Remark.* — It is easy to see that if $X_t$, $0 \leqslant t < \infty$ is a local martingale, then we can take $T'_n \uparrow \infty$ such that $X_{t \wedge T'_n}$ is a martingale on $[0, \infty]$.

The aim of this section is to prove the

FACTORIZATION THEOREM FOR POSITIVE SUPERMARTINGALES. — A positive (= nonnegative) supermartingale $X_t$ with $P(T_X > 0) = 1$ is factorized as

$$\text{(2.6)} \qquad X_t = X_t^{(0)} \, . \, X_t^{(1)}$$

with a positive local martingale $X_t^{(0)}$ and a natural decreasing process $X_t^{(1)} (X_0^{(1)} = 1)$. If there are two such factorizations, then they are identical in $0 \leqslant t < T_X$.

Before proving this we shall prove some preliminary facts.

Let $X_t$ be a positive supermartingale for which there exist a constant M and an almost surely finite Markov time T such that

$$\text{(2.7 a)} \qquad \frac{1}{M} \leqslant X_t \leqslant M \qquad \text{for every } t < T,$$

$$\text{(2.7 b)} \qquad X_t = X_T \qquad \text{for every } t \geqslant T.$$

Let $X_t = Y_t - A_t$ be the Meyer decomposition of $X_t$ which is possible because $X_t$ is a supermartingale of class D by virtue of (2.7 *a, b*). We shall further impose the following conditions:

$$\text{(2.7 c)} \qquad A_t \leqslant M \qquad \text{for every } t < T.$$

Now, writing $A_t^c$ for the continuous part of $A_t$ and setting

$$\text{(2.8 a)} \qquad X_t^{(0)} = X_{t \wedge T} \exp\left(\int_0^{t \wedge T} \frac{dA_u^c}{X_u}\right) . \prod_{\substack{0 \leqslant u \leqslant t \wedge T \\ \Delta A_u > 0}} \left(1 + \frac{\Delta A_u}{X_u}\right)$$

$$(2.8\,b)\quad X_t^{(1)} = \exp\left(-\int_0^{t\wedge T}\frac{dA_u^c}{X_u}\right)\prod_{\substack{0\leqslant u\leqslant t\wedge T\\ \Delta A_u>0}}\left(1+\frac{\Delta A_u}{X_u}\right)^{-1}.$$

If $t \geqslant T$ and $X_T = 0$, then the last factor $1 + \dfrac{\Delta A_T}{X_T}$ is meaningless and is to be interpreted by the following convention:

$$(2.8\,c)\quad X_T\left(1+\frac{\Delta A_T}{X_T}\right) = X_T + \Delta A_T \quad \text{in } (2.8\,a)$$

$$\left(1+\frac{A_T}{X_T}\right)^{-1} = 0 \quad \text{in } (2.8\,b)$$

Then $X_t^{(i)}$, $i = 0, 1$, are well-defined by virtue of (2.7 *a, b, c*).

LEMMA 2.1. — $X_t = X_t^{(0)}X_t^{(1)}$ is the unique factorization of $X_t$ into a martingale $X_t^{(0)}$ and a natural decreasing process $X_t^{(1)}$ with

$$X_t^{(i)} = X_{t\wedge T}^{(i)}\ (i = 1, 2),\qquad X_0^{(1)} = 1 \qquad \text{and} \qquad X_t^{(0)} \leqslant Z$$

for some integrable Z.

*Proof.* — Let $\delta = (0 = u_0 < u_1 < \cdots < u_n \cdots \to \infty)$ be a division of $[0, \infty)$ with $|\delta| \equiv \sup_i |u_i - u_{i-1}| < \infty$ and define

$$(2.9)\quad X_{t,\delta}^{(0)} = \prod_{i\leqslant m}\left(1+\frac{A_{u_i}-A_{u_{i-1}}}{X_{u_i}}\right)(X_{t\wedge T} + A_{t\wedge T} - A_{u_m})$$

where $m = m(t, T)$ is the maximum number for which $u_m < t \wedge T$. Then it is easy to see that

$$(2.10)\qquad \lim_{|\delta|\to 0} X_{t,\delta}^{(0)} = X_t^{(0)}, \qquad 0 \leqslant X_{t,\delta}^{(0)} < e^{M^2}(X_T + A_T).$$

Denoting with $\phi(x)$ a function equal to 0 or $x^{-1}$ according as $x = 0$ or $x > 0$, and noticing

$$(2.11)\qquad X_{u_k,\delta}^{(0)} = X_{u_{k-1},\delta}^{(0)}(1 + \phi(X_{u_{k-1}})(Y_{u_k} - Y_{u_{k-1}})),$$

we can see

$$E(X_{u_k,\delta}^{(0)}|F_{u_{k-1}}) = X_{u_{k-1},\delta}^{(0)}$$

and so

$$(2.12)\qquad E(X_{u_k,\delta}^{(0)}|F_{u_l}) = X_{u_l,\delta}^{(0)} \qquad (l < k).$$

Given any pair $s < t$, let us consider only $\delta \ni s, t$ such that $|\delta| \to 0$ and notice (2.10) to derive from (2.12)

$$E(X_t^{(0)}|\mathbf{F}_s) = X_s^{(0)}, \tag{2.13}$$

which proves that $X_t^{(0)}$ is a martingale.

Since $X_t^{(1)}$ has the same discontinuity points as $A_t$ by the definition, $X_t^{(1)}$ is natural.

It is now easy to see that the factorization $X_t = X_t^{(0)} X_t^{(1)}$ satisfies the conditions stated in the lemma.

Now we shall prove the uniqueness.

Let $X_t = \Phi_t^{(0)} . \Phi_t^{(1)}$ be any such factorization and consider a decomposition of $X_t$:

$$X_t = Y'_t - A'_t \tag{2.14}$$

$$Y'_t = X_t - \int_0^t \Phi_u^{(0)} \, d\Phi_u^{(1)}, \qquad A'_t = -\int_0^t \Phi_u^{(0)} \, d\Phi_u^{(1)}.$$

Then it is clear that $A'_t$ is a natural increasing process. Now we shall prove that $Y'_t$ is a martingale. Using the division

$$\delta = (0 = u_0 < u_1 < \cdots \to \infty)$$

used above, we shall define

$$Y_t^\delta = X_t - \sum_{i \leqslant m} \Phi_{u_i}^{(0)}(\Phi_{u_i}^{(1)} - \Phi_{u_{i-1}}^{(1)}) - \Phi_t(\Phi_t^{(1)} - \Phi_{u_m}^{(11)}) \tag{2.15}$$

$$(u_m < t \leqslant u_{m+1}).$$

Writing $Y_{u_k}^\delta$ as

$$Y_{u_k}^\delta = X_0 + \sum_{i=1}^{k} \Phi_{u_{i-1}}^{(1)}(\Phi_{u_i}^{(0)} - \Phi_{u_{i-1}}^{(0)}) \tag{2.16}$$

to see

$$E(Y_{u_k}^\delta|\mathbf{F}_u) = Y_{u_l}^\delta \qquad (l < k), \tag{2.17}$$

and noticing

$$\lim_{|\delta| \to 0} Y_t^\delta = Y'_t, \qquad |Y_t^\delta| \leqslant X_t + Z, \tag{2.18}$$

we can see

$$E(Y'_t|\mathbf{F}_s) = Y'_s \qquad (s < t), \tag{2.19}$$

which proves that $Y'_t$ is a martingale.

Therefore $X_t = Y'_t - A'_t$ is the Meyer decompositions of $X_t$, so that $Y'_t = Y_t$ and $A'_t = A_t$. Thus we have

$$(2.15) \qquad A_t = -\int_0^t \Phi_u^{(0)}\, d\Phi_u^{(1)} \qquad \text{i.e.} \qquad dA_t = -\Phi_u^{(0)}\, d\Phi_u^{(1)}$$

where $dF_t$ means the Lebesgue-Stieltjes measure induced by $F_t$.

Let $A_t = A_t^c + A_t^d$ be the decomposition of $A_t$ into the continuous part $A_t^c$ and the discontinuous part $A_t^d$ and $\Phi_t^{(1)} = \Phi_{t,c}^{(1)}\Phi_{t,d}^{(1)}$ be the factorization of $\Phi_t^{(1)}$ into the continuous factor $\Phi_{t,c}^{(1)}$ and the discontinuous factor $\Phi_{t,d}^{(1)}$. Then (2.15) is written as

$$(2.16) \qquad dA_t^c + dA_t^d = -\Phi_t^{(0)}\Phi_{t,d}^{(1)}\, d\Phi_{t,c}^{(1)} - \Phi_t^{(0)}\Phi_{t,c}^{(1)}\, d\Phi_{t,d}^{(1)}$$

Equating the continuous parts and the discontinuous parts of both sides of (2.16) respectively, we get

$$(2.17) \qquad dA_t^c = -\Phi_t^{(0)}\Phi_{t,d}^{(1)}\, d\Phi_{t,c}^{(1)}, \qquad dA_t^d = -\Phi_t^{(0)}\Phi_{t,c}^{(1)}\, d\Phi_{t,d}^{(1)}.$$

Consider now the case $t < T$, so that $X_t > 0$. Recalling

$$X_t = \Phi_t^{(0)}\Phi_t^{(1)} = \Phi_t^{(0)}\Phi_{t,c}^{(1)}\Phi_{t,d}^{(1)}$$

we can derive from (2.17)

$$(2.18\,a) \qquad \frac{d\Phi_{t,c}^{(1)}}{\Phi_{t,c}} = -\frac{dA_t^c}{X_t}$$

$$(2.18\,b) \qquad \frac{d\Phi_{t,d}^{(1)}}{\Phi_{t,d}} = -\frac{dA_t^d}{X_t}.$$

Since $\Phi_{t,c}^{(1)}$ is continuous and $\Phi_{0,c}^{(1)} = 1$, we get from (2.18 *a*)

$$\Phi_{t,c}^{(1)} = \exp\left(-\int_0^t \frac{dA_u^c}{X_u}\right).$$

Since $\Phi_{t,d}^{(1)}$ is purely discontinuous and $\Phi_{0,d}^{(1)} = 1$, we get from (2.18 *b*)

$$\Phi_{t,d}^{(1)} = \prod_{\substack{\Delta A_u > 0 \\ 0 \leqslant u \leqslant t}} \frac{\Phi_{u,d}^{(1)}}{\Phi_{u-,d}^{(1)}} = \prod_{\substack{\Delta A_u > 0 \\ 0 \leqslant u \leqslant t}} \left(1 + \frac{\Delta A_u}{X_u}\right)^{-1}.$$

Thus we have proved $\Phi_t^{(1)} = \Phi_{t,c}^{(1)}\Phi_{t,d}^{(1)} = X_t^{(1)}$ and so $\Phi_t^{(0)} = X_t^{(0)}$, as far as $t < T$.

Since $\Phi_{T-}^{(i)} = X_{T-}^{(i)}$ is now evident, it is sufficient for the proof of $\Phi_T^{(i)} = X_T^{(i)}$ to recall the special care we took when we defined $X_T^{(i)}$. If $t \geqslant T$, then $\Phi_t^{(i)} = \Phi_T^{(i)} = X_T^{(i)} = X_T^{(i)}$, which completes the proof of Lemma 1.

LEMMA 2.—A positive supermartingale $X_t$ is decomposed uniquely as $X_t = Y_t - A_t$, where $Y_t$ is a local martingale and $A_t$ is a natural increasing process with $A_0 = 0$.

*Proof.*— Define $T_n$ by

$$T_n = \min[\inf(t : X_t \geqslant n), n].$$

Then $X_{t \wedge T_n}$ is a supermartingale of the class (D) and has a Meyer decomposition

$$X_{t \wedge T_n} = Y_t^{(n)} - A_t^{(n)}.$$

By the uniqueness of the Meyer decomposition we have

$$Y_t^{(m)} = Y_t^{(n)} \quad \text{and} \quad A_t^{(m)} = A_t^{(n)}$$

for $m < n$ and $t \leqslant T_m$, so that we have a decomposition of $X_t$: $X_t = Y_t - A_t$ such that $Y_t = Y_t^{(n)}$ and $A_t = A_t^{(n)}$ for $t < T_n$. This decomposition of $X_t$ satisfies the conditions in Lemma 2. The uniqueness is easy to see.

LEMMA 3. — Let $X_t$ be a supermartingale and T a Markov time. Suppose that $X_t = X_T$ for $t \geqslant T$ and that

$$\Omega_T \cap \{T < \infty\} \underset{\text{a.s.}}{\subset} \{\Delta X_T = 0, T < \infty\}. \tag{2.19}$$

If there exists an increasing sequence $T_n$ with $\lim T_n \geqslant T$ such that $X_{t \wedge T_n}$ is a martingale, then $X_t$ is a local martingale.

*Proof.* — Let $X_t = Y_t - A_t$ be the decomposition of Lemma 2. It is clear that $A_t = 0$ for $t \leqslant T_n$ and so for $t < T$. Since $A_t = A_T$ ($t \geqslant T$) follows from $X_t = X_T$ ($t \geqslant T$), $A_t$ is a step function with a single jump at $t = T$. Since $Y_t$ is a local martingale, we have a sequence of Markov times $S_n \uparrow \infty$ such that $Y_{t \wedge S_n}$ is a martingale. By a theorem due to Meyer [6], we have

$$\Omega_T \cap \{T < \infty\} \underset{\text{a.s.}}{\subset} \bigcup_n \{\Delta Y_T = 0, T < S_n\} \underset{\text{a.s.}}{\subset} \{\Delta Y_{T=0}, T < \infty\},$$

and so

$$\Omega_T \cap \{T < \infty\} \underset{\text{a.s.}}{\subset} \{\Delta Y_T = 0, \Delta X_T = 0, T < \infty\} \underset{\text{a.s.}}{\subset} \{\Delta A_T = 0, T < \infty\}$$

by (2.19), while

$$\{\Delta A_T > 0, T < \infty\} \underset{\text{a.s.}}{\subset} \Omega_T$$

because $A_t$ is natural. Thus

$$\{\Delta A_T > 0, T < \infty\} \underset{\text{a.s.}}{\subset} \{\Delta A_T = 0, T < \infty\},$$

which implies $\Delta A_T = 0$ almost surely and so $X_t \equiv Y_t$.

Now we shall come back to the

*Proof of the theorem.* — Let $X_t$ be a positive supermartingale and let $X_t = Y_t - A_t$ be the decomposition of Lemma 2 and define $T'_n$ by

$$\begin{aligned} T'_n &= \inf\left( t : X_t \geqslant n \quad \text{or} \quad X_t \leqslant \frac{1}{n} \quad \text{or} \quad A_t \leqslant n \right) \\ &= \infty \end{aligned}$$

if there exists no such $t$ and set $T_n = \min(T'_n, n)$. Then it is clear $T_n \uparrow T_\infty \equiv \lim T_n \geqslant T_X$.

Define $X_t^{(0)}$ by

$$X_t^{(0)} = X_t \exp\left[\int_0^t \frac{dA_u^c}{X_u}\right] \prod_{\substack{\Delta A_u > 0 \\ 0 \leqslant u \leqslant t}} \left(1 + \frac{\Delta A_u}{X_u}\right)$$

$$(A_t^c = \text{the continuous part of } A_t)$$

for $t < T_X$ and

$$X_t^{(0)} = X_{T_X}^{(0)} = \begin{cases} X_{T_X-}^{(0)} & \text{on} \quad \Omega_X \\ 0 & \text{on} \quad \Omega - \Omega_X \end{cases}$$

for $t \geqslant T_X$. Then $X_{t \wedge T_n}^{(0)}$ is a martingale by Lemma 1 and therefore a local martingale by Lemma 3.

Define $X_t^{(1)}$ by

$$X_t^{(1)} = \exp\left[-\int_0^t \frac{dA_u^{(c)}}{X_u}\right] \prod_{\substack{\Delta A_u > 0 \\ 0 \leqslant u \leqslant t}} \left(1 + \frac{\Delta B_u}{X_u}\right)^{-1}$$

for $t < T_X$ and

$$X_t^{(1)} = X_{T_X}^{(1)} = \begin{cases} 0 & \text{on} \quad \Omega_X \\ X_{T_X-}^{(1)} & \text{on} \quad \Omega - \Omega_X \end{cases}$$

if $t \geqslant T_X$. It is then clear that $X_t^{(1)}$ is natural.

Thus $X_t = X_t^{(0)} . X_t^{(1)}$ is a decomposition which satisfies all conditions of our theorem. The uniqueness follows at once from the uniqueness part of Lemma 1.

### 3. Factorization theorem for multiplicative functionals.

Let us recall the definition and notations on Markov processes. Let S be a locally compact Hausdorff space with a countable open base and $\bar{S} = S \vee \{\Delta\}$ the one-point compactification of S in case S is not compact. If S is already compact, $\Delta$ is adjointed to S as an isolated point. The topological $\sigma$-algebra $\mathbf{B}(S)$ on S, i.e., the least $\sigma$-algebra containing all open subsets of S is denoted by $\mathbf{B}(S)$. A function $w:[0, \infty) \to S$ is called a *path* if it satisfies the following conditions:

(3.1 *a*) $w(t) = \Delta \Rightarrow w(s) = \Delta \quad$ for $\quad s \geqslant t$,

(3.1 *b*) $w(t)$ is right-continuous,

(3.1 *c*) *either* $\lim_{s \uparrow t-} w(s)$ exists *or* $w(t) = \Delta$.

The space of all paths is denoted by W. To emphasize the fact that $w(t)$ is a function of $w \in W$ for each $t$, we shall write $X_t(w)$ for $w(t)$. The *terminal time* $T_\Delta(w)$ of the path $w$ is defined as

$$(3.2) \qquad T_\Delta(w) = \begin{cases} \inf(t : X_t(w) = \Delta) & \text{if } X_t(w) = \Delta \text{ for some } t \\ \infty & \text{if otherwise} \end{cases}$$

The shift *transformation* $\theta_t$ in W is defined by

$$(3.3) \quad \theta_t w(s) = w(s+t) \qquad \text{i.e.} \qquad X_s(\theta_t w) = X_{t+s}(w).$$

$\mathbf{B}_t$ is the least $\sigma$-algebra on W for which $X_s(w)$ is measurable for every $s \leqslant t$ and $\mathbf{B}$ denotes the lattice sum of all $\mathbf{B}_t$, $t \geqslant 0$. $P_x$, $x \in S$, is a system of probability measures on $W(\mathbf{B})$ such that

(3.4 *a*) $P_x(B)$ is $\mathbf{B}(S)$-measurable in $x$ for every $B \in \mathbf{B}$

(3.4 *b*) $P_x(X_0(w) = x) = 1$

(3.4 *c*) (Markov property)

$$E_x[f(\theta_t w), B \wedge (T_\Delta > t)] = E_x[E_{X_t}(f), B \wedge (T_\Delta > t)]$$

for every $B \in \mathbf{B}_t$, for every $\mathbf{B}$-measurable $f$, every $t \geqslant 0$ and every $B \in \mathbf{B}_t$.

The triple $(X_t, P_a, S)$ is called a Markov process and it is also denoted by $X_t(P_a, S)$ or simply by $X_t$. Given a probability measure $\mu$

on $S(\mathbf{B}(S))$, let $P_\mu$ be the probability measure defined by

$$P_\mu(B) = \int_S P_a(B)\, d\mu(a),$$

$B \in \mathbf{B}$. Let $\mathbf{N}$ be the class of all sets N such that $P_\mu(N) = 0$ for every $\mu$, and $\mathbf{F}_t$ be the least $\sigma$-algebra containing $\mathbf{B}_t$ and $\mathbf{N}$. $\mathbf{F}$ is defined similarly as the least $\sigma$-algebra containing $\mathbf{B}$ and $\mathbf{N}$. A function $T: W \to [0, \infty)$ is called a *Markov time* if $(T < t) \in \mathbf{F}_t$ for every $t \geqslant 0$. Given a Markov time T, $\mathbf{F}_T$ is defined as

(3.5) $$\mathbf{F}_T = (B \in \mathbf{F} : (T < t) \wedge B \in \mathbf{F}_t \text{ for every } t).$$

A Markov process $X_t(P_a, S)$ is called a *standard process* if the following two conditions are satisfied:

(3.6) (strict Markov property). For any Markov time

$$E_x[f(\theta_T w), B \wedge (T < T_\Delta)] = E_x[E_{X_T}(f), B \wedge (T < T_\Delta)]$$

for every $B \in \mathbf{F}_T$.

(3.7) (quasi-left continuity before $T_\Delta$). Given any Markov time T if $T_n$ is a sequence of Markov times such that $T_n \uparrow T$, then

$$P_x(X_{T_n} \to X_T / T < T_\Delta) = 1$$

A standard process is called a *Hunt Process* if the following two conditions are satisfied

(3.8) (existence of left-limit).

$$P_x(\lim_{t \uparrow T-} X_t \text{ exists } | T < \infty) = 1$$

for every Markov time T,

(3.9) (quasi-left continuity). Given any Markov time T, if $T_n$ is a sequence of Markov times such that $T_n \uparrow T$, then

$$P_x(X_{T_n} \to X_T | T < \infty) = 1.$$

For a Hunt process $X_t$ we have

(3.10) Given any Markov time T, if $T_n$ is a sequence of Markov time such that $T_n \uparrow T$, then $\bigvee_n \mathbf{F}_{T_n} = \mathbf{F}_T$.

Let $X_t$ be a standard process.

DEFINITION.—$\alpha_t(w)$, $t \in (0, \infty) w \in W$, is called a *multiplicative functional* (= m.f.), if it satisfies the following conditions

(3.11) $\alpha_t(w)$ is $\mathbf{F}_t$-measurable in $w$ for each fixed $t$,

(3.12) Except on a subset N of W such that $P_x(N) = 0$ for every $x$, we have

(i) $\alpha_0(w) = 1$

(ii) $\alpha_t(w)$ is right-continuous in $t$ and has finite left limits in $0 \leqslant t < T_\Delta$

(iii) $\alpha_t(w) = 0,\ t \geqslant T_\Delta$

(iv) $\alpha_{t+s}(w) = \alpha_t(w)\alpha_s(\theta_t w)$, for every pair $t, s \geqslant 0$.

Given a m.f. $\alpha_t$, we shall denote $\inf(t : \alpha_t = 0)$ by $T_\alpha$.

DEFINITION.—A m.f. $\alpha_t$ is called *regular* (superregular, subregular) if there exists an increasing sequence of Markov times $T_n$ such that $\lim_n T_n \geqslant T_\alpha$ and that $\alpha_{t \wedge T_n}$ is a martingale (supermartingale, submartingale) for each $P_x$.

It is easy to see by Fatou's lemma on integrals that the following conditions are equivalent to each other.

*a*) $\alpha_t$ is superregular

*b*) $\alpha_t$ is a supermartingale for each $P_x$

*c*) $E_x(\alpha_t) \leqslant 1$ for every pair $(t, x)$.

Suppose $X_t$ be a Hunt process and $\alpha_t$ a superregular m.f. Since $\alpha_t$ is a supermartingale by (*b*), we can apply the results of Section 1 to see that there exist a local martingale $\beta_t^{(x)}$ and a natural decreasing process $\gamma_t^{(x)}(\gamma_0^{(x)} = 1)$ such that $\alpha_t = \beta_t^{(x)} \cdot \gamma_t^{(x)}$. Recalling how the natural increasing part was constructed in the Meyer decomposition, we can get a version $\beta_t$ of $\beta_t^{(x)}$ and a version $\gamma_t$ of $\gamma_t^{(x)}$, both independent of $x$. $\beta_t$ and $\gamma_t$ satisfy

$$\text{(3.13)} \qquad \begin{aligned} \beta_{t+s}(w) &= \beta_t(w)\beta_s(\theta_t w) \\ \gamma_{t+s}(w) &= \gamma_t(w)\gamma_s(\theta_t w) \end{aligned} \qquad t + s < T_\alpha.$$

We shall discuss the details in a separate paper.

Now we shall define $\alpha_t^{(0)}$ and $\alpha_t^{(1)}$ by

$$\text{(3.14)} \qquad \alpha_t^{(0)} = \beta_t e_{[T_\alpha > t]}, \qquad \alpha_t^{(1)} = \gamma_t e_{[T_\alpha > t]}$$

where $e_{[T_\alpha > t]}$ takes 1 or 0 according as $T_\alpha > t$ or $T_\alpha \leqslant t$. It is easy to see that $T_{\alpha^{(0)}} = T_{\alpha^{(1)}} = T_\alpha$; $\alpha_t^{(0)}$ is regular. $X_t$ is continuous at every jump point of $\alpha_t^{(1)}$ in $[0, T_\alpha)$ since $\gamma_t$ is natural. Thus we have

FACTORIZATION THEOREM FOR M.F.—A m.f. $\alpha_t$ of a Hunt process $X_t(P_a, S)$ such that $E_x[\alpha_t] \leqslant 1$ for every $(t, x)$ can be factorized uniquely as

$$\alpha_t = \alpha_t^{(0)}\alpha_t^{(1)}$$

with a regular m.f. $\alpha_t^{(0)}$ and a decreasing m.f. $\alpha_t^{(1)}$ such that

$$T_\alpha = T_{\alpha^{(0)}} = T_{\alpha^{(1)}}$$

and that $\alpha_t^{(1)}$ has no common discontinuity points with $X_t$ on $[0, T_\alpha)$.

### 4. Construction of the $\alpha$-subprocess of a Hunt process $X_t$.

Let $X_t(P_a, S)$ be a Hunt process and $\alpha_t$ be a multiplicative functional for $X_t$. We shall construct the $\alpha$-subprocess of $X_t$.

*Regular case.* — Let $\alpha_t$ be regular. Then we have a sequence of Markov times $T_n \uparrow T_\alpha$ such that for each $n$, $\alpha_{T_n \wedge t}$ be a martingale on $[0, \infty]$ for every $P_x$. We shall prove that *there exists one and only one standard process* $X_t^{(0)}(P_a^{(0)}, S)$ that satisfies

(4.1) $$P_x^{(0)}(B) = E_x[\alpha_{T_n}, B] \qquad B \in \mathbf{F}_{T_n}, n = 1, 2, \ldots$$

and

(4.2) $$P_x^{(0)}(T_\Delta = \lim T_n) = 1.$$

*This process is semi-conservative, in fact, we have*

(4.3) $$P_x^{(0)}(T_n < T_\Delta) = 1, \qquad n = 1, 2, \ldots.$$

(4.1) and (4.2) are written symbolically as

(4.4) $$dP_x = \lim_{n \to \infty} \alpha_{T_n} \, dP_a | \mathbf{F}_{T_n}.$$

Let $\alpha_t = \beta_t . \gamma_t$ be the factorization that we introduced in Section 3. Since $\alpha_t$ is regular, it is also the regular part of the factorization of the m.f. $\alpha_t$, so that we have

(4.5) $$\alpha_t = \beta_t e_{[T_\alpha > t]} = \beta_t \qquad \text{for} \qquad t < T_\alpha.$$

Setting

(4.6) $$\tau = \min[\inf(t : \beta_t > 2), 1],$$

we shall obtain

(4.7) $$E_x[\beta_\tau | \mathbf{F}_t] = \beta_{\tau \wedge t};$$

in fact, since $\beta_t$ is a local martingale, there exists a sequence of Markov times $S_n \uparrow \infty$ such that $\beta_{S_n \wedge t}$ is a martingale on $0 \leqslant t \leqslant \infty$ and so we get

(4.8) $$E_x[\beta_{S_n \wedge \tau} | \mathbf{F}_t] = \beta_{S_n \wedge \tau \wedge t},$$

but since $\beta_{t\wedge\tau} \leqslant 2 + \beta_\tau$ follows from (4.6), we get (4.7) by letting $n$ tend to $\infty$ in (4.8).

Define $\tau_n, n = 1, 2, \ldots$ recursively by

(4.9) $$\tau_1 = \tau, \qquad \tau_2(w) = \tau_{n-1}(w) + \tau(\theta_{\alpha_{n-1}(w)} w)$$

and set

(4.10) $$\begin{aligned} \tilde{\beta}_t &= \beta_t & (0 \leqslant t \leqslant \tau_1) \\ &= \beta_{\tau_1} \cdot \beta_{t-\tau_1}(\theta_{\tau_1} w) & (\tau_1 \leqslant t \leqslant \tau_2) \\ &= \beta_{\tau_2} \cdot \beta_{t-\tau_2}(\theta_{\tau_2} w) & (\tau_2 \leqslant t \leqslant \tau_3) \end{aligned}$$

*etc.*

It is clear that

(4.11) $$\tilde{\beta}_t = \alpha_t \qquad \text{for} \qquad t < T_\alpha.$$

Using (4.9), we have

(4.12) $$E_x[\tilde{\beta}_{\tau_n} | F_s] = \tilde{\beta}_{\tau_n \wedge s}.$$

It is easy to see that $q(x, B) = E_x[\beta_\tau, B]$ is a probability measure in $B \in \mathbf{F}_\tau$ for each $x$ by (4.7) ($t = 0$) and measurable in $x$ for each $B \in \mathbf{F}_\tau$, so that $p(w, B) = q(X_{\tau(w)}(w, B)$ is also a probability measure in $B \in \mathbf{F}_\tau$ for each $x$ and $\mathbf{F}_\tau$-measurable in $\omega$ for each $B \in \mathbf{F}_\tau$. Using Ionescu Tulcea's theorem [2], we can define a probability measure on the direct product space $\Omega(\mathbf{B}_\Omega) = W(\mathbf{B}) \times \mathbf{W}(\mathbf{B}) \times \ldots$ such that

$$\begin{aligned} &P(\omega \in \Omega : \omega_1 \in B_1, \omega_2 \in B_2, \ldots, \omega_n \in B_n) \\ &= \int_{B_1} \cdots \int_{B_n} q(x, dw_1) p(w_1, dw_2) \ldots p(w_{n-1}, dw_n), \end{aligned}$$

where $\omega_i$ denote the $n$-th component of $\omega \in \Omega$.

Define $\tilde{\xi}_t(\omega)$ by.

(4.13) $$\begin{aligned} \tilde{\xi}_t &= X_t(\omega_1) & 0 \leqslant t \leqslant \tau(\omega_1), \\ &= X_{t-\tau(\omega_1)}(\omega_2) & 0 < t \leqslant \tau(\omega_2), \\ &= X_{t-\tau(\omega_1)-\tau(\omega_2)}(\omega_3) & 0 < t \leqslant \tau(\omega_3), \\ &\ \ \vdots \\ &= \Delta & t \geqslant \sum_i \tau(\omega_i), \end{aligned}$$

denote by $\tilde{P}_x^{(0)}$ the probability law (on $\mathbf{W}(\mathbf{B})$) of the stochastic process $\xi_t$, $0 \leqslant t < \infty$, and define

$$\begin{aligned} \xi_t(w) &= X_t(w) \qquad t < T_\alpha \\ &= \Delta \qquad t \geqslant T_\alpha \end{aligned} \tag{4.14}$$

on the probability space $W(\mathbf{B}, P_x^{(0)})$. Using the properties of $\alpha_t$ and $\beta_t$ and noticing (4.11) and (4.12), we can prove that the probability law $P_x^{(0)}$ of the stochastic process $\xi_t$, $0 \leqslant t < \infty$ satisfies (4.1) and (4.2). (4.3) follows from

$$P_x^{(0)}(T_n < T_\Delta) = E_x(\alpha_{T_n}) = 1.$$

*Decreasing case.* — In this case the construction of the $\alpha$-process is the usual terminating procedure. Consider a probability measure space $\Omega = W \times [0, \infty]$ associated with the probability measure $\tilde{P}_x^{(1)}$ such that

$$\tilde{P}_x^{(1)}([s, \infty] \times B) = \int_B \alpha_s(w)\, dP(w), \qquad B \in \mathbf{B}, \tag{4.15}$$

and define a stochastic process $\xi_t(\omega)$, $0 \leqslant t < \infty$, on this probability measure space by

$$\xi_t(\omega) = \begin{cases} X_t(w) & (t < s) \\ \Delta & (t \geqslant s) \end{cases} \tag{4.16}$$

for $\omega = (w, s)$. Let $P_x^{(1)}$ be the probability law of the process $\xi_t(\omega)$, $0 \leqslant t < \infty$. Then $X_t(P_x^{(1)}, S)$ is the $\alpha$-subprocess.

*General case.* — Let $\alpha_t = \alpha_t^{(0)}\alpha_t^{(1)}$ be the factorization of $\alpha_t$ into the regular part $\alpha_t^{(0)}$ and the decreasing part $\alpha_t^{(1)}$. The $\alpha$-subprocess can be constructed by the superposition of the transformations by $\alpha_t^{(0)}$ and by $\alpha_t^{(1)}$ in this order, each having been explained above.

*Remark.* — It is to be noted that the $\alpha$-subprocess is not always a Hunt process but a standard process in general, even if the original one is a Hunt process. In order to construct the $\alpha$-subprocess of a standard process as Dynkin and Kunita-Watanabe did, we should overcome some technical difficulties. We would like to discuss this in a separate paper.

## 5. Examples.

*a*) Let the original process be the Brownian motion in the *n*-dimensional domain D terminated at the boundary and let $\alpha_t$ be defined as

$$(5.1) \qquad \alpha_t = \begin{cases} u(X_t)/u(X_0) & t < T_\Delta \\ 0 & t \geqslant T_\Delta, \end{cases}$$

where $u$ is a positive superharmonic function of class $C^2$ in D. Note that $T_\Delta$ is nothing but the exit time from D. In this case the regular and decreasing parts of $\alpha_t$ are expressed in $t < T_\Delta$ as

$$(5.2\ a) \qquad \alpha_t^{(0)} = \exp\left[\int_0^t \frac{\operatorname{grad} u(X_s)}{u(X_s)}\, dX_s - \frac{1}{2}\int_0^t \frac{|\operatorname{grad} u|^2(X_s)}{u^2(X_s)}\, ds\right]$$

$$(5.2\ b) \qquad \alpha_t^{(1)} = \exp\left[\frac{1}{2}\int_0^t \frac{\Delta u(X_s)}{u(X_s)}\, ds\right],$$

since a formula of stochastic integrals shows

$$(5.3) \qquad \log \alpha_t = \log u(X_t) - \log u(X_0)$$

$$= \int_0^t \frac{\operatorname{grad} u(X_s)}{u(X_s)}\, dX_s + \frac{1}{2}\int_0^t \left(-\frac{|\operatorname{grad} u|^2(X_s)}{u(X_2)^2} + \frac{\Delta u(X_s)}{u(X_s)}\right) ds.$$

In particular, $\alpha_t$ is regular if and only if $u$ is harmonic.

*b*) Let $\tilde{X}_t(\tilde{P}_a, S)$ be a conservative Hunt process and $X_t(P_a, S)$ be the subprocess of $X_t$ by the multiplicative functional:

$$(5.4) \qquad \tilde{\alpha}_t = \exp\left(-\int_0^t f(\tilde{X}_s)\, ds\right).$$

Then

$$(5.5) \qquad \alpha_t = e_{[T_\Delta > t]}$$

is a multiplicative functional of $X_t$, whose regular and decreasing parts are

$$(5.6\ a) \qquad \alpha_t^{(0)} = \alpha_t \,.\, \exp\left(\int_0^t f(X_s)\, ds\right)$$

(5.6 *b*) $$\alpha_t^{(1)} = \alpha_t \exp\left(-\int_0^t f(X_s)\,ds\right).$$

To prove this, observe

$$\begin{aligned} E_x(\alpha_t - \alpha_0) &= \tilde{E}_x\left(\exp\left(-\int_0^t f(\tilde{X}_s)\,ds\right)\right) - 1 \\ &= -\tilde{E}_x\left(\int_0^t \exp\left(-\int_0^s f(\tilde{X}_u)\,du\right) f(\tilde{X}_s)\,ds\right) \\ &= -E_x\left(\int_0^t \alpha_s \,.\, f(X_s)\,ds\right) \end{aligned}$$

to see that the increasing process $A_t$ appearing in the Meyer decomposition of $\alpha_t$ is $\int_0^t \alpha_s f(X_s)\,ds$. It is now easy to verify (5.6 *a*, *b*) by

$$\alpha_t^{(1)} = \exp\left(-\int_0^t \frac{dA_s}{\alpha_s}\right) = \exp\left(\int_0^t f(X_s)\,ds\right).$$

The $\alpha^{(0)}$-subprocess of $X_t$ is exactly the process $\tilde{X}_t$.

*c*) In the construction discussed in Section 4, the $\alpha^{(0)}$-subprocess was semi-conservative. We shall prove that *it is conservative if* $E_x(\alpha_t)$ *tends to* 1 *uniformly in x as t tends to* 0.

Let $P_x^{(0)}$ be the probability law of the $\alpha^{(0)}$-subprocess. Then

$$P_x^{(0)}(T_\Delta > t) = E_x[\alpha_t^{(0)}] \geqslant E_x[\alpha_t].$$

As we saw in Section 4, we have $T_n \uparrow T$ such that

$$P_x^{(0)}(X_{T_n} \in S) = P_x^{(0)}(T_n < T_\Delta) = 1.$$

To complete the proof, we need only observe

$$\begin{aligned} P_x^{(0)}(T_\Delta < \infty) &= \lim_{n\to\infty} P_x^{(0)}(T_\Delta < T_n + t) \\ &= \lim_{n\to\infty} E_x^{(0)}[P_{xT_n}^{(0)}(T_\Delta < t)] \\ &\leqslant 1 - \inf_{x\in S} E_x(\alpha_t) \to 0 \qquad (t \to 0). \end{aligned}$$

*d*) Using the results of (*c*) we shall prove that *a Hunt process on a compact state space* S *whose semi-group transforms* C(S) *into* C(S) *can be obtained from a conservative Hunt process by terminating procedure.*

Since $P_x(T_\Delta > t) = T_t 1(x)$ is continuous in $x \in S$ and $P_x(T_\Delta > t) \to 1$ $(t \to 0)$, it is clear by Dini's theorem that the convergence is uniform, so that the m.f. $\alpha_t = e_{(T_\Delta > t)}$ satisfies the condition of (*c*). Let $\alpha_t = \alpha_t^{(0)}\alpha_t^{(1)}$ be the factorization of $\alpha_t$. Then $X_t$ is obtained by terminating procedure from its $\alpha^{(0)}$-subprocess which is conservative by (*c*).

*Remark.* — V. A. Volkonsky [8] discussed the same problem by a different method.

## 6. Subregular multiplicative functionals and creation.

Let $\alpha_t$ be a subregular multiplicative functional of a Hunt process $X_t(P_a, S)$. Under reasonable conditions $\alpha_t$ can be factorized as $\alpha_t = \alpha_t^{(0)}\alpha_t^{(1)}$ with a regular $\alpha_t^{(0)}$ and an increasing $\alpha_t^{(1)}$. Construct first the $\alpha^{(0)}$-subprocess $X_t^{(0)}$ which is semi-conservative and then the *branching process* of the particle subject to the same probability law as $X_t^{(0)}$ except the particle will create a new particle of the same probabilistic character by the rate $d\alpha_t^{(1)}/\alpha_t^{(1)}$, each performing the same random motion and creation. This branching process is called the $\alpha$-subprocess of $X_t$, because $E(\alpha_t, X_t \in E)$ *is the expected number of the particles in* $E(\in S)$ *at time t*. Several interesting results concerning creation are found in [3] and [7], in case the original process is a Brownian motion.

BIBLIOGRAPHY

[1] E. B. DYNKIN, *Markov processes,* Moscow (1963).

[2] C. T. IONESCU TULCEA, Mesures dans les espaces produits, *Atti Acad. Naz. Lincei Rend. Cl. Sci. Fis. Mat. Nat.* (8), 7 (1949), 208-211.

[3] K. ITÔ and H. P. MCKEAN, Jr., *Diffusion processes and their sample paths, Grundlehre d. Math. Wiss. Bd.* 125, Springer, Berlin, 1965.

[4] H. KUNITA and T. WATANABE, Notes on transformations of Markov processes connected with multiplicative functionals, *Memoirs of Fac. Sci. Kyushu Univ. Ser. A. Math.*, 17 (1963), 181-191.

[5] P. A. MEYER, A decomposition theorem for supermartingales, *Ill. Jour. Math.*, 6, 2 (1962), 193-205.

[6] P. A. MEYER, Decomposition of supermartingales: The uniqueness theorem, *Ill. Jour. Math.*, 7-1 (1963), 1-17.

[7] B. A. SEVAST'YANOV, Extinction conditions for branching stochastic processes with diffusion, *Th. Prob. Appl.*, 5-3 (1961).

[8] V. A. VOLKONSKY, Additive functionals of Markov processes, *Trudy Moskov. Mat. Obs.*, 9 (1960), 143-189.

Kiyosi ITÔ et Shinzo WATANABE,
Department of Mathematics,
Kyoto University,
Kyoto (Japan).

J. Math. Soc. Japan
Vol. 20, Nos. 1-2, 1968

# The canonical modification of stochastic processes

Dedicated to Professor S. Iyanaga for his sixtieth birthday

By Kiyosi ITÔ

(Received Sept. 11, 1967)

## § 1. Introduction and summary.

In his book [1], J. L. Doob proved that every stochastic process continuous in probability has a standard separable measurable modification. This theorem plays a fundamental role in the sample path approach of stochastic processes. The aim of our paper is to give a more concrete formulation to this important fact to make it easier to visualize the probability law of the sample path.

In Section 2 we shall introduce the space $M\equiv M(T)$ of *canonical measurable functions* on the time interval $T$. The space $\tilde{M}\equiv\tilde{M}(T)$ contains bad functions such as the Dirichlet function that takes 1 on rationals and 0 elsewhere. Since we have a good function $f\equiv 0$ equivalent to the Dirichlet function, this can be discarded from $\tilde{M}$ without any essential loss. We shall pick up at least one good function, called canonical measurable function here, from among each equivalent class in $\tilde{M}$ and consider the space $M\equiv M(T)$ of all canonical functions in behalf of $\tilde{M}$. By definition a canonical function takes one of its general approximate limits at each point. All continuous functions are in $M$ and if a function in $M$ is equal to a continuous function almost everywhere on $T$, they are equal everywhere on $T$. A similar fact holds for functions with no discontinuities of the second kind. These facts suggest that $M$ is suitable for the function space in which the path of a reasonable stochastic process is ranging.

In Section 3 we shall define a $\sigma$-algebra $\mathcal{B}=\mathcal{B}(M)$ of subsets of $M$ which will determine a measurable structure in $M$. $\mathcal{B}$ is generated by all sets of the following types

(i) $$\{f\in M: f(t)<a\},$$

(iii) $$\{f\in M: \int_I \arctan f(t)dt<a\}.$$

where $a$ ranges over reals, $t$ over $T$ and $I$ over all compact intervals in $T$. The scaling "arctan" was used in the integral to make it converge. We shall write $\mathcal{B}_K\equiv\mathcal{B}_K(M)$ and $\mathcal{B}_\rho\equiv\mathcal{B}_\rho(M)$, respectively, for the $\sigma$-algebra generated by the sets (i) only and that generated by the sets (ii) only. The space $C \equiv C(T)$ of all continuous functions, the space $D \equiv D(T)$ of all functions with no discontinuities of the second kind and the space $L^p \equiv L^p(T)$ of all canonical functions with finite $p$-th norm are $\mathcal{B}_\rho$-measurable and so $\mathcal{B}$-measurable.

In Section 4 we shall discuss *canonical stochastic processes*. Let $\{X_t(\omega), t \in T\}$ be a stochastic process defined on $(\Omega, \mathcal{F}, P)$. The sample path $X.(\omega)$ is a function of $\omega$ ranging in the function space $\bar{R}^T$ in general. $\{X_t\}$ is called *canonical* if $X.(\omega) \in M$ for every $\omega$ and if the map: $\omega \to X.(\omega)$ from $\Omega$ into $M$ is measurable $(\mathcal{F}, \mathcal{B})$. In other words, the sample path $X.(\omega)$ is an $(M, \mathcal{B})$-valued random variable. We shall prove that every canonical process continuous in probability is *measurable* in the pair $(t, \omega)$ with respect to the product measure $dt\, dP$ and also *separable* relative to closed sets with respect to every countable dense subset of $T$. The *(standard) canonical modification* is defined in the same way as Doob's separable measurable modification, but our meaning of "standard" is more strict than Doob's. It will be proved that every stochastic process continuous in probability has one and *only one* (in a reasonable sense) canonical modification.

In Section 5 we shall discuss probability measures on $M$. Let $\{X_t(\omega)\}$ be a canonical process continuous in probability. Then the sample path $X.(\omega)$ is an $(M, \mathcal{B})$-valued random variable on $(\Omega, \mathcal{F}, P)$. The probability law of $X.(\omega)$ defined as usual will be a complete $\mathcal{B}$-regular probability measure $\mu$ on $M$ which satisfies

$$(\Gamma.\ 1) \qquad \mu\{f \in M : |f(t)| < \infty\} = 1 \qquad t \in T,$$

$$(\Gamma.\ 2) \qquad \lim_{s \to t} \mu\{f \in M : |f(s) - f(t)| > \varepsilon\} = 0 \qquad \varepsilon > 0, \quad t \in T.$$

The finite-dimensional marginal distribution $m_{t_1,\cdots,t_n}$ of over $\{t_1, \cdots, t_n\}$ is defined by

$$m_{t_1,\cdots,t_n}(E) = \mu\{f \in M : (f(t_1), \cdots, f(t_n)) \in E\}.$$

The system of all such marginal distributions will satisfy

$$(m.\ 1) \qquad m_t\{(-\infty, \infty)\} = 1 \qquad t \in T$$

$$(m.\ 2) \qquad \lim_{s \to t} m_{st}\{(x, y) : |x - y| > \varepsilon\} = 0 \qquad \varepsilon > 0, \quad t \in T$$

in addition to Kolmogorov's consistency condition. Conversely, if we are given such a system of finite dimensional probability measures $\{m_{t_1,\cdots,t_n}\}$, we can construct *one and only one* $\mu$ with $(\Gamma.\ 1)$ and $(\Gamma.\ 2)$ whose marginal distributions are the given $\{m_{t_1,\cdots,t_n}\}$. It is to be noted that the finite-dimensional marginal distributions determine the probability measure $\mu$ not only on $\mathcal{B}_K$ but also on $\mathcal{B}$, provided the conditions $(m.\ 1)$ and $(m.\ 2)$ are satisfied.

REMARKS. We shall list some notations and definitions which will be used repeatedly in this paper.

$T$ stands for a real interval, bounded or unbounded and open, closed or semi-closed. It indicates the time interval. The $\sigma$-algebra of all Borel subsets of $T$ is denoted by $\mathcal{B}(T)$.

An open subset of $T$ means a subset of $T$ open in $T$, not in $(-\infty, \infty)$. Similarly for a closed subset.

$R$ and $\bar{R}$ stand respectively for the open real line $(-\infty, \infty)$ and the closed one $[-\infty, \infty]$. $\mathcal{B}(R)$ denotes the system of all Borel subsets of $R$. Similarly for $\mathcal{B}(\bar{R})$ and $\mathcal{B}(R^n)$.

A subset $E$ of $T$ is always assumed to be measurable and $|E|$ denotes the Lebesgue measure of $E$.

$I$, $I_n$ etc. stand for bounded subintervals of $T$.

$\mathcal{I}_{\text{rat}}$ stand for the system of all compact subintervals of $T$ expressible as the intersection of $T$ with a rational interval $\subset R$, and $\mathcal{F}_{\text{rat}}$ for the system of all non-empty sets expressible as a finite sum of compact intervals in $\mathcal{I}_{\text{rat}}$. Both are countable systems. Every $E$ with $|E|<\infty$ can be approximated in measure by sets in $\mathcal{F}_{\text{rat}}$.

A function $f$ is always assumed to be a real measurable function defined on $T$. $f$ may take $\pm\infty$ on a null set.

$\{f<c\}$ denotes the set $\{t\in T: f(t)<c\}$.

"arctan" is abbreviated as "atn". This gives a homeomorphism from $\bar{R}$ onto $[-\pi/2, \pi/2]$.

$(X, \mathcal{B})$ stands for a *measurable space*.

Let $(X, \mathcal{B})$ and $(Y, \mathcal{C})$ be measurable spaces. A map $f: X\to Y$ (into) is said to be *measurable* $(\mathcal{B}, \mathcal{C})$ if $f^{-1}(\mathcal{C})\subset\mathcal{B}$ i.e. if $f^{-1}(C)\in\mathcal{B}$ for every $C\in\mathcal{C}$. $f$ is said to be *measurable* $(\mathcal{B})$ or $\mathcal{B}$*-measurable* in case $Y=R$ or $\bar{R}$ and $\mathcal{C}=\mathcal{B}(R)$ or $\mathcal{B}(\bar{R})$.

The *basic probability space* is denoted by $(\Omega, \mathcal{F}, P)$. $\mathcal{F}$ is always assumed to be $P$-complete. In other words, every subset of a set with $P$-measure 0 belongs to $\mathcal{F}$. The *generic element* in $\Omega$ is denoted by $\omega$.

Let $\mathcal{F}_1$ be a $\sigma$-subalgebra of $\mathcal{F}$. The *P-completion* of $\mathcal{F}_1$, denoted by $\mathcal{F}_1^P$ is defined to be the $\sigma$-algebra that consists of all $A\in\mathcal{F}$ with the property:

$$\exists A_1, A_2\in\mathcal{F}_1 \text{ such that } A_1\subset A\subset A_2,\ P(A_2-A_1)=0.$$

Notice that not every $P$-null set belongs to $\mathcal{F}_1^P$.

An $(X, \mathcal{B})$-valued *random variable* $X$ is a function of $\omega$ such that the map $\omega\to X(\omega)$ is $(\mathcal{F}, \mathcal{B})$-measurable. The *probability law* of $X$, denoted by $P_X$, is defined by

$$P_X(B)=P\{\omega: X(\omega)\in B\}=P(X^{-1}(B)) \quad \text{if } X^{-1}(B)\in\overline{X^{-1}(\mathcal{B})}^P.$$

$P_X$ is a *complete* $\mathcal{B}$*-regular probability measure* on $X$. The $P_X$-completion of $\mathcal{B}$

is denoted by $\overline{\mathcal{B}}^{P_X}$. $P\{\omega : X \in B\}$ may be meaningful for a set $B \in \overline{\mathcal{B}}^{P_X}$, but we do *not* define $P_X$ for such $B$.

An $(\bar{R}, \mathcal{B}(\bar{R}))$-valued random variable $X(\omega)$ is called a *real random variable* if $P(X \in R) = 1$. $X_t(\omega)$ is written as $X(t, \omega)$ without explanation. The $\sigma$-algebra generated by the sets $\{\omega : X_t(\omega) < a,\ t \in T,\ a \in R\}$ is denoted by $\mathcal{B}[X_t, t \in T]$ or $\mathcal{B}[X]$.

## § 2. Canonical measurable function.

Let $\tilde{M} \equiv \tilde{M}(T)$ stand for the space of all real measurable functions defined on $T$. We shall introduce some notions following Saks [2]. The parameter of regularity of $E$, denoted by $\alpha(E)$, is defined to be the supremum of $|E|/|I|$, $I$ ranging over all intervals $\supset E$. We shall write $E_n \underset{r}{\to} t$ ($E_n$ *tends to* $t$ *regularly*), if $t \in E_n$ for every $n$, if the diameter of $E$ tends to 0 and if $\inf_n \alpha(E_n) > 0$. $a \in \bar{R}$ is called a *general approximate limit* of $f \in \tilde{M}$ at $t$ if for every neighborhood $U$ of $a$, we have $E_n \underset{r}{\to} t$ such that

$$\lim_n \frac{|\{f \in U\} \cap E_n|}{|E_n|} > 0 .$$

The set of all general approximate limits of $f$ at $t$ is denoted by $L(f, t)$. The *approximate upper limit* of $f$ at $t$, denoted here by $\bar{f}(t)$, is defined to be the infimum of $b \in \bar{R}$ such that for every $E_n \underset{r}{\to} t$ we have

$$\overline{\lim_n} \frac{|\{f > b\} \cap E_n|}{|E_n|} = 0 ;$$

we set $\bar{f}(t) = -\infty$, if there is no such $b$. The *approximate lower limit* $\underline{f}(t)$ is defined similarly. It is a well-known important fact [2] that $\bar{f}(t) = \underline{f}(t)$ *a.e.* on $T$. It is easy to see

PROPOSITION (2.1). $\bar{f}(t)$ ($\underline{f}(t)$) *is the largest (least) element in* $L(f, t)$ *and so* $L(f, t)$ *is non-empty.* $L(f, t)$ *consists of a single point for almost every* $t \in T$.

Let $\mathcal{E}_{mn}(t)$ denote the class of all sets $E \subset T$ such that $E \subset (t - 1/m, t + 1/m)$ and that $\alpha(E) > 1/n$ and $\mathcal{F}_{mn}(t)$ the class of all sets in $\mathcal{F}_{\text{rat}}$ with the same property. The following proposition that will be useful later can be proved by a routine.

PROPOSITION (2.2).

(i) $a \in L(f, t)$ *if and only if for every* $\varepsilon > 0$, *we can find* $E_n \underset{r}{\to} t$ *such that*

$$\frac{1}{|E_n|} \int_{E_n} |\operatorname{atn} f(s) - \operatorname{atn} a|\, ds < \varepsilon, \qquad n = 1, 2, \cdots$$

(ii) $$\operatorname{atn} \bar{f}(t) = \sup_n \inf_m \sup_{E \in \mathcal{E}_{mn}(t)} \frac{1}{|E|} \int_E \operatorname{atn} f(s)\, ds$$

$\mathcal{E}_{mn}(t)$ *can be replaced by* $\mathcal{F}_{mn}(t)$. atn $\underline{f}(t)$ *is also expressed similarly.*

DEFINITION. $f \in \tilde{M}$ is called a canonical measurable function, if $f(t) \in L(f, t)$ for every $t \in T$. $M \equiv M(T)$ denotes the space of all canonical functions.

We shall use $M$ in behalf of $\tilde{M}$. This is justified by the following

PROPOSITION (2.3). *For every* $f \in M$ *we have at least one* $g \in M$ *equal to* $f$ *a.e. on* $T$. *All such* $g$*'s can be obtained by taking any point in* $L(f, t)$ *for the value of* $g$ *at each* $t$.

"$f = g$ a.e." is a strong condition in $M$ as is seen in the following propositions (2.4), (2.5) and (2.6).

PROPOSITION (2.4). *The following conditions on* $f, g \in M$ *are equivalent, where* $U$ *is an open subset of* $T$.

(i) $f = g$ *a.e. on* $U$,

(ii) $L(f, t) = L(g, t)$ *for every* $t \in U$,

(iii) $\bar{f} = \bar{g}$ *on* $U$,

(iv) $\underline{f} = \underline{g}$ *on* $U$.

Let $C = C(T)$ denote the space of all continuous functions. Then it is obvious that $C \subset M$.

PROPOSITION (2.5). *Suppose* $f \in C$, $g \in M$ *and* $U$ *is open in* $T$. *Then we have*

(i) $\bar{f} = \underline{f} = f$: $L(f, t)$ *is a single point for every* $t \in U$,

(ii) $g \leqq (\geqq)\, f$ *a.e. on* $U \Rightarrow g \leqq (\geqq) f$ *everywhere on* $U$,

(iii) $g = f$ *a.e. on* $U \Rightarrow g = f$ *everywhere on* $U$.

The case $f \equiv$ const in (ii) will be useful later.

Let $D \equiv D(T)$ denote the space of all functions having no discontinuities of the second kind. We assume $f \in D$ to be right or left continuous at every jump point and continuous at the end points (if any). It is clear that $D \subset M$.

PROPOSITION (2.6). *Suppose* $f \in D$ *and* $g \in M$. *If* $f = g$ *a.e. on* $T$, *then*

(i) $g \in D$,

(ii) *The set of continuity points of* $g$ *is the same as that for* $f$: $g = f$ *on that set and* $g(t) = f(t+)$ *or* $f(t-)$ *elsewhere.*

## § 3. The $\sigma$-algebras $\mathcal{B}_\rho$, $\mathcal{B}_K$ and $\mathcal{B}$.

The space $M$ of all canonical functions on $T$ is topologized by the following pseudo-metric:

$$\rho(f, g) = \int_T |\text{atn} f(t) - \text{atn}\, g(t)| \frac{dt}{1+t^2} \qquad (\text{atn} = \text{arc tan});$$

$\rho(f, g) = 0$ does not always imply $f = g$ but only $f = g$ a.e. (see Propositions (2.4), (2.5) and (2.6)). The *$\rho$-convergence* is equivalent to the convergence in measure on every compact subset of $T$. The space $C$ of all continuous functions is $\rho$-dense in $M$. It is easy to see that the pseudo-metric space $M(\rho)$ is complete and separable. Referring to the *$\rho$-topology*, we can define *$\rho$-open sets, $\rho$-Borel sets, $\rho$-continuous functions, lower semi-continuous* ($\rho$) functions *etc.*

Let $\mathcal{B}_\rho \equiv \mathcal{B}_\rho(M)$ denote the $\sigma$-algebra of all $\rho$-Borel sets in $M$. Another characterization for $\mathcal{B}_\rho$ will be given in Theorem (3.11).

A second $\sigma$-algebra $\mathcal{B}_K \equiv \mathcal{B}_K(M)$ is defined to be the $\sigma$-algebra generated by

$$\{f \in M : f(t) < a\}, \qquad t \in T, \qquad a \in R.$$

There is no inclusion relation between these two $\sigma$-algebras, as we can see in Theorems (3.12) and (3.13). Therefore we shall introduce a third $\sigma$-algebra $\mathcal{B} \equiv \mathcal{B}(M)$. It is the join $\mathcal{B}_\rho \vee \mathcal{B}_K$, the least $\sigma$-algebra containing both, so that every $\mathcal{B}_\rho$- or $\mathcal{B}_K$-measurable set or function is $\mathcal{B}$-measurable. $\mathcal{B}$ is the $\sigma$-algebra we refer to in discussing probability measures on $M$.

The following notations will be used in this section.

$$
\begin{aligned}
&A(E,f) = \int_E \operatorname{atn} f(s)ds && E \text{ bounded}\\
&S(E,f) = \sup f(s), \quad s \in E &&\\
&L(E,f) = \inf f(s), \quad s \in E &&\\
&a(E) = \sup(1+s^2), \quad s \in E && E \text{ bounded.}
\end{aligned}
$$

PROPOSITION (3.1). *$A(E,f)$ is $\rho$-continuous in $f$ for $E$ fixed.*

PROOF. $|A(E,f)-A(E,g)| \leqq a(E)\rho(f,g)$.

THEOREM (3.2). *Let $U$ be a non-empty open subset of $T$, $\Gamma$ the class of all $E \subset U$ with $|E| < \infty$ and $\Gamma' = \Gamma \cap \mathcal{J}_{\text{rat}}$. Then we have*

$$\operatorname{atn} S(U,f) = \sup_{E \in \Gamma} \frac{A(E,f)}{|E|} = \sup_{I \in \Gamma'} \frac{A(I,f)}{|I|}$$

*and a similar identity for $L(U,f)$. $S(U,f)$ ($L(U,f)$) is lower (upper) semi-continuous ($\rho$) and so $\mathcal{B}$-measurable in $f$ for $U$ fixed.*

PROOF. We shall discuss $S(U,f)$ only. First we shall prove that the following conditions are equivalent, where $c$ is a constant.

(a) $\operatorname{atn} f \leqq c$ *on* $U$,

(b) $A(E,f) \leqq c|E|$ *for every* $E \in \Gamma$,

(c) $A(I,f) \leqq c|I|$ *for every* $I \in \Gamma'$.

(a)$\Rightarrow$(b)$\Rightarrow$(c) is obvious. Assuming (c), we have $f \leqq c$ a.e. on $U$ by the density theorem and so $f \leqq c$ on $U$ by Proposition (2.5) (ii). Hence (c) implies (a). Thus (a), (b) and (c) are equivalent. This proves the identity in the theorem. As $A(I,f)$ is $\rho$-continuous in $f$ by (3.1), $S(U,f)$ is lower semi-continuous ($\rho$).

COROLLARY (3.3). *Let $U$ be a non-empty open subset of $T$. Then $\{f \in M : f \leqq c$ on $U\}$ and $\{f \in M : f \geqq c$ on $U\}$ are $\rho$-closed and so $\mathcal{B}$-measurable.*

THEOREM (3.4). $C \in \mathcal{B}_\rho$.

PROOF. We shall discuss the case $T=[0,1]$ only. Let $U_{ni}$ be the intersection of the interval $((i-1)/n, (i+1)/n)$ with $T$, where $i$ ranges over $-1<i<n+1$ so that $U_{ni}$ is non-empty. As each $f\in C$ is bounded and uniformly continuous on $T$, we have

$$C=\bigcap_m \bigcup_n \bigcap_i \bigcup_{k=-m^2}^{m^2} \left\{f\in M: S(U_{ni}, f)\leqq \frac{k+1}{m},\ L(U_{ni}f)\geqq \frac{k-1}{m}\right\}.$$

The set in the bracket is $\rho$-closed by (3.3) and so $C\in\mathcal{B}_\rho$.

THEOREM (3.5). $D\in\mathcal{B}_\rho$.

PROOF. We shall discuss the case $T=[0,1]$ only. As every $f\in D$ is bounded on $T$, we have

$$D\subset \bigcup_n B_n,\qquad B_n=\{f\in M: |f|\leqq n \text{ on } T\}.$$

$B_n$ is $\rho$-closed by (3.3). If we can prove that the set $\Gamma_k(r, r')$ defined by the condition on $f\in M$:

(c) $$\exists t_1<t_2<\cdots<t_{2k}\quad \forall i\quad f(t_{2i-1})<r,\quad f(t_{2i})>r'$$

is $\rho$-open for $r<r'$, then we have

$$B_n-D=B_n\cap\bigcup_{r<r'}\bigcap_k \Gamma_k(r, r')\qquad r, r' \text{ rational}$$

and so

$$D=\bigcup_n B_n\cap D=\bigcup_n [B_n-(B_n-D)]\in\mathcal{B}_\rho.$$

It remains only to prove that $\Gamma_k(r, r')$ is $\rho$-open. Consider the following condition on $f\in M$:

(c′) $\exists E_1, E_2, \cdots, E_{2k}\subset T$ such that

(i) $|E_i|>0\qquad i=1,2,\cdots,2k$

(ii) $A(E_{2i-1}, f)<\operatorname{atn} r\cdot|E_{2i-1}|,\quad A(E_{2i}, f)>\operatorname{atn} r\cdot|E_{2i}|\qquad i=1,2,\cdots,k$

(iii) $E_1<E_2<\cdots<E_{2k}$, namely

$$a_1<a_2<\cdots<a_{2k} \text{ for every } a_i\in E_i,\ 1<i<2k.$$

As $A(E, f)$ is continuous in $f\in M$ by (3.1), (c′) determines a $\rho$-open set as a union of $\rho$-open sets. To prove that $\Gamma_k(r, r')$ is open, it suffices to show the equivalence of (c) and (c′). As $A(f, E)\gtrless \operatorname{atn} r\cdot|E|$ implies the existence of $t\in E$ with $f(t)\gtrless r$, (c′) implies (c). Suppose (c) holds. Recalling $f(t_i)\in L(f, t_i)$ for $f\in M$, we have a neighborhood $U_i$ of $t_i$ for each $i$ such that

$$|\{f<r\}\cap U_{2i-1}|>0,\ |\{f>r'\}\cap U_{2i}|>0\qquad i=1,2,\cdots,k.$$

Write $E_{2i-1}$ and $E_{2i}$ for these $t$-sets. Then the sets $\{E_i\}$ satisfy. (i) and (ii) in (c′). By taking all $U_i$ small enough, we can assume $U_1 < U_2 < \cdots < U_{2k}$. Therefore $E_i$, $i = 1, \cdots, 2k$ satisfy (iii) in (c′). Thus (c) implies (c′). This completes the proof.

THEOREM (3.6). *The p-th norm* $\|f\|_p$ $(1 \leqq p \leqq \infty)$ *is lower semi-continuous* ($\rho$) *in* $f \in M$.

PROOF. The set $\{f \in M : \|f\|_\infty \leqq c\}$ is $\{f \in M : -c \leqq f \leqq c \text{ on } T\}$ and so $\rho$-closed by (3.3). Hence $\|f\|_\infty$ is lower-semi-continuous ($\rho$) in $f$. Suppose $p < \infty$. Consider first the functions

$$F_n(f) = \int_{T \wedge [-n, n]} \min(|f(t)|^p, n) dt \qquad n = 1, 2, \cdots .$$

Since the $\rho$-convergence is equivalent to the convergence in measure on every compact subset of $T$, $F_n(f)$ is $\rho$-continuous in $f$ for each $n$. But $F_n(f) \uparrow \|f\|_p$ as $n \to \infty$ and so $\|f\|_p$ is lower semi-continuous ($\rho$).

COROLLARY (3.7). $L^p \equiv \{f \in M : \|f\|_p < \infty\} \in \mathcal{B}_\rho$.

THEOREM (3.8). *For t fixed,* $\bar{f}(t)$ *and* $\underline{f}(t)$ *are* $\mathcal{B}_\rho$*-measurable in* $f \in M$.

PROOF. This follows at once from Propositions (2.2) (ii) and (3.1).

Now we shall discuss the measurability of $\bar{f}(t)$ and $\underline{f}(t)$ in the pair $(t, f) \in T \times M$. Let $\mathcal{B}(T) \otimes \mathcal{B}_\rho$ denote the product $\sigma$-algebra. It is the same as the $\sigma$-algebra $\mathcal{B}(T \times M)$ of all Borel subsets of $T \times M$ with respect to the product topology.

PROPOSITION (3.9). *Suppose* $I \equiv [u - \varepsilon, v + \varepsilon] \subset T$ *and* $E \subset (-\varepsilon, \varepsilon)$. *Then* $E_t \equiv \{s + t : s \in E\} \subset I$ *for* $t \in [u, v]$ *and* $A(E_t, f)$ *is continuous in* $(t, f) \in [u, v] \times M$.

PROOF.

$$\begin{aligned}
&|A(E_t, f) - A(E_s, g)| \\
&\leqq |A(E_t, f) - A(E_s, f)| + |A(E_s, f) - A(E_s, g)| \\
&\leqq \int_{E_t \triangle E_s} |\operatorname{atn} f(\theta)| d\theta + \int_I |\operatorname{atn} f(\theta) - \operatorname{atn} g(\theta)| d\theta \\
&\leqq \frac{\pi}{2} |E_t \triangle E_s| + a(I)\rho(f, g) \\
&= \frac{\pi}{2} |E_{t-s} \triangle E| + a(I)\rho(f, g) \\
&\to 0, \text{ as } |t - s| + \rho(f, g) \to 0 .
\end{aligned}$$

THEOREM (3.10). *Both* $\bar{f}(t)$ *and* $\underline{f}(t)$ *are measurable* ($\mathcal{B}(T) \otimes \mathcal{B}_\rho$) *in the pair* $(t, f) \in T \times M$.

PROOF. We shall discuss the case $T = [u, v)$ only. Let $\{I_p \equiv [u_p, v_p]\}$ be a sequence of intervals such that

$$u_1 > u_2 > \cdots \to u , \qquad v_1 < v_2 < \cdots \to v .$$

It suffices to show that

$$\{(t, f) \in T \times M : \bar{f}(t) < c\} \in \mathcal{B}(T) \otimes \mathcal{B}_\rho .$$

This set is the union of the following sets

$$\{u\} \times \{f \in M : \bar{f}(u) < c\}, \quad \{(t, f) \in I_p \times M : \bar{f}(t) < c\}, \qquad p = 1, 2, \cdots .$$

The first set is measurable ($\mathcal{B}(T) \otimes \mathcal{B}_p$) by (3.8). It is therefore enough to prove that $\bar{f}(t)$ is measurable ($\mathcal{B}(I_p) \otimes \mathcal{B}_p$) in $(t, f) \in I_p \times M$ for each $p$. Noticing that $[u_p - 1/m_0, v_p + 1/m_0] \subset T$ for some big $m_0$, we have

$$\operatorname{atn} \bar{f}(t) = \sup_n \inf_{m < m_0} \sup_{E \in \mathcal{F}_{nm}(0)} A(E_t, f)/|E| \qquad t \in I_p, f \in M$$

by Proposition (2.2) (ii). As $A(E_t, f)$ is continuous in $(t, f) \in I_p \times M$ by (3.9), $\bar{f}(t)$ is measurable ($\mathcal{B}(I_p) \otimes \mathcal{B}_p$) in $(t, f) \in I_p \times M$.

Now we shall prove some facts that will show the difference between $\mathcal{B}_\rho$ and $\mathcal{B}_K$. First we shall prove

THEOREM (3.11). *$\mathcal{B}_p$ is generated by the sets*

$$\{f \in M : \int_I \operatorname{atn} f(t) dt < a\}$$

*where $I$ ranges over all compact intervals $\subset T$ and $a$ ranges over $R$.*

PROOF. Let $\mathcal{B}'$ denote the $\sigma$-algebra generated by the sets above. It is obvious by (3.1) that $\mathcal{B}' \subset \mathcal{B}_\rho$. To prove the opposite inclusion relation, it is enough to prove that $\rho(f, g)$ is $\mathcal{B}'$-measurable in $f$ for $g$ fixed. For this purpose it suffices to show that

$$F(f) \equiv \int_I |\operatorname{atn} f(t) - \operatorname{atn} g(t)| \frac{dt}{1+t^2}$$

is $\mathcal{B}'$-measurable for every compact interval $I \subset T$, because $\rho(f, g)$ is the limit of a sequence of functions of this form. Let $I = I_{n1} \cup I_{n2} \cup \cdots \cup I_{nn}$ be a non-overlapping decomposition of $I$ into $n$ compact intervals with equal length and set

$$\varphi_n(t) = \frac{1}{|I_{ni}|} \int_{I_{ni}} \operatorname{atn} f(s) ds ,$$

$$\psi_n(t) = \frac{1}{|I_{ni}|} \int_{I_{ni}} \operatorname{atn} g(s) ds , \qquad t \in I_{ni}, \quad i = 1, 2, \cdots, n .$$

By the density theorem we have, as $n \to \infty$,

$$\varphi_n(t) \to \operatorname{atn} f(t), \qquad \psi_n(t) \to \operatorname{atn} g(t) \qquad \text{a.e. on } I .$$

$$F_n(f) \equiv \int_I |\varphi_n(s) - \psi_n(s)| \frac{ds}{1+s^2} \to F(f) \quad \text{as} \quad n \to \infty .$$

Observing

$$F_n(f)=\sum_i\left|\frac{1}{|I_{ni}|}\int_{I_{ni}}\operatorname{atn} f(s)ds-\frac{1}{|I_{ni}|}\int_{I_{ni}}\operatorname{atn} g(s)ds\right|\int_{I_{ni}}\frac{dt}{1+t^2},$$

we can see that $F_n(f)$ is $\mathcal{B}'$-measurable, so that $F(f)$ is also $\mathcal{B}'$-measurable.

THEOREM (3.12). *$\mathcal{B}_K-\mathcal{B}_\rho$ is not empty.*

PROOF. Take a point $t_0$ strictly inside of $T$ and consider the set

$$A=\{f\in M: f(t_0)=1\}.$$

$A\in\mathcal{B}_K$ is clear. We shall prove $A\notin\mathcal{B}_\rho$. Consider two functions $f_1, f_2\in M$

$$f_1(t)=1 \quad\text{for}\quad t\leqq t_0,\quad =0\quad\text{for}\quad t>t_0$$

$$f_2(t)=1 \quad\text{for}\quad t<t_0,\quad =0\quad\text{for}\quad t\geqq t_0.$$

As $\rho(f_1, f_2)=0$, *either* both of $f_1, f_2$ are in $A$ *or* none of $f_1, f_2$ is in $A$, if $A\in\mathcal{B}_\rho$. But $f_1\in A$ and $f_2\notin A$. Therefore $A\notin\mathcal{B}_\rho$.

THEOREM (3.13). *$\mathcal{B}_\rho-\mathcal{B}_K$ is not empty.*

PROOF. Take a compact interval $I\subset T$ and consider the set

$$A=\{f\in M: \int_I \operatorname{atn} f(s)ds=|I|\}.$$

It is obvious by (3.11) that $A\in\mathcal{B}_\rho$. We shall prove $A\notin\mathcal{B}_K$. Suppose $A\in\mathcal{B}_K$. Then we have a countable subset $Q$ of $T$ with the following property

$$g\in M,\quad f\in A,\quad g=f\quad\text{on}\quad Q\ \Rightarrow\ g\in A. \tag{3.14}$$

Since $|Q|=0$, we have an open neighborhood $U$ of $Q$ with $|U|<|I|/2$. Consider two functions $f_1, f_2\in M$

$$f_1(t)=\alpha\quad\text{on}\quad T \qquad (\alpha=\tan 1)$$

$$f_2(t)=\alpha\quad\text{on}\quad U,\quad =0\qquad\text{elsewhere}.$$

$f_1\in A$ is clear. Since $f_2\in\tilde{M}$, we have $f_3\in M$ such that $f_3=f_1$ a. e. Since $f_3=f_1=\alpha$ a. e. on $U$, $f_3=\alpha$ on $U$ by Proposition (2.5) (iii). Therefore $f_1=f_3=\alpha$ on $Q$. But $f_1\in A$. Therefore $f_2\in A$ by (3.14). This is a contradiction, because

$$\int_I \operatorname{atn} f_3(s)ds=\int_I \operatorname{atn} f_2(s)ds=|U\cap I|\leqq|U|<|I|/2.$$

## § 4. Canonical stochastic process.

A stochastic process $\{X_t(\omega), t\in T, \omega\in\Omega, \mathcal{F}, P)\}$ is defined to be a family of real random variables indexed by the time parameter $t$ ranging over $T$. Fixing $\omega$ and changing $t$ in $X_t(\omega)$, we have a function of $t\in T$ which is an element of $\bar{R}^T$. This is denoted by $X.(\omega)$ and is called the *sample path* of the process corresponding to $\omega$. $X.(\omega)$ is considered a function of $\omega$ ranging in $\bar{R}^T$ in general. Let $\mathcal{B}[X]$ be the $\sigma$-algebra (of subsets of $\Omega$) that is generated by the sets

$$\{\omega : X_t\omega) < a\}, \qquad t \in T, \qquad a \in R.$$

$\mathcal{B}[X] \subset \mathcal{F}$ is obvious.

$\{X_t\}$ is said to be *continuous in probability* if

$$\lim_{s \to t} P\{|X_s - X_t| > \varepsilon\} = 0 \qquad \varepsilon > 0, \ t \in T,$$

or equivalently if

$$\lim_{\tau \to t} \boldsymbol{E}\{|\operatorname{atn} X_s - \operatorname{atn} X_t|\} = 0 \qquad t \in T,$$

where $\boldsymbol{E}(Y) = \int_\Omega Y(\omega) P(d\omega)$. This can also be stated as follows:

atn $X_t$ is continuous in $t$ with respect to the norm in $L^1(\Omega, \mathcal{F}, P)$.

If this condition is satisfied, then atn $X_t$ is uniformly continuous on every compact $E \subset T$ with respect to this norm. Therefore we have $\delta = \delta(E, \varepsilon)$ for $\varepsilon > 0$ such that

$$(4.1) \qquad |t - s| < \delta, \qquad t, s \in E \Rightarrow \boldsymbol{E}\{|\operatorname{atn} X_s - \operatorname{atn} X_t|\} < \varepsilon.$$

Definition (4.2). A stochastic process $\{X_t(\omega), t \in T\}$ is called *canonical*, if the following two conditions are satisfied:

(C. 1) $X.(\omega) \in M$ for every $\omega \in \Omega$

(C. 2) The map $\omega \to X.(\omega)$ from $\Omega$ into $M$ is measurable $(\mathcal{F}, \mathcal{B})$, i. e. $X.^{-1}(\mathcal{B}) \subset \mathcal{F}$.

($M$ and $\mathcal{B}$ were defined in the previous sections.) In other words, $X.(\omega)$ is an $(M, \mathcal{B})$-valued random variable, so that we can define the probability law of $X.(\omega)$ which is a complete $\mathcal{B}$-regular probability measure on $M$ (see Remarks in Section 1). As it is easy to see

$$X.^{-1}(\mathcal{B}_K) = \mathcal{B}[X_t, t \in T],$$

(C. 2) can be replaced by a weaker condition:

$$(\text{C. } 2') \qquad X.^{-1}(\mathcal{B}_\rho) \subset \mathcal{F}.$$

Theorem (4.3). (*Measurability of canonical processes*). *Let $\{X_t(\omega)\}$ be a canonical process and m the product (complete) measure of the Lebesgue measure on T and the measure P on $\Omega$. Then $X_t(\omega)$ is measurable $(\overline{\mathcal{B}(T) \otimes \mathcal{F}^m})$ in the pair $(t, \omega) \in T \times \Omega$.*

Proof. First we shall prove the measurability of $\overline{X.(\omega)}(t)$, where the top bar means the approximate upper limit introduced in Section 2. As the map $\omega \to X.(\omega)$ is measurable $(\mathcal{F}, \mathcal{B})$ by (C. 2), the map $(t, \omega) \to (t, X.(\omega))$ is measurable $(\mathcal{B}(T) \otimes \mathcal{F}, \mathcal{B}(T) \times \mathcal{B})$. By Theorem (3.10), the map $(t, f) \to \bar{f}(t)$ is measurable $(\mathcal{B}(T)\otimes\mathcal{B}_\rho)$ and so measurable $(\mathcal{B}(T)\otimes\mathcal{B})$. Composing these two maps, we can see that $(t,\omega)\to\overline{X.(\omega)}(t)$ is measurable $(\mathcal{B}(T)\otimes\mathcal{F})$ and so measurable $(\overline{\mathcal{B}(T)\otimes\mathcal{F}^m})$, namely that $\overline{X.(\omega)}(t)$ is measurable $(\overline{\mathcal{B}(T)\otimes\mathcal{F}^m})$ in the pair $(t,\omega)$. Similarly for $\underline{X.(\omega)}(t)$. As it is obvious that $\underline{X.(\omega)}(t)\leq X_t(\omega)\leq\overline{X.(\omega)}(t)$ for every pair $(t,\omega)$, we get the measurability $(\overline{\mathcal{B}(T)\otimes\mathcal{F}^m})$ of $X_t(\omega)$ in the pair $(t,\omega)$, observing

$$m\{(t,\omega):\underline{X.(\omega)}(t)<\overline{X.(\omega)}(t)\}$$

$$=\int_\Omega|\{t\in T:\underline{X.(\omega)}(t)<\overline{X.(\omega)}(t)\}|P(d\omega)$$

by Fubini's theorem

$$=0 \qquad \text{by } X.(\omega)\in M.$$

THEOREM (4.4). (*Separability of canonical processes continuous in probability*). *Every canonical process $\{X_t(\omega)\}$ continuous in probability is separable relative to closed sets with respect to every countable dense subset $Q$ of $T$.*

PROOF. Let us write $Y_t(\omega)$ for $\operatorname{atn} X_t(\omega)$. Let $\mathcal{F}_{\mathrm{rat}}$ be the set system introduced in Remarks in Section 1 and $\Theta$ a countable dense subset of $[-\pi/2,\pi/2]$.

First we shall find, for each pair $(E,\theta)\in\mathcal{F}_{\mathrm{rat}}\times\Theta$, a sequence of non-overlapping decompositions of $E$ into intervals

$$E=I_{n1}\cup I_{n2}\cup\cdots\cup I_{np(n)},\qquad n=1,2,\cdots,$$

and a system of points

$$r_{ni}\in I_{ni}\cap Q,\qquad i=1,2,\cdots,p(n),$$

with the following property:

$$\int_E|Y_s(\omega)-\theta|\,dt=\lim_n\sum_i|Y(r_{ni},\omega)-\theta|\,|I_{ni}|\qquad\text{a. e. on }\Omega. \tag{4.5}$$

Since $E$ is compact, we can take $\delta(\varepsilon)=\delta(E,\varepsilon)$ in (4.1) to get

$$|t-s|<\delta(\varepsilon),\ t,s\in E\Rightarrow \boldsymbol{E}(|Y_t-Y_s|)<\varepsilon. \tag{4.1'}$$

As $E$ is a finite sum of compact intervals, $E$ can be decomposed into non-overlapping intervals with length $<\delta(\varepsilon)$, say

$$E=I_1\cup I_2\cup\cdots\cup I_p.$$

Since $Q$ is dense in $T$, $Q\cap I_i$ is non-empty. Take a point $r_i$ from $Q\cap I_i$. Then we have

$$E\Big[\Big|\int_E |Y_s-\theta|\,ds-\sum_i |Y(r_i)-\theta|\,|I_i|\Big|\Big]$$
$$\leqq \sum_i E\Big[\int_{I_i}\Big||Y_s-\theta|-|Y(r_i)-\theta|\Big|ds\Big]$$
$$\leqq \sum_i E\Big[\int_{I_i}|Y_s-X(r_i)|\,ds\Big].$$

As $Y_s(\omega)$ is measurable $(\overline{\mathcal{B}(T)\otimes\mathcal{F}^m})$ in $(s,\omega)$ by Theorem (4.3), we can exchange the order of $E$ and $\int$. Using (4.1′), we get

$$E\Big[\Big|\int_E |Y_s-\theta|\,ds-\sum_i |Y(r_i)-\theta|\,|I_i|\Big|\Big]<\varepsilon|E|.$$

Writing $I_i$ and $r_i$ for $\varepsilon=2^{-n}$ as $I_{ni}$ and $r_{ni}$ respectively, we have

$$E\Big[\sum_n\Big|\int_E |Y_s-\theta|\,ds-\sum_i |Y(r_{ni})-\theta|\,|I_{ni}|\Big|\Big]<\infty,$$

which implies (4.5).

Writing $\Omega(E,\theta)$ for the set of all $\omega$ for which (4.5) holds and setting

$$\Omega_1=\cap\,\Omega(E,\theta),\qquad E\in\mathcal{F}_{\mathrm{rat}},\ \theta\in\Theta$$

we have $P(\Omega_1)=1$, because $\mathcal{F}_{\mathrm{rat}}\times\Theta$ is countable. To prove our theorem, it suffices to show that for every $\omega\in\Omega_1$, every closed $F\subset[-\pi/2,\pi/2]$, every open $U\subset T$ and every $t\in U$, if $Y_r(\omega)\in F$ for $r\in Q\cap U$ then $Y_t(\omega)\in F$; notice here that atn gives a homeomorphism from $[-\infty,\infty]$ onto $[-\pi/2,\pi/2]$. For this purpose it is enough to find, for every $\varepsilon>0$, $r\in Q\cap U$ such that

$$|Y_r(\omega)-Y_t(\omega)|<\varepsilon,$$

because $F$ is closed. As $X.(\omega)\in M$, we have $X_t(\omega)\in L(X.,t)$ and so we can find $E\in\mathcal{F}_{\mathrm{rat}}$ such that $t\in E\subset U$ and that

$$\frac{1}{|E|}\int_E |Y_s(\omega)-Y_t(\omega)|\,ds<\varepsilon/3$$

by virtue of Proposition (2.2) (i). Since $Y_t(\omega)\in[-\pi/2,\pi/2]$ and $\Theta$ is dense in this interval, we can find $\theta\in\Theta$ such that

$$|Y_t(\omega)-\theta|<\varepsilon/3. \tag{4.6}$$

These two inequalities will imply

$$\frac{1}{|E|}\int_E |Y_s(\omega)-\theta|\,ds<2\varepsilon/3.$$

As $\omega\in\Omega_1$ and $E\in\mathcal{F}_{\mathrm{rat}}$, we can use (4.5) to get $n$ such that

$$\frac{1}{|E|}\sum_i |Y(r_{ni},\omega)-\theta|\,|I_{ni}|<2\varepsilon/3.$$

As $|E|=\sum_i |I_{ni}|$, we can find $i$ such that

$$|Y(r_{ni},\omega)-\theta|<2\varepsilon/3,$$

which, combined with (4.6), implies

$$|Y(r_{ni},\omega)-Y_t(\omega)|<\varepsilon.$$

As $r_{ni}\in Q\cap I_{ni}\subset Q\cap E\subset Q\cap U$, $r=r_{ni}$ is what we wanted to find out. This completes the proof.

Let $\{X_t(\omega)\}$ be a stochastic process. A canonical stochastic process $\{S_t(\omega)\}$ (defined on the some probability space) is called a *canonical modification* of $\{X_t(\omega)\}$ if

$$P\{X_t=S_t\}=1 \qquad \text{for every } t\in T.$$

CANONICAL MODIFICATION THEOREM (4.7). *Every stochastic process $\{X_t(\omega)\}$ continuous in probability has a canonical modification. If we have two such modifications $\{S_t\}$ and $\{S_t'\}$ for the same process, then we have*

$$P\{S_t=S_t' \quad \text{for almost every } t\}=1$$

*in addition to the automatic property:*

$$P\{S_t=S_t'\}=1 \qquad \text{for every } t\in T.$$

PROOF. We shall discuss the case $T=[0,1]$ only, because a small change in the proof will take care of the other cases. The metric $\rho$ in $M=M(T)$ can now be replaced by a simpler one

$$\rho(f,g)=\int_T |\operatorname{atn} f(t)-\operatorname{atn} g(t)|\,dt,$$

which induces the same topology and therefore the same $\sigma$-algebras $\mathscr{B}_\rho$ and $\mathscr{B}(\equiv\mathscr{B}_\rho\vee\mathscr{B}_K)$. Let us write $Y_t$, $Y_t^n$ and $V_t$ respectively for atn $X_t$, atn $X_t^n$ and atn $U_t$. By (4.1) we can take $\delta=\delta(\varepsilon)$ for $\varepsilon>0$ such that

$$\text{(4.1}'')\qquad |s-t|<\delta,\quad s,t\in T\Rightarrow \boldsymbol{E}(|Y_s-Y_t|)<\varepsilon.$$

It is harmless to assume that $\delta(\varepsilon)\downarrow 0$ as $\varepsilon\downarrow 0$.

Let us consider for each $n=1,2,\cdots$, a finite decomposition of $I$ into non-overlapping compact intervals:

$$I=I_{n1}\vee I_{n2}\vee\cdots\vee I_{np(n)},\quad |I_{ni}|<\delta(2^{-n})$$

and a corresponding step processes

$$X_t^n(\omega)=X(t_{ni},\omega) \qquad \text{for } I_{ni},\ i=1,2,\cdots,p(n),$$

where $t_{ni}$ is a point in $I_{ni}$, say the left end point.

$\rho(X_\cdot^n(\omega), X_\cdot^{n+1}(\omega))$ is a Borel measurable function of $X(t_{ni},\omega)$, $i=1,2,\cdots,p(n)$ and $X(t_{nj},\omega)$, $j=1,2,\cdots,p(n+1)$ and so measurable $(\mathscr{F})$. $X_t^n(\omega)$ is also measurable $(\mathcal{B}(T)\otimes\mathcal{B}[X])$ in $(t, \omega)$ and

$$E[\rho(X_\cdot^n, X_\cdot^{n+1})]$$

$$\leqq \int_E E[|Y_t^n - Y_t^{n+1}|]dt < 2^{-n}.$$

Therefore we have

$$E[\sum_n \rho(X_\cdot^n, X_\cdot^{n+1})] < \infty,$$

It follows from this that there exists an $\bar{R}$-valued function $W_t(\omega)$ measurable in $(t, \omega)\in T\times\Omega$ such that

(i) $$\int_T |\text{arc tan } X_t^n(\omega) - \text{arc tan } W_t(\omega)|\,dt \to 0$$

for almost every $\omega\in\Omega$,

(ii) $X_t^n(\omega)\to W_t(\omega)$ for almost every $(t, \omega)\in T\times\Omega$.

Since $X_s(\omega)$ is continuous in probability, it follows from (ii) that

(iii) for almost every $t$ fixed, we have

$$W_t(\omega) = X_t(\omega) \quad \text{for almost every} \quad \omega\in\Omega.$$

By Fubini's theorem we get

$$\int_\Omega |\{t: |W_t(\omega)| = \infty\}|\,P(d\omega)$$

$$= \int_T P\{\omega: |W_t(\omega)| = \infty\}dt$$

$$= \int_T P\{\omega: |X_t(\omega)| = \infty\}dt$$

$$= 0.$$

Therefore $W_\cdot(\omega)\in\tilde{M}$ for almost every $\omega$. Let

$$U_t(\omega) = \overline{W}_t(\omega) \qquad (= \text{the approximate upper limit of } W_\cdot(\omega) \text{ at } t).$$

Then $U_\cdot(\omega)\in M$ for almost every $\omega$ and (i) implies that

$$P(\Omega') = 1 \quad \text{for} \quad \Omega' = \{\omega: U_\cdot(\omega)\in M,\ \rho(X_\cdot^n(\omega), U_\cdot(\omega))\to 0\}.$$

As $\rho(X_\cdot^n(\omega), f)$ is equal to a Borel measurable function of $X(t_{ni}, \omega)$, $i=1, 2, \cdots, p(n)$ for each $f\in M$, it is measurable $(\mathcal{F})$ in $\omega$. Therefore $\rho(U_\cdot(\omega), f)$ is also measurable $(\mathcal{F})$ in $\omega$, because

$$\rho(U_\cdot(\omega), f) = \lim_n \rho(X_\cdot^n(\omega), f) \quad \text{for } \omega\in\Omega'$$

$$\Omega'\in\mathcal{F}, \qquad P(\Omega') = 1.$$

This shows that the set $\{\omega : U.(\omega) \in N_\rho(f, r)\}$ $(N_\rho(f, r) = \{g \in M : \rho(g, f) < r\})$ is measurable $(\mathcal{F})$. As $\mathcal{B}_\rho$ is generated by the sets $N_\rho(f, r), r > 0, f \in M$, $\{\omega : U.(\omega) \in B\}$ is measurable $(\mathcal{F})$ for $B \in \mathcal{B}_\rho$.

We shall modify $\{U_t(\omega)\}$ at each point $t$ to get a canonical modification $\{S_t(\omega)\}$. Fix the time point $t$ for the moment. Observing

$$\int_I |V_s(\omega) - Y_t(\omega)| ds \lesseqgtr \int_I |Y_s^n(\omega) - Y_t(\omega)| ds \pm \rho(U.(\omega), X.^n(\omega)),$$

we have

$$\int_I |V_s(\omega) - Y_t(\omega)| ds = \lim_n \int_I |Y_s^n(\omega) - Y_t(\omega)| ds \qquad \text{for } \omega \in \Omega',$$

where $I$ is a compact interval $\subset T$. This shows that the left side is measurable $(\mathcal{F})$. Noticing $P(\Omega') = 1$, we have

$$\boldsymbol{E}\left[\frac{1}{|I|}\int_I |V_s - Y_t| ds\right] = \lim_n \boldsymbol{E}\left[\frac{1}{|I|}\int_I |Y_s^n - Y_t| ds\right].$$

Since $X_s^n(\omega)$ (and so $Y_s^n(\omega)$) is measurable $(\mathcal{B}(T) \otimes \mathcal{F})$ in $(s, \omega)$ and $X_t(\omega)$ (and so $Y_t(\omega)$) is measurable $(\mathcal{F})$ in $\omega$ and so measurable $(\mathcal{B}(T) \otimes \mathcal{F})$ in $(s, \omega)$, we can use Fubini's theorem to get

$$\boldsymbol{E}\left[\frac{1}{|I|}\int_I |V_s - Y_t| ds\right] = \frac{1}{|I|}\int_I \boldsymbol{E}[|V_s - V_t|] ds < \varepsilon, \qquad \text{if } |I| < \delta(\varepsilon).$$

Taking a sequence $\{I_n\}$ with $t \in I_n$ and $|I_n| < \delta(2^{-n})$, we have

$$\boldsymbol{E}\left[\sum_n \frac{1}{|I_n|}\int_{I_n} |V_s - Y_t| ds\right] < \sum_n 2^{-n} < \infty.$$

Write $\Omega_t$ for the set of all $\omega$ for which the infinite series converges. Since each term is measurable $(\mathcal{F})$ as we mentioned above, $\Omega_t$ is measurable $(\mathcal{F})$ and $P(\Omega_t) = 1$. It is easy to see by Proposition (2.2) (i) that

$$X_t(\omega) \in L(U., t) \qquad \text{for } \omega \in \Omega_t.$$

Set

$$S_t(\omega) = \begin{cases} X_t(\omega) & \text{for } \omega \in \Omega_t \\ U_t(\omega) & \text{for } \omega \in \Omega - \Omega_t. \end{cases}$$

Define $S_t(\omega)$ at each point $t \in T$ in such a way.

Now fix $\omega$. By our construction we have

$$U.(\omega) \in M \quad \text{and} \quad S_t(\omega) \in L(U., t) \qquad \text{for every } t \in T.$$

By Proposition (2.3) we have

$$S.(\omega) \in M \quad \text{and} \quad S_t(\omega) = U_t(\omega) \qquad \text{for almost all } t.$$

This implies $\rho(S.(\omega), U.(\omega)) = 0$. Since that is true for every $\omega$, we have

$$\{\omega : S.(\omega) \in B\} = \{\omega : U.(\omega) \in B\} \qquad \text{for } B \in \mathcal{B}_\rho .$$

But the right side belongs to $\mathcal{F}$ as we mentioned above. Therefore we have

$$S.^{-1}(\mathcal{B}_\rho) \subset \mathcal{F} .$$

Since $\Omega_t \in \mathcal{F}$, $S_t(\omega)$ is measurable ($\mathcal{F}$) for each $t$ fixed by the definition. This shows

$$\{\omega : S. \in B_{t,a}\} \in \mathcal{F} \qquad \text{for } B_{t,a} = \{f : f(t) < a\} .$$

As $\mathcal{B}_K$ is generated by $B_{t,a}$, $t \in T$, $a \in R$, we have

$$\{\omega : S. \in B\} \in \mathcal{F}, \qquad B \in \mathcal{B}_K$$

namely

$$S.^{-1}(\mathcal{B}_K) \subset \mathcal{F} .$$

Therefore

$$S.^{-1}(\mathcal{B}) = S.^{-1}(\mathcal{B}_\rho) \vee S.^{-1}(\mathcal{B}_K) \subset \mathcal{F} .$$

Thus we have proved that $\{S_t(\omega)\}$ is a canonical process. It is obvious by the construction that $P(S_t = X_t) = 1$ for each $t$. Therefore $\{S_t\}$ is a canonical modification of $\{X_t\}$.

Now we shall prove the uniqueness. Take any arbitrary canonical modification $\{\tilde{S}_t(\omega)\}$. $\{X_t^n(\omega)\}$ is measurable ($\mathcal{B}(T) \otimes \mathcal{F}$) in $(t, \omega)$ by the definition. $\{\tilde{S}_t(\omega)\}$ is measurable ($\mathcal{B}(T) \otimes \mathcal{F}$) in $(t, \omega)$ by Theorem (4.3). Therefore we can use Fubini's theorem to get

$$\begin{aligned} \boldsymbol{E}[\rho(S., X.^n)] &= \int_T \boldsymbol{E}[|\operatorname{atn} S_t - \operatorname{atn} X_t^n|]dt \\ &= \int_T \boldsymbol{E}[|\operatorname{atn} X_t - \operatorname{atn} X_t^n|]dt \\ &< 2^{-n} \to 0 \quad \text{as} \quad n \to \infty . \end{aligned}$$

Since $\rho(X.^n(\omega), U.(\omega)) \to 0$ a. e. on $\Omega$, we can use the bounded convergence theorem to get

$$\boldsymbol{E}[\rho(X.^n, U.(\omega))] \to 0 \qquad \text{as } n \to \infty ;$$

Therefore

$$\boldsymbol{E}[\rho(\tilde{S}., U.)] = 0 .$$

But $\rho(S., U.) = 0$ for every $\omega$. Thus we get

$$\boldsymbol{E}[\rho(\tilde{S}., S.)] = 0$$

i. e. $P[\tilde{S}_t(\omega) = S_t(\omega)$ for almost every $t] = 1$. This completes the proof.

A canonical modification $\{S_t\}$ of $\{X_t\}$ is called standard if

$$S.^{-1}(\mathcal{B}) \subset \overline{\mathcal{B}[X_t\ t \in T]}^P ,$$

where $\mathcal{B}[X_t, t \in T]$ is the $\sigma$-algebra of subsets of $\Omega$ generated by the sets

$\{\omega : X_t < a\}$, $a \in R$, $t \in T$.

STANDARD CANONICAL MODIFICATION THEOREM (4.8). *Every stochastic process $X_t(\omega)$ continuous in probability has a standard canonical modification. It is unique in the same sense as in Theorem (4.7).*

PROOF. Let $\mathcal{F}_1$ denote $\overline{\mathcal{B}[X_t, t \in T]}^P$. Then $\{X_t\}$ is considered a stochastic process continuous in probability on $(\Omega, \mathcal{F}_1, P)$. Therefore we have a canonical process $\{S_t(\omega)\}$ on $(\Omega, \mathcal{F}_1, P)$ such that $P(S_t = X_t) = 1$ for each $t$. $\{S_t(\omega)\}$ can be considered a canonical process on $(\Omega, \mathcal{F}, P)$. $S.^{-1}(\mathcal{B}) \subset \mathcal{F}_1 = \overline{\mathcal{B}[X_t, t \in T]}^P$ is now obvious, namely $\{S_t\}$ is a standard canonical modification of $\{X_t\}$. The uniqueness part is contained in theorem (4.7).

REMARK. Doob's definition of the standard property is

$$\{\omega : S_t \neq X_t\} \in \overline{\mathcal{B}[X_t, t \in T]}^P \qquad \text{for every } t.$$

This means

$$S.^{-1}(\mathcal{B}_R) \subset \overline{\mathcal{B}[X_t, t \in T]}^P$$

in our case. We required more than this.

## § 5. Probability measure on $M$.

Let $\{X_t(\omega), t \in T, \omega \in \Omega, \mathcal{F}, P)\}$ be a canonical process. Then the probability law of the sample path $X.(\omega)$ viewed as an $(M, \mathcal{B})$-valued random variable (see Remarks in Section 1), is a complete $\mathcal{B}$-regular probability measure on $M$ satisfying

$$(\Gamma.1) \qquad \mu\{f : f(t) \in R\} = 1, \qquad t \in T.$$

Conversely, if we have such a probability measure $\mu$ on $M$, then $\{\xi_t(f) \equiv f(t), t \in T, f \in M, \overline{\mathcal{B}}^\mu, \mu)\}$ is a canonical process for which the probability law of the sample path is the measure $\mu$.

If $\{X_t\}$ is *continuous in probability*, then the probability law of the sample path satisfies

$$(\Gamma.2) \qquad \lim_{s \to t} \mu\{f : |f(s) - f(t)| > \varepsilon\} = 0, \qquad \varepsilon > 0,\ t \in T.$$

It is obvious that $\{\xi_t(f)\}$ is also continuous in probability under the condition $(\Gamma.2)$.

Let $m_{t_1 \cdots t_n}$ be the marginal distribution of $\mu$ over the time points $t_1 \cdots t_n$, namely

$$(5.1) \qquad m_{t_1 \cdots t_n}(E) = \mu\{f : (f(t_1), \cdots, f(t_n)) \in E\}, \qquad E \in \mathcal{B}[\overline{R}^n].$$

This is the joint distribution of the process $\{\xi_t\}$ (or $\{X_t\}$) over $(t_1 \cdots t_n)$. Let $\mathcal{M}$ denote the system of all marginal distributions. It is obvious that (m.0) $\mathcal{M}$ satisfies Kolmogorov's consistency condition.

$(\Gamma.1)$ and $(\Gamma.2)$ are formulated in terms of $\mathcal{M}$ as follows.

$$(m.1) \qquad m_t(R)=1\,, \quad t\in T\,,$$

$$(m.2) \qquad \lim_{s\to t} m_{st}\{(x,y):|x-y|>\varepsilon\}=0\,, \quad \varepsilon>0\,, \quad t\in T\,.$$

Our problem is to determine $\mu$ from $\mathcal{M}$.

THEOREM (5.2). *For every $\mathcal{M}$ with (m.0), (m.1) and (m.2), there exists one and only one complete $\mathcal{B}$-regular probability measure $\mu$ with* (5.1).

PROOF OF EXISTENCE. Let $\Omega$ be $\bar{R}^T$ and $\mathcal{B}(\bar{R}^T)$ the $\sigma$-algebra generated by the sets $\{\omega\in\bar{R}^T:\omega(t)<a\}$, $t\in T$, $a\in R$. Since (m.0) is assumed, we can construct a complete regular $(\mathcal{B}(\bar{R}^T))$ probability measure $P$ on $\Omega$ such that

$$P\{\omega:(\omega(t_1)\cdots\omega(t_n))\in E\}=m_{t_1\cdots t_n}(E)$$

by Kolmogorov's theorem. Then $\{X_t(\omega)\equiv\omega(t), t\in T, \omega\in(\Omega,\mathcal{F},P)\}$ $(\mathcal{F}=\overline{\mathcal{B}(\bar{R}^T)}^P)$, is a stochastic process continuous in probability by (m.1) and (m.2). By the canonical modification theorem (4.7) we have a canonical process $\{S_t(\omega)\}$ such that $P\{S_t=X_t\}=1$, $t\in T$.

Then the probability law $P_{S.}$ of the sample path $S.(\omega)$ is a complete $\mathcal{B}$-regular probability measure on $M$ such that

$$\begin{aligned} P_{S.}\{f:(f(t_1),\cdots f(t_n))\in E\} &= P\{\omega:(S(t_1,\omega)\cdots S(t_n,\omega))\in E\}\\ &= P\{\omega:(X(t_1,\omega)\cdots X(t_n,\omega))\in E\}\\ &= m_{t_1\cdots t_n}(E)\,. \end{aligned}$$

PROOF OF UNIQUENESS. We shall consider the case $T=[0,1]$ only; the other cases can be treated similarly. Suppose that two complete $\mathcal{B}$-regular measures on $M$ (say $\mu$ and $\nu$) satisfy

$$(5.3) \qquad \begin{aligned} \mu\{f:(f(t_1),\cdots,f(t_n))\in E\} &= \nu\{f:(f(t_1),\cdots,f(t_n))\in E\}\\ &= m_{t_1\cdots t_n}(E)\,. \end{aligned}$$

This is equivalent to

$$(5.3') \qquad \int_M F(f(t_1),\cdots,f(t_n))\mu(df)=\int_M F(f(t_1),\cdots,f(t_n))\nu(df)$$

where $F$ ranges over all continuous functions on $\bar{R}^n$.

Let us write $\hat{f}(I)$ for $\int_I \operatorname{atn} f(t)dt$. Then $\mathcal{B}_\rho$ is generated by $\{f:\hat{f}(I)<a\}$, $a\in R$, $I$ being a compact interval, by virtue of theorem (3.11). $\mathcal{B}_K$ is generated by $\{f:f(t)<a\}$, $a\in R$, $t\in T$. Therefore $\mathcal{B}(=\mathcal{B}_\rho\vee\mathcal{B}_K)$ is generated by

all such sets. (5.3) or (5.3′) shows that $\mu$ and $\nu$ are equal on $\mathcal{B}_K$. To prove that they are equal on $\mathcal{B}$, it is enough to prove that

$$\int_M G[\hat{f}(I_1), \cdots, \hat{f}(I_p), \ f(t_1), \cdots, f(t_q)]\mu(df) \tag{5.4}$$
$$= \int_M G[\hat{f}(I_1), \cdots, \hat{f}(I_p), \ f(t_1), \cdots, f(t_q)]\nu(df)$$

where $p, q = 1, 2, \cdots$, $\{I_i\}$ range over all compact intervals, $\{t_i\}$ range over $T$ and $F$ ranges over all continuous functions on $\bar{R}^{p+q}$. The stochastic processes

$$X_t(f) = f(t), \qquad f \in (M, \overline{\mathcal{B}}^\mu, \mu)$$
$$Y_t(f) = f(t), \qquad f \in (M, \overline{\mathcal{B}}^\nu, \nu)$$

are canonical processes continuous in probability. Therefore we have $\delta(\varepsilon) > 0$ for every $\varepsilon > 0$ such that

$$|t-s| < \delta(\varepsilon) \tag{5.5}$$
$$\Rightarrow \int_M |\operatorname{atn} f(t) - \operatorname{atn} f(s)| \mu(df) < \varepsilon$$
$$\Rightarrow \int_M |\operatorname{atn} f(t) - \operatorname{atn} f(s)| \nu(df) < \varepsilon \qquad \text{by (5.3′)}.$$

Fix a compact interval $I$, consider its decomposition into non-overlapping intervals

$$I = I_{n1} \cup I_{n2} \cup \cdots \cup I_{np(n)} \qquad |I_{ni}| < \delta(2^{-n})$$

for each $n$ and let $t_{ni}$ be the left and points of $I_{ni}$. Using the measurability of $X_t(f)$ in $(t, f)$ (Theorem (4.3)), we can get

$$\int_M \sum_n \left| |f(I) - \sum_i \operatorname{atn} f(t_{ni}) | I_{ni}| \right| \mu(df) < \sum_n 2^{-n} |I| < \infty .$$

Therefore

$$\mu\{f : \hat{f}(I) = \lim_n \sum_i \operatorname{atn} f(t_{ni}) |I_{ni}|\} = 1 .$$

Applying this to $I = I_1, I_2, \cdots, I_p$ we have

$$\mu\{f : \hat{f}(I_k) = \lim_n \sum_i \operatorname{atn} f(t^k{}_{ni}) |I^k{}_{ni}|\} = 1 .$$

We have the same result for $\nu$, i.e.

$$\nu\{f : \hat{f}(I_k) = \lim_n \sum_i \operatorname{atn} f(t^k{}_{ni}) |I^k{}_{ni}|\} = 1 .$$

Notice here that we can take the same $t^k{}_{ni}$ and $I^k{}_{ni}$ for $\mu$ and $\nu$ by virtue of (5.5). By (5.3′) we have

$$\int_M G[\sum_i \operatorname{atn} f(t^k{}_{ni})|I^k{}_{ni}|,\ f(t_h),\ k=1,\cdots,p,\ h=1,\cdots,q]\mu(df)$$

$$=\int_M G[\sum_i \operatorname{atn} f(t^k{}_{ni})|I^k{}_{ni}|,\ f(t_h),\ k=1,\cdots,p,\ h=1,\cdots,q]\nu(df)$$

Letting $n\to\infty$, we have (5.4). This completes the proof.

Aarhus University, Denmark

## Bibliography

[1] J. L. Doob, Stochastic Processes, New York, 1952.
[2] S. Saks, Theory of the Integral, 2nd rev. ed., English translation by L. C. Young, New York, 1933.

Itô, K. and Nisio, M.
Osaka J. Math.
5 (1968), 35–48

# ON THE CONVERGENCE OF SUMS OF INDEPENDENT BANACH SPACE VALUED RANDOM VARIABLES

KIYOSI ITÔ AND MAKIKO NISIO

(Received February 26, 1968)

## 1. Introduction

The purpose of this paper is to discuss the convergence of sums of independent random variables with values in a separable real Banach space and to apply it to some problems on the convergence of the sample paths of stochastic processes.

For the real random variables, we have a complete classical theory on the convergence of independent sums due to P. Lévy, A. Khinchin and A. Kolmogorov. It can be extended to finite dimensional random variables without any change. In case the variables are infinite dimensional, there are several points which need special consideration. The difficulties come from the fact that bounded subsets of Banach space are not always conditionally compact.

In Section 2 we will discuss some preliminary facts on Borel sets and probability measures in Banach space. In Section 3 we will extend P. Lévy's theorem. In Section 4 we will supplement P. Lévy's equivalent conditions with some other equivalent conditions, in case the random variables are symmetrically distributed. Here the infinite dimensionality will play an important role. The last section is devoted to applications.

## 2. Preliminary facts

Throughout this paper, $E$ stands for a separable real Banach space and the topology in $E$ is the norm topology, unless stated otherwise. $E^*$ stands for the dual space of $E$, $\mathcal{B}$ for all Borel subsets of $E$ and $\mathcal{P}$ for all probability measures on $(E, \mathcal{B})$.

The basic probability measure space is denoted by $(\Omega, \mathcal{F}, P)$ and the generic element of $\Omega$ by $\omega$. An $E$-valued random variable $X$ is a map of $\Omega$ into $E$ measurable $(\mathcal{F}, \mathcal{B})$. The probability law $\mu_X$ of $X$ is a probability measure in $(E, \mathcal{B})$ defined by

$$\mu_X(B) = P(X \in B)\,, \qquad B \in \mathcal{B}\,.$$

According to Prohorov [5], every $\mu \in \mathcal{P}$ is tight, i.e.

$$\forall \varepsilon>0 \quad \exists K \text{ compact} \subset E \quad \mu(K)>1-\varepsilon .$$

A subset $\mathcal{M}$ of $\mathcal{P}$ is called *uniformly tight* if

$$\forall \varepsilon>0 \quad \exists K \text{ compact} \subset E \quad \forall \mu \in \mathcal{M} \quad \mu(K)>1-\varepsilon .$$

$\mathcal{P}$ is a complete metric space with respect to the Prohorov metric [5]. $\mathcal{M} \subset \mathcal{P}$ is *conditionally compact*, if and only if $\mathcal{M}$ is uniformly tight.

Let $C$ denote the algebra of all cylinder sets;

$$\{x \in E: (\langle z_1, x\rangle, \langle z_2, x\rangle, \cdots, \langle z_n, x\rangle) \in \Gamma\}$$
$$n = 1, 2, \cdots, z_i \in E^*, \Gamma \in \mathcal{B}(R^n)$$

and $\mathcal{B}[C]$ the $\sigma$-algebra generated by $C$.

**Proposition 2.1.** $\mathcal{B}[C]=\mathcal{B}$.

Proof. Let $\{b_n\}$ be a countable dense subset of $E$ and take $\{z_n\} \subset E^*$ such that

$$||z_n|| = 1 , \quad \langle z_n, b_n\rangle = ||b_n|| , \quad n = 1, 2, \cdots .$$

Such $z_n$ exists by Hahn-Banach's extension theorem for each $n$. Now we shall prove that

$$\{x: ||x|| \le r\} = \bigcap_n \{x: \langle z_n, x\rangle \le r\} .$$

Write $B_1$ and $B_2$ for the sets on both sides. $B_1 \subset B_2$ is obvious. To prove $B_1^c \subset B_2^c$, take an arbitrary point $b$ in $B_1^c$. Then we have $||b||>r$. As $\{b_n\}$ is dense, we can find $b_n$ such that

$$||b-b_n||<\tfrac{1}{2}(||b||-r) .$$

Then we get

$$||b_n|| \ge ||b||-||b-b_n||>\tfrac{1}{2}(||b||+r) ,$$
$$|\langle z_n, b\rangle-||b_n||| = |\langle z_n, b\rangle-\langle z_n, b_n\rangle| \le ||z_n|| \, ||b-b_n||<\tfrac{1}{2}(||b||-r)$$

and so

$$\langle z_n, b\rangle>||b_n||-\tfrac{1}{2}(||b||-r)>\tfrac{1}{2}(||b||+r)-\tfrac{1}{2}(||b||-r) = r .$$

This shows $b \in B_2^c$. Thus $B_1=B_2$ is proved. Since $B_2 \in \mathcal{B}[C]$, $B_1 \in \mathcal{B}[C]$. Since $\mathcal{B}[C]$ is translation invariant, $\{x: ||x-a||<r\}$ also belongs to $\mathcal{B}[C]$. Therefore $\mathcal{B} \subset \mathcal{B}[C]$. Since $\mathcal{B}[C] \subset \mathcal{B}$ is obvious, we have $\mathcal{B}=\mathcal{B}[C]$.

The *characteristic functional* of $\mu \in \mathcal{P}$ is defined by

$$C(z: \mu) = \int_E e^{i\langle z,x\rangle} \mu(dx) , \qquad z \in E^* .$$

It is clear that

$$C(z: \mu_X) = E[e^{i\langle z,X\rangle}] .$$

$\mu_n \to \mu$ (Prohorov metric) implies $C(z: \mu_n) \to C(z: \mu)$ for every $z \in E^*$. $C(z: \mu)$ is continuous in the norm topology in $E^*$.

**Proposition 2.2.**

$$C(z: \mu) = C(z: \nu) \Rightarrow \mu = \nu .$$

Proof. Setting $z = \sum_{j=1}^{n} t_j z_j$ and using the one-to-one correspondence between the probability measures and the characteristic functions in $R^n$, we can easily see that $\mu = \nu$ on $\mathcal{C}$. Since $\mathcal{C}$ is an algebra which generates $\mathcal{B}$ by Proposition 2.1, we have $\mu = \nu$ on $\mathcal{B}$.

**Proposition 2.3.** *If we have $r > 0$ such that*

$$C(z: \mu) = 1 \qquad \textit{for} \qquad \|z\| < r ,$$

*then $\mu$ is concentrated at* 0, i.e. $\mu = \delta$.

Proof. Let $\varphi(t) = C(tz: \mu)$; $t$ real, $z \neq 0$. Then $\varphi(t)$ is a characteristic function in $R^1$ and

$$\varphi(t) = 1 \qquad \text{for} \qquad |t| < \frac{r}{\|z\|} .$$

Using the inequality

$$|\varphi(t) - \varphi(s)| \leq \sqrt{2|1 - \varphi(t-s)|}$$

we can get $\varphi(t) = 1$ for every $t$. Setting $t = 1$ we have

$$C(z: \mu) = 1 = C(z: \delta) \qquad \text{for} \qquad z \neq 0 .$$

This is obvious for $z = 0$. Hence $\mu = \delta$ follows by Proposition 2.2.

## 3. Sums of independent random variables

Let $X_n(\omega)$, $n = 1, 2, \cdots$ be a sequence of independent $E$-valued random variables and set

$$S_n = \sum_{1}^{n} X_i , \qquad \mu_n = \text{the probability law of } S_n .$$

Then we have

**Theorem 3.1.**[1] (*Extension of P. Lévy's theorem*). *The following conditions are equivalent.*

(a) *$S_n$ converges a.s. (=almost surely),*

(b) *$S_n$ converges in probability,*

(c) *$\mu_n$ converges (Prohorov metric).*

---

(1) In the course of printing the authors noticed that a more general fact was proved by A. Tortrat [7].

Proof. (a)⇒(b)⇒(c) is obvious. We can prove (b)⇒(a) in the same way as in the real case, using the inequality that can be verified also as in the real case:

$$P(\max_{m<k\leq n}\|X_{m+1}+X_{m+2}+\cdots+X_k\|>2c)$$
$$\leq \frac{P(\|X_{m+1}+\cdots+X_n\|>c)}{1-\max_{m<k\leq n} P(\|X_{k+1}+\cdots+X_n\|\geq c)}$$

To prove (c)⇒(b), let us denote the probability law of $S_n-S_m=\sum_{m+1}^n X_i$ by $\mu_{mn}$, $m<n$. As $\mu_n$ tends to a probability measure $\mu$ on $E$ by the assumption (c), $\{\mu_n\}$ is conditionally compact and so uniformly tight, i.e.

$$\forall \varepsilon>0 \quad \exists K \text{ compact} \quad \forall n \quad \mu_n(K)>1-\varepsilon\,.$$

Let $K_1$ denote the set $\{x-y\colon x, y\in K\}$. $K_1$ is also compact by the continuity of the map $(x, y)\to x-y$. As $S_n$, $S_m\in K$ implies $S_n-S_m\in K_1$, we have

$$\begin{aligned}\mu_{mn}(K_1)&\geq P(S_n\in K, S_m\in K)\\ &\geq 1-P(S_n\in K^c)-P(S_m\in K^c)\\ &=1-\mu_n(K^c)-\mu_m(K^c)\\ &>1-2\varepsilon\,.\end{aligned}$$

This shows that $\{\mu_{mn}\colon m<n\}$ is also conditionally compact. We shall now prove (b), i.e.

(1) $$\forall \varepsilon>0 \quad \exists N \quad \forall m<n<N \quad \mu_{mn}(U_\varepsilon)>1-\varepsilon\,,$$

where $U_\varepsilon$ denotes the $\varepsilon$-neighbourhood of the origin 0 in $E$. Suppose to the contrary that

(2) $$\exists \varepsilon>0 \quad \forall N \quad \exists n(N)>m(N)>N \quad \mu_{m(N)n(N)}(U_\varepsilon)\leq 1-\varepsilon\,.$$

As $\{\mu_{mn}\}$ is conditionally compact, we can assume that $\mu_{m(N)n(N)}$ converges to a probability measure $\nu$ on $(E, \mathcal{B})$, then

(3) $$\nu(U_\varepsilon)\leq \varliminf_{N\to\infty} \mu_{m(N)n(N)}(U_\varepsilon)\leq 1-\varepsilon\,.$$

On the other hand we have

$$\begin{aligned}C(z\colon \mu_{n(N)}) &= E[e^{i<z,S_{n(N)}>}]\\ &= E[e^{i<z,S_{m(N)}>}]\,E[e^{i<z,S_{n(N)}-S_{m(N)}>}]\\ &= C(z\colon \mu_{m(N)})\,C(z\colon \mu_{m(N)n(N)})\end{aligned}$$

by the independence of $X_n$, $n=1, 2, \cdots$. Letting $N\to\infty$, we have

$$C(z\colon \mu) = C(z\colon \mu)\,C(z\colon \nu)\,.$$

Since $C(0: \mu)=1$, we have $r>0$ such that

$$C(z: \mu)\neq 0 \quad \text{for} \quad ||z||\leq r .$$

Then

$$C(z: \nu) = 1 \quad \text{for} \quad ||z||\leq r ,$$

so that $\nu=\delta$ by Proposition 2.3. This contradicts (3).

**Theorem 3.2.** *The uniform tightness of $\{\mu_n\}$ implies that we have a sequence $c_n\in E$, $n=1, 2, \cdots$ such that $S_n-c_n$ converges a.s.*

Proof. Let $(Y_n, n=1, 2, \cdots)$ be a copy of $(X_n, n=1, 2, \cdots)$ independent of this random sequence. Then $X_1, X_2, \cdots, Y_1, Y_2, \cdots$ are independent. Now set

$$T_n = \sum_1^n Y_i , \qquad U_n = \sum_1^n (X_i - Y_i) = S_n - T_n$$

$$\nu_n = \text{the probability law of } U_n .$$

Then $X_1-Y_1, X_2-Y_2, \cdots$ are independent.

By our assumption we have

$$\forall \varepsilon>0 \quad \exists K = K(\varepsilon) \quad \text{compact} \quad \forall n \quad \mu_n(K)>1-\varepsilon .$$

Write $K_1=K_1(\varepsilon)$ for the set $\{x-y: x, y\in K\}$. Then $K_1$ is also compact and we have

$$\begin{aligned} \nu_n(K_1) &= P(S_n-T_n\in K_1) \\ &\geq P(S_n\in K, T_n\in K) \\ &\geq 1-P(S_n\in K^c)-P(T_n\in K^c) \\ &= 1-2\mu_n(K^c)>1-2\varepsilon . \end{aligned}$$

Therefore $\{\nu_n\}$ is also uniformly tight and so conditionally compact.

Since $X_1, X_2, \cdots Y_1, Y_2 \cdots$ are independent and $X_n$ and $Y_n$ have the same distribution, we have

$$\begin{aligned} C(z: \nu_n) &= E[e^{i<z,U_n>}] \\ &= \prod_{j=1}^n E[e^{i<z,X_j>}]E[e^{-i<z,Y_j>}] \\ &= \prod_{j=1}^n |E(e^{i<z,X_j>})|^2 . \end{aligned}$$

Since $0\leq |E[e^{i<z,X_n>}]|^2\leq 1$, $\lim_{n\to\infty} C(z: \nu_n)$ exists for every $z\in E^*$.

Now we shall prove that $\{\nu_n\}$ is convergent. Since it is conditionally compact, it is enough to prove that two arbitrary convergent subsequences $\{\nu_n'\}$, $\{\nu_n''\}$ of $\{\nu_n\}$ have the same limit. Since $\lim_{n\to\infty} C(z: \nu_n)$ exists, we have, for

$\nu'=\lim \nu'_n$ and $\nu''=\lim \nu''_n$,

$$C(z:\nu')=\lim_n C(z:\nu'_n)=\lim_n C(z:\nu''_n)=C(z:\nu''),$$

and so $\nu'=\nu''$ by Proposition 2.3.

By Theorem 3.1 the convergence of $\nu_n$ implies the a.s. convergence of $U_n\equiv S_n-T_n$. Since $\{S_n\}$ and $\{T_n\}$ are independent, we can use Fubini's theorem to see that for almost every sample sequence $(c_1, c_2, \cdots)$ of $(T_1, T_2, \cdots)$, $S_n-c_n$ converges a.s. This completes the proof.

## 4. Sum of independent random variables with symmetric distributions

Let $\{X_n\}$ be independent $E$-valued random variables, $S_n$ denote $\sum_1^n X_i$, $n=1, 2, \cdots$ and $\mu_n$ the probability law of $S_n$, $n=1, 2, \cdots$ as before. In this section we shall impose an additional condition:

(SD) Each $X_n$ is symmetrically distributed.

**Theorem 4.1.** *The conditions* (*a*), (*b*) *and* (*c*) *in Theorem* 3.1 *and the following conditions are all equivalent.*

(d) $\{\mu_n\}$ *is uniformly tight.*

(e) *There exists an E-valued random variable S such that* $\langle z, S_n\rangle\to\langle z, S\rangle$ *in probability for every* $z\in E^*$.

(f) *There exists a probability measure* $\mu$ *on E such that*

$$E[e^{i\langle z,S_n\rangle}]\to C(z:\mu)$$

*for every* $z\in E^*$.

REMARK. In the finite dimensional case, (SD) is not necessary for the proof of the equivalence of all conditions except (d). But (f) does not always imply (c) in the infinite dimensional case without (SD). For example, let $E$ be a Hilbert space and $\{e_n\}$ be an orthonormal base. Now set

$$X_1(\omega)\equiv e_1, \qquad X_n(\omega)\equiv e_n-e_{n-1}, \qquad n=1, 2, \cdots.$$

Then $S_n(\omega)\equiv e_n$ and

$$\langle z, S_n\rangle=\langle z, e_n\rangle\to 0=\langle z, S\rangle, \qquad S\equiv 0.$$

But $S_n$ does not converge to $S$.

Proof. (a)$\Leftrightarrow$(b)$\Leftrightarrow$(c)$\Rightarrow$(d) and (a)$\Rightarrow$(e)$\Rightarrow$(f) are both obvious. Therefore it remains only to prove (f)$\Rightarrow$(e)$\Rightarrow$(d)$\Rightarrow$(a).

Suppose that (d) holds. By Theorem 3.2 we have $\{c_n\}$ such that $S_n-c_n$

converges a.s. Since each $X_n$ is symmetrically distributed and $X_1, X_2, \cdots$ are independent, the random sequence $(X_1, X_2, \cdots)$ has the same probability law as $(-X_1, -X_2, \cdots)$. Since $S_n - c_n \equiv \sum_1^n X_i - c_n$ converges a.s., $-S_n - c_n = \sum_1^n (-X_i) - c_n$ also converges a.s. and so does $S_n = [(S_n - c_n) - (-S_n - c_n)]/2$. Thus (d)$\Rightarrow$(a) is proved.

Suppose that (e) holds. Take $z_1, z_2, \cdots z_p \in E^*$ arbitrarily and fix them. Then the sequence of random vectors

$$\sigma_m(\omega) \equiv (\langle z_1, S_m(\omega)\rangle, \cdots, \langle z_p, S_m(\omega)\rangle), \qquad m = 1, 2, \cdots.$$

converges in probability to the random vector

$$\sigma(\omega) \equiv (\langle z_1, S(\omega)\rangle, \cdots, \langle z_p, S(\omega)\rangle).$$

Since $\sigma_n$ and $\sigma_m - \sigma_n$, $(m>n)$, prove to be independent by the assumption, $\sigma_n$ and $\sigma - \sigma_n$ are also independent, i.e.

$$P(\sigma_n \in \Gamma_1, \sigma - \sigma_n \in \Gamma_2) = P(\sigma_n \in \Gamma_1) P(\sigma - \sigma_n \in \Gamma_2), \quad \text{for} \quad \Gamma_1, \Gamma_2 \in \mathcal{B}[R^p].$$

Writing this in terms of $S_n$ and $S - S_n$, we have

$$P(S_n \in C_1, S - S_n \in C_2) = P(S_n \in C_1) P(S - S_n \in C_2)$$

for $C_1, C_2 \in \mathcal{C}$. Since $\mathcal{C}$ is an algebra which generates $\mathcal{B}$ by Proposition 2.1, we have

$$P(S_n \in B_1, S - S_n \in B_2) = P(S_n \in B_1) P(S - S_n \in B_2)$$

for $B_1, B_2 \in \mathcal{B}$, namely $S_n$ and $S - S_n$ are independent. Thus we have

$$P(S \in K) = \int_E P(S_n + x \in K) P(S - S_n \in dx)$$

and so we can find $x_0 = x_0(K, n)$ such that

$$P(S_n + x_0 \in K) \geq P(S \in K).$$

Since $S_n$ proves to be symmetrically distributed by our assumption, we have

$$P(-S_n + x_0 \in K) = P(S_n + x_0 \in K) \geq P(S \in K).$$

Writing $K_1$ for the set $\{(x-y)/2 : x, y \in K\}$, we get

$$\begin{aligned} \mu_n(K_1) = P(S_n \in K_1) &\geq P(S_n + x_0 \in K, -S_n + x_0 \in K) \\ &\geq 1 - P(S_n + x_0 \in K^c) - P(-S_n + x_0 \in K^c) \\ &\geq 1 - 2P(S \in K^c). \end{aligned}$$

By taking a compact set $K = K(\varepsilon)$ for $\varepsilon > 0$, we can make the right hand side greater than $1-\varepsilon$. Since $K_1$ is compact with $K$, $\{\mu_n\}$ proves to be uniformly tight. This proves (e)⇒(d).

We shall now prove (f)⇒(e).

Take $z\in E^*$ and fix it for the moment. $\langle z, X_n\rangle$, $n=1, 2, \cdots$ are independent real random variables and

$$\langle z, S_n\rangle = \sum_1^n \langle z, X_i\rangle .$$

By our assumption, we have

$$E[e^{it\langle z,S_n\rangle}] = E[e^{i\langle tz,S_n\rangle}] \to C(tz\colon \mu)$$

for every real $t$. Since the right hand side is a characteristic function of $t$ for the probability measure on $R^1$ induced from $\mu$ by the map $x\to\langle z, x\rangle$, the probability law of $\langle z, S_n\rangle$ converges to this measure. Therefore $\langle z, S_n\rangle$ converges a.s. to a real random variable, say $Y_z$, by Lévy's theorem. Notice that the exceptional $\omega$-set depends on $z$. Since a countable sum of null sets also is a null set, we have a P-null $\omega$-set $N=N(z^1, z^2, \cdots)$ such that

$$\langle z^{(k)}, S_n\rangle \to Y_{z^{(k)}} \qquad k=1, 2, \cdots$$

for every $\omega\in N^c$.

Now we shall compare two systems of real random variables:

$$Y_z(\omega), \qquad \omega\in(\Omega, \mathcal{F}, \mathcal{P}), \qquad z\in E^*$$

and

$$\langle z, x\rangle, \qquad x\in(E, \mathcal{B}, \mu), \qquad z\in E^*.$$

They have the same finite jont distributions. To prove this, take $z^{(1)}, z^{(2)}, \cdots, z^{(p)}\in E^*$. Then

$$E\Big[e^{i\sum_j t_j Y_{z^{(j)}}}\Big] = \lim_{n\to\infty} E\Big[e^{i\sum_j t_j\langle z^{(j)}, S_n\rangle}\Big] = \lim_{n\to\infty} E\Big[e^{i\langle\sum_j t_j z^{(j)}, S_n\rangle}\Big]$$
$$= C\Big(\sum_j t_j z^{(j)}\colon \mu\Big) = E_\mu\Big[e^{i\langle\sum_j t_j z^{(j)}, x\rangle}\Big]$$
$$= E_\mu\Big[e^{i\sum_j t_j\langle z^{(j)}, x\rangle}\Big]$$

by our assumption (f), where $E_\mu$ is the expectation sign based on the measure $\mu$.

Let $R^\infty$ be a countable product space $R^1\times R^1\times\cdots$ and $\mathcal{B}(R^\infty)$ the $\sigma$-algebra generated by all cylindrical Borel sets in $R^\infty$. $\mathcal{B}(R^\infty)$ is also the $\sigma$-algebra of all Borel subsets of $R^\infty$ with respect to the product topology.

Let $z', z'', \cdots$ be any sequence in $E^*$. Then we have

$$(1) \qquad P[(Y_{z'}, Y_{z''}, \cdots)\in B] = \mu[(\langle z', x\rangle, \langle z'', x\rangle\cdots)\in B]$$

for $B\in\mathcal{B}(R^\infty)$; in fact, if $B$ is a cylindrical Borel set, this identity holds because of the same joint distributions mentioned above and so it proves to hold for every $B\in\mathcal{B}(R^\infty)$ by usual argument.

Now we shall prove the existence of an $E$-valued random variable $S(\omega)$ with $Y_z=\langle z, S\rangle$ a.s. for each $z\in E^*$, which will complete the proof of (e). To find out such $S(\omega)$, we shall use the facts mentioned above.

Since $\mu$ is tight, we can find an increasing sequence of compact sets $K_1, K_2, \cdots\subset E$ such that $\mu(K_n)\to 1$, so that $\mu(K_\infty)=1$ for $K_\infty\equiv\bigcup_n K_n$. $K_\infty$ is clearly a Borel subset of $E$. Take $z_1, z_2, \cdots\in E^*$ in Proposition 2.1. The map $\theta: E\to R^\infty$ denfined by

$$\theta x = (\langle z_1, x\rangle, \langle z_2, x\rangle\cdots)$$

is continuous and one-to-one from $E$ onto $\theta E$; in fact, if $\theta x=(0, 0, \cdots)$, then

$$x\in \bigcap_n \{x: \langle z_n, x\rangle\leq r\} = \{x: \|x\|\leq r\}$$

for every $r>0$ (see the proof of Proposition 2.1) and so $x=0$. Therefore $\theta K_n$ is compact and the restriction $\theta|K_n$ has a continuous inverse map. Since $\theta K_\infty=\bigcup_n\theta K_n$ is a Borel subset of $R^\infty$, the restriction of $\theta$ to $K_\infty$ has an inverse which can be extended to a map $\varphi: R^\infty\to E$ measurable $(\mathcal{B}(R^\infty), \mathcal{B})$. It is clear that

$$x = \varphi(\theta x) = \varphi(\langle z_1, x\rangle, \langle z_2, x\rangle, \cdots) \quad \text{on} \quad K_\infty\,. \tag{2}$$

Since

$$\mu((\langle z_1, x\rangle, \langle z_2, x\rangle\cdots)\in\theta K_\infty)\geq\mu(K_\infty) = 1\,,$$

we have

$$P((Y_{z_1}, Y_{z_2}, \cdots)\in\theta K_\infty) = 1\,.$$

Now we set

$$S(\omega) = \varphi(Y_{z_1}(\omega), Y_{z_2}(\omega), \cdots)\,.$$

Take an arbitrary $z\in E^*$. Then

$$\begin{aligned}\mu(\langle z, x\rangle &= \langle z, \varphi(\langle z_1, x\rangle, \langle z_2, x\rangle\cdots)\rangle)\\ &\geq\mu(x = \varphi(\langle z_1, x\rangle, \langle z_2, x\rangle, \cdots))\\ &\geq\mu(x = \varphi(\langle z_1, x\rangle, \langle z_2, x) \cdots\rangle, x\in K_\infty)\\ &= \mu(K_\infty) = 1 \qquad \text{by} \quad (2)\,.\end{aligned} \tag{3}$$

Since $\varphi$ is measurable $(\mathcal{B}(R^\infty),\mathcal{B})$ and $x\to\langle z, x\rangle$ is continuous; the condition on $(\xi_0, \xi_1, \xi_2, \cdots)$: $\xi_0=\langle z, \varphi(\xi_1, \xi_2, \cdots)\rangle$ is given by a Borel subset of $R^\infty$. By (1) and (3) we have

$$P(Y_z = \langle z, \varphi(Y_{z_1}, Y_{z_2}, \cdots)\rangle) = 1$$

and so

$$P(Y_z = \langle z, S\rangle) = 1\,,$$

which completes our proof.

## 5. Examples.

### a. Generalized Wiener expansion of Brownian motion

Let $B(t)$, $0 \leq t \leq 1$, be a Brownian motion with $B(0) \equiv 0$. Then we have an isomorphism between the real Hilbert Space $L^2[0, 1]$ and the real Hilbert Space $\boldsymbol{M}(B)$ spanned by $B(t)$, $0 \leq t \leq 1$ in $L^2(\Omega, \mathcal{F}, P)$ by

$$\varphi \to \int_0^1 \varphi(u)dB(u)\,.$$

The indicator $e_{0t}$ of the interval $(0, t)$ corresponds to $B(t)$. Let $\{\varphi_n\}_n$ be an orthonormal base in $L^2[0, 1]$. Then the $\{\xi_n\}_n$ that correspond to $\{\varphi_n\}_n$ form an orthogonal base in $\boldsymbol{M}(B)$. $\{\xi_n\}_n$ are independent since $B(t)$ is a Gaussian process. As $e_{0t}$ has an orthogonal expansion

$$e_{0t} = \sum_n b_n(t)\varphi_n\,,$$

where

$$b_n(t) = \int_0^1 e_{0t}(u)\varphi_n(u)du = \int_0^t \varphi_n(u)du\,,$$

we have an orthogonal expansion

$$B(t) = \sum_n b_n(t)\xi_n(\omega) = \sum_n \xi_n(\omega)\int_0^t \varphi_n(u)du\,. \tag{1}$$

For each $t$, this series converges in the mean square and so converges a.s. by Lévy's theorem. We shall make use of Theorem 4.1 ((e)⇒(a)) to prove

**Theorem 5.1.** *The right hand side of* (1) *converges uniformly in $t$ to $B(t)$ a.s.*

Proof. Let us introduce a sequence of stochastic processes

$$X_n(t, \omega) = \xi_n(\omega)\int_0^t \varphi_n(u)du\,, \qquad n = 1, 2, \cdots$$

and write the sample paths of $X_n(t, \omega)$, $n=1, 2, \cdots$ and $B(t, \omega)$ as $\boldsymbol{X}_n(\omega)$, $n=1, 2, \cdots$ and $\boldsymbol{B}(\omega)$ respectively. Then these are symmetrically distributed random variables with values in the Banach space $E \equiv C[0, 1]$ of continuous functions on $[0, 1]$. $\boldsymbol{X}_n$, $n=1, 2, \cdots$ are independent. Set

$$\boldsymbol{S}_n(\omega)=\sum_1^n \boldsymbol{X}_i(\omega)\,,\qquad n=1,2,\cdots.$$

$\boldsymbol{S}_n(\omega)$ is the sample path of the process $S_n(t,\omega)=\sum_1^n X_i(t,\omega)$. Our theorem claims that $\boldsymbol{S}_n\to\boldsymbol{B}$ a.s. To prove this it is enough by Theorem to prove that for every $z\in E^*$ i.e. for every signed measure $z(dt)$ on $[0,1]$, $\langle z,\boldsymbol{S}_n\rangle$ converges in probability to $\langle z,\boldsymbol{B}\rangle$. It is now enough to observe

$$\begin{aligned}
&E[|\langle z,\boldsymbol{S}_n\rangle-\langle z,\boldsymbol{B}\rangle|]\\
&\quad=E\left[\left|\int_0^1 z(dt)(S_n(t)-B(t))\right|\right], \qquad |z|=\text{total variation of } z\,,\\
&\quad\leq\int_0^1 |z|(dt)E[|S_n(t)-B(t)|]\to 0\,,
\end{aligned}$$

because

$$(E[|S_n(t)-B(t)|])^2\leq E[|S_n(t)-B(t)|^2]=\sum_{i=n+1}^{\infty} b_i(t)^2$$

and so

$$E[|S_n(t)-B(t)|]\begin{cases}\leq\left(\int_0^1 e_{0t}(u)^2du\right)^{1/2}\leq\sqrt{t} & \text{for } \forall t\\ \to 0 & \text{as } n\to\infty\,.\end{cases}$$

**b. Definition of Brownian motion**

Theorem 5.1 suggests that we can define Brownian motion as follows. Let $\xi_n(\omega)$, $n=1,2,\cdots$ be an independent sequence of real random variables with the distribution $N(0,1)$, (whose existence is guaranteed by Kolmogorov's extension theorem) and an orthonormal base $\varphi_n$, $n=1,2,\cdots$ in $L^2[0,1]$.

**Theorem 5.2.**

$$\sum_n \xi_n(\omega)\int_0^t \varphi_n(u)du\,,\qquad 0\leq t\leq 1 \tag{2}$$

*converges uniformly in t a.s. The limit process $S(t)$ is a stochastic process with independent increaments and continuous paths such that $S(t)S(s)$ is $N(0,t-s)$-distributed for $t>s$, i.e. $S(t)$ is a Brownian motion.*

REMARK. N. Wiener [6] defined Brownian motion in this way by taking $\varphi_n(u)=\sqrt{2}\sin n\pi t$, $n=1,2,\cdots$. He proved the a.s. uniform convergence of the grouped sums

$$\sum_{n=0}^{\infty}\sum_{k=2^{n-1}}^{2^n-1}\xi_k(\omega)\int_0^t\sqrt{2}\sin k\pi u du\,,$$

which was sufficient for his purpose. G. Hunt [4] proved a theorem that ensures the a.s. uniform convergence of the sum

$$\sum_{n=1}^{\infty} \xi_n(\omega) \int_0^t \sqrt{2} \sin n\pi u du .$$

We claim that this holds for a general orthonormal base $\{\varphi_n\}$, (Delporte. [2], Walsh [8]).

Proof. The only difficult part of our theorem is the a.s. uniform convergence. We shall use $\boldsymbol{X}_n(\omega)$, $S_n(t, \omega)$ and $\boldsymbol{S}_n(\omega)$ as before. By Theorem 3.2 it is enough to prove that the probability laws of $\boldsymbol{S}_n$, $n=1, 2, \cdots$ are uniformly tight.

Observing

$$E[|S_n(t)-S_n(s)|^2] = \sum_{i=1}^{n} |b_i(t)-b_i(s)|^2 \leq \int_0^1 |e_{0t}(u)-e_{0s}(u)|^2 du = |t-s| ,$$

we have

$$E[|S_n(t)-S_n(s)|^4] = 3E[|S_n(t)-S_n(s)|^2]^2 \leq 3(t-s)^2$$

because $S_n(t)-S_n(s)$ is Gauss distributed with the mean 0. Using the same technique of diadic expansions as in the proof of Kolmogorov's theorem, we can prove that for $\varepsilon>0$, there exists $\delta=\delta(\varepsilon)$ such that

$$P(\boldsymbol{S}_n \in K) > 1-\varepsilon, \; n = 1, 2, \cdots$$

where

$$\begin{aligned} K &= K(\delta) = K(\varepsilon) \\ &= \{f \in C[0, 1]: f(0) = 0, \; |f(t)-f(s)| \leq 5|t-s|^{1/8} \quad \text{for} \quad |t-s| \leq \delta(\varepsilon)\} , \end{aligned}$$

It is easy to see that $K$ is an equi-continuous and equibounded family. Therefore $K$ is conditionally compact in $C[0, 1]$ by Ascoli-Arzela's theorem. This completes the proof.

### c. Gaussian stationary processes

Let $S(t)$ be a Gaussian stationary process continuous in the square mean such that $E(S(t))=0$. Let us consider the sample path $\boldsymbol{S}$ of $S(t)$ on a bounded time interval $a \leq t \leq b$. By taking a measurable separable version we have $\boldsymbol{S} \in L^p[a, b]$ a.s. for every $p \geq 1$; we shall identify two functions on $[a, b]$ equal to each other a.e. In fact we have

$$\begin{aligned} E[\|\boldsymbol{S}\|_p^p] &= E[\int_a^b |S(t)|^p dt] \\ &= \int_a^b E[|S(t)|^p] dt = E[|S(0)|^p](b-a) \quad < \infty , \end{aligned}$$

noticing that $S(0, \omega)$ is Gauss distributed and so

$$P[\|S\|_p<\infty]=1.$$

Consider the spectral decomposition of $S(t)$:

$$S(t,\omega)=\int_{-\infty}^{\infty}e^{i\lambda t}\Phi(d\lambda,\omega)$$

and set

$$X_n(t,\omega)=\int_{n-1\leq|\lambda|<n}e^{i\lambda t}\Phi(d\lambda,\omega)\,,\quad n=1,2,\cdots.$$

Then $X_n(t)$ has a version whose sample path (on $a\leq t\leq b$) is continuous a.s. and so belongs to $L^p[a,b]$ a.s. Now set

$$S_n(t)=\sum_1^n X_k(t)=\int_{|\lambda|<n}e^{i\lambda t}\Phi(d\lambda)$$

Then it is easy to see that

$$E[|S_n(t)-S(t)|^2]=\int_{|\lambda|\geq n}F(d\lambda)\to 0\quad \text{as } n\to\infty$$

where $F(d\lambda)$ is the spectral measure of the covariance function of $S(t)$.

**Theorem 5.3.**

$$\int_a^b|S_n(t)-S(t)|^p dt\to 0\quad \text{a.s. } (p\geq 1)\,.$$

Proof. Using the same notation for the sample paths we can see that $X_n$, $n=1,2,\cdots$ are independent random variables with values in $L^p[a,b]$.

$$E[\|S_n-S\|_p^p]=E\left[\int_a^b|S_n(t)-S(t)|^p dt\right]$$
$$=\int_a^b E[|S_n(t)-S(t)|^p]\,dt=c_p\int_a^b(E[|S_n(t)-S(t)|^2])^{p/2}dt$$
$$=c_p(b-a)\left[\int_{|\lambda|\geq n}F(d\lambda)\right]^{p/2}\to 0\quad \text{as } n\to\infty\,,$$

where $c_p=\int_{-\infty}^{\infty}\frac{1}{\sqrt{2\pi}}e^{-(\xi^2/2)}|\xi|^p d\xi$. By Theorem 3.1 ((b)$\Rightarrow$(a)) for $E=L^p[a,b]$, this implies $\|S_n-S\|_p\to 0$ a.s.

**Theorem 5.4.** *If the sample path of* $S(t,\omega)$ *is continuous a.s., then*

$$\max_{a\leq t\leq b}|S_n(t)-S(t)|\to 0\quad \text{a.s.}$$

REMARK. Sufficient conditions for the a.s. continuity of the sample path of $S(t)$ in terms of the correlation function of $S(t)$ were given by G. Hunt [4], Belayev [1] and X. Fernique [3].

Proof. Using the same idea as above, we can apply Theorem 4.1 ((e)⇒(a)) for $E=C[0, 1]$ by observing

$$\begin{aligned} &E\left[\left|\int_a^b z(dt)\,S_n(t, \omega)-\int_a^b z(dt)\,S(t, \omega)\right|\right] \\ &\leq \int_a^b |z|(dt)E[|S_n(t, \omega)-S(t, \omega)|] \\ &\leq \int_a^b |z|(dt)[E(|S_n(t)-S(t)|^2)]^{1/2} \\ &= \int_a^b |z|(dt)\left[\int F(d\lambda)\right]^{1/2} \to 0\,, \quad \text{as } n\to\infty\,. \end{aligned}$$

Aarhus University
Kobe University

## References

[1] Yu. K. Belyaev: *Continuity and Hölder's conditions for sample function of stationary processes*, Proc. 4th Berkeley Symp. Math. Stat. Prob. II, University of Calif. Press, 1961, 23–33.

[2] J. Delporte: *Fonctions aléatoires presque sûrement continues sur un intervalle fermé*, Ann. Inst. Henri-Poincaré **1** (1964), 111–215.

[3] X. Fernique: *Continuité des processus gaussiens*, C.R. Acad. Paris **258** (1964), 6058–6060.

[4] G.A. Hunt: *Random Fourier transforms*, Trans. Amer. Math. Soc. **71** (1951), 38–69.

[5] Yu. V. Prohorov: *Convergence of random processes and limit theorems in probability theory*, Theor. Probability Appl. **1** (1956), 157–214.

[6] R. Paley and N. Wiener: Fourier transforms in the complex domain, Amer. Math. Soc. Coll. Publ. XIX, 1934.

[7] A. Tortrat: *Lois de probabilité sur un espace topologique complètement régulier et produits infinis à terms indépendants dans un groupe topologique*, Ann. Inst. Henri-Poincaré **1** (1965), 217–237.

[8] J.B. Walsh: *A note on uniform convergence of stochastic processes*, Proc. Amer. Math. Soc. **18** (1967), 129–132.

# GENERALIZED UNIFORM COMPLEX MEASURES IN THE HILBERTIAN METRIC SPACE WITH THEIR APPLICATION TO THE FEYNMAN INTEGRAL

Dedicated to Professor Charles Loewner

KIYOSI ITÔ
KYOTO UNIVERSITY and STANFORD UNIVERSITY

## 1. Introduction and summary

As we pointed out in the Fourth Berkeley Symposium [4], an infinite dimensional version of the complex measure in the $k$-space $E_k$,

$$F_k(dx) = \lambda_k(dx)/(\sqrt{2\pi hi})^k, \qquad \lambda_k = \text{Lebesgue measure} \tag{1}$$

is useful for a mathematical formulation of the Feynman integral [1]; $h$ is a positive constant which is supposed to indicate the Planck constant in its application to quantum mechanics and $\sqrt{z}$, $(z \neq 0)$ denotes the branch for which $-\pi/2 < \arg \sqrt{z} < \pi/2$ throughout this paper. Since neither $\lambda_k$ nor $(\sqrt{2\pi hi})^k$ has any meaning when $k = \infty$, we cannot directly extend this measure to the infinite dimensional space $E_\infty$ (Hilbert space). Therefore, we shall consider a linear functional $F_k(f)$ induced by the measure $F_k$ in (1):

$$F_k(t) = \int_{E_k} f(x) \frac{\lambda_k(dx)}{(\sqrt{2\pi hi})^k}, \tag{2}$$

and extend this by putting convergent factors as

$$F_k(f) = \lim_{n\to\infty} \int_{E_k} f(x) \exp\left[-\frac{1}{2n}(V^{-1}(x-a), (x-a))\right] \frac{\lambda_k(dx)}{(\sqrt{2\pi hi})^k} \tag{3}$$

where $a$ is any element of $E_k$ and $V$ is a strictly positive-definite symmetric operator. The domain $\mathfrak{D}(F_k)$ of definition of $F_k$ is the space of all Borel measurable functions for which the limit in (3) exists for every $(a, V)$ and has a finite value independent of $(a, V)$. We shall rewrite (3) as

$$F_k(f) = \lim_{n\to\infty} \prod_{\nu=1}^{k} \sqrt{1 + \frac{nv_\nu}{hi}} \int_{E_k} f(x) N(dx\colon a, nV), \tag{3'}$$

where $\{v_\nu\}$ are the eigenvalues of $V$ and $N(dx: a, V)$ denotes the Gauss measure with the mean vector $a$ and the covariance operator $V$. Since the Gauss measure $N(dx: a, V)$ can be defined in the real Hilbert space $E_\infty$ if $V$ is positive-definite symmetric operator with the sum of eigenvalues $< \infty$ [3], [7], we can define $F_\infty$ by putting $k = \infty$ in (3); notice that the infinite product in (3′) is convergent by virtue of $\sum_\nu v_\nu < \infty$.

Following an important suggestion of L. Gross we shall modify (3′) to make the approximating measures more uniform and define $F_k$ as follows. We introduce a directed semiorder in the class $\mathfrak{V}$ of all strictly positive-definite symmetric operators of finite trace by

$$V_1 < V_2 \qquad \text{if and only if} \quad V_2 - V_1 \in \mathfrak{V}. \tag{4}$$

DEFINITION. *Denoting with* $\lim_V$ *the limit along this directed system* $\mathfrak{V}$, *we shall define* $F_k(f)$ *as follows:*

$$F_k(f) = \lim_V \prod_{\nu=1}^{k} \sqrt{1 + \frac{v_\nu}{hi}} \int_{E_k} f(x) N(dx: a, V), \tag{5}$$

*where the domain* $\mathfrak{D}(F_k)$ *of definition of* $F_k$ *is the class of all Borel measurable functions for which this limit exists for every* $a$ *and has a finite value independent of* $a$.

In order to discuss this functional we shall introduce some notions.

Let $A$ be a bounded positive-definite symmetric operator. Then $\sum_n (Ae_n, e_n)$, $\{e_n\}$ being an orthonormal basis, is independent of the special choice of the basis $\{e_n\}$ and is called the *trace* of $A$, in symbol $\operatorname{Tr} A$. If $\operatorname{Tr} A < \infty$, then $A$ is completely continuous and $\operatorname{Tr} A$ is equal to the sum of all eigenvalues of $A$.

Let $A$ be a bounded operator. We shall define the trace norm of class $\alpha$ ($> 0$) of $A$ by

$$\|A\|_\alpha = [\operatorname{Tr}((A^*A)^{\alpha/2})]^{1/\alpha}. \tag{6}$$

If $A$ is symmetric, then $\|A\|_\alpha = [\operatorname{Tr}(|A|^\alpha)]^{1/\alpha}$. The uniform norm of $A$ is defined by $\|A\| = \sup_{|x|\leq 1} |Ax|$. It is easy to see that

$$\|A\|_\alpha < \infty \Rightarrow \|A\|_\beta < \infty \qquad \text{if} \quad \alpha < \beta \tag{7.a}$$

and

$$\|A\|_\alpha < \infty \Rightarrow \|A\| < \infty. \tag{7.b}$$

We shall call $A$ a *trace operator* if $\|A\|_1 < \infty$ and call it a *Hilbert-Schmidt operator* if $\|A\|_2 < \infty$. Because of (7.a) we can see that if $\|A\|_\alpha < \infty$ for some $\alpha < 1$, then $A$ is a trace operator, but not vice versa.

If $\|A\|_1 < \infty$, then $\sum (Ae_n, e_n)$ is also convergent for every orthonormal basis $\{e_n\}$ and has a value independent of $\{e_n\}$. This is also called the *trace* of $A$, in symbol $\operatorname{Tr} A$.

A bounded operator is called *nearly orthogonal* if $A$ is one-to-one and if $\|(A^*A)^{1/2} - I\|_\alpha < \infty$ for some $\alpha < 1$. The first condition is equivalent to the condition that $(A^*A)^{1/2}$ is strictly positive-definite.

A one-to-one transformation from $E_k$ onto itself is called *nearly isometric* if $C$ is expressed as $Cx = a + A \cdot x$, where $A$ is nearly orthogonal.

Writing the eigenvalues of $(A^*A)^{1/2} - I$ as $\{\alpha_\nu\}$, we shall define $J(C)$ by

$$J(C) = \prod_\nu (1 + \alpha_\nu). \tag{8}$$

This is well-defined and does not vanish because $A$ is nearly orthogonal. It is needless to say that every isometric transformation from $E_\infty$ onto $E_\infty$ is nearly isometric. In case $k < \infty$, every one-to-one linear transformation $C$ from $E_k$ onto $E_k$ is nearly isometric and $J(C)$ turns out to be the absolute value of the Jacobian of $C$.

We are now in a position to state the main properties of $F_k$.

**Theorem 1.** *The linear space $\mathfrak{D}(F_k)$ is invariant under nearly isometric transformations, and we have*

(i) $f = \alpha_1 f_1 + \alpha_2 f_2 \Rightarrow F_k(f) = \alpha_1 F_k(f_1) + \alpha_2 F_k(f_2)$;

(ii) *if $g(x) = f(Cx)$ and $C$ is nearly isometric, then $F_k(g) = J(C)^{-1}F_k(f)$.*

The functional $F_k$ is not trivial. In other words $\mathfrak{D}(F_k)$ is a fairly large class of functions, as the following theorem shows.

**Theorem 2.** *If $f(x)$ is of the form*

$$f(x) = \exp\left[\frac{i}{2h}|x|^2\right]\int_{E_k} e^{i(x,y)}\mu(dy) \tag{9}$$

*where $\mu$ is a complex measure of bounded absolute variation defined for all Borel subsets of $E_k$, then*

$$f \in \mathfrak{D}(F_k) \quad \text{and} \quad F_k(f) = \int_{E_k} \exp\left[\frac{h}{2i}|y|^2\right]\mu(dy). \tag{10}$$

Combining this with theorem 1, we can see that all functions $f(Cx)$, $f$ being of the form (9) and $C$ being nearly isometric, and their linear combinations belong to $\mathfrak{D}(F_k)$.

In view of these facts, we shall call the functional $F_k$ a *generalized uniform complex measure* and write it as

$$F_k(f) = \int_{E_k} f(x)F_k(dx). \tag{11}$$

In case $k < \infty$, every Borel measurable function $f$ that is $\lambda_k$-summable on $E_k$ belongs to $\mathfrak{D}(F_k)$, and we have (1). Therefore, we call $F_k$, $(k = 1, 2, \cdots, \infty)$ a generalized uniform measure with *index* $\sqrt{2\pi hi}$.

We shall write $\mathcal{E}_k$ for the class of all functions $f$ of the form in theorem 2. It is easy to see that *$\mathcal{E}_k$ is a linear space invariant under isometric transformations.*

The space $E_{k+\ell}$, $(k, \ell = 1, 2, \cdots, \infty)$ is considered as the Cartesian product of $E_k$ and $E_\ell$. Any point $x$ of $E_{k+\ell}$ is written as the pair $(y, z)$, $y \in E_k$, $z \in E_\ell$. Then we can easily prove the following theorem.

**Theorem 3.** *Suppose that $f \in \mathcal{E}_{k+\ell}$. Then $f(x) = f(y, z)$ belongs to $\mathcal{E}_k$ as a function of $y$ for each $z$ and $\int f(y, z)F_k(dy)$ belongs to $\mathcal{E}_\ell$ as a function of $z$. Furthermore,*

$$\int_{E_{k+l}} f(x)F_{k+l}(dx) = \int_{E_l} \left[\int_{E_k} f(y, z)F_k(dy)\right] F_l(dz). \tag{12}$$

Similarly, we have

$$\int_{E_{k+l}} f(x)F_{k+l}(dx) = \int_{E_k} \left[\int_{E_l} f(y, z)E_l(dz)\right] F_k(dz). \tag{13}$$

If $g \in \mathcal{E}_k$ and $h \in \mathcal{E}_l$, then $f(y, z) = g(y)h(z)$ belongs to $\mathcal{E}_{k+l}$ and

$$\int_{E_{k+l}} f(x)F_{k+l}(dx) = \int_{E_k} g(y)F_k(dy) \int_{E_l} h(z)F_l(dz). \tag{14}$$

A metric space $M_k$ is called *Hilbertian* with dimension $k$ if there exists an isometric mapping $\Phi$ from $M_k$ onto $E_k$. If there are two such mappings $\Phi_1$ and $\Phi_2$, then $\Phi_2\Phi_1^{-1}$ will be an isometric mapping from $E_k$ onto itself. Because of this fact, all notions in $E_k$ invariant under isometric transformations can be transplanted into the Hilbertian metric space $M_k$: for example, Gaussian measure, nearly isometric transformations, $J(C)$, the function class $\mathcal{E}_k$, the generalized uniform complex measure $F_k$, and so on.

Theorems 1, 2, and 3 can be restated in terms of the Hilbertian metric space. The statement does not change for theorem 1. Theorem 2 should be stated as follows.

THEOREM 2′. *If $f(x)$ is of the form*

$$f(x) = \int \exp\left[\frac{i}{2h}\overline{xy}^2\right] \nu(dy), \qquad \overline{xy} = \text{distance } (x, y), \tag{9′}$$

*then we have*

$$f \in \mathfrak{D}(F_k) \quad \text{and} \quad F_k(f) = \nu(E_k). \tag{10′}$$

In order to state theorem 3 for $M_k$ we need some notions. The concept of a linear mapping with a translation from $E_{k+l}$ onto $E_l$ is invariant under isometric transformations, so that we can define such a mapping from $M_{k+l}$ onto $M_k$, which we shall call a *linear mapping* from $M_{k+l}$ onto $M_k$. Let $C$ be such a mapping. Then

$$d(y_1, y_2) = \inf \{\overline{x_1x_2}: x_1 \in C^{-1}(y_1), x_2 \in C^{-1}(y_2)\} \tag{15}$$

defines a new metric on $M_l$, so that there exists a linear mapping $T$ from $M_l$ onto itself such that

$$d(y_1, y_2) = \overline{(Ty_1)(Ty_2)}. \tag{16}$$

We shall call $C$ *normal*, or *nearly normal*, according to whether $T$ is isometric or nearly isometric. If $C$ is nearly normal, then we shall define $J(C)$ to be $J(T)$. There are many $T$'s for a single $C$, but these notions are independent of the choice of $T$. Now we shall state theorem 3 for the Hilbertian metric space, omitting the statements about the domain of definition.

THEOREM 3′. *Let $C$ be a normal linear mapping from $M_{k+l}$ onto $M_l$. Then $C^{-1}(y)$ is a $k$-dimensional Hilbertian metric subspace of $M_{k+l}$, and we have, for $f \in \mathcal{E}_{k+l}$,*

(17) $$\int_{M_{k+\ell}} f(x)F_{k+\ell}(dx) = \int_{M_\ell}\int_{C^{-1}(y)} f(x)F_k(dx)F_\ell(dy).$$

Combining theorem 3′ with theorem 2, we shall obtain a slight generalization of theorem 3′.

THEOREM 3″. *Let $C$ be nearly normal from $M_{k+\ell}$ onto $M_\ell$. Then we have*

(18) $$\int_{M_{k+\ell}} f(x)F_{k+\ell}(dx) = J(C)\int_{M_\ell}\int_{C^{-1}(y)} F_k(dx)F_\ell(dy).$$

Let us now formulate the Feynman integral in terms of $F_\infty$.

Consider a classical mechanical system of a particle of mass $m$ moving on the real line $-\infty < q < \infty$ where the field of force is given by a potential $u(q)$. The *Lagrangian* of this system is

(19) $$L(q, \dot{q}) = \frac{m}{2}\dot{q}^2 - mu(q)$$

and its action integral $\mathfrak{A}(\gamma)$ along a motion $\gamma = \gamma(\tau)$, $s \leq \tau \leq t$, is given by

(20) $$\mathfrak{A}(\gamma) = \int_s^t L(\gamma(\tau), \gamma'(\tau))\, d\tau = \tfrac{1}{2}\int_s^t m\gamma'(\tau)^2\, d\tau - \int_s^t mu(\gamma(\tau))\, d\tau.$$

Let $\Gamma = \Gamma(t, b|s, a)$ be the space of all motions $\gamma = \gamma(\tau)$, $s \leq \tau \leq t$, starting at $\gamma(s) = a$ and ending with $\gamma(t) = b$ such that the velocity function $\gamma'(\tau)$ is square summable on $s \leq \tau \leq t$. The space $\Gamma$ is a Hilbertian metric space $M_\infty$ with the metric

(21) $$\overline{\gamma_1\gamma_2}^2 = \int_s^t m(\gamma_1'(\tau) - \gamma_2'(\tau))^2\, d\tau.$$

Thus we can define the generalized measure on $F_\infty$ on $\Gamma$. The *Feynman principle of quantization* of this mechanical system is that the function

(22) $$G(t, b|s, a) \equiv \frac{\sqrt{m}}{\sqrt{2\pi hi(t-s)}}\int_\Gamma \exp\left[\frac{i}{h}\mathfrak{A}(\gamma)\right] F_\infty(d\gamma)$$

is the *Green function* of the corresponding quantum mechanical system, namely that $\exp[(i/h)\mathfrak{A}(\gamma)]$ belongs to $\mathfrak{D}(F_\infty)$, and the function $G(t, b|s, a)$ defined above is the elementary solution of the *Schrödinger equation*

(23) $$\frac{h}{i}\frac{\partial\varphi}{\partial t} = \frac{h^2}{2m}\frac{\partial^2\varphi}{\partial q^2} - m\cdot u\cdot\varphi$$

for the quantum mechanical system; the right side of (22) is called the *Feynman integral*. We shall prove this for the following cases.

*Case 1*: $u$ is the Fourier transform of a complex measure of bounded absolute variation on $(-\infty, \infty)$.

*Case 2*: $u(q) \equiv c_1\cdot q$, where $c_1$ is a real constant.

*Case 3*: $u(q) \equiv c_2q^2$, where $c_2$ is a positive constant.

Let us mention one word about the index $c = \sqrt{2\pi hi}$. As a matter of fact, we can carry out the same argument in the case where $c \neq 0$ and $\operatorname{Re} c^2 \geq 0$ by replacing $hi$ in (5) with $c^2/2\pi$, and the case where $c = \sqrt{2\pi}$ turns out to be in close connection with the *Wiener measure*.

In the course of writing this paper we found a gap in our argument based on definition (3′). We also found that we were able to overcome the difficulty by adopting definition (5) suggested by Professor L. Gross, and by using Kuroda's theory on infinite determinants [5] to which Professor T. Kato drew our attention. We would like to express our hearty thanks to them.

## 2. Properties of the generalized uniform complex measure $F_k$

We shall start with some preliminary facts.

LEMMA 1. *Suppose that $V$ is a positive-definite symmetric trace operator with the eigenvalues $\{v_\nu\}$ and the eigenvectors $\{e_\nu\}$. Then $\xi_\nu(x) \equiv (x, e_\nu)$, $\nu = 1, 2, \cdots$ are independent random variables on the probability space $(E_k, N(dx{:}\ a, V))$, each $(x, e_\nu)$ having a Gaussian distribution with the mean $a_\nu = (a, e_\nu)$ and the variance $v_\nu$.*

PROOF. It is enough to observe that

$$(24)\qquad \int \exp\left[i \sum_{\nu=1}^{n} z_\nu(x, e_\nu)\right] N(dx, a, V) = \int \exp\,[i(x, \textstyle\sum z_\nu e_\nu]N(dx, a, V)$$
$$= \exp\,\{i(a, \textstyle\sum z_\nu e_\nu) - \frac{1}{2}(V \sum z_\nu e_\nu, \sum z_\nu e_\nu)\}$$
$$= \exp\,[\textstyle\sum (iz_\nu a_\nu - \frac{1}{2}v_\nu z_\nu^2)] = \displaystyle\prod_{\nu=1}^{n} \exp\,(iz_\nu a_\nu - \tfrac{1}{2}v_\nu z_\nu^2).$$

LEMMA 2. *For $-\infty < y < +\infty$ and $\mathrm{Re}\,\alpha > 0$, one has*

$$(25)\qquad \int_{-\infty}^{\infty} e^{ixy}e^{-(\alpha/2)x^2}\,dx = \sqrt{\frac{2\pi}{\alpha}}\,e^{-y^2/2\alpha}.$$

PROOF. This is well known for $\alpha > 0$. By analytic continuation we can see that it holds for $\mathrm{Re}\,\alpha > 0$.

LEMMA 3. *Let $H$ be a real or complex Hilbert space and suppose that $V_1$ and $V_2$ are bounded symmetric linear operators. If*

$$(26)\qquad |V_1x|^2 \geq |V_2x|^2 \geq c|x|^2, \qquad c \text{ is a positive constant,}$$

*then*

$$(27)\qquad |V_1^{-1}x|^2 \leq |V_2^{-1}x|^2 \leq c^{-1}|x|^2.$$

PROOF. It follows from the assumption that $|V_2V_1^{-1}x|^2 \leq |x|^2$, that is, $\|V_2V_1^{-1}\| \leq 1$, so that $\|V_1^{-1}V_2\| = \|(V_1^{-1}V_2)^*\| = \|V_2V_1^{-1}\| \leq 1$. Thus we have $|V_1^{-1}V_2x| \leq |x|^2$, that is, $|V_1^{-1}x| \leq |V_2^{-1}x|$. It is obvious that $|V_1^{-1}x|^2 \leq c^{-1}|x|^2$.

LEMMA 4. *For any bounded Borel measurable function $f(x)$ defined on $E_k$, any linear operator $A{:}\ E_k \rightarrow E_k$ and any $b \in E_k$, we have*

$$(28)\qquad \int_{E_k} f(b + Ax)N(dx{:}\ a, V) = \int_{E_k} f(x)N(dx{:}\ b + Aa, AVA^*).$$

PROOF. If $f(x)$ is of the form $e^{i(x,y)}$, $y$ being any fixed element in $E_k$, then this is true by virtue of

$$(29)\qquad \int e^{i(x,y)}N(dx{:}\ a, V) = \exp\,\{i(a, y) - \tfrac{1}{2}(Vy, y)\}.$$

Taking linear combinations and limits, we can see that it is true in general.

In section 1 we defined $\operatorname{Tr} A$, $\|A\|_\alpha$, and $\|A\|$ for the real separable Hilbert space $E_k$. We have analogous concepts for the complex separable Hilbert space $H_k$ of dimension $k$. Any linear operator $A$ from $E_k$ into itself induces an operator $\tilde{A}$ on $H_k$ by $\tilde{A}(x + iy) = Ax + iAy$. It is easy to see $\operatorname{Tr} A = \operatorname{Tr} \tilde{A}$, $\|A\|_\alpha = \|\tilde{A}_\alpha\|$, and $\|A\| = \|\tilde{A}\|$. Therefore, we write $A$ for $\tilde{A}$ without any ambiguity. We summarize some known facts about norms and determinants. Kuroda's paper [5] is to be referred to concerning the definition and the properties of the determinant of a linear operator $A$ such that $\|A - I\|_1 < \infty$.

LEMMA 5. *The following relations hold:*

$$\|A\| \leq \|A\|_1, \tag{30.a}$$

$$\|AB\|_1 = \|(AA^*)^{1/2}(B * B)^{1/2}\|_1 \leq \|A\| \, \|B\|_1 \leq \|A\|_1 \|B\|_1, \tag{30.b}$$

$$|\operatorname{Tr} A| = \|A\|_1, \tag{30.c}$$

$$\det A = \lim_{n\to\infty} \det (Ae_p, e_q)^n_{p,q=1}, \tag{30.d}$$

*where* $\|A - I\|_1 < 1$ *and* $\{e_p\}$ *is any orthonormal basis;*

$$\det AB = \det A \det B \quad \text{if} \quad \|A - I\|_1 < 1 \quad \text{and} \quad \|B - I\|_1 < 1, \tag{30.e}$$

$$\det (I + D) = \exp\left[\sum_{n=1}^{\infty} \frac{(-1)^n}{r} \operatorname{Tr} (D^n)\right] \quad \text{if} \quad \|D\|_1 < 1. \tag{30.f}$$

With these facts in mind we shall discuss the properties of $F_k$.

*Proof of theorem 1.* It is clear by the definition that $\mathfrak{D}(F_k)$ is a linear space and $F_k$ is a linear functional on $\mathfrak{D}(F_k)$.

Using lemma 4 and the fact that $OVO^*$ has the same eigenvalues as $V$ for any orthogonal transformation $O$, we can easily verify the invariance of $\mathfrak{D}(F_k)$ and $F_k$ under translations and orthogonal transformations.

Let $C$ be nearly isometric. Then we have $Cx = a + A \cdot x$, where $A$ is nearly orthogonal. Applying Neumann's decomposition to $A$, we can write $C$ as

$$Cx = a + BOx \tag{31}$$

where $O$ is orthogonal and $B$ is a strictly positive definite symmetric operator with

$$\|B - I\|_\alpha < \infty \qquad \text{for some } \alpha \text{ such that } 0 < \alpha < 1. \tag{32}$$

Since we have already proved the invariance under translations and orthogonal transformations, it is enough to discuss the case $C = B$ in (31).

Let $\{\beta_\nu\}$ be the eigenvalues of $B - I$. Then we have $\beta_\nu > -1$ and $\sum_\nu |\beta_\nu|^\alpha < \infty$. The second inequality implies that $0 < \gamma_* < 1 + \beta_\nu < \gamma^* < \infty$ with $\gamma_*$ and $\gamma^*$ independent of $\nu$. Thus we have

$$\|B^{1/n} - I\|_\alpha \leq \|B - I\|_\alpha < \infty \qquad \text{for every} \quad n = 1, 2, \cdots, \tag{33}$$

and

$$\|B^{-2/n} - I\|_1 = \sum_\nu |(1 + \beta_\nu)^{-2/n} - 1| < 1 \qquad \text{for } n \text{ sufficiently large.} \tag{34}$$

If we can prove our theorem for $C = B^{1/n}$, then we can verify it for $C = B$ by applying $B^{1/n}$ $n$ times and noticing $J(B) = \prod (1 + \beta_\nu) = (\prod (1 + \beta_\nu)^{1/n})^n = J(B^{1/n})^n$. Therefore, in order to prove that

$$f \in \mathfrak{D}(F_k) \Rightarrow f_B \equiv f(Bx) \in \mathfrak{D}(F_k) \quad \text{and} \quad F_k(f_B) = J(B)^{-1}F_k(f), \tag{35}$$

we can assume, in addition to (32), that

$$\|B^{-2} - I\| < 1. \tag{36}$$

Let $\{v_\nu\}$, $\{\beta_\nu\}$, and $\{w_\nu\}$ be the eigenvalues of $V$, $B - I$, and $BVB$ respectively. By the definition of $F_k$, we can derive (35) easily from the fact that

$$\lim_V \frac{\prod_\nu \left[1 + \frac{w_\nu}{hi}\right]^{1/2}}{\prod_\nu \left[1 + \frac{v_\nu}{hi}\right]^{1/2}} = \prod_\nu (1 + \beta_\nu), \quad (= \det B). \tag{37}$$

To prove this, we shall consider the complexified Hilbert space $H_k = E_k + iE_k$ and denote the complexification of $V$ and that of $B$ with the same notations as we remarked before.

First, we shall derive the identity

$$\frac{\prod_\nu \left[1 + \frac{tw_\nu}{hi}\right]^{1/2}}{\prod_\nu \left[1 + \frac{tv_\nu}{hi}\right]^{1/2}} = \det B \exp\left\{\frac{1}{2} \sum_{n=1}^{\infty} \frac{(-1)^n}{n} \operatorname{Tr}(D_t^n)\right\}, \tag{38}$$

where

$$D_t = \left(I + \frac{tV}{hi}\right)^{-1} (B^{-2} - I). \tag{39}$$

Using lemma 7, for $0 \leq t \leq 1$ we have

$$\begin{aligned} \frac{\prod_\nu \left(1 + \frac{tw_\nu}{hi}\right)}{\prod_\nu \left(1 + \frac{tv_\nu}{hi}\right)} &= \frac{\det\left(I + \frac{tBVB}{hi}\right)}{\det\left(I + \frac{tV}{hi}\right)} = \frac{\det B \det\left(B^{-2} + \frac{tV}{hi}\right) \det B}{\det\left(I + \frac{tV}{hi}\right)} \\ &= (\det B)^2 \det\left[\left(I + \frac{tV}{hi}\right)^{-1} \left(B^{-2} + \frac{tV}{hi}\right)\right] \\ &= (\det B)^2 \det[I + D_t] \\ &= (\det B)^2 \exp\left\{\sum_{n=1}^{\infty} \frac{(-1)^n}{n} \operatorname{Tr}(D_t^n)\right\}, \end{aligned} \tag{40}$$

where we should notice

$$\left\|B^{-2} + \frac{tV}{hi} - I\right\|_1 = \|B^{-2} - I\|_1 + \left\|\frac{tV}{hi}\right\|_1 < \infty, \tag{41}$$

$$\text{(42)}\qquad \|D_t\|_1 = \left\| \left(I + \frac{t^2V^2}{h^2}\right)^{-1/2} |B^{-2} - I| \right\|_1, \qquad \text{by (30.b)},$$

$$\leq \left\| \left(I + \frac{t^2V^2}{h^2}\right)^{-1/2} \right\| \|B^{-2} - I\|_1, \qquad \text{by (30.b)},$$

$$\leq \|B^{-2} - I\|_1 < 1,$$

and so

$$\text{(43)}\qquad |\mathrm{Tr}\,(D_t^n)| \leq \|D_t^n\|_1 \leq \|D_t\|_1^n \leq \|B^{-2} - I\|_1^n, \qquad \text{by (30.b)}.$$

It follows from (40) that (38) holds with the sign $+$ or $-$ in front of det $B$. By our convention for $\sqrt{z}$ (section 1), the left side of (38) is continuous in $t \in [0, 1]$. The right side of (38) is also continuous in $t \in [0, 1]$, because the infinite series is convergent uniformly on $0 \leq t \leq 1$ by virtue of (42) and (43). Since the right side of (38) never vanishes, the $\pm$ sign remains unchanged as $t$ moves from 0 to 1. But both sides of (38) are positive at $t = 0$ and so (38) holds for every $t \in [0, 1]$.

Setting $t = 1$, we have

$$\text{(44)}\qquad \frac{\prod_\nu \left[1 + \frac{w_\nu}{hi}\right]^{1/2}}{\prod_\nu \left[1 + \frac{v_\nu}{hi}\right]^{1/2}} = \det B \exp\left\{\tfrac{1}{2} \sum_{n=1}^{\infty} \frac{(-1)^n}{n} \mathrm{Tr}\,(D(V)^n)\right\},$$

where $D(V) = (I + V/hi)^{-1}(B^{-2} - I)$.

Since we have, by (43),

$$\text{(45)}\qquad \left|\sum_{n=1}^{\infty} \frac{(-1)^n}{n} \mathrm{Tr}\,(D(V)^n)\right| \leq \sum_{n=1}^{\infty} \|D(V)\|_1^n$$

$$\leq \|D(V)\|_1 \sum_{n=1}^{\infty} \|B^{-2} - I\|_1^{n-1}$$

$$\leq \frac{\|D(V)\|_1}{1 - \|B^{-2} - I\|_1},$$

the proof will be completed, if we prove

$$\text{(46)}\qquad \lim_V \|D(V)\|_1 = 0.$$

Using (30.b), we shall evaluate $D(V)$ obtaining

$$\text{(47)}\qquad \|D(V)\|_1 = \left\| \left(I + \frac{V^2}{h^2}\right)^{-1/2} |B^{-2} - I| \right\|_1$$

$$\leq c_1 \left\| \left(I + \frac{V^2}{h^2}\right)^{-1/2} |B - I| \right\|_1 \quad (c_1 = \|B^{-2}(B + I)\| < \infty)$$

$$\leq c_1c_2 \left\| \left(I + \frac{V^2}{h^2}\right)^{-1/2} |B - I|^{1-\alpha} \right\| \quad (c_2 = \|\, |B - I|^\alpha\|_1 = \|B - I\|_\alpha < \infty).$$

Using the spectral decomposition of $V$ and noticing $2(1 + \lambda^2) \geq (1 + \lambda)^2$, we have

$$\left\| \left(I + \frac{V^2}{h^2}\right)^{-1/2} x \right\| \leq \sqrt{2} \left\| \left(I + \frac{V}{h}\right)^{-1} x \right\|, \tag{48}$$

and so

$$\|D(V)\|_1 < \sqrt{2}\, c_1 c_2 \left\| \left(I + \frac{V}{h}\right)^{-1} |B - I|^{1-\alpha} \right\|. \tag{49}$$

Hence, it follows by lemma 3 that $V > \ell|B - I|$ implies that

$$\|D(V)\|_1 \leq \sqrt{2}\, c_1 c_2 \left\| \left(I + \frac{\ell}{h} |B - I|\right)^{-1} |B - I|^{1-\alpha} \right\|. \tag{50}$$

Using the spectral decomposition of $B$, we have

$$\|D(V)\|_1 \leq \sqrt{2}\, c_1 c_2 \sup_{\lambda \geq 0} \left(1 + \frac{\ell\lambda}{h}\right)^{-1} \lambda^{1-\alpha} \leq \sqrt{2}\, c_1 c_2 \left(\frac{h}{\ell}\right)^{1-\alpha}. \tag{51}$$

Taking $\ell = \ell(\epsilon)$ large enough, we have $\|D(V)\|_1 < \epsilon$ for $V > \ell(\epsilon)|B - I|$, which proves (46).

PROOF OF THEOREM 2. Setting

$$F(a, V) = \prod_\nu \sqrt{1 + \frac{v_\nu}{hi}} \int_{E_k} \exp\left(\frac{i}{2h} |x|^2\right) \int_{E_k} e^{i(x,y)} \mu(dy) N(dx: a, V), \tag{52}$$

and changing the order of integration, we have

$$F(a, V) = \int_{E_k} I(y: a, V) \mu(dy), \tag{53}$$

where

$$\begin{aligned} I(y: a, V) &= \prod_\nu \sqrt{1 + \frac{v_\nu}{hi}} \int_{E_k} \exp\left[\frac{i}{2h} |x|^2 + i(x, y)\right] N(dx: a, V) \\ &= \prod_\nu \sqrt{1 + \frac{v_\nu}{hi}} \int_{E_k} \exp\left[\frac{i}{2h} |x + a|^2 + i(x + a, y)\right] N(dx: 0, V). \end{aligned} \tag{54}$$

Let $\{e_\nu\}$ be the eigenvectors of $V$ corresponding to $\{\lambda_\nu\}$ and write $x_\nu$, $y_\nu$, and $a_\nu$ for $(x, e_\nu)$, $(y, e_\nu)$, and $(a, e_\nu)$, respectively. Noticing that

$$\exp\left[\frac{i}{2h} |x + a|^2 + i(x + a, y)\right] = \prod_\nu \exp\left[\frac{i}{2h} (x_\nu + a_\nu)^2 + i(x_\nu + a_\nu) y_\nu\right] \tag{55}$$

and using lemmas 1 and 2, we can get

$$I(y: a, V) = \exp\left[\frac{h}{2i} \sum_\nu \frac{1}{v_\nu + hi} \left(y_\nu + \frac{a_\nu}{hi}\right)^2\right] \exp\left[\frac{h}{2i} |y|^2\right] \tag{56}$$

by usual computation. Therefore,

$$\begin{aligned} &\left|F(a, V) - \int \exp\left(\frac{h}{2i} |y|^2\right) \mu(dy)\right| \\ &\qquad \leq \int \left| \exp\left[\frac{h}{2i} \sum_\nu \frac{1}{v_\nu + hi} \left(y_\nu + \frac{a_\nu}{h}\right)^2 - 1\right] \right| |\mu|(dy), \end{aligned} \tag{57}$$

where $|\mu|$ is the absolute variation of $\mu$.

Writing $A(y: a, V)$ for the inside of the bracket, we have

(58.a) $\quad \operatorname{Re} A(y: a, V) \leq 0, \quad \text{and so} \quad |e^{A(y:a,V)} - 1| \leq 2,$

and

(58.b) $$|A(y: a, V)| \leq h \sum_\nu \frac{1}{v_\nu + h}\left(y_\nu + \frac{a_\nu}{h}\right)^2, \quad (\text{by } v_\nu + h \leq 2|v_\nu + hi|),$$
$$\leq h\left|(V + hI)^{-1/2}\left(y + \frac{a}{h}\right)\right|^2.$$

In order to prove our theorem, it is enough to find $V_0 = V_0(\epsilon)$ for every $\epsilon > 0$ such that the integral in (35) is less than $\epsilon$ for every $V > V_0$.

Since $|\mu|$ is a measure with $|\mu|(E_\infty) < \infty$, we have a compact subset $K$ of $E_\infty$ such that

(59) $$|\mu|(E_\infty - K) < \frac{\epsilon}{4}$$

by a theorem due to Prohorov [6]. Take $\delta > 0$ such that $e^{2\delta h} - 1 < \epsilon/2$, or find a finite set $\{y', y'', \cdots, y^{(n)}\}$ such that every $y \in K$ is within the distance $h^{1/2}\delta$ from some $y^{(\nu)}$, and choose $V_0 \in \mathfrak{V}$ such that

(60) $$\left|V_0^{-1/2}\left(y^{(\nu)} + \frac{a}{h}\right)\right| < \delta, \qquad \nu = 1, 2, \cdots, n.$$

Such a $V_0$ can be constructed easily.

If $|y - y^{(\nu)}| < \delta h$, we have

(61) $$\left|(V_0 + hI)^{-1/2}\left(y + \frac{a}{h}\right)\right| \leq \left|(V_0 + hI)^{-1/2}\left(y^{(\nu)} + \frac{a}{h}\right)\right| + |(V_0 + hI)^{-1/2}(y - y^{(\nu)})|$$
$$\leq \left|V_0^{-1/2}\left(y^{(\nu)} + \frac{a}{h}\right)\right| + h^{-1/2}|y - y^{(\nu)}| < 2\delta.$$

Hence, we have

(62) $$\left|(V_0 + hI)^{-1/2}\left(y + \frac{a}{h}\right)\right| \leq 2\delta \qquad \text{for} \quad y \in K.$$

Therefore,

(63) $$\left|(V + hI)^{-1/2}\left(y + \frac{a}{h}\right)\right| \leq 2\delta \qquad \text{for} \quad y \in K \quad \text{and} \quad V > V_0$$

by lemma 3, observing that $V > V_0$ implies

(64) $$(Vx, x) + h(x, x) \geq (V_0x, x) + h(x, x) \geq h(x, x),$$
$$|(V + hI)^{1/2}x|^2 \geq |(V_0 + hI)^{1/2}x|^2 \geq h|x|^2.$$

Using (58.a) and (63), we have, for $V > V_0$,

(65) $$\int |e^{A(y:a,V)} - 1|\mu(dy)$$
$$\leq \int_K + \int_{E_\infty - K} \leq (e^{2\delta h} - 1)\mu(K) + 2\mu(E_\infty - K) \leq (e^{2\delta h} - 1) + \frac{\epsilon}{2} < \epsilon,$$

which completes the proof.

Since theorems 3, 3′, and 3″ can be easily proved, we shall omit the proof.

### 3. Application to the Feynman integral

In section 1 we formulated the Feynman integral in terms of our generalized measure $F_k$. We shall now carry out the computation in the three cases mentioned there.

Let $L_0^2[s, t]$ be the space of all square summable functions $x(\tau)$ on the interval $s \leq \tau \leq t$,

$$\int_s^t x(\tau)\, d\tau = 0. \tag{66}$$

$L_0^2[s, t]$ is a Hilbert space $E_\infty$ with the usual norm in the space $L^2[s, t]$. Recalling the definition of the metric in $\Gamma$ in section 1, we can see that

$$\begin{aligned} &\Phi: \Gamma \to L_0^2[s, t], \\ &(\Phi\gamma)(\tau) = \sqrt{m}\left[\gamma'(\tau) - \frac{b-a}{t-s}\right], \qquad s \leq \tau < t \end{aligned} \tag{67}$$

defines an isometric mapping from $\Gamma$ onto $L_0^2[s, t]$. Introducing $e_{s,t,\tau} \in L_0^2[s, t]$ by

$$e_{s,t,\tau}(\sigma) = \begin{cases} \dfrac{t-\tau}{t-s}, & s \leq \sigma \leq \tau \leq t, \\ -\dfrac{\tau - s}{t-s}, & s \leq \tau \leq \sigma \leq t, \end{cases} \tag{68}$$

for $x \in L_0^2[s, t]$, we have the following relations:

$$\int_s^\tau x(\sigma)\, d\sigma = (e_{s,t,\sigma}, x), \tag{69.a}$$

$$\gamma(\tau) = \gamma_0(\tau) + \frac{1}{\sqrt{m}}(e_{s,t,\tau}, x), \qquad \gamma_0(\tau) = \frac{(t-\tau)a + (\tau - s)b}{t-s}, \tag{69.b}$$

$$\int_s^t m\gamma'(\tau)^2\, d\tau = |x|^2 + \frac{m(b-a)^2}{t-s}. \tag{69.c}$$

Therefore,

$$\mathfrak{A}(\gamma) = \tfrac{1}{2}|x|^2 + \tfrac{1}{2}\frac{m(b-a)^2}{t-s} - \int_s^t m\cdot u\left[\gamma_0(\tau) + \frac{1}{\sqrt{m}}(e_{s,t,\tau}, x)\right] d\tau. \tag{69.d}$$

Let $\mathfrak{M}_\infty$ denote the class of complex measures of bounded absolute variation on $L_0^2[s, t]$ and $\mathfrak{F}\mathfrak{M}_\infty$ the class of all functions

$$g(x) = \mathfrak{F}\mu(x) \equiv \int_{L_0^2[s,t]} e^{i(x,y)}\mu(dy), \qquad \mu \in \mathfrak{M}_\infty \tag{70}$$

where $\mathfrak{F}$ denotes the Fourier transform. The following fact will be useful here:

$$g \in \mathfrak{F}\mathfrak{M} \Rightarrow \varphi(g) \in \mathfrak{F}\mathfrak{M}_\infty \qquad \text{for every entire function } \varphi; \tag{71}$$

in fact, if $g = \mathfrak{F}\mu$ and $\varphi(\xi) = \sum \alpha_n \xi^n$, then $\nu = \sum \alpha_n \mu^{*n}$, ($\mu^{*n} = n$ times convolution of $\mu$) converges in the norm of total absolute variation and $\varphi(t) = \mathfrak{F}\cdot\nu$.

*Case 1. Assume that* $u(q) = \int_{-\infty}^{\infty} e^{iq\xi}\theta(d\xi)$, *where* $\theta$ *is a complex measure of bounded absolute variation on* $(-\infty, \infty)$.

First we shall prove that

$$\exp\left(\frac{i}{h}\,\mathcal{A}(\gamma)\right) \in \mathcal{E}_\infty. \tag{72}$$

According to theorem 2, it is sufficient to prove that

$$\exp\left\{-\frac{im}{h}\int_s^t u\left[\gamma_0(\tau) + \frac{1}{\sqrt{m}}\,(e_{s,t,\tau}, x)\right]d\tau\right\} \in \mathcal{FM}_\infty. \tag{73}$$

By virtue of (71), it is also enough to show that

$$\int_s^t u\left[\gamma_0(\tau) + \frac{1}{\sqrt{m}}\,(e_{s,t,\tau}, x)\right]d\tau \in \mathcal{FM}_\infty. \tag{74}$$

The left side is

$$\int_s^t \int_{-\infty}^{\infty} \exp\left[i\left(\frac{\xi}{\sqrt{m}}\,e_{s,t,\tau}, x\right)\right] e^{i\xi\gamma_0(\tau)}\theta(d\xi)\cdot d\tau, \tag{75}$$

and so it is the Fourier transform of the measure $\mu \in \mathcal{M}_\infty$,

$$\mu(\cdot) = \int_s^t \int_{-\infty}^{\infty} \delta\left(\cdot, \frac{\xi}{\sqrt{m}}\,e_{s,t,\tau}\right) e^{i\xi\gamma_0(\tau)}\theta(d\xi)\,d\tau, \tag{76}$$

$\delta(\cdot, y)$ being the $\delta$-measure concentrated at $y$. Therefore, we shall have (74).

Using theorem 2 we can see that

$$\begin{aligned} G(t, b|s, a) &= \frac{\sqrt{m}}{\sqrt{2\pi hi(t-s)}}\int_\Gamma \exp\left(\frac{i}{h}\,\mathcal{A}(\gamma)\right)F_\infty(d\gamma) \\ &= \frac{\sqrt{m}}{\sqrt{2\pi hi(t-s)}}\exp\left[-\frac{m(b-a)^2}{2hi(t-s)}\right]\sum_n \frac{1}{n!}\left(-\frac{im}{h}\right)^n \int_s^t \cdots \int_s^t Q_n\,d\tau_1\cdots d\tau_n \end{aligned} \tag{77}$$

where

$$\begin{aligned} Q_n &= Q_n(\tau_1, \tau_2, \cdots, \tau_n; t, b, s, a) \\ &= \int_{-\infty}^{\infty}\cdots\int \exp\left\{i\sum_{\nu=1}^n \gamma_0(\tau_\nu)\xi_\nu + \frac{h}{2i}\int_s^t \left|\sum_{\nu=1}^n \frac{\xi_\nu}{\sqrt{m}}\,e_{s,t,\tau_\nu}(\sigma)\right|^2 d\sigma\right\}\theta(d\xi_1)\cdots\theta(d\xi_n). \end{aligned} \tag{78}$$

If $t - s$ is small, then

$$\begin{aligned} &G(t, b|s, a) \\ &= \frac{\sqrt{m}}{\sqrt{2\pi hi(t-s)}}\exp\left[-\frac{m(b-a)^2}{2hi(t-s)}\right]\left[1 - \frac{im}{h}(t-s)\,\frac{u(a)+u(b)}{2} + o(t-s)\right] \end{aligned} \tag{79}$$

where $o$ does not depend on $(a, b)$ as long as they stay in a compact set. Hence, it follows that if $\varphi$ is a $C_2$-function with compact support, we have

$$\lim_{t\downarrow s}\frac{1}{t-s}\left[\int G(t, b|s, a)\varphi(a)\,da - \varphi(b)\right] = \frac{hi}{2m}\varphi''(b) - \frac{im}{h}\,u(b)\varphi(b). \tag{80}$$

We have a *composition rule*

$$G(t, b|s, a) = \int_{-\infty}^{\infty} G(t, b|u, c)G(u, c|s, a)\, dc \tag{81}$$

where the integral is to be understood in an improper sense, namely

$$\sqrt{2\pi hi} \int_{E_1} G(t, b|u, c)G(u, c|s, a)F_1(dc) \tag{82}$$

or

$$\lim_{n\to\infty} \int_{-\infty}^{\infty} G(t, b|u, c)G(u, c|s, a)e^{-c^2/2n}\, dc. \tag{83}$$

The composition rule can be written in terms of $F_\infty$:

$$\begin{aligned} &\int_{\Gamma(t,b|s,a)} \exp\left(\frac{i}{h}\,\mathfrak{A}(\gamma)\right) F_\infty(d\gamma) \\ &\quad = \sqrt{\frac{m(t-s)}{(u-s)(t-u)}} \int_{E_1} \int_{\Gamma(t,b|u,c)} \exp\left(\frac{i}{h}\,\mathfrak{A}(\gamma_1)\right) F_\infty(d\gamma_1) \\ &\quad\quad \times \int_{\Gamma(u,c|s,a)} \exp\left(\frac{i}{h}\,\mathfrak{A}(\gamma_2)\right) F_\infty(d\gamma_2)F_1(dc), \end{aligned} \tag{81'}$$

which we can get by applying theorem 3″ first to

$$C\colon \Gamma \to E_1; \qquad C(\gamma) = \gamma(u); \qquad J(C) = \sqrt{\frac{m(t-s)}{(u-s)(t-u)}}, \tag{84}$$

and then to

$$\begin{aligned} &C_1\colon \Gamma_c \equiv \{\gamma \in \Gamma\colon \gamma(u) = c\} \to \Gamma(u, c|s, a); \\ &C_1(\gamma) = \text{restriction of } \gamma \text{ onto } [s, u], \qquad J(C_1) = 1. \end{aligned} \tag{85}$$

*Case 2.* This case, $u(q) = c\cdot q$, $-\infty < c < \infty$, is not included in the first case, because $u(q)$ is unbounded. By a simple computation, we have

$$\mathfrak{A}(\gamma) = \tfrac{1}{2}|x|^2 + \tfrac{1}{2}\,\frac{m(b-a)^2}{t-s} - \tfrac{1}{2}mc(a+b)(t-s) - \sqrt{m}\, c(y, x), \tag{86}$$

where $y(\sigma) \equiv (1/2)(t+s) - \sigma$. Therefore, it is obvious that $\exp\{(i/h)\mathfrak{A}(\gamma)\} \in \mathcal{E}$, and we get

$$\begin{aligned} &G(t, b|s, a) \\ &= \frac{\sqrt{m}}{\sqrt{2\pi hi(t-s)}} \int_\Gamma \exp\left(\frac{i}{h}\,\mathfrak{A}(\gamma)\right) F_\infty(d\gamma) \\ &= \frac{\sqrt{m}}{\sqrt{2\pi hi(t-s)}} \exp\left\{-\frac{m}{2hi}\left[\frac{(b-a)^2}{t-s} - c(t-s)(a+b) - \frac{c^2}{12}(t-s)^3\right]\right\} \end{aligned} \tag{87}$$

observing that

$$|y|^2 = \int_s^t \left(\frac{t+s}{2} - \sigma\right)^2 d\sigma = \tfrac{1}{12}(t-s)^3. \tag{88}$$

*Case 3.* In the case where $u(q) = cq^2$, $c > 0$, we can prove that

$$(89) \qquad e^{\frac{i}{\hbar}\mathfrak{a}(\gamma)} = f(Ax), \qquad \in \mathcal{E}_\infty; \quad A \text{ is nearly isometric,}$$

and so this belongs to $\mathfrak{D}(F_\infty)$.

Applying (69.d) to our case, we have

$$(90) \qquad \mathfrak{a}(\gamma) = \tfrac{1}{2}|x|^2 + \tfrac{1}{2}\frac{m(b-a)^2}{t-s} - \int_s^t mc\left[\gamma_0(\tau) + \frac{1}{\sqrt{m}}(e_{s,t,\tau}, x)\right]^2 d\tau,$$

$$= \tfrac{1}{2}|x|^2 + \tfrac{1}{2}\frac{m(b-a)^2}{t-s} - mcI_1 - 2\sqrt{m}\,cI_2 - cI_3,$$

where

$$(91) \qquad I_1 = \int_s^t \gamma_0(\tau)^2\, d\tau = \tfrac{1}{3}(t-s)(a^2 + ab + b^2),$$

$$(92) \qquad I_2 = \int_s^t \gamma_0(\tau)(e_{s,t,\tau}, x)\, d\tau = \int_s^t \gamma_0(\tau) \int_s^\tau x(\sigma)\, d\sigma\, d\tau,$$

$$(93) \qquad I_3 = \int_s^t (e_{s,t,\tau}, x)^2\, d\tau = \int_s^t \left(\int_s^\tau x(\sigma)\, d\sigma\right)^2 d\tau.$$

Let us set

$$(94) \qquad C_n(\tau) = \sqrt{\frac{2}{t-s}} \cos\frac{n\pi(\tau - s)}{t-s}, \qquad n = 0, 1, 2, \cdots,$$

$$(95) \qquad S_n(\tau) = \sqrt{\frac{2}{t-s}} \sin\frac{n\pi(\tau - s)}{t-s}, \qquad n = 1, 2, \cdots.$$

Then each of $\{C_n(\tau), n = 0, 1, 2, \cdots\}$ and $\{S_n(\tau), n = 1, 2, \cdots\}$ is a complete orthonormal system in $L^2[s, t]$ and $\{C_n(\tau), n = 1, 2, \cdots\}$ is a complete orthonormal system in $L_0^2[s, t]$. Expanding $x$ as

$$(96) \qquad x(\tau) = \sum_{n=1}^\infty (x, C_n)\cdot C_n(\tau),$$

we can express $I_2$ and $I_3$ as follows.

$$(97) \qquad I_2 = (y, x), \qquad y = \sum_{n=1}^\infty \frac{\sqrt{2}(t-s)^{3/2}(a + (-1)^{n-1}b)}{n^2\pi^2}\cdot C_n,$$

$$(98) \qquad I_3 = (Bx, x), \qquad Bx = \sum_{n=1}^\infty \frac{(t-s)^2}{n^2\pi^2}(x, C_n)\cdot C_n.$$

Thus we have

$$(99) \qquad \mathfrak{a}(\gamma) = \tfrac{1}{2}|x|^2 + \tfrac{1}{2}\frac{m(b-a)^2}{t-s} - \frac{mc}{3}(t-s)(a^2 + ab + b^2)$$

$$- 2\sqrt{m}\,c(y, x) - c(Bx, x) = \tfrac{1}{2}|Ax|^2 + \tfrac{1}{2}\frac{m(b-a)^2}{t-s}$$

$$- \frac{mc}{3}(t-s)(a^2 + ab + b^2) + c(z, Ax),$$

where

$$(100) \qquad Ax = (I - 2cB)^{1/2}x = \sum_{n=1}^\infty \left(1 - \frac{2c(t-s)^2}{n^2\pi^2}\right)^{1/2}(x, C_n)\cdot C_n,$$

(101) $$z = (I - 2cB)^{-1/2}y$$
$$= \sum_{n=1}^{\infty} \left(1 - \frac{2c(t-s)^2}{n^2\pi^2}\right)^{-1/2} \frac{\sqrt{2}(t-s)^{3/2}(a + (-1)^{n-1}b)}{n^2\pi^2} C_n,$$

as far as $t - s < \pi/\sqrt{2c}$. Since we have

(102) $$(\|A - I\|_\alpha)^\alpha = \mathrm{Tr}\,[|A - I|^\alpha]$$
$$= \sum_{n=1}^{\infty} \left|\left(1 - \frac{2c(t-s)^2}{n^2\pi^2}\right)^{-1/2} - 1\right|^\alpha \sim \sum_{n=1}^{\infty} n^{-2\alpha},$$

we get $\|A - I\|_\alpha < \infty$ for every $\alpha > 1/2$. Therefore, it is obvious $\|A - I\|_{2/3} < \infty$, which shows that $A$ is nearly isometric because $A$ is a bounded symmetric operator. By theorems 1 and 2 we have $\exp((i/h)\mathfrak{A}(\gamma) \in \mathfrak{D}(F_\infty))$ and

(103) $$G(t, b|s, a) = \frac{\sqrt{m}}{\sqrt{2\pi hi(t-s)}} \int_\Gamma \exp\left(\frac{i}{h}\,\mathfrak{A}(\gamma)F_\infty(d\gamma)\right)$$
$$= \frac{\sqrt{m}}{\sqrt{2\pi hi(t-s)}} J(A)^{-1}$$
$$\times \exp\left\{\frac{i}{2h}\frac{m(b-a)^2}{t-s} - \frac{imc}{3h}(t-s)(a^2 + ab + b^2) + \frac{h}{2i}\left|\frac{ic}{h}\right|^2 |z|^2\right\}.$$

Using

(104) $$\frac{\sin z}{z} = \prod_{n=1}^{\infty} \left(1 - \frac{z^2}{n^2\pi^2}\right),$$

we have

(105) $$J(A) = \prod_n \left(1 - \frac{2c(t-s)}{n^2\pi^2}\right)^{1/2} = \sqrt{\frac{\sin[\sqrt{2c}(t-s)]}{\sqrt{2c}(t-s)}}.$$

Observing that

(106) $$|z|^2 = \sum_{n=1}^{\infty} \left(1 - \frac{2c(t-s)^2}{n^2\pi^2}\right)^{-1} \frac{2(t-s)^3(a + (-1)^{n-1}b)^2}{n^4\pi^4}$$
$$= \frac{t-s}{c} \sum_{n=1}^{\infty} \left(\frac{1}{n^2\pi^2 - 2c(t-s)^2} - \frac{1}{n^2\pi^2}\right)(a^2 + b^2)$$
$$+ 2\,\frac{t-s}{c} \sum_{n=1}^{\infty} \left(\frac{(-1)^{n-1}}{n^2\pi^2 - 2c(t-s)^2} - \frac{(-1)^{n-1}}{n^2\pi^2}\right) ab$$

and using

(107) $$\sum_{n=1}^{\infty} \frac{1}{n^2\pi^2 - x^2} = \frac{1}{2x}\left(\frac{1}{x} - \cot x\right),$$

(108) $$\sum_{n=1}^{\infty} \frac{(-1)^{n-1}}{n^2\pi^2 - x^2} = \frac{1}{2x}\left(\frac{1}{\sin x} - \frac{1}{x}\right),$$

(109) $$\sum_{n=1}^{\infty} \frac{1}{n^2} = \frac{\pi^2}{6}, \qquad \sum_{n=1}^{\infty} \frac{(-1)^{n-1}}{n^2} = \frac{\pi^2}{12},$$

we have

$$(110) \quad |z|^2 = \frac{t-s}{c}\left[\frac{1}{2\sqrt{2c}(t-s)}\left(\frac{1}{\sqrt{2c}(t-s)}\cot(\sqrt{2c}(t-s))\right) - \tfrac{1}{6}\right](a^2+b^2)$$
$$+ 2\frac{t-s}{c}\left[\frac{1}{2\sqrt{c}(t-s)}\left(\frac{1}{\sin[2c(t-s)]} - \frac{1}{\sqrt{2c}(t-s)}\right) - \tfrac{1}{12}\right]ab.$$

Putting (105) and (110) in (103), we get

$$(111) \quad G(t, b|s, a)$$
$$= \sqrt{\frac{m\sqrt{2c}}{2\pi hi \sin[\sqrt{2c}(t-s)]}} \exp\left\{\frac{im\sqrt{2c}}{2h}\cdot\frac{(a^2+b^2)\cos[\sqrt{2c}(t-s)] - 2ab}{\sin[\sqrt{2c}(t-s)]}\right\},$$

for $t - s < \pi/\sqrt{2c}$.

## REFERENCES

[1] R. P. FEYNMAN, "Space-time approach to non-relativistic quantum mechanics," *Rev. Mod. Phys.*, Vol. 20 (1948), pp. 368–387.

[2] I. M. GELFAND and A. M. YAGLOM, "Integration in function space and its applications in quantum physics," *Uspehi Mat. Nauk.*, Vol. II (1956), pp. 77–114. (English translation, *J. Mathematical Phys.*, Vol. 1 (1960), pp. 48–69.)

[3] L. GROSS, "Harmonic analysis on Hilbert space," *Mem. Amer. Math. Soc.*, Vol. 46 (1963), ii and 62 pp.

[4] K. ITÔ, "Wiener integral and Feynman integral," *Proceedings of the Fourth Berkeley Symposium on Mathematical Statistics and Probability*, Berkeley and Los Angeles, University of California Press, 1961, Vol. 2, pp. 227–238.

[5] S. S. KURODA, "On a generalization of the Weinstein-Aronszajn formula and the infinite determinant," *Sci. Papers College Gen. Ed. Univ. Tokyo*, Vol. 11 (1961), pp. 1–12.

[6] YU. V. PROHOROV, "Convergence of random processes and limit theorems in probability theory," *Teor. Verojatnost. i Primenen.*, Vol. 1 (1946), pp. 177–238. (English translation, pp. 156–214.)

[7] V. SAZONOV, "A remark on characteristic functions," *Teor. Verojatnost. i Primenen.*, Vol. 3 (1958), pp. 201–205. (English translation, pp. 188–192.)

Reprinted from
*Proc. Fifth Berkeley Symp. Math. Stat. Probab.* **II,** 145–161 (1965)

MATH. SCAND. 22 (1968), 209–223

# ON THE OSCILLATION FUNCTIONS OF GAUSSIAN PROCESSES

KIYOSI ITÔ and MAKIKO NISIO

## 1. Introduction and results obtained.

By a *Gaussian process* we shall understand a separable, measurable, jointly Gauss distributed process with the time parameter on $[0,1]$, continuous in the (second order) mean.

As to the regularity of the sample path of a stationary Gaussian process, we have Yu. K. Belayev's theorem of alternatives [1] which reads as follows: the sample function (path) of a stationary Gaussian process is either continuous with probability one or unbounded on every interval with probability one.

What will happen for a non-stationary Gaussian process? Our purpose is to answer this question.

Given a Gaussian process $x=x(t,\omega)$, $0 \leqq t \leqq 1$, $\omega \in \Omega(\mathscr{B},P)$, we shall define the *oscillation function* of $x$ by

$$W_x(t,\omega) = \lim_{\varepsilon \downarrow 0} \sup_{u,v \in (t-\varepsilon,\, t+\varepsilon) \cap [0,1]} |x(v,\omega)-x(u,\omega)| ,$$

where we apply the usual convention $(+\infty)-(+\infty)=(-\infty)-(-\infty)=0$, etc. Because of the separability of our process $x$, the supremum can be taken only for the $u$ and $v$ in the separant $Q$ of $x$ which is countable, so that $W_x(t,\omega)$ is measurable in $\omega$ for each $t \in [0,1]$.

An interesting fact is that the oscillation function of a Gaussian process is a deterministic function. Precisely speaking, we have the following theorem which will be proved in Section 2.

THEOREM 1. *There exists a function* $\alpha = \alpha_x(t)$, $0 \leqq t \leqq 1$, *which does not depend on* $\omega$, *such that*

$$P[W_x(t,\omega) = \alpha(t) \text{ for every } t \in [0,1]] = 1 .$$

In view of this theorem we call $\alpha$ the *oscillation function* of the Gaussian process $x$.

Received August 25, 1967.

Math. Scand. 22 — 14

Using the fact that the probability law of the process $x(t)-E(x(t))$, $0\leqq t\leqq 1$, is invariant under reflection, we shall prove the following theorem in Section 3:

THEOREM 2. *For each* $t\in[0,1]$, *we have*

$$P\left[\overline{\lim_{s\to t}}\,x(s)=x(t)+\tfrac{1}{2}\alpha(t),\ \underline{\lim_{s\to t}}\,x(s)=x(t)-\tfrac{1}{2}\alpha(t)\right]=1.$$

It is clear that, with probability one, the sample function of $x(t)$ is continuous at every point $t$ where $\alpha(t)$ vanishes.

We have also the following theorem which will be proved in Section 4.

THEOREM 3. *If, for some constant* $a$, $\alpha(t)\geqq a>0$ *on a dense subset* $D$ *of an open interval* $I\subset[0,1]$, *then*

$$P\left[\overline{\lim_{s\to t}}\,x(s)=\infty,\ \underline{\lim_{s\to t}}\,x(s)=-\infty \text{ for every } t\in I\right]=1.$$

In Section 5 and 6 we shall prove the following properties that characterize oscillation functions.

THEOREM 4. (a) *The oscillation function of a Gaussian process satisfies*

$(\alpha,1)$ $\qquad\alpha(t)$ *is upper semi-continuous,*

$(\alpha,2)$ $\qquad\{t:\ a\leqq\alpha(t)<\infty\}$ *is nowhere dense for every* $a>0$.

(b) *Conversely, given a function* $\alpha:[0,1]\to[0,\infty]$ *satisfying* $(\alpha,1)$ *and* $(\alpha,2)$, *we can construct a (not necessarily unique) Gaussian process whose oscillation function is* $\alpha$.

Let us derive Belayev's theorem of alternatives from Theorems 1 and 3. Suppose $x(t)$ is a stationary Gaussian process. The oscillation function $\alpha(t)$ of the restriction of $x$ to $0\leqq t\leqq 1$ is constant because of the stationarity. If the constant is 0, then almost all sample functions of $x$ are continuous; if it is positive, then almost all sample functions are, by Theorem 3, unbounded both below and above on every interval.

## 2. Proof of Theorem 1.

We can assume that $Ex(t)=0$, because $Ex(t)$ is continuous in $t$. Let $R(s,t)$, $t,s\in[0,1]$, be the covariance function of $x$. Then $R$ is real, symmetric, positive-definite and continuous on $I\times I$. Using Mercer's theorem, we can expand $R$ as follows:

$$R(t,s) = \sum_n \varphi_n(t)\,\varphi_n(s)/\lambda_n\,, \tag{2.1}$$

where $\{\lambda_n\}$ and $\{\varphi_n\}$ are, respectively, the positive eigenvalues and the corresponding real (normalized) eigenfunctions for the integral operator with the kernel $R(t,s)$, that is,

$$\varphi_n(t) = \lambda_n \int_0^1 R(t,s)\,\varphi_n(s)\,ds, \qquad 0 \leqq \lambda_1 \leqq \lambda_2 \leqq \ldots\,, \tag{2.2}$$

and the sum in (2.1) converges absolutely and uniformly on $I \times I$.

We shall define a sequence of random variables $x_n$, $n=1,2,\ldots$, as the Fourier coefficients of the sample function of $x(t)$ with respect to $\{\varphi_n\}$:

$$x_n = \int_0^1 x(t)\,\varphi_n(t)\,dt\,. \tag{2.3}$$

Observing that

$$E\left[\int_0^1 x(s)^2\,ds\right] = \int_0^1 R(s,s)\,ds < \infty\,,$$

we have

$$P\left[\int_0^1 x(s)^2\,ds < \infty\right] = 1\,,$$

so that $x_n$ is well-defined.

A simple computation shows that

$$E(x_n x_m) = 0,\ n \neq m, \qquad E(x_n{}^2) = \lambda_n{}^{-1}\,,$$

which implies that $x_n$, $n=1,2,\ldots$, are independent, each having the Gauss distribution with mean 0 and variance $\lambda_n{}^{-1}$. Observing that

$$E\left[\left|x(t) - \sum_1^N \varphi_n(t)\,x_n\right|^2\right] = R(t,t) - \sum_1^N \lambda_n{}^{-1}\varphi_n(t)^2 \to 0$$

as $n \to \infty$, we have

$$P\left[x(t) = \sum_{n=1}^{\infty} \varphi_n(t)\,x_n\right] = 1 \quad \text{for each } t; \tag{2.4}$$

the infinite series converges to $x(t)$ in the mean for each $t$ and so it converges with probability one for each $t$ because of the independence of $x_n$, $n=1,2,\ldots$.

Let us define the *maximum oscillation* $W_y(s,t,\omega)$ of a separable process on the interval $[s,t]$ by

(2.5) $$W_y(s,t,\omega) = \lim_{n\uparrow\infty} \lim_{p\uparrow\infty} \sup_{\substack{u,v\in(s-n^{-1},\, t+n^{-1})\cap[0,1] \\ |u-v|<p^{-1}}} |y(u,\omega)-y(v,\omega)| ,$$

where the interval $[0,1]$ can be replaced by the separant $Q$ of $y$.

We shall now prove that $W_x(s,t,\omega)$ is a constant; more precisely, that there exists a function $\alpha(s,t)$ independent of $\omega$ such that

(2.6) $$P[W_x(s,t,\omega)=\alpha(s,t)] = 1$$

for each pair $s \leqq t$.

Since $\varphi_j(t)$ is continuous in $t$, we have

(2.7) $$P[W_x(s,t,\omega)= W_{y_n}(s,t,\omega)] = 1 ,$$

where

$$y_n(t) = x(t) - \sum_{j=1}^{n} \varphi_j(t)x_j = \sum_{j=n+1}^{\infty} \varphi_j(t)x_j .$$

The separant $Q$ of $x$ is also that of $y_n$ because of the continuity of $\varphi_j(t)$. By virtue of (2.4), $y_n(t)$ is measurable with respect to $\mathscr{B}(x_k, k \geqq n)$ for each $t$. But $W_{y_n}(s,t,\omega)$ is measurable with respect to $\mathscr{B}(y_n(t), t \in Q)$. Since $Q$ is countable, $W_{y_n}(s,t,\omega)$ is measurable with respect to $\mathscr{B}(x_k, k \geqq n)$ and so $W_x(s,t,\omega)$ is also measurable with respect to $\mathscr{B}(x_k, k \geqq n)$ for every $n$, by (2.7). Since the $x_n$, $n=1,2,\ldots$, are independent, Kolmogorov's zero-one law shows that $W_x(s,t,\omega)$ is a constant with probability one.

We shall now strengthen (2.6) to get

(2.6′) $$P[W_x(s,t,\omega)=\alpha(s,t) \text{ for every pair } s \leqq t] = 1 .$$

It follows from (2.6) that

(2.6″) $$P[W_x(s,t,\omega)=\alpha(s,t) \text{ for every rational pair } s \leqq t] = 1 .$$

Since $W_x(s,t,\omega)$ is left-continuous in $s$ and right-continuous in $t$, (2.6″) implies (2.6′). Writing $W_x(t,\omega)$ and $\alpha(t)$ for $W_x(t,t,\omega)$ and $\alpha(t,t)$, respectively, we get from (2.6′)

(2.7) $$P[W_x(t,\omega)=\alpha(t) \text{ for every } t] = 1 .$$

## 3. Proof of Theorem 2.

We shall use the same notation as in Section 2. Using Kolmogorov's zero-one law in the same way as before, we can see that $\overline{\lim}_{s\to t}(x(s,\omega) - x(t,\omega))$ is a constant, say $\beta(t)$, with probability one. Since the process $y(t) \equiv -x(t)$ has the same Gaussian probability law as the process $x$, we have

$$\overline{\lim_{s\to t}}(-x(s)+x(t)) = \beta(t) ,$$

that is,

$$\varliminf_{s\to t}(x(s)-x(t)) = -\beta(t)$$

with probability one. By definition we have

$$\begin{aligned} W_x(t,\omega) &= \varlimsup_{s\to t} x(s) - \varliminf_{s\to t} x(s) \\ &= \varlimsup_{s\to t}(x(s)-x(t)) - \varliminf_{s\to t}(x(s)-x(t)) . \end{aligned}$$

Therefore, $\alpha(t)$ must be equal to $2\beta(t)$. This completes the proof of Theorem 2.

## 4. Proof of Theorem 3.

In the following lemma and throughout this paper an open subset of $[0,1]$ means a subset open in $[0,1]$. For example, $[0,u)$ is open and 0 is an interior point of this interval.

LEMMA 4.1. *Let $x(t)$, $0 \leqq t \leqq 1$, be a separable process continuous in probability, and $D$ a dense subset of an open subinterval $I$ of $[0,1]$. For each $t \in I$, we can then find a sequence $s_n \in D$ such that $s_n \to t$ and that*

$$P\left[\varlimsup_{n\to\infty} x(s_n) = \varlimsup_{s\to t} x(s)\right] = 1, \tag{4.1}$$

*and hence a fortiori*

$$P\left[\varlimsup_{\substack{s\to t\\ s\in D}} x(s) = \varlimsup_{s\to t} x(s)\right] = 1 . \tag{4.1'}$$

PROOF. Let $Q=\{t_n\}$ be a separant of $x(t)$ and let $\{U_n\}$ be a sequence of neighborhoods of $t$ converging to $t$. Let $\{y_n(\omega)\}$ be a sequence converging to $\bar{x}(t)=\varlimsup_{s\to t}x(t)$ strictly from below, for example

$$y_n(\omega) = \min(\bar{x}(t), n) - 1/n .$$

Since $Q$ is a separant of $x(t)$, we can find $u_{n1}, u_{n2}, \ldots, u_{np_n} \in U_n \cap Q$ such that

$$P[\max_k x(u_{nk}) > y_n(\omega)] > 1-2^{-n} .$$

By Borel–Cantelli's lemma we have

$$P[\lim_n \max_k x(u_{nk}) \geqq \bar{x}(t)] = 1 .$$

Writing $\{u_n\}$ for $\{u_{11},\ldots,u_{1p_1},u_{21},\ldots,u_{2p_2},\ldots\}$, we have

$$P\left[\overline{\lim_{n\to\infty}}\, x(u_n) \geqq \bar{x}(t)\right] = 1. \tag{4.2}$$

Since $x(t)$ is continuous in probability, we can find

$$s_n \in (u_n - 1/n, u_n + 1/n) \cap D$$

such that

$$P[|x(u_n) - x(s_n)| > 2^{-n}] < 2^{-n}, \qquad n = 1, 2, \ldots .$$

Using Borel–Cantelli's lemma again, we have

$$P[\overline{\lim}_n x(u_n) = \overline{\lim}_n x(s_n)] = 1, \tag{4.3}$$

which, combined with (4.2), implies (4.1) and so (4.1′).

We shall now prove Theorem 3. Using (4.1) and Theorem 2, we can see that the event

$$\Omega_1 = \left\{\omega : \overline{\lim_{\substack{s\to t\\ s\in D}}}\, x(s,\omega) = x(t,\omega) + \tfrac{1}{2}\alpha(t) \text{ for every } t \in D\right\}$$

has probability one, since $D$ is a countable dense subset of $I$. For every $\omega \in \Omega_1$ and every $t \in D$, we have

$$\begin{aligned}\overline{\lim_{\substack{s\to t\\ s\in D}}}\, x(s,\omega) &= \overline{\lim_{\substack{s\to t\\ s\in D}}}\ \overline{\lim_{\substack{u\to s\\ u\in D}}}\, x(u,\omega)\\ &= \overline{\lim_{\substack{s\to t\\ s\in D}}}\, (x(s,\omega) + \tfrac{1}{2}\alpha(s)) \geqq \overline{\lim_{\substack{s\to t\\ s\in D}}}\, x(s,\omega) + \tfrac{1}{2}a .\end{aligned}$$

But this is impossible, unless

$$\overline{\lim_{\substack{s\to t\\ s\in D}}}\, x(s,\omega) = \infty ,$$

that is, unless $\overline{\lim}_{s\to t}\, x(s,\omega) = \infty$.

Therefore

$$P\left[\overline{\lim_{s\to t}}\, x(s,\omega) = \infty \text{ for every } t \in D\right] = 1.$$

Similarly we have

$$P\left[\underline{\lim_{s\to t}}\, x(s,\omega) = -\infty \text{ for every } t \in D\right] = 1.$$

This completes our proof, since $D$ is dense in $I$.

**5. Proof of Theorem 4(a).**

Let $\alpha(t)$ be the oscillation function of a Gaussian process $x(t)$, $0<t<1$. Then we have

$$P[W_x(t,\omega)=\alpha(t) \text{ for every } t] = 1$$

by the definition of $\alpha(t)$ in Section 2. By the definition of $W_x(t,\omega)$, it holds that

$$\overline{\lim_{s\to t}}\, W_x(s,\omega) \leqq W_x(t,\omega),$$

so that we have

$$\overline{\lim_{s\to t}}\, \alpha(s) \leqq \alpha(t),$$

which shows that $\alpha$ is upper semi-continuous. Hence it follows that $T_a=\{t: \alpha(t)\geqq a\}$ is a closed subset of $[0,1]$. If $T_a-T_\infty$ contains a dense subset $D$ of an open interval for some $a>0$, then $D\subset T_\infty$ by Theorem 3, in contradiction with $D\subset T_a-T_\infty$. Therefore, $T_a-T_\infty$ is nowhere dense for $a>0$. This completes the proof of Theorem 4(a).

**6. Proof of Theorem 4(*b*).**

We shall denote the mean square norm of a random variable $x$ by $\|x\|$,

$$\|x\|^2 = E(x^2).$$

Consider a Brownian motion $B(t)$, $0\leqq t\leqq 1$, and a stationary Gaussian process $S(t)$, $-\infty<t<\infty$, with $ES(t)=0$, $ES(t)^2=1$ and $\alpha(t,S)\equiv\infty$. The existence of the latter process was proved by Belayev [1]. We can assume that these two processes are independent. Let $L$ be the $\|\ \|$-closure of all finite linear combinations of $B(t)$, $0\leqq t\leqq 1$, and $S(t)$, $0\leqq t\leqq 1$. It is clear that any process $x(t)$, $0\leqq t\leqq 1$, such that $x(t)\in L$ for each $t$ is jointly Gauss distributed.

We shall prove Theorem 4(b) by constructing a Gaussian process $x(t)\in L$, $0\leqq t\leqq 1$, with $\alpha(t,x)=\alpha(t)$ for any given function

$$\alpha: [0,1]\to[0,\infty] \text{ satisfying } (\alpha,1) \text{ and } (\alpha,2).$$

Let us start with some lemmas.

LEMMA 6.1. *Given* $I=[u,v]\subset[0,1]$ *and* $\varepsilon>0$, *we can construct a Gaussian process* $x(t)\in L$, $0\leqq t\leqq 1$, *satisfying*

(a) $x(t,\omega)=0$ *for* $t\in[0,1]-I^\circ$ ($I^\circ$ = the interior of $I$),
(b) $\alpha(t,x)=\infty$ *for* $t\in I$,
(c) $\|x(t)\|\leqq\varepsilon$ *for* $t\in[0,1]$.

**Proof.** Take a continuous function $f(t)$ such that $0<f(t)<\varepsilon$ in $I^\circ$ and $f(t)=0$ elsewhere. Then $x(t,\omega)\equiv f(t)S(t,\omega)$ is a Gaussian process satisfying our conditions.

**Lemma 6.2.** *Given* $0<a<\infty$, $\varepsilon>0$, *and* $I=[u,v]\subset[0,1]$, *we can construct a Gaussian process* $x(t)\in L$, $0\leqq t\leqq 1$, *satisfying the following conditions:*

(a) $x(t,\omega)$ *has continuous paths,*
(b) $x(t,\omega)=0$ *for* $t\in[0,1]-I^\circ$,
(c) $Ex(t)=0$, $\|x(t)\|<\varepsilon$ *for every* $t$,
(d) $P(|\sup_I x(t)-a|>\varepsilon)<\varepsilon$.

*Such a process will be denoted by* $x(t;I,a,\varepsilon)$.

**Proof.** Let $B(t)$ be a Brownian motion and define a process $y(t)$ by

$$\begin{aligned} y(t) &= 0, & 0\leqq t\leqq u\,, \\ &= \frac{a(B(t)-B(u))}{(2(t-u)\log\log(t-u)^{-1})^{\frac{1}{2}}}, & u\leqq t\leqq v'\equiv\min(v,u+9)\,, \\ &= \frac{a(B(v')-B(u))}{(2(v'-u)\log\log(v'-u)^{-1})^{\frac{1}{2}}}, & v'\leqq t\leqq 1\,. \end{aligned}$$

Then $y(t)$ is jointly Gauss distributed with $Ey(t)=0$ and the sample path of $y(t)$ is continuous except at $t=u$. The continuity in the mean follows from

$$E[y(t)^2] = \frac{a^2}{2\log\log(t-u)^{-1}} \to 0 \quad \text{as } t\downarrow u\,.$$

By the law of the iterated logarithm we have

$$P\left[\overline{\lim_{t\downarrow u}}\, y(t)=a\right] = 1\,.$$

We now determine $u<s_1<s_2<s_3<s_4<v$ as follows. By taking $s_4$ sufficiently close to $u$, we have

$$E[y(t)^2] < \varepsilon^2\,,$$

$$P\left[\sup_{u<t<s_4} y(t)<a+\varepsilon\right] > 1-\tfrac{1}{2}\varepsilon\,,$$

$$P\left[\sup_{u<t<s_4} y(t)>a-\varepsilon\right] > 1-\tfrac{1}{2}\varepsilon\,.$$

By taking $s_2$ sufficiently close to $u$ and $s_3$ sufficiently close to $s_4$, we have

$$P\left[\sup_{s_2 \leqq t \leqq s_3} y(t) > a - \varepsilon\right] > 1 - \tfrac{1}{2}\varepsilon .$$

We shall take $s_1$ in $(u, s_2)$ arbitrarily.

Let $f(t)$ be a polygonal function of $t$ vanishing on $[0, s_1] \cup [s_4, 1]$, equal to 1 on $[s_2, s_3]$ and linear in each of $[s_1, s_2]$ and $[s_3, s_4]$. Then $x(t) = f(t)\, y(t)$ is a Gaussian process satisfying our conditions.

**Lemma 6.3.** *Suppose that $\alpha_1(t)$ and $\alpha_2(t)$ satisfy $(\alpha, 1)$ and the following condition (stronger than $(\alpha, 2)$):*

$(\alpha, 2')$ $\quad\{t : \alpha_2(t) > 0\}$ is nowhere dense .

*For any Gaussian process $x_1(t)$ with $\alpha(t, x_1) = \alpha_1(t)$ and any $\varepsilon > 0$, we can construct a Gaussian process $x_2(t) \in L$, $0 \leqq t \leqq 1$, satisfying the conditions:*

(a) $\alpha(t, x_1 + x_2) = \alpha_1(t) + \alpha_2(t)$,
(b) $\|x_2(t)\| < \varepsilon$,
(c) $P[\sup_t |x_2(t)| > \sup_t \alpha_2(t)] < \varepsilon$.

**Proof.** We can assume that $c = \sup \alpha_2(t) > 0$. If otherwise, $x_2(t) \equiv 0$ will satisfy our conditions trivially.

Write $\alpha(t)$ for $\alpha_1(t) + \alpha_2(t)$. The set $\{(t, \alpha(t)) : \alpha(t) > 0\}$ is a subset of $[0, 1] \times [0, \infty]$. Let $\{(t_n, \alpha(t_n))\}_n$ be a countable dense subset of $\{(t, \alpha(t)) : \alpha(t) > 0\}$. Then the sets $\{t_n\}_n$ are dense in $\{t : \alpha(t) > 0\}$ and so dense in its closure $F$. By our assumptions $(\alpha, 2')$, $F$ is nowhere dense and its complement $G$ is a dense open subset of $[0, 1]$. Since $\alpha(t) \geqq \alpha(t, x_1)$, we have $\alpha(t, x_1) = 0$ for $t \in G$, so that the sample path of $x_1(t)$ is continuous in $t \in G$ with probability one.

Using Theorem 2 and Lemma 4.1, we can find $\{t_{1n}\}_n$ in $G$ tending to $t_1$ as $n \to \infty$ such that

$$P\left[\overline{\lim_{n \to \infty}}\, x_1(t_{1n}) = x(t_1) + \tfrac{1}{2}\alpha_1(t_n)\right] = 1 .$$

Since the path of $x_1(t)$ is continuous in $t \in G$, we can find, for each $t_{1n}$, a closed interval $I_{1n} \subset G$ containing $t_{1n}$ in its interior such that

$$P\left[x_1(t_{1n}) \geqq \inf_{I_{1n}} x_1(s) \geqq x_1(t_{1n}) - 2^{-n}\right] < 2^{-n} .$$

This implies

$$P\left[\overline{\lim_{n}} \inf_{I_{1n}} x_1(s) = \overline{\lim_{n \to \infty}}\, x_1(t_{1n})\right] = 1$$

by Borel–Cantelli's lemma. Thus we have

$$P\left[\overline{\lim_{n\to\infty}}\ \inf_{I_{1n}} x_1(s) = x(t_1) + \tfrac{1}{2}\alpha_1(t_1)\right] = 1.$$

By taking $I_{1n}$ sufficiently small, we can achieve that the $\{I_{1n}\}_n$ are disjoint. Then $\sum_n |I_{1n}| \leqq 1$ where $|\cdot|$ denotes length. Therefore $|I_{1n}| \to 0$ and so $I_{1n}$ tends to $t_1$ by $t_{1n} \to t_1$.

It is clear that $\bigcup_n I_{1n} \cup \{t_1\}$ is a closed set which does not contain $t_2$. Therefore we can find a neighborhood $U$ of $t_2$ which does not intersect this closed set. In the same way as above, we can find a sequence of disjoint intervals $I_{2n} \subset G \cap U$, $n = 1, 2, \ldots$, tending to $t_2$ such that

$$P\left[\overline{\lim_{n\to\infty}}\ \inf_{I_{2n}} x_1(s) = x_1(t_2) + \tfrac{1}{2}\alpha_1(t_2)\right] = 1.$$

Continuing this procedure we can get a double sequence of disjoint closed intervals $I_{kn} \subset G$, $k, n = 1, 2, \ldots$, such that $I_{kn}$ tends to $t_k$ as $n \to \infty$ for each $k$ and that

$$P\left[\overline{\lim_{n\uparrow\infty}}\ \inf_{I_{kn}} x_1(s) = x_1(t_k) + \tfrac{1}{2}\alpha_1(t_k)\right] = 1, \quad k = 1, 2, \ldots. \tag{6.1}$$

By removing a finite number of intervals from $\{I_{kn}\}_n$ we can achieve that $I_{kn} \subset (t_k - 1/k, t_k + 1/k)$ for each $k$.

Take $\varepsilon_{kn} > 0$ such that

$$\sum_{kn} \varepsilon_{kn} < \min\left(\tfrac{1}{2}\varepsilon, \tfrac{1}{2}c\right)$$

and set

$$x_{kn}(t) = x(t; I_{kn}, a_{kn}, \varepsilon_{kn}) \quad \text{(see Lemma 6.2)},$$

where $a_{kn} = \frac{1}{2}\alpha_2(t_{kn})$ if this is finite and $= n$ otherwise, so that $a_{kn} \uparrow \frac{1}{2}\alpha_2(t_{kn})$ as $n \to \infty$ for each $k$. We shall prove that

$$x_2(t) = \sum_{kn} x_{kn}(t)$$

is a Gaussian process satisfying our conditions.

Since this implies

$$\begin{aligned} x_2(t) &= x_{kn}(t) \quad \text{on} \quad I_{kn}, \quad k, n = 1, 2, \ldots, \\ &= 0 \qquad\quad \text{elsewhere}, \end{aligned}$$

$x_2(t)$ is well-defined and $x_2(t) \in L$. Therefore $x_2(t)$ is jointly Gauss distributed, separable and measurable.

By $\|x_{kn}(t)\| < \varepsilon_{kn}$ and $\sum \varepsilon_{kn} < \infty$, the series $\sum_{kn} x_{kn}(t)$ converges in the mean uniformly in $t$ and so $x_2(t)$ is continuous in the mean and we have

$$\|x_2(t)\| < \sum_{kn} \|x_{kn}(t)\| < \varepsilon ,$$

which proves (b).

Observing that

$$\begin{aligned} P\left[\sup_t x_2(t) > c\right] &\leqq \sum_{kn} P\left[\sup_{t\in I_{kn}} x_2(t) > c\right] \\ &\leqq \sum_{kn} P\left(\sup_t x_{kn}(t) > a_{kn} + \varepsilon_{kn}\right) \\ &\leqq \sum_{kn} \varepsilon_{kn} < \tfrac{1}{2}\varepsilon , \end{aligned}$$

we have

$$P(\sup_t |x_2(t)| > c) < \varepsilon ,$$

since the probability law of the sample path of $x_2(t)$ is symmetric by $Ex_2(t) = 0$. Thus (c) is proved.

Now we shall prove that $\alpha(t, x_2) \leqq \alpha_2(t)$.

If $t_0 \in G$, then $t_0$ has a positive distance from $F$. Since $I_{kn}$ is in the $1/k$-neighborhood of $t_k \in F$, a small neighborhood $U$ ($\subset G$) of $t_0$ does not intersect $I_{kn}$ with $k \geqq k_0$, $n = 1, 2, \ldots$, for some $k_0$. Since $I_{kn} \to t_k \in F$ for each $k$, $U$ can intersect only a finite number of intervals among $\{I_{kn}\}_{kn}$. Let us denote these intervals by $I_{k(i), n(i)}$, $i = 1, 2, \ldots, m$. Then $x_2(t)$ is the sum of $x_{k(i), n(i)}$, $i = 1, 2, \ldots, m$, as far as $t$ lies in $U$. Therefore the path of $x_2(t)$ is continuous in $U$ and so $\alpha(t_0, x) = 0 \leqq \alpha_2(t_0)$.

If $t_0 \in F$, then $x_2(t_0) = 0$ by our construction. Take an arbitrary $\delta > 0$. Then there exists a neighborhood $U_1$ of $t_0$ such that

$$\sup_{s\in U_1} \alpha_2(s) < \alpha_2(t_0) + \delta .$$

Take a neighborhood $U_2$ of $t_0$ such that $\overline{U}_2 \subset U_1$. Then the distance $\varrho(U_2, U_1{}^c)$ is positive. Since the $\{I_{kn}\}_{kn}$ are disjoint and $\sum_{kn} |I_{kn}| \leqq 1$, we have only a finite number of intervals among $\{I_{kn}\}_{kn}$ with the length $\geqq \varrho(U_2, U_1{}^c)$ and only such intervals can intersect both $U_2$ and $U_1{}^c$. Since each $I_{kn}$ has positive distance from $t_0$, we have a neighborhood $U_3$ ($\subset U_2$) of $t_0$ such that $I_{kn} \subset U_1$ as far as $I_{kn}$ intersects $U_3$.

Write $\sum'$ for the summation over those indices $(k,n)$ for which $I_{kn}$ intersects $U_3$. Then any interval $I_{kn}$ with the index $(k,n)$ appearing in $\sum'$ is contained in $U_1$. By taking $U_3$ small enough, we can achieve that

$$\sum_{kn}' \varepsilon_{kn} < \delta .$$

We then have

$$P\left[\sup_{s\in U_3} x_2(s) > \tfrac{1}{2}\alpha_2(t_0)+2\delta\right]$$

$$\leqq \sum{}' P\left[\sup x_{kn}(s) > \sup_{s\in U_1}\alpha_2(s)+\delta\right]$$

$$\leqq \sum{}' P(\sup x_{kn}(s) > a_{kn}+\varepsilon_{kn}) < \sum{}'\varepsilon_{kn} < \delta\,,$$

by the construction in Lemma 6.2. Therefore we get

$$P\left[\overline{\lim_{s\to t_0}}\, x_2(s) > \tfrac{1}{2}\alpha_2(t_0)+2\delta\right] < \delta$$

for every $\delta > 0$. Letting $\delta \downarrow 0$, we have

$$P\left[\overline{\lim_{s\to t_0}}\, x_2(s) \geqq x_2(t_0)+\tfrac{1}{2}\alpha_2(t_0)\right] = 0. \qquad (\text{Note } x_2(t_0)=0.)$$

This implies that $\alpha(t_0, x_2) \leqq \alpha_2(t_0)$ by Theorem 2.

Thus $\alpha(t, x_2) \leqq \alpha_2(t)$ is proved for every $t$. By the definition of the oscillation function we have

$$\alpha(t, x_1+x_2) \leqq \alpha(t, x_1)+\alpha(t, x_2) \leqq \alpha_1(t)+\alpha_2(t) = \alpha(t)\,.$$

We shall now prove that

$$\alpha(t, x_1+x_2) \geqq \alpha(t)\,.$$

Consider first the case $t = t_k$. It holds that

$$(6.2)\quad P\left[\overline{\lim_{s\to t_k}}\,(x_1(s)+x_2(s)) \geqq x_1(t_k)+x_2(t_k)+\tfrac{1}{2}\alpha(t_k)\right]$$

$$\geqq P\left[\overline{\lim_{n\to\infty}}\, \sup_{I_{kn}}[x_1(s)+x_2(s)] \geqq x_1(t_k)+x_2(t_k)+\tfrac{1}{2}\alpha(t_k)\right]$$

$$\geqq P\left[\overline{\lim_{n\to\infty}}\left(\inf_{I_{kn}} x_1(s)+\sup_{I_{kn}} x_2(s)\right) \geqq x_1(t_k)+x_2(t_k)+\tfrac{1}{2}\alpha(t_k)\right]$$

$$\geqq P\left[\underline{\lim_{n\to\infty}}\, \inf_{I_{kn}} x_1(s)+\overline{\lim_{n\to\infty}}\, \sup_{I_{kn}} x_2(s) \geqq x_1(t_k)+x_2(t_k)+\tfrac{1}{2}\alpha(t_k)\right].$$

Since we have

$$P[\sup_{I_{kn}} x_2(s) < a_{kn}-\varepsilon_{kn}] < \varepsilon_{kn}$$

by virtue of $x_2(s) = x_{kn}(s)$ on $I_{kn}$, we get

$$(6.3)\qquad P\left[\overline{\lim_{n\to\infty}}\, \sup_{I_{kn}} x_2(s) \geqq x_2(t_k)+\tfrac{1}{2}\alpha_2(t_k)\right] = 1$$

by $x_2(t_k)=0$ and $\sum_n \varepsilon_{kn} < \infty$. By (6.1), (6.2) and (6.3) we get

$$P\left[\overline{\lim_{s\to t_k}}\,(x_1(s)+x_2(s)) \geqq x_1(t_k)+x_2'(t_k)+\tfrac{1}{2}(t_k)\right] = 1\,,$$

that is, by Theorem 2,

$$\alpha(t_k, x_1+x_2) \geqq \alpha(t_k)\,.$$

For $t \neq t_k$, $k=1,2,\ldots$, the point $(t,\alpha(t))$ is an accumulation point of $(t_n,\alpha(t_n))$, $n=1,2,\ldots$. Therefore we have a subsequence $\{s_n\}$ of $\{t_n\}$ such that

$$\alpha(t) = \lim_n \alpha(s_n)\,.$$

By Theorem 4 (a) $(\alpha, 1)$ we have

$$\alpha(t, x_1+x_2) \geqq \overline{\lim}_n \alpha(s_n, x_1+x_2) = \overline{\lim}_n \alpha(s_n) = \alpha(t)\,,$$

which completes the proof of Lemma 6.3.

Now we shall come back to the proof of Theorem 4 (b). We shall first assume

$(\alpha, 2'')$ $\{t : \alpha(t) \geqq c\}$ is nowhere dense for every $c>0$;

this is stronger than $(\alpha, 2)$ but weaker than $(\alpha, 2')$. Set

$$\begin{aligned}\alpha_0(t) &= 0\,,\\ \alpha_1(t) &= \max(\alpha(t), \tfrac{1}{2}) - \tfrac{1}{2}\,,\\ \alpha_n(t) &= \max\left(\min(\alpha(t), 2^{-n+1}), 2^{-n}\right) - 2^{-n}, \qquad n=2,3,\ldots\,.\end{aligned}$$

It is then easy to verify the following properties of $\alpha_n(t)$:

$$\begin{aligned}&0 \leqq \alpha_n(t) \leqq 2^{-n-1}, \qquad n=2,3,\ldots\,,\\ &\textstyle\sum_0^n \alpha_i(t) \uparrow \alpha(t)\\ &\{t : \alpha_n(t) > 0\} \text{ is nowhere dense}\,.\end{aligned}$$

Starting with the Gaussian process $x_0(t) \equiv 0$ whose oscillation function is $\alpha_0(t) \equiv 0$, we can use Lemma 6.3 to define a sequence of Gaussian processes $x_n(t) \in L$, $0 \leqq t \leqq 1$, $n=1,2,\ldots$, satisfying

(a) $\alpha(t, \sum_0^n x_i) = \sum_0^n \alpha_i(t)$,
(b) $\|x_n(t)\| < 2^{-n}$,
(c) $P[\sup_t |x_n(t)| > \sup_t \alpha_n(t)] < 2^{-n}$.

Now set

$$x(t) = \textstyle\sum_n x_n(t)\,.$$

By Borel–Cantelli's lemma it follows from (c) and $\alpha_n(t) \leqq 2^{-n-1}$, $n \geqq 2$, that this infinite series converges uniformly in $t$ with probability one.

Therefore $x(t)$ is well-defined, separable, measurable and jointly Gauss distributed, and we have

$$\alpha(t,x) = \lim_n \alpha(t, \textstyle\sum_0^n x_i) = \lim_n \sum_0^n \alpha_i(t) = \alpha(t) .$$

It follows from (b) that this infinite series converges also in the mean, uniformly in $t$, so that $x(t)$ is continuous in the mean.

We shall now remove the assumption that $T_c = \{t: \alpha(t) \geqq c\}$ is nowhere dense.

Let $I_1, I_2, \ldots$ be the maximal intervals contained in the set $T_\infty = \{t: \alpha(t) = \infty\}$. Since, by $(\alpha, 2)$, $\{t: c \leqq \alpha(t) < \infty\}$ is nowhere dense, $I_1, I_2, \ldots$ are also the maximal intervals contained in the set $T_c$ $(c>0)$.

We now define

$$\begin{aligned} \beta(t) &= 0, \qquad t \in I_n^\circ, \quad n = 1,2,\ldots, \\ &= \alpha(t) \quad \text{elsewhere} . \end{aligned}$$

Then $\beta(t)$ satisfies $(\alpha, 1)$ and $(\alpha, 2'')$. Therefore we can construct a Gaussian process $y(t) \in L$, $0 \leqq t \leqq 1$, such that $\alpha(t,y) = \beta(t)$, as we proved above.

By Lemma 6.1, we can construct a sequence of Gaussian processes $y_n(t) \in L$, $0 \leqq t \leqq 1$, $n = 1,2,\ldots$, such that $\alpha(t, y_n) = \infty$ on $I_n$, $y_n(t) = 0$ on $[0,1] - I_n^\circ$ and $\|y_n(t)\| < 2^{-n}$. Now consider

$$x(t) = y(t) + \textstyle\sum_n y_n(t) .$$

Then

$$\begin{aligned} x(t) &= y_n(t) + y(t), \qquad t \in I_n^\circ, \quad n = 1,2,\ldots, \\ &= y(t), \qquad \text{elsewhere}, \end{aligned}$$

and $x(t) \in L$, $0 \leqq t \leqq 1$. Therefore $x(t)$ is jointly Gaussian, measurable and separable. Its continuity in the mean follows from $\|y_n(t)\| < 2^{-n}$.

To complete our proof, we need only to show that $\alpha(t,x) = \alpha(t)$. Since $x(t) = y(t)$ on the set $G = [0,1] - \bigcup_n I_n$ which is open in $[0,1]$, we have

$$\alpha(t,x) = \alpha(t,y) = \beta(t) = \alpha(t), \quad t \in G .$$

Since $x(t) = y_n(t) + y(t)$ and $y(t)$ is continuous in $I_n^\circ$, we have

$$\alpha(t,x) = \alpha(t,y_n) = \infty = \alpha(t), \quad t \in \textstyle\bigcup_n I_n^\circ .$$

If $t \in \overline{\bigcup_n I_n} - \bigcup_n I_n^\circ$, then $t$ is an accumulation point of $\bigcup_n I_n^\circ$ and so we get

$$\alpha(t,x) \geqq \overline{\lim_{\substack{s \to t \\ s \in \cup I_n^\circ}}} \alpha(s,x) = \infty$$

by Theorem 4 (a) $(\alpha, 1)$.

## REFERENCE

1. Yu. K. Belayev, *Continuity and Hölder's conditions for sample functions of stationary Gaussian processes*, Proc. 4th Berkeley Sympos. Math. Statist. and Prob., Vol. 2., 23–33. Univ. Calif. Press, 1961.

AARHUS UNIVERSITY, DENMARK

AND

KOBE UNIVERSITY, JAPAN

# Canonical measurable random functions

By Kiyosi ITÔ

## § 1. Introduction.

The basic concepts in probability theory are measurable spaces and complete regular probability measures [1], [2]. We will impose the following assumption.

(A) All measurable spaces appearing here are standard.
A measurable space is called *standard* if it is Borel (point-)isomorphic with a Borel subset of the real line $R^1$.

This standpoint appears quite restricted but is in fact general enough to cover all interesting problems in probability theory. For example the following spaces and their Borel subsets are standard, where all topological spaces below are assumed to be endowed with their topological $\sigma$-algebras.

(1.1) a finite or countable set.

(1.2) the $n$-space $R^n$.

(1.3) the sequence space $R^\infty = R^1 \times R^1 \times \cdots$.

(1.4) the space of all continuous functions on [0, 1] with the maximum metric: $C[0, 1]$.

(1.5) the space of all functions with no discontinuities of the second kind endowed with the Skorokhod metric: $D[0, 1]$.

(1.6) the space of all measurable functions on [0, 1] with topology of convergence in measure: $\tilde{M}[0, 1]$, where two equivalent measurable functions are identified.

(1.7) a complete separable metric space.

(1.8) the space of all signed measures on a standard measurable space endowed with the $\sigma$-algebra generated by the cylinder sets.

(1.9) the space of all distributions in the sense of Schwartz: $\mathcal{D}'$.

(1.10) the product of a countable number of standard measurable spaces.

Let $(\Omega, \mathcal{B}, P)$ denote the basic probability space, the space of elementary events, where $(\Omega, \mathcal{B})$ is a standard measurable space and $P$ is a complete $\mathcal{B}$-regular probability measure on $\Omega$. The domain of definition of $P$ is denoted by $\mathcal{M}_P$. Let $(E, \mathcal{E})$ be a standard measurable space. An $(E, \mathcal{E})$*-valued random variable* on $(\Omega, \mathcal{B}, P)$ is defined to be a map from $\Omega$ into $E$ measurable $\mathcal{M}_P/\mathcal{E}$. We identify two random variables if they are equal almost surely. A random variable is rather an equivalence class of measurable functions and each measurable function in a class is regarded as a version of the random variable. Let $\mathcal{R}(\Omega, E)$ denote the space of all $(E, \mathcal{E})$-valued random variables on $(\Omega, \mathcal{B}, P)$. $\mathcal{R}(\Omega, R^1)$ is denoted simply by $\mathcal{R}$. It is a complete separable metric space with the metric

$$\rho(\xi, \eta) = E(|\xi-\eta| \wedge 1)\,.$$

The topological $\sigma$-algebra on $(\mathcal{R}, \rho)$ is denoted by $\mathcal{B}(\mathcal{R})$. The *probability distribution* $P_\xi$ of an $(E, \mathcal{E})$-valued random variable $\xi$ is defined to be

$$P_\xi(A) = P(\omega:\ \xi(\omega) \in A) = P(\xi^{-1}(A))$$

as far as $\xi^{-1}(A) \in \mathcal{M}_P$. Then

(i) $P_\xi$ is a complete $\mathcal{E}$-regular probability measure on $E$.

It is easy to verify the following.

(ii) The conditional probability distribution $P(\cdot\,|C)$ is well-defined for every sub-$\sigma$-algebra $C$ of $\mathcal{M}_P$.

(iii) The Kolmogorov extension theorem [1] holds for a countable product of standard measurable spaces.

Now we consider *joint variables.* No special consideration is necessary for the joint variable of a countable number of random variables. Let $\xi_n$ be an $(E_n, \mathcal{E}_n)$-valued random variable for $n=1, 2, \cdots$ (finite or countably infinite). Then the joint variable is a random variable with values in $(\prod_n E_n, \prod_n \mathcal{E}_n)$. If $\xi_n = \eta_n$ a. s., for each $n$, we have

$$(\xi_1, \xi_2, \cdots) = (\eta_1, \eta_2, \cdots) \qquad \text{a. s.}$$

Since an uncountable union of null sets is not always a null set, the joint variable of an uncountable number of random variables is not always well-defined. This situation has produced a difficult problem on the definition of sample functions of stochastic processes. The notions of *separable modification, measurable modification* and *separable measurable modification* have been introduced in this connection by J. L. Doob [2], E. Slutsky [3] and others. However, these notions do not lie in our framework, so that we will present a slightly different method.

A stochastic process has so far been defined in the following three different ways, the time interval and the state space being assumed to be respectively $[0, 1]$ and $R^1$.

(a) A family of random variables $\xi_t$ indexed by the time parameter $t$.

(b) A random variable with values in a function space.

(c) A function of the pair $(t, \omega)$.

(c) is outside of our framework. (b) is in our framework if the function space in consideration is endowed with a $\sigma$-algebra which makes the function space standard. (a) is obviously in our framework. To distinguish (a) and

(b) from each other, we call (a) a *stochastic process* and (b) a *random function.*

To discuss a random function we should specify a function space. The conceivable function spaces are for example

$$C[0,1]\subset D[0,1]\subset \tilde{L}^p[0,1]\subset \tilde{M}[0,1].$$

We put $\sim$ on $L^p$ and $M$ to note that the elements in these spaces are not functions but equivalent classes, so that $f$ in $D[0,1]$ should be identified with the equivalence class containing $f$ when we interpret the inclusion relation $D[0,1]\subset \tilde{L}^p[0,1]$. To avoid such a trouble of identification, we will pick up a nice function, called *canonical measurable function* from each equivalence class in $\tilde{M}[0,1]$ and form the spaces $M[0,1]$ and $L^p[0,1]$ (see Section 2). Then we have

$$C[0,1]\subset D[0,1]\subset L^p[0,1]\subset M[0,1].$$

$M=M[0,1]$ is topologized by convergence in measure and so is a standard measurable space with its topological $\sigma$-algebra $\mathcal{B}(M)$. An $(M,\mathcal{B}(M))$-valued random variable is called a *canonical measurable random function* or simply a *random M-function.* Similarly we define a *random C-function,* a *random D-function* and a *random $L^p$-function.* However, it is enough to consider only random $M$-functions, for we can define a random $C$-, $D$- or $L^p$-function to be a random $M$-function whose probability distribution gives 1 to $C$, $D$ or $L^p$ in view of the fact that $C[0,1]$, $D[0,1]$ and $L^p[0,1]$ belong to $\mathcal{B}(M)$. We will discuss canonical measurable random functions in Section 3.

A stochastic process $\xi_t$, $t\in[0,1]$ is called *strongly measurable* if the map $t\to\xi_t$ is measurable $\mathcal{M}_\lambda/\mathcal{B}(\mathcal{R})$, where $\mathcal{M}_\lambda$ is the family of all measurable sets in $[0,1]$ with respect to the Lebesgue measure. According to a theorem due to J. Hoffmann-Jørgensen [4], strong measurability is equivalent to the following simple condition:

> $P(\xi_t\in A,\ \xi_s\in B)$ is measurable in $t$ for every pair $A, B\in\mathcal{B}(R^1)$ and every $s\in T$.

If $\xi_t$ is strongly measurable, then there exists a unique (up to $P$-measure 0) canonical measurable random function whose value at $t$ is equal to $\xi_t$ a. s. for almost every $t$. Such a random function is called the *M-regularization* or the *regularization of the process.* It is also called the *regularized sample function of the process* and it is a natural extension of the notion of joint variables.

We can define the *C-regularization* of a stochastic process with some conditions, but such a notion is unnecessary, for if it exists, then the $M$-regularization exists obviously and is continuous in $t$ almost surely. Similarly for the $D$- or $L^p$-regularization.

The notion of regularization can be defined in the case that the parameter space $T$ is a $C^1$-manifold and the state space is a complete separable metric space, as we will explain it in Section 5.

## § 2. Canonical measurable functions.

A measurable set $A\subset T=[0,1]$ is called *d-open* if every point in $A$ is a density point of $A$. The family of all $d$-open sets in $T$ defines a Hausdorff topology in $T$, called the *d-topology*, which is stronger than the original topology in $T$. The limit of $f(t)$ as $t \underset{d}{\to} t_0$ is known in the name of *approximate limit* [5]. It is easy to see

$$\lim_{\substack{t\to t_0\\ d}} f(t)=a$$

if and only if

$$\lim_{I\to t_0} \frac{1}{|I|}\int_I |f(t)-a|\wedge 1\,dt=0,$$

where $I$ denotes an interval and $|\,|$ denotes the Lebesgue measure.

Let $f\in \tilde{M}=\tilde{M}(T)$. Then it is known [5] that

$$\lim_{\substack{s\to t\\ d}} f(s)=f(t) \qquad \text{for almost every } t\in T.$$

The *canonical modification* of $f\in\tilde{M}$, $f^c$ in notation, is defined by

$$f^c(t)=\lim_{\substack{s\to t\\ d}} f(s) \qquad \text{if this limit exists and is finite,}$$

$$=\Delta \qquad \text{if otherwise,}$$

where $\Delta$ denotes an extra point not in $R^1$ and $f^c(t)=\Delta$ means that $f^c$ is not defined at $t$. This definition is slightly different from that in our previous paper [6]. It is easy to see that (i) $f^{cc}=f^c$, (ii) $f^c=f$ a.e., (iii) $f=g$ a.e. implies $f^c=g^c$. The space $M=M(T)=\{f^c: f\in\tilde{M}\}$ is a complete separable metric space with the metric

$$\delta(f,g)=\int_T |f(t)-g(t)|\wedge 1\,dt.$$

A member of $M$ is called a *canonical measurable function* on $T$. $\mathcal{D}(f)$ denotes the set of all $t$ for which $f(t)\neq\Delta$. It is obvious that if $f, g\in M$, then $f=g$ a.e. is equivalent to $f=g$. The topology induced by the metric $\delta$ is that of convergence in measure. Let $\mathcal{B}(M)$ denote the topological $\sigma$-algebra in $M$. Then $(M, \mathcal{B}(M))$ is a standard measurable space.

Let $C=C(T)$ be the space of all continuous functions on $T$. Then it is obvious that $C\subset M$.

THEOREM 2.1. $C\in\mathcal{B}(M)$.

PROOF. Let us introduce a modified oscillation function

$$\alpha(f, u, v) = \sup \{ |f(t)-f(s)| \wedge 1 : t, s \in \mathcal{D}(f) \cap (u, v) \cap T \} .$$

Recalling the definition of $M$, we can easily verify for $f \in M$

$$\alpha(f, u, v) = \sup \frac{1}{|I||J|} \int_J \int_I |f(t)-f(s)| \wedge 1 \, dt ds ,$$

where $I$ and $J$ move over all rational intervals $\subset (u, v)$. As the double integral above is $\mathcal{B}(M)$-measurable in $f \in M$, so is $\alpha(f, u, v)$. Then we have

$$C = \bigcap_m \bigcup_p \left\{ f \in M : \operatorname*{Max}_{q=1,2,\cdots,p} \alpha\left(f, \frac{q-1}{p}, \frac{q+1}{p}\right) \leqq \frac{1}{m} \right\} \in \mathcal{B}(M) .$$

Let $D = D(T)$ be the space of all functions on $T$ with no discontinuities of the second kind such that

$$f(0+) = f(0) , \qquad f(1-) = f(1)$$

and that $f = \Delta$ at its jump points. It is obvious that $D \subset M$.

THEOREM 2.2. $D \in \mathcal{B}(M)$.

PROOF. Let $N(f, m, n)$ be the number of $k \in \{1, 2, \cdots, n\}$ for which

$$\alpha\left(f, \frac{k-1}{n}, \frac{k+1}{n}\right) \geqq \frac{1}{m} .$$

Then $N(f, m, n)$ is $\mathcal{B}(M)$-measurable in $f \in M$, so that we have

$$D = \bigcap_m \{ f \in M : \sup_n N(f, m, n) < \infty \} \in \mathcal{B}(M) .$$

Since the $p$-th order norm of $f \in M$ is $\mathcal{B}(M)$-measurable in $f$, we have

THEOREM 2.3. $L^p(T) \in \mathcal{B}(M)$.

## § 3. Canonical measurable random functions.

A *canonical measurable random function* is defined to be an $(M, \mathcal{B}(M))$-valued random variable on $(\Omega, \mathcal{B}, P)$. Let $X(\omega)$ be a canonical measurable random function. The value of $X(\omega)$ at time $t \in T$ is denoted by $X_t(\omega)$.

THEOREM 3.1. *$X_t(\omega)$ is defined for almost every pair $(t, \omega)$ and is measurable in $(t, \omega)$.*

PROOF. Let us define $\overline{X}(t, \omega)$ to be

$$\sup_n \limsup_{I \downarrow a} \frac{1}{|I|} \int_I X_t(\omega) \wedge n \, dt ,$$

where $n$ moves over 1, 2, 3, ⋯ and $I$ moves on all rational intervals. $\overline{X}(t, \omega)$ may take $\pm\infty$. Replacing sup, lim sup and $X_t(\omega) \wedge n$ respectively, by inf, lim inf and $X_t(\omega) \vee (-n)$, we define $\underline{X}(t, \omega)$. Since $X(\omega) \in M$ for every $\omega$, we have, for every $\omega$,

$$\bar{X}(t, \omega)=\underline{X}(t, \omega)=X_t(\omega)\in R^1 \qquad \text{a. e. on } T.$$

It is easy to see that both $\bar{X}$ and $\underline{X}$ are measurable in the pair $(t, \omega)$. Hence it follows by Fubini's theorem that

$$\bar{X}(t, \omega)=\underline{X}(t, \omega)\in R^1 \qquad \text{for almost every pair } (t, \omega).$$

For such pair $(t, \omega)$ $X_t(\omega)$ is well-defined and equals $\bar{X}(t, \omega)$. Therefore $X_t(\omega)$ is measurable in the pair $(t, \omega)$.

Applying Fubini's theorem again, we have

THEOREM 3.2. *For almost every $t$, $X_t(\omega)$ is defined for almost every $\omega$ and is measurable in $\omega$.*

We will denote the exceptional $t$-set in the theorem by $N=N(X)$.

THEOREM 3.3. *$X_t(\omega)$ is separable in the following sense. There exist $\Omega_1\subset\Omega$ with $P(\Omega_1)=1$ and a countable set $C\subset T$ such that*

(i) *$C\subset\mathcal{D}(X(\omega))$ for $\omega\in\Omega_1$ and that*

(ii) *for every $\omega\in\Omega_1$, every closed set $F\subset R^1$ and every set $G\subset T$ open in $T$, we have*

$$X_t(\omega)\in F \text{ for } t\in C\cap G \;\Rightarrow\; X_t(\omega)\in F \text{ for } t\in G\cap\mathcal{D}(X(\omega)).$$

REMARK. The separability above is slightly different from Doob's [2] and it may be better to call it *essential separability.*

PROOF OF THE THEOREM. Let $\{a_n\}$ be a countable dense subset of $R^1$ and set

$$\begin{aligned} g_n(t, \omega) &= |X(t, \omega)-a_n|\wedge 1 \qquad &&\text{for } t\in T-N \\ &=0 &&\text{for } t\in R^1-(T-N). \end{aligned}$$

Then $g_n(t, \omega)$ is defined everywhere on $R^1\times\Omega$ and measurable in $(t, \omega)$. By a device due to Doob [2] (p. 441) we can find $s_n\in R^1$ and a sequence of integers $a(1)<a(2)<\cdots$ depending on $n$ such that

$$s_n+2^{-a(p)}k\in R^1-N \qquad \text{for } p, k=1, 2, \cdots$$

and that for almost every $\omega$,

$$\int_T |g_n(t, \omega)-g_n^p(t, \omega)|\,dt\to 0,$$

where $g_n^p(t, \omega)$ is a step function equal to $g_n(s_n+2^{-a(p)}(k-1), \omega)$ on $s_n+2^{-a(p)}(k-1)\leqq t<s_n+2^{-a(p)}k$.

Write $C$ for the set $(T-N)\cap\{s_n+2^{-p}k: k, p, n=1, 2, \cdots\}$. Then it is easy to see that on a set $A\subset\Omega$ with $P(A)=1$, for every interval $I\subset T$ and every $n$, the integral $\int_I g_n(t, \omega)dt$ can be approximated by a finite sum of the form

$\sum_i g_n(u_{i-1}, \omega)(u_i - u_{i-1})$ where $u_i \in C$ and $\sum_i (u_i - u_{i-1}) \to |I|$.

Since $C \subset T-N$ and since $C$ is countable, we have a set $B \subset \Omega$ with $P(B)=1$ such that

$$C \subset \mathcal{D}(X(\omega)) \qquad \text{for every } \omega \in B.$$

Let $\Omega_1 = A \cap B$. Then $P(\Omega_1)=1$.

Now we will verify (i) and (ii) for $C$ and $\Omega_1$. (i) is obvious. To prove (ii), fix any $\omega \in \Omega_1$, any closed set $F \subset R^1$ and any open $G \subset T$ and suppose that

$$X_t(\omega) \in F \text{ for every } t \in C \cap G \text{ but } X_{t_0}(\omega) \notin F \text{ for some } t_0 \in G \cap \mathcal{D}(X(\omega)).$$

Then there exist $a_n$ and $\varepsilon \in (0, 1)$ such that

$$(1) \qquad |X_{t_0}(\omega) - a_n| < \varepsilon$$

and

$$(2) \qquad \inf\{|a_n - a| : a \in F\} > \varepsilon,$$

because $\{a_n\}$ is dense in $R^1$. By $\omega \in \Omega_1 \subset A$ we have

$$\int_I |X_s(\omega) - a_n| \wedge 1 \, ds \geqq \varepsilon |I|$$

as far as $I \subset G$. Since $t_0 \in \mathcal{D}(X(\omega)) \cap G$, we have

$$\lim_{I \downarrow t_0} \frac{1}{|I|} \int_I |X_s(\omega) - X_{t_0}(\omega)| \wedge 1 \, ds = 0.$$

But

$$\begin{aligned} \varepsilon &\leqq \frac{1}{|I|} \int_I |X_s(\omega) - a_n| \wedge 1 \, ds \\ &\leqq \frac{1}{|I|} \int_I |X_s(\omega) - X_{t_0}(\omega)| \wedge 1 \, ds + |X_{t_0}(\omega) - a_n| \end{aligned}$$

Therefore we have $|X_{t_0}(\omega) - a_n| \geqq \varepsilon$ in contradiction with (1). This completes the proof.

## § 4. The regularized sample functions of strongly measurable processes.

Let $X(\omega)$ be a canonical random function. Then we get a stochastic process $\xi_t = X_t(\omega)$, $t \in T-N(X)$. By setting $\xi_t = 0$ for $t \in N(X)$, we have $\xi_t$ for every $t \in T$. This extension is not essential because $N(X)$ is a null set in $T$ by Theorem 3.2. The stochastic process $\xi_t$ is strongly measurable.

THEOREM 4.1. *Given a strongly measurable process $\xi_t$, $t \in T$, we have a unique canonical measurable random function such that*

$$P(X_t = \xi_t) = 1 \qquad \textit{for} \quad t \in T - N(X).$$

PROOF. By a theorem due to J. Hoffmann-Jørgensen [4] we can find a function $f(t, \omega)$ measurable in the pair $(t, \omega)$ such that

$$P(f(t, \omega)=\xi_t(\omega))=1 \qquad \text{for every } t\in T.$$

$f(t, \omega)$ is a measurable function of $t$ for almost every $\omega$. Define $X(\omega)$ to be the canonical modification of $f(\cdot, \omega)$ if $f(t, \omega)$ is measurable in $t$ and to be 0 if otherwise. Then it is easy to see by Fubini's theorem and Theorem 3.1 that $X(\omega)$ is a canonical measurable random function which satisfies our condition. Recalling the fact that $f=g$ a. e. is equivalent to $f=g$ for $f, g\in M$, we can prove the uniqueness by Fubini's theorem.

The canonical measurable random function $X$ in the theorem above is called the *regularization* or the *regularized sample function* of $\xi_t$.

EXAMPLE 1. Let $\xi_t$ be an additive process continuous in probability. Then the regularized sample function $X$ of $\xi_t$ lies in $D$ a. s. Since $D\in\mathcal{B}(M)$, $P(X\in D)$ is well-defined. We can prove that it is one by the same technique as Doob used for his separable version.

EXAMPLE 2. Let $\xi_t$ be a strongly measurable and strictly stationary process where the time interval $T$ is $R^1$. We can carry out all arguments above for this case with trivial modifications. Note that strong measurability is equivalent to the existence of a measurable (in $(t, \omega)$) version by virtue of a theorem due to J. Hoffmann-Jørgensen [4]. Let $X$ be the regularized sample function of this process. We have $N(X)=\emptyset$ by stationarity. The probability distribution $P_x$ is shift invariant. Let $\{T_t\}$ be the group of shift transformations in $M$. Then the map

$$(t, f)\to T_t f$$

is continous from $R^1\times M$ into $M$ and therefore measurable. The individual ergodic theorem can apply to $\{T_t\}$ on $(M, \mathcal{B}(M), P_X)$. Thus the theory of strictly stationary processes corresponds completely to that of measure-preserving transformation groups. If we use Doob's separable measurable modification, the situation will be rather involved in this point.

EXAMPLE 3. Let $\xi_t$ be a Gaussian stationary process continuous in probability. Yu. K. Belayev [5] proved that its measurable separable version is either continuous a. s. or unbounded on every interval a. s. This fact can be stated also in terms of the regularized sample function. Let $P_X$ denote the probability distribution of the regularized sample function $X$ of the process and $U$ the set of all functions in $M$ unbounded on every interval in $R^1$. Then this theorem is stated as follows. Either $P_X(C)=1$ or $P_X(U)=1$. Notice that $P_X$ is completely determined by the mean and the covariance function of the process.

A corresponding fact for the non-stationary case by Itô and Nisio [8] is also stated in this way.

## § 5. Generalizations.

It is easy to see all results above can be extended to the case that the state space is a complete metric space and that the time interval may be infinite.

The generalization to the case that the time parameter space $T$ is a $C^1$-manifold with or without boundaries will be necessary for discussion of random scalar or vector fields on a manifold. Because of continuity of the functional determinants between two local coordinates it is easy to find a measure $\lambda$ on $T$ such that for every local coordinate $(U, \varphi)$ and for every open $V$ with $\bar{V} \subset U$, the restriction $\lambda | V$ is abolutely continuous with respect to the measure induced from the Lebesgue measure in $\varphi(V)$ and that the density function is bounded below and above by positive numbers depending on $V$. This measure $\lambda$ will be used in place of the Lebesgue measure in the discussion above and all notions defined through $\lambda$ are well-defined independently of the special choice of $\lambda$.

## References


[1] A. Kolomogorov, Grundbegriffe der Wahrscheinlichkeitsrechnung, Ergeb. Math., 2 Heft 3, Springer, Berlin, 1933.

[2] J.L. Doob, Stochastic Processes, J. Wiley, New York, 1952.

[3] E. Slutsky, Sur les fonctions aléatoires presques périodiques et sur la décomposition des fonctions aléatoires en composants, dans S. Bernstein, E. Slutsky et A. Steinhaus: Les Fonctions Aléatoires, Act. Sci. Ind. 738, Hermann, Paris, 1938, pp. 35-55.

[4] J. Hoffmann-Jørgensen, Measurable versions of stochastic processes, Aarhus Univ. Preprint Series 1967/68, No. 9.

[5] S. Saks, Theory of the Integral, Warszawa, 1937.

[6] K. Itô, The canonical modification of stochastic processes, J. Math. Soc. Japan, 20 (1968), 130-150.

[7] Yu. K. Belayev, Continuity and Hölder condition for sample functions of stationary Gaussian processes, Proc. Fourth Berkeley Symp. Math. Stat. and Prob. Vol. 2, Univ. Calif. Press, 1961, pp. 23-33.

[8] K. Itô and M. Nisio, On the oscillation function of Gaussian processes, Math. Scand., 22 (1968), 209-223.



Matematisk Institut
Aarhus Universitet
Aarhus, Denmark



Reprinted from
*Proc. Int. Conf. Funct. Anal. Rel. Topics.* 369-377 (1969)

K. Ito
Nagoya Math. J.
Vol. 38 (1970), 181-183

# THE TOPOLOGICAL SUPPORT OF GAUSS MEASURE ON HILBERT SPACE

KIYOSI ITO

***dedicated to Professor K. Ono for his sixtieth birthday***

## 1. Introduction

Let $X$ be a Hilbert space. The *topological support* of a Radon probability measure $P$ on $X$ is the least closed subset $M$ of $X$ that carries the total measure 1. A closed subset $M$ of $X$ is called a *linear subvariety* if

$x, y \in M$ implies $\alpha x + (1-\alpha)y \in M$ for every $\alpha \in R^1$,

or equivalently if $M = a + Y$ for some $a \in X$ and some closed linear subspace $Y$ of $X$. A Radon probability measure $P$ on $X$ is called a *Gauss measure* if for every $a \in X$, the image measure of $P$ by the map

$$f_a(x) = (a, x): X \longrightarrow R^1$$

is a Gauss measure on $R^1$.

The purpose of this note is to prove

THEOREM. *Let $P$ be a Gauss measure on a Hilbert space $X$. Then the topological support $S(P)$ of $P$ is a linear subvariety of $X$.*

This fact is obvious in case $X$ is finite dimensional but we need a small trick to discuss the infinite dimensional case as we shall see below.

## 2. Proof of the theorem.

Since $P$ is a Gauss measure, its characteristic functional

$$C(z) = \int_X e^{i(z,x)} P(dx)$$

is expressed as

$$C(z) = \exp\left\{i\,(z, m) - \frac{1}{2}\sum_k v_k (z, e_k)^2\right\}$$

Received July 3, 1969

where $\{e_k\}$ is an orthonormal sequence (finite or countable) and

$$z \in X, \quad v_k > 0 \quad \sum_k v_k < \infty.$$

By the translation $x \longrightarrow x + m$, we can assume that $m = 0$, namely that

$$C(z) = \exp\left\{-\frac{1}{2}\sum_k v_k(z, e_k)^2\right\}.$$

Let $Y$ be the closed linear subspace spanned by $\{e_k\}$. If $z \perp Y$, then

$$E(e^{it(z,x)}) = C(tz) = 1, \quad E(f(x)) = \int_X f(x)P(dx),$$

for every $t \in R^1$. Therefore we get

$$P(L_z) = 1, \quad L_z = \{x : (z, x) = 0\}.$$

Since $Y = \bigcap_z L_z$, we obtain

(1) $P(Y) = 1$,

because $L_z$ is closed and $P$ is Radon.

Now we will prove that $Y = S(P)$. For this purpose it is enough to prove that

$$P\{x \in X : \| x - a \| < r\} > 0$$

for every $a \in Y$ and every $r > 0$. Suppose to the contrary that we have $a \in Y$ and $r > 0$ such that

$$P\{x \in X : \| x - a \| < r\} = 0.$$

Then we have

(2) $E(e^{-\alpha\|x-a\|^2/2}) \leqq e^{-\alpha r^2/2}, \quad \alpha > 0.$

On the other hand we have by (1)

$$E(e^{-\alpha\|x-a\|^2/2}) = E(e^{-\alpha\sum_k(x_k-a_k)^2/2}), \quad x_k = (x, e_k), \quad a_k = (a, e_k).$$

Since

$$E(e^{i\sum_{k=1}^n z_k x_k}) = \exp\{-\sum_{k=1}^n v_i z_k^2/2\}, \quad n = 1, 2, \cdots,$$

$x_k$, $k = 1, 2, \cdots$ are independent and each $x_k$ is $N(0, v_k)$-distributed on the probability space $(X, P)$. Thus we have

(3) $$E(e^{-\alpha\|x-a\|^2/2}) = \prod_k E(e^{-\alpha(x_k-a_k)^2/2})$$

$$= \prod_k \exp - \frac{\alpha a_k^2}{2(1-v_k\alpha)}(1+v_k\alpha)^{-1/2}.$$

Comparing (2) with (3) we have

(4) $$\prod_k \exp \frac{a_k^2\alpha}{1+v_k\alpha}(1+v_k\alpha) \geq e^{\alpha r^2}.$$

Writing $I_1$ and $I_2$ for the products corresponding to $k \leq N$ and $k > N$ respectively, we have

$$I_2 \leq \prod_{k>N} e^{a_k^2\alpha} e^{v_k\alpha} = e^{\alpha \sum_{k>N}(v_k+a_k^2)}.$$

Since $\sum v_k$ and $\sum a_k^2$ are both finite, we have

(5) $$I_2 \leq e^{\alpha r^2/2}$$

for some large $N$ which is independent of $\alpha$. Fix such $N$. From (4) and (5). we have

$$\prod_{k=1}^N \exp \frac{a_k^2\alpha}{1+v_k\alpha}(1+v_k\alpha) \geq e^{\alpha r^2/2}$$

namely

$$\prod_{k=1}^N \exp \frac{a_k^2\alpha}{1+v_k\alpha} \cdot \frac{\prod_{k=1}^N (1+v_k\alpha)}{e^{\alpha r^2/2}} \geq 1.$$

Letting $\alpha \uparrow \infty$, we have

$$\prod_{k=1}^N e^{a_k^2/v_k} \cdot 0 \geq 1,$$

which is a contradiction. This completes the proof.

*Cornell University*

# POISSON POINT PROCESSES ATTACHED TO MARKOV PROCESSES

KIYOSI ITÔ
CORNELL UNIVERSITY

## 1. Introduction

The notion of point processes with values in a general space was formulated by K. Matthes [4]. A point process is called *Poisson*, if it is $\sigma$-discrete and is a renewal process. We will prove in this paper that such a process can be characterized by a measure on the space of values, called the *characteristic measure*.

Let $X$ be a standard Markov process with the state space $S$. Fix a state $a \in S$ and suppose that $a$ is recurrent state for $X$. Let $S(t)$ be the inverse local time of $X$ at $a$. By defining $Y(t)$ to be the excursion of $X$ in $(S(t-), S(t+))$ for the $t$ value such that $S(t+) > S(t-)$, we shall obtain a point process called the *excursion point process* with values in the space of paths. Using the strong Markov property of $X$, we can prove that $Y$ is a Poisson point process. Its characteristic measure, called the *excursion law*, is a $\sigma$-finite measure on the space of paths. Although it may be an infinite measure, the conditional measure, when the values of the path up to time $t$ is assigned, is equal to the probability law of the path of the process $X$ starting at the value of the path at $t$ and stopped at the hitting time for $a$. Using this idea, we can determine the class of all possible standard Markov processes whose stopped process at the hitting time for $a$ is a given one.

We presented this idea in our lecture at Kyoto in 1969 [3] and gave the *integral representation* of the excursion law to discuss the jumping-in case in which the excursion starts outside $a$. P. A. Meyer [5] discussed the general case in which continuous entering may be possible by introducing the *entrance law*. In our present paper, we will prove the *integral representation* of the excursion law in terms of the *extremal excursion laws* for the general case. It is not difficult to parametrize the extremal excursion laws by the entrance Martin boundary points for the stopped process and to determine the generator of $X$, though we shall not discuss it here.

E. B. Dynkin and A. A. Yushkevich [1], [2] discussed a very general extension problem which includes our problem as a special case. We shall deal with their case from our viewpoint. The excursion point process defined similarly is no longer Poisson but will be called Markov. It seems useful to study point processes of Markov type in general.

This work was supported by a National Science Foundation Grant GP-19658.

## 2. Point functions

Let $U$ be an abstract Borel space associated with a $\sigma$-algebra $B(U)$ on $U$ whose member is called a Borel subset of $U$. Let $T_+$ be the open half line $(0, \infty)$ which is called the *time interval*. The product space $T_+ \times U$ is considered as a Borel space associated with the product $\sigma$-algebra $B(T_+ \times U)$ of the topological $\sigma$-algebra $B(T_+)$ on $T_+$ and the $\sigma$-algebra $B(U)$ on $U$.

A *point function* $p: T_+ \to U$ is defined to be a map from a countable set $D_p \subset T$ into $U$. The space of all point functions: $T_+ \to U$ is denoted by $\Pi = \Pi(T_+^-, U)$. For $p \in \Pi$ and $E \in B(T_+ \times U)$, we denote by $N(E, p)$ the number of the time points $t \in D_p$ for which $(t, p(t)) \in E$. The space $\Pi$ is regarded as a Borel space associated with the $\sigma$-algebra $B(\Pi)$ generated by the sets: $\{p \in \Pi: N(E, p) = k\}$, $E \in B(T_+ \times U)$, $k = 0, 1, 2, \cdots$.

We will introduce several operations in $\Pi$. Let $E \in B(T_+ \times U)$. The *restriction* $p|E$ of $p$ to $E$ is defined to be the point function $g$ such that

$$D_g = \{t \in D_p: (t, p(t)) \in E\} \tag{2.1}$$

and $g(t) = p(t)$ for $t \in D_g$.

Let $V \in B(U)$. The *range restriction* $p|_r V$ of $p$ to $V$ is defined to be the restriction $p|T_+ \times V$. Let $S \in B(T_+)$. The *domain restriction* $p|_d S$ is defined to be the restriction $p|S \times U$. Let $s \in T_+$. The *stopped point function* $\alpha_s p$ of $p$ at $s$ is defined to be the domain restriction of $p$ to $(0, s]$ and the *shifted point function* $g = \theta_s p$ is defined by $D_g = \{t: t + s \in D_p\}$ and $g(t) = p(t + s)$.

Let $p_n$ for $n = 1, 2, \cdots$, be a sequence of point processes. If the $D_{p_n}$, $n = 1, 2, \cdots$, are disjoint, then the direct sum $p = \Sigma_n p_n$ is defined by $D_p = \cup_n D_{p_n}$ and $p(t) = p_n(t)$ for $t \in D_{p_n}$.

## 3. Point processes

Let $\Pi$ be the space of all point functions: $T_+ \to U$ as in Section 2. Let $(\Omega, P)$ be a probability space, where $P$ is a complete probability measure on $\Omega$. A map $Y: \Omega \to \Pi$ is called a *point process* if it is measurable. The image of $\omega$ by $Y$ is denoted $Y_\omega$ and is called the sample point function of $Y$ corresponding to $\omega$. The value of $Y_\omega$ at $t$ if $t$ belongs to the domain of $Y_\omega$ is denoted by $Y_\omega(t)$. It follows from the definition that $N(E, Y_\omega)$ is measurable in $\omega$ and is therefore a random variable on $(\Omega, P)$ if $E \in B(T \times U)$. The probability law of $Y$ is clearly a probability measure on $(\Pi, B(\Pi))$.

The process $Y$ is called *discrete* if $N((0, t) \times U, Y) < \infty$ a.s. for every $t < \infty$. It is called *$\sigma$-discrete* if for every $t$ we can find $U_n = U_n(t) \in B(U)$, $n = 1, 2, \cdots$ such that $N((0, t) \times U_n, Y) < \infty$ a.s. for every $n$ and that $U = \cup_n U_n$.

The process $Y$ is called "*renewal*" if for every $t < \infty$ we have that $\alpha_t Y$ and $\theta_t Y$ are independent and that $\theta_t Y$ has the same probability law as $Y$ for every $t$. By virtue of the following theorem, we call a $\sigma$-discrete and renewal point process a *Poisson point process*.

THEOREM 3.1. *Let $Y$ be $\sigma$-discrete and renewal. Then $N(E_i, Y), i = 1, 2, \cdots, k$ are independent and Poisson distributed for every finite disjoint system $\{E_i\} \subset B(T_+ \times U)$.*

REMARK. A probability measure concentrated at $\infty$ is regarded as a special case of Poisson measure with mean $\infty$.

PROOF. Because of the $\sigma$-discreteness of $Y$ and the second condition of the renewal property, we can take $\{U_n\}$ independent of $t$ such that $N((0, t) \times U_n, Y) < \infty$ a.s. for every $t$ and such that $U = \bigcup_n U_n$. It can be also assumed that $U_n$ increases with $n$. Since

$$N(E, Y) = \lim_n N(E \cap (T_+ \times U_n), Y), \tag{3.1}$$

and since the Poisson property and the independence property are inherited by the limit of random variables, we can assume that all $E_i$ are included in $T_+ \times U_n$. By the renewal property of $Y$, we can easily see that $Z(t) \equiv N((0, t) \times U_n, Y)$ is a stochastic process with stationary independent increments and that its sample function increases only with jumps $=1$, a.s. This $Z(t)$ is therefore a Poisson process. This implies that $Z(t)$ is continuous in probability.

Now set $Z_i(t) = N(E_i, (0, t) \times U_n)$ and $Z^*(t) = \Sigma_i\, iZ_i(t)$. Since $Z^*(t) - Z^*(s) \leqq k(Z(t) - Z(S))$ for $s < t$, $Z^*$ is continuous in probability with $Z$. The process $Z^*$ has independent increments and its sample function increases with jumps. Thus, $Z^*$ is a Lévy process and $Z_i(t)$ is the number of jumps $= i$ of $Z^*$ up to time $t$. The $\{Z_i(t)\}_i$ are therefore independent and Poisson distributed. Since

$$N(E_i, Y) = \lim_{t \to \infty} Z_i(t), \tag{3.2}$$

the proof is completed.

As an immediate result of Theorem 2.1 we have the following theorem.

THEOREM 3.2. *In order for $Y$ to be a Poisson point process, it is necessary and sufficient that it be the sum of independent discrete Poisson point processes.*

## 4. Characteristic measure

Let $Y$ be a Poisson point process defined in Section 3 and set $m(E) = E_P(N(E, Y))$, where $E_P$ = expectation. Then $m$ is a measure on $T_+ \times U$ which is shift invariant in the time direction because of the second condition in the renewal property of $Y$. By the $\sigma$-discreteness of $Y$, we have $m((0, t) \times U_n) < \infty$ for the $U_n$ introduced in the proof of Theorem 3.1. This implies that $m$ is $\sigma$-finite. Therefore, $m$ is the product measure of the Lebesgue measure on $T_+$ and a unique $\sigma$-finite measure $n$ on $U$. The measure $n$ is called the *characteristic measure* $n$ of $Y$ by virtue of the following theorem.

THEOREM 4.1. *The probability law of a Poisson point process $Y$ is determined by its characteristic measure $n$.*

PROOF. The measure $n$ determines the joint distribution of $N(E_i, Y)$, $i = 1, 2, \cdots, k$ for disjoint $E_i \in B(T_+ \times U)$ by Theorem 3.1. Since $N(E, Y)$ is additive in $E$, it is also true for nondisjoint $E_i$. This completes the proof.

The following theorem shows that any arbitrary $\sigma$-finite measure on $U$ induces a Poisson point process.

THEOREM 4.2. *Let $n$ be a $\sigma$-finite measure on $U$. Then there exists a Poisson point process whose characteristic measure is $n$.*

Before proving this, we will study the structure of a Poisson point process whose characteristic measure is finite.

THEOREM 4.3. *A Poisson point process is discrete if and only if its characteristic measure is finite.*

PROOF. Observe that the number of $s \in D_Y \cap (0, t)$ is $N\big((0, t) \times U, Y\big)$.

THEOREM 4.4A. *Let $Y$ be a discrete Poisson point process with characteristic measure $n$. (Then $n$ is a finite measure by the previous theorem.) Let $D_Y$ be $\tau_1(\omega)$, $\tau_2(\omega), \cdots$, and let $\xi_i(\omega) = Y_\omega\big(\tau_i(\omega)\big)$. Then we have the following:*

(i) $\tau_i - \tau_{i-1}$, $i = 1, 2, \cdots$, $(\tau_0 = 0)$, $\xi_1, \xi_2, \cdots$ *are independent;*

(ii) $\tau_i - \tau_{i-1}$ *is exponentially distributed with mean* $1/n(U)$, *that is,* $P(\tau_i - \tau_{i-1} > t) = e^{-tn(U)}$;

(iii) $P(\xi_i \in A) = n(A)/n(U)$, $A \in B(U)$.

PROOF. Let $\alpha_i > 0$ and $V_i \in B(U)$, $i = 1, 2, \cdots, k$. Write $\phi_p(t)$ for $\big([pt] + 1\big)/p$, $[t]$ being the greatest integer $\leqq t$. Then we have

$$E_P\left[\exp\left\{-\sum_{i=1}^{k} \alpha_i \tau_i\right\}, \xi_i \in V_i, i = 1, 2, \cdots, k\right] \tag{4.1}$$

$$= \lim_{p\to\infty} E_P\left[\exp\left\{-\sum_{i=1}^{k} \alpha_i \phi_p(\tau_i)\right\}, \xi_i \in V_i, \tau_i - \tau_{i-1} > \frac{1}{p}, \right.$$

$$\left. i = 1, 2, \cdots, k\right]$$

$$= \lim_{p\to\infty} \sum_{0<\nu_1<\cdots<\nu_k} \exp\left\{-\sum_{i=1}^{k} \frac{\alpha_i \nu_i}{p}\right\} P\left(\xi_i \in V_i, \frac{\nu_i - 1}{p} < \tau_i \leqq \frac{\nu_i}{p},\right.$$

$$\left. i = 1, 2, \cdots, k\right).$$

The event in the second factor can be expressed as

$$\left.\begin{aligned} &N\left(Y, \left(\frac{\nu_i - 1}{p}, \frac{\nu_i}{p}\right] \times V_i\right) = 1 \\ &N\left(Y, \left(\frac{\nu_i - 1}{p}, \frac{\nu_i}{p}\right] \times (U - V_i)\right) = 0 \end{aligned}\right\} \quad \text{for } i = 1, 2, \cdots, k, \tag{4.2}$$

$$N\left(Y, \left(\frac{\nu - 1}{p}, \frac{\nu}{p}\right] \times U\right) = 0 \quad \text{for } \nu \neq \nu_1, \cdots, \nu_k, \nu \leqq \nu_k. \tag{4.3}$$

Using Theorem 3.1, we can easily see that its probability is

$$\exp\left\{\frac{-v_k n(U)}{p}\right\}\left(\frac{1}{p}\right)^k \prod_{i=1}^{k} n(V_i). \tag{4.4}$$

Therefore, the above limit is expressed as an integral form

$$\begin{aligned} &E_P\left[\exp\left\{-\sum_{i=1}^{k}\alpha_i\tau_i\right\}, \xi_i \in V_i, i = 1, 2, \cdots, k\right] \\ &\quad = \prod_{i=1}^{k} n(V_i) \int\cdots\int_{0<t_1<\cdots<t_k} \exp\left\{-\sum_{i=1}^{k}\alpha_i t_i - t_k n(U)\right\} dt_1 \cdots dt_k. \end{aligned} \tag{4.5}$$

Changing variables in the integral, we obtain

$$\begin{aligned} &E_P\left(\exp\left\{-\sum_{i=1}^{k}\beta_i(\tau_i - \tau_{i-1})\right\}, \xi_i \in V_i, i = 1, 2, \cdots, k\right) \\ &\quad = \prod_{i=1}^{k} n(V_i) \int\cdots\int_{s_1\cdots s_k>0} \exp\left\{-\sum_{i=1}^{k}\beta_i s_i - \left(\sum_{i=1}^{k} s_i\right) n(U)\right\} ds_1 \cdots ds_k \end{aligned} \tag{4.6}$$

for $\beta_1 > \beta_2 > \cdots > \beta_k > 0$. This is true for $\beta_1, \beta_2, \cdots, \beta_k > 0$ by analytic continuation in $\beta_i$. It is now easy to complete the proof.

This theorem suggests a method to construct a Poisson point process whose characteristic measure is a given finite measure.

THEOREM 4.4B. *Let $n$ be a finite measure on $U$. Suppose that $\sigma_1, \sigma_2, \cdots, \xi_1, \xi_2, \cdots$ are independent and that*

$$P(\sigma_i > t) = \exp\{-tn(U)\}, \qquad P(\xi_i \in A) = \frac{n(A)}{n(U)}. \tag{4.7}$$

Define a point process $Y$ by

$$D_Y = \{\sigma_1, \sigma_1 + \sigma_2, \cdots\}, \qquad Y(\sigma_1 + \cdots + \sigma_k) = \xi_k. \tag{4.8}$$

Then $Y$ is a Poisson point process with characteristic measure $= n$.

Now we will prove Theorem 4.2. The case $n(U) < \infty$ has been discussed above. If $n(U) = \infty$, then we have a disjoint countable decomposition of $U$: $U = \cup_i U_i$, $n(U_i) < \infty$. Let $n_i(A) = n(A \cap U_i)$. Then $n_i$ is a finite measure on $U$ and we have a Poisson point process $Y_i$ with characteristic measure $n_i$. We can assume that $Y_1, Y_2, \cdots$, are independent. First we will remark that

$D_{Y_i}$, $i = 1, 2, \cdots$, are disjoint a.s. In fact for $i \neq j$,

(4.9)

$$P(D_{Y_i} \cap D_{Y_j} \cap (0, t) \neq \varnothing)$$
$$\leqq \sum_{k=1}^{p} P\left(N\left(U \times \left(\frac{(k-1)t}{p}, \frac{kt}{p}\right], Y_i\right) \neq 0, N\left(U \times \left(\frac{(k-1)t}{p}, \frac{kt}{p}\right], Y_j\right) \neq 0\right)$$
$$\leqq t^2 n_i(U) n_j(U) \left(\frac{1}{k}\right)^2 k \to 0,$$

as $k \to \infty$. By letting $t \to \infty$, we have $P(D_{Y_i} \cap D_{Y_j} \neq \varnothing) = 0$ for $i, j$ fixed. Therefore, $D_{Y_1}, D_{Y_2}, \cdots$, are disjoint a.s.

Let $Y$ be the sum of $Y_1, Y_2, \cdots$, (see the end of Section 1). It is easy to show that $Y$ is a Poisson point process whose characteristic measure is $n$.

Let $\varphi: T \times U \twoheadrightarrow [0, \infty)$ be measurable $B(T \times U)/B[0, \infty)$. Then we have the following result.

THEOREM 4.5. *With the convention that* $\exp\{-\infty\} = 0$, *we have*

$$E_P[\exp\{-\alpha \sum_{t \in D_Y} \varphi(t, Y_t)\}] = \exp\left\{\iint_{T_+ \times U} (\exp\{-\alpha\varphi(t, u)\} - 1)\, dt\, n(du)\right\}. \tag{4.10}$$

PROOF. If $\varphi$ is a simple function, this follows from Theorem 3.1 and the definition of $n$. We can derive the general case by taking limits.

Let $\Phi$ be a random variable with values in $[0, \infty]$. The condition $\Phi < \infty$ a.s. is equivalent to

$$\lim_{\alpha \to 0+} E(\exp\{-\alpha\Phi\}) = 1. \tag{4.11}$$

Using this fact, we get the following theorem from Theorem 4.5.

THEOREM 4.6. *The condition*

$$\sum_{t \in D_Y} \varphi(t, Y_t) < \infty \text{ a.s.} \tag{4.12}$$

*is equivalent to*

$$\iint_{T \times U} \varphi(t, u) \wedge 1\, dt\, n(du) < \infty. \tag{4.13}$$

REMARK. This condition is also equivalent to

$$\iint_{T \times U} (1 - \exp\{-\varphi(t, u)\})\, dt\, n(du) < \infty. \tag{4.14}$$

THEOREM 4.7. *Let $Y$ be a Poisson point process with values in $U$. The range restriction $Y^*$ of $Y$ to a set $U^* \in B(U)$ is also a Poisson point process whose characteristic measure is the restriction of that of $Y$ to $U^*$.*

## 5. The strong renewal property of Poisson point processes

Let $Y$ be a Poisson point process and $B_t(Y)$ be the $\sigma$-algebra generated by the stopped process $\alpha_t Y$. It is easy to see that $B_t(Y)$ is right continuous, that is, $B_t(Y) = \bigcap_{s>t} B_s(Y)$ a.s., where two $\sigma$-algebras are said to be equal a.s. if every member of one $\sigma$-algebra differs from a member of the other by a null set.

Let $\sigma$ be a stopping time with respect to the increasing family $B_t(Y)$, $t \in T_+$. Suppose that $\sigma < \infty$ a.s. Then we have the following.

THEOREM 5.1 (Strong Renewal Property). *The process $Y$ has the strong renewal property:*

(i) *$\alpha_\sigma Y$ and $\theta_\sigma Y$ are independent;*

(ii) *$\theta_\sigma Y$ has the same probability law as $Y$.*

The idea of the proof is the same as that of the proof of the strong Markov property in the theory of Markov processes and is omitted.

## 6. The recurrent extension of a Markov process at a fixed state

Let $S$ be a locally compact metric space. Let $\delta$ stand for the *cemetery*, an extra point to be added to $S$ as an isolated point. Denote the topological $\sigma$-algebra on $S$ by $B(S)$ and let $T$ stand for the closed half line $[0, \infty)$ with the topological $\sigma$-algebra $B(T)$.

Let $U$ stand for the space of all right continuous functions: $T \to S \cup \{\delta\}$ with left limits. Let $B(U)$ denote the $\sigma$-algebra on $U$ generated by the cylinder Borel subsets of $U$. It is the same as the topological $\sigma$-algebra with respect to the Skorohod topology in $U$. A member of $U$ is often called a *path*. The hitting time for $a$, the stopped path at $t$, and the shifted path at $t$ are denoted by $\sigma_a(u)$, $\alpha_t(u)$, and $\theta_t(u)$ as usual. To avoid typographical difficulty, we write $\alpha_a(u)$ for the stopped path at $\sigma_a(u)$.

Let $X = (X_t, P_b)$ be a standard Markov process with the state space $S$, where $P_b$ denotes the probability law of the path starting at $b$ which is clearly a probability measure on $(U, B(U))$, completed if necessary.

The process $X$ stopped at the hitting time $\sigma_a$ for $a$ is also a standard Markov process which is denoted by $\alpha_a X$. The state $a$ is a trap for $\alpha_a X$.

Let $X^0 = (X_t^0, P_b^0)$ be a standard Markov process with a trap at $a$. Any standard Markov process $X = (X_t, P_b)$ with the state space $S$ is called a *recurrent extension* of $X^0$ at $a$ if $\alpha_a X$ is equivalent to $X^0$ and if $a$ is a *recurrent state* for $X$, that is, $P_a(\sigma_a < \infty) = 1$. The process $X^0$ itself is a recurrent extension of $X^0$, but there are many other extensions. Our problem is to determine all possible recurrent extensions of $X^0$. We will exclude the trivial extension $X^0$ from our discussion.

Let $X$ be a recurrent extension of $X^0$. We will exclude the trivial extension. Since $P_a(\sigma_a < \infty) = 1$, we have two cases.

*Case* 1 (Discrete Visiting Case). $P_a(0 < \sigma_a < \infty) = 1$. In this case the visiting times of the path of $X$ at $a$ form a discrete set.

*Case* 2. $P_a(\sigma_a = 0) = 1$. In this case we have exactly one additive functional

$A(t)$ such that

$$E_b(e^{-\sigma_a}) = E_b\left(\int_0^\infty e^{-t}\, dA(t)\right). \tag{6.1}$$

This function $A(t)$ is called the Blumenthal–Getoor *local time* of $X$ at $a$. The path of $A(t)$ increases continuously from 0 to $\infty$ with $t$. This case is divided into two subcases. Let $\tau_a$ denote the exit time from $a$.

*Case* 2(*a*) (Exponential Holding Case). $\tau_a$ is exponentially distributed with finite and positive mean.

*Case* 2(*b*). (Instantaneous Case). $P_a(\tau_a = 0) = 1$.

We will define the excursion process $Y$ of $X$ with respect to $P_a$. In case 1, $Y$ is a sequence of random variables with values in $U$, $Y_k = \alpha_{\sigma(k)}(\theta_{\sigma(k-1)}X)$, where $\sigma(0) = 0$ and $\sigma(k)\,(k > 0)$ is the $k$th hitting time for $a$. Since $Y_k$ is a random variable with values in $U$ for each $k$, it is a stochastic process and

$$Y_k(t) = \begin{cases} X(\sigma(k-1) + t), & 0 \leqq t < \sigma(k) - \sigma(k-1), \\ a, & t \geqq \sigma(k) - \sigma(k-1). \end{cases} \tag{6.2}$$

By the strong Markov property of $X$ at $\sigma(k)$, we can easily prove that all $Y_k$ have the same probability law. In Case 2, $Y$ is a point process with values in $U$:

$$\begin{aligned} D_Y &= \{s : S(s+) - S(s-) > 0\}, & S(t) &= A^{-1}(t), \\ Y_s &= \alpha_a(\theta_{S(s-)}X), & s &\in D_Y; \end{aligned} \tag{6.3}$$

the second equation can be written as

$$Y_s(t) = \begin{cases} X(S(s-) + t), & 0 \leqq t < S(s+) - S(s-), \\ a, & t \geqq S(s+) - S(s-). \end{cases} \tag{6.4}$$

The process $Y$ is discrete in Case 2(a), but not in Case 2(b). Even in Case 2(b) $Y$ is $\sigma$-discrete, because $S(s+) - S(s-) > 1/k$ is possible only for a finite number of $s$ values in every finite time interval. $Y$ is also renewal, as we can prove by the strong Markov property of $X$. Therefore, the excursion process is a Poisson point process in Case 2.

The *excursion law* of $X$ at $a$ is defined to be the common probability law of $Y_k$ in Case 1 and the characteristic measure of $Y$ in Case 2.

THEOREM 6.1. *The excursion law $n$ of $X$ and the probability laws $\{P_b^0\}_{b \neq a}$ determine the probability laws $\{P_b\}$ of $X$.*

PROOF. Because of the strong Markov property, it is enough to prove that $P_a$ is determined by $n$ and $\{P_b^0\}_{b \neq a}$. Since $n$ determines $Y$, this is obvious in Case 1. We will discuss Case 2. Since $S(s)$ is a subordinator, that is, an increasing homogeneous Lévy process, we have

$$S(s-) = ms + \sum_{t \leqq s} (S(t-) - S(t-)) = ms + \sum_{t \leqq s, t \in D_y} \sigma_a(Y_t), \tag{6.5}$$

$m$ being a nonnegative constant, called the *delay coefficient* of $X$ at $a$. Therefore, $S$ is determined by $m$ and $Y$. Since $S(s)$ is increasing and since we have

$$X_t = \begin{cases} Y_s(t - S(s-)), & S(s-) \leqq t < S(s+), \\ a, & S(s-) = t = S(s-), \end{cases} \tag{6.6}$$

a.s. $(P_a)$, the probability law $P_a$ is determined by $m$ and the probability law of $Y$. But the latter is determined by $n$. Since $m$ is also determined by $n$ by the theorem below, $P_a$ is determined by $n$.

THEOREM 6.2. *In Case 2, we have*

$$m = 1 - \int_U (1 - \exp\{-\sigma_a(u)\}) n(du). \tag{6.7}$$

PROOF. By the definition of the local time, we have

$$\begin{aligned} 1 &= E_a(\exp\{-\sigma_a\}) = E_a\left(\int_0^\infty \exp\{-t\}\, dA(t)\right) \\ &= E_a\left(\int_0^\infty \exp\{-S(s)\}\, ds\right) = \int_0^\infty E_a(\exp\{-S(s)\})\, ds \\ &= \int_0^\infty \exp\{-ms\}\, E_a\left[\exp\left\{-\sum_{t \leqq s, t \in D_y} \sigma_a(Y_t)\right\}\right] ds \\ &= \int_0^\infty \exp\{-ms\} \exp\left\{-s\int_U (1 - \exp\{-\sigma_a(u)\}) n(du)\right\} ds \end{aligned} \tag{6.8}$$

by Theorem 4.5, which equals $[m - \int_U (1 - \exp\{-\sigma_a(u)\}) n(du)]^{-1}$. This completes the proof.

THEOREM 6.3. *The excursion law satisfies the following conditions:*

(i) *$n$ is concentrated on the set $U' \equiv \{u : 0 < \sigma_a(u) < \infty,\ u(t) = a$ for $t \geqq \sigma_a(u)\}$;*

(ii) *$n\{u \in U : u(0) \notin V(a)\} < \infty$ for every neighborhood $V(a)$ of $a$;*

(ii′) *$\int_U (1 - \exp\{-\sigma_a(u)\}) n(du) \leqq 1$;*

(iii) *$n\{u : \sigma_a(u) > t,\ u \in \Lambda_t,\ \theta_t u \in M\} = \int_{\Lambda_t \cap (\sigma_a > t)} P^0_{u(t)}(M) n(du)$ for $t > 0$, $\Lambda_t \in B_t(U)$ and $M \in B(U)$;*

(iii′) *$n(u\colon u(0) \in B,\ u \in M) = \int_{u(0) \in B} P^0_{u(0)}(M) n(du)$ for $B \in B(S - \{a\})$ and $M \in B(U)$.*

PROOF. Condition (i) is obvious. Condition (ii′) is obvious in Case 1 and it follows at once from Theorem 6.2 in Case 2. Condition (ii) is obvious in Case 1. To prove it in Case 2, consider the restriction $Y^*$ of $Y$ to $U^* = \{u : u(0) \notin V(a)\}$. Then $t \in D_{Y^*}$ implies $X(S(t-)-) = a$ and $X(S(t-)+) \notin V(a)$. Since the set of such $t$ values is discrete, $Y^*$ is discrete. But $Y^*$ is a Poisson point process with the characteristic measure $= n|U^*$ (Theorem 4.7). Therefore, the total measure of $n|U^*$ is finite and so we have $n(U^*) < \infty$. Condition (iii) is obvious and condition (iii′) is trivial in Case 1.

To prove (iii) in Case 2, consider the measures (Meyer's entrance law):

$$r_t(B) = n\{u : \sigma_a(u) > t,\ u(t) \in B\} \qquad B \in B(S),\ t > 0. \tag{6.9}$$

First we will prove that for $B \in B(S)$ and $M \in B(U)$,

$$n\{n : \sigma_a(u) > t,\ u(t) \in B,\ \theta_t u \in M\} = \int_B r_t(db) P_b(M). \tag{6.10}$$

Let $V = \{u \in U : \sigma_a(u) > t\}$ and $Z = Y|_r V$. Since $s \in D_Z$ implies $S(s+) - S(s-) > t$, the set of such $s$ values is discrete. Therefore, $Z$ is a discrete Poisson point process with the characteristic measure $= n \mid V$. This implies that $n(V) < \infty$. Let $\tau$ be the first time in $D_Y$. By Theorem 4.4A we have

$$P_a(\alpha_a(\theta_{S(\tau-)}X) \in M') = \frac{n(M' \cap V)}{n(V)}, \qquad M' \in B(U). \tag{6.11}$$

Setting $M' = \{u : \sigma_a(u) > t,\ u(t) \in B,\ \theta_t u \in M\}$, $M \in B(U)$, we have

$$\begin{aligned} &n\{u : \sigma_a(u) > t,\ u(t) \in B,\ \theta_t u \in M\} \\ &\quad = n(V) P_a(\sigma_a(\theta_{S(\tau-)}X) > t,\ X_{S(\tau-)+t} B,\ \alpha_a(\theta_{S(\tau-)+t}(X)) \in M). \end{aligned} \tag{6.12}$$

Since $S(\tau-) + t = \inf\{\alpha > 0 : X_s \neq a \text{ for } \alpha - t < s \leqq \alpha\}$, $S(\tau-) + t$ is a stopping time for $X$. Since $\sigma_a(\theta_{S(\tau-)}X) > t$ is the same as $X_{S(\tau-)+t} \neq a$, this event is measurable $(B_{S(\tau-)+t})$. Therefore, we have

$$\begin{aligned} &n\{u : \sigma_a(u) > t,\ u(t) \in B,\ \theta_t u \in M\} \\ &\quad = n(V) \int_S P_a(\sigma_a(\theta_{S(\tau-)}X) > t,\ X_{S(\tau-)+t} \in db) P_b(X \in M). \end{aligned} \tag{6.13}$$

Setting $M = \{u : u(0) \in B\}$, we have

$$\begin{aligned} r_t(B) &\equiv n\{u : \sigma_a(u) > t,\ u(t) \in B\} \\ &= n(V) P_a(\sigma_a(\theta_{S(\tau-)}X) > t,\ X_{S(\tau-)+t} \in B). \end{aligned} \tag{6.14}$$

Putting this in the above formula, we obtain (6.9). Equation (6.9) can be written as

$$n\{u : \sigma_a(u) > t,\ u(t) \in B,\ \theta_t u \in M\} = \int_{(u(t) \in B) \cap (\sigma_a(u) > t)} P_{u(t)}(M) n(du). \tag{6.15}$$

Using this and the Markov property of $X$ and noticing that $\sigma_a(u) > t$ implies $\sigma_a(u) > s$ and $\sigma_a(u) = s + \sigma_a(\theta_s u)$ for $s \leqq t$, we get

$$\begin{aligned} &n\{u : \sigma_a(u) > t,\ u(t_1) \in B_1, \cdots, u(t_k) \in B_k,\ \theta_t u \in M\} \\ &\quad = \int_{u(t_1) \in B_1, \cdots, u(t_k) \in B_k, \sigma_a(u) > t} P_{u(t)}(M) n(du), \end{aligned} \tag{6.16}$$

which implies (iii). A similar and even simpler argument shows (iii′).

THEOREM 6.4. *In order for a $\sigma$-finite measure $n$ on $U$ to be the characteristic measure of a recurrent extension of $X^0$ it is necessary that $n$ satisfies the following conditions in addition to* (i), (ii), (ii′), (iii) *and* (iii′) *in Theorem* 6.3:

*Case* 1 (*Discrete Visiting Case*). *$n$ is a probability measure concentrated on* $U^a \equiv \{u \in U' : u(0) = a\}$;

*Case* 2(*a*) (*Exponential Holding Case*). *$n$ is a finite but not identically zero measure concentrated on* $U^+ \equiv \{u \in U' : u(0) \neq a\}$ *such that* $\int_U (1 - \exp\{-\sigma_a(u)\}) n(du) < 1$;

*Case* 2(*b*) (*Instantaneous Case*). *$n$ is an infinite measure such that* $n(U^a) = 0$ or $\infty$.

PROOF. First we will prove the necessity of the conditions. In Case 1, $n$ is the probability law of the path of $X^0$ starting at $a$ and is therefore concentrated in $U^a$. In Case 2, $n$ is proportional to the probability law (with respect to $P_a$) of $\alpha_a(\theta_\tau X)$, when $\tau$ is the exit time from $a$. Since $\infty > \tau > 0$ a.s., $X(\tau) \equiv \alpha_a(\theta_\tau X)(0) \neq a$ by the strong Markov property of $X$. This implies that $n$ is concentrated in $U^+$. Since the local time of $X$ at $a$ is proportional to the actual visiting time in this case, the delay coefficient $m$ must be positive. Thus, we have the inequality. In Case 3, $n$ must be an infinite measure, because the excursion process $Y$ is not discrete. If $0 < n(U^a) < \infty$, then $n(U^+) = \infty$ and $Y^a \equiv Y|_r U^a$ would be a discrete Poisson point process. Let $\tau$ be the first time in $D_{Y^a}$. Then $S(\tau-)$ would be a stopping time for $X$, that is,

$$S(\tau-) = \inf\{t : X_t = a \text{ and there exists } t' > t \text{ for all } s \in (t, t') X_s \neq a\} \tag{6.17}$$

and the strong Markov property of $X$ would be violated at $S(\tau-)$.

For the proof of the sufficiency, it is enough to construct the path of $X$ starting at $a$ with the excursion measure $= n$. First we construct the Poisson point process $Y$ with the characteristic measure $= n$ by Theorem 4.2. It is easy to construct the path of $X$ starting at $a$ whose excursion process has the same probability law as $Y$ by reversing the procedure of deriving the excursion process from a Markov process.

## 7. The integral representation of the excursion law

Let $X^0$ be a standard process on $S$ with a trap at $a \in S$ and $\mathscr{E}(X^0)$ be the set of all $\sigma$-finite measures on $U$ satisfying the five conditions (i), (ii), (ii′), (iii) and (iii′) in Theorem 6.3. Define the norm $n \in \mathscr{E}(X^0)$ by

$$\|n\| = \int_U (1 - \exp\{-\sigma_a(u)\}) n(du). \tag{7.1}$$

Let $\mathscr{E}_1(X^0)$ denote the set of all $n \in \mathscr{E}(X^0)$ with $\|n\| = 1$. Clearly, $\mathscr{E}_1(X^0)$ is a convex set. This suggests that any $n \in \mathscr{E}(X^0)$ has an integral representation in terms of extremal ones. The measure $n \in \mathscr{E}_1(X^0)$ is called *extremal* if

$$n = c_1 u_1 + c_2 u_2 (c_1, c_2 > 0, c_1 + c_2 = 1, u_1, u_2 \in \mathscr{E}_1(X)) \tag{7.2}$$

implies $u_1 = u_2 = u$.

Suppose that $n$ is concentrated on $U^+ = \{u \in U' : u(0) \neq a\}$. Then condition (iii′) implies

$$n(\cdot) = \int_{b \neq a} k(db) P_b(\cdot),\ k(B) = n\{u : u(0) \in B\}. \tag{7.3}$$

Since $P_b(\cdot)$, $b \neq a$, satisfies all conditions (i) to (iii′). it belongs to $\mathscr{E}(X^0)$, and therefore we have

$$n_b \equiv \frac{P_b(\cdot)}{\|P_b\|} = \frac{P_b(\cdot)}{E_b(1 - \exp\{-\sigma_a\})} \in \mathscr{E}_1(X). \tag{7.4}$$

Therefore (7.3) can be written as

$$n(\cdot) = \int_{b \in S - \{a\}} \lambda(db) n_b(\cdot). \qquad \lambda(db) = E_b(1 - \exp\{-\sigma_a\}) k(db). \tag{7.5}$$

Since $k(S - V(a)) < \infty$ for every neighborhood $V(a)$ of $a$ and

$$\int_{S - \{a\}} E_b(1 - \exp\{\sigma_a\}) k(db) = 1 \tag{7.6}$$

by (ii) and $\|u\| = 1$, $\lambda$ satisfies

$$\int_{S - V(a)} \frac{\lambda(db)}{E_b(1 - \exp\{-\sigma_a\})} < \infty \qquad \text{for every } V(a) \tag{7.7}$$

and $\lambda(S - \{a\}) = 1$.

Using (7.5) and noticing that $n_b = c_1 n_1 + c_2 n_2$, $c_1, c_2 > 0$, implies that both $n_1$ and $n_2$ are concentrated on $U^b \equiv \{u \in U' : u(0) = b\}$, we can easily prove that $n_b$, $b \neq a$, is extremal.

Suppose that $u \in \mathscr{E}_1(X^0)$ and $B \in B(S)$ and set

$$U^B = \{u \in U : u(0) \in B\}, \qquad n^B = n \,|\, U^B. \tag{7.8}$$

If $n^B \neq 0$, then $n^B / \|n^B\| \in \mathscr{E}_1(X^0)$. Using this fact, we can easily see that if $n \in \mathscr{E}_1(X^0)$ is extremal, then $n$ is concentrated in $U^b$ for some $b \in S$. If $b \neq a$, then $n = n_b$ by (7.5). However, there are many extremal ones concentrated in $U^a$. Let $N^a$ be the set of all such extremal ones and write $B(N^a)$ for the $\sigma$-algebra on $N^a$ generated by the sets $\{v \in N^a : v(\Lambda) < c\}$, $\Lambda \in B(U)$, $c > 0$.

If $n$ is concentrated on $U^a$, then $n(\cdot) = \int_{N^a} v(\cdot) \lambda(dv)$, where $\lambda$ is a probability measure on $N^a$. Once this is proved, we can easily obtain the following theorem.

THEOREM 7.1. *The measure $n \in \mathscr{E}_1(X^0)$ can be expressed uniquely as*

$$n(\cdot) = \int_{b \neq a} n_b(\cdot) \lambda(db) + \int_{N^a} v(\cdot) \lambda(dv) \tag{7.9}$$

*with a probability measure $\lambda$ on $(S - \{a\}) \cup N^a$.*

PROOF. Since $n$ is concentrated on $U^a$, we will regard $n$ as a measure on $U^a$ from now on. Let $\Omega$ be the product space $T \times U^a$ associated with the product $\sigma$-algebra. We introduce a probability measure $Q$ on $\Omega$ by

$$Q(d\tau du) = 1_{\tau<\sigma_a(u)} \exp\{-\tau\}\, d\tau\, n(du), \tag{7.10}$$

where $1_{\tau<\sigma_a(u)}$ denotes the indicator of the set $\{(\tau, u) \in \Omega : \tau < \sigma_a(u)\}$. It is easy to see that $Q(\Omega) = \|n\| = 1$. We will use the same notation for a measure and the integral based on it, for example $Q(f) = \int_\Omega f(\omega) Q(d\omega)$. We also use the same notation for a $\sigma$-algebra and for the class of all bounded real functions measurable with respect to it.

Consider a stochastic process $Z_t(\omega)$ on $(\Omega, Q)$ defined by

$$Z_t(\omega) = Z_t(\tau, u) = \begin{cases} u(t) & \text{for } t < \tau, \\ a & \text{for } t \geqq \tau, \end{cases} \tag{7.11}$$

and let $B_t(Z)$ denote the $\sigma$-algebra generated by $Z_s$, $s \leqq t$. We will first prove that

$$Q[1_{t<\tau} g(\theta_t u) | B_t(Z)] = 1_{t<\tau} n_{u(t)}((1 - \exp\{-\sigma_a\})g) \quad \text{a.s. } (Q) \tag{7.12}$$

for every $g \in B(U^a)$. For this purpose it is enough to prove that

$$\begin{aligned} &Q[f_1(Z(t_1)) \cdots f_n(Z(t_n)) 1_{t<\tau} g(\theta_t u)] \\ &\quad = Q[f_1(Z(t_1)) \cdots f_n(Z(t_n)) 1_{t<\tau} n_{u(t)}((1 - \exp\{-\sigma_a\})g)], \end{aligned} \tag{7.13}$$

for $t_1 < t_2 < \cdots < t_n \leqq t$ and $f_i \in B(S_i)$. The $Z(t_i)$ can be replaced by $u(t_i)$ in the above equation because of the factor $1_{t<\tau}$. It is therefore enough to prove that

$$Q[f(u) 1_{t<\tau} g(\theta_t u)] = Q[f(u) 1_{t<\tau} n_{u(t)}(1 - \exp\{-\sigma_a)g] \tag{7.14}$$

for every $f \in B_t(U)$. Integrating by $d\tau$ and using property (iii), we can prove that both sides are equal to

$$\int f(u) 1_{t<\sigma_a} \exp\{-t\}\, E^0_{u(t)}((1 - \exp\{-\sigma_a\})g) n(du). \tag{7.15}$$

The set $U^a$ is a Borel subset of $U$ with respect to the Skorohod topology whose topological $\sigma$-algebra is the same as $B(U)$. Therefore letting $B_{0+}(Z) = \bigcap_{t>0} B_t(Z)$, we can define on $(U, B(U))$ the *conditional probability measure* $\tilde{Q}(\cdot | B_{0+}(Z))$ of the random variable $\omega = (\tau, u) \rightsquigarrow u$. Define a measure $\nu_\omega$ on $(U, B(U))$ depending on $\omega$ by

$$\nu_\omega(du) = \frac{1}{1 - \exp\{-\sigma_a(u)\}} \tilde{Q}(du | B_{0+}). \tag{7.16}$$

Then we have

$$\nu_\omega(g) = Q\left(\frac{g(u)}{1 - \exp\{-\sigma_a(u)\}} \middle| B_{0+}\right) \quad \text{a.s. } (Q). \tag{7.17}$$

The measure $\nu_\omega$ is clearly an $N^a$ valued function on $\Omega$, measurable with respect to $B_{0+}(Z)$. Let $\lambda$ be the probability law of this random variable.

Let $g \in B(U^a)$. Then we have

$$\begin{aligned}(7.18)\qquad \int_{N^a} \nu(g)\lambda(d\nu) &= Q(\nu_\omega(g)) = Q\left(\frac{g(u)}{1 - \exp\{-\sigma_a(u)\}}\right)\\ &= \int_{U^a}\int_0^{\sigma_a(u)} e^{-\tau}\, d\tau \frac{g(u)}{1 - \exp\{-\sigma_a(u)\}} n(du)\\ &= n(g).\end{aligned}$$

To complete the proof, we need only prove $\nu_\omega \in \mathscr{E}_1(X^0)$. The only difficult condition we have to verify is (iii), that is,

$$(7.19)\qquad \nu_\omega(f(u)g(\theta_t u)) = \nu_\omega(f(u)E^0_{u(t)}(g))$$

for $t > 0$, $f \in B_t(U^a)$ and $g \in B(U^a)$. We have to prove this except on a null set which is independent of $t$, $f$ and $g$. First fix $t > 0$. Since $\nu_\omega$ and $P^0_b$ are measures, it is enough to prove (7.19) in case $f$ and $g$ are of the following form:

$$(7.20)\qquad \begin{aligned} f(u) &= f_1(u(t_1))f_2(u(t_2)) \cdots f_k(u(t_k)), &\qquad 0 < t_1 < \cdots < t_k,\\ g(u) &= g_1(u(s_1))g_2(u(s_2)) \cdots g_m(u(s_m)), &\qquad 0 < s_1 < \cdots < s_m,\end{aligned}$$

where $t_i$ and $s_j$ are taken from a fixed countable dense subset of $(0, \infty)$ and $f_i$ and $g_j$ are taken from a fixed countable number of bounded continuous functions on $S$ which form a ring generating the $\sigma$-algebra on $S$. Take $\delta < t_1$ and write $f_\delta$ for $f \circ \theta_{-\delta}$. Then we have

$$\begin{aligned}(7.21)\qquad \nu_\omega(f(u)g(\theta_t u)) &= Q\left(\frac{f(u)g(\theta_t u)}{1 - \exp\{-\sigma_a(u)\}}\middle| B_{0+}(Z)\right)\\ &= \lim_{\delta \to 0+} Q\left(1_{\delta<\tau}\frac{f_\delta(\theta_\delta u)g(\theta_{t-\delta}\theta_\delta u)}{1 - \exp\{-\sigma_a(\theta_\delta(u))\}}\middle| B_{0+}(Z)\right).\end{aligned}$$

Noticing that $Q(\cdot|B_{0+}(Z)) = Q(Q(\cdot|B_t(Z))|B_{0+}(Z))$ and using (7.14), we can see that the expression above is

$$\begin{aligned}(7.22)\qquad &\lim_{\delta \to 0+} Q(1_{\delta<\tau}n_{u(\delta)}(f_\delta(u)g(\theta_{t-\delta}u))|B_{0+}(Z))\\ &= \lim_{\delta \to 0+} Q(1_{\delta<\tau}n_{u(\delta)}(f_\delta(u)\cdot E_{u(t-\delta)}(g))|B_{0+}(Z)).\end{aligned}$$

By a similar argument, we have

$$\begin{aligned}(7.23)\qquad \nu_\omega(f(u)E_{u(t)}(g)) &= Q\left(\frac{f(u)E_{u(t)}(g)}{1 - \exp\{-\sigma_a(u)\}}\middle| B_{0+}\right)\\ &= \lim_{\delta \to 0+} Q(1_{\delta<\tau}n_{u(\delta)}(f_\delta(u)\cdot E_{u(t-\delta)}(g))|B_{0+}(Z)).\end{aligned}$$

Thus, (7.19) is proved for such $f$ and $g$ except on a null $\omega$ set. Since there are a countable number of possible pairs of $(f, g)$, equation (7.19) holds for every

$(f, g)$ except on a null $\omega$ set which may depend on $t$. If (7.19) is true for $t$, then it is true for every $t' > t$ by the Markov property of $X$. It is therefore enough to verify (7.19) only for a sequence $t_k \downarrow 0$. Thus, the exceptional $\omega$ set can be taken independently of $t$.

◇ ◇ ◇ ◇ ◇

We wish to express our gratitude to K. L. Chung and R. K. Getoor for their valuable suggestions.

## REFERENCES

[1] E. B. DYNKIN, "On excursions of Markov processes," *Theor. Probability Appl.*, Vol. 13 (1968), pp. 672–676.

[2] E. B. DYNKIN and A. A. YUSHKEVICH, "On the starting points of incursions of Markov processes," *Theor. Probability Appl.*, Vol. 13 (1968), pp. 469–470.

[3] K. ITÔ, "Poisson point processes and their application to Markov processes," Department of Mathematics, Kyoto University, mimeographed notes, 1969.

[4] K. MATTHES, "Stationäre zufällige Punktmengen, I," *Jber. Deutsch. Math.-Verein.*, Vol. 66 (1963), pp. 69–79.

[5] P. A. MEYER, "Processus de Poisson ponctuels, d'après K. Itô," *Séminaire de Probabilités*, No. 5, 1969–1970, Lecture Notes in Mathematics, Berlin–Heidelberg–New York, Springer-Verlag, 1971.

Reprinted from
*Proc. Sixth Berkeley Symp. Math. Stat. Probab.* **III,** 225–239 (1970)

# Stochastic Differentials

K. Itô*

Department of Mathematics
Cornell University
Ithaca, New York 14850
Communicated by A. V. Balakrishnan

By a stochastic differential we understand a random interval function induced by a continuous local quasi-martingale [11]. Hence the stochastic integral in the usual sense, the stochastic integral in the Stratonovich sense and the quadratic variation in the Kunita-Watanabe sense are interpreted as operations in the space of stochastic differentials. In this note we will present a simple relation among these operations and its application.

## *1. The Space of Stochastic Differentials.*

Let $\{\mathscr{F}_t\}_{t\in[0,\infty)}$ be an increasing family of $\sigma$-algebras of measurable events and $\mathscr{B}$ denote the measurable processes adapted to $\{\mathscr{F}_t\}$ and having locally bounded sample functions. Among the sub-classes of $\mathscr{B}$ the following classes are important for our purpose.

$$\mathscr{C} = \{X \in \mathscr{B}\colon \text{ the sample function of } X \text{ is continuous}\}$$

$$\mathscr{M} = \{X \in \mathscr{C}\colon\ X \text{ is a local martingale relative to } \{\mathscr{F}_t\}\}$$

$$\mathscr{A} = \{X \in \mathscr{C}\colon \text{ the sample function of } X \text{ is locally bounded variation and } X(0) = 0\}$$

$$\mathscr{Q} = \{X = M + A\colon\ M \in \mathscr{M} \text{ and } A \in \mathscr{A}\}$$

$$= \{X \in \mathscr{C}\colon\ X \text{ is a local quasi-martingale relative to } \{\mathscr{F}_t\}\}.$$

Every $X \in \mathscr{Q}$ is expressed uniquely as $X = M_X + A_X$, $M_X \in \mathscr{M}$ and $A_X \in \mathscr{A}$. $M_X$ and $A_X$ are called the *martingale part* and the *bounded variation part* of $X$ respectively.

With each $X \in \mathscr{Q}$ we associate a random interval function

$$dX(I) = X(t) - X(s), \ I = (s.t],$$

which is continuous in the sense that

$$dX(I) \to 0 \text{ as } I \downarrow t \text{ for every } t.$$

---

* Supported by NSF GP-33136X, Cornell University.

374

Applied Mathematics & Optimization, Vol. 1, No. 4

Let $d\mathcal{Q}$, $d\mathcal{M}$ and $d\mathcal{A}$ denote the classes

$$\{dX : X \in \mathcal{Q}\}, \ \{dM : M \in \mathcal{M}\} \text{ and } \{dA : A \in \mathcal{A}\}$$

respectively.

Now we introduce two operations in $d\mathcal{Q}$.

*A. Addition*:

$$dX + dY = d(X+Y)$$

*M. $\mathcal{B}$-Multiplication*:

$$\Phi \cdot dX(I) = \int_I \Phi \cdot dX \ (\textit{Stochastic integral } [1], [2], [3], [7], [8]).$$

In case $\Phi \in \mathcal{C}$, we have

$$(\Phi \cdot dX)(I) = \underset{|\Delta| \to 0}{\text{l.i.p.}} \sum_{i=1}^{n} \Phi(t_{i-1}) dX(I_i) \tag{1.1}$$

where $\Delta = \{I_i \equiv (t_{i-1}, t_i]\}_{i=1}^n$ is a subdivision of $I$ and $|\Delta| = \max_i(t_i - t_{i-1})$.

*P. Product*:

$$(dX \cdot dY)(I) = d[M_X, M_Y], \quad X, Y \in \mathcal{Q}$$

where $[M_X, M_Y]$ is the *quadratic variation* of the pair $(M_X, M_Y)$. For each $I$ we have

$$(dX \cdot dY)(I) = \underset{|\Delta| \to 0}{\text{l.i.p.}} \sum_{i=1}^{n} dX(I_i) dY(I_i).$$

where $\Delta$ and $|\Delta|$ are the same as above.

Observing

$$dX(I_i) \cdot dY(I_i) = d(XY)(I_i) - Y(t_{i-1}) dX(I_i) - X(t_{i-1}) dY(I_i),$$

we can show that

$$dX \cdot dY = d(X \cdot Y) - Y dX - X dY. \tag{1.2}$$

Hence

$$d(XY) = Y \cdot dX + X \cdot dY + dX \cdot dY, \text{ i.e. } XY \in \mathcal{Q}.$$

Therefore *$\mathcal{Q}$ is a commutative ring.*

It is easy to prove the following:

***Theorem 1.*** *The space $d\mathcal{Q}$ with the operations A, M and P is a commutative $\mathcal{B}$-ring. $d\mathcal{A}$ is a subring of $d\mathcal{Q}$, and $d\mathcal{M}$ is a submodule of the $\mathcal{B}$-module $d\mathcal{Q}$ with A and M. Also we have*

$$d\mathcal{Q} \cdot d\mathcal{Q} \subset d\mathcal{A}, \ d\mathcal{A} \cdot d\mathcal{Q} = 0 \textit{ and } d\mathcal{Q} \cdot d\mathcal{Q} \cdot d\mathcal{Q} = 0.$$

Let us consider a third operation:

*SM. Symmetric $\mathcal{Q}$-multiplication.*

$$Y \circ dX = Y \cdot dX + \tfrac{1}{2} dX \cdot dY$$

For each $I$ we have

$$Y \circ dX(I) = \underset{|\Delta| \to 0}{\text{l.i.p.}} \sum_{i=1}^{n} \frac{Y(t_{i-1}) + Y(t_i)}{2} dX(I_i), \quad Y \in \mathcal{Q}, dX \in d\mathcal{Q} \tag{1.3}$$

where $\Delta$ and $|\Delta|$ are the same as above. This is obvious by

$$\frac{Y(t_{i-1}) + Y(t_i)}{2} \cdot dX(I_i) = Y(t_{i-1}) \cdot dX(I_i) + \tfrac{1}{2} dY(I_i) \cdot dX(I_i)$$

and the formula (1.1). $Y \circ dX(I)$ coincides with the *symmetric stochastic integral of Stratonovich*: $(s) \int_I Y \cdot \delta X$ in the special case discussed by him [9]. The stochastic integral of this type was also discussed by Wong and Zakai [4] and most extensively by McShane [5]. It is trivial that

$$Y \circ dX = Y \cdot dX \text{ for } Y \in \mathcal{A} \text{ or } dX \in d\mathcal{A}$$

We can easily prove the following:

***Theorem 2.*** *The space $d\mathcal{Q}$ with operations A, SM and P is a $\mathcal{Q}$-ring.*

Since the product $dX \cdot dY$ is expressed by (2) in terms of $\mathcal{B}$-multiplication, the symmetric $\mathcal{Q}$-multiplication can be expressed in terms of $\mathcal{B}$-multiplication as follows:

$$Y \circ dX = Y \cdot dX + \tfrac{1}{2}(d(XY) - X \cdot dY - Y \cdot dX).$$

However, there is no way of expressing $\mathcal{B}$-multiplication in terms of the symmetric $\mathcal{Q}$-multiplication even in case $Y \in \mathcal{Q}$. The only conceivable way is

$$Y \cdot dX = p_{d\mathcal{M}}(Y \circ dX) + Y \cdot (dX - p_{d\mathcal{M}} dX),$$

where $p_{d\mathcal{M}}$ is the projection from $d\mathcal{Q}$ to $d\mathcal{M}$, i.e.

$$p_{d\mathcal{M}} dZ = dM_Z,$$

but this needs an extra operation $p_{d\mathcal{M}}$.

In the stochastic calculus the following transformation formula plays an important role. Let $f: \mathbf{R}^n \to \mathbf{R}$ be of class $C^2$. If $X_1, X_2, \ldots, X_n \in \mathcal{Q}$, then

$$f(X) = f(X_1, X_2, \ldots, X_n) \in \mathcal{Q} \text{ and } \partial_i f(X), \partial_i \partial_j f(X) \in \mathcal{C}$$

and

$$(T) \quad df(X) = \sum_{i=1}^{n} \partial_i f(X) \cdot dX_i + \tfrac{1}{2} \sum_{i,j=1}^{n} \partial_i \partial_j f(X) \cdot dX_i \cdot dX_j.$$

For the index $i$ with $X_i \in A$, we need only assume $\partial_i f \in C^1$ by removing the terms involving all such indexes from the second summation.

If $f$ is of class $C^3$, then $\partial_i f(X) \in \mathcal{Q}$ and

$$d(\partial_i f(X)) = \sum_j \partial_j \partial_i f(X) \cdot dX_j + \tfrac{1}{2} \sum_{j,k} \partial_k \partial_j \partial_i f(X) dX_j dX_k.$$

Since $dX_k \cdot dX_j \cdot dX_i = 0$ by Theorem 1, the formula $(T)$ implies that

$$\sum_{i=1}^{n} \partial_i f(X) \circ dX_i = \sum_{i=1}^{n} \partial_i f(X) \cdot dX_i + \tfrac{1}{2} \sum_{i=1}^{n} d(\partial_i f(X)) \cdot dX_i$$

$$= \sum_{i=1}^{n} \partial_i f(X)\cdot dX_i + \tfrac{1}{2} \sum_{i=1}^{n} \partial_i \partial_j f(X)\cdot dX_j dX_i,$$

$$= df(X),$$

i.e.

$$(T_s) \quad df(X) = \sum_{i=1}^{n} \partial_i f(X) \circ dX_i.$$

This is the *Stratonovich transformation formula* [9]. This formula can be proved for $f \in C^2$ if we extend the definition of the symmetric multiplication but we will not get into it, because in practical application the function $f$ can be assumed to be continuously differentiable as many times as we need.

Because of simplicity of the formula $(T_s)$, the symmetric multiplication is very convenient for some purposes, as will be illustrated in Section 3.

## 2. *Stochastic Differential Equations.*

Let $B(t) = (B_1(t), B_2(t), \ldots, B_n(t))$, $t \in [0,1)$, be an $n$-dimensional Brownian motion and let $\{\mathscr{F}_t\}$ be an increasing family of $\sigma$-algebras of measurable events. We assume that (1) $(B(t))$ is adapted to $\{\mathscr{F}_t\}$ and (2) for every $t$ $(B(s+t)-B(t))_{s\in[0,\infty)}$ and $\mathscr{F}_t$ are independent; for example, the $\sigma$-algebras

$$\mathscr{F}_t = \sigma[X_u, u \le t], \ t \in [0,\infty)$$

have these properties.

Let us consider the space $d\mathscr{Q}$ of stochastic differentials relative to this increasing family $\{\mathscr{F}_t\}$ and its subspaces $d\mathscr{M}$ and $d\mathscr{A}$. Then

$$dB_i \in d\mathscr{M}, \ dt \in d\mathscr{A} \tag{2.1}$$

$$dB_i dB_j = \delta_{ij} dt, \ dB_i dt = 0 \text{ and } dt\cdot dt = 0. \tag{2.2}$$

Now we consider two stochastic differential equations:

$$(E) \quad dX_\alpha = a_\alpha(t,X)dt + \sum_{i=1}^{n} \sigma_{\alpha i}(t,X)dB_i, \ \alpha = 1,2,\ldots,m,$$

$$(E_s) \quad dY_\alpha = a_\alpha(t,Y)\circ dt + \sum_{i=1}^{n} \sigma_{\alpha i}(t,Y)\circ dB_i, \ \alpha = 1,2,\ldots,m.$$

where $a_\alpha(t,\xi)$ and $\sigma_{\alpha i}(t,\xi)$ are smooth.

From now on we often omit the summation sign $\sum$ on the indexes appearing twice as in differential geometry.

It is known that $(E)$ determines a continuous Markov process (i.e. a diffusion) with generator at $t$:

$$\mathscr{G}_t f = a_\alpha \partial_\alpha f + \tfrac{1}{2}\sigma_{\alpha i}\sigma_{\beta i}\partial_\alpha\partial_\beta f.$$

We can derive this by using the transformation formula $(T)$ and (2) as follows.

$$\begin{aligned} df(X) &= \partial_\alpha f dX + \tfrac{1}{2}\partial_\alpha\partial_\beta f\, dX_\alpha dX_\beta \\ &= \partial_\alpha f(a_\alpha dt + \sigma_{\alpha i} dB_i) + \tfrac{1}{2}\partial_\alpha\partial_\beta f(a_\alpha dt + \sigma_{\alpha i} dB_i)(a_\beta dt + \sigma_{\beta i} dB_i) \\ &= (\partial_\alpha f a_\alpha + \tfrac{1}{2} a_{\alpha i} a_{\beta i}\partial_\alpha\partial_\beta f)dt + \partial_\alpha f \sigma_{\alpha i} dB_i, \end{aligned}$$

so
$$\begin{aligned} \mathscr{G}_t f(X(t))) &= \lim_{\epsilon\downarrow 0} \frac{1}{\epsilon} E(df(X)(t, t+\epsilon] | \mathscr{F}_t) \\ &= \lim_{\epsilon\downarrow 0} E\left(\frac{1}{\epsilon}\int_t^{t+\epsilon} (a_\alpha\partial_\alpha f + \tfrac{1}{2}\sigma_{\alpha i}\sigma_{\beta i}\partial_\alpha\partial_\beta f + \partial_\alpha f\sigma_{\alpha i} dB_i)d\tau | \mathscr{F}_t\right) \\ &= (a_\alpha\partial_\alpha f + \tfrac{1}{2}\sigma_{\alpha i}\sigma_{\beta i} f)(X(t)). \end{aligned}$$

The equation $(E_s)$ is the *symmetric stochastic differential equation due to Stratonovich* [4]. By the definition $(E_s)$ is written as

$$\begin{aligned} dY_\alpha &= a_\alpha dt + \sigma_{\alpha i} dB_i + \tfrac{1}{2} d\sigma_{\alpha i} dB_i \\ &= a_\alpha dt + \sigma_{\alpha i} dB_i + \tfrac{1}{2}\partial_\beta\sigma_{\alpha i} dY_\beta dB_i \text{ (note } d\mathscr{Q}\cdot d\mathscr{Q}\cdot d\mathscr{Q} = 0) \\ &= (a_\alpha + \tfrac{1}{2}(\partial_\beta\sigma_{\alpha i})\sigma_{\beta i})dt + \sigma_{\alpha i} dB_i \text{ by (2.2)} \end{aligned}$$

Therefore $(E_s)$ determines a Markov process $Y$ different from the process $X$ determined by $(E)$. The generator of $Y$ is given by

$$\begin{aligned} \mathscr{G}_t f &= a_\alpha\partial_\alpha f + \tfrac{1}{2}\partial_\beta(\sigma_{\alpha i})\sigma_{\beta i}\partial_\alpha\partial_\beta f + \tfrac{1}{2}\sigma_{\alpha i}\sigma_{\beta i}\partial_\alpha\partial_\beta f \\ &= a_\alpha\partial_\alpha f + \tfrac{1}{2}\sigma_{\beta i}\partial_\beta(\sigma_{\alpha i}\partial_\alpha f). \end{aligned}$$

We can derive this more directly from $(E_s)$ by using the Stratonovich transformation formula $(T_s)$ as follows:

$$\begin{aligned} df(Y) &= \partial_\alpha f \circ dY \\ &= \partial_\alpha f \circ (a_\alpha \circ dt + \sigma_{\alpha i} \circ dB_i) \\ &= a_\alpha\partial_\alpha f dt + \sigma_{\alpha i}\partial_\alpha f dB_i + \tfrac{1}{2}\partial_\beta(\sigma_{\alpha i}\partial_\alpha f) dY_\beta dB_i \\ &= (a_\alpha\partial_\alpha f + \tfrac{1}{2}\partial_\beta(\sigma_{\alpha i}\partial_\alpha f)\sigma_{\beta i})dt + \sigma_{\alpha i}\partial_\alpha f dB_i, \end{aligned}$$

so $\mathscr{G}_t f(Y(t)) = (a_\alpha\partial_\alpha f + \tfrac{1}{2}\sigma_{\beta i}\partial_\beta(\sigma_{\alpha i}\partial_\alpha f))(Y(t))$.

Consider the ordinary differential equation

$$(C) \quad \frac{dx_\alpha}{dt} = a_\alpha(t,x) + \sigma_{\alpha i}(t,x)u_i(t), \quad \alpha = 1,2,\ldots,m,$$

where $u = (u_1(t), u_2(t), \ldots, u_n(t))$ is a square integrable vector function. We can regard $(C)$ as a control equation of the equation

$$\frac{dx_\alpha}{dt} = a_\alpha(t,x), \quad \alpha = 1,2,\ldots,m,$$

under the control $u$. The *Stroock-Varadhan support theorem* [6] claims that the solutions of $(C)$, with initial vector $X(0) = (X(0), \ldots, X_n(0))$ given, is dense in the topological support of the probability law of the sample function of the solution of the Stratonovich symmetric stochastic differential equation $(E_s)$ with the same initial vector. In this sense $(E_s)$ *is more closely connected to* $(C)$ *than* $(E)$.

### 3. *Examples of Stochastic Calculus.*

*Example* 1. The ordinary differential equation

$$dy = ydx$$

has a unique solution $y = y(0)e^x$. We have two stochastic versions of this equation:

$$(E_s) \qquad dY = Y \circ dX,$$

$$(E) \qquad dY = Y \cdot dX.$$

Using the transformation formula $(T_s)$ we can prove that $(E_s)$ has unique solution:

$$Y(t) = Y(0)\exp\left(X(t)-X(0)\right) (= Y(0)\exp(dX(0,t]) = Y(0)\exp\int_0^t dX). \quad (1)$$

The equation $(E)$ can be written as

$$\begin{aligned} dY &= Y \circ dX - \tfrac{1}{2} dY \cdot dX \\ &= Y \circ dX - \tfrac{1}{2} Y \cdot (dX)^2 \\ &= Y \circ (dX - \tfrac{1}{2}(dX)^2) \quad (\text{Note: } Y \cdot (dX)^2 = Y \circ (dX)^2) \\ &= Y \circ dZ,\ Z(t) = \int_0^t (dX - \tfrac{1}{2}(dX)^2). \end{aligned}$$

Hence we have

$$Y = Y(0)e^{Z(t)} = Y(0)\exp\left\{\int_0^t (dX - \tfrac{1}{2}(dX)^2\right\} \quad (2)$$

by (1).

As a special case: $dX = \Phi \cdot dB$ ($B$: Brownian motion), we obtain the following results:

$$dY = Y \circ (\Phi \cdot dB) \Rightarrow Y(t) = Y(0)\exp\left(\int_0^t \Phi dB\right) \quad (1')$$

$$dY = Y \cdot \Phi \cdot dB \Rightarrow Y(t) = Y(0)\exp\left(\int_0^t \Phi dB - \tfrac{1}{2}\Phi^2 ds\right). \quad (2')$$

*Example* 2. Consider the Stroock equation [10] determining the spherical Brownian motion:

$$dX = a(X) \circ dB,\ a(x) = I - \frac{x \cdot x'}{|x|^2}, \quad (3)$$

$$\alpha,\ i = 1,2,\ldots,n,$$

where $B = (B_1, B_2, \ldots, B_n)$ is an $n$-dimensional Brownian motion, all $n$-vectors in (1) are column vectors and $x'$ denotes the transpose of the vector $x$ (regarded as an $n \times 1$ matrix). The matrix $a(x)$ gives the projection to the hyperplane normal to the direction $x$; in fact,

$$x' \cdot a(x) = 0.$$

By applying the formula $(T_s)$ component-wise and using Theorem 2, we have

$$\begin{aligned} d|X|^2 &= 2X' \circ dX (= 2\textstyle\sum_\alpha X_\alpha \circ dX_\alpha) \\ &= 2X' \cdot a(X) \circ dB \\ &= 0. \end{aligned}$$

Hence the solution of (3) lies on a sphere with center 0. This fact is also regarded as an example of the Stroock-Varadhan support theorem. Thus the solution $X$ of (3) is a diffusion process on a sphere with centre 0. Since $dB$ is rotation invariant, and since

$$a(0x) = 0a(x)0' \qquad (0 = \text{an orthogonal matrix}),$$

we have

$$d(0X) = 0 \cdot dX = 0a(X) \cdot dB = 0a(X)0'0dB = a(0X)0B,$$

showing that $X$ is a rotation invariant diffusion on a sphere with centre 0, i.e. a Brownian motion on the sphere.

Writing (3) (dimension $n = 3$) in polar coordinates we have

$$\begin{pmatrix} \sin\Theta\cos\Phi & R\cos\Theta\cos\Phi & -R\sin\Theta\sin\Phi \\ \sin\Theta\sin\Phi & R\cos\Theta\sin\Phi & R\sin\Theta\cos\Phi \\ \cos\Theta & -R\sin\Theta & 0 \end{pmatrix} \circ \begin{pmatrix} dR \\ d\Theta \\ d\Phi \end{pmatrix}$$

$$= \begin{pmatrix} 1-\sin^2\Theta\cos^2\Phi & -\sin^2\Theta\cos\Phi\sin\Phi & -\cos\Theta\sin\Theta\cos\Phi \\ -\sin^2\Theta\cos\Phi\sin\Phi & 1-\sin^2\Theta\sin^2\Phi & -\cos\Theta\sin\Theta\sin\Phi \\ -\cos\Theta\sin\Theta\cos\Phi & -\cos\Theta\sin\Theta\sin\Phi & \sin^2\Theta \end{pmatrix} \circ \begin{pmatrix} dB_1 \\ dB_2 \\ dB_3 \end{pmatrix}$$

By Theorem 2 in Section 1 we can use the Cramer formula to solve this for $dR$, $d\Theta$ and $d\Phi$ and we obtain

$$dR = 0$$

$$d\Theta = \frac{1}{R}\cos\Theta\cos\Phi \circ dB_1 + \frac{1}{R}\cos\Theta\sin\Phi \circ dB_2 - \frac{1}{R}\sin\Theta \circ dB_3,$$

$$d\Phi = -\frac{1}{R}\frac{\sin\Phi}{\sin\Theta} \circ dB_1 + \frac{1}{R}\frac{\cos\Phi}{\sin\Theta} \circ dB_2.$$

Thus the generator of the diffusion $X$ in polar coordinates is

$$\frac{1}{r^2}\left[\frac{1}{\sin\theta}\,\partial_\theta(\sin\theta\,\partial_\theta) + \frac{1}{\sin^2\theta}\,\partial_\varphi^2\right];$$

note that the expression in the square bracket equals the spherical Laplacian on the unit sphere.

## References

[1] D. L. Fisk, Quasi-martingales, *Trans. Amer. Math. Soc.*, **120** (1965), 369–389.
[2] P. Courrege, Intégrals stochastiques et martingales de carré integrable, *Sem. Th. Potent.* 7. Inst. Henri Poincaré, Paris, 1963.

[3] H. KUNITA and S. WATANABE, On square integrable martingales, *Nagoya Math. J.* **30** (1967), 209–245.

[4] E. WONG and M. ZAKAI, On the relation between ordinary and stochastic differential equations, *Internat. J. Engrg. Sci.* **3** (1965), 213–229.

[5] E. J. MCSHANE, Stochastic differential equations and models of random processes, *Prc. 6th Berkeley Symp. on Math. Stat. and probability* **3** (1970), 263–294.

[6] D. W. STROOCK and S. R. S. VARADHAN, On the support of diffusion processes with applications to strong maximum principle, *Proc. 6th Berkeley Symp. on Math. Stat. and probability* **3** (1970), 333–360.

[7] P. A. MEYER, *Intégrals stochastiques I-IV*. Lecture Notes in Math. (Springer) 39 Semi. Prob. (1967), 72–162.

[8] P. W. MILLAR, Martingale intégrals, *Trans. Amer. Math. Soc.* **133** (1968), 145–168.

[9] R. L. STRATONOVICH, *Conditioned Markov processes and their application to optimal control*, R. Elsevier, New York (1968).

[10] D. W. STROOCK, On the growth of stochastic integrals, *Z. Wakr. and Vew. Geb.* **18** (1971), 340–344.

[11] K. ITÔ, Stochastic differentials of continuous local quasi-martingales, Stability of stochastic dynamical systems. Lecture Notes in Math. (Springer), **294** (1972), 1–7.

# Stochastic parallel displacement

Kiyosi Itô*

1. Introduction. In our previous paper [1] we have introduced a stochastic differential $dX$ as a random interval function induced from a continuous local quasi-martingale $X$. If $F(x_1, x_2, \dots, x_n)$ is sufficiently smooth, we have a chain rule

$$(C) \qquad dF(X_1, X_2, \dots, X_n) = \sum_i \partial_i F\, dX_i + \frac{1}{2} \sum_{i,j} \partial_i \partial_j F\, dX_i\, dY_j$$

where

$$(A\, dX)(I) = \int_I A\, dX \qquad \text{(stochastic integral)}$$

and $dX\, dY = d(XY) - XdY - YdX$ (quadratic covariation).
If we use the symmetric multiplication

$$A \circ dX = AdX + \frac{1}{2}\, dAdX$$

which corresponds to the Fisk-Stratonovich symmetric stochastic integral [5][6], the chain rule (C) can be written as

$$(C_s) \qquad dF(X_1, X_2, \dots, X_n) = \sum_i \partial_i\, F \circ dX_i.$$

Since $(C_s)$ takes the same form as in the ordinary calculus, the symmetric multiplication is convenient for some purpose. We have given such examples in our previous paper [1]. In the present paper we will discuss stochastic parallel displacement as another interesting example.

*Supported by NSF GP-33136X, Cornell University.

2. Stochastic parallel displacement. Let us review some notation in differential geometry following Dynkin [3]. Let $S = (S, \Gamma^i_{jk})$ be an affinely connected $\ell$-dimensional $C^3$-manifold and $T^n_m = (T^n_m(x), x \in S)$ the bundle of tensors of type $(m,n)$. $T^n_m(x)$ is dual to $T^m_n(x)$ relative to the invariant bilinear form:

(1) $$(u,w) := u^{j_1 \cdots j_n}_{i_1 \cdots i_m} w^{i_1 \cdots i_m}_{j_1 \cdots j_n},$$

where the summation sign is omitted as is common in differential geometry. Using the symbol

(2) $$(\Gamma_i u)^{j_1 \cdots j_n}_{i_1 \cdots i_m} = \Gamma^k_{i i_\mu} u^{j_1 \cdots j_n}_{i_1 \cdots i_{\mu-1} k\, i_{\mu+1} \cdots i_m} - \Gamma^{j_\nu}_{ik} u^{j_1 \cdots j_{\nu-1} k j_{\nu+1} \cdots j_n}_{i_1 \cdots i_m},$$

we can express the covariant derivative $\nabla_i$ as follows:

(3) $\nabla_i = \partial_i - \Gamma_i$, where $\partial_i = \dfrac{\partial}{\partial x_i}$.

Let $C: y(t)$, $t_0 \leq t \leq t_1$, be a smooth curve on $S$. The tensor $u_1 \in T^n_m(y(t_1))$ is called parallel to $u_0 \in T^n_m(y(t_0))$ along $C$, if $u_0$ and $u_1$ are connected by a family of tensors $u(t) \in T^n_m(y(t))$, $t_0 \leq t \leq t_1$ satisfying

(4) $\dot{u} = (\Gamma_i u)\dot{y}^i$, i.e. $du = (\Gamma_i u)dy^i$, $i = 1,2,\ldots,\ell$.

Now we want to define the stochastic parallelism along a random curve $C(\omega)$: $Y_t(\omega)$, $t_0 \leq t \leq t_1$. Since the sample curve of $C(\omega)$ is not smooth in general, we cannot apply the above definition to each sample curve, but we can define stochastic parallelism by replacing the equation (4) with its stochastic analogue:

$$(5) \qquad dU = (\Gamma_i U) \circ dY^i,$$

where the small circle denotes the symmetric multiplication of stochastic differentials [1], so (5) is also expressed as

$$(5') \qquad dU = (\Gamma_i U) dY^i + 1/2 d(\Gamma_i U) dY^i.$$

We can equivalently define stochastic parallelism in the following geometric way. Let $U_1(\omega)$ be a random tensor in $T^n_m(Y(t_0,\omega))$ and $\Delta = (t_0 = s_0 < s_1 < \ldots < s_r = t_1)$ a subdivision of the internal $[t_0, t_1]$. By connecting $Y(t_{i-1}, \omega)$ with $Y(t_i,\omega)$ by a geodesic curve for $i = 1,2,\ldots,n$, we obtain a piece-wise smooth curve $C_\Delta(\omega)$ approximating $C(\omega)$. Take a random tensor $U_2^\Delta(\omega) \in T^n_m(Y(t_1,\omega))$ parallel to $U_1(\omega)$ along $C_\Delta(\omega)$. Then the random tensor in $T^n_m(Y(t_1,\omega))$:

$$U_2(\omega) = \underset{|\Delta|\to 0}{\text{l.i.p.}}\, U_2^\Delta(\omega), \qquad |\Delta| = \max_i (t_i - t_{i-1}),$$

is said to be parallel to $U_1(\omega)$ along $C(\omega)$.

We will give two interesting examples of application of stochastic parallelism.

<u>1. Diffusions of tensors</u>. The diffusion of tensors was introduced by K. Itô [2] for the diffusion of covariant tensors on a Riemannian manifold induced by the Brownian motion on the manifold and extended to by E.B. Dynkin [3] to the general case we are dealing with below. Let $\{x_t(\omega)\}$ be a diffusion on $S$ with generator

$$(6) \qquad \mathfrak{G} = 1/2 a^{ij} \nabla_i \nabla_j \neq b^i \nabla_i.$$

If $u(x)$ is a scaler field, we define a semi-group

(7) $$H_t u(a) = E_a(u(X_t))$$

and we have

(8) $$\lim_{t \downarrow 0} \frac{H_t u(a) - u(a)}{t} = (1/2a^{ij} \nabla_i \nabla_j + b^i \nabla_i)u(a) \quad \text{for } u \text{ smooth.}$$

If $u(x)$ is a tensor field of type $(m,n)$, we have

$$u(X_t) \in T^n_m(X_t) \quad \text{but} \quad u(X_t) \notin T^n_m(X_0) \equiv T^n_m(a),$$

so the right hand side of (7) has no meaning. Hence we interpret (3) as follows. Take a tensor $\hat{U}_t(\omega)$ in $T^n_m(a)$ $(= T^n_m(X_0))$ such that $u(X_t)$ is parallel to $\hat{U}_t(\omega)$ along the curve $\{X_t(\omega)\}$. Then

(7') $$H_t u(a) = E_a(\hat{U}_t(\omega))$$

defines a semi-group with generator $\mathcal{G}$ of the form (6). The family of tensors $V_t(\omega) \in T^m_n(X_t)$ parallel to $v(a) \equiv v(X_0)$ along the curve $(X_t(\omega))$ is a diffusion on the bundle $T^m_n$ defined by Dynkin [3]. Since $(X_t(\omega))$ is determined by a stochastic differential equation:

(9) $$dX^i_t = \sigma^i_\alpha \, d\xi^\alpha + \rho^i \, dt,$$

where $(\xi_t(\omega))$ is a Brownian motion and

$$a^{ij} = \sum_\alpha \sigma^i_\alpha \sigma^j_\alpha \quad \text{and} \quad b^i = \rho^i + a^{jk} \Gamma^i_{jk} \,,$$

$V_t(\omega)$ is determined by the stochastic differential equation:

(10) $$dV_t = (\Gamma_i V_t) \cdot dX^i_t \,,$$

i.e.

$$dV_t = (\Gamma_i V_t) \cdot (\sigma^i_\alpha \, d\xi^\alpha + \rho^i \, dt).$$

To obtain (7') it is crucial to note that the equation

$$(V_t, u(X_t)) = (v(a), \hat{U}_t(\omega)) \tag{11}$$

holds by virtue of the invariance of the bilinear form (1) under parallel displacement.

## 2. Rolling along Brownian motion.

Let $S$ and $\tilde{S}$ be $\ell$-dimensional Riemannian manifolds with metric tensors $\{g_{ij}\}$ and $\{\tilde{g}_{ij}\}$ respectively. Let

$$C: x(t), \ t_0 \leq t \leq t_2.$$

be a curve on $S$. If we roll $\tilde{S}$ on $S$ along $C$ without slipping, the touching point on $\tilde{S}$ describes a curve

$$\tilde{C}: \tilde{x}(t), \ t_0 \leq t \leq t_2.$$

Let $\{v_\alpha(t)\}_{\alpha=1,2,\ldots,\ell}$ denote the family of covariant vectors in $T_1(x(t))$ obtained from an orthonormal basis in $T_1(x(t_0))$ by parallel displacement along $C$. Then $\{v_\alpha(t)\}$ is an orthonormal basis in $T_1(x(t))$ for every $t$. Similarly we obtain a family of covariant vectors $\{\tilde{v}_\alpha(t)\}_{\alpha=1,2,\ldots,\ell}$ in $\tilde{T}_1(\tilde{x}(t))$ by parallel displacement along $\tilde{C}$. $\{\tilde{v}_\alpha(t)\}$ is also an orthonormal basis in $\tilde{T}_1(x(t))$, as we can prove by using the symmetric chain rule $(C_s)$. We choose the initial orthonormal basis $\{\tilde{v}_\alpha(t_0)\}$ so that $\{\tilde{v}_\alpha(t_0)\}$ coincides with $\{v_\alpha(t_0)\}$ when $S$ starts rolling. As $S$ rolls, $\tilde{x}(t)$ coincides with $x(t)$, $\tilde{T}_1(\tilde{x}(t))$ with $T_1(x(t))$, $\{\tilde{v}_\alpha(t)\}$ with $\{v_\alpha(t)\}$ and $d\tilde{x}(t)/dt$ with $dx(t)/dt$ at every $t$. Thus we have

$$\tilde{v}_{\alpha,i} \frac{d\tilde{x}^i}{dt} = v_{\alpha,i} \frac{dx^i}{dt} \tag{12}$$

$$\frac{dv_{\alpha,i}}{dt} = \Gamma^k_{ji} v_{\alpha,k} \frac{dx^j}{dt} \tag{13}$$

$$\frac{d\tilde{v}_{\alpha,i}}{dt} = \tilde{\Gamma}^k_{ji} \tilde{v}_{\alpha,k} \frac{d\tilde{x}^j}{dt} \tag{14}$$

Since $x(t)$ is given, $v_{\alpha,i}(t)$ is determined by (13). By solving (12) and (14) for $\tilde{x}^i(t)$ and $\tilde{v}_{\alpha,i}(t)$, we obtain the curve $\tilde{C}$: $\tilde{x}^i(t)$, $t_0 \leq t \leq t_1$.

Let $C(\omega)$: $\{x_t(\omega)\}t \in [0,\infty)$ be Brownian motion on S. Since the sample curve of $C(\omega)$ is not smooth, we cannot apply the above argument to each sample curve to obtain the random curve $\tilde{C}(\omega)$ on $\tilde{S}$ by rolling $\tilde{S}$ on S along $C(\omega)$. However, the geodesic approximation method will work as for stochastic parallelism. Then $\tilde{C}(\omega)$: $\{X_t(\omega)\}t \in [0,\infty)$ is determined by the following stochastic differential equations analogous to (12), (13) and (14):

$$(12') \qquad \tilde{V}_{\alpha,i} \circ d\tilde{X}^i = V_{\alpha,i} \circ dX^i$$

$$(13') \qquad dV_{\alpha,i} = \Gamma^k_{ji} V_{\alpha,k}) \circ dX^j$$

$$(14') \qquad d\tilde{V}_{\alpha,i} = (\tilde{\Gamma}^k_{ji} \tilde{V}_{\alpha,k}) \circ d\tilde{X}^j ,$$

from which we can prove that $\{\tilde{X}_t(\omega)\}$ is a diffusion on S with generator

$$(15) \qquad \tilde{\mathcal{J}} = 1/2\, \tilde{g}^{ij} \tilde{\nabla}_i \tilde{\nabla}_j ,$$

i.e. Brownian motion on S. Using the symmetric chain rule $(C_s)$, we can easily verify the formula (15) in terms of the geodesic coordinates on S and $\tilde{S}$ at the starting points of C and $\tilde{C}$. The rolling problem along Brownian motion was first discussed by H.P. McKean [4] in the case that $S_1$ is a plane and $S_2$ is a sphere.

## Bibliography

[1] K. Itô, Stochastic differentials, to appear in Applied Mathematics and optimization.

[2] K. Itô, The Brownian motion and tensor fields on Riemannian manifold, Proc. Int. Congress Math. 1962 (Stockholm), 536-539.

[3] E.B. Dynkin, Diffusions of tensors, Soviet Math. Dokl. Vol. 9 (1968), No. 2, 532-535.

[4] H.P. McKean, Jr., Brownian motions on the 3-dimensional rotation group, Mem. Coll. Sci., Univ. Kyoto, Ser. A, 33, Math. No. 1 (1960), 25-38.

[5] R.L. Stratonovich, Conditional Markov processes and their application to optimal control, Elsevier, New York (1968).

[6] D.L. Fisk, Quasi-martingales and stochastic integrals, Technical Report No. 1, Dept. of Math., Michigan State University, 1963.

Reprinted from
*Probabilistic Methods in Stochastic Differential Equations*
(Lecture Notes in Mathematics 451) Springer-Verlag, 1974, pp. 1–7

Proc. of Intern. Symp. SDE
Kyoto 1976, pp. 95–109

# Extension of Stochastic Integrals

Kiyosi Itô

## § 1. Introduction

Let $B = \{B_t, 0 \leq t < \infty\}$ be a Brownian motion. From the intuitive meaning of integrals we want to have

$$\int_s^t B_1 dB_u = B_1(B_t - B_s) . \tag{1.1}$$

Unfortunately the left hand side has no meaning in the sense of Brownian stochastic integrals [1] because $Y_t \equiv B_1$ depends on the future development of $B$ for $t < 1$. However, we can interpret the integral changing our view from Brownian stochastic integrals into quasimartingale integrals. Let $\mathscr{F} = \{\mathscr{F}_t, 0 \leq t < \infty\}$ be the reference family (i.e. the filtration) generated by two processes $B$ and $Y_t \equiv B_1$. Then $Y = \{Y_t, 0 \leq t < \infty\}$ is $\mathscr{F}$-adapted and $B$ is an $\mathscr{F}$-quasimartingale with the canonical $\mathscr{F}$-decomposition:

$$B_t = M_t + A_t , \quad A_t = \int_0^{t \wedge 1} \frac{B_1 - B_u}{1 - u} du , \quad M_t = B_t - A_t . \tag{1.2}$$

Hence the left hand side of (1) is meaningful in the sense of quasimartingale stochastic integrals and

$$\begin{aligned} \int_s^t B_1 dB_u &= \int_s^t B_1 dM_u + \int_s^t B_1 dA_u \\ &= B_1(M_t - M_s) + B_1(A_t - A_s) \\ &= B_1(B_t - B_s) , \end{aligned}$$

as we desire. The purpose of this paper is to develop this idea and present some of its applications.

In Section 2 we will review the theory of quasimartingale stochastic integrals from our view-point. This theory was first initiated by D. L. Fisk [2] and later developed extensively by P. Courrège [3], H. Kunita and S. Watanabe [4] and P. A. Meyer [5]. In Section 3 we will observe different aspects of Stratonovich's symmetric stochastic integrals [6]. The forward symmetric stochastic integrals discussed in this section was introduced first by Fisk [2]; see also K. Itô [7]. In the last section we will discuss stochastic paralled displacement on a Riemannian manifold introduced by K. Itô [8], [9]. E. B. Dynkin [10] extended it to the case of a manifold with connection. This case can be discussed in the same way as we do here.

## § 2. Stochastic integrals

Let $\Omega = (\Omega, \mathscr{F}_\Omega, P)$ be a complete probability space. A stochastic process $X = \{X_t(\omega), 0 \leqq t < \infty\}$ on $\Omega$ is called *continuous* (resp. *locally bounded, locally finite variation*) if its sample function is continuous a.s. (resp. bounded on every finite interval a.s., of finite absolute variation on every finite interval a.s.).

Let $\mathscr{F} = \{\mathscr{F}_t, 0 \leqq t < \infty\}$ be a right-continuous increasing family of sub-$\sigma$-algebras of $\mathscr{F}_\Omega$ such that $\mathscr{F}_t$ contains all $P$-null sets. Such a family is simply called a *reference family* in this paper. A process adapted to $\mathscr{F}$ is called an *$\mathscr{F}$-adapted process* or an *$\mathscr{F}$-process*. Similarly we use the terms *$\mathscr{F}$-martingales, $\mathscr{F}$-stopping times, $\mathscr{F}$-well-measurable processes*, etc.. We may omit the prefix $\mathscr{F}$ in case we refer to a fixed reference family. Let $X$ be a stochastic process. The $\sigma$-algebra generated by the sets

$$\{\omega\colon X_t(\omega) < a\}\,, \quad t \in [0, \infty)\,, \quad a \in (-\infty, \infty)$$

and the $P$-null sets is denoted by $\bar{\sigma}_t[X]$. The reference family

$$\bar{\sigma}_{t+}[X] \equiv \bigcap_n \bar{\sigma}_{t+1/n}[X]$$

is called the *reference family generated by* $X$, $\mathscr{F}[X] = \{\mathscr{F}_t[X]\}$ in notation. $\mathscr{F}[X]$ is the least of all reference families to which $X$ is adapted. Similarly for the reference family $\mathscr{F}[X, Y, \cdots]$ generated by a family of processes $X, Y, \cdots$.

A processes $X = \{X_t\}$ is called a *continuous local $\mathscr{F}$-martingale* if $X$ is continuous and if there exists a sequence of $\mathscr{F}$-stopping times $\theta_1 \leqq \theta_2 \leqq \cdots \to \infty$ a.s. such that the stopped process $\{X_{t\wedge\theta_n}\}$ is an $\mathscr{F}$-martingale for every $n$. $X$ is called a *continuous local $\mathscr{F}$-quasimartingale* if $X$ is expressible as

$$X = M + A\,, \qquad A(0) = 0 \tag{2.1}$$

where $X$ is a continuous local $\mathscr{F}$-martingale and $A$ is a continuous, locally bounded variation $\mathscr{F}$-process. Since the expression (2.1) is uniquely determined by $X$ and $\mathscr{F}$ if exists, it it is called the *canonical $\mathscr{F}$-decomposi-*

*tion* of $X$. The family of all continuous local $\mathcal{F}$-quasimartingales is denoted by $\mathcal{Q}_{\mathcal{F}}$.

Let $\mathcal{W}_{\mathcal{F}}$ denote the $\mathcal{F}$-well-measurable processes and $\mathcal{B}_{\mathcal{F}}$ the locally bounded $\mathcal{F}$-well-measurable processes. For every $X \in \mathcal{Q}_{\mathcal{F}}$ with the canonical $\mathcal{F}$-decomposition $X = M + A$ we denote by $\mathcal{L}_{\mathcal{F}}(dX)$ the family of all $Y \in \mathcal{W}_{\mathcal{F}}$ such that

$$\text{(2.2)} \qquad \int_0^t Y^2(dM)^2 < \infty \quad \text{and} \quad \int_0^t |Y||dA| \qquad \text{for every } t$$

a.s., where $(dM)^2$ is the Lebesgue-Stieltjes measure induced by the quadratic variation $\{\langle M\rangle_t\}$ of $\{M_t\}$ and $|dA|$ is that induced by the absolute variation $\{|A|_t\}$ of $\{A_t\}$.

Suppose that $X \in \mathcal{Q}_{\mathcal{F}}$. For every $Y \in \mathcal{L}_{\mathcal{F}}(dx)$ we define

$$\text{(2.3)} \qquad \mathcal{F}\text{-}\int_0^t YdX = \mathcal{F}\text{-}\int_0^t YdM + \int_0^t YdA\,, \qquad 0 < t < \infty\,,$$

where $\mathcal{F}\text{-}\int YdM$ is the stochastic (martingale) $\mathcal{F}$-integral and $\int YdA$ is the Lebesgue-Stieltjes integral (for each sample). The process

$$\mathcal{F}\text{-}\int_0^t YdX\,, \qquad 0 \leq t < \infty\,,$$

is called the *stochastic (quasimartingale) $\mathcal{F}$-integral* of $Y$ based on $dX$. It is obviously a continuous local $\mathcal{F}$-quasimartingale whose canonical decomposition is given by (2.1). The properties of quasi-martingale integrals will follow immediately from those of martingale integrals and Lebesgue-Stieltjes integrals. For example,

$$\text{(2.4)} \qquad \begin{aligned} &\int_0^t (Y_n - Y)^2(dM)^2 + \int_0^t |Y_n - Y||dA| \to 0, \ \forall t, \text{ a.s. }.\\ &\qquad \Longrightarrow \mathcal{F}\text{-}\int_0^t Y_n dX \to \mathcal{F}\text{-}\int_0^t YdX, \ \forall t, \text{ a.s.} \end{aligned}$$

Since $\mathcal{B}_{\mathcal{F}} \subset \mathcal{L}_{\mathcal{F}}(dX)$ for every $X \in \mathcal{Q}_{\mathcal{F}}$, the integral $\mathcal{F}\text{-}\int YdX$ is defined for every $Y \in \mathcal{B}_{\mathcal{F}}$.

Thus far we have fixed the reference family $\mathcal{F}$. Now we will discuss what happens on when we replace $\mathcal{F}$ by another reference family $\tilde{\mathcal{F}}$ finer than $\mathcal{F}$; $\tilde{\mathcal{F}}$ is called *finer* than $\mathcal{F}$, $\tilde{\mathcal{F}} \succ \mathcal{F}$ in notation, if $\tilde{\mathcal{F}}_t \supset \mathcal{F}_t$ for every $t$. Suppose that $\tilde{\mathcal{F}} \succ \mathcal{F}$. Then $\mathcal{W}_{\tilde{\mathcal{F}}} \supset \mathcal{W}_{\mathcal{F}}$ and $\mathcal{B}_{\tilde{\mathcal{F}}} \supset \mathcal{B}_{\mathcal{F}}$, but there is no inclusion relation between $\mathcal{Q}_{\tilde{\mathcal{F}}}$ and $\mathcal{Q}_{\mathcal{F}}$. Even if $X \in \mathcal{Q}_{\tilde{\mathcal{F}}} \cap \mathcal{Q}_{\mathcal{F}}$, there is no inclusion relation between $\mathcal{L}_{\mathcal{F}}(dX)$ and $\mathcal{L}_{\tilde{\mathcal{F}}}(dX)$, but we have

$$\mathcal{B}_{\mathcal{F}} \subset \mathcal{L}_{\mathcal{F}}(dX) \cap \mathcal{L}_{\tilde{\mathcal{F}}}(dX) .$$

**Theorem 2.1.** *Suppose that $\tilde{\mathcal{F}} \succ \mathcal{F}$. If $X \in \mathcal{Q}_{\mathcal{F}} \cap \mathcal{Q}_{\tilde{\mathcal{F}}}$, then*

$$\mathcal{F}\text{-}\int_0^t YdX = \tilde{\mathcal{F}}\text{-}\int_0^t YdX , \quad 0 \leqq t < \infty, \text{ a.s.} \tag{2.5}$$

*for every $Y \in \mathcal{L}_{\mathcal{F}}(dX) \cap \mathcal{L}_{\tilde{\mathcal{F}}}(dX)$ (and hence for every $Y \in \mathcal{B}_{\mathcal{F}}$).*

*Proof.* Let $\mathcal{L}$ denote the family of all $Y$'s in $\mathcal{L}_{\mathcal{F}}(dX) \cap \mathcal{L}_{\tilde{\mathcal{F}}}(dX)$ for which the equation (2.5) holds. It is obvious that $\mathcal{L}$ is closed under linear combinations. Using the property (2.4) we can easily prove that if (i) $Y^{(n)} \in \mathcal{L}$, $n = 1, 2, \cdots$, (ii) for each $\omega$, $Y_t^{(n)}(\omega)$ is bounded on every finite $t$-interval and (iii) $Y_t^{(n)}(\omega) \to Y_t(\omega)$ for every $(t, \omega)$, then $Y \in \mathcal{L}$. By the definition $\mathcal{L}$ contains all right-continuous step $\mathcal{F}$-processes with possible jumps on a discrete subsets of $[0, \infty)$ independent of $\omega$. Hence $\mathcal{L} \supset \mathcal{B}_{\mathcal{F}}$. If $Y \in \mathcal{L}_{\mathcal{F}}(dX) \cap \mathcal{L}_{\tilde{\mathcal{F}}}(dX)$, then the sequence of the truncated processes

$$Y_t^{(n)}(\omega) = Y_t(\omega) 1_{[-n,n]}(Y_t(\omega)) , \qquad n = 1, 2, \cdots$$

satisfies the assumption of (2.4) and

$$Y^{(n)} \in \mathcal{B}_{\mathcal{F}} \subset \mathcal{L} , \qquad n = 1, 2, \cdots$$

as was proved above. Hence we have $Y \in \mathcal{L}$, completing the proof.

*Example 2.1.* (The example mentioned in the introduction). Let $B = \{B_t, 0 \leqq t < \infty\}$ be a Brownian motion. We consider two reference families

$$\mathcal{F} = \mathcal{F}[B] \quad \text{and} \quad \tilde{\mathcal{F}} = \mathcal{F}[B_1, B]$$

where $B_1$ denotes the process equal to $B_1(\omega)$ for every $t$. Since $B = \{B_t\}$ is a continuous $\mathcal{F}$-martingale (so $B \in \mathcal{Q}_{\mathcal{F}}$), we can define

$$\mathcal{F}\text{-}\int_0^t YdB$$

for every $Y$ in the family

$$\mathcal{L}_{\mathcal{F}}(dB) = \left\{ Y \in \mathcal{W}_{\mathcal{F}} : \int_0^t Y_s^2 ds < \infty, \forall t, \text{ a.s.} \right\} ,$$

noting that $(dB)^2 = dt$. In fact this $\mathcal{F}$-integral is essentially the same as the classical Brownian stochastic integral. Since $B = \{B_t\}$ is a continuous $\tilde{\mathcal{F}}$-quasi-martingale with the canonical decomposition

$$B_t = M_t + A_t\,, \quad A_t = \int_0^{t\wedge 1} \frac{B_1 - B_s}{1-s} ds\,, \quad M_t = B_t - A_t\,,$$

we can define

$$\tilde{\mathscr{F}}\text{-}\int_0^t Y dB$$

for every $Y$ in the family

$$\mathscr{L}_{\tilde{\mathscr{F}}}(dB) = \left\{ Y \in \mathscr{W}_{\tilde{\mathscr{F}}} : \int_0^t Y_s^2 ds + \int_0^{t\wedge 1} |Y_s| \frac{|B_1 - B_s|}{1-s} ds < \infty, \forall t, \text{ a.s.} \right\}.$$

Noting that

$$\langle M \rangle_t = \langle B \rangle_t = t$$

and

$$E(|A|_1) = E\left( \int_0^1 \left| \frac{B_1 - B_s}{1-s} \right| ds \right) = \sqrt{\frac{2}{\pi}} \int_0^1 \frac{\sqrt{1-s}}{1-s} ds < \infty\,,$$

so

$$|A|_t \leqq |A|_1 < \infty \qquad \text{a.s.}\,.$$

It is obvious that

$$(2.6) \qquad Y = \{Y_t \equiv B_1,\ 0 \leqq t < \infty\} \in \mathscr{L}_{\tilde{\mathscr{F}}}(dB) \backslash \mathscr{L}_{\mathscr{F}}(dB)\,.$$

If we could prove that $\mathscr{L}_{\tilde{\mathscr{F}}}(dB) \supset \mathscr{L}_{\mathscr{F}}(dB)$, the $\tilde{\mathscr{F}}$-integral would be a proper extension of the $\mathscr{F}$-integral by virtue of Theorem 2.1, but we can neither prove nor disprove this inclusion relation at present.

*Example 2.2.* This is another example which is slightly more complicated than the previous one. Let $B = \{B_t\}$ be a Brownian motion as above. We want to define the stochastic integral

$$(2.7) \qquad \int_0^t f_s\left( \int_{s\wedge 1}^1 B_u du \right) dB_s$$

where $f_s(x)$ is a locally bounded Borel function of $(s, x) \in [0, \infty) \times (-\infty, \infty)$. It is obvious that this integral is meaningless in the sense of the classical Brownian stochastic integral. Let $\tilde{\mathscr{F}} = \{\tilde{\mathscr{F}}_t, 0 \leqq t < \infty\}$ be the reference family generated by the processes

$$B = \{B_t, 0 \leqq t < \infty\} \quad \text{and} \quad I = \left\{I_t \equiv \int_{t \wedge 1}^1 B_u du, 0 \leqq t < \infty\right\}.$$

Observing that for every $t \in (0, 1)$

$$\begin{aligned} \sigma[B_s, I_s, s \leqq t] &= \sigma\left[B_s, \int_s^1 B_u du, s \leqq t\right] \\ &= \sigma\left[B_s, s \leqq t, \int_t^1 B_u du\right] \\ &= \sigma\left[B_s, s \leqq t, \int_t^1 (B_u - B_t) du\right] \end{aligned}$$

and recalling that $B$ is a Gaussian process with $E(B_t) = 0$ and $E(B_t B_s) = s \wedge t$, we can prove that $B$ is a continuous local $\tilde{\mathscr{F}}$-quasimartingale with the canonical $\tilde{\mathscr{F}}$-decomposition

$$B_t = M_t + A_t$$

where

$$A_t = \int_0^{t \wedge 1} \frac{3}{(1-s)^2} \int_s^1 |B_u - B_s| \, duds \quad \text{and} \quad M_t = B_t - A_t \, .$$

It should be noted that

$$\begin{aligned} E(|A|_1) &\leqq E\left(\int_0^1 \frac{3}{(1-s)^2} \int_s^1 |B_u - B_s| \, duds\right) \\ &= \int_0^1 \frac{3}{(1-s)^2} \int_s^1 \sqrt{u-s} \, duds \\ &= \int_0^1 \frac{2}{\sqrt{1-s}} ds < \infty \, , \end{aligned}$$

so $|A|_t \leqq |A|_1 < \infty$ for every $t$ a.s.. It is obvious that

$$\langle M \rangle_t = \langle B \rangle_t = t \, .$$

Hence

$$\begin{aligned} \mathscr{L}_{\mathscr{F}}(dB) = \Big\{ Y \in \mathscr{W}_{\mathscr{F}} : &\int_0^t Y^2 dt \\ &+ \int_0^{t \wedge 1} |Y| \frac{3}{(1-s)^2} \left| \int_s^1 (B_u - B_s) du \right| ds < \infty, \forall t, \text{a.s.} \Big\} \, . \end{aligned}$$

Since the integrand process of (2.7) belongs to $\mathscr{B}_{\mathscr{F}} \subset \mathscr{L}_{\mathscr{F}}(dX)$, the stochastic integral (2.7) is well-defined.

*Example 2.3.* The above examples show that we may define the stochastic integral of an integrand depending on the future development of the base process by taking a finer reference family to which the integrand and the base process are adapted. But this is not always possible. Let $B = \{B_t\}$ be a Brownian motion as above. Since the space $C$ of all continuous functions on $[0, \infty)$ is a Polish space with non-denumerable points, there exists a Borel isomorphism $f: C \to [0, 1]$. Let

$$Y_t(\omega) = f(B_u(\omega), 0 \leqq u < \infty) \qquad \text{for every } t\,.$$

In order to define the stochastic integral $\int YdB$ we must choose a reference family $\mathscr{F}$ finer than $\mathscr{F}[B, Y]$, but $B$ cannot be a continuous local $\mathscr{F}$-quasimartingale for such an $\mathscr{F}$. Suppose that it had a canonical $\mathscr{F}$-decomposition

$$B = M + A\,.$$

Since $\mathscr{F}_t \supset \sigma[Y_t] = \sigma[B_u, 0 \leqq u < \infty]$ for every $t$, we have

$$(B_{t+\Delta} - B_t) - E(B_{t+\Delta} - B_t | \mathscr{F}_t) = (B_{t+\Delta} - B_t) - (B_{t+\Delta} - B_t) = 0\,,$$

which implies $M_t \equiv 0$, so

$$B_t \equiv A_t\,.$$

Hence it follows that $dt = (dB_t)^2 = (dA_t)^2 = 0$, which is a contradiction.

Let $\mathscr{F} = \{\mathscr{F}_t, 0 \leqq t < \infty\}$ be a reference family. Then so is the family $\mathscr{F}^{(s)} = \{\mathscr{F}_{s+t}, 0 \leqq t < \infty\}$. If $X \in \mathscr{Q}_{\mathscr{F}}$ and $Y \in \mathscr{L}_{\mathscr{F}}(dX)$, then

$$X^{(s)} = \{X_{s+t}, 0 \leqq t < \infty\} \in \mathscr{Q}_{\mathscr{F}^{(s)}}$$

and

$$Y^{(s)} = \{Y_{s+t}, 0 \leqq t < \infty\} \in \mathscr{L}_{\mathscr{F}^{(s)}}(dX^{(s)})\,,$$

so the stochastic integral

$$\mathscr{F}^{(s)}\text{-}\int_0^t Y^{(s)} dX^{(s)}$$

is well-defined and is denoted by

$$\mathscr{F}\text{-}\int_t^{s+t} YdX\,.$$

If $Z \in \mathscr{Q}_{\mathscr{F}}$ and if

$$Z_t - Z_s = \mathscr{F}\text{-}\int_s^t YdX\,, \qquad 0 \leqq s \leqq t < \infty\,,$$

then we denote this relation by

$$dZ = \mathscr{F}\text{-}YdX\,;$$

the prefix will often be omitted if there is no possibility of confusion.

If $dZ_i = Y_i dX_i$, $i = 1, 2, \cdots, n$ and if $F = f(Z)$ where

$$Z = (Z_1, Z_2, \cdots, Z_n) \quad \text{and} \quad f \in C^2(R^n)\,,$$

then the following stochastic chain rule holds:

$$dF = \sum_i \partial_i f(Z) dZ_i + \frac{1}{2} \sum_{i,j} \partial_i \partial_j f(Z) dZ_i dZ_j\,, \tag{2.8}$$

where $\partial_i$ denotes the partial derivative with respect to the $i$-th variable and

$$dZ_i dZ_j = d\langle Z_i, Z_j \rangle$$
$$(\langle Z_i, Z_j \rangle = \tfrac{1}{4}(\langle Z_i + Z_j \rangle - \langle Z_i - Z_j \rangle))\,.$$

Let us mention one word about the case where the time interval is the whole real line $(-\infty, \infty)$. Let $\mathscr{F} = \{\mathscr{F}_t, -\infty < t < \infty\}$ be a right-continuous increasing family of sub-$\sigma$-algebras of $\mathscr{F}_\Omega$ such that $\mathscr{F}_t \supset \mathbf{2}$ for every $t$. Such a family is called a reference family on the time interval $(-\infty, \infty)$. A stochastic process $X = \{X_t, -\infty < t < \infty\}$ is called a *continuous local $\mathscr{F}$-quasimartingale*, $X \in \mathscr{Q}_{\mathscr{F}}$ in notation, if

$$X^{(s)} = \{X_{s+t}, 0 \leqq t < \infty\}$$

is a continuous local quasimartingale relative to the reference family

$$\mathscr{F}^{(s)} = \{\mathscr{F}_{s+t}, 0 \leqq t < \infty\}\,.$$

A process $Y = \{Y_t, -\infty < t < \infty\}$ is said to belong to $\mathscr{L}_{\mathscr{F}}(dX)$, if

$$Y^{(s)} = \{Y_{s+t}, 0 \leqq t < \infty\} \in \mathscr{L}_{\mathscr{F}^{(s)}}(dX^{(s)})$$

for every $s \in (-\infty, \infty)$. For $X \in \mathscr{Q}_{\mathscr{F}}$ and $Y \in \mathscr{L}_{\mathscr{F}}(dX)$ we define

$$\mathscr{F}\text{-}\int_s^{s+t} YdX = \mathscr{F}^{(s)}\text{-}\int_0^t Y^{(s)} dX^{(s)}\,.$$

This stochastic integral on $(-\infty, \infty)$ has the same properties as the above stochastic integral on $[0, \infty)$. It is obvious that

$$\mathscr{F}\text{-}\int_s^t YdX + \mathscr{F}\text{-}\int_t^u YdX = \mathscr{F}\text{-}\int_s^u YdX\,, \qquad s \leqq t \leqq u\,.$$

## § 3. Symmetric stochastic integrals

In the last section we have defined the stochastic integral

$$\mathscr{F}\text{-}\int_s^t YdX\,, \quad -\infty < s \leqq t < \infty\,, \quad X \in \mathscr{Q}_{\mathscr{F}}\,, \quad Y \in \mathscr{L}_{\mathscr{F}}(dX)\,.$$

By reversing the time order we can define *time reversed stochastic integrals*. The family

$$\mathscr{G} = \{\mathscr{G}_t,\ \infty > t > -\infty\}$$

is called a *time-reversed reference family*, if

$$\mathscr{G}^- = \{\mathscr{G}_t^- \equiv \mathscr{G}_{-t},\ -\infty < t < \infty\}$$

is a reference family in the sense of the last section. For a time-reversed reference family $\mathscr{G} = \{\mathscr{G}_t,\ \infty > t > -\infty\}$ we define a process $X = \{X_t,\ \infty > t > -\infty\}$ to be a *time reversed continuous local $\mathscr{G}$-quasimartingale, $X \in \mathscr{Q}_{\mathscr{G}}$* in notation, if

$$X^- = \{X_t^- \equiv X_{-t},\ -\infty < t < \infty\}$$

is a continuous local $\mathscr{G}^-$-quasimartingale. Similarly we define

$$Y = \{Y_t,\ \infty > t > -\infty\} \in \mathscr{L}_{\mathscr{G}}(dX)$$

if $Y^- \in \mathscr{L}_{\mathscr{G}^-}(dX^-)$. The (*time-reversed*) *stochastic $\mathscr{G}$-integral* is defined as follows:

$$\mathscr{G}\text{-}\int_t^s YdX = \mathscr{G}^-\text{-}\int_{-s}^{-t} Y^-dX^-\,, \qquad \infty > t \geqq s > -\infty$$

for $X \in \mathscr{Q}_{\mathscr{G}}$ and $Y \in \mathscr{L}_{\mathscr{G}}(dX)$.

Let $(\mathscr{F}, \mathscr{G})$ be a pair consisting of a reference family $\mathscr{F}$ and a time reversed reference family $\mathscr{G}$. We define

$$\mathscr{Q}_{(\mathscr{F},\mathscr{G})} = \mathscr{Q}_{\mathscr{F}} \cap \mathscr{Q}_{\mathscr{G}}$$

and

$$\mathscr{L}_{(\mathscr{F},\mathscr{G})}(dX) = \mathscr{L}_{\mathscr{F}}(dX) \cap \mathscr{L}_{\mathscr{G}}(dX) \qquad \text{for } X \in \mathscr{Q}_{(\mathscr{F},\mathscr{G})}\,.$$

For $X \in \mathscr{Q}_{(\mathscr{F},\mathscr{G})}$ and $Y \in \mathscr{L}_{(\mathscr{F},\mathscr{G})}(dX)$ we define

$$(\mathscr{F},\mathscr{G})\text{-}\int_s^t Y\circ dX = \frac{1}{2}\left[\mathscr{F}\text{-}\int_s^t YdX - \mathscr{G}\text{-}\int_t^s YdX\right]$$

$$\text{or} \qquad = \frac{1}{2}\left[\mathscr{G}\text{-}\int_s^t YdX - \mathscr{F}\text{-}\int_t^s YdX\right]$$

according to whether $s \leqq t$ or $s \geqq t$ and call this integral the *symmetric stochastic integral* of $Y$ based on $dX$ relative to $(\mathscr{F},\mathscr{G})$. We will omit the prefix $(\mathscr{F},\mathscr{G})$ if there is no possibility of confusion.

For symmetric integrals we have

$$\int_s^t Y\circ dX + \int_t^u Y\circ dX = \int_s^t Y\circ dX$$

irrespectively of the order of $s$, $t$ and $u$, so

$$\int_s^t Y\circ dX = -\int_t^s Y\circ dX\,.$$

**Theorem 3.1.** *Let $X \in \mathscr{Q}_{(\mathscr{F},\mathscr{G})}$ and $Y \in \mathscr{L}_{(\mathscr{F},\mathscr{G})}(dX)$. If $Y$ is a continuous process, then*

$$(3.1)\qquad (\mathscr{F},\mathscr{G})\text{-}\int_s^t Y\circ dX = \operatorname*{l.i.p.}_{|\Delta|\to 0}\sum_{i=1}^n \frac{1}{2}(Y_{t_{i-1}} + Y_{t_i})(X_{t_i} - X_{t_{i-1}})\,,\qquad s\leqq t$$

*where $\Delta = \{s = t_0 < t_1 < \cdots < t_n = t\}$ is an arbitrary partition of the interval $[s, t]$ and*

$$|\Delta| = \max_{1\leq i\leq n}|t_i - t_{i-1}|\,.$$

*Proof.* Since $Y$ is continuous,

$$Y_u(\omega) = \lim_{|\Delta|\to 0}\sum_i Y_{t_{i-1}}(\omega)1_{[t_{i-1},t_i]}(u)\,,\qquad s\leqq u<t \quad \text{a.s.},$$

we obtain

$$\mathscr{F}\text{-}\int_s^t YdX = \operatorname*{l.i.p.}_{|\Delta|\to 0}\sum_{i=1}^n Y_{t_{i-1}}(X_{t_i} - X_{t_{i-1}})\,.$$

Similarly

$$\mathscr{G}\text{-}\int_t^s YdX = \operatorname*{l.i.p.}_{|\Delta|\to 0}\sum_{i=1}^n Y_{t_i}(X_{t_{i-1}} - X_{t_i})\,.$$

Substracting and dividing by 2, we obtain (3.1).

We denote by $dZ = Y \circ dX$ the relation

$$Z_t - Z_s = \int_s^t Y \circ dX\,, \qquad -\infty < s, t < \infty \qquad (Z \in \mathcal{Q}_{(\mathcal{F},\mathcal{G})})\,.$$

Using Theorem 2.1 we obtain

**Theorem 3.2** (*The chain rule for symmetric stochastic integrals*). *If* $dZ_i = Y_i \circ dX_i$, $i = 1, 2, \cdots, n$ *and if* $F = f(Z)$ *where* $Z = (Z_1, Z_2, \cdots, Z_n)$ *and* $f \in C^3(R^n)$, *then*

$$dF = \sum_i \partial_i f(Z) \circ dZ_i\,. \tag{3.2}$$

*Proof*. Observing that

$$\begin{aligned} f(b) - f(a) &\qquad (b = (b_i),\ a = (a_i)) \\ &= \sum_i \frac{1}{2}(\partial_i f(a) + \partial_i f(b))(b_i - a_i) + (|b - a|^3) \end{aligned}$$

we can prove (3.2) by the argument used in proving (2.8).

*Example 3.1*. Let $B = \{B_t, 0 \leqq t < \infty\}$ be a Brownian motion with $B_0 = 0$. By defining $B_t = 0$ for $t < 0$ we have $B = \{B_t, -\infty < t < \infty\}$. Let

$$\mathcal{F}_t = \bigcap_n \sigma\left[B_s, s \leqq t + \frac{1}{n}\right] \vee \mathbf{2} \qquad (\mathbf{2} = \{A \in \mathcal{F}_\Omega : P(A) = 0 \text{ or } 1\})$$

and

$$\mathcal{G}_t = \bigcap_n \sigma\left[B_s, s \geqq t - \frac{1}{n}\right] \vee \mathbf{2}\,.$$

Then $\mathcal{F} = \{\mathcal{F}_t, -\infty < t < \infty\}$ is a reference family and $\mathcal{G} = \{\mathcal{G}_t, \infty > t > -\infty\}$ is a time-reversed reference family. $B = \{B_t\}$ is a continuous $\mathcal{F}$-martingale. Observing that

$$E(B_s - B_t \mid \mathcal{G}_t) = \frac{s-t}{t} B_t\,, \qquad 0 < s < t < \infty\,,$$

we can prove that $B = \{B_t\}$ is a continuous time reversed $\mathcal{G}$-quasimartingale with the canonical decomposition

$$B = M + A\,,$$

where

$$A_t = \int_{\infty}^{t\vee 0} \frac{1}{s} B_s ds\,, \qquad M = B - A\,.$$

Hence $B$ belongs to $\mathscr{Q}_{(\mathscr{F},\mathscr{G})}$. Let $h$ be any Borel function on $\boldsymbol{R}^1$. Then $Y$ belongs to $\mathscr{L}_{(\mathscr{F},\mathscr{G})}(dB)$ (so the symmetric stochastic integral

$$\int_s^t h(B_u) \circ dB_u\,, \qquad -\infty < s,\, t < \infty$$

is well-defined), if and only if

$$\int_0^n h(B_t)^2 dt + \int_0^n |h(B_t)| \frac{1}{t} |B_t|\, dt < \infty,\ \forall n\,. \tag{3.3}$$

This condition is satisfied if $h$ is a locally bounded Borel function. By Theorem 3.2 we have

$$f(B_t) - f(B_s) = \int_s^t f'(B_u) \circ dB_u \tag{3.4}$$

for $f \in C^3(\boldsymbol{R}^1)$, but we can prove this (identity) for every absolutely continuous function $f$ whose Lebesgue derivative $f'$ satisfies (3.3), observing that the symmetric stochastic integral commutes with limits by virtue of the definition.

*Example 3.2.* Let $X = (X_t, 0 \leqq t < \infty)$ be a diffusion process with generator

$$A = \frac{1}{2} a\partial^2 + b\partial\,, \quad a > 0, \quad \partial = \frac{d}{dx}\,. \tag{3.5}$$

Then $X$ is determined by the stochastic differential equation:

$$dX = \sqrt{a(X)} dB + b(X) dt\,, \qquad t \geqq 0\,. \tag{3.6}$$

By defining $X_t = X_0$ for $t \leqq 0$ we have a diffusion process on the time interval $(-\infty, \infty)$. Let

$$\mathscr{F}_t = \bigcap_n \sigma\left[X_s,\, s \leqq t + \frac{1}{n}\right] \vee \mathbf{2}$$

and

$$\mathscr{G}_t = \bigcap_n \sigma\left[X_s,\, s \geqq t - \frac{1}{n}\right] \vee \mathbf{2}\,.$$

Using (3.6) we can easily prove that $X \in \mathcal{Q}_{\mathcal{F}}$. $X$ is a diffusion process by reversing the time order. Under some reasonable conditions we can prove that the generator of this backward diffusion is also a second order differential operator. Hence $X \in \mathcal{Q}_{\mathcal{G}}$, so $X \in \mathcal{Q}_{(\mathcal{F},\mathcal{G})}$. As in Example 3.1 we can define a symmetric stochastic integral of the type

$$\int_s^t h(X_u) \circ dX_u$$

and prove the chain rule

$$f(x_t) - f(x_s) = \int_s^t f'(X_u) \circ dX_u$$

as in Example 3.1.

In the above discussions we considered symmetric stochastic integrals relative to a pair $(\mathcal{F}, \mathcal{G})$ of a reference family $\mathcal{F}$ and a time-reversed reference family $\mathcal{G}$. In view of Theorem 3.1 we may define *symmetric stochastic integral* as follows:

$$(3.7) \qquad \int_s^t Y \circ dX = \underset{|\Delta| \to 0}{\text{l.i.p.}} \sum_{i=1}^n \frac{1}{2}(Y_{t_i} + Y_{t_{i-1}})(X_{t_i} - X_{t_{i-1}})$$

if this limit exists for every $(s, t) \in \boldsymbol{R}^2$ and has a version continuous in $(t, s)$, where $X$ is a continuous process with

$$(3.8) \qquad \lim_{|\Delta| \to 0} \sum_{i=1}^n |X_{t_i} - X_{t_{i-1}}|^3 \to 0$$

and $Y$ is a continuous process. This definition has nothing to do with reference families. The proof of Theorem 3.2 actually proves that the chain rule holds for such a symmetric stochastic integral. Writing the integral of (3.7) as $Z_t - Z_s$, we can easily show that $Z = \{Z_t\}$ satisfies (3.8).

As a special case we can define the *forward symmetric stochastic integral* as follows. If $\mathcal{F}$ is a reference family and if $X, Y \in \mathcal{Q}_{\mathcal{F}}$, the stochastic integral (3.6) is well-defined and

$$(3.9) \qquad \int_s^t Y \circ dX = \mathcal{F}\text{-}\int_s^t Y dX + \frac{1}{2}\int_s^t dY dX .$$

Similarly for the *backward symmetric stochastic integral* relative to a time-reversed reference family $\mathcal{F}$.

To distinguish the symmetric stochastic $(\mathcal{F}, \mathcal{G})$-integral from others we call it a *two-sided symmetric stochastic integral*.

## § 4. Stochastic parallel displacement

Let $E$ be a $d$-dimensional Riemannian manifold with metric tensor $(g_{ij})$ and $X = \{X_t, -\infty < t < \infty\}$ be a Brownian motion on $E$, i.e., a diffusion process whose generator is $\Delta/2$ where $\Delta$ is the Laplace-Beltrami operator on $E$. The sample function of $X$ can be constructed from the Brownian motion $B$ on $R^d$ by a stochastic differential equation:

$$dX^i = \sum_\alpha \sigma^i_\alpha dB^\alpha + b^i dt \tag{4.1}$$

where

$$\sum_\alpha \sigma^i_\alpha \sigma^j_\alpha = g^{ij}\,, \qquad (g^{ij}) = (g_{ij})^{-1}$$

and

$$b^i = \frac{1}{2}\sum_{\alpha,\beta} g^{\alpha\beta}\Gamma^i_{\alpha\beta}\,.$$

Hence $X^i$ is a continuous quasimartingale relative to the reference family $\mathscr{F}$ generated by $X$ (if it is stopped at the exit time from the neighborhood of the local coordinates). Since the tensor $(g^{ij})$ is strictly positive definite and belongs to the class $C^\infty$, $X$ is a continuous, time-reversed quasimartingale relative to the time reversed reference family $\mathscr{G}$ generated by $X$ in the same way as in Example 3.2. Hence the symmetric stochastic integral of the type

$$(\mathscr{F},\mathscr{G})\text{-}\int_s^t h_i(X_t)\circ dX^i_t\,, \qquad i = 1, 2, \cdots, n$$

is well-defined. Since symmetric stochastic differentials are subject to the chain rule of Theorem 3.2 that takes the same form as in the ordinary calculus, we can easily prove that the above integral along the sample curve of $X$ is independent of the choice of local coordinates whenever $h = (h_1, h_2, \cdots, h_d)$ is a covariant vector field.

The stochastic parallel displacement along the Brownian curve $X$ is determined by the symmetric stochastic differential equation:

$$dU^i_t = -\sum_{j,k}\Gamma^i_{jk}(X_t)U^j\circ dX^k_t\,, \qquad i = 1, 2, \cdots, d\,. \tag{4.2}$$

For simplicity we will discuss such a parallel displacement along the conditional Brownian motion starting at $X_0 = a$ and ending at $X_1 = b$. Such a conditional Brownian motion is also a diffusion whose generator is an elliptic differential operator. Hence $X^i \in \mathscr{Q}_\mathscr{F} \cap \mathscr{Q}_\mathscr{G}$ where $\mathscr{F}$ and $\mathscr{G}$ are defined relative to the conditional Brownian motion $X$. Interpreting the above differential equation in the forward (resp. backward) sense we obtain $U_1$ from $U_0$ (resp. $U_0$ from $U_1$). Noticing that the chain rule for symmetric stochastic differentials takes the same form as in ordinary calculus we can prove that both the map $\Phi: U_0 \to U_1$ and the map $\Psi: U_1 \to U_0$ is orthogonal and depends on the curve $X$ (forward for $\Phi$ and backward for $\Psi$). Since the equation (4.2) is invariant under the time reversal, we can also prove that $\Phi$ and $\Psi$ are inverse to each other.

## References


[1] McKean, H. P., Jr., Stochastic integrals, Academic Press, 1969.

[2] Fisk, D. L., Quasi-martingales and stochastic integrals, Technical Report No. 1, Dept. Math. Michigan State Univ., 1963.

[3] Courrège, P., Intégrales stochastiques et martingales de carré integrable, Sem. Th. Potent. 7 Inst. Henri Poincare, Paris, 1963.

[4] Kunita, H. and Watanabe, S., On square integrable martingales, Nagoya Math. J., **30** (1967), 209–245.

[5] Meyer, P. A., Intégrales stochastiques, I–IV Sém. de Prob., Lecture Notes in Math., **39** (1967), 72–162, Springer.

[6] Stratonovich, R. L., Conditional Markov processes and their application to optimal control, English Translation, Elsevier, New York, 1968.

[7] Itô, K., Stochastic differentials, Appl. Math. and Optimization, **1** (1975), 347–381.

[8] ——, The Brownian motion and tensor fields on Riemannian manifold, Proc. Internat. Congress of Math. Stockholm, 1962.

[9] ——, Stochastic parallel displacement, Lecture Notes in Math., **451** (1975), 1–7, Springer.

[10] Dynkin, E. B., Diffusion of tensors, Dokl. Akad. Nauk SSSR, **179**-6 (1968), 532–535.



Research Institute
for Mathematical Sciences
Kyoto University
Kyoto 606, Japan

Taniguchi Symp. SA
Katata 1982, pp. 197–224

# Infinite Dimensional Ornstein-Uhlenbeck Processes

Kiyosi Itô

## § 1. Introduction

The purpose of this paper is to prove an infinite dimensional version of the following well-known fact:

An Ornstein-Uhlenbeck process, a centered Gaussian, Markov, stationary and mean-continuous process $\{X_t\}$ satisfies the Langevin equation:

$$dX_t = \beta \cdot dB_t + \alpha X_t dt \quad (\alpha < 0) \tag{1.1}$$

and is expressible as follows:

$$X_t = e^{\alpha(t-s)} X_s + \int_s^t e^{\alpha(t-u)} \beta \cdot dB_u \tag{1.2}$$

$$X_t = \int_{-\infty}^t e^{\alpha(t-u)} \beta dB_u\,. \tag{1.3}$$

We consider a stochastic process of linear random functionals on a vector space $E$:

$$X_t : E \to L_2(\Omega, \mathscr{F}, P)\,, \qquad t \in (-\infty, \infty)\,. \tag{1.4}$$

We define an *Ornstein-Uhlenbeck process of linear random functionals* in the same way as in the one dimensional case. With such a process we can associate a semigroup of operators on a spearable Hilbert space $\bar{E}$ (§ 2):

$$\{S_t, t > 0\}\,, \tag{1.5}$$

which corresponds to the semigroup $\{e^{\alpha t}, t > 0\}$ in (1.2) and (1.3). The constant $\alpha$ turns out to be the infinitesimal generator $A$ of the semigroup $\{S_t\}$ in our infinite dimensional case. There is a nice parallelism between the one-dimensional case and the infinite dimensional case, but we obtain an additional term, called the deterministic part in (1.3) in our case (§ 6, § 8). In the last section we will discuss the continuous regular versions of the processes in consideration.

For the Langevin equation in infinite dimensions there are several interesting papers related to the general theory, Y. Okabe [3], J.T. Lewis-L.C. Thomas [2] and to special problems, B. Gaveau [1].

## § 2. Definitions

Let $(\Omega, \mathscr{F}, P)$ be a probability space and $L$ denote the real $L_2$-space on $(\Omega, \mathscr{F}, P)$ endowed with the $L_2$-norm, denoted by $\| \ \|_L$. We assume $(L, \| \ \|_L)$ to be separable. Let $E$ be a vector space over $\boldsymbol{R}$. A linear map of $E$ into $L$ is called a *linear random functional* (abbr. LRF) on $E$. We denote by $\mathscr{L}(E)$ the vector space that consists of all LRF's.

The time interval $T$ is $(-\infty, \infty)$ unless stated otherwise. A map $X: T \to \mathscr{L}(E)$, $t \mapsto X_t$ is called a *process of LRF's* on $E$ or an $\mathscr{L}(E)$-*process*.

Let $X$ be an $\mathscr{L}(E)$-process. $X$ is called *centered Gaussian* if $\{X_t(f): t \in T, f \in E\}$ is a Gaussian system with mean 0. The finite joint distributions of a centered Gaussian $\mathscr{L}(E)$-process are determined by the *covariance functional*:

$$V_{s,t}(f, g) \equiv V^X_{s,t}(f, g) := E(X_s(f) \cdot X_t(g)) = (X_s(f), X_t(g))_L . \tag{2.1}$$

For any set $I \subset T$ we denote by $\sigma_I(X)$ the $\sigma$-algebra generated by $X_t(f)$, $t \in I$, $f \in E$ and all $P$-null sets. For $I = (-\infty, t]$, $[t, \infty)$ and $\{t\}$ (a singleton) we denote $\sigma_I$ by $\sigma_t^-$, $\sigma_t^+$ and $\sigma_t$ respectively. We denote by $M_I(X)$ the closed linear span (abbr. CLS) of $X_t(f)$, $t \in I, f \in E$. $M_t^-$, $M_t^+$ and $M_t$ are defined from $M_I$ in the same way as for $\sigma$-algebras.

$X$ is called *Markov* if

$$P(B|\sigma_t^-(X)) = P(B|\sigma_t(X)), \quad B \in \sigma_t^+(X), \quad t \in T, \tag{2.2}$$

and *weakly Markov* if

$$\hat{E}(X_t(f)|M_s^-(X)) = \hat{E}(X_t(f)|M_s(X)), \quad f \in E, \quad t > s, \tag{2.3}$$

where $\hat{E}(\cdot|\cdot)$ denotes the orthogonal projection in $L$. These two properties are equivalent to each other for a centered Gaussian $\mathscr{L}(E)$-process.

$X$ is called *stationary* if the finite joint distributions of $\{X_t(f)\}_{t,f}$ are invariant under time shift, and *weakly stationary* if the covariance functional of $X$ is invariant under time shift. These two properties are equivalent to each other for a centered Gaussian $\mathscr{L}(E)$ process.

$X$ is called *mean-continuous* if

$$\lim_{h \to 0} E((X_{t+h}(f) - X_t(f))^2) = 0, \quad f \in E, \quad t \in T. \tag{2.4}$$

$X$ is said to have *independent increments* if the $\sigma$-algebras generated by $\{X_{t_i}(f) - X_{s_i}(f), f \in E\}$, $i = 1, 2, \cdots, n$, are independent whenever $(s_i, t_i]$, $i = 1, 2, \cdots, n$, are disjoint. Furthermore if the finite joint distributions of $X_t(f) - X_s(f), f \in E$ are invariant under time shift, then $X$ is said to have *stationary independent increments.* $X$ is said to have *orthogonal increments* if $X_t(f) - X_s(f)$ and $X_v(g) - X_u(g)$ are orthogonal to each other whenever $s < t \leqq u < v$ and $f, g \in E$. Furthermore if the mean functional and the covariance functional of $X_t(f) - X_s(f)$, $f \in E$ are invariant under time shift, then $X$ is said to have *stationary orthogonal increments.* The property of having (stationary) independent increments and that of having (stationary) orthogonal increments are equivalent to each other for a centered Gaussian $\mathscr{L}(E)$-process.

**Definition 2.1.** An $\mathscr{L}(E)$-process is called a *Wiener $\mathscr{L}(E)$-process* if it is centered Gaussian and has stationary independent increments.

The following theorem follows easily from the definitions:

**Theorem 2.1.** *A centered Gaussian $\mathscr{L}(E)$-process $B = \{B_t, t \in T\}$ is a Wiener $\mathscr{L}(E)$-process if and only if the covariance functional of its increments*

$$\{B_{st} = B_t - B_s\}_{s<t}$$

*is represented in the form:*

$$E(B_{st}(f) \cdot B_{uv}(g)) = |(s, t] \cap (u, v]| b(f, g) , \tag{2.5}$$

*where $b(f, g)$ is a general inner product in $E$ ($b(f, f) = 0$ does not imply $f = 0$ in general) and $|\ |$ indicates the Lebesgue measure.*

Every general inner product $b(f, g)$ induces a Hilbertian seminorm $b(f) = b(f, f)^{1/2}$ which determines $b(f, g)$ in the obvious way. The Hilbertian seminorm $b(f)$ corresponding to $b(f, g)$ in (2.5) is determined by $B$ as follows:

$$b(f)^2 = E(B_{01}(f)^2) . \tag{2.6}$$

Hence $b$ is denoted by $b_B$.

**Definition 2.2.** $b_B$ is called the *characteristic seminorm* of $B$.

**Theorem 2.2.** *For every Hilbertian seminorm $b$ on a vector space $E$ there exists a unique (in the finite joint distributions) Wiener $\mathscr{L}(E)$-processes with characteristic seminorm $b$.*

*Proof.* $V((s, t, f), (u, v, g)) := |(s, t] \cap (u, v]| b(f, g)$ is non-negative-definite, because it is represented as follows:

$$\int_T b(1_{(s,t]}(\tau)f,\ 1_{(u,v]}(\tau)g)d\tau\,,$$

so $V$ is symmetric and

$$\sum_{ij} x_i x_j V((s_i, t_i, f_i), (s_j, t_j, f_j))$$
$$= \int_T b(\sum_i x_i 1_{s_i t_i}(\tau) f_i)^2 d\tau \geqq 0\,.$$

The rest of the proof is routine.

**Definition 2.3.** An $\mathscr{L}(E)$-process is called an *Ornstein-Uhlenbeck* (abbr. OU) *$\mathscr{L}(E)$-process* if it is centered Gaussian, Markov, stationary and mean-continuous.

Let $X = \{X_t, t \in T\}$ be an OU $\mathscr{L}(E)$-process, Then

$$p(f) := \|X_t(f)\|_L$$

is a Hilbertian seminorm on $E$. Because of the stationarity of $X$, $p$ does not depend on $t$ but is determined only by $X$. Hence it is denoted by $p_X$.

**Definition 2.4.** $p_X$ is called the *characteristic seminorm* of the OU $\mathscr{L}(E)$-process.

By identifying two vectors in $E$ with $p$-distance $0 (p = p_X)$, we obtain a pre-Hilbert space $(\tilde{E}, \tilde{p})$ and by completing $(\tilde{E}, \tilde{p})$ we obtain a Hilbert space $(\bar{E}, \bar{p})$. Using the natural surjection $\sigma: E \to \tilde{E}$ we define an $\mathscr{L}(E)$-process

$$\tilde{X}_t(\sigma f) := X_t(f)\,.$$

This is clearly an OU $\mathscr{L}(\tilde{E})$-process with characteristic norm $p_{\tilde{X}} = \tilde{p}$. Since $X_t: (\tilde{E}, \tilde{p}) \to (L, \|\ \|_L)$ is isometric, it can be uniquely extended to an isometric map $\bar{X}_t: (\bar{E}, \bar{p}) \to (L, \|\ \|_L)$. It is easy to check that $\bar{X} = \{X_t, t \in T\}$ is an OU $\mathscr{L}(E)$-process with charactristic norm $p_{\bar{X}} = \bar{p}$. Since $(\bar{E}, \bar{p})$ is isomorphic to $(\bar{X}_t(\bar{E}), \|\ \|_L)$, our separability assumption of $(L, \|\ \|_L)$ implies that $(\bar{E}, \bar{p})$ is a separable Hilbert space.

Using the natural injection $\theta: E \to \bar{E}$ we obtain $X$ from $\bar{X}$ as follows:

$$X(f) = \bar{X}(\theta f)\,. \tag{2.7}$$

**Definition 2.5.** $\bar{X}$ is called the *completion* of $X$. If $\bar{X} = X$, then $X$ is called *complete*.

**Theorem 2.3.** *If $X$ is a complete OU $\mathscr{L}(E)$-process, then*

(2.8) $$X_t(E) = M_t(X).$$

*Proof.* Since $\bar{X} = X$ and so $(E, p) = (\bar{E}, \bar{p})$, the argument above implies that $(E, p)$ is a separable Hilbert space. Being isomorphic to $(E, p)$, $(X_t(E), \|\ \|_L)$ is also a separable Hilbert space. Hence $X_t(E)$ is a closed linear subspace of $(L, \|\ \|_L)$. Thus we obtain

$$M_t(X) = CLS(X_t(E)) = X_t(E).$$

In the subsequent sections except in section 9 we assume that $X$ is a complete $OU\,\mathscr{L}(E)$-process, so that $(E, p)$ is a separable Hilbert space and $X_t(E) = M_t(X)$.

## § 3. The characteristic operator

Let $X = \{X_t\}$ be a complete $OU\,\mathscr{L}(E)$-process, and $p$ its characteristic norm. Then $(X_t(E), \|\ \|_L)$ is isomorphic to $(E, p)$ and $X_t(E) = M_t(X)$, as we mentioned in the previous section.

Since $X$ is centered Gaussian and Markov, we obtain

$$\hat{E}(X_t(f) \mid M_0^-(X)) = \hat{E}(X_t(f) \mid M_0(X)) \in M_0(X) = X_0(E),$$

for $t \geqq 0$, so we can find a unique $g = g(t, f) \in E$

$$\hat{E}(X_t(f) \mid M_0^-(X)) = X_0(g(t, f)),$$

which implies that

(3.1) $$\hat{E}(X_{t+s}(f) \mid M_s^-(X)) = X_s(g(t, f)), \qquad t, s \geqq 0,$$

by the stationarity of $X$. Defining an operator $S_t \colon E \to E$ by $S_t f = g(t, f)$ for each $t \geqq 0$, we can rewrite (3.1) as follows:

(3.2) $$\hat{E}(X_{t+s}(f) \mid M_s^-(X)) = X_s(S_t f), \qquad t \geqq 0,\ s \in T.$$

**Theorem 3.1.** $\{S_t,\ t \geqq 0\}$ *is a strongly continuous semigroup of linear contractions on the separable Hilbert space* $(E, p)$.

*Proof.* Since $f \mapsto \hat{E}(X_t(f) \mid M_0^-(X))$ is linear in $f$, (3.2) ($s=0$) implies that $S_t$ is linear. Since

$$\begin{aligned} p(S_t f) &= \|X_0(S_t f)\|_L = \|\hat{E}(X_t(f) \mid M_0^-(X))\|_L \\ &\leqq \|X_t(f)\|_L = p(f), \end{aligned}$$

$S_t$ is a linear contraction in $(E, p)$. To prove the semigroup property, observe

$$\begin{aligned}X_0(S_{s+t}f) &= \hat{E}(X_{s+t}(f)|M_0^-) \quad (M_\theta^- = M_\theta^-(X)) \\ &= \hat{E}(\hat{E}(X_{s+t}(f)|M_s^-)|M_0^-) \quad \text{(by } M_s^- \supset M_0^-) \\ &= \hat{E}(X_s(S_t f)|M_0^-) \quad \text{(by (3.2))} \\ &= X_0(S_s S_t f) \quad \text{(by 3.2))}\,.\end{aligned}$$

$S_0 = I$ is obvious. Hence $\{S_t, t \geqq 0\}$ is a semigroup of linear contractions in $(E, p)$, which is strongly continuous, because

$$\begin{aligned}p(S_t f - f) &= \|X_0(S_t f) - X_0(f)\|_L \\ &= \|\hat{E}(X_t(f) - X_0(f)|M_0^-(X))\|_L \\ &\leqq \|X_t(f) - X_0(f)\|_L \to 0 \quad (t \downarrow 0)\,.\end{aligned}$$

Now we will apply the Hille-Yosida theory [5] to $\{S_t\}$. Let $A$ be the infinitesional generator of $\{S_t\}$. Since $(E, p)$ is a separable Hilbert space, $A$ is characterized by the following three conditions:

(A.1) $A$ is a closed linear operator in $E$ and $D = \mathscr{D}(A)$ is dense in $E$.
(A.2) $\mathscr{R}(I - A) = E$.
(A.3) (*dissipativity*) $p(Af, f) \leqq 0, \quad f \in D$.

Also

$$f \in D \Longrightarrow S_t f \in D\,, \qquad \frac{d}{dt} S_t f = A S_t f = S_t A f\,,$$

where the differentiation is taken in the sense of norm convergence in $(E, p)$ for each $f$.

**Definition 3.1.** $A$ is called the *characteristic operator* of the $OU\ \mathscr{L}(E)$-process $X$.

Since $X = \{X_t\}$ is centered Gaussian, the finite joint distributions $\{X_t(f)\}_{t,f}$ are determined by its covariance function, which is given as follows:

**Theorem 3.2.**

$$E(X_s(f)X_t(g)) = p(f, S_{t-s}g) \quad (t \geqq s)\,, \qquad p(S_{s-t}f, g) \quad (s \geqq t)\,.$$

*Proof.* In case $t \geqq s$ we obtain

$$\begin{aligned}E(X_s(f)X_t(g)) &= E(X_s(f)E(X_t(g)|\sigma_s^-(X))) \\ &= E(X_s(f)\hat{E}(X_t(g)|M_s^-(X))) \\ &= E(X_s(f)X_s(S_{t-s}g)) \\ &= p(f, S_{t-s}g)\,.\end{aligned}$$

Similarly for the case $s \geqq t$.

## § 4. Three Hilbertian seminorms on $D$

We define two Hilbertian norms $q$ and $r$ and a Hilbertian seminorm $b$ on $D = \mathscr{D}(A)$ as follows:

$$q(f, g) = p(f, g) + p(Af, Ag) . \tag{4.1}$$

$$r(f, g) = p((I - A)f, (I - A)g) . \tag{4.2}$$

$$b(f, g) = - p(Af, g) - p(f, Ag) . \tag{4.3}$$

It is obvious that $q$ is a Hilbertian norm and $r$ and $b$ are Hilbertian seminorms by the dissipativity of $A$. We can see that $r$ is a Hilbertian norm by observing

$$r(f)^2 = p(f)^2 + p(Af)^2 - 2p(Af, f) = q(f)^2 + b(f)^2 .$$

Thus we obtain the following:

**Theorem 4.1.**

$$b(f) \leqq q(f) \leqq r(f) \leqq \sqrt{2}\, q(f) . \tag{4.4}$$

**Theorem 4.2.** *$(D, q)$ and $(D, r)$ are separable Hilbert spaces.*

*Proof.* Let $G$ be the graph of $A$, i.e.

$$G := \{(f, Af): f \in D\} . \tag{4.5}$$

Since $A$ is a closed linear operator, $G$ is a closed linear subspace of the separable Hilbert space $(E, p) \oplus (E, p)$. Hence $(G, p \oplus p)$ is also a separable Hilbert space. Being isomorphic to this space by the projection $(f, Af) \mapsto f$, $(D, q)$ is a separable Hilbert space and so is $(D, r)$ by (4.4).

## § 5. The innovation process

Let $D := \mathscr{D}(A)$ as in the previous sections. We define an $\mathscr{L}(D)$-process $B = \{B_t\}$ whose increments $\{B_{st}\}$ are given as follows:

$$B_{st}(f) := X_t(f) - X_s(f) - \int_s^t X_u(Af)du , \qquad f \in D , \tag{5.1}$$

where the integral is the Bochner integral.

**Definition 5.1.** The $\mathscr{L}(D)$-process $B = \{B_t\}$ is called the *innovation process* of $X$. This name will be justfied at the end of § 8.

**Theorem 5.1.** *The innovation process is a Wiener $\mathscr{L}(D)$-process with characteristic norm b* $(b(f)^2 = -2p(Af, f)$, § 4).

*Proof.* $B = \{B_t\}$ is obviously centered Gaussian. It is obvious that

$$B_{st}(D) \subset M_s^+(X) \cap M_t^-(X), \qquad s < t. \tag{5.2}$$

We can check that

$$\hat{E}(B_{uv}(f) | M_t^-(X)) = 0, \qquad f \in D, \quad t \leqq u < v, \tag{5.3}$$

observing that

$$\hat{E}(B_{uv}(f) | M_t^-(X)) = X_t(S_{v-t}f) - X_t(S_{u-t}f) - \int_u^v X_t(S_{\theta-t}Af)d\theta$$

(by (5.1) and (3.2))

$$= X_t\left(S_{v-t}f - S_{u-t}f - \int_u^v S_{\theta-t}Af d\theta\right) = 0.$$

Hence if $s < t \leqq u < v$, then $B_{st}(f) \in M_t^-(X)$ by (5.2) and

$$(B_{st}(f), B_{uv}(g))_L = \hat{E}(B_{st}(f)\hat{E}(B_{uv}(g) | M_t^-(X))) = 0. \tag{5.4}$$

Next we will check that

$$\| B_{st}(f) \|_L^2 = (t - s)b(f)^2. \tag{5.5}$$

To do this we will use Theorem 3.2:

$$(X_s(f), X_t(g))_L = p(f, S_{t-s}g), \qquad t \geqq s.$$

Since there is no possibility of confusion, we will omit $L$ and $p$ in the proof below.

$$\begin{aligned}
\| B_{st}(f) \|^2 &= \| X_t(f) \|_s^2 + \| X_s(f) \|^2 + \int_s^t \int_s^t (X_u(Af), X_v(Af))dudv \\
&\quad - 2(X_t(f), X_s(f)) - 2\int_s^t (X_t(f), X_u(Af))du \\
&\quad + 2\int_s^t (X_v(Af), X_s(f))dv \\
&= 2\|f\|^2 + 2\int_s^t \int_u^t (S_{v-u}Af, Af)dvdu - 2(S_{t-s}f, f) \\
&\quad - 2\int_s^t (S_{t-u}f, Af)du + 2\int_s^t (S_{v-s}Af, f)dv \\
&= 2\|f\|^2 + 2\int_s^t ((S_{t-u}f, Af) - (f, Af))du - 2(S_{t-s}f, f)
\end{aligned}$$

$$- 2\int_s^t (S_{t-u}f, Af)du + 2((S_{t-s}f, f) - (f, f))$$
$$= - 2(f, Af)(t - s) = (t - s)b(f)^2.$$

This proves (5.5). Now it is easy to derive the following from (5.4) and (5.5):

(5.6) $$(B_{st}(f), B_{uv}(g))_L = |(s, t] \cap (u, v]| b(f, g), f, g \in D .$$

Hence $B = \{B_t\}$ is a Wiener $\mathscr{L}(D)$-process with characteristic seminorm $b$ by Theorem 2.1.

We denote by $M_t^-(dB)$(resp. $M_t^+(dB)$) the following subspace of $(L, \| \|_L)$:

$$\text{CLS}\{B_{uv}(f) : u < v \leqq t, f \in D\}$$
$$(\text{resp. CLS}\{B_{uv}(f) : t \leqq u < v, f \in D\}) .$$

Similarly we define the $\sigma$-algebras $\sigma_t^-(dB)$ and $\sigma_t^+(dB)$.

**Theorem 5.2.**

(i) $M_t^-(dB) \subset M_t^-(X)$, $\quad \sigma_t^-(dB) \subset \sigma_t^-(X)$,

(ii) $M_t^+(dB) \perp M_t^-(X)$, $\quad \sigma_t^+(dB) \perp\!\!\!\perp \sigma_t^-(X)$,

*where $\perp$ means orthogonality in $(L, \| \|_L)$ and $\perp\!\!\!\perp$ means independence.*

*Proof.* The statements for the subspaces of $L$ are obvious by (5.2) and (5.3). Hence the statements for the $\sigma$-algebras follow at once, because $X$ is centered Gaussian.

## § 6. The determistic part

Define

$$\sigma_{-\infty}(X) := \bigcap_t \sigma_t^-(X) = \lim_{t\to\infty} \sigma_{-t}^-(X) ,$$
$$M_{-\infty}(X) := \bigcap_t M_t^-(X) = \lim_{t\to\infty} M_{-t}^-(X) .$$

Since $X$ is centered Gaussian,

(6.1) $$E(X_t(f)|\sigma_s^-(X)) = \hat{E}(X_t(f)|M_s^-(X)) .$$

Since $\sigma_{-s}^-(X)$ and $M_{-s}^-(X)$ decrease as $s$ increases, we obtain

(6.2) $$E(X_t(f)|\sigma_{-\infty}(X)) = \lim_{s\to\infty} E(X_t(f)|\sigma_{-s}^-(X))$$

$$\hat{E}(X_t(f)|M_{-\infty}(X)) = \lim_{s\to\infty} \hat{E}(X_t(f)|M_s^-(X)) \tag{6.3}$$

and so

$$E(X_t(f)|\sigma_{-\infty}(X)) = \hat{E}(X_t(f)|M_{-\infty}(X)), \tag{6.4}$$

where 'lim' indicates 'limit in $(L, \| \|)$'. We denote (6.4) by $X_t^d(f)$. Then $X^d = \{X_t^d, t \in T\}$ is an $\mathscr{L}(E)$-process.

**Definition 6.1.** $X^d$ is called the *deterministic part* of $X$.

**Proposition 6.1.**

$$X_t^d(f) = \lim_{s\to\infty} X_{-s}(S_{s+t}f). \tag{6.5}$$

Hence $X^d$ is centered Gaussian.

*Proof.* Use (6.3) and (3.2).

**Proposition 6.2.**

$$E(X_t^d(f)X_t^d(g)) = \lim_{u\to\infty} p(S_u f, S_u g). \tag{6.6}$$

Hence this expectation is independent of $t$ and is denoted by $p^d(f, g)$.

*Proof.* Use (6.5).

**Proposition 6.3.**

$$X_t^d(f) = X_s^d(S_{t-s}f), \quad t \geqq s. \tag{6.7}$$

*Proof.*

$$\begin{aligned} X_t^d(f) &= \hat{E}(X_t(f)|M_{-\infty}(X)) \\ &= \hat{E}(\hat{E}(X_t(f)|M_s^-(X))|M_{-\infty}(X)) \\ &= \hat{E}(X_s(S_{t-s}f)|M_{-\infty}(X)) \\ &= X_s^d(S_{t-s}f). \end{aligned}$$

**Proposition 6.4.**

$$E(X_s^d(f)X_t^d(g)) = \begin{cases} p^d(f, S_{t-s}g), & t \geqq s \\ p^d(S_{s-t}f, g), & s \geqq t. \end{cases} \tag{6.8}$$

Hence $X^d$ is stationary by the centered Gaussian property of $X^d$.

*Proof.* Obvious from (6.7) and the definition of $p^d$.

**Proposition 6.5.**

(6.9) $$M_t(X^d) \subset M(X^d) = M_{-\infty}(X) = M_s^-(X^d) = M_{-\infty}(X^d).$$

*Proof.* The first inclusion relation is trivial. $M_t(X^d) \subset M_{-\infty}(X)$ by the definition of $X^d$. Hence $M(X^d) \subset M_{-\infty}(X)$, so

$$M_{-\infty}(X^d) \subset M_s^-(X^d) \subset M(X^d) \subset M_{-\infty}(X).$$

Suppose that $Y \in M_{-\infty}(X)$. Then $Y \in M_{-n}^-(X)$, so we have

$$\left\| Y - \sum_{i=1}^{n} a_i X_{u_i}(f_i) \right\|_L < n^{-1}, \qquad u_i \leqq -n .$$

Projecting this vector to $M_{-\infty}(X)$, we obtain

$$\left\| Y - \sum_{i=1}^{n} a_i X_{u_i}^d(f_i) \right\|_L < n^{-1}.$$

This implies that we can find $Y_n \in M_{-n}^-(X^d)$ such that $\|Y - Y_n\| < n^{-1}$. Since $Y_{k+n} \in M_{-n}^-(X^d)$ for every $k$,

$$Y = \lim_{k\to\infty} Y_{k+n} \in M_{-n}^-(X^d)$$

for every $n$, so $Y \in M_{-\infty}(X^d)$. Hence we have $M_{-\infty}(X) \subset M_{-\infty}(X^d)$, which completes the proof.

**Proposition 6.6.** *$X^d$ is Markov.*

*Proof.* Because of the centered Gsussian property of $X^d$ it suffices to observe that for $t > s$

$$\begin{aligned} \hat{E}(X_t^d(f)\,|\,M_s^-(X^d)) &= X_t^d(f) && \text{(by (6.9))} \\ &= X_s^d(S_{t-s}f) && \text{(by (6.7))} \\ &\in M_s(X^d). \end{aligned}$$

**Proposition 6.7.**

(6.10) $$p^d(S_t f, S_t g) = p^d(f, g)$$

(6.11) $$E(X_s^d(f) X_t^d(g)) = p^d(S_{s+u} f, S_{t+u} g)$$

*for every* $u \geqq \max(-s, -t)$.

*Proof.* By the definition of $p^d$ (Proposition 6.2) we obtain the first equality, which, combined with Proposition 6.4, implies the second equality.

**Proposition 6.8.**

(6.12) $\|X^d_{t+h}(f) - X^d_t(f)\|_L \leqq \|X_{t+h}(f) - X_t(f)\|_L \to 0 \quad (h \to 0).$

*Proof.* Obvious by the definition of $X^d$.

By Proposition 6.1, 6.4, 6.6, 6.8 and 6.2 we obtain the following:

**Theorem 6.1.** *$X^d$ is an OU $\mathscr{L}(E)$-process with charactreistic seminorm $p^d$ and $X^d_t(f)$ is $\sigma_{-\infty}(X)$-measurable for every $t \in T$ and every $f \in E$. And so $\sigma(X^d)$ is independent of $\sigma(dB)$ (Theorem 5.2 (ii)).*

## § 7. A special Wiener integral

Let $B = \{B_t\}$ be the Wiener $\mathscr{L}(D)$-process with characteristic norm $b$ introduced in § 5. We denote its increments by $\{B_{st}\}_{s<t}$.

Suppose that we are given a family of bounded linear operators on $(D, r)$: $Q_t$, $t \in T$. (See § 4 for the definition of $r$). Let $Q'_t$ be a linear operator in $\mathscr{L}(D)$ such that

(7.1) $$(Q'_t Y)(f) = Y(Q_t f).$$

We want to define an integral of the type:

(7.2) $$I(Q, f) = \int_T Q'_t \, dB_t(f).$$

First consider the case where $Q_t$ is a step operator-valued function of $t$ vanishing outside of a bounded interval $[a, b)$. Then there exists a decomposition of $[a, b)$

(7.3) $$\Delta\colon a = a_0 < a_1 < \cdots < a_n = b$$

such that

(7.4) $$Q_t = Q_{a_i} \quad \text{on } [a_i, a_{i+1}),\ i = 1, 2, \cdots, n.$$

For such $\{Q_t\}$ we define

(7.5) $$\begin{aligned} I(Q, f) :&= \sum_{i=0}^{n-1} Q'_{a_i} B_{a_i, a_{i+1}}(f) \\ &= \sum_{i=0}^{n-1} B_{a_i, a_{i+1}}(Q_{a_i} f). \end{aligned}$$

Then $I(Q, f)$ is bilinear in $(Q, f)$. Since $\{B_t\}$ has independent increments with mean 0, we obtain

$$\|I(Q,f)\|_L^2 = \sum_{i=0}^{n-1} \|B_{a_i a_{i+1}}(Q_{a_i}f)\|^2$$

$$= \sum_{i=0}^{n-1} (a_{i+1} - a_i) b(Q_{a_i} f)^2$$

$$= \int_{-\infty}^{\infty} b(Q_t f)^2 dt .$$

We can use the extension theorem for bounded linear operators to define $I(Q,f)$ under the following condition on $Q = \{Q_t\}$:
(I) There exists a sequence of operator-valued step functions vanishing outside of a bounded interval:

$$Q^{(n)} = \{Q_t^{(n)}\} \tag{7.6}$$

such that

$$\lim_{n\to\infty} \int_T b(Q_t^{(n)} f - Q_t f)^2 dt = 0, \qquad f \in E . \tag{7.7}$$

Let $J$ be a finite or infinite interval. If $Q$ satisfies the condition (I), then $\{I_J(t)Q_t\}_t$ also satisfies (I) and we define

$$I(Q,f,J) \equiv \int_J Q'_t dB_t(f) := \int_T 1_J(t) Q'_t dB_t(f) . \tag{7.8}$$

In this case $Q_t$ need not be defined outside of $J$.
By routine we can prove the following:

**Theorem 7.1.**
(i) *$I(Q,f,J)$ is bilinear in $(Q,f)$ and additive in $J$.*
(ii) $E(I(Q,f,J)) = 0$.
(iii) $\|I(Q,f,J)\|_L^2 = \int_J b(Q_t f)^2 dt$.
(iv) *If $\{Q_t, t \in [a,b)\}$ is a family of operators on $(D,r)$ such that $t \mapsto Q_t$ is piece-wise strongly continuous and if*

$$\int_J b(Q_t f)^2 dt < \infty \tag{7.9}$$

*then $I(Q,f,J)$ is well-defined and*

$$I(Q,f,J) = \lim_{|\Delta|\to 0} \sum_i B_{a_i a_{i+1}}(Q_{a_i} f) \tag{7.10}$$

*where* $\Delta = (a = a_0 < a_1 < \cdots < a_n = b)$, $|\Delta| = \max_i (a_{i+1} - a_i)$ *and* 'lim' *indicates* 'limit *in* $(L, \|\ \|)$'.

(v) *If* $[a_i, b_i)$, $i = 1, 2, \cdots, n$ *are disjoint, then the* $\sigma$*-algebras*:

$$\sigma\{I(Q, f, a_i, b_i)) : f \in D\},\ i = 1, 2, \cdots, n$$

*are independent.*

(vi) $\sigma\{I(Q, f, (a, b)) : f \in D, -\infty < a < b < \infty\}$ *is included in* $\sigma(dB)$ *and so it is independent of* $\sigma(X^d)$.

## § 8. The canonical representation

Rewriting (5.1) we obtain

$$X_t - X_s = B_{st} + \int_s^t A' X_u du \quad \text{in } \mathscr{L}(D) \tag{8.1}$$

or

$$dX_t = dB_t + A' X_t dt \tag{8.2}$$

where $A' : \mathscr{L}(E) \to \mathscr{L}(D)$ is defined by the condition $A'X(f) = X(Af)$, $f \in D$. This equation is called the *Langevin equation* for $X_t$. Now we want to express $X$ in terms of $B$, using the Wiener integral defined in the provious section.

**Theorem 8.1.**

$$X_t = S'_{t-s} X_s + \int_s^t S'_{t-u} dB_u \quad \textit{in } \mathscr{L}(E), \tag{8.3}$$

i.e.

$$X_t(f) = X_s(S_{t-s} f) + \left(\int_s^t S'_{t-u} dB_u\right)(f), \qquad f \in E. \tag{8.4}$$

*Remark.* By the definition given in the last section we can define

$$I(f) = \left(\int_s^t S'_{t-u} dB_u\right)(f)$$

for $f \in D$ only. But $I(f)$ is linear in $f \in D$ and

$$\begin{aligned} \|I(f)\|_L^2 &= \int_s^t b(S_{t-u} f)^2 du = -2 \int_s^t p(AS_{t-u} f, S_{t-u} f) du \\ &= \int_s^t \frac{d}{du} p(S_{t-u} f)^2 du = p(f)^2 - p(S_{t-s} f)^2 \\ &\leqq p(f)^2, \end{aligned} \tag{8.5}$$

so $I: (D, p) \to (L, \| \|_L)$ is linear and bounded. Hence we can extend $I$ to a bounded linear operator of $(E, p) \to (L, \| \|_L)$. Thus (8.3) and (8.4) are meaningful.

*Proof of the Theorem* 8.1. By Theorem 7.1 (iv), the integral is equal to

$$\lim_{|\Delta|\to 0} \sum_{i=0}^{n-1} B_{s_i, s_{i+1}}(S_{t-s_i}f)$$

for $f \in D$, where $\Delta = \{s_i\}$ is a decomposition of $[s, t)$.

First we prove (8.4) in case $f \in \mathscr{D}(A^2)$. To do this, observe the following:

$$X_t(f) - X_s(S_{t-s}f) = \sum_{i=0}^{n-1} (X_{s_{i+1}}(S_{t-s_{i+1}}f) - X_{s_i}(S_{t-s_i}f)),$$

$$X_v(S_{t-v}f) - X_u(S_{t-u}f) = X_v(S_{t-v}f - S_{t-u}f) + (X_v - X_u)(S_{t-u}f)$$
$$= :U_1 + U_2,$$

$$U_1 = X_v\left(-\int_u^v S_{t-\alpha} Afd\alpha\right) = -\int_u^v X_v(S_{t-\alpha}Af)d\alpha,$$

$$U_2 = B_{uv}(S_{t-u}f) + \int_u^v X_\alpha(AS_{t-\alpha}f)d\alpha \qquad (by\ (8.1))$$

$$= B_{uv}(S_{t-u}f) + U_3, \tag{8.6}$$

$$U: = U_1 + U_3 = -\int_u^v (X_v(S_{t-\alpha}g) - X_\alpha(S_{t-u}g))d\alpha,$$

$$\|U\| \leqq \int_u^v \|X_v(S_{t-v}g) - X_\alpha(S_{t-u}g)\| d\alpha \quad (g = Af,\ \| \| = \| \|_L),$$

$$\|X_v(S_{t-v}g) - X_\alpha(S_{t-u}g)\|$$
$$\leqq \|X_v(S_{t-\alpha}g - S_{t-u}g)\| + \|(X_v - X_\alpha)(S_{t-u}g)\|$$
$$\leqq \int_u^\alpha \|X_v(S_{t-\beta}Ag)\|d\beta + \|B_{\alpha v}(S_{t-u}g)\| + \int_\alpha^v \|X_\beta(AS_{t-u}g)\|d\beta$$
$$\leqq (\alpha - u)p(Ag) + \sqrt{2}(v - \alpha)p(Ag)p(g) + (v - \alpha)p(Ag)$$

(Note that $b(S_\theta g) \leqq 2p(AS_\theta g)p(S_\theta g) = 2p(S_\theta Ag)p(S_\theta g) \leqq 2p(Ag)p(g)$)

$$\leqq 4(v - u)(p(A^2f) + p(Af)) = : c(v - u),$$

$$\|U\| \leqq c(v - u)^2, \quad \|X_t(f) - X_s(S_{t-s}f) - \sum_i B_{s_i s_{i+1}}(S_{t-s_i}f)\|$$
$$\leqq c\sum_i (s_{i+1} - s_i)^2 \leqq c|\Delta|\cdot(t - s).$$

Letting $|\Delta| \to 0$, we obtain (8.4) in case $f \in \mathscr{D}(A^2)$.

Since $\mathscr{D}(A^2)$ is dense in $(E, p)$, we obtain (8.4) for $f \in E$, observing that

$$\|X_t(f)\| = p(f),\ \|X_s(S_{t-s}f)\| \leqq p(f)$$

and

$$\left\|\left(\int_s^t S'_{t-u}\,dB_u\right)(f)\right\| \leqq p(f)\,.$$

Because of the inequality (8.5) we can let $s \to -\infty$ in (8.4). Also

$$X_s(S_{t-s}f) = \hat{E}(X_t(f)\,|\,M_s^-(X)) \to X_t^d(f) \tag{8.7}$$

as $s \to -\infty$. Thus we obtain in the following:

**Theorem 8.2.** (*The canonical representation of an OU $\mathscr{L}(E)$-process*)

$$X_t(f) = X_t^d(f) + \left(\int_{-\infty}^t S'_{t-u}\,dB_u\right)(f), \qquad f \in E \tag{8.8}$$

*where $X^d = \{X_t^d\}$ is the deterministic part of $X$ and $B = \{B_t\}$ is the innovation process of $X$.*

In Theorem 8.1 $S'_{t-s}X_s$ is $\sigma_s^-(X)$-measurable and the integral is independent of $\sigma_s^-(X)$. Theorem 8.1 asserts that $X_t$ is decomposed into these two parts and explains why $B = \{B_t\}$ is called the innovation process.

## § 9. The characteristics of Ornstein-Uhlenbeck processes

In the previous sections we derived several concepts such as $p$, $\{S_t\}$, $A$, $X^d$, $p^d$, $B$, etc. related to a given $OU\ \mathscr{L}(E)$-processes. However, the following two concepts, called the *characterstics* of the process are basic and the others can be derived from these characteristics:

1. the characteristic Hilbertian seminorm $p$ on$E$,
2. the characteristic operator $A$ on $(\bar{E}, \bar{p})$ satisfying the conditions (A.1), (A.2) and (A.3) (§ 3).

**Theorem 9.1.** *Suppose that we are given an arbitrary Hilbertian seminorm $p$ on $E$ and an arbitrary operator $A$ in $(\bar{E}, \bar{p})$ satisfying $(A.1)$ $(A.2)$ and $(A.3)$. Then there exists a unique (in the finite joint distributions) $OU$ $\mathscr{L}(E)$-process whose charateristics are $p$ and $A$.*

*Proof of existence.* From $p$ we can define the completion $(\bar{E}, \bar{p})$ as in Section 2. To avoid notational complications, we omit the bar in $\bar{E}$ and $\bar{p}$ for the moment. Since $A$ satisfies $(A.1)$. $(A.2)$ and $(A.3)$, it generates a strongly continuous semigroup of linear contractions $\{S_t,\ t \geqq 0\}$ on $(E, p)$.

Since $p(S_t f)$ is decreasing in $t$,

$$p^d(f) := \lim_{t\to\infty} p(S_t f) \tag{9.1}$$

is well-defined. Hence

$$p^d(f, g) := \lim_{t\to\infty} p(S_t f, S_t g)$$

(9.2)
$$= \lim_{t\to\infty} \frac{1}{4}(p(S_t(f+g))^2 - p(S_t(f-g))^2)$$

is well-defined and

(9.3)
$$p^d(f, f) = p^d(f)^2.$$

This implies that $p^d$ is a Hilbertian seminorm in $E$. By (9.1) it holds that

(9.4)
$$p^d(S_t f, S_t g) = p^d(f, g), \qquad t \geqq 0.$$

Let

(9.5)
$$V_{s,t}(f, g) := p^d(S_{s+u} f, S_{t+u} g), \qquad s, t \in T$$

where $u$ is any positive number larger than $\max(-s, -t)$. $V_{s,t}(f, g)$ is well-defined independently of the choice of $u$ by virtue of (9.4). $V_{st}(f, g)$ is symmetric and nonnegative-definite in $(s, f)$ and $(t, g)$, because

$$V_{s_i s_j}(f_i, f_j) = p^d(S_{s_i+u} f_i, S_{s_j+u} f_j), \qquad i = 1, 2, \cdots, n$$

for a sufficiently large $u$ and

$$\sum_{i,j} x_i x_j V_{s_i s_j}(f_i, f_j) = p^d(\sum_i x_i S_{s_i+u} f_i)^2 \geqq 0.$$

Hence we can construct a centered Gaussian $\mathscr{L}(E)$-process $X^d$ with covariance functional $V_{s,t}(f, g)$.

Since $A$ satisfies ($A$.1), ($A$.2) and ($A$.3),

(9.6)
$$b(f, g) := -(Af, g) - (f, Ag)$$

defines a Hilbertian seminorm on $D = \mathscr{D}(A)$. Hence we can construct a Wiener $\mathscr{L}(D)$-process $B$ with characteristic seminorm $b$ (Theorem 2.1). Also we can assume that $B$ and $X^d$ are independent.

Define

(9.7)
$$X_t := X_t^d + \int_{-\infty}^t S'_{t-u}\, dB_u.$$

We can easily prove that $X = \{X_t\}$ is an $OU$ $\mathscr{L}(E)$-process whose deterministic part is $X^d$ and whose innovation process is $B$.

Restoring the bar we denote this process by $\bar{X}$. $\bar{X}$ is an $OU$ $\mathscr{L}(\bar{E})$-process. Defining

$X(f) := \bar{X}(\theta f)$ ($\theta$: the canonical injection: $E \to \bar{E}$), we obtain the process $X$ which is to be constructed.

*Proof of the uniqueness.* Let $X$ be any $OU\ \mathscr{L}(E)$-process with characteristic norm $p$ and characteristic operator $A$. $p$ determines $(\bar{E}, \bar{p})$. $A$ determines the semigroup $\{S_t\}$. $\{S_t\}$ determines the covariance functional $V_{s,t}$ of the completetion $\bar{X}$. $V_{st}$ determines the finite joint distributions of $\bar{X}$ and so those of $X$.

## § 10. Continuous regular versions

In the previous sections we observed an $OU\ \mathscr{L}(E)$-process $X = \{X_t(f)\}$ and its innovation process $B = \{B_t(f)\}$ which is a Wiener $\mathscr{L}(D)$-process. In this section we define their continuous regularizations to discuss the properties of these processes more thoroughly. $c, c_1, c_2, \cdots$ stand for positive constants throughout this section.

(a) *A Gelfand triple* $\{(D_K, r_K), (D, r), (H, \|\ \|)\}$.

In § 4 we introduced a separable Hilbert space $(D, r)$:

$$D = \mathscr{D}(A), \quad r(f) = p((I - A)f)\,. \tag{10.1}$$

Let $\{d_n\}$ be an orthonormal base (abbr. ONB) in $(D, r)$ such that $d_n \in \mathscr{D}(A^2)$ for every $n$. Note that such an ONB exists because $D(A^2)$ is dense in $(D, r)$. Let $\{\lambda_n\}$ be a sequence such that

$$\lambda_n > 0\,, \qquad |\lambda|^2 := \sum_n \lambda_n^2\,. \tag{10.2}$$

Then

$$K := \sum_n \lambda_n d_n \otimes d_n \tag{10.3}$$

is a strictly positive definite Hilbert Schmidt operator in $(D, r)$. Define a separable Hilbert space $(D_K, r_K)$ as follows:

$$D_K := K(D)\,,$$
$$r_K(f) := r(K^{-1}f) = \Big(\sum_n r(f, d_n)^2 \lambda_n^{-2}\Big)^{1/2}\,,$$
$$r_K(f, g) := r(K^{-1}f, K^{-1}g) = \sum_n r(f, d_n) r(g, d_n) \lambda_n^{-2}\,.$$

It is easy to check that

$$r(f) \leqq c\, r_K(f)\,, \qquad c = \sup\{\lambda_n\} < \infty\,. \tag{10.4}$$

The sequence

$$e_n := \lambda_n d_n, \qquad n = 1, 2, \cdots$$

is an ONB in $(D_K, r_K)$.

The dual space of $(D_K, r_K)$ is denoted by $(H, \| \|)$. $H$ consists of all bounded linear functionals on $(D_K, r_K)$ and $\| \|$ is the usual supremum. The value of $y \in H$ evaluated at $f \in D_K$ is denoted by $\langle y, f \rangle$. $(H, \| \|)$ is a separable Hilbert space isomorphic to $(D_K, r_K)$ under the correspondence:

$$y \mapsto f_y, \langle y, \cdot \rangle = r_K(f_y, \cdot).$$

The triple

$$\{(D_K, r_K), (D, r), (H, \| \|)\}. \tag{10.5}$$

is a *Gelfand triple of Hilbert spaces.* The ONB in $(H, \| \|)$ dual to $\{e_n\}$ is denoted by $\{e'_n\}$, i.e.

$$\langle e'_m, e_n \rangle = \delta_{mn}.$$

The $\sigma$-algebra on $H$ generated by the $\| \|$-open subsets (equivalently the $\sigma$-algebra generated by the weakly open subsets or by the half spaces: $\{x \in H: \langle x, f \rangle \leqq a\}, f \in D_K, a \in \boldsymbol{R}$) is denoted by $\mathscr{B}(H)$ and is called *the Borel system* on $H$. We define *Borel measurable functions* on $H$, *Borel measurable maps* from $H$ into $H$ and *H-valued random variables* (abbr. $H$-variables) on $(\Omega, \mathscr{F}, P)$ with respect to $\mathscr{B}(H)$. $\langle x, y \rangle$ is Borel measurable in $x \in H$ for every $f \in D_K$. If $Y(\omega)$ is an $H$-valued random variable, then $\langle Y(\omega), f \rangle$ is a real random variable.

Let $\{S_t\}$ be the semigroup of linear contractions on $(E, p)$ we have introduced in § 3, and let $A$ be its generator. When we restrict $\{S_t\}$ to $D$, we obtain a family of linear operators $\{S_t^D\}$. Since $S_t D \subset D$ and $S_t Af = AS_t f$ for $f \in D$, $S_t^D$ carries $D$ into $D$. Since

$$\begin{aligned} r(S_t^D f) &= p((I - A)S_t f) = p(S_t(I - A)f) \\ &\leqq p((I - A)f) = r(f), \qquad f \in D, \end{aligned}$$

$S_t^D$ is a linear contraction on $(D, r)$. Similarly we can prove the following:

**Theorem 10.1.** *$\{S_t^D\}$ is a strongly continuous semigroup of linear contractions on $(D, r)$ with generator $A^D$: = the $A$ restricted to $\mathscr{D}(A^D) := \mathscr{D}(A^2)$. The resolvent operator $(I - A)^{-1}: E \to D$ gives an isomorphic map from $(E, p)$ to $(D, r)$ which transforms $S_t$ to $S_t^D$ and $A$ to $A^D$.*

From now on we restrict our consideration to $(D, r)$ and omit the index $D$ in $S_t^D$ and $A^D$ for notational simplicity.

(*b*) *Stochastically r-bounded H-variables and r-bounded LRF's.* Let $Y(\omega)$ be an $H$-variable. $Y$ is called *stochastically r-bounded* if

$$E(\langle Y, f\rangle^2) \leqq c_1^2 r(f)^2, \qquad f \in D_K .$$

$Y$ is called *centered Gaussian* if $\langle Y, f\rangle$, $f \in D_K$ form a centered Gaussian system. The family of all stochastically $r$-bounded $H$-variables is denoted by $\mathscr{R}(H, r)$. $\mathscr{R}(H, r)$ is a vector space with the usual linear operations.

Let $X(f)$ be an LRF on $D$. $X$ is called *r-bounded* if

$$E(X(f)^2) \leqq c_2^2 r(f)^2 , \qquad f \in D .$$

The family of all $r$-bounded LRF's on $D$ is denoted by $\mathscr{L}(D, r)$. $\mathscr{L}(D, r)$ is also a vector space.

For $Y \in \mathscr{R}(H, r)$ we define

$$X\colon D_K \longrightarrow L, \qquad f \longmapsto X(f)(\omega) = \langle Y(\omega), f\rangle .$$

Since $Y$ is stochastically $r$-bounded, $X$ is a bounded linear map from $(D_K, r)$ into $(L, \| \ \|)$ and so it is extended to a bounded linear map from $(D, r)$ into $(L, \| \ \|)$ because $D_K$ contains $e_n$, $n = 1, 2, \cdots$ and so is dense in $(D, r)$. The $X$, thus extended, belongs to $\mathscr{L}(D, r)$. The map $Y \mapsto X$ is a linear map from $\mathscr{R}(H, r)$ into $\mathscr{L}(D, r)$. This map, denoted by $\varphi$, is injective. Suppose that $\varphi Y = 0$. Then $\langle Y, e_n\rangle = 0$ a.s. for every $n$. Hence

$$P\{\langle Y, e_n\rangle = 0 \ \ n = 1, 2, \cdots\} = 1 .$$

Since $Y(\omega) \in H$, this implies that $P(Y = 0) = 1$, proving the injectivity of $\varphi$. In fact we have the following:

**Theorem 10.2.** $\varphi\colon \mathscr{R}(H, r) \to \mathscr{L}(D, r)$ *is bijective.*

*Proof.* It suffices to prove that $\varphi$ is surjective, i.e. that for every $X \in \mathscr{L}(D, r)$ we can find an $H$-variable $Y$ such that $\varphi Y = X$; then $Y \in \mathscr{R}(H, r)$ follows automatically. Observing that

$$\begin{aligned} E\Big(\sum_n X(e_n)^2\Big) &= \sum_n E(X(e_n)^2) \leqq c_2^2 \sum_n r(e_n)^2 \\ &= c_2^2 \sum_n \lambda_n^2 = c_2^2 |\lambda|^2 , \end{aligned}$$

we obtain

$$\sum_n X(e_n)^2 < \infty \qquad \text{a.s.}$$

Hence

$$Y:=\sum_n X(e_n)e'_n \quad \text{(in norm convergence in } (H, \|\ \|))$$

defines an $H$-variable. Observing that

$$E(\langle Y,f\rangle^2)=E((\sum_n X(e_n)\langle e'_n,f\rangle)^2)$$
$$\leqq E(\sum_n X(e_n)^2)\sum_n \langle e'_n,f\rangle^2 \leqq c_2^2|\lambda|^2 r_K(f)^2$$

for every $f \in D_K$, we will check that $\varphi Y = X$, i.e.

$$\langle Y,f\rangle = X(f) \quad \text{a.s.,} \quad f \in D_K .$$

This is obvious by the definition of $Y$ for $f=e_n$ and so for every $f \in F$: $= LS\ (e_1, e_2, \cdots)$. For every $f \in D_K$ we can find $f_n \in F$ such that

$$r_K(f-f_n) \longrightarrow 0 \quad \text{(and so } r(f-f_n) \longrightarrow 0 \text{ by (10.4))} .$$

Hence $\langle Y,f\rangle = \|\ \|_L\text{-}\lim_n \langle Y,f_n\rangle = \|\ \|_L\text{-}\lim_n X(f_n) = X(f)$ a.s.

**Definition 10.1.** For $X \in \mathscr{L}(D,r)$ the $H$-variable $Y:=\varphi^{-1}X \in \mathscr{R}(H,r)$ is called a *regular version* of $X$.

**Theorem 10.3.** *Suppose that* $Y \in \mathscr{R}(H,r)$.
(i) $E(\|Y\|^2) < \infty$,
(ii) *If $Y$ is centered Gaussian, then*

(10.6) $$E(\|Y\|^{2\alpha}) \leqq c(\alpha)E(\|Y\|^2)^\alpha < \infty , \qquad \alpha \geqq 1$$

*where $c(\alpha)$ id the $2\alpha$-order absolute moment of the standard Gauss distribution $N(0,1)$.*

*Proof of* (i). Representing $Y \in H$ in the form:

$$Y=\sum_n \langle Y,e_n\rangle e'_n$$

we obtain

$$E(\|Y\|^2)=\sum_n E(\langle Y,e_n\rangle^2) \leqq c_1^2|\lambda|^2 < \infty .$$

*Proof of* (ii) (due to D. Stroock). Let $\{a_n(t), n=1,2,\cdots\}$ be the Rademacher ONB on $L_2[0,1]$ and let

$$p_Y(f)^2 := E(\langle Y,f\rangle^2), \qquad f \in D_K .$$

Then $p_Y(f)$ is a Hilbertian seminorm in $D_K$ and

$$\sum_n p_Y(e_n)^2 = E(\|Y\|^2) < \infty .$$

Hence we can find a Hilbert-Schmidt operator $B$ in $(D_K, r_K)$ such that

$$p_Y(f) = r_K(Bf) , \qquad f \in D_K .$$

Let $\{\mu_n\}$ and $\{\varepsilon_n\}$ be the eigenvalues and the eigenvectors of $B$. Then $\{\varepsilon_n\}$ is an ONB in $(D_K, r_K)$ and

$$B\varepsilon_n = \mu_n \varepsilon_n , \qquad |\mu|^2 := \sum_n \mu_n^2 < \infty .$$

Let

$$Y_n := \langle Y, \varepsilon_n \rangle , \qquad n = 1, 2, \cdots .$$

Then $\{Y_n\}$ is a centered Gaussian system, and

$$E(Y_m Y_n) = p_Y(\varepsilon_m, \varepsilon_n) = r_K(B\varepsilon_m, B\varepsilon_n) = \mu_m \mu_n \delta_{mn} .$$

Therefore $\{Y_n\}$ is independent and Gauss distributed, each $Y_n$ having mean 0 and variance $\mu_n^2$, and so are $\{a_n(t) Y_n\}$ for every $t$, because $a_n(t) = \pm 1$ for every $(t, n)$. Since $|\mu|^2 = \sum_n \mu_n^2 < \infty$,

$$S(t) := \sum_{n=1}^{\infty} a_n(t) Y_n$$

is a centered Gaussian variable with variance $|\mu|^2$. Therefore

$$E(S(t)^{2\alpha}) = c(\alpha) |\mu|^{2\alpha} , \qquad t \in [0, 1] .$$

Since $\{a_n\}$ is an ONB in $L_2[0, 1]$ and $\{\varepsilon_n\}$ is an ONB in $(D_K, r_K)$,

$$\int_0^1 S(t)^2 dt = \sum_{n=1}^{\infty} Y_n^2 = \|Y\|^2 .$$

Since $s^\alpha$ is convex in $s \in [0, \infty]$ for $\alpha \geqq 1$, we obtain

$$\begin{aligned} E(\|Y\|^{2\alpha}) &= E\left(\left(\int_0^1 S(t)^2 dt\right)^\alpha\right) \leqq E\left(\int_0^1 S(t)^{2\alpha} dt\right) \\ &= \int_0^1 E(S(t)^{2\alpha}) dt = c(\alpha) |\mu|^{2\alpha} \\ &= c(\alpha) E(\|Y\|^2)^\alpha \end{aligned}$$

because

$$E(\|Y\|^2) = \sum_n E(Y_n^2) = \sum_n \mu_n^2 = |\mu|^2 .$$

(c) *The generalized coupling.* The coupling $\langle y, f\rangle$, $y \in H, f \in D_K$ is bilinear and continuous in $(y,f) \in (H, \|\ \|) \times (D_K, r_K)$. For the later use we will define a generalized coupling

$$[y,f], \qquad y \in H, \qquad f \in D$$

in such a way that $[y,f]$ is Borel measurable in $(y,f) \in H \times D$ with respect to $\mathscr{B}(H) \otimes \mathscr{B}(D)$ and coincides with $\langle y, f\rangle$ for $y \in H$ and $f \in D_K$. There are many such extensions but we will fix a particular one below. Such a specification does not matter for the later use: only the Borel measurability of $[y,f]$ in $y \in H$ does matter.

Let $C = \{c_n\}$ be a countable dense subset of $(D_K, r_K)$. As $D_K$ is dense in $(D, r)$, so is $C$. With each $f \in D$ we associate a sequence $\{f_n\}$ in $C$ as follows. Suppose that $f \in D - D_K$. If $r(c_i - f) > 2^{-n}$, $i = 1, 2, \cdots, m-1$ and $r(c_m - f) \leqq 2^{-n}$, we set $f_n = c_m$. For $f \in D_K$ we define in the same way by replacing $r$ by $r_K$. Since $D_K = \{f \in D: \sum_n r(f, d_n)^2 \lambda_n^{-2} < \infty\}$, $D_K$ is a Borel subset of $D$. Hence the map $f \mapsto (f_n, n = 1, 2, \cdots)$ is Borel measurable from $D$ into $D_K^\infty = D_K \times D_K \times \cdots$ and

$$\sum_n r(f_n - f) < \infty, \quad \sum_n r(f_{n+1} - f_n) < \infty, \quad r(f_n - f) \longrightarrow 0$$

obviously.

**Definition 10.2.**

$$[y,f] := \begin{cases} \lim_n \langle y, f_n\rangle & \text{if } \{\langle y, f_n\rangle\}_n \text{ is convergent} \\ 0 & \text{if otherwise} \end{cases}$$

and $[y,f]$ $(y \in H, f \in D)$ is called a *generalized coupling*.

From the definition it is easy to check that $[y,f]$ is Borel measurable in $(y,f)$ and coincided with $\langle y, f\rangle$ for $y \in H$ and $f \in D_K$.

If $Y$ is an $H$-variable, then $\langle Y, f\rangle$ is obviously a real random variable for $f \in D_K$ and $[Y,f]$ is a real random variable for $f \in D$ because of Borel measurability of $[y,f]$ in $y$. If $Y \in \mathscr{R}(H, r)$, then $\varphi Y \in \mathscr{L}(D, r)$ (Theorem 10.2). But we have the following:

**Theorem 10.4.** *If $Y \in \mathscr{R}(H, r)$ is a regular version of $X \in \mathscr{L}(D, r)$, then*

$$X(f) = [Y,f] \quad \text{a.s.}, \qquad f \in D.$$

*Proof.* Take a sequence $\{f_n\}$ associated with $f \in D$ above. Then

$$\begin{aligned} E\Big(\sum_n |\langle Y, f_{n+1}\rangle - \langle Y, f_n\rangle|\Big) &\leqq \sum_n \|\langle Y, f_{n+1} - f_n\rangle\|_L \\ &\leqq c_1 \sum r(f_{n+1} - f_n) < \infty. \end{aligned}$$

Hence $\langle Y, f_n \rangle$ is convergent a.s. Hence

$$[Y, f] = \lim_{n\to\infty} \langle Y, f_n \rangle \quad \text{a.s.}$$

by Definition 10.2. Since $X \in \mathscr{L}(D, r)$, we obtain

$$\|X(f) - X(f_n)\|_L \leqq c_2 r(f_n - f) \longrightarrow 0\,.$$

But $X(f_n) = \langle Y, f_n \rangle$ a.s. by the definition of regular versions, so $X(f) = [Y, f]$ a.s.

(*d*) *The dual operator of a linear operator $Q$ in $D$.* Let $Q$ be a bounded linear operator in $(D_K, r_K)$. Then, for every $y \in H$ $\langle y, Qf \rangle$ is bounded and linear in $f \in (D_K, r_K)$, so there exists a unique element $z \in H$ such that $\langle z, f \rangle = \langle y, Qf \rangle$. Setting $Q'y = z$, we obtain a bounded linear operator $Q'$: $H \to H$, the dual operator of $Q$, which is represented as

$$Q'y = \sum_n \langle y, Qe_n \rangle e'_n\,.$$

In view of this fact we will extend this notion to the case where $Q$ is a linear operator in $D$ with $\mathscr{D}(Q) \supset \{e_n, n = 1, 2, \cdots\}$ as follows:

**Definition 10.3.**

$$Q'y := \begin{cases} \sum_n [y, Qe_n]e'_n & \text{if } \sum_n [y, Qe_n]^2 < \infty \\ 0 & \text{if otherwise.} \end{cases}$$

It is obvious that $Q'$: $H \to H$ is Borel measurable. In the previous section we defined $Q'$ as a linear operator for LRF's but the dual operator $Q'$ defined here is a Borel measurable operator in $H$. Hence if $Y$ is an $H$-variable, then $Q'Y$ is also an $H$-variable.

This definition of $Q'$ is artificial but is useful for our discussion because of the following:

**Theorem 10.5.** *If $Y \in \mathscr{R}(H, r)$ is a regular version of $X \in \mathscr{L}(D, r)$ and if*

$$\|X(Qf)\|_L \leqq cr(f) \tag{10.7}$$

*then $Q'Y \in \mathscr{R}(H, r)$ and $Q'Y$ (in the sense of $H$-variables) is a regular version of $Q'X$ (in the sense of LRF's).*

*Proof.* Using Theorem 10.4 and assumption (10.7) we obtain

$$E\left(\sum_n [Y, Qe_n]^2\right) \leqq c^2 \sum_n r(e_n)^2 = c^2 |\lambda|^2 < \infty\,.$$

This implies that

$$\sum_n [Y, Qe_n]^2 < \infty \qquad \text{a.s.},$$

so

$$Q'Y = \sum [Y, Qe_n]e'_n \ (\in H) \qquad \text{a.s.} \tag{10.8}$$

Hence for $f \in D_K$ we have

$$\begin{aligned} E(\langle Q'Y, f\rangle^2) &\leqq E(\sum_n [Y, Qe_n]^2) \sum_n \langle e'_n, f\rangle^2 \\ &\leqq c^2|\lambda|^2 r_K(f)^2 . \end{aligned}$$

But $\langle Q'Y, e_n\rangle = [Y, Qe_n] = X(Qe_n)$ a.s. by (10.8) and Theorem 10.4. Hence $\langle Q'Y, f\rangle = X(Qf) = (Q'X)(f)$ a.s. for $f \in D_K$. Thus $Q'Y$ is a regular version of $Q'X$. Since $Q'X \in \mathscr{L}(D, r)$ by (10.7), we obtain $Q'Y \in \mathscr{R}(H, r)$.

(e) *The continuous regular versions of* $\{X_t\}$ *and* $\{B_t\}$. We will apply the results obtained above to the $OU\ \mathscr{L}(E)$-process $X = \{X_t\}$ and its innovation process $B = \{B_t\}$. $X_t(f)$ was defined as an LRF on $E$, but here we regard it an LRF on $D$, as it is harmless because $D$ is dense in $(E, p)$. Thus all processes of LRF's can be regarded as $\mathscr{L}(D)$-processes. Also we consider $S_t$ on $D$ and $A$ on $\mathscr{D}(A^2)$ in view of Theorem 10.1.

Since

$$\begin{aligned} &\|X_t(f)\|_L = p(f) \leqq r(f), \qquad f \in D \\ &\|B_t(f)\|_L = |t|b(f) \leqq |t|r(f), \qquad f \in D, \end{aligned}$$

$X_t, B_t \in \mathscr{L}(D, r)$. Hence $X_t$ and $B_t$ have regular versions $\tilde{X}_t$ and $\tilde{B}_t$ respectively, both belong to $\mathscr{R}(H, r)$ by Theorem 10.2 and Definition 10.1. But

$$\begin{aligned} &\|X_t(S_u f)\|_L = p(S_u f) \leqq p(f) \leqq r(f), \qquad f \in D \\ &\|B_t(S_u f)\|_L = |t|b(S_u f) \leqq |t|r(S_u f) \leqq |t|r(f), \qquad f \in D \\ &\|X_t(Af)\|_L = p(Af) \leqq r(f), \qquad f \in D(A^2). \end{aligned}$$

Since both $D$ and $\mathscr{D}(A^2)$ contain every $e_n$, we can define $S'_u$ and $A'\colon H \to H$ and check that $S'_u\tilde{X}_t$, $S'_u\tilde{B}_t$ and $A'\tilde{X}_t$ are regular versions of $S'_uX_t$, $S'_uB_t$ and $A'X_t$ respectively. Hence $S'_u\tilde{X}_t - S'_u\tilde{X}_s$ is a regular version of $S'_uX_t - S'_uX_s = S'_u(X_t - X_s)$ (a.s.) and so

$$S'_u\tilde{X}_t - S'_u\tilde{X}_s = S'_u(\tilde{X}_t - \tilde{X}_s) \quad \text{a.s.}$$

In fact

$$\langle S'_u\tilde{X}_t - S'_u\tilde{X}_s, f\rangle = \langle S'_u(\tilde{X}_t - \tilde{X}_s), f\rangle \quad \text{a.s.}$$

holds for $f = e_n$ and so for every $f \in D_K$, because

$$S'_u\tilde{X}_t - S'_u\tilde{X}_s, \quad S'_u(\tilde{X}_t - \tilde{X}_s) \in H.$$

Similarly

$$S'_u\tilde{B}_{st} = S'_u\tilde{B}_t - S'_u\tilde{B}_t \quad \text{a.s.}$$

**Definition 10.4.** A faimly of $H$-variables $\{Y_t, t \in T\}$ is called an *$H$-valued process* or an *$H$-process.* An $H$-process $\{Y_t\}$ is called *sample continuous* if $Y_t(\omega)$ is $\|\ \|$-continuous in $t$ for every $\omega$. Let $\{Y_t\}$ be an $H$-process. A sample continuous $H$-process $\{Y'_t\}$ is called a *continuous version* of $\{Y_t\}$ if $Y'_t = Y_t$ a.s. for every $t$. Let $\{Y_t\}$ be an $\mathscr{L}(D)$-process and let $\tilde{Y}_t$ be a regular version of $Y_t$ for each $t$. A continuous version of $\{\tilde{Y}_t\}$, if it exists, is called a *continuous regular version* of $\{Y_t\}$.

Let $\{Y_t\}$ be an $H$-process. It is called *centered Gaussian* if $\{\langle Y_t, f\rangle, f \in D_K, t \in T\}$ form a centered Gaussian system. The property of being centered Gaussian is preserved by taking versions.

**Theorem 10.6.** *Let $\{Y_t\}$ be a centered Gaussian $\mathscr{L}(D, r)$ process. If*

$$E((Y_t(f) - Y_s(f))^2) \leqq c\cdot|t-s|r(f)^2 \tag{10.9}$$

*then $\{Y_t\}$ has a continuous regular version.*

*Proof.* Let $\tilde{Y}_t$ be a regular version of $Y_t$. $\tilde{Y}_t$ exists and belongs to $\mathscr{R}(H, r)$. Using (10.9), we obtain

$$\begin{aligned} E(\|\tilde{Y}_t - \tilde{Y}_s\|^2) &= \sum_n E(\langle Y_t - Y_s, e_n\rangle^2) \\ &\leqq c|t-s|\sum_n r(e_n)^2 = c_1|t-s| \\ &\qquad (c_1 := c|\lambda|^2 < \infty) \end{aligned}$$

and so

$$E(\|\tilde{Y}_t - \tilde{Y}_s\|^4) \leqq 3c_1^2|t-s|^2. \quad \text{(Theorem 10.3 (ii)).}$$

Using Kolmogorov's continuous version theorem, we obtain a continuous version of $\{\tilde{Y}_t\}$, which turns out to be a continuous regular version of $\{Y_t\}$.

**Theorem 10.7.** *The OU $\mathscr{L}(D)$-process $\{X_t\}$ and the Wiener $\mathscr{L}(D)$-process $\{B_t\}$ have continuous regular versions.*

*Proof.* If suffices only to check (10.9) for $\{X_t\}$ and $\{B_t\}$. This condition is obvious for $\{B_t\}$, because

$$E((B_t(f) - B_s(f))^2) = |t - s| b(f)^2 \leqq |t - s| r(f)^2 .$$

Since $E(X_t(f)X_s(f)) = p(S_{|t-s|}f, f)$ by Theorem 3.2, we obtain

$$\begin{aligned} E((X_t(f) - X_s(f))^2) &= 2p(f, f) - 2p(S_u f, f) \ (u: = |t - s|) \\ &= -2\int_0^u p\left(\frac{d}{dv} S_v f, f\right) dv = -2\int_0^u p(S_u A f, f) du \\ &\leqq \int_0^u (p(S_v A f)^2 + p(f)^2) dv \leqq u(p(Af)^2 + p(f)^2) \\ &\leqq u r(f)^2 = |t - s| r(f)^2, \end{aligned}$$

so (10.9) holds for $\{X_t\}$.

Similarly we can prove that $\{S'_{t-s} X_s, t \geqq s\}$, $\{X_t^d\}$ and the stochastic integrals observed in § 8 have continuous regular versions.

Denoting all such regularizations by the same notation as their original ones, we obtain

$$dX_t = dB_t + A' X_t \, dt , \tag{10.10}$$

$$X_t = S'_{t-s} X_s + \int_s^t S'_{t-u} \, dB_u , \tag{10.11}$$

$$X_t = X_t^d + \int_{-\infty}^t S'_{t-u} \, dB_u , \tag{10.12}$$

and all these equations are regarded equations for sample continuous $H$-processes. Now it is easy to prove the following:

**Theorem 10.8.** *$\{X_t\}$ is a sample continous H-Process which is centered Gaussian, Markov and stationary with invariant measure*

$$\mu: = N(0, \{p(f, g)\}_{f, g \in D_K})$$

*and with transition probability*

$$\mu_t(x, \cdot): = N(S'_{t-u} x, \{p(f, g) - p(S_t f, S_t g)\}_{f, g \in D_K}) ,$$

*where $N(m, \{v(f, g)\})$ is the Gaussian measure on $H$ with mean vector $m$ and variance functional $v$.*

*Acknowledgement.* Professor Y. Okabe read our original manuscript with great care. We are very grateful to him for his valuable comments.

## References

[ 1 ] B. Gaveau, Noyaux des probabilités de transition de certains opérateur d'Ornstein-Uhlenbeck dans L'éspace de Hilbert, C.R. Acad. Sci Paris. Série I, **293** (1981), 460–472.

[ 2 ] J. T. Lewis and L. C. Thomas, A characterization of regular solutions of a linear stochastic differential equation, Z. Wahr, verw. Gebiete, **30** (1974), 45–55.

[ 3 ] Y. Okabe, On a stationary Gaussian process with T-positivity and its associated Langevin equation and S-matrix, J. Fac. Sci. Univ. Tokyo, Soc. IA, **26** (1979), 115–116.

DEPARTMENT OF MATHEMATICS
GAKUSHUIN UNIVERSITY
MEJIRO, TOKYO 171, JAPAN

# REGULARIZATION OF LINEAR RANDOM FUNCTIONALS

Kiyosi Itô and Masako Nawata

## §1. INTRODUCTION

In probability theory we are often concerned with regularization problems ; for example, Doob's separable versions, Kolmogorov's continuous version theorem, Cramer's D-version theorem ( D: the right continuous functions with left limits), Doob's D-version theorem for martingales, the continuous version theorems of Gaussian processes, measurable versions, regular conditional probabilities, and so on. In this paper we discuss the regularization problems of linear random functionals.

The base probability space is denoted by

$$\Omega = (\Omega, \mathcal{F}, P).$$

We denote by $L_0 = L_0(\Omega, \mathcal{F}, P)$ the Frechet space consisting of all random variables ( $\mathcal{F}$- measurable real-valued functions on $\Omega$ ) endowed with the Frechet norm :

$$\|X\|_0 = E(|X| \wedge 1).$$

The $\| \|_0$ - convergence is equivalent to the convergence in probability (abbr.i.p.). Two equivalent ( = equal a.s.) random variables are identified in $L_0$ .

Let $E$ be an abstract space. A family of random variables indexed by elements in $E$ : $\{X(f) = X(f,w),\ f \in E\}$ is called a <u>random functional</u> on E. When we fix w, we obtain a functional $X_w : E \to \mathbb{R}$, $f \mapsto X(f,w)$, called the <u>sample functional</u> of X corresponding to w.

Let X and Y be two random functionals on E. If

$$P\,(X(f) = Y(f)) = 1 \quad \text{for every} \quad f,$$

then X is said to be <u>equivalent</u> to Y. If

$$P\,(X(f) = Y(f) \quad \text{for every} \quad f) = 1, \quad \text{i.e. } P\{X_w = Y_w\} = 1,$$

then X is said to be strictly equivalent to X.

Let E be a vector space (with real coefficients) and t a linear topology in E which makes E be a topological vector space $E_t$. A random functional X on E is called <u>linear</u> ( or <u>linear in the stochastic sense</u>), if

$$X(a_1f_1 + a_2f_2) = a_1X(f_1) + a_2X(f_2) \quad \text{a.s.}$$

for every choice of $(a_1, a_2, f_1, f_2)$; the exceptional w-set obviously dependes on the choice in general. X is linear if and only if $f \mapsto X(f)$ is a linear map from E into $L_0$.

A random functional X on E is called <u>t-continuous</u> (or <u>t-continuous i.p.</u>), if X(f) is t-continuous i.p. in f. A random functional X is linear and t-continuous if and only if $f \mapsto X(f)$ is a continuous linear map from $E_t$ into $(L_0, \| \|_0)$.

A random functional X on E is called <u>sample linear</u> if the sample functional $X_w : f \to X(f,w)$ is linear a.s., and <u>sample t-continuity</u> is defined similarly. A sample linear and sample t-continuous random functional is called a <u>t-regular random functional.</u> X is t-regular if and only if

$$X_w \in E_t' \quad \text{a.s.},$$

where $E_t'$ is the dual of $E_t$, the continuous linear functionals on $E_t$. Let X be a linear random functional on E. A t-regular random functional equivalent to X is called a <u>t-regularization</u> of X.

In this paper we will discuss the problem of existence and

uniqueness of the $t_r$-regular version of a linear random functional on a vector space E in case the linear topology t is determined by a family of separable Hilbertian seminorms. The regularization theorem is a probabilistic formulation of the generalized Bochner theorem due to Sazonov [1], Minlos [2] and Kolmogorov [3] and we can easily derive one of these theorems from another.

## §2. MULTI-HILBERTIAN TOPOLOGIES.

Let E be a vector space with real coefficients. In this paper a non-trivial (= not identically 0) separable Hilbertian seminorms on E is called an H-norm for simplicity. If p is an H-norm, then so is $a \cdot p$ $(a > 0)$. If $p_1, p_2, \cdots, p_n$ are H-norms, then so is

$$p_1 \vee p_2 \vee \cdots \vee p_n := \left(\sum_{k=1}^{n} p_k^2\right)^{1/2} .$$

Let p and q be two H-norms. $p \leqq q$ meens that $p(x) \leqq q(x)$ for every $x \in E$. $p \prec q$ means that $p \leqq aq$ for some $a > 0$. In this case p is said to be weaker than q. Let $\{e_n\}$ be a q-orthonormal base (abbr. q-ONB). Then

$$(p : q)_{HS} := \left(\sum_k p(e_k)^2\right)^{1/2}$$

is independent of $\{e_n\}$. If this is finite, then p is HS-weaker (weaker in the Hilbert-Schmidt sense) than q.

A linear topology on E determined by a directed (with respect to $\prec$) family of H-norms is called a multi-Hilbertian topology. If it is determined by a countable family, then it is called countably Hilbertian. A countably Hilbertian topology is determined by an increasing sequence of H-norms.

From now on we denote H-norms by $p, p_k, q, q_k, \cdots$ and multi-Hilbertian topologies by $s, t, u, \cdots$. Also the p-topology is denoted by the same notation p. Hence $p \prec t$ means that the p-topology is weaker than the topology t. It is obvious that the topology $t$ is determined by the family of all $p$'s such that $p \prec t$.

$s$ is said to be <u>HS-weaker</u> than $t$, written $s \prec_{HS} t$ if for every $p \prec s$ there exists $q \prec t$ such that $p \prec_{HS} q$. Suppose that $s$ and $t$ are determined by $\{p_\alpha\}$ and $\{q_\beta\}$ respectively. Then $s \prec_{HS} t$ if and only if for every $\alpha$ there exists $\beta$ such that $p_\alpha \prec_{HS} q_\beta$.

The Kolmogorov I-topology of $t$, written $I(t)$, is the strongest of all multi-Hilbertian topologies HS-weaker than $t$. If $t$ is determined by $\{q_\beta\}$, then $I(t)$ is determined by all H-norms $p$'s such that $p \prec_{HS} q_\beta$ for some $\beta$.

## §3. REGULARIZATION THEOREM.

Let $E$ be a vector space (with real coefficients) and $t$ a multi-Hilbertian topology on $E$. Let $X = \{X(f),\ f \in E\}$ be a linear ranodm functional and $Y$ a $t$-regular version of $X$. If there exists a countably Hilbertian topology $s \prec t$ such that

$$P\left\{Y_w \in E'_s\right\} = 1 \qquad (\text{Note that } E'_s \subset E'_t),$$

then $Y$ is called a <u>separable $t$-regular version</u> of $X$. In this case $Y$ turns out to be an $s$-regular version of $X$.

<u>REGULARIZATION THEOREM</u>. Let $t$ be a multi-Hilbertian topology on a vector space $E$ and $X = \{X(f),\ f \in E\}$ a linear ranodm functional on $E$. $X$ has a separable $t$-regular version if and only if $X$ is $I(t)$-continuous i.p. Any two separable $t$-regular versions are strictly equivalent to each other.

<u>Proof of the 'only if' part</u>. Let $Y$ be a separable $t$-regular version of $X$. Then there exists a countably Hilbertian topology $s \prec t$ such that $Y_w \in E'_s$ a.s. Let $\{q_n\}$ be an increasing sequence determining $s$. Then

$$E'_n := E'_{q_n} \uparrow E'_s \text{ and so } P(Y_w \in E'_n) \uparrow 1 \qquad (1).$$

$E_n'$ is a Hilbert space with the dual norm $\| \|_{-q_n}$. If $\{e_k\}$ is any $q_n$-ONB, then

$$\sum_k x(e_k)^2 = \|x\|^2_{-q_n} < \infty \quad , \quad x \in E_n' \tag{2}$$

Let $\varepsilon > 0$. Then we can find $n = n(\varepsilon)$ such that

$$P(\, Y_w \in E_n' \,) > 1-\varepsilon \ ,$$

so

$$P(\sum_k Y(e_k)^2 < \infty) > 1-\varepsilon$$

by (2). Taking $r = r(\varepsilon)$ sufficiently large so that

$$P(\sum_k Y(e_k)^2 < r) > 1-\varepsilon \,.$$

This implies that

$$\|Y(f)\|_0^2 \le E(Y(f)^2 \wedge 1)$$

$$= E(Y(f)^2, \ \sum_k Y(e_k)^2 < r) + \varepsilon \tag{3}$$

Define $p(f) = p_\varepsilon(f)$ to be the square root of the last $E(\ ,\ )$. Then $p$ is an Hilbertian seminorm. Since

$$\sum_k p(e_k)^2 = E(\sum_k Y(e_k)^2 \ , \ \sum_k Y(e_k)^2 < r)$$

$$< r \ ,$$

$p$ is a separable Hilbertian seminorm HS-bounded by $q_n$, and so $p \prec I(t)$. Since $Y$ is equivalent to $X$, (3) implies that

$$\|X(f)\|^2 \le p(f)^2 + \varepsilon$$

Hence

$$p(f) < \varepsilon \quad \Rightarrow \quad \|X(f)\|_0 < \varepsilon^2 + \varepsilon.$$

This proves that $X$ is $I(t)$-continuous i.p. We used Sazonov's idea [1] to define $p \prec I(t)$.

<u>Proof of the 'if' part</u>. Suppose that $X(f)$ is $I(t)$-continuous i.p. in $f$. Take a sequence $\varepsilon_n > 0$ such that

$$\sum_n \varepsilon_n < \infty .$$

Then we can find $\delta_n > 0$ and $p_n \prec I(t)$ such that

$$p_n(f) < \delta_n \quad \Rightarrow \quad \|X(f)\|_0 < \varepsilon_n \tag{4}$$

We can assume that $\{p_n\}$ is increasing, by replacing $p_n$ by $p_1 \vee p_2 \vee \cdots \vee p_n$ . This implies that

$$\mathrm{Re}E(e^{iX(f)}) > 1-2\varepsilon_n - 2p(f)^2 \delta_n^{-2} \tag{5}$$

In fact, if $p_n(f) < \delta_n$ , then

$$\begin{aligned} |E(e^{iX(f)}) - 1| &\leqq E(|X(f)| \wedge 2) \\ &\leqq 2\|X(f)\|_0 < 2\,\varepsilon_n , \end{aligned}$$

so the left hand side of (4) is larger than $1-2\,\varepsilon_n$ , and if $p_k(f) \geqq \delta_n$ , then the right hand side of (4) is less than $-1$.

Denote by $s$ the multi-Hilbertian topology determined by $\{p_n\}$. Since $p_n \prec I(t)$, we can find $q_n \prec t$ such that $p_n \prec_{HS} q_n$ . We can assume that $\{q_n\}$ is increasing, by replacing $q_n$ by $q_1 \vee q_2 \vee \cdots \vee q_n$ . Let $u$ be the multi-Hilbertian topology determined by $\{q_n\}$ . Then $E$ has a countable $u$-dense subset

$$D = \{d_1, d_2, \cdots \} .$$

Since $s$ is weaker than $u$, $D$ is also $s$-dense in $E$.

Appling the Schmidt $q_n$-orthogonalization to $D$, we obtain an $q_n$- ONB in $E$, say $\{e_{kn}, k=1,2,\cdots\}$ , for which $d_j$ is expressible in the form :

$$d_j = \sum_{k=1}^{j} \alpha_{jkn}\, e_{kn} + r_{jn} , \tag{6}$$

where

$$q_n(r_{jn}) = 0 \tag{7}$$

Since $p_n \prec_{HS} q_n$, this implies

$$p_n(r_{jn}) = 0$$

Fix n for the moment and suppress the suffix n for simplicity of notation. Setting

$$f = \sum_{j=1}^{J} a_j r_j$$

in (5) and noting that p(f) = 0 by (7), we obtain

$$\mathrm{Re}\, E(\exp(i \sum_{j=1}^{J} a_j X(r_j))) > 1-2\varepsilon$$

Integrating both sides by $\prod_{j=1}^{J} N_v(da_j)$ where $N_v$ is the centered Gaussian measure on $\mathbb{R}$ with variance $v > 0$, we obtain

$$E(\exp(-\frac{v}{2} \sum_{j=1}^{J} X(r_j)^2)) > 1-2\varepsilon .$$

Letting $J \to \infty$ and then $v \to \infty$, we obtain

$$P( \sum_{j=1}^{\infty} X(r_j)^2 = 0) \geq 1-2\varepsilon \qquad (8).$$

Setting

$$f = \sum_{k=1}^{J} a_k e_k$$

in (5), we obtain

$$\mathrm{Re}\, E(\exp i \sum_{k=1}^{J} a_k X(e_k))$$
$$\geq 1 - 2\varepsilon - 2\delta^{-2} \sum_{j,k=1}^{J} a_k a_j p(e_k,e_j)$$

Integrating both sides by the same measure as above, we obtain

$$E(\exp(-\frac{v}{2} \sum_{k=1}^{J} X(e_k)^2)$$
$$> 1 - 2\varepsilon - 2v\delta^{-2} \sum_{k=1}^{J} p(e_k)^2$$
$$1 - 2\varepsilon - 2v\delta^{-2} (p : q)_{HS}^2$$

where $(p:q)_{HS} < \infty$ by $p \prec_{HS} q$ . Letting $J \to \infty$ and $v \downarrow 0$, we obtain

$$P(\sum_k X(e_k)^2 < \infty) \geq 1-2\varepsilon \qquad (9).$$

Now restoring the suffix $n$, we can deduce from (8) and (9) that

$$P(\Omega_n) \geq 1-4\varepsilon_n , \quad n=1,2,\cdots \qquad (10)$$

where

$$\Omega_n = \{X(r_{jn}) = 0, \ j=1,2,\cdots, \quad \sum_k X(e_{kn})^2 < \infty\} \qquad (11).$$

Now define

$$Y_n(f,w) = \begin{cases} \sum_k q_n(f,e_{kn}) X(e_{kn},w) & \text{on } \Omega_n \\ 0 & \text{elsewhere} \end{cases} \qquad (12).$$

$Y_n(f,w)$ is well-defined by the Schwarz inequality and we obtain

$$Y_{n,w} \in E'_{q_n} \subset E'_u \qquad (13).$$

The expression (12) implies that

$$Y_n(e_{kn}) = X(e_{kn}) , \quad k=1,2,\cdots \quad \text{on } \Omega_n \qquad (14).$$

Since

$$|q_n(r_{jn}, e_{kn})| \leq q_n(r_{jn}) = 0 ,$$

(11) and (12) imply that

$$Y_n(r_{jn}) = X(r_{jn}) = 0, \quad j=1,2,\cdots \quad \text{on } \Omega_n \qquad (15)$$

But $d_j$ is a linear combination of $(e_{1n}, e_{2n},\cdots, e_{jn},r_{jn})$, $Y_n(f,w)$ is linear in $f$ for every $w$, and $X(f)$ is stochastically linear in $f$. Hence (14) and (15) imply that

$$Y_n(d_j) = X(d_j) , \quad j=1,2,\cdots \quad \text{on } \Omega'_n := \Omega_n - N_n \qquad (16)$$

where $P(N_n) = 0$. Therefore

$$P(\Omega'_n) = P(\Omega_n) > 1-4\varepsilon_n$$

by (10). Since $\sum_k \varepsilon_n < \infty$ , the Borel-Cantelli lemma implies that

$$P(\Omega') = 1, \quad (\Omega' = \bigcup_N \bigcap_{n \geq N} \Omega'_n) \tag{17}$$

Since $Y_n(d_j,w) = X(d_j,w)$ for every $j$ on $\Omega'_n$ , for every $w \in \Omega'$ we can find $N(w)$ such that

$$Y_m(d_j,w) = Y_n(d_j,w) = X(d_j,w), \quad j=1,2,\cdots \tag{18}$$

whenever $w \in \Omega'$ and $m,n \geq N(w)$. Since $\{d_j\}$ is u-dense in E and $Y_k(f,w)$ is u-continuous in f by (13), we can deduce from (18) that

$$Y_m(f,w) = Y_n(f,w), \quad f \in E$$

whenever $w \in \Omega'$ and $m,n \geq N(w)$. Define

$$Y(f,w) := \begin{cases} Y_n(f,w) & \text{on } \Omega' \quad (n \geq N(w)) \\ 0 & \text{elsewhere} \end{cases} \tag{19}$$

It is obvious that $Y_w \in E'_u \subset E'_t$ for every w.

It remains only to prove that

$$Y(f) = X(f) \quad \text{a.s.}$$

Since $Y_n(d_j) = X(d_j)$ on $\Omega'_n$ and $P(\Omega'_n) \to 1$,

$$Y_n(d_j) \to X(d_j) \quad \text{i.p.}$$

Since $Y_n(d_j) = Y(d_j)$ whenever $n \geq N(w)$ and $w \in \Omega'$ ,

$$Y_n(d_j) \to Y(d_j) \quad \text{a.s.} \quad \text{by (19).}$$

Hence

$$Y(d_j) = X(d_j) \quad \text{a.s.} \tag{20}$$

Let f be any point in E. Since $\{d_j\}$ is u-dense in E we can find a subsequence $\{f_n\}$ which u-converges to f. Since $Y_w \in E'_{\kappa}$ ,

$$Y(f_n) \to Y(f) \quad \text{for every w.}$$

Since s is weaker than u and X(f) is s-countinuous i.p. by (4),

$$X(f_n) \to X(f) \quad \text{i.p.}$$

Since $X(f_n) = Y(f_n)$ a.s. by (20), $Y(f) = X(f)$ a.s.

We used Sazonov's idea [1] of noting the inequality (5) and Yamazaki's idea [4] of applying integration by the Gauss measure to obtain (8) and (9).

Proof of the last part (uniqueness) Let Y and Z be two separable t-regular versions of X. Then we can easily find a countably Hilbertian topology $u \prec t$ such that

$$P(Y_w \in E_u', \ Z_w \in E_u) = 1 \tag{21}$$

E has a countable u-dense set, say D. Since both Y and Z are equivalent to X

$$P\{Y(d) = X(d) = Z(d)\} = 1 \quad \text{for every } d \in D.$$

Since D is countable,

$$P\{Y(d) = Z(d) \ \text{ for every } d \in D\} = 1 \tag{22}$$

Since D is u-dense, (21) and (22) implies that

$$P(Y(f) = Z(f) \ \text{ for every } f) = 1$$

i.e. $Y_w = Z_w$ a.s.

REFERENCES

[1]. Sazonov, A.V., A remark on characteristic functions, Th. Prob. and Appl. III 3 (1958) 188-192.

[2]. Minlos, R.A., On the extension of a generalized stochastic process to an additive measure, Th. Prob. and Appl. III 3 (1958) 199. (See Gelfand-Vilenkin, Generalized functions IV,

Academic Press (1964) (chap. 4) for details.)

[3]. Kolmogorov, A.N., A note on the papers of R.A. Minlos and A. Sazanov, Th. Prob. and Appl. IV 3 (1959) 221-223.

[4]. Yamazaki, Y., Measures on infinite dimensional spaces I (in Japanese), Kinokuniya-shoten, Tokyo, 1978.

Department of Mathematics, Gakushuin University
Mejiro, Toshimaku, Tokyo 171, Japan
and
Department of Mathematics, Sophia University
Kioicho, Chiyodaku, Tokyo 102, Japan

Reprinted from
*Probability Theory and Mathematical Statistics*
(Lecture Notes in Mathematics 1021) Springer-Verlag, 1982, pp. 257–267

Math. Z. 182, 17–33 (1983)

Mathematische Zeitschrift


# Distribution-Valued Processes Arising from Independent Brownian Motions

Kiyosi Itô

Department of Mathematics, Gakushuin University, Mejiro, Tokyo 171, Japan

## § 1. Introduction

Let $B_k(t)$, $k=1, 2, 3, \ldots$ be a sequence of independent 1-dimensional Brownian motions with a common initial distribution $\mu$ and $N_n(t, A)$ denote the number of $k \leqq n$ for which $B_k(t) \in A$. Normalizing $N_n(t, A)$, we define

$$X_n(t, A) := n^{-\frac{1}{2}}(N_n(t, A) - E N_n(t, A)).$$

Since $X_n(t, A)$ is a (signed) measure in $A$,

$$(X_n(t) = X_n(t, \cdot),\ t \in [0, \infty))$$

is a measure-valued stochastic process.

Let us define the limit process $X(t) := \lim_{n \to \infty} X_n(t)$. Since $X_n(t, A)$ turns out to be extremely irregular as a measure in $A$ when $n$ gets large, we cannot expect the limit process $X(t)$ to be a measure-valued process. Hence we observe the following distribution-valued process corresponding to $X_n(t, A)$:

$$X_n(t) = X_n(t, \varphi) := \int_{\mathbb{R}} \varphi(x)\, X_n(t, dx), \qquad \varphi \in \mathscr{D},$$

i.e.

$$X_n(t, \varphi) = n^{-\frac{1}{2}} \sum_{k=1}^{n} (\varphi(B_k(t)) - E\varphi(B_k(t))).$$

$X_n(t)$ is a continuous $\mathscr{D}'$-valued process for each $n$.

Using the central limit theorem in several dimensions we can construct a continuous centered Gaussian $\mathscr{D}'$-valued process $X_t = X(t, \varphi)$, $\varphi \in \mathscr{D}$ such that every finite dimensional distribution of $\{X_n(t, \varphi)\}_{t, \varphi}$ converges to the corresponding one of $\{X(t, \varphi)\}_{t, \varphi}$. $X_t$ satisfies the following stochastic differential equation:

$$\text{(S)} \qquad dX_t = (\partial \circ \sqrt{\mu_t})\, db_t + \tfrac{1}{2} \partial^2 X_t\, dt$$

where $\{b_t\}$ is a standard Wiener $\mathscr{S}_2'$ process (see §4), $\mu_t = \mu * g_t (g_t$: the Gauss density with mean 0 and variance $t)$, $\sqrt{\mu_t} (\in C^\infty(\mathbb{R}))$ is viewed as a multiplication operator in $\mathscr{D}'$. $\partial$ is the differentiation in $\mathscr{D}'$ and $\partial \circ \sqrt{\mu_t}$ is the composition of these operators.

In our previous paper [2] we studied the process $X_t$ and derived the equation (S) in case $B_k(t)$, $k = 1, 2, \ldots$ are independent $n$-dimensional Brownian motions starting at the origin. But we used heuristic arguments in several parts of the paper. The first purpose of this paper is to make rigorous those arguments.

The probability distribution of the sample function of $X_t$ is a centered Gaussian measure on the $\mathscr{D}'$-valued continuous functions on $[0, \infty)$ whose covariance functional is determined by $\mu$, so we denote it by $X_t^\mu$. Since the operator-valued coefficient of $db_t$ in the equation (S) contains $\mu$ explicitely, the equation (S) should be denoted by $(S^\mu)$. The second purpose of this paper is to find a stochastic differential equation independent of $\mu$ of which $X_t^\mu$ is a solution for every $\mu$. In §8 we will prove that for every $\mu$, $X_t = X_t^\mu$ satisfies the following equation:

$$(S_D) \qquad dX_t = \sigma(X_t)\, db_t + \tfrac{1}{2}\partial^2 X_t,$$

where $\sigma$ is a Borel measurable map from $\mathscr{D}'$ into $\mathscr{L}' = \mathscr{L}'(\mathscr{D}')$, the space of all continuous linear operators from $\mathscr{D}'$ into itself. The essential part of the proof is the construction of the map $\sigma: \mathscr{D}' \to \mathscr{L}'$ independent of $(\mu, t)$ such that

$$\sigma(X_t^\mu) = \partial \circ \sqrt{\mu_t} \quad \text{a.s.} \quad \text{for every } (\mu, t).$$

One might wonder how the sample value $(\in \mathscr{D}')$ of $X_t^\mu$ determines the deterministic operator $\partial \circ \sqrt{\mu_t}$. As a matter of fact this is due to the ergodic property of $X_t^\mu$ that follows from the fact that for each $t$, $\{X_t^\mu(\varphi), \varphi \in \mathscr{D}\}$ is similar to the white noise on $\mathbb{R}$.

$(S_D)$ is a time homogeneous stochastic differential equation in $\mathscr{D}'$. It has many solutions other than $X_t^\mu$. Assume, as we can, that the base probability space $(\Omega, \mathscr{F}, P)$ is standard. Then, for almost every $f \in \mathscr{D}'$ with respect to the measure $P(X_0^\mu \in \cdot)$, we can define the regular conditional probability under $X_0^\mu = f$, written $P_f^\mu$. For such $f$ the process $X_t^\mu$, when observed on $(\Omega, \mathscr{F}, P_f^\mu)$, is a solution of $(S_D)$ with the initial value $f$. At present we have no idea of how to discuss the uniqueness of the solution.

## §2. A Family of Hilbert Spaces of Functions and Distributions

In this section we mention some known facts on the Schwartz distributions and fix the notation. Most of the facts were mentioned in my previous papers [1, 2], but we change the notation slightly for convenience; for example, the $\mathscr{S}_p$ here corresponds to the $\mathscr{S}_{p/2}$ there. All functions and functionals are assumed to be real valued. $c, c', c'', \ldots, c_1, c_2, \ldots$ denote positive constants unless explicitly stated otherwise.

The $L_2$ space on $\mathbb{R}$ is denoted by $L_2$ and the inner product and the norm in $L_2$ are denoted by $( \, , \, )$ and $\| \; \|$ resepectively. $\mathscr{D}$ and $\mathscr{D}'$ denote the $C^\infty$-

functions of compact support and the Schwartz distributions respectively. $H_n(x)$ is the Hermite polynomial of degree $n$ and $h_n(x)$ is the corresponding Hermite functions, i.e.

$$h_n(x)=c_n H_n(x)\, e^{-x^2/2}, \qquad n=0,1,2,\dots$$

where $c_n$ is a constant for which $\|h_n\|=1$. The Hermite functions form an orthonormal base (abbrev. ONB) in $L_2$. The $p$-norm $\|\ \|_p$ in $L_2$ and the $(-p)$-norm $\|\ \|_{-p}$ in $\mathscr{D}'$ are defined as follows:

$$\|\varphi\|_p^2=\sum_{n=0}^{\infty}(\varphi,h_n)^2(2n+1)^p\in[0,\infty]$$

$$\|f\|_{-p}^2=\sum_{n=0}^{\infty}f(h_n)^2(2n+1)^{-p}\in[0,\infty]\ (f\in\mathscr{S}') \quad \text{and}$$

$$=\infty(f\in\mathscr{D}'-\mathscr{S}'),$$

where $p=0,1,2,\dots$ and $f(\varphi)$ denotes the value of $f$ evaluated at $\varphi\in\mathscr{D}$. It is obvious that

$$\|f\|_{-p}=\sup\{|f(\varphi)|:\varphi\in\mathscr{D},\ \|\varphi\|_p\leqq 1\},$$

which implies

$$|f(\varphi)|\leqq\|f\|_{-p}\|\varphi\|_p, \quad f\in\mathscr{D}', \quad \varphi\in\mathscr{D}.$$

Define $\mathscr{S}_p$, $(\varphi,\psi)_p$, $\mathscr{S}_p'$ and $(f,g)_{-p}$ as follows:

$$\mathscr{S}_p=\{\varphi\in L_2:\ \|\varphi\|_p<\infty\},$$

$$\mathscr{S}_p'=\{f\in\mathscr{D}':\ \|f\|_{-p}<\infty\},$$

$$(\varphi,\psi)_p=\sum_{n=0}^{\infty}(\varphi,h_n)(\psi,h_n)(2n+1)^p,$$

$$(f,g)_{-p}=\sum_{n=0}^{\infty}f(h_n)\,g(h_n)(2n+1)^{-p}.$$

Then $(\mathscr{S}_p,(\ ,\ )_p)$ and $(\mathscr{S}_p',(\ ,\ )_{-p})$ are Hilbert spaces with the norms $\|\ \|_p$ and $\|\ \|_{-p}$ respectively. As $p$ increases, $\|\ \|_p$ increases and $\|\ \|_{-p}$ decreases, so $\mathscr{S}_p$ decreases and $\mathscr{S}_p'$ increases. The intersection $\bigcap_p \mathscr{S}_p$ coincides with the rapidly decreasing functions $\mathscr{S}$ and the union $\bigcup_p \mathscr{S}_p'$ with the tempered distributions $\mathscr{S}'$. It follows from the definition that $\mathscr{S}_0=L_2$, $\|\ \|_0=\|\ \|$ and $(\ ,\ )_0=(\ ,\ )$. By identifying $\psi\in L_2$ with $i\psi\in\mathscr{D}'$ where $(i\psi)(\varphi):=(\psi,\varphi)$, we have $L_2=\mathscr{S}_0'$, $\|\ \|=\|\ \|_{-0}$ and $(\ ,\ )=(\ ,\ )_{-0}$. Hence

$$\mathscr{D}\subset\mathscr{L}\subset\dots\subset\mathscr{S}_2\subset\mathscr{S}_1\subset\mathscr{S}_0=L_2=\mathscr{S}_0'\subset\mathscr{S}_1'\subset\dots\subset\mathscr{S}'\subset\mathscr{D}'.$$

Let $\mathscr{B}_K(\mathscr{D}')$ denote the Kolmogorov $\sigma$-algebra on $\mathscr{D}'$, the least $\sigma$-algebra that makes measurable the map $f\mapsto f(\varphi)$ for every $\varphi\in\mathscr{D}$. The $\mathscr{S}_p'\in\mathscr{B}_K(\mathscr{D}')$ and the topological $\sigma$-algebra on $\mathscr{S}_p'$, written $\mathscr{B}(\mathscr{S}_p')$, is the restriction of $\mathscr{B}_K(\mathscr{D}')$ to $\mathscr{S}_p'$.

Because of the inequality $|f(\varphi)| \leqq \|f\|_{-p} \|\varphi\|_p$ every $f \in \mathscr{S}_p'$ can be extended to a unique bounded linear functional on $\mathscr{S}_p$ with the same norm. With every ONB $\{\varepsilon_n\}$ in $\mathscr{S}_p$ we associate a unique ONB $\{e_n\}$ in $\mathscr{S}_p'$, called the ONB in $\mathscr{S}_p'$ dual to $\{\varepsilon_n\}$, such that $e_m(\varepsilon_n) = \delta_{mn}$, $m, n = 0, 1, 2, \ldots$. In this case every $f \in \mathscr{S}_p'$ has an orthogonal expansion

$$f = \sum_n f(\varepsilon_n)\, e_n$$

and

$$\|f\|_{-p}^2 = \sum_n f(\varepsilon_n)^2.$$

The operators in $\mathscr{D}$:

$$\partial\colon \varphi \mapsto \varphi' \quad \text{and} \quad \alpha\colon \varphi \mapsto \alpha\varphi \; (\alpha \in C^\infty(\mathbb{R}))$$

are extended onto $\mathscr{D}'$ in the Schwartz theory of distributions. In case $\alpha(x) \equiv x$ we denote $\alpha\varphi$ by $x\varphi$. These extended operators have the following properties:

$$\|xf\|_{-p-1} \leqq c\,\|f\|_{-p}, \qquad \|\partial f\|_{-p} \leqq c'\,\|f\|_{-p}$$

which we can easily check by observing that

$$\|x\varphi\|_p \leqq c\,\|\varphi\|_{p+1}, \qquad \|\varphi'\|_p \leqq c'\,\|\varphi\|_{p+1},$$

which in turn follow from the fact that both $xh_n$ and $h_n'$ are represented as linear combinations of $h_{n+1}$ and $h_{n-1}$ with coefficients of order $\sqrt{n}$.

## § 3. $\mathscr{D}'$-Valued Random Variables

Let $\Omega = (\Omega, \mathscr{F}, P)$ be a probability space. A $\mathscr{D}'$-valued function on $\Omega$ measurable $\mathscr{F}/\mathscr{B}_K(\mathscr{D}')$ is called a *$\mathscr{D}'$-valued random variable* or a *$\mathscr{D}'$ variable*. Let $X = X(\omega)$ be a $\mathscr{D}'$ variable. The value of $X(\omega)$ evaluated at $\varphi \in \mathscr{D}$ is denoted by $X(\omega)(\varphi)$, $X(\varphi, \omega)$ or $X(\varphi)$. This is a real random variable. Since $\mathscr{B}_K(\mathscr{D}')$ is the Kolmogorov $\sigma$-algebra on $\mathscr{D}'$, the finite dimensional distributions of $\{X(\varphi), \varphi \in \mathscr{D}\}$ determines the *probability distribution* of $X$:

$$P^X(A) := P\{X \in A\}, \qquad A \in \mathscr{B}_K(\mathscr{D}').$$

The *characteristic functional* of a $\mathscr{D}'$ variable $X$, written $C_X(\varphi)$, is defined by

$$C_X(\varphi) = E e^{iX(\varphi)}, \qquad \varphi \in \mathscr{D}.$$

Since $C(t_1, t_2, \ldots, t_n) := C_X(\sum_k t_k \varphi_k)$ is the characteristic function of the joint variable $(X(\varphi_k), k = 1, 2, \ldots, n)$, $C_X$ determines $P^X$.

A $\mathscr{D}'$ variable $X$ is called *centered Gaussian* if the family $\{X(\varphi), \varphi \in \mathscr{D}\}$ is a centered Gaussian system, a Gaussian system with mean 0. The positive quadratic functional

$$V_X(\varphi) := E(X(\varphi)^2), \qquad \varphi \in \mathscr{D}$$

is called the *variance functional* of $X$. If $X$ is centered Gaussian, then

$$C_X(\varphi)=e^{-V_X(\varphi)/2}.$$

The probability distribution of a centered Gaussian $\mathscr{D}'$ variable is determined by its variance functional.

A $\mathscr{D}'$ variable $X$ is called an $\mathscr{S}_p'$ variable if $X(\omega)\in\mathscr{S}_p'$ for every $\omega\in\Omega$. If $X$ is an $\mathscr{S}_p'$ variable, then we can extend $X(\omega)$ to a bounded linear functional on $\mathscr{S}_p(\supset\mathscr{D})$, as we do always. If $X$ is a centered Gaussian $\mathscr{S}_p'$ variable, then $\{X(\varphi),\varphi\in\mathscr{S}_p\}$ form a centered Gaussian system.

Let $Y=\{Y(\varphi),\varphi\in\mathscr{D}\}$ be a family of real random variables and $X$ a $\mathscr{D}'$ (resp. $\mathscr{S}_p'$) variable. If $X(\varphi)=Y(\varphi)$ a.s. for every $\varphi\in\mathscr{D}$, then $X$ is called a $\mathscr{D}'$ (resp. $\mathscr{S}_p'$) *regularization* of $Y$. If $X_1$ and $X_2$ are two $\mathscr{D}'$ (or $\mathscr{S}_p'$) regularizations, then $X_1=X_2$ a.s., because there exists a countable sequence $\{\psi_n\}$ in $\mathscr{D}$ such that for every $\varphi\in\mathscr{D}$ we can find a subsequence $\{\psi_n\}$ of $\{\varphi_n\}$ with the property that $\psi_n^{(k)}$ converges to $\varphi^{(k)}(x)$ uniformly in $x$ for every $k$.

**Theorem 3.1.** *Let $Y=\{Y(\varphi),\ \varphi\in\mathscr{D}\}$ be a family of real random variables. If $Y$ has the properties:*

(L) *(almost linear)*

$$Y(c_1\varphi_1+c_2\varphi_2)=c_1Y(\varphi_1)+c_2Y(\varphi_2)\quad a.s.,$$

*where the exceptional $\omega$-set may depend on the choice of $(c_1,\varphi_1,c_2,\varphi_2)\in\mathbb{R}\times\mathscr{D}\times\mathbb{R}\times\mathscr{D}$,*

(B) *(p-bounded)*

$$E(Y(\varphi)^2)\leqq c\|\varphi\|_p^2,$$

*then $Y$ has an $\mathscr{S}_{p+2}'$ regularization $X$ satisfying*

(C)

$$E(\|X\|_{-p-2}^2)\leqq\alpha c,\qquad \alpha=\pi^2/8.$$

*Proof.* The conditions (L) and (B) imply that the map $Y\colon\varphi\mapsto Y(\varphi)$ is a bounded linear operator from the pre-Hilbert space $(\mathscr{D},\|\ \|_p)$ into the Hilbert space $L_2=L_2(\Omega,\mathscr{F},P)$. This map can be extended to a unique linear operator from $\mathscr{S}_p$ into $L_2$, denoted by the same notation $Y$, because $\mathscr{D}$ is dense in $\mathscr{S}_p$. Let $\{h_n\}$ be the Hermite functions (§ 2). Then

$$\varepsilon_n=(2n+1)^{-(p+2)/2}h_n\in\mathscr{S}_p,\qquad n=0,1,2,\dots$$

form an ONB in $\mathscr{S}_{p+2}$. The ONB in $\mathscr{S}_{p+2}'$ dual to $\{\varepsilon_n\}$ is denoted by $\{e_n\}$. Denoting $Y(\varepsilon_n)$ by $X_n$, we obtain

$$E(\sum_n X_n^2)\leqq c\sum_n\|\varepsilon_n\|_p^2$$

by (B). But $\|\varepsilon_n\|_p^2=(2n+1)^{-2}$ by the definition of the $p$-norm. Hence

$$E(\sum_n X_n^2)\leqq c\sum_n(2n+1)^{-2}=c\pi^2/8=\alpha c.$$

This implies that

$$P(\Omega_1)=1, \qquad \Omega_1:=\{\omega\in\Omega: \sum_n X_n(\omega)^2<\infty\}.$$

We define

$$X(\omega):=\sum_n X_n(\omega)\, e_n \text{ on } \Omega_1, \text{ and } =0 \text{ elsewhere.}$$

Then $X(\omega)\in\mathscr{S}'_{p+2}$ for every $\omega$ and $X(\omega)$ is measurable $\mathscr{F}/\mathscr{B}(\mathscr{S}'_{p+2})$ in $\omega$. Hence $X$ is an $\mathscr{S}'_{p+2}$ variable and

$$E(\|X\|^2_{-p-2})=E(\sum_n X_n^2)\leqq\alpha c.$$

Since $e_m(\varepsilon_n)=\delta_{mn}$ and $P(\Omega_1)=1$, $X(\varphi)=Y(\varphi)$ a.s. for $\varphi=\varepsilon_n$, $n=0,1,2,\ldots$ and so for every finite linear combination of $\{\varepsilon_n\}$ by (L). For a general $\varphi\in\mathscr{S}_{p+2}(\subset\mathscr{S}_p)$ we have

$$\|\varphi-\varphi_n\|_{p+2}\to 0(n\to\infty), \qquad \varphi_n:=\sum_{k=0}^{n}(\varphi,\varepsilon_n)_{p+2}\,\varepsilon_n.$$

Since $X\in\mathscr{S}'_{p+2}$,

$$|X(\varphi)-X(\varphi_n)|\leqq\|X\|_{-p-2}\|\varphi-\varphi_n\|_{p+2}\to 0$$

for every $\omega$. Also we can use (B) to check that

$$E((Y(\varphi)-Y(\varphi_n))^2)\leqq c\|\varphi-\varphi_n\|_p^2\leqq c\|\varphi-\varphi_n\|_{p+2}^2\to 0.$$

Thus $X(\varphi)=Y(\varphi)$ a.s. for every $\varphi\in\mathscr{S}_p(\supset\mathscr{D})$. □

**Theorem 3.2.** *Let $Y$ be a $\mathscr{D}'$ variable. If $Y$ has the property*

$$\text{(B)} \quad E(Y(\varphi)^2)\leqq c\|\varphi\|_p^2,$$

*then*

$$E(\|Y\|^2_{-p-2})\leqq\alpha c, \qquad \alpha=\pi^2/8.$$

*Proof.* The family $\{Y(\varphi), \varphi\in\mathscr{S}\}$ has the properties (L) and (B) of Theorem 3.1. Hence this family has an $\mathscr{S}'_{p+2}$ regularization $X$ satisfying the condition (C) in the theorem. Since $Y$ is a $\mathscr{D}'$ variable, $Y$ itself is a $\mathscr{D}'$ regularization of this family. Being an $\mathscr{S}'_{p+2}$ regularization, $X$ is a $\mathscr{D}'$ regularization obviously. Therefore $Y=X$ a.s. Hence the conclusion follows immediately. □

**Theorem 3.3.** *Let $X$ be a centered Gaussian $\mathscr{D}'$ variable. If $X\in\mathscr{S}'_p$ a.s., i.e. $\|X\|_{-p}<\infty$ a.s., then*

(i) $E(\|X\|^2_{-p})<\infty$,

(ii) $E(\|X\|^{2m}_{-p})\leqq\alpha_m E(\|X\|^2_{-p})^m$, $\alpha_m=1.3\ldots(2m-1)$.

*Proof.* Let $\{\varepsilon_n, n=0,1,2,\ldots\}$ be any ONB in $\mathscr{S}_p$ and $L_n$ the subspace of $\mathscr{S}_p$ spanned by $\{\varepsilon_i, i\leqq n\}$. Define

$$V(\varphi,\psi):=E(X(\varphi)\,X(\psi)), \qquad \varphi,\psi\in L_n.$$

Then $V$ is a positive definite symmetric bilinear form on $L_n$. Let $\{\delta_i = \delta_i^n,\ i = 0, 1, 2, \ldots, n\}$ be an ONB in $L_n$ such that

$$V(\varphi, \psi) = \sum_{i=1}^{n} v_i(\varphi, \psi_i)(\psi, \delta_i), \qquad v_i \geqq 0.$$

Then we have

$$E(X(\delta_i)X(\delta_j)) = V(\delta_i, \delta_j) = 0 \qquad (i \neq j),$$

$$E(X(\delta_i)^2) = V(\delta_i, \delta_i) = v_i.$$

Since $X$ is centered Gaussian, $X(\delta_i), i = 0, 1, 2, \ldots, n$ are independent and each $X(\delta_i)$ is centered Gaussian with variance $v_i$. Since both $(\varepsilon_i, i \leqq n)$ and $(\delta_i, i \leqq n)$ are ONB's in $L_n$, they are transformed from one to another by an orthogonal transformation, so $(X(\varepsilon_i), i \leqq n)$ and $(X(\delta_i), i \leqq n)$ have the same property. Hence we have

$$\sum_{i=0}^{n} X(\varepsilon_i)^2 = \sum_{i=0}^{n} X(\delta_i)^2 := S_n,$$

so

$$ES_n = \sum_{i=0}^{n} E(X(\delta_i)^2) = \sum_{i=0}^{n} v_i.$$

Recalling that $X(\delta_i), i \leqq n$ are independent and Gauss distributed, we have

$$Ee^{-S_n} = \prod_i Ee^{-X(\delta_i)^2} = \prod_i (1 + 2v_i)^{-\frac{1}{2}}$$

$$\leqq (1 + 2\sum_i v_i)^{-\frac{1}{2}} = (1 + 2ES_n)^{-\frac{1}{2}}.$$

But

$$S_n \leqq \sum_{i=0}^{\infty} X(\varepsilon_i)^2 = \|X\|_{-p}^2 < \infty \qquad \text{(by the assumption)}$$

and so

$$(1 + 2ES_n)^{-\frac{1}{2}} \geqq Ee^{-\|X\|_{-p}^2} > 0$$

which implies that $\{ES_n\}_n$ is bounded. Therefore.

$$E(\|X\|_{-p}^2) = E\left(\sum_{i=0}^{\infty} X(\varepsilon_i)^2\right) = \lim_{n \to \infty} ES_n < \infty.$$

This completes the proof of (i).

To prove (ii) we will observe the case where $m = 2$; the proof of the general case is similar. Using the same notation as above, we obtain

$$E(S_n^2) = \sum_i E(X(\delta_i)^4) + 2\sum_{i<j} E(X(\delta_i)^2)\,E(X(\delta_j)^2)$$

$$\text{(by the independence of } X(\delta_i) \text{ and } X(\delta_j))$$

$$= \sum_i 3v_i^2 + 2\sum_{i<j} v_i v_j$$

$$\leqq 3(\sum_i v_i)^2 = 3E(S_n^2).$$

Letting $n \to \infty$, we can check that (ii) holds for $m=2$. □

*White noise.* A central Gaussian $\mathscr{S}_2'$ variable with variance functional $c\|\varphi\|^2 (c \geqq 0)$ is called a *white noise with intensity c*. It is obvious that any two white noise with intensity $c$ have the same probability distribution. We will prove that for every $c \geqq 0$ there exists a white noise with intensity $c$. Since $V(\varphi, \psi) := c(\varphi, \psi)$ is positive definite in $(\varphi, \psi)$, there exists a centered Gaussian system $w = \{w(\varphi), \varphi \in \mathscr{D}\}$ satisfying

$$E(w(\varphi)\, w(\psi)) = c(\varphi, \psi).$$

$w(\varphi)$ is almost linear in $\varphi$, because

$$E((w(\varphi) - c_1 w(\varphi_1) - c_2 w(\varphi_2))^2) = 0.$$

Also $E(w(\varphi)^2) = c\|\varphi\|^2$. Hence Theorem 3.1 assures that $w$ has an $\mathscr{S}_2'$ regularization, denoted by $w$ again. Then $w$ is a white noise with intensity $c$. Since $\varepsilon_n := h_n/(2n+1)$, $n = 0, 1, 2, \ldots$ form an ONB in $\mathscr{S}_2$, we obtain

$$\begin{aligned} E(\|w\|_{-2}^2) &= E(\sum_n w(\varepsilon_n)^2) = \sum_n c\|\varepsilon_n\|^2 \\ &= \sum_n c(2n+1)^{-2} = c\pi^2/8. \end{aligned}$$

## § 4. $\mathscr{D}'$-Valued Stochastic Processes

A family of $\mathscr{D}'$ variables $X = \{X_t, t \in [0, \infty)\}$ is called a *$\mathscr{D}'$-valued stochastic process* or simply an *$\mathscr{D}'$ process.* A $\mathscr{D}'$ process $X$ is called *centered Gaussian* if the real random variables $X_t(\varphi), \varphi \in \mathscr{D}, t \in [0, \infty)$ form a centered Gaussian system.

A $\mathscr{D}'$ process $X$ is called (sample) *continuous* if $X_t(\omega)$ is continuous in $t$ for every $\omega$. Since a $\mathscr{D}'$-valued function on $[0, \infty)$ is strongly continuous if and only if it is weakly continuous, we do not have to specify the topology in $\mathscr{D}'$ in the definition of continuous $\mathscr{D}'$ processes. The function $t \mapsto X_t(\omega)$, denoted by $X_\cdot(\omega)$, is called the sample function of $X$ for $\omega$. Let Let $\mathbb{C}$ denote the space of all $\mathscr{D}'$-valued continuous functions on $[0, \infty)$. We endow $\mathbb{C}$ with the Kolmogorov $\sigma$-algebra $\mathscr{B}_K(\mathbb{C})$, the least $\sigma$-algebra on $\mathbb{C}$ that makes measurable the map $\gamma \in \mathbb{C} \mapsto \gamma(t) \in \mathscr{D}'$ with respect to the Kolmogorov $\sigma$-algebra $\mathscr{B}_K(\mathscr{D}')$. If $X$ is a continuous $\mathscr{D}'$ process, then its sample function $X_\cdot(\omega)$ is regarded as a $\mathbb{C}$-valued random variable whose probability idstribution is defined on $\mathscr{B}_K(\mathbb{C})$. If $X$ is a centered Gaussian continuous $\mathscr{D}'$ process, then its probability distribution is determined by the *covariance functional* of $\{X_t(\varphi)\}_{t,\varphi}$:

$$V(s, \varphi, t, \psi) = E(X_s(\varphi)\, X_t(\psi)),\ s, t \in [0, \infty);\ \varphi, \psi \in D.$$

A $\mathscr{D}'$ process $X = \{X_t\}_t$ is called an $\mathscr{S}_p'$ process, if $X_t$ is an $\mathscr{S}_p'$ variable for every $t$. Let $X$ be an $\mathscr{S}_p'$ process. If $X$ is centered Gaussian, then $X_t(\varphi), \varphi \in \mathscr{S}_p$, $t \in [0, \infty)$ form a centered Gaussian system. If $X$ is continuous, then every sample function of $X$ is $\|\ \|_{-p}$-continuous.

Let $X$ be a $\mathscr{D}'$ process and $Y$ a continuous $\mathscr{D}'$ process. $Y$ is called a *continuous $\mathscr{D}'$ version* of $X$, if $X_t = Y_t$ a.s. for every $t$. Similarly we define a continuous $\mathscr{S}_p'$ version of a $\mathscr{D}'$ process. The following theorem is an infinite dimensional version of *the Kolmogorov continuous version theorem.*

**Theorem 4.1.** *Let $X$ be an $\mathscr{S}_p'$ process. If there exist positive constants $\alpha$, $\beta$ and $\gamma$ such that*

$$\text{(K)} \qquad E(\|X_t - X_s\|_{-p}^{\alpha}) \leqq \beta(t-s)^{1+\gamma}, \quad 0 \leqq s < t < \infty$$

*then $X$ has a continuous $\mathscr{S}_p'$ version.*

**Theorem 4.2.** *Let $X$ be a centered Gaussian $\mathscr{S}_p'$ process. If there exist positive constants $\beta$ and $\gamma$ such that*

$$E(\|X_t - X_s\|_{-p}^2) \leqq \beta(t-s)^{\gamma}, \quad 0 \leqq s < t < \infty,$$

*then $X$ has a continuous $\mathscr{S}_p'$ version.*

*Proof.* Take an integer $m$ such that $m\gamma > 1$. Since $X_t - X_s$ is centered Gaussian, we can use Theorem 3.3 (ii) to obtain

$$E(\|X_t - X_s\|_{-p}^{2m}) \leqq \alpha_m \beta^m (t-s)^{m\gamma}, \quad s < t,$$

which implies that the condition (K) holds. Hence $X$ has a continuous $\mathscr{S}_p'$ version. □

*Standard Wiener $\mathscr{S}_2'$ Process.* A continuous centered Gaussian $\mathscr{S}_2'$ process with covariance functional

$$V(s, \varphi, t, \psi) = (s \wedge t)(\varphi, \psi)$$

is called a standard Wiener $\mathscr{S}_2'$ process [1]. It is obvious that if $b$ and $b'$ are two standard Wiener $\mathscr{S}_2'$ processes, then the probability distribution of the sample function of $b$ is equal to that of $b'$. We will prove the existence of a standard Wiener $\mathscr{S}_2'$ process. Denoting by $e_a(u)$ the indicator of the interval $[0, a]$ we obtain

$$V(s, \varphi, t, \psi) = \int_0^\infty e_s(u)\, e_t(u)\, (\varphi, \psi)\, du,$$

which implies that $V$ is positive definite in $((s, \varphi), (t, \psi))$. Hence there exists a centered Gaussian system $\{b(s, \varphi)\}_{s,\varphi}$ such that

$$E(b(s, \varphi)\, b(t, \psi)) = (s \wedge t)(\varphi, \psi).$$

Then $b(t, \varphi)$ is almost linear in $\varphi$ and $E(b(t, \varphi)^2) = t\|\varphi\|^2$. Hence Theorem 3.1 ensures that for every $t$, $\{b(t, \varphi)\}_\varphi$ has an $\mathscr{S}_2'$ regularization, written $b_t$. $b_t$ is a white noise with intensity $t$. Also we have

$$E((b_t(\varphi) - b_s(\varphi))^2) = |t-s|\, \|\varphi\|^2$$

and so

$$E(\|b_t - b_s\|_{-2}^2) \leqq \alpha |t-s|, \quad \alpha = \pi^2/8 \quad \text{(Theorem 3.2)},$$

which implies the existence of a continuous $\mathscr{S}_2'$ version of $\{b_t\}$ by Theorem 4.2. This version, denoted by $b$, is a standard Wiener $\mathscr{S}_2'$ process.

## § 5. Stochastic Integrals in $\mathscr{D}'$

There are many references on the general theory of stochastic integrals in infinite dimensions (see [3–5]). Here we will explain the stochastic integral of the type we are going to use in the next two sections.

We endow the base probability space $(\Omega, \mathscr{F}, P)$ with a reference family of $\sigma$-subalgebras of $\mathscr{F}, \mathscr{F}_t, 0 \leqq t < \infty$. A standard Wiener $\mathscr{S}_2'$ process $b$ is called a *standard Wiener $\mathscr{S}_2'$ matringale* (relative to $\{\mathscr{F}_t\}$) if (i) $\{b_t(\varphi)\}$ is adapted to $\{\mathscr{F}_t\}$ for every $\varphi \in \mathscr{D}$ and (ii) for $s < t, b_t - b_s$ is independent of $\mathscr{F}_s$.

Let $\mathscr{L} = \mathscr{L}(\mathscr{D})$ (resp. $\mathscr{L}' = \mathscr{L}'(\mathscr{D}')$) denote the continuous linear operators from $\mathscr{D}$ (resp. $\mathscr{D}'$) into itself. For every $\alpha \in \mathscr{L}'$ we define the *conjugate operator* $\alpha^* \in \mathscr{L}$ by the following relation:

$$(\alpha f)(\varphi) = f(\alpha^* \varphi), \qquad f \in \mathscr{D}', \qquad \varphi \in \mathscr{D}.$$

Let $\alpha(\omega)$ be an $\mathscr{L}'$-valued function of $\omega$. If for every $(f, \varphi) \in \mathscr{D}' \times \mathscr{D}$, $(\alpha(\omega) f)(\varphi)$ is $P$-measurable in $\omega$, then $\alpha(\omega)$ is called an *$\mathscr{L}'$-valued random variable* or simply an *$\mathscr{L}'$- variable.* In this case we define the $(-p)$-norm of $\alpha^*$ as follows:

$$|\alpha^*|_{-p}^2 = \sup\{E(\|\alpha^*(\omega)\varphi\|^2) : \|\varphi\|_p \leqq 1\}.$$

This norm may be infinite.

Let $\alpha_t(\omega)$ be an $\mathscr{L}'$-valued function of $(t, \omega)$. Then, for every $(f, \varphi) \in \mathscr{D}' \times \mathscr{D}$, $(\alpha_t(\omega) f)(\varphi)$ is a real-valued function of $(t, \omega)$. If this function is a predictable process (relative to $\{\mathscr{F}_t\}$) for every $(t, \varphi)$, then $\{\alpha_t(\omega)\}$ is called a *predictable $\mathscr{L}'$-valued process* or simply a *predictable $\mathscr{L}'$ process* (relative to $\{\mathscr{F}_t\}$). In this case $\alpha_t(\omega)$ is an $\mathscr{L}'$ variable for almost every $t$, $|\alpha_t^*|_{-p}$ is well-defined for almost every $t$, and $|\alpha_t^*|_{-p}$ is measurable in $t$.

Keeping these preliminary facts in mind, we define the *stochastic integral*

$$\text{(I)} \qquad M_t = \int_0^t \alpha_s db_s$$

under the following conditions:

$(A_1)$ $\qquad \{\alpha_r(\omega)\}$ is a predictable $\mathscr{L}'$ process,

$(A_2)$ $\qquad \int_0^t |\alpha_s^*|_{-p}^2 ds < \infty, \quad 0 \leqq t < \infty,$

where $p$ is fixed. The process $\{M_t\}$ has the following properties:

$(I_1)$ $M_t$ is a continuous $\mathscr{S}_{p+2}'$ process,

$(I_2)$ $M_t(\varphi)$ is a martingale (relativ to $\{\mathscr{F}_t\}$) for every $\varphi \in \mathscr{D}$.

$$(\mathrm{I}_3)\qquad \begin{aligned} E(M_t(\varphi)^2) &= \int_0^t E(\|\alpha_s^*\varphi\|^2)\,ds \\ &\leqq \|\varphi\|_p^2 \int_0^t \|\alpha_s^*\|_{-p}^2 ds < \infty \end{aligned}$$

$$(\mathrm{I}_4)\qquad E(M_t(\varphi)\,M_s(\psi)) = \int_0^{t\wedge s} E(\alpha_u^*\varphi, \alpha_u^*\psi)\,du.$$

The proof is similar to the one-dimensional case.

In the special case where $\alpha_s$ is deterministic, $M_t$ turns out to be centered Gaussian. Hence $M_t - M_s$ is independent of $\mathscr{F}_s$ whenever $t > s$. $M_t$ has the property:

$$E(M_s(\varphi)\,M_t(\psi)) = \int_0^{s\wedge t} (\alpha_u^*\varphi, \alpha_u^*\psi)\,du.$$

Conversely every centered Gaussian $\mathscr{D}'$ process $M_t$ with this property where $\{\alpha_t\}$ satisfies $(\mathrm{A}_2)$ has the *stochastic integral representation* (I) except on a $P$-null set independent of $t$ by extending the base probability space $(\Omega, \mathscr{F}, P)$ and the reference family $\{\mathscr{F}_t\}$.

## § 6. The Limit Process $\{X_t\}$

In this section we will discuss the limit process $\{X_t\}$ mentioned in the introduction. Let $B_k(t)$, $n = 1, 2, \ldots$ be a sequence of independent 1-dimensional Brownian motions with a common initial distribution $\mu$. Using the notation

$$F_k(t, \varphi) = \varphi(B_k(t)) - E\varphi(B_k(t)), \qquad k = 1, 2, \ldots,$$

we can express the $\mathscr{D}'$ process $X_n(t, \varphi)$ introduced in § 1 as follows:

$$X_n(t, \varphi) = n^{-\frac{1}{2}} \sum_{k=1}^{n} F_k(t, \varphi), \qquad \varphi \in \mathscr{D}.$$

Since one of $(B_k, F_k)$, $k = 1, 2, \ldots$, say $(B_1, F_1)$, will be used frequently, we denote it by $(B, F)$. The limit process $X_t = X(t, \varphi)$ is defined in view of the following theorem.

**Theorem 6.1.** (i) *There exists a continuous centered Gaussian* $\mathscr{S}_3'$ *process* $X_t$ *satisfying the condition:*

(C) *Every finite dimensional joint distribution of* $\{X_n(t, \varphi)\}_{t,\varphi}$ *converges to the corresponding one of* $\{X(t, \varphi)\}_{t,\varphi}$.

(ii) *Such process* $X_t$ *is uniquely determined by* $\mu$ *up to equivalence in distributions.*

*In view of this theorem we should denote* $X_t$ by $X_t^\mu$, when we consider the limit process for different $\mu$'s.

$\{X_t\}$ is called the *limit process* of $\{X_n(t, \varphi)\}$, $n = 1, 2, \ldots$.

*Proof.* The systems of real random variables:

$$F_k = \{F_k(t, \varphi)\}_{t,\varphi}, \qquad k = 1, 2, \ldots$$

are independently identically distributed and

$$E(F_k(t, \varphi)) = 0 \quad \text{and } |F_k(t, \varphi)| \leqq 2 \sup_x |\varphi(x)|.$$

Applying the central limit theorem in several dimensions and using the Kolmogorov extension theorem, we can find a centered Gaussian system $\{X(t, \varphi)\}$ for which the condition (C) holds. $\{F_k\}$ and $\{X(t, \varphi)\}$ may be defined on the different probability spaces, say $(\Omega_i, \mathscr{F}_i, P_i)$, $i = 1, 2$. But we take their product probability space $(\Omega, \mathscr{F}, P)$ on which both systems are defined. It is obvious that

$$E(X(t, \varphi)\, X(s, \psi)) = E(F(t, \varphi)\, F(s, \psi)).$$

Let $H_X$ and $H_F$ denote the Hilbert subspaces of $L^2(\Omega)$ spanned by $\{X(t, \varphi)\}_{t,\varphi}$ and $\{F(t, \varphi)\}_{t,\varphi}$ respectively. In view of the relation above we can find an isomorphism $\theta$: $H_X \to H_F$ which carries $F(t, \varphi)$ to $X(t, \varphi)$ for every $(t, \varphi)$. As $\{X(t, \varphi)\}_{t,\varphi}$ is a centered Gaussian system, so is $H_X$. $F_t(\varphi) := F(t, \varphi)$ is a continuous $\mathscr{D}'$ process, but $X_t(\varphi) := X(t, \varphi)$ is not. Hence we will take a reasonable modification of $X_t(\varphi)$.

Let us fix $t > 0$ for the moment. Since $F_t(\varphi)$ is linear in $\varphi$, the isomorphism $\theta$: $H_X \to H_F$ ensures that $X_t(\varphi)$ is almost linear in $\varphi$. Let

$$g_t := e^{-x^2/2t}/\sqrt{2\pi t} \quad \text{and } \mu_t := \mu * g_t.$$

$\mu_t$ is the probability density of $B_t := B(t)$. Observing that

$$E(X_t(\varphi)^2) = E(F_t(\varphi)^2) \leqq E(\varphi(B_t)^2)$$
$$= \int_{-\infty}^{\infty} \varphi(x)^2 \mu_t(x)\, dx \leqq \|\varphi\|^2 t^{-\frac{1}{2}}$$

and using Theorem 3.1, we can see that $\{X_t(\varphi)\}_\varphi$ has an $\mathscr{S}_2'$ regularization (written $X_t$ again) satisfying

$$E(\|X_t\|_{-2}^2) \leqq \alpha t^{-\frac{1}{2}}, \qquad \alpha = \pi^2/8.$$

Next we observe $\{X_0(\varphi)\}$. The isomorphism $\theta$ carries $X_t(\varphi) - X_0(\varphi)$ to $F_t(\varphi) - F_0(\varphi)$. Hence $X_t(\varphi) - X_0(\varphi)$ is almost linear in $\varphi$ and

$$E[(X_t(\varphi) - X_0(\varphi))^2] = E[(F_t(\varphi) - F_0(\varphi))^2]$$
$$\leqq E[(\varphi(B_t) - \varphi(B_0))^2] = E\left[\left(\int_{B_0}^{B_t} \varphi'(x)\, dx\right)^2\right]$$
$$\leqq E\left(|B_t - B_0| \int_{-\infty}^{\infty} \varphi'(x)^2\, dx\right)$$
$$= t^{\frac{1}{2}} \|\varphi'\|^2 \leqq c t^{\frac{1}{2}} \|\varphi\|_1^2.$$

Therefore $\{X_t(\varphi)-X_0(\varphi)\}_\varphi$ has an $\mathscr{S}_3'$ regularization by Theorem 3.1. Since $\{X_t(\varphi)\}_\varphi$ has an $\mathscr{S}_2'$ regularization, it has an $\mathscr{S}_3'$ regularization. Therefore $\{X_0(\varphi)\}_\varphi$ has an $\mathscr{S}_3'$ regularization, written $X_0$, because $X_0=X_t-(X_t-X_0)$. Thus $\{X_t\}_t$ is regarded as an $\mathscr{S}_3'$ process.

In the same way as above we obtain

$$E[(X_t(\varphi)-X_s(\varphi))^2]\leqq c\,|t-s|^{\frac{1}{2}}\,\|\varphi\|_1^2,$$

so Theorem 3.1 ensures that

$$E(\|X_t-X_s\|_{-3}^2)\leqq c'|t-s|^{\frac{1}{2}}.$$

Hence $\{X_t\}$ has a continuous $\mathscr{S}_3'$ version, written $\{X_t\}$ again, by Theorem 4.2. This completes the proof of (i). The assertion (ii) is obvious. □

From the proof above we can see that the following theorem holds.

**Theorem 6.2.** (i) $X_t$ *is a continuous centered Gaussian* $\mathscr{S}_3'$ *procees,*

(ii) $E(X_t(\varphi)\,X_s(\psi))=E(\varphi(B_t)\,\psi(B_s))-E\varphi(B_t)\,E\psi(B_s)$,

(iii) $E(X_t(\varphi)^2)\leqq\|\varphi\|^2\,t^{-\frac{1}{2}}(t>0)$,

(iv) $E((X_t(\varphi)-X_s(\varphi))^2)\leqq c'|t-s|^{\frac{1}{2}}\,\|\varphi\|_1^2$,

(v) $X_t\in\mathscr{S}_2'$ *a.s.* $(t>0)$ *and* $X_0\in\mathscr{S}_3'$ *a.s.*,

(vi) $E(\|X_t\|_{-2}^2)\leqq\alpha t^{-1/2}$ $(t>0)$,

(vii) $E(\|X_t-X_s\|_{-3}^2)\leqq c'(t-s)^{\frac{1}{2}}$.

## § 7. The Stochastic Differential Equation (S)

In this section we will prove that the limit process $X_t$ obtained in § 6 satisfies the stochastic differential equation:

$$\text{(S)}\qquad dX_t=(\partial\circ\sqrt{\mu_t})\,db_t+\tfrac{1}{2}\partial^2 X_t dt$$

where $\{b_t\}$ is a standard Wiener $\mathscr{S}_2'$ martingale (see § 5), and $\mu_t=\mu * g_t$ (see § 6).

Observing that

$$\varphi(B_t)=\varphi(B_0)+\int_0^t\varphi'(B_s)\,dB_s+\tfrac{1}{2}\int_0^t\varphi''(B_s)\,ds$$

and recalling that

$$F_t(\varphi)=\varphi(B_t)-E(\varphi(B_t)),$$

we obtain

$$F_t(\varphi)=F_0(\varphi)+S_t(\varphi)+\tfrac{1}{2}\int_0^t F_s(\varphi'')\,ds \tag{7.1}$$

where

$$S_t^{(\varphi)}:=\int_0^t\varphi'(B_s)\,dB_s. \tag{7.2}$$

Since $F_t(\varphi)\in H_F$ for every $(t, \varphi)$, (7.1) ensures that $S_t(\varphi)$ belongs to $H_F$ and corresponds to

$$M_t(\varphi):=X_t(\varphi)-X_0(\varphi)-\tfrac{1}{2}\int_0^t X_s(\varphi'')\,ds \tag{7.3}$$

by the isomorphism $\theta$: $H_X \to H_F$. This formula defines a $\mathscr{D}'$ process $M_t$, which is also expressed as follows:

$$M_t=X_t-X_0-\tfrac{1}{2}\int_0^t \partial^2 X_s ds. \tag{7.4}$$

This implies that $M_t$ is a continuous centered Gaussian $\mathscr{D}'$ process. We can find the covariance functional of $M_t$ by virtue of the isomorphism $\theta$ as follows:

$$\begin{aligned} E(M_s(\varphi)\,M_t(\psi)) &= E(S_s(\varphi)\,S_t(\psi)) \\ &= \int_0^{s\wedge t} E(\varphi'(B_u)\,\psi'(B_u))\,du \\ &= \int_0^{s\wedge t} (\sqrt{\mu_u}\,\varphi', \sqrt{\mu_u}\,\psi')\,du \qquad \text{(by (7.2))} \\ &= \int_0^{s\wedge t} ((\partial\circ\sqrt{\mu_u})^*\,\varphi, (\partial\circ\sqrt{\mu_u})^*\,\psi)\,du; \end{aligned}$$

note that

$$(\partial\circ\sqrt{\mu_u})\,f(\varphi)=(\sqrt{\mu_u}\,f)\,(-\varphi')=f(-\sqrt{\mu_u}\,\varphi').$$

Since

$$\|(\partial\circ\sqrt{\mu_s})^*\,\varphi\|^2=\|\sqrt{\mu_s}\,\varphi'\|^2 \leqq c s^{-\frac{1}{2}}\|\varphi\|_1^2,$$

we obtain

$$\int_0^t |(\partial\circ\sqrt{\mu_s})^*|_{-1}^2\,ds<\infty.$$

As we mentioned in § 5, $M_t$ has the following stochastic integral representation by extending the base probability space and the reference family:

$$M_t=\int_0^t (\partial\circ\sqrt{\mu_s})\,db_s \qquad \text{(an } \mathscr{S}_3 \text{ process)} \tag{7.5}$$

(7.4) and (7.5) imply

$$X_t=X_0+\int_0^t (\partial\circ\sqrt{\mu_s})\,db_s+\tfrac{1}{2}\int_0^t \partial^2 X_s ds,$$

which proves (S).

Since $\{X_t\}$ is an $\mathscr{S}_3'$ process, $\partial^2 X_t$ is an $\mathscr{S}_5'$ process and so the integral

$$I_t=\tfrac{1}{2}\int_0^t \partial^2 X_s ds$$

appears to be an $\mathscr{S}_5'$ processes. But it is an $\mathscr{S}_3'$ process, being equal to $X_t - M_t$. This suggests that the irregularities of $\{\partial^2 X_s\}_s$ must have cancelled each other by integration.

## § 8. The Stochastic Differential Equation ($S_D$)

In this section we will construct a Borel measurable map

$$\sigma: \mathscr{D}' \to \mathscr{L}' := \mathscr{L}'(\mathscr{D}')$$

such that

$$\sigma(X_t^\mu) = \partial \circ \sqrt{\mu_t} \quad \text{a.s.} \quad \text{for every } \mu \text{ and every } t>0$$

(see Theorem 6.1 for $X_t^\mu$). Once this is done, for every $\mu$ $X_t = X_t^\mu$ is a solution of the following stochastic differential equation of the temporally homogeneous diffusion type independent of $\mu$:

$$(S_D) \qquad dX_t = \sigma(X_t)\, db_t + \tfrac{1}{2}\partial^2 X_t\, dt.$$

Also, for almost all $f$ with respect to the initial distribution of $X_t^\mu$ (i.e. $P\{X_0^\mu \in \cdot\}$), the process

$$X_t^\mu(\omega), \; t \in [0, \infty), \; \omega \in (\Omega, F, P\{\cdot | X_0^\mu = f\}$$

is a solution of ($S_D$) starting at $f$.

Let $C_0 = C_0(\mathbb{R})$ denote the continuous functions on $\mathbb{R}$ vanishing at 0. It is obvious that $C_0 \subset \mathscr{D}'$. Let $\partial C_0$ denote the set $\{\partial\eta : \eta \in C_0\}$. Then $C_0 \subset \partial C_0 \subset \mathscr{D}'$. If $\xi \in C_0$, then there exists one and only one $\eta \in C_0$ such that $\xi = \partial\eta$, which is denoted by $I\xi$. Define $I\xi := 0$ on $\mathscr{D}' - \partial C_0$. Then $I$ maps $\mathscr{D}'$ into $C_0$.

We will prove that

$$X_t^\mu \in \partial C_0 \quad \text{a.s.} \tag{8.1}$$

for every $t>0$ and every $\mu$. We will delete $\mu$ from $X_t^\mu$ for the moment. Since

$$E[X_t(\varphi)^2] \leqq \|\varphi\|^2 t^{-\frac{1}{2}} \quad \text{(Theorem 6.2 (iii))},$$

the map $\varphi \mapsto X_t(\varphi)$ is a bounded linear operator from $(\mathscr{D}, \| \ \|)$ into $L_2(\Omega, P)$. Since $\mathscr{D}$ is dense in $L_2 = L_2(\mathbb{R})$, we can extend the map to a bounded linear operator from $L_2$ into $L_2(\Omega, P)$, which we denote by $\bar{X}_t$. Define a family of real random variables $Y_t(x)$, $x \in \mathbb{R}$ by

$$Y_t(0) = 0, \quad Y_t(y) - Y_t(x) = \bar{X}_t(e_y - e_x) \tag{8.2}$$

for each $t$, where $e_a$ denotes the indicator of the interval $(-\infty, a)$. As $\{X_t(\varphi)\}_\varphi$ is a centered Gaussian system, so is $\{Y_t(x)\}_x$ for every $t$. Hence $Y_t(y) - Y_t(x)$ is Gauss distributed with mean 0. Also we have

$$\begin{aligned} E[(X_t(y) - Y_t(x))^2] &= E[\bar{X}_t(e_y - e_x)^2] \\ &\leqq \|e_y - e_x\|^2 t^{-\frac{1}{2}} \leqq |y-x| t^{-\frac{1}{2}}, \\ E[(Y_t(y) - Y_t(x))^4] &\leqq 3|y-x|^2/t. \end{aligned}$$

Hence $\{Y_t(x)\}_x$ has a continuous version for every $t$, which we denote by the same notation $Y_t(x)$. Then $Y_t \in C_0$ a.s. for every $t$. On one hand

$$\partial Y_t(\varphi) = Y_t(-\varphi') = -\int_{-\infty}^{\infty} Y_t(x)\,\varphi'(x)\,dx.$$

Approximating this integral by Riemann sums and using (8.2) we can see that the integral is equal to $\bar{X}_t(\varphi)$ a.s. On the other hand $\bar{X}_t(\varphi) = X_t(\varphi)$, $\varphi \in \mathscr{D}$ by the definition of $\bar{X}_t$. Therefore $\partial Y_t(\varphi) = X(\varphi)$ a.s. for every $\varphi \in \mathscr{D}$. Using the argument used in the uniqueness proof of the $\mathscr{D}'$ regularization, we can check that $\partial Y_t = X_t$ a.s., which implies that $X_t \in \partial C_0$ and $I X_t = Y_t$ a.s. for every $t > 0$.

Let $J_i = (a_i, b_i)$, $i = 1, 2$ be disjoint intervals, and let

$$\alpha_i := \int_{J_i} \mu_t(x)\,dx, \qquad X_i := Y_t(b_i) - Y_t(a_i).$$

Recalling that

$$Y_t(b_i) - Y_t(a_i) = \bar{X}_t(e_{b_i} - e_{a_i}) \quad \text{a.s.,}$$

we can check that $X_1$ and $X_2$ form a Gaussian system and that

$$E(X_i) = 0, \qquad E(X_i^2) = \alpha_i - \alpha_i^2,$$
$$E(X_i X_j) = -\alpha_i \alpha_j \quad (i \neq j).$$

Then the density of the joint distribution of $(X_i, X_j)$ can be computed concretely. After elementary but lengthy computation we can check that

$$E[(X_i^2 - \alpha_i)^2] = \alpha_i^2 (2 - 4\alpha_i + 3\alpha_i^2), \tag{8.3}$$
$$E[(X_i^2 - \alpha_i)(X_j^2 - \alpha_j)] = 3\alpha_i^2 \alpha_j^2. \tag{8.4}$$

Let $J$ be any interval $(a, b]$. Divide $J$ into $2^n$ subintervals of equal length, $J_{n,i} = (a_{n,i-1}, a_{n,i}]$, $i = 1, 2, \ldots, 2^n$ and let

$$X_{ni} := Y(a_{n,i}) - Y(a_{n,i-1})$$

and

$$\alpha_{ni} := \int_{J_{n,i}} \mu_t(x)\,dx, \qquad \alpha := \int_J \mu_t(x)\,dx.$$

Using (8.3) and (8.4) and noting that $\alpha_{ni} \leqq (b-a)\,2^{-n} t^{-\frac{1}{2}}$, we obtain

$$\begin{aligned} E[(\sum_i X_{ni}^2 - \alpha)^2] &= E[(\sum_i (X_{ni}^2 - \alpha_{ni}))^2] \\ &= \sum_i E[(X_{n,i}^2 - \alpha_{n,i})^2] + 2 \sum_{i<j} E[(X_{n,i}^2 - \alpha_{n,i})(X_{n,j}^2 - \alpha_{n,j})] \\ &= \sum_i \alpha_{n,i}^2 (2 - 4\alpha_{n,i} + 3\alpha_{n,i}^2) + 6 \sum_{i<j} \alpha_{n,i}^2 \alpha_{n,j}^2 \\ &= O(2^{-n}), \end{aligned}$$

which implies by the Borel-Cantelli lemma that

$$\sum_i X_{n,i}^2 \to \alpha \quad \text{a.s.} \tag{8.5}$$

Keeping this in mind we define $e_J\colon C_0 \to [0, \infty]$ as follows:

$$e_J(\eta) = \limsup_{n\to\infty} \sum_{i=1}^{2^n} (\eta(a_{n,i}) - \eta(a_{n,i-1}))^2.$$

Then (8.5) implies that

$$e_J(Y_t) = \int_J \mu_t(x)\,dx \quad \text{a.s.} \tag{8.6}$$

for each interval $J$. Hence this holds for every rational interval $J$ simultaneously a.s.. Denoting the positive $C^\infty$ functions by $C_+^\infty$, the rational interval of the form $(k/n, (k+1)/n]$ containing $x$ by $J_n(x)$ and the Lebesgue measure of $J$ by $|J|$, we define $e\colon C_0 \to C_+^\infty$ as follows:

$$(e\eta)(x) = \limsup_{n\to\infty} e_{J_n(x)}(\eta)/|J_n(x)|$$

if this function belongs to $C_0^+$ and $(e\eta)(x) = 1$ otherwise. Then (8.6) implies that

$$eY_t = \mu_t \quad \text{a.s.} \qquad \text{for every } t > 0. \tag{8.7}$$

Define

$$\theta\colon C_+^\infty \to \mathscr{L}' = \mathscr{L}'(\mathscr{D}'), \qquad \gamma \mapsto \partial \circ \sqrt{\gamma}.$$

and $\sigma := \theta \circ e \circ I$. Then

$$\sigma(X_t) = \partial \circ \sqrt{\mu_t} \quad \text{a.s.} \qquad \text{for every } t.$$

The measurability of $I$, $e$, $\theta$ and $\sigma$ can be proved by routine.

## References


1. Itô, K.: Continuous additive $\mathscr{S}'$-processes. In: Stochastic Differential Systems - Filtering and Control. Proceedings (Vilnius, Lithuania 1978), pp. 143–151. Lecture Notes in Control and Information Sciences **25**. Berlin-Heidelberg-New York: Springer 1980
2. Itô, K.: Stochastic analysis in infinite dimensions. In: Stochastic Analysis (A. Frieman and M. Pinsky, eds.). Proceedings (Evanston 1978), pp. 187–197. New York-San Francisco-London: Academic Press 1980
3. Kunita, H.: Stochastic integrals based on martingales taking values in Hilbert spaces. Nagoya Math. J. **38**, 41–52 (1970)
4. Kuo, H.H.: Gaussian measures in Banach spaces. Lecture Notes in Mathematics **463**, Berlin-Heidelberg-New York: Springer 1975
5. Métivier, M.: Reelle and vektorwertige Quasimartingale und die Theorie der stochastischen Integration. Lecture Notes in Mathematics **607**. Berlin-Heidelberg-New York: Springer 1977


Received June 7, 1982

# Permissions

Springer-Verlag would like to thank the original publishers of Itô's papers for granting permissions to reprint specific papers in his collection. The following list contains the credit lines for those articles.

[1] Reprinted from *Japan. Journ. Math* **18,** © 1942 by Science Council of Japan.
[2] Reprinted from *Journ. Pan-Japan Math. Coll.* No. 1077, © 1942 by Osaka University.
[3] Reprinted from *Proc. Imp. Acad. Tokyo* **20,** © 1944 by The Japan Academy.
[4] Reprinted from *Proc. Imp. Acad. Tokyo* **20,** © 1944 by The Japan Academy.
[5] Reprinted from *Proc. Imp. Acad. Tokyo* **20,** © 1944 by The Japan Academy.
[7] Reprinted from *Proc. Imp. Acad. Tokyo* **20,** © 1944 by The Japan Academy.
[9] Reprinted from *Proc. Japan. Acad.* **22,** © 1946 by The Japan Academy.
[10] Reprinted from *Nagoya Math. Journ.* **1,** © 1950 by Nagoya University.
[11] Reprinted from *Proc. Japan. Acad.* **26,** © 1950 by The Japan Academy.
[12] Reprinted from *Mem. Amer. Math. Soc.* 4, © 1951 by the American Mathematical Society.
[13] Reprinted from *Nagoya Math. Journ.* **3,** © 1951 by Nagoya University.
[14] Reprinted from *Journ. Math. Soc. Japan* **3,** © 1951 by Journ. Math. Soc. Japan.
[15] Reprinted from *Mem. Coll. Science, Univ. Kyoto, Ser. A,* **28,** © 1953 by Kyoto University.
[16] Reprinted from *Mem. Coll. Science, Univ. Kyoto, Ser. A,* **28,** © 1953 by Kyoto University.
[17] Reprinted from *Japan. Journ. Math.* **22,** © 1953 by Science Council of Japan.
[18] Reprinted from *Proc. Third Berkeley Symp. on Math., Stat. and Prob.* **2,** © 1955 by University of California Press.
[19] Reprinted from *Trans. Amer. Math. Soc.* **81,** © 1956 by Amer. Math. Soc.
[20] Reprinted from *Illinois Journ. Math.* **4,** © 1960 by University of Illinois Press.
[21] Reprinted from *Proc. Fourth Berkeley Symp. on Math., Stat. and Prob.* **2,** © 1960 by University of California Press.
[22] Reprinted from *Ann. Fac. Sci. Univ. Clermont* **2,** © 1962 by University of Clermont.
[23] Reprinted from *Proc. Intern. Congress of Math.,* © 1962 by Mittag-Leffler.
[24] Reprinted from *Illinois Journ. Math.* **7,** © 1963 by University of Illinois Press.
[25] Reprinted from *Journ. Math. Kyoto Univ.* **3,** © 1964 by Kyoto University.
[26] Reprinted from *Journ. Math. Kyoto Univ.* **4,** © 1964 by Kyoto University.
[27] Reprinted from *Ann. Inst. Fourier, Univ. Grenoble,* © 1965 by University of Grenoble.
[28] Reprinted from *Journ. Math. Soc. Japan* **20,** © 1968 by Journ. Math. Soc. Japan.
[29] Reprinted from *Osaka Journ. Math.* **5,** © 1968 by Osaka University.
[30] Reprinted from *Proc. Fifth Berkeley Symp. on Math., Stat. and Prob.* **2,** © 1965 by University of California Press.
[31] Reprinted from *Math. Scand.* **22,** © 1968 by Math. Scand.
[32] Reprinted from *Proc. Int. Conf. on Funct. Anal. and Rel. Topics* (Toyko), © 1969 by Tokyo University.
[33] Reprinted from *Nagoya Math. Journ.* **38,** © 1970 by Nagoya University.
[34] Reprinted from *Proc. Sixth Berkeley Symp. on Math., Stat. and Prob.* **3,** © 1970 by University of California Press.
[40] Reprinted from *Proc. Int. Symp. on Stoch. Diff. Equat.* (Kyoto), © 1976 by K. Ito.